Physics and Chemistry of Micro-Nanotribology

Jianbin Luo, Yuanzhong Hu, and Shizhu Wen

ASTM Stock Number: MONO7

ASTM International
100 Barr Harbor Drive
PO Box C700
West Conshohocken, PA 19428–2959

Printed in the U.S.A.

Library of Congress Cataloging-in-Publication Data

Luo, Jianbin, 1961–
Physics and chemistry of micro-nanotribology/Jianbin Luo, Yuanzhong Hu, Shizhu Wen.
p. cm.
"ASTM Stock Number: MONO7."
ISBN 978-0-8031-7006-3
1. Tribology. I. Hu, Yuanzhong, 1946– II. Wen, Shizhu, 1932– III. Title

TJ1075.L825 2008
621.8'9—dc22

2008023825

Copyright © 2008 ASTM International, West Conshohocken, PA. All rights reserved. This material may not be reproduced or copied, in whole or in part, in any printed, mechanical, electronic, film, or other distribution and storage media, without the written consent of the publisher.

Photocopy Rights
Authorization to photocopy items for internal, personal, or educational classroom use of specific clients, is granted by ASTM International provided that the appropriate fee is paid to ASTM International, 100 Barr Harbor Drive, P.O. Box C700, West Conshohocken, PA 19428-2959, Tel: 610–832–9634; online:http://www.astm.org/copyright/
The Society is not responsible, as a body, for the statements and opinions advanced in this publication.
ASTM International does not endorse any products represented in this publication.

Printed in Baltimore, Maryland
November, 2008

Dedication

This Monograph is dedicated to the 20th anniversary of the State Key Laboratory of Tribology.

Jianbin Luo
Yuanzhong Hu
Shizhu Wen

Acknowledgment

This book was brought to fruition by the efforts of many individuals. We would like to thank all of them, beginning with the editor and the publication staff of ASTM International, especially Dr. George Totten who has encouraged us to publish our research achievements in this monograph, and Kathy Dernoga and Monica Siperko who have given us guidance and assistance from the outset of the venture. In addition, we wish to convey appreciation to the authors who have devoted considerable time, energy, and resources to support this endeavor. We are also grateful to the reviewers of the various chapters who, through their suggestions, permitted good manuscripts to be made better.

Finally, we are grateful for the support from government and industry through various research programs, including National Basic Research Program of China, National Natural Science Foundation of China, and international joint researches. Their support of our research activities has led to this publication.

Jianbin Luo
Yuanzhong Hu
Shizhu Wen

Foreword

THIS PUBLICATION, *Physics and Chemistry of Micro-Nanotribology*, was sponsored by Committee D02 on Petroleum Products and Lubricants. This is Monograph 7 in ASTM International's manual series.

Contents

Preface xi

Chapter 1: Introduction, *Shizhu Wen, Jianbin Luo, and Yuanzhong Hu* 1
    The Measurement and Investigation of Thin Film Lubrication (TFL) 2
    Surface Coatings 2
    Applications of Micro/Nanotribology 3
    Summary 4

Chapter 2: Measuring Techniques, *Dan Guo, Jianbin Luo, and Yuanzhong Hu* 7
    Introduction 7
    Optical Measuring Techniques 8
    Surface Force Apparatus 14
    Scanning Probe Microscope 18
    Nanoindentation and Nanoscratching 22
    Other Measuring Techniques 26

Chapter 3: Thin Film Lubrication—Experimental Study, *Jianbin Luo and Shizhu Wen* 37
    Introduction 37
    Properties of Thin Film Lubrication 39
    The Failure of Lubricant Film 53
    Thin Film Lubrication of Ionic Liquids 54
    Gas Bubble in Liquid Film under External Electric Field 55
    Summary 60

Chapter 4: Thin Film Lubrication—Theoretical Modeling, *Chaohui Zhang* 63
    Introduction 63
    Spatial Average and Ensemble Average 64
    Velocity Field of Lubricants with Ordered Molecules 65
    Simulations via Micropolar Theory 67
    Rheology and Viscosity Modification 72
    Other Approaches Related to TFL Theories 74
    Conclusions 77

Chapter 5: Molecule Films and Boundary Lubrication, *Yuanzhong Hu* 79
    Introduction 79
    Mechanisms of Boundary Lubrication 80
    Properties of Boundary Films as Confined Liquid 82
    Ordered Molecular Films 88
    Discussions on Boundary Friction 93
    Summary 94

Chapter 6: Gas Lubrication in Nano-Gap, *Meng Yonggang* 96
    History of Gas Lubrication 96
    Theory of Thin Film Gas Lubrication 97
    Application of Gas Lubrication Theory 103
    Summary 114

Chapter 7: Mixed Lubrication at Micro-scale, *Wen-zhong Wang, Yuanzhong Hu, and Jianbin Luo* 116
    Introduction 116
    Statistic Approach of Mixed Lubrication 116
    A DML Model Proposed by the Present Authors 118
    Validation of the DML Model 125
    Performance of Mixed Lubrication—Numerical and Experimental Studies 130
    Summary 144

Chapter 8: Thin Solid Coatings, *Chenhui Zhang and Tianmin Shao* ... 147
    Introduction ... 147
    Diamond-like Carbon (DLC) Coatings ... 147
    CNx Films ... 151
    Multilayer Films ... 153
    Superhard Nanocomposite Coatings ... 157

Chapter 9: Friction and Adhesion, *Yuanzhong Hu* ... 167
    Introduction ... 167
    Physics and Dynamics of Adhesion ... 167
    Models of Wearless Friction and Energy Dissipation ... 171
    Correlations Between Adhesion and Friction ... 178
    The Nature of Static Friction ... 181
    Summary ... 184

Chapter 10: Microscale Friction and Wear/Scratch, *Xinchun Lu and Jianbin Luo* ... 187
    Introduction ... 187
    Differences Between Macro and Micro/Nano Friction and Wear ... 188
    Calibration of the Friction Force Obtained by FFM ... 189
    Microscale Friction and Wear of Thin Solid Films ... 191
    Microscale Friction and Wear of Modified Molecular Films ... 194
    Microscale Friction and Scratch of Multilayers ... 200
    Summary ... 208

Chapter 11: Tribology in Magnetic Recording System, *Jianbin Luo, Weiming Lee, and Yuanzhong Hu* ... 210
    Introduction ... 210
    Surface Modification Films on Magnetic Head ... 211
    Lubricants on Hard Disk Surface ... 226
    Challenges from Developments of Magnetic Recording System ... 231

Chapter 12: Tribology in Ultra-Smooth Surface Polishing, *Jianbin Luo, Xinchun Lu, Guoshun Pan, and Jin Xu* . 237
    Introduction ... 237
    Nanoparticles Impact ... 237
    Chemical Mechanical Polishing (CMP) ... 245
    The Polishing of Magnetic Head Surface ... 262

Subject Index ... 270

Preface

The roots of micro/nanotribology can be found deep in conventional concepts of tribology. The recognition in the last century of elasto-hydrodynamic lubrication (EHL) as the principal mode of fluid-film lubrication in many machine components enabled reliable design procedures to be developed for both highly stressed and low elastic modulus machine elements. Towards the end of the last century submicron film thicknesses were recognized in many EHL applications. It is now being asked how EHL concepts can contribute to understanding the behavior of even thinner lubricating films. The answer is to be found in the subject widely known as micro/nanotribology.

As early as 1929 Tomlinson considered the origin of friction and the mechanism of energy dissipation in terms of an independent oscillator model. This important approach provided the foundation for many present studies of atomic scale friction. The rapid development of micro/nanotribology in recent decades is certainly a significant and fascinating aspect of modern tribology. New scientific instruments, impressive modeling, and computer simulations have contributed to the current fascination with nanotechnology.

A remarkable indication of these developments is evident in the boom of publications. Nevertheless, the knowledge and understanding of micro/nanotribology remains incomplete, although several books related to the subject have now been published. The interdisciplinary nature of tribology persists in studies of microscopic scale tribology. Individual investigators contribute to specific aspects of the field as they help to develop a general picture of the new field of micro/nanotribology, thus adding additional bricks to the house of truth.

The present book is written by authors whose backgrounds are mainly in mechanical engineering. They present individual contributions to the development of microscopic tribology, with significant effort being made to form a bridge between fundamental studies and applications.

I am confident that readers in both academic and industrial sectors will find the text interesting and beneficial to their understanding of an exciting aspect of modern tribology.

Duncan Dowson
Leeds, U.K. June 2008

1

Introduction

Shizhu Wen,[1] Jianbin Luo,[1] and Yuanzhong Hu[1]

IN 1966, "TRIBOLOGY," AS A NEW WORD IN SCIENCE, was first presented in a report by the U.K. Department of Education and Science, which has been usually known as the Jost report. Tribology is defined in this report as the science and technology of interacting surfaces in relative motion and of related subjects and practices. The report emphasized the importance and a great potential power of tribology as an individual branch of science in the development of modern national economy. In the history of science, however, research activities on tribology can be traced back to the 15th century, when Leonardo da Vinci (1452–1519) presented a scientific deduction on solid surface friction.

As a practice-based subject, the formation and development of tribology have always been associated with the requirement from society and technology development. Tribology experienced several different stages in its history. Its developing process indicates an obvious trend of integration and combination of multi-scientific subjects in a multi-scale nature from macroscopic dimension to nanometre.

The most remarkable character of tribology is the integration, combination, and interaction between multi-scientific subjects. This not only broadens the scope of tribology research, but also enriches the research mode and methodology. An early research was typically of Amontons and Coulomb's work on solid surface friction before the 18th century. Based on experimental observations, they concluded an empirical formula of sliding friction. An experiment-based research mode represented a characteristic of this stage. At the end of the 19th century, Reynolds [1] revealed load carrying mechanics of lubricating films and established a foundation of the fluid lubrication theory based on viscous hydrodynamics. A new theoretical research mode was then initiated, which is associated with the continuous medium mechanics.

After the 1920s, the multi-subject nature of tribology research was enhanced due to rapid development of economy and relative technologies. During this period, Hardy [2] proposed a model of boundary lubrication. He explained that the polar molecules in lubricant had a physicochemical interaction with metal surfaces, from which the boundary lubricating films were formed. At the same time, Ostwald [3] presented a conception of mechano-chemistry, referring to the physicochemical change and effect induced by the energy alternation in the friction and impact process. Subsequently, Heinicke [4] published his monograph in a book titled *Tribochemistry*, emphasizing an integration of tribology and chemistry. Bio-tribology emerged in the 1970s and is another example of the integration of multi-scientific subject research that bridges tribology, biology, and iatrology. In development, it integrated with bionics and nanotechnology, and created a new research field [5]. Clearly, modern tribology, in the process of its maturity, has combined different scientific subjects into an integrated science and technology.

A new stage of tribology started in the 1980s because of an awareness of 21st century-oriented nanotechonology, which resulted in a series of new scientific branches, such as nanoelectronics, nanomaterials, and nanobiology, etc. Micro/nanotribology [6], or molecular tribology as some prefer to call [7], is one of the most important branches that emerged during that period. Nanotechnology studies behaviors and interactions of atoms and molecules in natural or technical phenomena at nanometre scale (0.1 nm ~ 100 nm) to improve and enhance our understanding of nature. This would enable us to deal with the existing world more effectively. In other words, micro/nanotribology creates a microscopic research mode of tribology.

Another remarkable aspect of tribology is the transition from macro scale to micro scale research known as scale-down development. The foundation of micro/nanotribology is not only a result of the integration of multi-scientific subjects, but also originates from the understanding that a tribology process can proceed across several scales. A reduction in the research scale from macro to micro metre is also determined by the nature of the tribology process itself. In a friction process, for example, the macro tribology property of sliding surfaces depends closely on micro structure or micro interactions on the interface. Micro/nanotribology provides a new insight and an innovative research mode. It reveals mechanisms of the friction, wear, and lubrication on atomic and molecular scale, or both, and establishes a relationship between the microstructure and macroscopic performance. This is very important for the further development of tribology.

In addition, micro/nanotribology also has a broad application foreground. A development of modern precision machinery, high technology equipment, and especially the newly born scientific areas promoted by nanotechnology, such as nano electronics, nano biology, and the micro electromechanical system, leads to an urgent demand on micro/nanotribology research for theoretical support.

It is clear that the emergence of micro/nanotribology marks a new stage in tribology progress. Winer [8] pointed out that a promising development of tribology is the micro or atom-scale tribology. In such an area, new instruments for surface observations with sub-nanometre resolution have been established, and computer simulations allow one to ex-

[1] *State Key Laboratory of Tribology, Tsinghua University, Beijing, China.*

plore a tribological process at atomic scale. These might bring a great breakthrough in the field of tribology.

As research progresses, a great many papers on the subject of micro/nanotribology appear in journals of science and engineering, and several books have been already published over the past decade. Examples include *Micro and Nanotribology* by N. Ohmae, J. M. Martin, and S. Mori [9], *Handbook of Micro/Nanotribology* edited by B. Bhushan [10], and many more. To the present authors, however, it is worthwhile to write a new monograph on this subject for good reasons such as: (1) The present book is contributed by a group of authors who have been working on various aspects of micro/nanotribology for more than a decade and yet closely cooperating in the same institution, the State Key Laboratory of Tribology (SKLT), which makes the book more systemic and more extensive. (2) The book focuses on physics/mechanics of micro/nanotribology, which may be found interesting and convenient to the readers with a background in mechanical engineering. The book was written on the basis of the new progress in micro/nanotribology since the 1990s. Since it is difficult for a single book to cover all related subjects, more attention is paid to the areas we have been working on, and in particular to the subjects briefly discussed in the following.

The Measurement and Investigation of Thin Film Lubrication (TFL)

Since the 1990s, significant progress has been made in this area, particularly three methods for the measurement of lubricant film at nanoscale. They are the spacer layer optical interferometry (SLOI) proposed by Johnston et al. in 1991 [11], the relative optical interference intensity (ROII) technique proposed by Luo et al. in 1994 [12,13], and the thin film colorimetric interferometry (TFCI) proposed by Hartl et al. in 1997 [14,15]. These instruments are powerful for an investigation of the properties and characteristics of oil films from a few nanometres to hundreds of nanometres in a contact region between a steel ball and a glass disk with a semi-reflected coating. Therefore, thin film lubrication (TFL) as a lubrication regime between elastohydrodynamic lubrication (EHL) and boundary lubrication has been proposed and well studied from the 1990s [12–14]. In this regime, the isoviscosities of liquid depends on multiple factors, such as the distance between two solid surfaces, the polarity of additives, the surface energy of the materials in contact, the external voltage applied, etc. [13,16,17]. The isoviscosity of pure hexadecane in a 7 nm gap, e.g., is about two times of its bulk viscosity, or, about three times to more than ten times of their bulk viscosities when the polarity additive is increased to a concentration of 2 %. The critical film thickness for the transition from EHL to TFL was proposed as [13]:

$$h_{ct} = a + be^{c/\eta_0} \qquad (1)$$

where h_{ct} is the critical film thickness; η_0 is the initial viscosity of lubricant; a, b, and c are coefficients that are related, respectively, to surface energy or surface tension, the electric-field intensity, and the molecular structure and polarity of the lubricant.

Another powerful tool for investigating a rheology of liquid films on nano-scale is Surface Force Apparatus (SFA) which was invented in 1969 by Tabor and Winterton [18] and further developed in 1972 by Israelachivili and Tabor [19]. In the 1980s, SFA was further improved by Israelachivili and McGuiggan [20], Prieve et al. [21], and Tonock et al. [22]. A great number of interesting results have been obtained by using SFA. It is indicated that as film thickness decreases to a molecule dimension, the confinement of walls would induce dramatic changes in rheological properties of thin films, including the viscosity enhancement, non-Newtonian shear response, formation of ordering structures, and solidification [23]. These have greatly improved our understanding of TFL and boundary lubrication as well.

Surface Coatings

In the past ten years, another significant progress in tribology is attributed to surface coatings and surface texture. New coating materials and technologies for preparing ultrathin solid films have been developed which has attracted great attention in the field of tribology.

Near Frictionless Coatings (NFC)

It has been a dream for a tribologist to create a motion with a super low friction or even no friction between two contact surfaces. In order to reduce friction, great efforts have been made to seek materials that can exhibit lower friction coefficients. It is well known that friction coefficients of high quality lubricants, e.g., polytetrafluoroethylene (PTFE), graphite, molybdenum disulphide (MoS_2), etc., are hardly reduced below a limit of 0.01.

Diamond-like carbon (DLC) coating has emerged as one of the most attractive coatings. It exhibits many excellent properties, for instance, a low friction coefficient, high hardness, good bio-consistence, etc. At the end of the last century, Erdemir et al. [24], from Argonne National Laboratory USA, reported a new type of DLC film called near frictionless carbon (NFC) coating. It is reported that super low friction coefficients in a range from 0.001 to 0.003 have been achieved between ball and disk, both coated with NFC films. Such a low friction coefficient was proposed mainly due to the elimination of the strong covalent and π-π^* interaction at sliding DLC interfaces plus good shielding of carbon atoms by dihydration [25]. The additional works have been tasked for an improvement of anti-wear property of these low friction coatings.

Superhard Nanocomposite Coatings

Superhard materials refer to the solids with Vickers hardness higher than 40 GPa. A great number of attempts have been made to synthesize these superhard materials with the hardness close to the diamond. Two approaches were adopted to achieve this objective. One is to synthesize intrinsic superhard material. In general, it is believed that the diamond is the hardest intrinsic material due to strong, nonpolar C-C covalent bonds which make the hardness as high as 70–100 GPa. Synthetic c-BN is another of the hardest bulk material with a hardness of about 48 GPa. Ta-C coatings with a sp^3 fraction of larger than 90 % show a superhardness of 60–70 GPa. The second approach is to make nanostructured superhard coatings. Their superhardness and other mechanical properties are determined by a proper design of microstructure. A typical example for nanostructured superhard coatings is the heterostructures or superlattices. TiN/VN superlattice coatings, for instance, can achieve a superhardness of 56 GPa as the lattice period is 5.2 nm [26]. Carbon nitrides are another type of coating material. It is claimed that their bulk modulus could probably be greater than diamond. This has attracted a great deal of attention since its first prediction in 1989 [27].

Applications of Micro/Nanotribology

The technologies resulting from the progress in micro/nanotribology have been successfully applied to the manufacture of high-tech products, such as hard disk drivers (HDD), integrated circuits (IC), particularly high density multilevel interconnected circuits, and microelectromechanical systems (MEMS). For example, the fast growth of areal recording density of HDD, continuous decrease in IC line width or the size of micro/nano-printing, and improvements of MEMS service life and performance are greatly attributed to recent advancements in micro/nanotribology. Two typical applications are given in the following.

Hard Disk Driver (HDD)

The HDD recording density has been increasing at a high rate of 100 % per year in the past ten years. It is expected that the recording density is to be increased to over 1,000 Gbit/in.^{2} and the fly height be decreased to about 3 nm in the near future. There are three major challenges for tribologists to face today: (1) to make a solid protective coating, i.e., a diamond-like carbon (DLC) layer, with a thickness of about 1 nm without any micro-pinholes; (2) to make a lubricant film about 1 nm thick on the disk and head surfaces, or both, to minimize the wear, friction, and erosion; and (3) to control vibration of the magnetic head and its impact on the surface of the disk.

Ultra-Thin Coatings

DLC coatings on both HDD head and disk surfaces become thinner with an increase of the storage area. Initially, sputtered DLC coatings with a thickness of 7 nm were used for an HDD with an area density of 10 Gbits/in.^{2} [28]. Later, PECVD was used to deposit a-C:H coatings less than 5 nm in thickness on the disk surface, giving rise to an area density of 20–30 Gbits/in.^{2} [29]. Recently, a-CNx coatings [28] and Si doped coatings were introduced to replace a-C:H. In order to achieve a storage density more than 200 Gbit/in.^{2}, only 2 nm is allowed for DLC protective coatings to remain on the head and disk surfaces, or both. To synthesize this thin and smooth DLC coating, it is required to grow the coating layer by layer to avoid island formation. As a result, a deposition method with a high fraction of energetic carbon ions is needed. Traditional deposition methods, such as magnetron sputtering and PECVD, cannot meet this requirement. Due to a nearly 100 % ions fraction, FCVA is a promising technique to deposit an ultra-thin DLC coating with a thickness of 2 nm or even less. Recent research indicates that a continuous DLC coating with a thickness of 2 nm could be synthesized by FCVA. Additional work has to be carried out to be successful in preparing continuous DLC coatings with a thickness of 1 nm.

Lubrication and Monolayers

Impressive advances in lubrication technology of HDDs have also made a great contribution to the theoretical development of boundary lubrication. Perfluoropolyethers (PFPE), particularly PFPE Z-DOL, is one of the synthetic lubricants that is widely applied due to its excellent performance, such as chemical inertness, oxidation stability, lower vapor pressure, and good lubrication properties [30]. Since the recording density is approaching more than 1 Tbit/in.^{2}, a flying height has to be reduced to about 2 nm. Sliding contacts between two surfaces of the head and disk would occur much more frequently than before. A good surface mobility of lubricant film could ensure the lubricant reflow and cover the area where the lubricant molecules are depleted after a head-disk interaction. Therefore, a proper combination of the mobile and surface grafted molecules would be preferred for the lubricant to be used in HDDs. Interaction of PFPE with a DLC coating and lubricant degradation are important subjects to be considered in HDD lubrication [31]. A more detailed discussion on this issue will be given in Chapter 11.

In addition, efforts have been made to explore the possibility of applying a monolayer as a lubricant to the surfaces of the magnetic head and disk, or both [32–35]. The results indicate:

1. The FAS SAMs on the magnetic heads lead to a considerable improvement on tribological and corrosion-resistant properties, a high water contact angle, and electron charge adsorption-resistant property of the magnetic head.
2. The monolayers of organic long-chain molecules are fairly well oriented relative to the substrates. Unlike the bulk crystals, however, the ordering found in the monolayers is short-range in nature, extending over a few hundred angstroms.
3. The short-range order gradually disappears as the temperature rises, and the structure becomes almost completely disordered near the bulk melting point of the monolayer materials.

Gas Lubrication Theory in HDD

In a hard disk drive, the read-write components attached to a slider are separated from the disk surface by an air-bearing force generated by a thin air layer squeezed into a narrow space between the slider and disk surfaces due to a high rotation speed of the disk. An increase in the HDD storage density requires a corresponding reduction of the smallest thickness of the air bearing between the slider and disk surfaces, estimated as low as 2 nm in the near future. At such a small spacing, many new problems, such as particle flow and contamination [36], surface force effect [37], surface texture effect [38], etc., emerge and have been extensively investigated in recent years. From a theoretical point of view, the most important problem is that the physical models that describe an air-bearing phenomenon well at larger spacing could no longer give any prediction close to the reality of 2 nm FH. So it is important to have an improved lubrication model to ensure the read-write elements attached at their trailing edge to fly at a desirable attitude.

In 1959, Burgdorfer [39] first introduced a concept of the kinetic theory to the field of gas film lubrication. This was to derive an approximation equation, called the modified Reynolds equation, using a slip flow velocity boundary condition for a small Knudsen number $Kn \ll 1$. For larger Knudsen numbers, Hisa [40] in 1983 proposed a higher order approximation equation by considering both the first- and the second-order slip flows. In 1993 Mitsuya [41] introduced a 1.5-order slip flow model which incorporated different order slippage boundary conditions into an integration of the traditional macroscopic continuum compressible Stokes equations under an isothermal assumption.

In 1985, Gans [42], who treated the linearized Boltzmann equation as a basic equation, derived an approximate lubrication equation analytically using a successive approximation method. Fuikui and Kaneko [43] started from a linearized Boltzmann equation with slip boundary conditions similar to Gan's but with different solution methods. Consequently, they derived the generalized Reynolds equation including thermal creep flow. Their results showed that the Burgdorfer's first-order slip model overestimated the load-carrying capacities, while the second-order slip model underestimated it. In 1990, Fuikui and Kaneko [44] proposed a polynomial fitting procedure to explicitly express Poseuille flow rate as a function of the inverse Knudsen number using cubic polynomials.

In 2003, Wu and Bogy [45] introduced a multi-grid scheme to solve the slider air bearing problem. In their approach, two types of meshes, with unstructured triangles, were used. They obtained the solutions with the minimum flying height down to 8 nm.

For a flying height around 2 nm, collisions between the molecules and boundary have a strong influence on the gas behavior and lead to an invalidity of the customary definition of the gas mean free path. This influence is called a "nanoscale effect" [46] and will be discussed more specifically in Chapter 6.

Chemical Mechanic Polishing (CMP)

Chemical Mechanical Polishing, also referred to as Chemical Mechanical Planarization (CMP), is commonly recognized to be the best method of achieving global planarization in a super-precision surface fabrication. From the technology point of view, the original work to develop CMP for semiconductor fabrication was done at IBM [47]. It also has been the key technology for facilitating the development of high density multilevel interconnected circuits [48], such as silicon, dielectric layer, and metal layers [49].

The material removal mechanisms in the CMP process involve abrasive action, material corrosion, electrochemistry, and the hydrokinetics process. They are closely related to tribology. To date, it is extremely difficult to clearly separate the key factors associated with a required removal and surface quality during CMP. There is still a lack of knowledge on the fundamental understanding of polishing common materials widely used in microelectronic industries, such as silicon, SiO_2, tungsten, copper, etc. The growing and widespread applications of CMP seem to exceed the advances of our scientific understanding. Modeling the material removal process is now an active area of investigation, which may help us to improve the understanding of CMP mechanisms. In the recent ten years, progress in the following aspects has contributed much to the development of CMP technology and the understanding of CMP mechanisms.

1. An investigation of interaction between individual particles and solid surface indicates that the atoms on the surface have been extruded out by the incident particles. This forms a pileup at the rim of the impacted region. Amorphous phase transition takes place and materials in the contact region are deformed due to plastic flow inside the amorphous zone [50,51].
2. The material removal in CMP is attributed to multi mechanisms of wear, including abrasive, adhesive, erosive, and corrosive wear.
3. An abrasive-free CMP is an enhanced chemically active process, which provides lower dishing, erosion, and less or no mechanical damage of low-k materials compared to conventional abrasive CMP processes [52].
4. Electric chemical polish (ECP) and electric chemical mechanical polish (ECMP) [53] have been developed as promising methods for global planarization of LSI fabrication and abrasive-free polish.
5. The surface stress free (SSF) approach for removal of the Cu layer and planarization without polishing is critical for manufacturing a new generation of IC wafer composed of soft low-k materials [54].

Summary

Micro/nanotribology emerges as a new area of tribology and has been growing very fast over the past ten years. Both the experimental measuring technique and the theoretical simulation method have been scaled down to atomic level, which provides powerful tools for us to explore new tribological phenomena or rules in a nano-scale. As we recognized, new challenges would emerge as a result of development in nanotechnology, particularly in MEMS, HDD, and nano-manufacturing, which are expected to be the most important and fastest developing areas in the 21st century. Therefore, micro/nanotribology will play a more significant role in the next 30 years.

References

[1] Reynolds, D., "On the Theory of Lubrication and Its Application to Mr. Beauchamp Tower's Experiments, Including an Experiment Determination of the Viscosity of Olive Oil," *Phil. Trans. Roy. Soc. A*, Vol. 177, 1866, pp. 159–234.

[2] Hardy, W. B., and Hardy, J. K., "Note on Static Friction and on the Lubricating Properties of Certain Chemical Substance," *Philos. Mag.*, Vol. 28, 1919, pp. 33–38.

[3] Ostwald, W., *Handbuch allg. Chemic.*, Leipzig, 1919.

[4] Heinicke, G., *Tribochemistry*, Berlin, Academic-Verlag, 1984.

[5] Scherge, M., and Gorb, S., *Biological Micro- and Nanotribology: Nature's Solutions*, Berlin, Springer-Verlag, 2001.

[6] Belak, J. F., "Nanotribology," *MRS Bull.*, Vol. 18, No. 5, 1993, pp. 55–60.

[7] Granick, S., "Molecular Tribology," *MRS Bull.*, Vol. 16, No. 10, 1991, pp. 33–35.

[8] Winer, W. O., "Future Trends in Tribology," *Wear*, Vol. 136, 1990, pp. 19–27.

[9] Ohmae, N., Martin, J. M., and Mori, S., *Micro and Nanotribology*, TJ1075.O36 2005, ASME Press, New York, 2005.

[10] Bhushan, B., *Handbook of Micro/Nanotribology*, Boca Raton, CRC Press, LLC, Vol. 1, 1998.

[11] Johnston, G. J., Wayte, R. H., and Spikes A., "The Measurement and Study of Very Thin Lubricant Films in Concentrate Contact," *STLE Tribol. Trans.*, Vol. 34, 1991, pp. 187–194.

[12] Luo, J. B., "Study on the Measurement and Experiments of Thin Film Lubrication," Ph.D. thesis, Tsinghua University, Beijing, China, 1994.

[13] Luo, J. B., Wen, S. Z., and Huang, P., "Thin Film Lubrication, Part I: The Transition Between EHL and Thin Film Lubrication," *Wear*, Vol. 194, 1996, pp. 107–115.

[14] Hartl, M., Krupka, I., and Liska, M., "Differential Colorimetry: Tool for Evaluation of Chromatic Interference Patterns," *Optical Engineering*, Vol. 36, No. 9, 1997, pp. 2384–2391.

[15] Hartl, M., Krupka, I., Poliscuk, R., and Liska, M., "An Automatic System for Real-Time Evaluation of EHD Film Thickness and Shape Based on the Colorimetric Interferometry," *STLE Tribol. Trans.*, Vol. 42, No. 2, 1999, pp. 303–309.

[16] Shen, M. W., Luo, J. B., Wen, S. Z., and Yao, J. B., "Investigation of the Liquid Crystal Additive's Influence on Film Formation in Nano Scale," *Lubr. Eng.*, Vol. 58, No. 3, 2002, pp. 18–23.

[17] Luo, J. B., Shen, M. W., and Wen, S. Z., "Tribological Properties of Nanoliquid Film under an External Electric Field," *J. Appl. Phys.*, Vol. 96, No. 11, 2004, pp. 6733–6738.

[18] Tabor, D., and Winterton, R. H., "The Direct Measurement of Normal and Retarded van der Waals Forces," *Proc. R. Soc. London, Ser. A*, Vol. 312, 1969, pp. 435–450.

[19] Israelachvili, J. N., and Tabor, D., "The Measurement of van der Waals Dispersion Forces in the Range 1.5 to 130 nm," *Proc. R. Soc. London, Ser. A*, Vol. 331, 1972, pp. 19–38.

[20] Israelachvili, J. N., and McGuiggan, P. M., "Adhesion and Short-Range Forces between Surfaces. Part I: New Apparatus for Surface Force Measurement," *J. Mater. Res.*, Vol. 5, No. 10, 1990, pp. 2223–2231.

[21] Prieve, D. C., Luo, F., and Lanni, F., "Brownian Motion of a Hydrosol Particle in a Colloidal Force Field," *Faraday Discuss. Chem. Soc.*, Vol. 83, 1987, pp. 297–308.

[22] Tonock, A., Georges, J. M., and Loubet, J. L., "Measurement of Intermolecular Forces and the Rheology of Dodecane between Alumina Surfaces," *J. Colloid Interface Sci.*, Vol. 126, 1988, pp. 1540–1563.

[23] Hu, Y. Z., Wang, H., and Guo, Y., "Simulation of Lubricant Rheology in Thin Film Lubrication, Part I: Simulation of Poiseuille Flow," *Wear*, Vol. 196, 1996, pp. 243–248.

[24] Erdemir, A., Eryilmaz, O. L., and Fenske, G., "Synthesis of Diamond-like Carbon Films with Superlow Friction and Wear Properties," *J. Vac. Sci. Technol. A*, Vol. 18, 2000, pp. 1987–1992.

[25] Erdemir, A., "The Role of Hydrogen in Tribological Properties of Diamond-Like Carbon Films," *Surf. Coat. Technol.*, Vol. 146/147, 2001, pp. 292–297.

[26] Veprek, S., Reiprich, S., and Li, S. Z., "Superhard Nanocrystalline Composite Materials: The TiN/Si3N4 System," *Appl. Phys. Lett.*, Vol. 66, 1995, pp. 2640–2642.

[27] Liu, A. Y., and Marvin, L., "Cohen, Prediction of New Compressibility Solids," *Science*, Vol. 245, 1989, p. 841.

[28] Robertson, J., "Requirements of Ultrathin Carbon Coatings for Magnetic Storage Technology," *Tribol. Int.*, Vol. 36, 2003, pp. 405–415.

[29] Goglia, P. R., Berkowitz, J., Hoehn, J., Xidis, A., and Stover, L., "Diamond-Like Carbon Applications in High Density Hard Disc Recording Heads," *Diamond Rel. Mat.*, Vol. 10, 2001, pp. 271–277.

[30] Demczyk, B., Liu, J., Chao, Y., and Zhang, S. Y., "Lubrication Effects on Head-Disk Spacing Loss," *Tribol. Int.*, Vol. 38, 2005, pp. 562–565.

[31] Kajdas, C., and Bhushan, B., "Mechanism of Interaction and Degradation of PFPEs with a DLC Coating in Thin-Film Magnetic Rigid Disks: A Critical Review," *J. Info. Storage Proc. Syst.*, Vol. 1, 1999, pp. 303–320.

[32] Hu, X. L., Zhang, C. H., Luo, J. B., and Wen, S. Z., "Formation and Tribology Properties of Polyfluoroalkylmethacrylate Film on the Magnetic Head Surface," *Chinese Science Bulletin*, Vol. 50, 2005, pp. 2385–2390.

[33] Ma, X., Gui, J., Smoliar, L., Grannen, K., Marchon, B., Jhon, M. S., and Bauer, C. L., "Spreading of Perfluoropolyalkylether Films on Amorphous Carbon Surfaces," *J. Chem. Phys.*, Vol. 110, 1999, pp. 3129–3137.

[34] Guo, Q., Izumisawa, S., Phillips, D. M., and Jhon, M. S., "Surface Morphology and Molecular Conformation for Ultrathin Lubricant Films with Functional End Groups," *J. Appl. Phys.*, Vol. 93, 2003, pp. 8707–8709.

[35] Li, X., Hu, Y. Z., and Wang, H., "A molecular Dynamics Study on Lubricant Perfluoropolyether in Hard Disk Driver," *Acta Phys. Sin.*, Vol. 54, 2005, pp. 3787–3792.

[36] Shen, X. J., and Bogy, D. B., "Particle Flow and Contamination in Slider Air Bearings for Hard Disk Drives," *ASME J. Tribol.*, Vol. 125, No. 2, 2003, pp. 358–363.

[37] Zhang, B., and Nakajima, A., "Possibility of Surface Force Effect in Slider Air Bearings of 100 Gbit/in.^{2} Hard Disks," *Tribol. Int.*, Vol. 36, 2003, pp. 291–296.

[38] Zhou, L., Kato, K., Vurens, G., and Talke, F. E., "The Effect of Slider Surface Texture on Flyability and Lubricant Migration under Near Contact Conditions," *Tribol. Int.*, Vol. 36, No. 4–6, 2003, pp. 269–277.

[39] Burgdorfer, A., "The Influence of the Molecular Mean Free Path on the Performance of Hydrodynamic Gas Lubricated Bearings," *ASME J. Basic Eng.*, Vol. 81, No. 3, 1959, pp. 94–100.

[40] Hisa, Y. T., and Domoto, G. A., "An Experimental Investigation of Molecular Rarefaction Effects in Gas Lubricated Bearings at Ultra-Low Clearances," *ASME J. Lubr. Technol.*, Vol. 105, 1983, pp. 120–130.

[41] Mitsuya, Y., "Modified Reynolds Equation for Ultra-Thin Film Gas Lubrication Using 1.5-Order Slip-Flow Model and Considering Surface Accommodation Coefficient," *ASME J. Tribol.*, Vol. 115, 1993, pp. 289–294.

[42] Gans, R. F., "Lubrication Theory at Arbitrary Knudsen Number," *ASME J. Tribol.*, Vol. 107, 1985, pp. 431–433.

[43] Fukui, S., and Kaneko, R., "Analysis of Ultra-Thin Gas Film Lubrication Based on Linearized Boltzmann Equation: First Report—Derivation of a Generalized Lubrication Equation Including Thermal Creep Flow," *ASME J. Tribol.*, Vol. 110, 1988, pp. 253–262.

[44] Fukui, S., and Kaneko, R., "A Database for Interpolation of Poiseuille Flow Rates for High Knudsen Number Lubrication Problems," *ASME J. Tribol.*, Vol. 112, 1990, pp. 78–83.

[45] Wu, L., and Bogy, D. B., "Numerical Simulation of the Slider Air Bearing Problem of Hard Disk Drives by Two Multidimensional Unstructured Triangular Meshes," *J. of Computational Physics*, Vol. 172, 2001, pp. 640–657.

[46] Peng, Y. Q., Lu, X. C., and Luo, J. B., "Nanoscale Effect on Ultrathin Gas Film Lubrication in Hard Disk Drive," *J. of Tribology-Transactions of the ASME*, Vol. 126, No. 2, 2004, pp. 347–352.

[47] Moy, D., Schadt, M., Hu, C. K., et al., *Proceedings, 1989 VMIC*

Conference, 1989, p. 26.

[48] Liu, R. C., Pai, C. S., and Martinez, E., "Interconnect Technology Trend for Microelectronics," *Solid-State Electron.*, Vol. 43, 1999, pp. 1003–1009.

[49] Zhou, C., Shan, L., Hight, J. R., Danyluk, S., Ng, S. H., and Paszkowski, A. J., "Influence of Colloidal Abrasive Size on Material Removal Rate and Surface Finish in SiO_2 Chemical Mechanical Polishing," *Lubr. Eng.*, Vol. 58, No. 8, 2002, pp. 35–41.

[50] Xu, J., Luo, J. B., Lu, X. C., Wang, L. L., Pan, G. S., and Wen, S. Z., "Atomic Scale Deformation in the Solid Surface Induced by Nanoparticle Impacts," *Nanotechnology*, Vol. 16, 2005, pp. 859–864.

[51] Duan, F. L., Luo, J. B., and Wen, S. Z., "Atomistic Structural Change of Silicon Surface under a Nanoparticle Collision," *Chinese Science Bulletin*, Vol. 50, No. 15, 2005, pp. 1661–1665.

[52] Balakumar, S., Haque, T., Kumar, A. S., Rahman, M., and Kumar, R., "Wear Phenomena in Abrasive-Free Copper CMP Process," *J. Electrochem. Soc.*, Vol. 152, No. 11, 2005, pp. G867–G874.

[53] Liu, F. Q., Du, T. B., Duboust, A., Tsai, S., and Hsu, W. Y., "Cu Planarization in Electrochemical Mechanical Planarization," *J. Electrochem. Soc.*, Vol. 153, No. 6, 2006, pp. C377–C381.

[54] Wang, D. H., Chiao, S., Afnan, M., Yih, P., and Rehayem, M., "Stress-free Polishing Advances Copper Integration with Ultralow-k Dielectrics," *Solid State Technol.*, Vol. 106, 2001, pp. 101–104.

2

Measuring Techniques

Dan Guo,[1] Jianbin Luo,[1] and Yuanzhong Hu[1]

1 Introduction

THE DEVELOPMENT OF NEW TECHNIQUES TO MEAsure surface topography, adhesion, friction, wear, lubricant film thickness, and mechanical properties on a micro- and nanometre scale has led to a new field referred to as micro/nanotribology, which is concerned with experimental and theoretical investigation of processes occurring from micro scales down to atomic or molecular scales. Such studies are becoming ever more important as moving parts and mating surfaces continue to be smaller. Micro/nanotribological studies are crucial to develop a fundamental understanding of interfacial phenomena occurring at such small scales and are boosted by the various industrial requirements.

The first apparatus for nanotribology research is the Surface Force Apparatus (SFA) invented by Tabor and Winterton [1] in 1969, which is used to study the static and dynamic performance of lubricant film between two molecule-smooth interactions.

The invention of the Scanning Tunneling Microscopy (STM) in 1981 by Binning and Rohrer [2] at the IBM Zurich Research Laboratory suddenly revolutionized the field of surface science and was awarded the Nobel Prize in 1986. This was the first instrument capable of directly obtaining three-dimensional images of a solid surface with atomic resolution and paved the way for a whole new family of Scanning Probe Microscopies (SPM), e.g., Atomic Force Microscopy (AFM), Friction Force Microscopy (FFM), and others. The AFM and FFM are widely used in nanotribological and nanomechanics studies for measuring surface topography and roughness, friction, adhesion, elasticity, scratch resistance, and for nanolithography and nanomachine.

As a major branch of nanotribology, Thin Film Lubrication (TFL) has drawn great concerns. The lubricant film of TFL, which exists in ultra precision instruments or machines, usually ranges from a few to tens of nanometres thick under the condition of point or line contacts with heavy load, high temperature, low speed, and low viscosity lubricant. One of the problems of TFL study is to measure the film thickness quickly and accurately. The optical method for measuring the lubricant film thickness has been widely used for many years. Goher and Cameron [3] successfully used the technique of interferometry to measure elastohydrodynamic lubrication film in the range from 100 nm to 1 μm in 1967. Now the optical interference method and Frustrated Total Reflection (FTR) technique can measure the film thickness of nm order.

Mechanical properties of solid surfaces and thin films are of interest as they may greatly affect the tribological performance of surfaces. Nanoindentation and nanoscratching are techniques developed since the early 1980s for probing the mechanical properties of materials at very small scales. Ultra-low load indentation and scratching employs high resolution sensors and actuators to continuously control and monitor the loads and displacements on an indenter as it is driven into and withdrawn from a material. In some systems, forces as small as a nano-Newton and displacements of about an Angstrom can be accurately measured. One of the great advantages of the technique is that many mechanical properties can be determined by analyses of the load-displacement data alone. The nanoindentation technique is usually used to measure the hardness, the elastic modulus, the fatigue properties of ultrathin films, the continuous stiffness, the residual stresses, the time dependent creep and relaxation properties, and the fracture toughness. The nanoscratching technique is used to measure the scratch resistance, film-substrate adhesion, and durability.

The brief history, operation principle, and applications of the above-mentioned techniques are described in this chapter. There are several other measuring techniques, such as the fluorometry technique, Scanning Acoustic Microscopy, Laser Doppler Vibrometer, and Time-of-flight Secondary Ion Mass Spectroscopy, which are successfully applied in micro/nanotribology, are introduced in this chapter, too.

Many technologies presented in this chapter were developed or improved by the authors of this book, or the institutes they belong to, as summarized as follows:

The Relative Optical Interference Intensity Method presented in Section 2.2 was first developed by Professor Luo (one of the authors of this book) and his colleagues in 1994 [4,5], for measuring the nanoscale lubricant film thickness.

In Section 2.4, we describe the principle of the Frustrated Total Reflection (FTR) Technique, which was first applied by Professor Wen's group at State Key Laboratory of Tribology, Tsinghua University, for measuring film thickness in mixed lubrication [6,7].

In Section 3, the contributions from one of the authors, Professor Hu, to a new version of Surface Force Apparatus (SFA) are included [8].

In Section 4.3, we introduce a Friction Force Microscopy (FFM), designed and made by one of the authors, Professor Lu et al. [9].

The research presented in Section 6.1 on two phase flow containing nano-particles was carried out using the Fluorometry Technique developed recently in Professor Luo's group.

---

[1] *State Key Laboratory of Tribology, Tsinghua University, Beijing, China.*

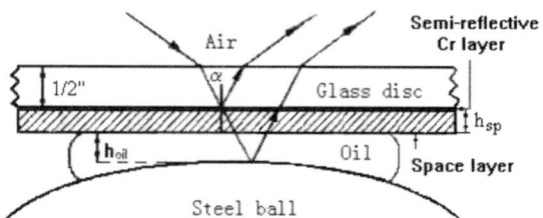

Fig. 1—Spacer layer method [10].

2 Optical Measuring Techniques

2.1 Wedged Spacer Layer Optical Interferometry [10]

This method was developed by Johnston et al. in 1991 and well described in Ref. [10], according to which the method is introduced as follows. The principle of optical interference is shown schematically in Fig. 1. A coating of transparent solid, typically silica, of known thickness, is deposited on top of the semi-reflecting layer. This solid thus permanently augments the thickness of any oil film present and is known as a "spacer layer." The destructive interference now obeys the equation:

$$n_{oil}h_{oil} + n_{sp}h_{sp} = \frac{\left(N + \frac{1}{2} - \phi\right)\lambda}{2\cos\theta} \quad N = 0,1,2 \quad (1)$$

and the first interference fringe occurs at a separation reduced by h_{sp}, where h_{sp} is the spatial thickness of the spacer layer. With a flat spacer layer, Westlake was able to measure an oil film of 10 nm thickness using optical interferometry [11]. Guangteng and Spikes [12] used an alumina spacer layer whose thickness varied in the shape of a wedge over the transparent flat surface in an optical rig. The method was able to detect oil film thickness down to less than 10 nm. A problem encountered using this technique was the difficulty of obtaining regular spacer layer wedges. Also, with low viscosity oils requiring high speeds to generate films, high speed recording equipment was needed to chart the continuously changing interference fringe colors.

The two limitations of optical interferometry, the one-quarter wavelength of light limit and the low resolution, have been addressed by using a combination of a fixed-thickness spacer layer and spectral analysis of the reflected beam. The first of these overcomes the minimum film thickness that can normally be measured and the second addresses the limited resolution of conventional chromatic interferometry.

A conventional optical test rig is shown in Fig. 2 [13]. A superfinished steel ball is loaded against the flat surface of a float-glass disk. Both surfaces can be independently driven. Nominally a pure rolling is used as shown in Fig. 2 that the disk is driven by a shaft and the ball is driven by the disk.

The disk was coated with a 20 nm sputtered chromium semi-reflecting layer, a silica spacer layer was sputtered on top of the chromium. This spacer layer varied in thickness in the radial direction, but was approximately constant circumferentially round the disk [10].

The reflected beam was taken through a narrow, rectangular aperture arranged parallel to the rolling direction, as

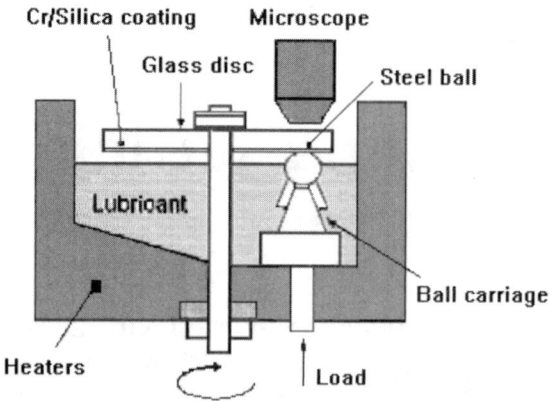

Fig. 2—Schematic representation of optical EHD rig [13].

shown in Fig. 3(a). It was then dispersed by a spectrometer grating and the resultant spectrum captured by a black and white video camera. This produced a band spectrum spread horizontally on a television screen, with the brightness of the

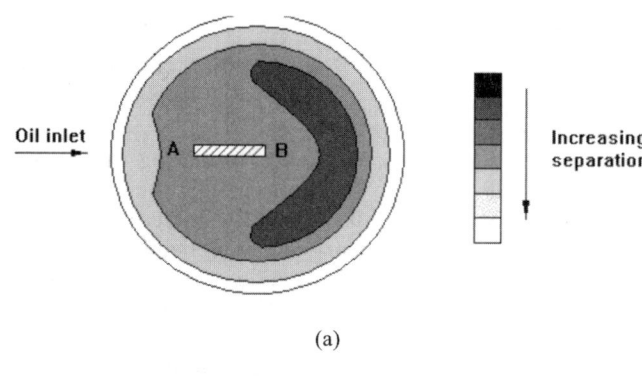

(a)

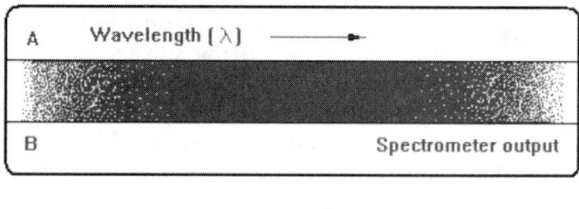

(b)

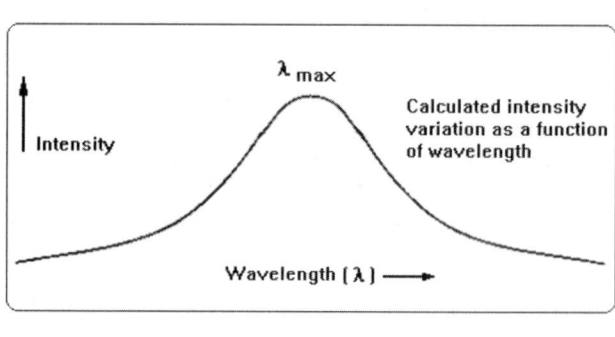

(c)

Fig. 3—Schematic representation of screen display showing calculated intensity profile.

CHAPTER 2 ■ MEASURING TECHNIQUES

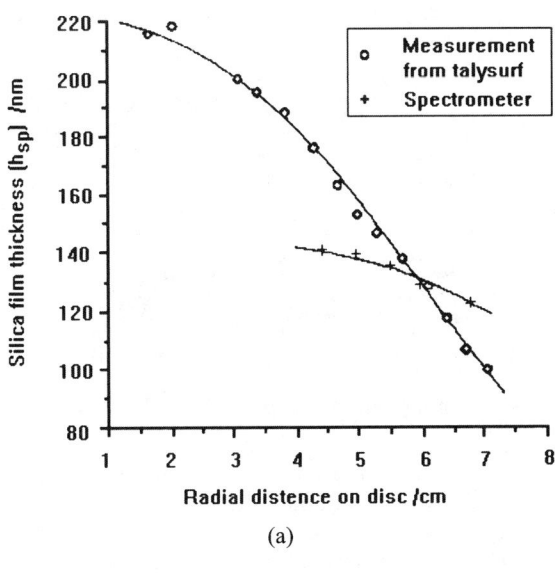

(a)

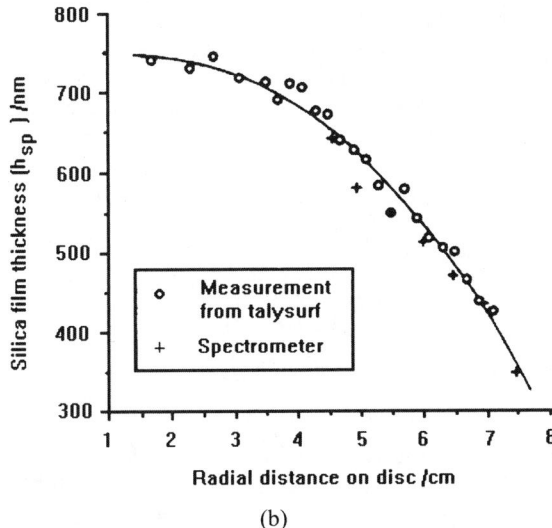

(b)

Fig. 4—Measured film thickness—thick spacer layer [10].

spectrum at each wavelength indicating the extent of interference, as shown schematically in Fig. 3(b). The vertical axis of the spectral band mapped across the center of the contact as illustrated in Fig. 3(a). This image was screen-dumped in digital form into a microcomputer, which drew an intensity profile of the spectrum as a function of wavelength. This is shown in Fig. 3(c). In practice, the spectrometer was arranged so that one digitized screen pixel corresponded to a wavelength change of 0.48 nm. A cold halogen white light source was employed and the angle of incidence was 0°.

By this means, the wavelength of light which constructively interfered could be determined accurately for separating film thickness, thus permitting a highly resolved film thickness measurement.

The silica film thickness was determined as a function of the disk radius by an optical interference method using the spectrometer. A steel ball was loaded against the silica surface to obtain an interference pattern of a central, circular Hertzian area with surrounding circular fringes due to the air gap between the deformed ball and flat. The thickness of

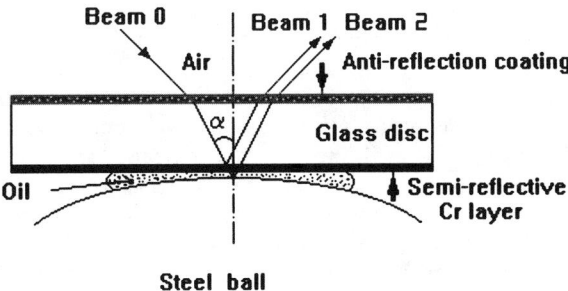

Fig. 5—Interference lights [4,5].

the silica layer in the Hertzian contact, h_{sp}, was calculated in terms of Eq (2), based on the measured wavelength, λ_{max}, at which maximum constructive interference occurred. N was known from the approximate thickness of the silica layer and Φ was taken to be 0.28 from the value measured previously in air. The refractive index of the separating medium is known to affect Φ by less than 2 % [14]. The refractive index of the spacer layer, n_{sp}, was measured as 1.476 ± 0.001 by the method of Kauffman [15].

$$h_{sp} = \frac{(N - \Phi)\lambda_{max}}{2n_{sp}} \qquad (2)$$

When using the thin silica spacer layer, however, it was found that the results from the above-mentioned methods did not agree with the direct measurements from the Talysurf profilemeter, as shown in Fig. 4(a). This was tentatively ascribed to the effect of penetration of the reflecting beam into the substrate. With a very thin silica layer, the depth of penetration and thus the phase change would depend upon the thickness of the silica spacer layer and also upon that of any oil film present.

The solution to this problem was to use a space layer of a thickness greater than the wavelength of visible light, which is above the limit of penetration of a reflected light beam. Due to space layer or oil, any variation above this value will have no further effect on phase change. Figure 4(b) shows that there is good agreement between Talysurf and optical calibration methods with a thick space layer.

2.2 Relative Optical Interference Intensity Method [4,5]

This method was proposed by Luo (one of the present authors) et al. in 1994 [4,5]. The principle of optical interference is shown in Fig. 5. On the upper surface of the glass disk there is an anti-reflective coating. There is an oil film between a super polished steel ball and the glass disk covered by a semi-transmitted Cr layer. When a beam of light reaches the upper surface of the Cr layer, it is divided into two beams—one reflected at the upper surface of the Cr layer and the other passing through the Cr layer and the lubricant film, and then reflected at the surface of the steel ball. Since the two beams come from the same light source and have different optical paths, they will interfere with each other. When the incident angle is 0°, the optical interference equation [16] is as follows:

$$I = I_1 + I_2 + 2\sqrt{I_1 I_2} \cos\left(\frac{4\pi\pi n}{\lambda} + \phi\right) \quad (3)$$

where I is the intensity of the interference light at the point where the lubricant film thickness h is to be measured, I_1 is the intensity of beam 1 and I_2 is that of beam 2 in Fig. 5, λ is the wavelength of the monochromatic light, ϕ is the system pure optical phase change caused by the Cr layer and the steel ball, and n is the oil refractive index.

I_1 and I_2 can be determined by the maximum interference intensity I_{max} and the minimum one I_{min} in the same interference order.

$$I_{max} = I_1 + I_2 + 2\sqrt{I_1 I_2}$$
$$I_{min} = I_1 + I_2 - 2\sqrt{I_1 I_2} \quad (4)$$

Therefore, Eq (3) can be written as:

$$I = \frac{1}{2}(I_{max} + I_{min}) + \frac{1}{2}(I_{max} - I_{min})\cos\left(\frac{4\pi n h}{\lambda} + \phi\right) \quad (5)$$

If I_a and I_d are expressed as follows:

$$I_a = \frac{I_{max} + I_{min}}{2}$$

$$I_d = \frac{I_{max} - I_{min}}{2}$$

We define the relative interference intensity as below:

$$\bar{I} = \frac{I - I_a}{I_d} = \frac{2I - (I_{max} + I_{min})}{I_{max} - I_{min}} \quad (6)$$

Hence, from Eqs (3)–(6), the lubricant film thickness can be determined as below:

$$h = \frac{\lambda}{4n\pi}[\arccos(\bar{I}) - \phi] \quad (7)$$

If the ball contacts the surface of the glass disk without oil, $h = 0$, and then the pure phase change ϕ of the system can be obtained as follows:

$$\phi = \arccos(\bar{I}_0)$$

$$\bar{I}_0 = \frac{I_0 - I_a}{I_d} \quad (8)$$

where I_0 is the optical interference intensity at the point where the film thickness is zero and should be determined by experiments. Then Eq (7) can be rewritten as:

$$h = \frac{\lambda}{4\pi n}[\arccos(\bar{I}) - \arccos(\bar{I}_0)] \quad (9)$$

A diagram of the measuring system is shown in Fig. 6 [4,5]. After the interference light beams reflected separately from the surfaces of the Cr layer and the steel ball passes through a microscope, they become interference fringes caught by a TV camera. The optical image is translated to a monitor and also sent to a computer to be digitized.

The experimental rig is shown in Fig. 7 [18]. The steel ball is driven by a system consisting of a motor, a belt, a shaft, a soft coupling, and a quill. The ball-mount is floating during the running process in order to keep the normal force constant. The micrometre enables the floating mount to move along the radial direction of the disk and to maintain a fixed position. The microscope can move in three dimensions.

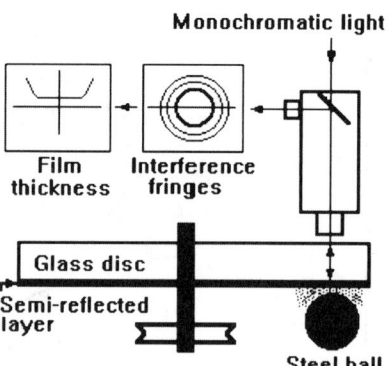

Fig. 6—Diagram of the measuring system [4,5].

The resolution of the instrument in the vertical direction depends upon the wavelength of visible light (450 to 850 nm), the oil refractive index, and the difference between the maximum and the minimum interference intensity as follows [5,18]:

$$\Delta h = \frac{\lambda}{4n\pi}[\arccos(\bar{I} + \Delta\bar{I}) - \arccos(\bar{I})] \quad (10)$$

$$\bar{I} = \frac{2I - (I_{max} + I_{min})}{I_{max} - I_{min}} \quad (11)$$

$$\Delta\bar{I} = \frac{2\Delta I}{I_{max} - I_{min}} \quad (12)$$

where n is oil refractive index, λ is the wavelength of the light and it is 600 nm in the normal experiment, I_{max} and I_{min} are the maximum and minimum interference intensity separately, which can be divided in 256 grades in the computer image card, ΔI is the resolution of optical interference intensity, which is one grade. The variation in the vertical resolution with respect to these factors is shown in Fig. 8. Among these factors, the wavelength is the most important in determining the vertical resolution. When effects of all these factors are considered, the vertical resolution is about 0.5 nm when wavelength is 600 nm. The horizontal resolution depends upon the distinguishing ability of CCD and the enlargement factor capacity of the micrometre. It is about 1 μm.

2.3 Thin Film Colorimetric Interferometry (TFCI) [19,20]

The method is proposed by Hartl et al. [19–21]. The colorimetric interferometry technique, in which film thickness is obtained by color matching between the interferogram and color/film thickness dependence obtained from Newton rings for static contact, represents an improvement of conventional chromatic interferometry.

The frame-grabbed interferograms with a resolution of 512 pixels by 512 lines are first transformed from RGB to CIELAB color space and they are then converted to the film thickness map using appropriate calibration and a color

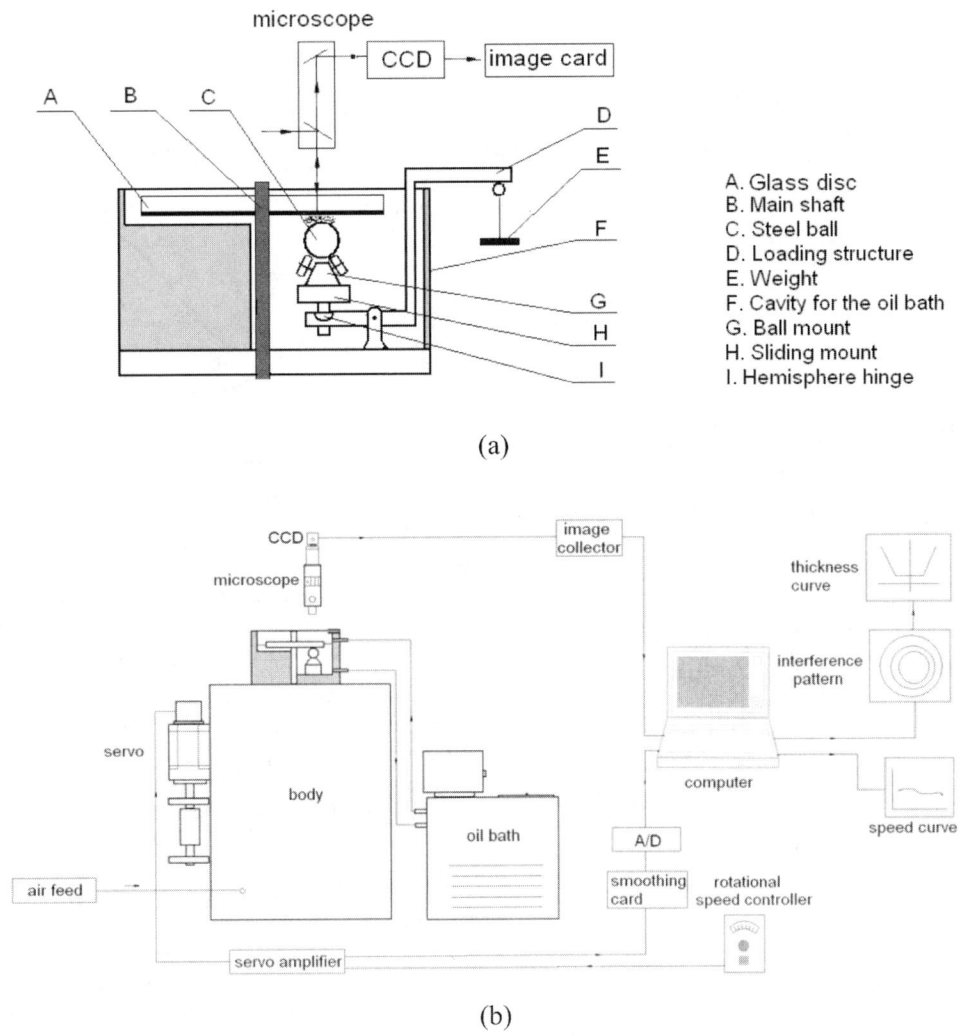

Fig. 7—Schematic representation of experiment rig [5,18], (a) measuring part, (b) whole structure.

matching algorithm. L*, a*, b* color coordinates/film thickness calibration is created from Newton rings for flooded static contact formed between the steel ball and the glass disk coated with a chromium layer. In the CIE curve as shown in Fig. 9, the wavelength can be determined by the ratio of R, G, B separately measured by color CCD. Therefore, this method has a much higher resolution than that of Gohar and Cameron [22] who measured the film thickness in terms of the colors of interference fringes observed by the eyes, which gives a resolution in the vertical direction of about 25 nm.

All aspects of interferogram and experimental data acquisition and optical test rig control are provided by a computer program that also performs film thickness evaluation. It is believed that the film thickness resolution of the colorimetric interferometry measurement technique is about 1 nm. The lateral resolution of a microscope imaging system used is 1.2 μm. Figure 10 shows a perspective view of the measurement system configuration. This is an even conventional optical test rig equipped with a microscope imaging system and a control unit.

The optical test rig consists of a cylindrical thermal isolated chamber enclosing the concentrated contact formed between a steel spherical roller and the flat surface of a glass disk. The underside of the glass disk is coated with a thin semi-reflective chromium layer that is overlaid by a silicon dioxide "spacer layer," as shown in Fig. 11. The contact is loaded through the glass disk that is mounted on a pivoted lever arm with a movable weight. The glass disk is driven, in nominally pure rolling, by the ball that is driven in turn by a servomotor through flexible coupling. The test lubricant is enclosed in a chamber that is heated with the help of an external heating circulator controlled by a temperature sensor. A heat insulation lid with a hole for a microscope objective seals the chamber and helps to maintain constant lubricant test temperature. Its stability is within ±0.2°C.

An industrial microscope with a long-working distance 20× objective is used for the collection of the chromatic interference patterns. They are produced by the recombination of the light beams reflected at both the glass/chromium layer and lubricant/steel ball interfaces. The contact is illuminated through the objective using an episcopic microscope illuminator with a fiber optic light source. The secondary beam splitter inserted between the microscope illuminator and an eyepiece tube enables the simultaneous use of a color video camera and a fiber optic spectrometer.

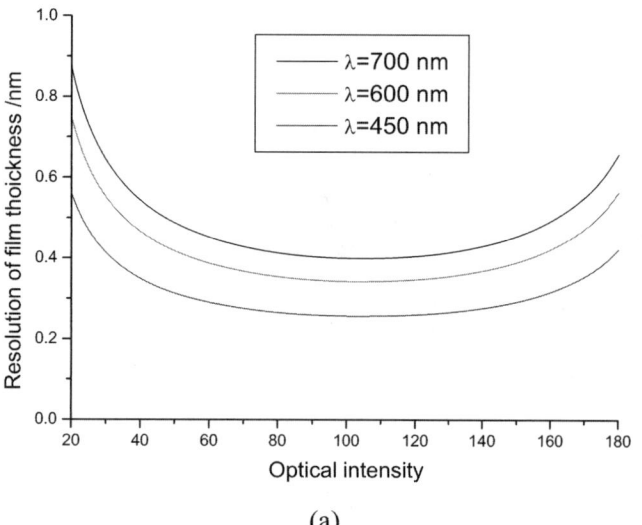

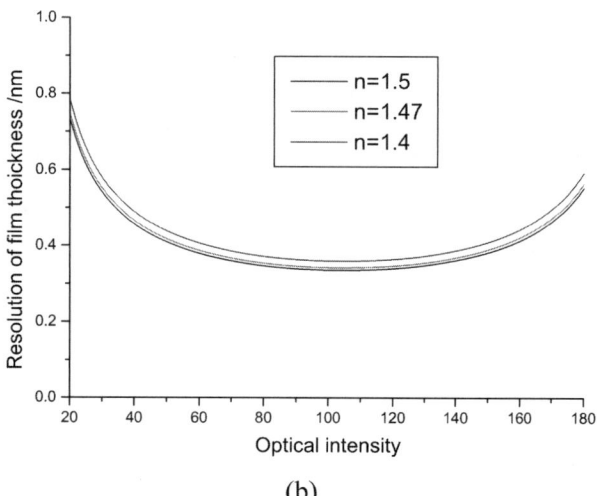

Fig. 8—Resolution of film thickness versus optical interference intensity [4,5], (a) different wavelengths, (b) different refractive indexes.

Both devices are externally triggered by an inductive sensor so that all measurements are carried out at the same disk position.

Spherical rollers were machined from AISI 52100 steel, hardened to a Rockwell hardness of Rc 60 and manually polished with diamond paste to RMS surface roughness of 5 nm. Two glass disks with a different thickness of the silica spacer layer are used. For thin film colorimetric interferometry, a spacer layer about 190 nm thick is employed whereas FECO interferometry requires a thicker spacer layer, approximately 500 nm. In both cases, the layer was deposited by the reactive electron beam evaporation process and it covers the entire underside of the glass disk with the exception of a narrow radial strip. The refractive index of the spacer layer was determined by reflection spectroscopy and its value for a wavelength of 550 nm is 1.47.

2.4 Frustrated Total Reflection (FTR) Technique

When a beam of light goes through a boundary between two dielectrics n_1 and n_2, with incident angle sufficiently large to

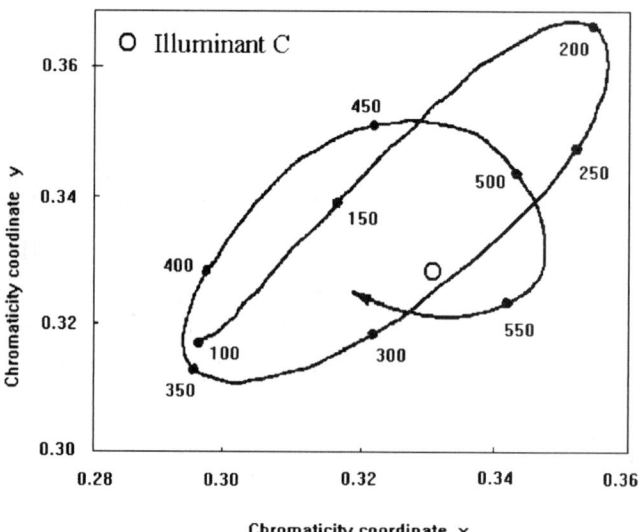

Fig. 9—Interference colors in CIE x, y chromaticity diagram [19].

the critical angle, total reflection will occur. There exists an inhomogeneous wave called the evanescent wave with its phase-normal parallel to the boundary and attenuated in the height direction. The behavior of light in this condition is changed remarkably if the second medium n_2 has a finite (and sufficiently small) thickness, and if the third medium n_3 behind the second boundary has a higher index of refraction ($n_3 > n_2$), the total reflection of energy originally occurring at the first boundary is now frustrated in that it becomes partial reflection, accompanied with some leakage of energy from the first to the third medium. This phenomenon is called frustrated total reflection (FTR). The principle of FTR was first applied for measuring the film thickness in mixed lubrication by Xian L., Kong and Wen of State Key Lab of Tribology, Tsinghua University in 1993 [6,7].

If a beam of light is incident upon Medium 1 from Medium 2 at an incident angle θ_1 as shown in Fig. 12, then according to the law of refraction:

$$n_1 \sin \theta_1 = n_2 \sin \theta_2 \quad (13)$$

where n_1 and n_2 are the refractive indexes of Media 1 and 2, respectively.

If $n_1 > n_2$, then, if $\theta_2 = 90°$, $\sin \theta_1 = n_2/n_1$. The incident angle at this moment is called the critical angle, designated by θ_c. If the incident angle is larger than the critical angle ($\theta_1 > \theta_c$), then the incident wave will be totally reflected back to Medium 1. This is known as total reflection.

According to the electromagnetic wave theory as applied to total reflection, the continuity condition at the boundary surface requires that an electromagnetic field in Medium 2 should persist in the form of a damped wave, called the evanescent wave. Its penetrating depth is the same order as the wavelength. If Medium 2 is unlimited, then the evanescent wave totally returns to Medium 1. The case is different if Medium 2 is only a layer bounded by Medium 3 and the latter is gradually brought closer to the boundary surface between Media 1 and 2. If the distance between Media 1 and 3 is within the penetration depth, then the total reflection is disturbed. Light energy no longer totally returns to Medium

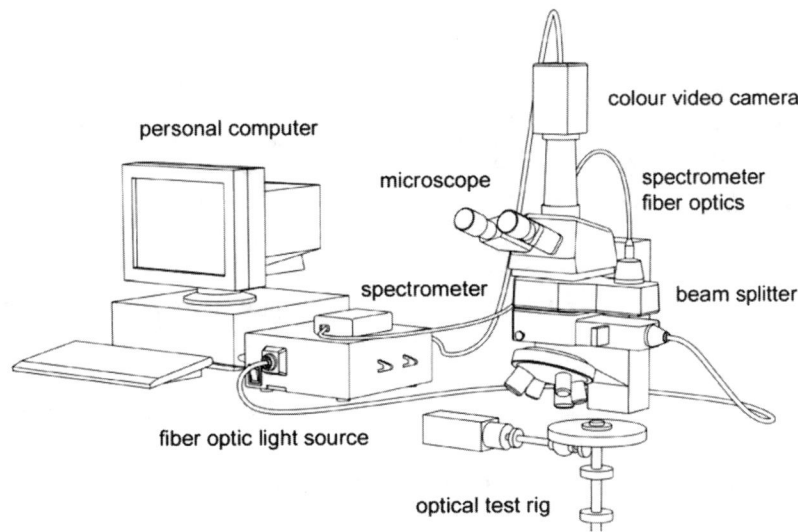

Fig. 10—Experimental apparatus [114].

1. Such a disturbed total reflection is FTR. In FTR, a definite relationship exists between the observed reflectivity and the distance between Media 1 and 3. If the reflectivity is measured, the distance can be calculated. This is the basic idea in using FTR to measure film thickness.

Figure 13 illustrates the principle of using FTR to measure the lubricating film thickness. A sapphire prism (Al_2O_3), the lubricant, and a steel specimen (steel ring, GCr_{15}) constitute the three media of FTR. The bottom surface of the sapphire prism and the cylindrical surface of the specimen form a line contact. They are separated by the lubricant. When the incident angle α is larger than the critical angle α_c, total reflection occurs at the boundary of the sapphire prism and lubricant. If the surface of the specimen approaches the bottom surface of the sapphire prism, and is within the penetrating depth, FTR will occur. The reflected image, which contains the information on film thickness at each point in the contact, can be recorded by a video camera and sent to a computer for processing. The film thickness can then be obtained at each point. In the apparatus, $n_1 = 1.77$, $n_2 = 1.5$, the critical angle is $\theta_c = 57.94°$, and the critical incident angle is $\alpha_c = 21.66°$.

The expression of film thickness deduced by Kong et al. takes the form of

$$h = \frac{\lambda_0 \ln[0.5(D \pm \sqrt{D^2 - 4\rho_{23}^2})]}{4\pi\sqrt{n_1^2 \sin^2 \theta_1 - n_2^2}} = \frac{\ln[0.5(D \pm \sqrt{D^2 - 4\rho_{23}^2})]}{C} \quad (14)$$

where

$$D = \frac{2\rho_{23}[R\cos(\varphi_{12} + \varphi_{23}) - \cos(\varphi_{12} - \varphi_{23})]}{1 - R}$$

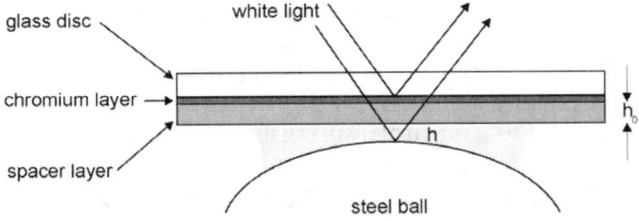

Fig. 11—Schematic representation of an interferometer [21] and rolling speed of 0.021 and 0.042 ms^{-1}.

$$C = \frac{4\pi\sqrt{n_1^2 \sin^2 \theta_1 - n_2^2}}{\lambda_0} \quad (15)$$

In Eqs (14) and (15), λ_0 is the wavelength of light in vacuum. ρ_{23} is the moduli of reflection coefficient when the light is propagated from Medium 2 to Medium 3, φ_{12} and φ_{23} are the phase angles of reflection coefficients when the lights are propagated from Medium 1 to Medium 2 and from Medium 2 to Medium 3, respectively.

FTR is an effective method for film thickness measurement in mixed lubrication. If the strength of incident light is properly adjusted, the resolution of film thickness by the FTR method can be limited within 5 nm. Theoretically, if the typical height of the surface asperities is less than the penetrating depth, the FTR method can be successfully used. In the tests, it is found that if Ra is greater than 0.15 µm, a con-

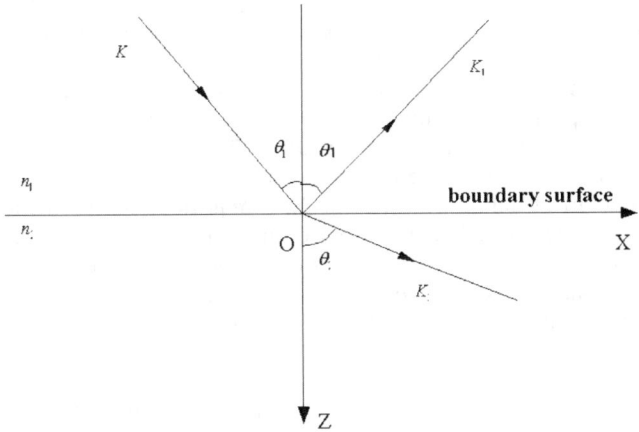

Fig. 12—Theory of refraction.

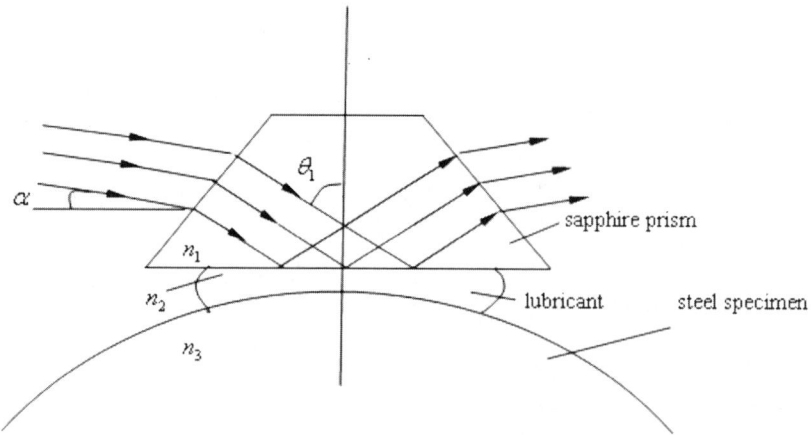

Fig. 13—Experimental arrangement for frustrated total reflection.

tinuous reflected image cannot be obtained. Therefore, for the FTR method to be successfully applied, the surface roughness should be better at less than 0.15 μm.

3 Surface Force Apparatus

3.1 A Brief Review

The surface force apparatus (SFA) is a device that detects the variations of normal and tangential forces resulting from the molecule interactions, as a function of normal distance between two curved surfaces in relative motion. SFA has been successfully used over the past years for investigating various surface phenomena, such as adhesion, rheology of confined liquid and polymers, colloid stability, and boundary friction. The first SFA was invented in 1969 by Tabor and Winterton [23] and was further developed in 1972 by Israelachivili and Tabor [24]. The device was employed for direct measurement of the van der Waals forces in the air or vacuum between molecularly smooth mica surfaces in the distance range of 1.5–130 nm. The results confirmed the prediction of the Lifshitz theory on van der Waals interactions down to the separations as small as 1.5 nm.

Afterward, Israelachivili and Adams [25] designed a new apparatus for measuring the forces between surfaces interacting in liquids and vapors, and the surface separation could be controlled and measured with a resolution of 0.1 nm by using the piezoelectric crystal. This was named SFA MK-I, the first mature apparatus that enabled the study of the two fundamental forces in colloid and biology science, namely the attractive van der Waals forces and the repulsive electrostatic double-layer forces between two charged surfaces immersed in an electrolyte solution. SFA MK-I was then developed into an improved version, SFA MK-II [26], in which various attachments were designed for extending the scope and versatility of the apparatus. For example, a small bath attachment inserted into the main chamber allows the experiments to be done in a much less quantity of liquid, the sample is supported on single or double cantilever with adjustable stiffness, and the range of forces measurement can be extended for about six orders of magnitude. In 1990, Israelachivili and McGuiggan [27] built the SFA MK-III, which overcame most of the limitations in the earlier models, e.g., the thermal drifts, difficulties of cleaning the chamber, insufficient spring stiffness, and the limited measurement range

required for more experiments. Since then, SFA MK-III has been widely applied to the experimental studies in polymer rheology, colloidal science, biology, and nanotribology.

Similar types of surface forces apparatus based on the same optical technique for the distance measurements were built during the same period, but with various modifications, for example, using samples made of different materials or in different shapes [8,28,29]. There were also other types of SFA, such as those developed by Prieve et al. [30] and Tonock et al. [31], in which different optical techniques or capacitive sensors were applied for measuring the gap between surfaces. In the apparatus developed by Tonock et al. [31], for example, a macroscopic ceramic sphere contacts against a plane, and three piezoelectric elements, combined with three capacitance sensors, permit accurate control and force measurement along three orthogonal axes. This design does not require optically transparent surfaces, and theoretically resolutions can achieve 10^{-8} N for the force measurements, and 10^{-3} nm in measuring the displacements.

3.2 Structure of SFA and Techniques in Measurement

3.2.1 Setup of the Apparatus

Figure 14 gives a sketch of an SFA consisting of several parts: a chamber, the samples, supporting/driving components, and an optical system for measuring the distance or gap between surfaces.

The steel or aluminum chamber is designed for providing a proper experimental environment. The samples are made of cylindrical glass of radius 10 or 20 mm, with their axes crossed in 90° to form a point contact. A cleaved mica sheet in a thickness about 1.5–2.5 μm is glued to the surface of each sample. The introduction of a mica sheet serves two purposes: to remove the effect of surface roughness for the mica surface is considered as molecularly smooth, and to facilitate the application of an optical technique of multiple-beam interference for distance measurement, which will be further discussed in the following section. The adhesion glue used for affixing the mica to the samples is sufficiently compliant, so the mica will flatten under applied load to produce a contact zone of a radius from 10 μm to 40 μm. The lower

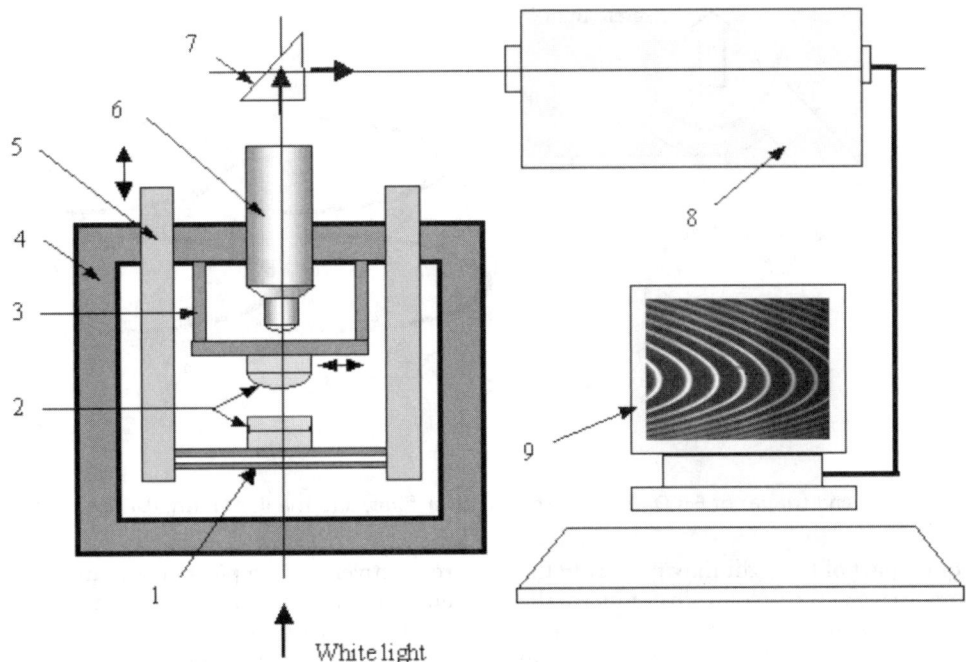

Fig. 14—A sketch of surface force apparatus. (1) cantilever, (2) samples, (3) supporter and driver for lateral motion, (4) chamber, (5) supporter and driver for normal displacement, (6) lens, (7) prism, (8) spectrometer, (9) computer for data collection.

sample is fixed on a cantilever of double spring with adjustable stiffness. The cantilever is connected to a supporter that drives the cantilever and the sample in the normal direction. The upper sample is fixed on another supporting frame that may serve in the meantime as a driver providing the upper sample a movement in the tangential direction.

In the following we will discuss the three key techniques involved in the SFA experiments: determining the distance or gap between contacting surfaces, positioning the sample in the normal direction and measuring normal surface forces, and driving the sample in the tangential direction and measuring friction forces.

3.2.2 Optical Technique for Gap Measurement

The most widely used technique in SFA for determining the distance or gap between the sample surfaces is based on the theory of multi-beam interference. A diagram of the optical system for the gap measurement is schematically shown in Fig. 15.

Before being glued to the glass sample, one face of each mica sheet has to be coated or spread with a thin layer of silver in 50~60 nm thick (reflectivity 96 % ~ 98 %) for facilitating the optical interference. As shown in Fig. 16(a), when a beam of white light goes up vertically through the lower sample and reaches the silver film, the beam is partly re-

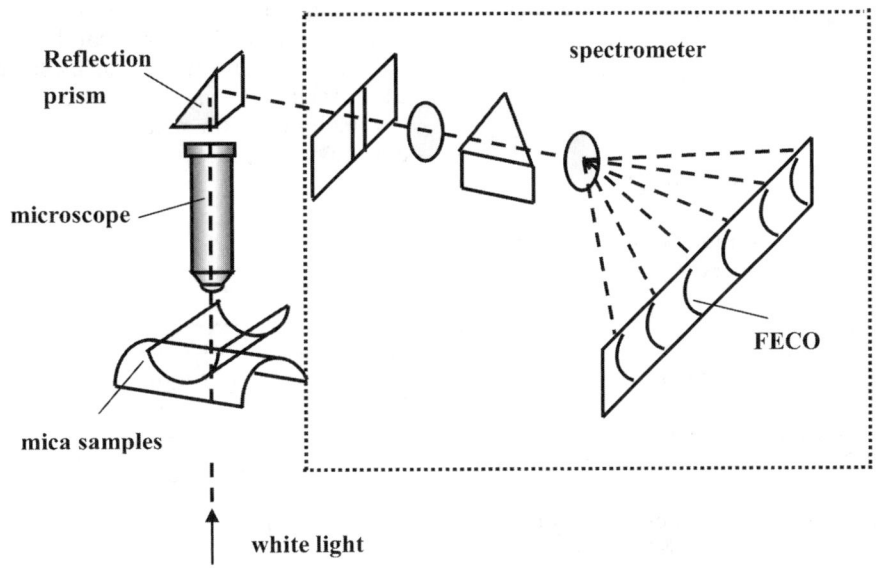

Fig. 15—A schematic diagram of the optical system based on FECO technique for the gap measurement.

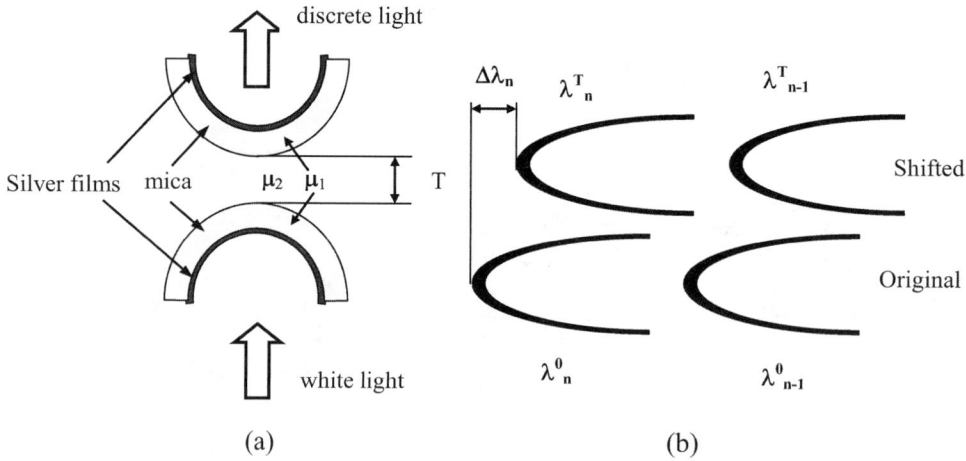

Fig. 16—A sketch of samples and fringes of FECO. (a) Mica sheets, silver films, and the light path. (b) Fringes before and after shift.

flected by the film, but a part of the light may penetrate the film, pass the gap, and arrive at the second silver film on the upper sample, where a similar process of reflection/penetration occurs and the beam reflected from the second film turns back to the first. In such a way, the white light beam reflects and oscillates between two silver films, which causes the multi-beam interference, and only a small portion of light in certain wavelengths can eventually pass through both films. The transmitted light therefore exhibits a discrete spectrum of wavelength. The light is then introduced into a spectrometer where the wavelengths are split up and an array of fringes appears in the screen of the spectrometer. The fringes are called the "Fringes of Equal Chromatic Order" (FECO), and have been studied extensively by Tolansky [32].

If there is a small increase in the distance between the two silver films, corresponding to a change in the gap between two mica sheets, the fringes will shift to the longer wavelengths (Fig. 16(b)), by a small interval of

$$\Delta\lambda_n = \lambda_n^T - \lambda_n^0 \qquad (16)$$

where n denotes the order of the fringes. The distance T between the mica surfaces can be estimated via the following expressions:

$$\begin{cases} T = \dfrac{n\Delta\lambda_n F_n}{2\mu_1} & (n = \text{odd integer}) \\ T\mu_2^2 = \dfrac{n\Delta\lambda_n \mu_1 F_n}{2} & (n = \text{even integer}) \end{cases} \qquad (17)$$

where μ_1 and μ_2 are the refractive index in the medium of mica and air, respectively (Fig. 16), and F_n stands for a correction factor. The surface distance can be measured in this approach with a resolution of 0.1–0.2 nm.

3.2.3 Positioning of the Sample and Measurement of Normal Force

There are several techniques available for SFA to drive a sample along the normal direction and to position it to a specified site. A common feature of these techniques is to employ a multi-stage driving system of increasing sensitivity. The apparatus developed by Israelachivili's team, for example, uses a two-stage screw as the coarse and medium control, which gives positioning accuracy of 1 μm and 1 nm, respectively, plus a piezocrystal tube that provides the finest adjustment of 0.1 nm. In another model of SFA developed by Zhou et al. [8], the driving system consists of two stepper motors and a differential elastic spring to achieve the positioning sensitivity within 0.13 nm. Now researchers are able to build the driving system in more flexible ways, by means of multi-stage piezocrystal of various combinations, which can provide long range positioning with the accuracy of 0.02 nm.

The accurate measurement of the normal force is a key target in the design of SFA. One way to do this is to measure the deflection of the cantilever spring. If the driving system gives a displacement D at the position where the cantilever is supported, as shown in Fig. 17, and the actual movement of the lower sample during the process can be measured according to the change in the gap between the lower and upper samples, $\Delta h = h_0 - h_1$, the difference in the two values, $d = D - \Delta h$, gives the deflection of the cantilever. As a result, the normal force can be calculated in terms of the displacement multiplied by the stiffness of the spring. In this method, the repulsive or attractive forces can be measured as a function of distance between two interaction surfaces and a wide range of interfacial forces can be detected by adjusting the stiffness of the force-measuring spring. In different types of SFA designed by various groups around the world, alternative techniques for measuring the normal force have been developed. For instance, the strain gage technique has been

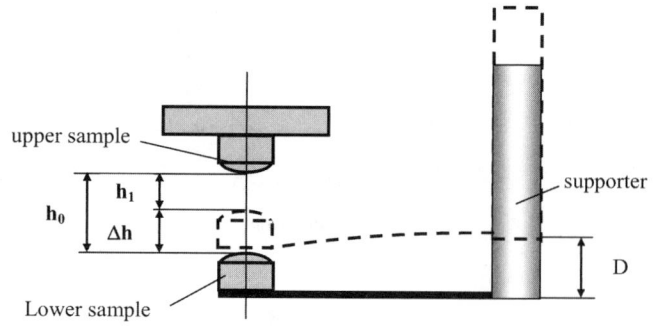

Fig. 17—Measurement of normal forces in terms of cantilever deflection.

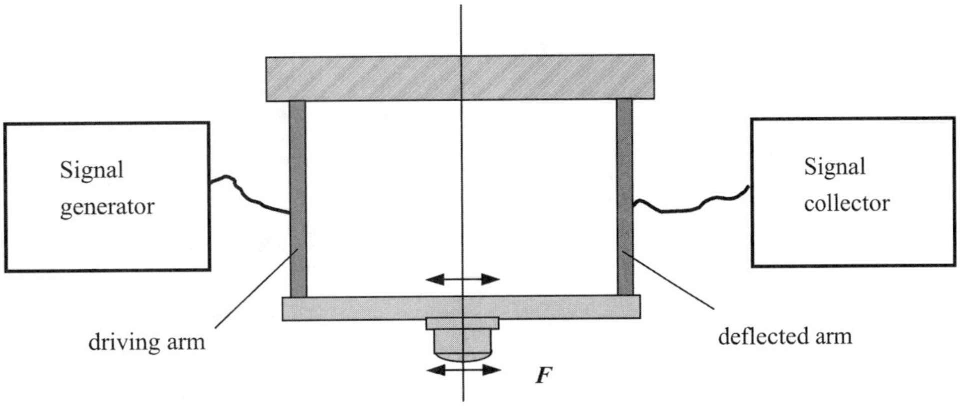

Fig. 18—Lateral driving system and measurement of friction force.

applied for the purpose of obtaining a real-time force curve during a process of approach and separation, although at the cost of reduction of resolution in the force measurement.

3.2.4 Measurement of Lateral Forces

A remarkable success of SFA in recent years has been found in applications to the field of tribology, particularly in the investigations of micro and nanotribology. To study friction or thin film rheology, one has to drive a sample in the tangential direction for producing a relative sliding, and to monitor the lateral forces during the process. There are two major approaches currently applied in SFA for this purpose: by driving the upper sample in a constant velocity [33,34], or by applying an excitation that makes the sample oscillating [8,35,36].

In the first approach, the translating system consists of a frame that supports the upper sample and a micrometre screw driven by a motor with adjustable speed. The rotation of the motor makes the screw move in the lateral direction and causes the frame deflection so that the upper sample translates at a steady rate. One arm of the frame, the vertical spring, acts at the same time as a force detector, using the strain gages adhered on it, so the friction force can be measured through a standard Wheatstone bridge. The translation of the upper sample results in a relative sliding between two surfaces, and the sliding speed can be adjusted continuously from 0.1 μm/s to 0.2 μm/s.

The second approach has been applied successfully for years in the SFA built by Granick et al. [36], in which a supporting/driving frame is composed of two vertical arms made of piezoelectric bimorphs, and a horizontal beam holding the upper sample, as illustrated in Fig. 18. One arm acts as a driver while the other serves as a force detector. The driving arm activated by the fluctuating voltage from a signal generator causes an oscillation in the frame. In the meantime, the second arm deflected by the oscillation produces a voltage signal proportional to its deflection. The deflection of the second arm in fact depends on two factors, the oscillation amplitude of the first arm and the lateral force acting on the upper sample and the frame. As a result, the lateral or friction forces can be evaluated by comparing the input of the first arm with the output signal from the second arm. The resolution of the force measurement can be in the order of micro-Newton, and the oscillation amplitude ranges from a few nanometres to 10 μm. A significant advantage of the design lies in the fact that it allows one to study the dynamic response of thin films under shear. A similar design has been widely adopted in recent models of SFA by other investigators.

Homola [37] compared the differences between the two approaches in measuring the shear performance. It is recognized that the first approach was suitable for examining the properties of sheared films composed of long-chain molecules, which requires a long sliding time to order and align and even a longer time to relax when sliding stops. In the second approach, on the other hand, there is not enough time to let molecules, especially those exhibiting a solid-like behavior, to respond sufficiently, thus the response of the sheared film will depend critically on the conditions of shearing, and this is maybe the main reason that the layering structure and "quantization" of the dynamic and static friction were not observed, in contrast to the results obtained when velocity was constant.

3.3 Applications of SFA

3.3.1 Surface Forces

Surface force apparatus has been applied successfully over the past years for measuring normal surface forces as a function of surface gap or film thickness. The results reveal, for example, that the normal forces acting on confined liquid composed of linear-chain molecules exhibit a periodic oscillation between the attractive and repulsive interactions as one surface continuously approaches to another, which is schematically shown in Fig. 19. The period of the oscillation corresponds precisely to the thickness of a molecular chain, and the oscillation amplitude increases exponentially as the film thickness decreases. This oscillatory solvation force originates from the formation of the layering structure in thin liquid films and the change of the ordered structure with the film thickness. The result provides a convincing example that the SFA can be an effective experimental tool to detect fundamental interactions between the surfaces when the gap decreases to nanometre scale.

3.3.2 Adhesion and Friction

It is observed from SFA experiments [38] that for two mica covered samples in contact, the load/contact-area relation

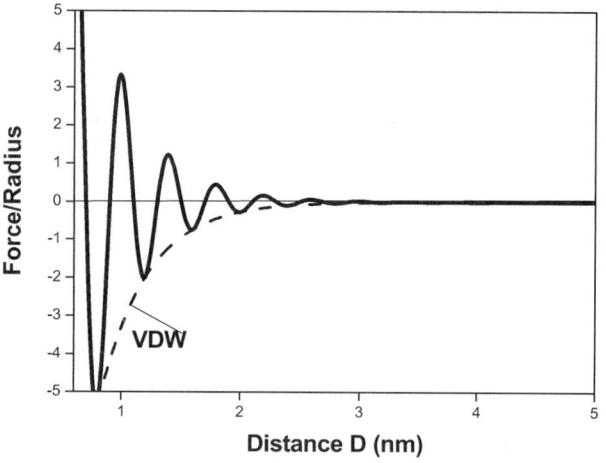

Fig. 19—A schematic force curve as a function of film thickness for the liquid of linear chain molecules.

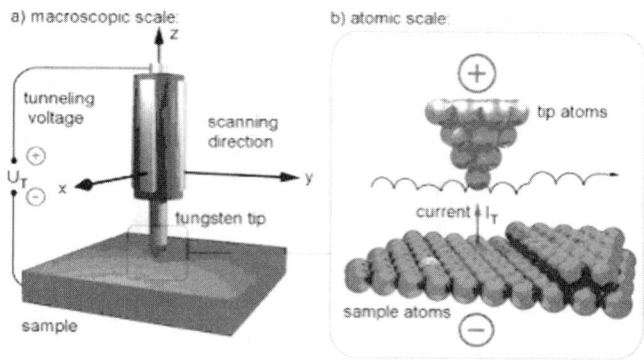

Fig. 20—Principle of scanning tunneling microscopy.

follows the JKR theory, and the friction between the smooth mica surfaces is much higher than that after the surfaces are damaged and wear takes place. If there are lubricants confined between smooth mica surfaces, the measurements of the friction forces on the film would give rise to the critical shear stress of the adsorbed boundary layers, which provides fundamental information for the study of boundary lubrication.

The adhesion hysteresis and its contribution to the friction have been studied extensively by means of SFA [39], which leads to an important conclusion that it is the adhesion hysteresis or the energy loss during the process of approach/separation, rather than the surface energy itself, that dominates the frictional behavior of boundary films.

Stick-slip motion is another issue that has been explored using SFA. It is found that the occurrence of stick-slip depends on the sliding velocity and the stiffness of the system, and the mechanism of the phenomenon can be interpreted in terms of periodic transition between liquid and solid states of the confined lubricant [40].

3.3.3 Thin Film Rheology

SFA has made a great contribution to the investigations of thin film rheology [41]. The measurements on SFA confirm that there is a significant enhancement of the effective viscosity in molecularly thin liquid films, and the viscosity grows constantly as the film thickness diminishes.

It is also observed in SFA experiments that the effective viscosity declines in a power law, as the shear rate increases. The observations of the dynamic shear response of confined liquid imply that the relaxation process in thin films is much slower and the time for the confined molecules to relax can increase by several orders.

The confined liquid is found to exhibit both viscous and elastic response, which demonstrates that a transition from the liquid to solid state may occur in thin films. The solidified liquid in the film deforms under shear, and finally yields when the shear stress exceeds a critical value, which results in the static friction force required to initiate the motion.

Much progress has been achieved so far, yet it can be expected that surface force apparatus will play a more significant role in future studies of micro/nanotribology, biology, and other fields of surface science.

4 Scanning Probe Microscope

The scanning tunneling microscope (STM) [42] has revolutionized the field of microscopy by stimulating an entire family of microscopes—generally referred to as scanning probe microscopes or SPMs [43,44], e.g., Scanning Force Microscope (SFM), Atomic Force Microscope (AFM), Friction Force Microscope (FFM), Scanning Near-field Optical Microscope (SNOM), Magnetic Force Microscope (MFM), and others, which are capable of measuring a range of physical and chemical properties, or both, on the nanometre scale. Although different types of SPMs have their own unique measuring ability, they are based on a common working feature: a mechanical probe sensor is scanned across an interface. During the scan, the probe sensor samples the signal which is interpreted in terms of structure, electronic, or force interaction information from the interface. A recently published book, edited by Bhushan et al. [45], gave an overview of new developments in the scanning probe method for both practical applications and basic research, and novel technical developments with respect to instrumentation and probes.

Three scanning probe techniques are described in more detail below: the scanning tunneling microscope, the atomic force microscope, and the friction force microscope.

4.1 Scanning Tunneling Microscope (STM)

In 1960, the principle of electron tunneling was first proposed by Giaever [42]. Binning et al. introduced vacuum tunneling combined with lateral scanning and successfully developed the first scanning tunneling microscope in 1982 [43], for which they were awarded the Nobel Prize in 1986. The STM allows one to image a surface with exquisite resolution, lateral 0.1 nm and vertical 0.01 nm, sufficient to define the position of single atoms.

The principle of the STM is based on the strong distance dependence of the quantum mechanical tunneling effect (Fig. 20) [44]. A thin metal tip is brought in close proximity to the sample surface. At a distance of only a few Angstroms, the overlap of tip and sample electron wave functions is large enough for a tunneling current I_t to occur, which is given by

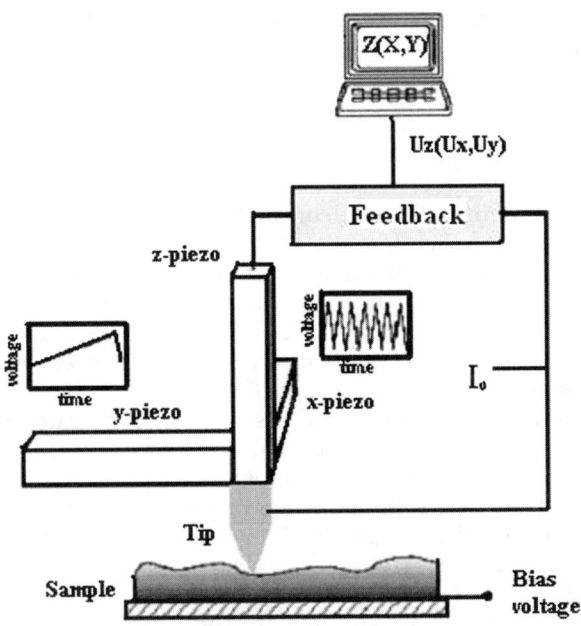

Fig. 21—Schematic drawing of scanning tunneling microscope.

$$I_t \sim e^{-2\kappa d} \qquad (18)$$

where d denotes the tip-sample distance and κ is a constant depending on the height of the potential barrier [46]. For metals with typical work functions of 4 eV~5 eV, the constant κ is of the order of 1 Å^{-1}. Hence, an increase of the tunneling distance of only 1 Å changes the tunneling currents by about an order of magnitude.

In Fig. 21 [47], a simplified schematic drawing of an STM is shown. The probe tip is attached to a piezodrive, which consists of three mutually perpendicular piezoelectric transducers. By applying a voltage fluctuating in a sawtooth form on the x-piezo and a voltage ramp on the y-piezo, the tip scans the xy plane. Scanning the tip over the sample surface while keeping the tunneling current constant by means of a feedback loop that is connected to the z-piezo (*constant current mode*), the tip will remain at a constant distance from the sample surface and will follow the surface contours. Monitoring the vertical position z of the tip as a function of the lateral position (x,y), one can get a two-dimensional array of z positions representing an equal tunneling-current surface.

Two working modes are used for the STM: first, the *constant height-mode*, in which the recorded signal is the tunneling current versus the position of the tip over the sample, and the initial height of the STM tip with respect to the sample surface is kept constant (Fig. 22(*a*)). In the *constant current-mode*, a controller keeps the measured tunneling current constant. In order to do that, the distance between tip and sample must be adjusted to the surface structure and to the local electron density of the probed sample via a feedback loop (Fig. 22(*b*)).

STM can be operated in a wide range of environments: a stable tunnel current can be maintained in almost any non-conducting medium, including air, liquid, or vacuum. It is also relatively forgiving for an STM operation to prepare a sample: the main requirement is that the sample conduct

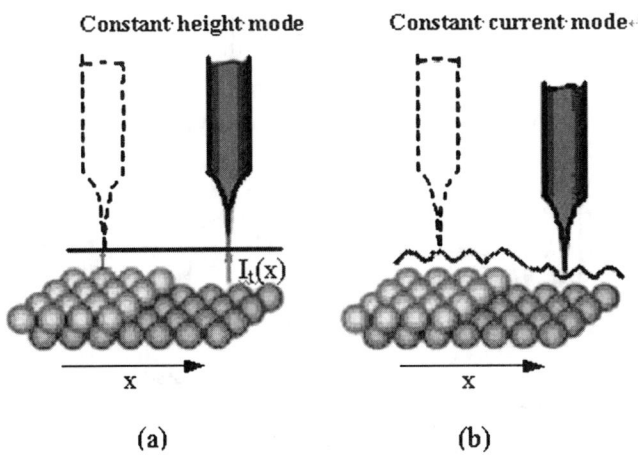

Fig. 22—Two working modes of STM, (a) constant height mode, (b) constant current mode.

~1 nA at ~1 V. This flexibility allows a wide range of applications. Due to the inherent surface sensitivity of STM, it is most widely applied in the field of surface science—the study of the structural, electronic, and chemical properties of surfaces, interfaces, and thin films, which is important to a wide range of technologies, including catalysis, semiconductor device fabrication, electrochemistry, tribology, and chemical sensors. The technique is especially useful for elucidating the properties of nanometre-sized surface structures. The operating flexibility of STM combined with the ability to acquire spectroscopic images has also led to its application in the study of novel electronic properties of materials, such as charge density waves and superconductivity. Moreover, the close proximity of the tip to the surface enables one to modify surfaces with atomic-scale precision.

4.2 Atomic Force Microscope (AFM)

The atomic force microscope was developed to overcome a basic drawback with the STM—that it can only image conducting or semiconducting surfaces. The AFM, however, has the advantage of imaging almost any type of surface, including polymers, ceramics, composites, glass, and biological samples. Unlike the STM, the physical magnitude monitored by the AFM is not tunneling current but the interaction force between the tip and sample. This is accomplished by attaching the tip to a cantilever-like spring and detecting its deflections due to forces acting on the tip. Since inter-atomic forces are always present when two bodies come into close proximity, the AFM is capable of probing surfaces of both conductors and insulators on an atomic scale.

Binnig et al. [48] invented the atomic force microscope in 1985. Their original model of the AFM consisted of a diamond shard attached to a strip of gold foil. The diamond tip contacted the surface directly, with the inter-atomic van der Waals forces providing the interaction mechanism. Detection of the cantilever's vertical movement was done with a second tip—an STM placed above the cantilever. Today, most AFMs use a laser beam deflection system, introduced by Meyer and Amer [49], where a laser is reflected from the back of the reflective AFM lever and onto a position-sensitive detector.

A schematic drawing of an AFM is in Fig. 23 [47]. A

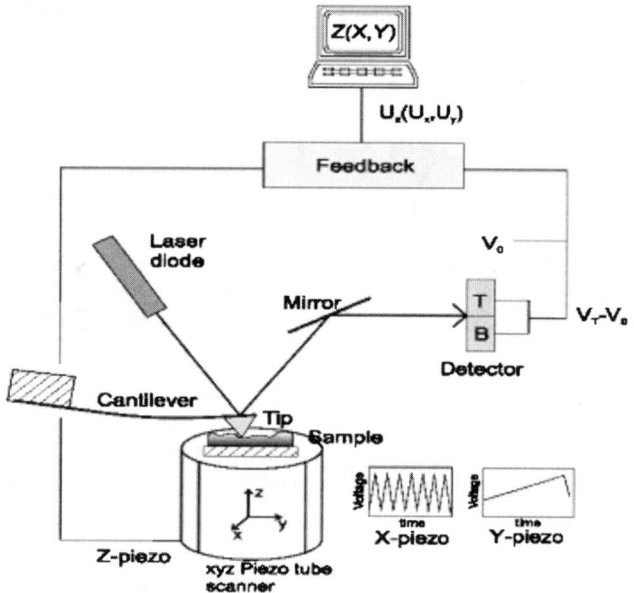

Fig. 23—Schematic drawing of the atomic force microscope.

sharp tip at the end of a cantilever is brought into contact with a sample surface via the z-piezo extension (either the sample or the tip can be scanned). The repulsive force F causes the cantilever to deflect vertically according to Hooke's law $F = k\Delta z$, where k is the spring constant and Δz the vertical displacement of the cantilever. The displacement Δz is monitored by the laser beam deflection technique. The back of the cantilever has to be a mirror-like reflecting surface. A laser beam (coming from a laser diode) is reflected off the rear side of the cantilever and the beam deflection is monitored with a position-sensitive detector (PSD) (split photodiode with two parts: top (T) and bottom (B)). The voltage difference from the top and bottom photodiodes, $V_T - V_B$, provides the AFM signal, which is a sensitive measure of the cantilever vertical deflection. During scanning via the x- and y-piezos, the z-piezo is connected to a feedback system. The feedback loop is used to keep the differential detector signal at a constant value V_0 by adjusting the vertical z position of the sample to achieve (almost) constant cantilever deflection Δz_0 even which corresponds to a constant force F_0 ($F_0 = k\Delta z_0$). The output signal of the feedback circuit U_z (z-piezo signal) is recorded as a function of (x,y) coordinates, which are determined by the voltages U_x and U_y applied to the x- and y-piezodrives. The two-dimensional array $U_z(U_x, U_y)$ can be transformed to "topography" $z(x,y)$, provided that the piezo coefficients are known. This mode of AFM operation is called constant force or constant cantilever deflection mode and is analogous to the constant current mode in an STM.

The most crucial component of an AFM is the cantilever. The deflection should be sufficiently large for ultra low forces (0.1 nN). Therefore, the spring constant should be as low as possible (lower than 1 N/m). On the other hand, the resonance frequency of the cantilever must be high enough (10 to 100 kHz) to minimize the sensitivity to mechanical vibrations (e.g., vibrational noise from the building ~100 Hz, frequency of the corrugation signal up to a few kHz). The resonant frequency of a spring loaded with an effective mass m is:

$$\omega_0 = \left(\frac{k}{m}\right)^{1/2} \quad (19)$$

Thus, in order to sustain a high resonance frequency, while reducing the spring constant, it is necessary to reduce the mass and therefore the geometrical dimensions of the cantilever. Microfabrication techniques are usually employed for the production of cantilever beams with integrated tips. Typically, AFM cantilevers are composed of single crystal silicon, or silicon nitride, with a reflective coating of gold or aluminum deposited on the top side. Cantilevers are generally single beams or v-shaped beams with lengths ranging from 50–200 μm, a thickness ranging from 0.5–2 μm, and spring constants of 0.1–100 N/m. The geometry of typical AFM tips is conical or square pyramidal, with a tip height of 3–15 μm, and an end radius of curvature ranging between 5–100 nm.

Because of the AFM's versatility, it has been applied to a large number of research topics. The AFM has also gone through many modifications for specific application requirements.

According to the distance from probe to the sample, three operation modes can be classified for the AFM. The first and foremost mode of operation is referred to as "contact mode" or "repulsive mode." The instrument lightly touches the sample with the tip at the end of the cantilever and the detected laser deflection measures the weak repulsion forces between the tip and the surface. Because the tip is in hard contact with the surface, the stiffness of the lever needs to be less than the effective spring constant holding atoms together, which is on the order of 1 ~ 10 nN/nm. Most contact mode levers have a spring constant of <1 N/m. The defection of the lever can be measured to within ±0.02 nm, so for a typical lever force constant at 1 N/m, a force as low as 0.02 nN could be detected [50].

The second operation mode is referred to as the "noncontact mode," in which the tip is brought in close proximity (with a few nanometres) to, and not in contact with the sample. A stiff cantilever is oscillated in the attractive regime. The forces between the tip and sample are quite low, on the order of pN (10^{-12} N). The detection scheme is based on measuring changes to the resonant frequency or amplitude of the cantilever [51,52]. In either mode, surface topography is generated by laterally scanning the sample under the tip while simultaneously measuring the separation dependent force or force derivative between the tip and the surface.

To minimize effects of friction and other lateral forces in the topography measurements in contact-modes AFMs and to measure topography of the soft surface, AFMs can be operated in so-called "tapping mode" [53,54]. It is also referred to as "intermittent-contact" or the more general term "Dynamic Force Mode" (DFM). A stiff cantilever is oscillated closer to the sample than in the noncontact mode. Part of the oscillation extends into the repulsive regime, so the tip intermittently touches or "taps" the surface. Very stiff cantilevers are typically used, as tips can get "stuck" in the water contamination layer. The advantage of tapping the surface is improved lateral resolution on soft samples. Lateral forces

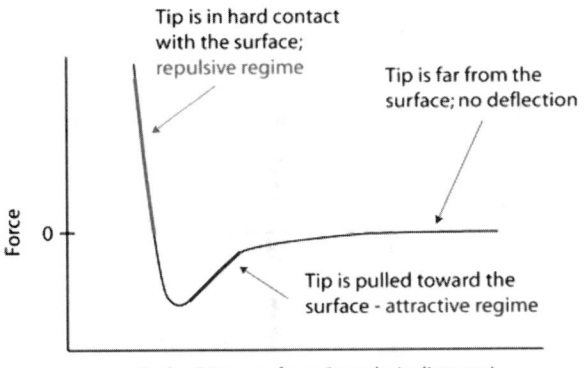

Fig. 24—The surface force field for different operation modes of AFM.

4.3 Friction Force Microscope (FFM)

The first FFM was developed by Mate et al. [55] at the IBM Almaden Research Center, San Jose, CA. In their setup, the tip and sample surface was in contact and in relative motion with respect to one another, the cantilever was expected to bend and twist, or both, in response to friction force. The principle of the FFM is shown in Fig. 25 [9]. It is similar to that of laser-AFM. The sample was mounted on a piezoelectrical tube (PZT), which scanned the X, Y plane and controlled the feedback of the Z axis. A laser beam from a laser diode was focused on the mirror of the free end of a cantilever with lens, and the reflected beam fell on the center of the position-sensitive detector (PSD), a four-segment photodiode. When the sample contacted with the tip and relatively moved in control of the computer, the reflected beam deflected and changed the position on PSD due to the twist and deflection of the cantilever caused by changes of surface roughness, friction force, and adhesive force between the sample and the tip. The extension and retraction of the PZT was feedback controlled by an electrical signal from the vertical direction of the PSD. Thereby the surface morphologies of the sample were obtained according to the movement of the PZT. At the same time, the electrical signal from the horizontal direction of the PSD caused by twist of the cantilever was transmitted to the computer and converted to a digital signal, and then the image of lateral force in the scanning area was obtained.

Figure 26 is the schematic diagram of the FFM designed by Lu (one of the present authors) et al. [9] in 1995. It consists of a stage system, circuitry controlling system, and computer controlling system.

The stage system consists of four parts: the light reflection, preamplifier, driver of scanning, and step motor controller. The part of light reflection consists of laser diode, lens, cantilever, PSD, and mechanical adjustment components. The preamplifier consists of precision instrument amplifiers to receive and amplify the electrical signal of the PSD, and convert the signal to a voltage signal at microvoltage scale. The driver of scanning refers to the cantilever,

such as drag, common in contact mode, are virtually eliminated. For poorly adsorbed specimens on a substrate surface the advantage is clearly seen. The surface force field for different operation modes of the AFM is shown in Fig. 24.

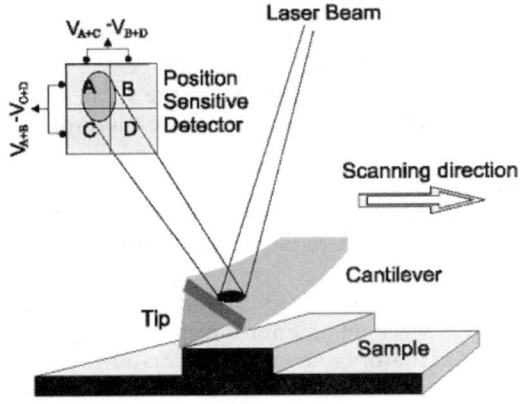

Fig. 25—Schematic drawing of friction force microscope.

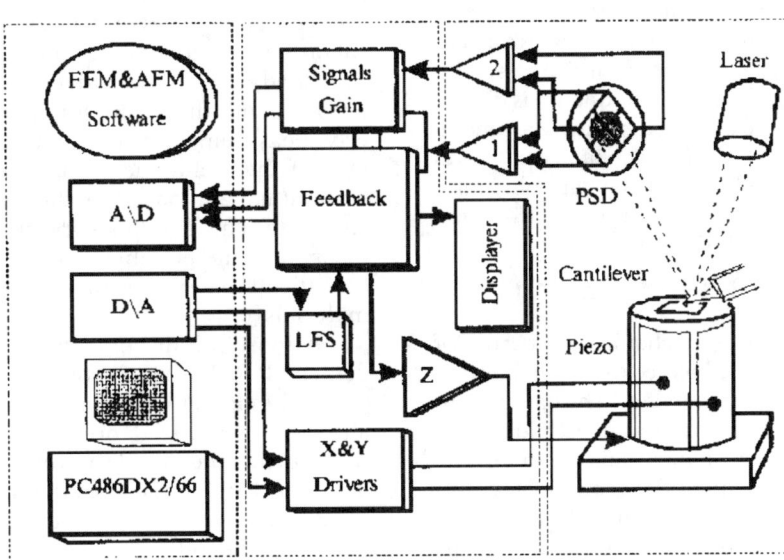

Fig. 26—The system schematic diagram of FFM designed by Lu et al. [9].

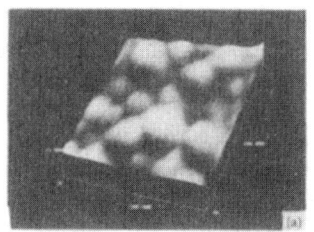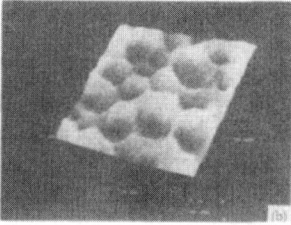

Fig. 27—The images of three-dimensional morphologies, (a) and lateral force (b) of Au film [9].

which can move at three-dimensions by charge. The step motor controller is used for moving the sample up and down. It can be automatically turned off when the tip-to-sample distance reaches the present distance.

The circuitry of the FFM is similar to that of the laser-AFM, in the aspect of feedback circuit, driver circuit, and the amplifier circuit of the vertical electrical signal. To facilitate the study of nanotribology, the unit of gain amplification of the electrical signal caused by lateral force, the unit of comparison and controlling of reference signal, and the unit of adjustment of the PZT were added to the circuitry. The output reference current signal (I_{ref}) can be controlled from the computer and compared with the signal from the PSD in order to change micro-load between the sample and the tip. The information of attractive force and adhesive force between the sample and the tip can also be obtained by turning off the feedback unit by computer and charge voltage to the PZT.

The software of the FFM mainly consists of three parts: (1) the program of scanning, data collecting, and image displaying, (2) the program of making force curve of cantilever and setting micro-load between the sample and the tip, and (3) the program of image processing. The setup aims not only for carrying out micro friction and wear test but also nano-scale processing. The maximum scanning range designed by Lu et al. is about 8 μm by 8 μm, and the resolution is about atomic scale.

The friction force microscope serves as an excellent tool to study friction and wear on nanometre scale. Figure 27 shows the image of three-dimensional morphologies and lateral force of an Au film by FFM [56]. The grain size of Au film is about 20–50 nanometres and the grain boundary is clearly observed. This result cannot be obtained by using a general scanning electronic microscope (SEM). The image of lateral force is different from morphology, but they are obviously correspondent to each other. The lateral force in the grain boundary is larger than that inside the grain.

5 Nanoindentation and Nanoscratching

Since the early 1980s, the study of mechanical properties of materials on the nanometre scale has received much attention, as these properties are size dependent. The nanoindentation and nanoscratch are the important techniques for probing mechanical properties of materials in small volumes. Indentation load-displacement data contain a wealth of information. From the load-displacement data, many mechanical properties such as hardness and elastic modulus can be determined. The nanoindenter has also been used to measure the fracture toughness and fatigue properties of ul-

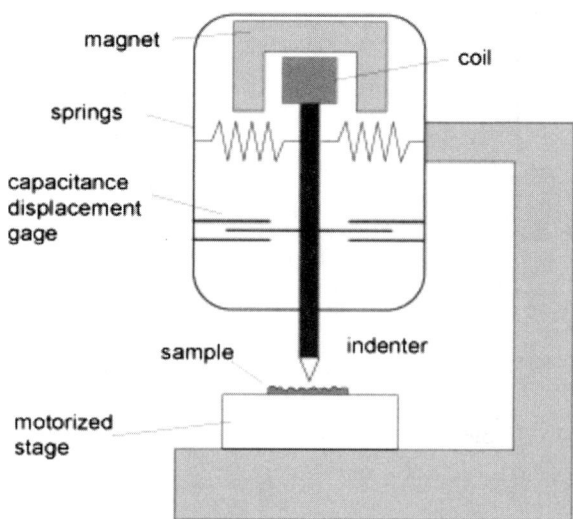

Fig. 28—Schematics of Nano Indenter.

tra thin films, which cannot be measured by conventional indentation tests. The continuous stiffness, residual stresses, time dependent creep, and relaxation properties can be measured by nanoindentation too. With a tangential force sensor, nanoscratch and wear tests can be performed at ramping loads. It can be used to measure the scratch resistance, film-substrate adhesion, and durability of thin solid films.

5.1 Nanoindentation

Nanoindentation is a technique developed over the past decade for probing the mechanical properties of materials at very small scales [57–63]. It is now used routinely in the mechanical characterization of thin films and thin surface layers [64]. Ultra low load indentation employs high resolution sensors and actuators to continuously control and monitor the loads and displacements on an indenter as it is driven into and withdrawn from a material [60,63]. In some systems, forces as small as a nano-Newton and displacements of about an Angstrom can be accurately measured. One of the great advantages of the technique is that many mechanical properties can be determined by analyses of the indentation load-displacement data alone, thereby avoiding the need to image the hardness impression and facilitating property measurement at the submicron scale. Another advantage of the technique is that measurements can be made without having to remove the film or surface layer from its substrate. This simplifies specimen preparation and makes measurements possible in systems which would otherwise be difficult to test, as is the case for most ion beam modified materials [65].

A most recent commercial Nano Indenter (Nano Indenter XP (MTS, 2001)) consists of three major components [66]: the indenter head, an optical/atomic force microscope, and x-y-z motorized precision table for positioning and transporting the sample between the optical microscopy and indenter (Fig. 28). The load on the indenter is generated using a voice coil in permanent magnet assembly, attached to the top of the indenter column. The displacement of the indenter is measured using a three plate capacitive displacement sensor. At the bottom of the indenter rod, a three-sided

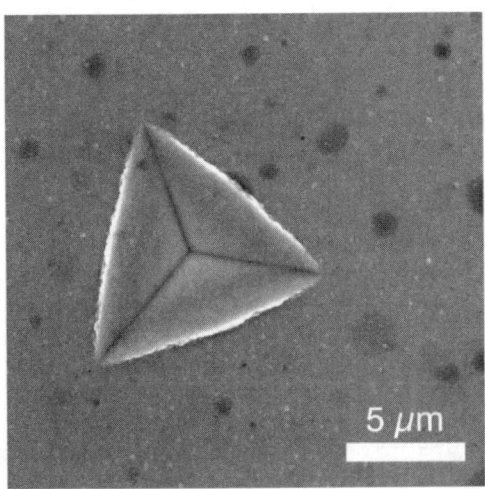

Fig. 29—Scanning elecron micrograph of a small nanoindentation made with a Berkovich indenter in a 500 nm aluminum film deposited on glass (from Ref. [64]).

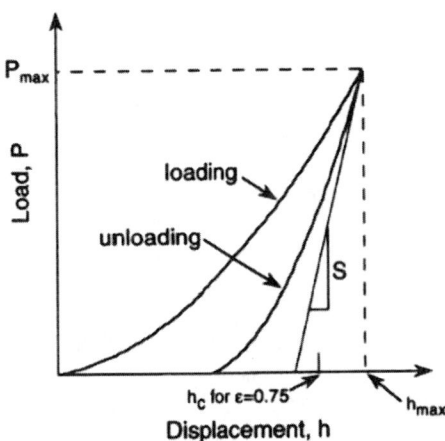

Fig. 30—Typical indentation load-displacement curve [64].

pyramidal diamond tip is generally attached. The indenter head assembly is rigidly attached to the U beam below which the x-y-z table rides. The optical microscope is also attached to the beam. The specimens are held on an x-y-z table whose position relative to the microscope or the indenter is controlled with a joystick. The three components are enclosed in a heavy wooden cabinet to ensure the thermal stability of the samples. The entire apparatus is placed on a vibration isolation table.

The main requirements for the indenter are high elastic modulus, no plastic deformation, low friction, smooth surface, and a well defined indentation impression. The first four requirements are satisfied by choosing diamond material for the tip. In general, for satisfying the last requirement, a sharp, geometrically-similar indenter such as the Berkovich triangular pyramid is useful. A scanning electron micrograph of a small nanoindentation made with a Berkovich indenter in a 500 nm aluminum film deposited on glass is shown in Fig. 29.

In practical indentation tests, multiple loading and unloading steps are performed to examine the reversibility of the deformation, ensuring that the unloading data used for analysis purposes are mostly elastic. A typical indentation experiment consists of a combination of several segments, e.g., approach, load, hold, and unload. Either constant loading or constant displacement experiments can be performed [59]. A typical constant loading indentation experiment consists of eight steps [67]: approaching the surface at 10 nm·s^{-1}; loading to peak load at a constant loading rate (10 % of peak load/s); unloading 90 % of peak load at a constant unloading rate (10 % of peak load/s); reloading to peak load; holding the indenter at peak load for 10 s; unloading 90 % of peak load; holding the indenter after 90 % unloading; and finally unloading completely.

The two mechanical properties measured most frequently using indentation techniques are the hardness, H, and the elastic modulus, E. A typical load-displacement curve of an elastic-plastic sample during and after indentation is presented in Fig. 30, which also serves to define some of the experimental quantities involved in the measurement.

The key quantities are the peak load, P_{max}, the displacement at peak load, h_{max}, and the initial unloading contact stiffness, $S = dP/dh$, i.e., the slope of the initial portion of the unloading curve.

The physical processes that occur during indentation are schematically illustrated in Fig. 31. As the indenter is driven into the material, both elastic and plastic deformation occurs, which results in the formation of a hardness impression conforming to the shape of the indenter to some contact depth, h_c. During indenter withdrawal, only the elastic portion of the displacement is recovered, which facilitates the use of elastic solutions in modeling the contact process.

The Oliver-Pharr data analysis procedure [59] begins by fitting the unloading curve to the power-law relation

$$P = B(h - h_f)^m \qquad (20)$$

where P is the indentation load, h is the displacement, B and m are empirically determined fitting parameters, and h_f is the final displacement after complete unloading (also determined by curve fitting). The unloading stiffness, S, is then established by differentiating Eq (20) at the maximum depth of penetration, $h = h_{max}$,

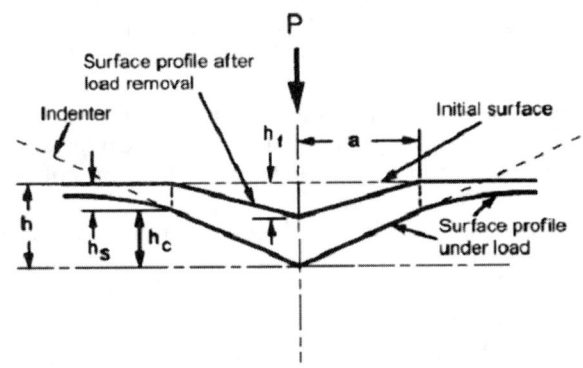

Fig. 31—The deformation pattern of an elastic-plastic sample during and after indentation [58].

$$S = \frac{dP}{dh}(h = h_{max}) = mB(h_{max} - h_f)^{m-1} \quad (21)$$

The contact depth is also estimated from the load-displacement data using:

$$h_c = h_{max} - \varepsilon \frac{P_{max}}{S} \quad (22)$$

where, P_{max} is the peak indentation load and ε is a constant, which depends on the indenter geometry. Empirical studies have shown that $\varepsilon \cong 0.75$ for a Berkovich indenter.

From the basic measurements contained in the load-displacement data, the projected contact area of the hardness impression, A, is estimated by evaluating an empirically determined indenter shape function at the contact depth, h_c; that is $A = f(h_c)$. The shape function, $f(d)$, relates the cross-sectional area of the indenter to the distance, d, from its tip. For a geometrically perfect Berkovich indenter, the shape function is given by $f(d) = 24.56d^2$, but for real Berkovich indenters, $f(d)$ is considerably more complex due to tip rounding. Even for the most carefully ground diamonds, mean tip radii are typically in the 10–100 nm range, and this must be accounted for in the analysis procedure if accurate results are to be obtained at small depths. A simple experimental procedure has been developed for determining shape functions without having to image the indenter or hardness impressions made with it [59].

Once the contact area is determined from the load-displacement data, the hardness, H, and effective elastic modulus, E_{eff}, follow from:

$$H = \frac{P_{max}}{A} \quad (23)$$

$$E_{eff} = \frac{1}{\beta} \frac{\sqrt{\pi}}{2} \frac{S}{\sqrt{A}} \quad (24)$$

where β is a constant, which depends on the geometry of the indenter ($\beta = 1.034$ for the Berkovich). The effective modulus, which accounts for the fact that elastic deformation occurs in both the specimen and the indenter, is given by

$$\frac{1}{E_{eff}} = \frac{1 - \nu^2}{E} + \frac{1 - \nu_i^2}{E_i} \quad (25)$$

where E and ν are the Young's modulus and Poisson's ratio for the specimen, and E_i and ν_i are the same quantities for the indenter. For diamond, $E_i = 1,141$ GPa and $\nu_I = 0.07$.

Equation (24) is originally derived for a conical indenter. Pharr et al. showed that Eq (24) holds equally well to any indenter, which can be described as a body of revolution of a smooth function [67]. Equation (24) also works well for many important indenter geometries, which cannot be described as bodies of revolution.

A recently developed technique, continuous stiffness measurement (CSM), offers a significant improvement in nanoindentation testing [59,68]. Using this technique, the contact stiffness, S, is measured continuously during the loading portion of the test. Continuous contact stiffness measurement is accomplished by imposing a small, sinusoidally varying signal on the output that drives the motion of the indenter and analyzes the resulting response of the system by means of a frequency specific amplifier [59]. The data obtained using this technique can be used to provide a continuous measurement of the hardness and elastic modulus as a function of depth in one simple experiment, which is useful for the evaluation of nanofatigue. Figure 32 shows the hardness of the DLC films as a function of indentation depth measured with a Nano Indenter XP system using continuous stiffness measurement (CSM) [69].

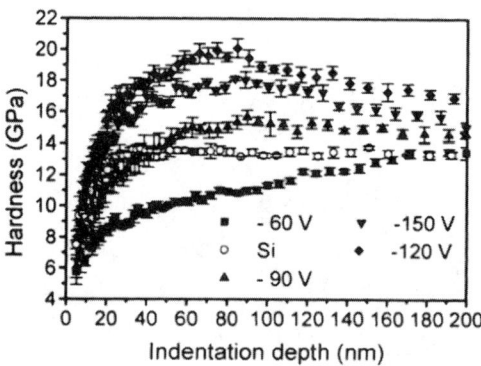

Fig. 32—The hardness of the DLC films as a function of indentation depth measured with a Nano Indenter XP system using CSM Technique.

With good experimental technique and careful analysis, the hardness and elastic modulus of many materials can be measured using these methods with accuracies of better than 10 % [59]. There are, however, some materials in which the methodology significantly overestimates H and E, specifically, materials in which a large amount of pile-up forms around the hardness impression. The reason for the overestimation is that Eqs (22) and (24) are derived from a purely elastic contact solution, which accounts for sink-in only [65].

It is widely accepted that to measure the true hardness of the films, the indentation depth should not exceed 10 % of the film thickness. Based on a finite element analysis, Bhattacharya [70] and Bhushan [71] conclude that the true hardness of the films can be obtained if the indentation depth does not exceed about 30 % of the film thickness.

The nanoindentation technique can also be used to measure fracture toughness at small scales, by using the radial crack, which occurs when brittle materials are indented by a sharp indenter [72,73]. Lawn et al. [74] have shown that a simple relationship exists between the fracture toughness, K_c, and the lengths of the radial cracks, c, in the form of:

$$K_c = \alpha \left(\frac{E}{H}\right)^{1/2} \left(\frac{P}{c^{3/2}}\right) \quad (26)$$

here, P is the peak indentation load and α is an empirical constant, which depends on the geometry of the indenter. E and H can be determined directly from analyses of the nanoindentation load-displacement data. Thus, provided one has a means for measuring crack lengths, the fracture toughness K_c can be obtained.

However, in order to probe the fracture toughness of thin films or small volumes using ultra low load indentation, it is necessary to use special indenters with cracking thresholds lower than those observed with the Vickers or Berkovich indenters (for Vickers and Berkovich indenters, cracking

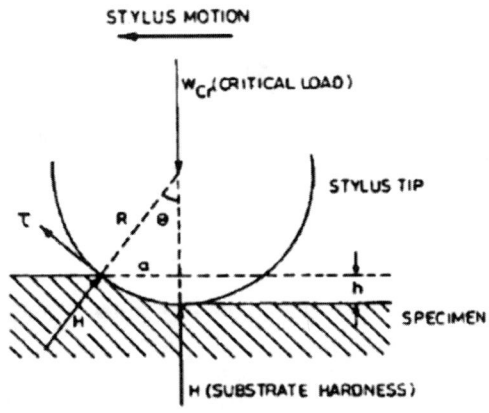

Fig. 33—Geometry of scratch for adhesion measurement technique [68].

thresholds in most ceramic materials are about 250 mN or more [75]). Significantly lower thresholds (less than 10 mN) can be achieved using the cube-corner indenter [72].

5.2 Nanoscratch

The nanoscratch technique is a very powerful tool for analyzing the wear resistance of bulk materials [71]. During the scratch test, normal load applied to the scratch tip is gradually increased until the material is damaged. Friction force is sometimes measured. After the scratch test, the morphology of the scratch region including debris is observed in an SEM. Based on the combination of changes in the friction force as a function of normal load and SEM observations, the critical load is determined and the deformation mode is identified. Any damage to the material surface as a result of scratching at a critical ramp-up load results in an abrupt or gradual increase in friction. The material may deform either by plastic deformation or fracture. Ductile materials (all metals) deform primarily by plastic deformation, resulting in significant plowing during scratching. Tracks are produced, the width and depth of which increase with an increase in the normal load. Plowing results in a continuous increase in the coefficient of friction along with an increase in the normal load during scratching. Debris is generally ribbon-like or curly. By contrast, brittle materials deform primarily by brittle fracture with plastic deformation to some extent. In the brittle fracture mode, the coefficient of friction increases very little until a critical load is reached, when the materials fail catastrophically and produce fine rounded debris, and the coefficient of friction increases rapidly afterward.

The other important application of nanoscratching is to measure adhesion strength and durability of ultrathin films and coatings. Scratch tests to measure adhesion of films were first introduced by Heavens in 1950 [76]. For nanoscratching, a conical diamond indenter is drawn across the coating surface. A normal load is applied to the scratch tip and is gradually increased during scratching until the coating is completely removed. The minimum or critical load at which the coating is detached or completely removed is used as a measure of adhesion. It is a most commonly used technique to measure adhesion of hard coatings with strong interfacial adhesion (>70 MPa).

For the scratch geometry shown in Fig. 33 [71], surface hardness H is given by

$$H = \frac{W_{cr}}{\pi a^2} \quad (27)$$

and adhesion strength τ is given by [77]

$$\tau = H \tan \theta = \frac{W_{cr}}{\pi a^2}\left[\frac{a}{(R^2 - a^2)^{1/2}}\right] \quad (28)$$

or

$$\tau = \frac{W_{cr}}{\pi a R} \text{ if } R >> a \quad (29)$$

where W_{cr} is the critical normal load, a is the contact radius, and R is the stylus radius.

Burnett and Rickerby [78] and Bull and Rickerby [79] analyzed the scratch test of a coated sample and derived a relation between the critical normal load W_{cr} and the work of adhesion W_{ad}

$$W_{cr} = \frac{\pi a^2}{2}\left(\frac{2EW_{ad}}{t}\right)^{1/2} \quad (30)$$

where E is Young's modulus of elasticity and t is the coating thickness. Plotting W_{cr} as a function of $a^2/t^{1/2}$ should give a straight line of slope $\pi(2EW_{ad}/t)^{1/2}/2$ from which W_{ad} can be calculated.

An accurate determination of critical load W_{cr} is sometimes difficult. Several techniques, such as (1) microscopic observation (optical or SEM) during the test, (2) chemical analysis of the bottom of the scratch channel (with electron microprobes), and (3) acoustic emission, have been used to obtain the critical load.

The Nano Indenter XP system was used for the nanoscratch tests to study the resistance of DLC films by Luo's (one of the authors of this book) group [69]. One edge of the diamond tip was aligned to the scratch direction. Before scratching, the surface profile of the sample was obtained via a "pre-scan" over a total scan length of 700 μm at a load of 20 μN. Then "scratch-scan" was carried out by ramping the load over a 500 μm length until a normal force of 60 mN was reached. The scratch depth and friction force between the tip and sample with increasing scratch length were measured during the process. Finally a "post-scan" was carried out at a load of 20 μN to measure the profile of the scratched surface. The morphologies of the scratch scars were observed with a scanning electron microscope (SEM).

Figure 34 shows a representative SEM micrograph and surface profiles of the scratch tracks and the evolutions of normal load and friction force between the tip and the DLC film. As shown in the SEM micrograph (Fig. 34(a)), fracture occurs at the end stage of scratching while the residual plastic deformation on the film surface is induced before approaching the initial point of fracture. In Fig. 34(b), the pre-scan profile is smooth and horizontal, which reveals that the original surface was smooth. However, both the post-scan and scratch-scan profiles include two regimes, namely, the smooth regime and zigzag regime. Moreover, the post-scan profile is located above the scratch-scan profile because of the elastic recovery of the film after scratching. Similar to the post-scan and scratch-scan profiles, the curve of friction force in Fig. 34(c) also exhibits a smooth and a zigzag regime.

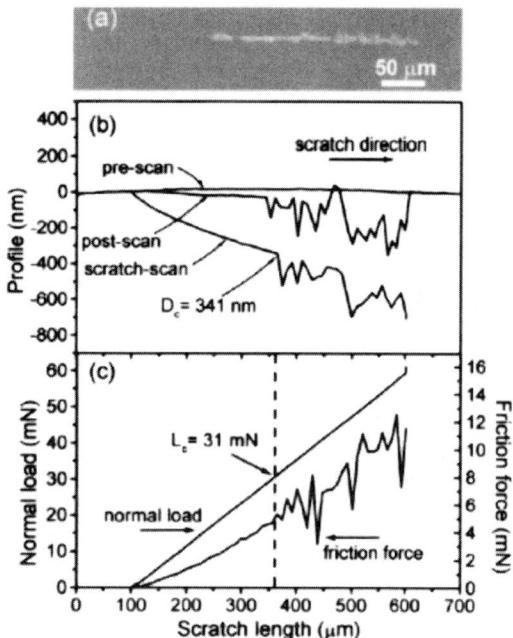

Fig. 34—(a) Representative SEM micrograph, (b) surface profiles of the scratch tracks, and (c) the evolutions of normal load and friction force between the tip and the film of the DLC film deposited at −90 V bias.

The abrupt increase in the penetration depth and friction force almost occurs simultaneously. Accordingly, the abrupt changes in the friction force represent the fracturing of films and the contact of the tip with the substrate. The normal load and the corresponding penetration depth associated with the abrupt changes are defined as the critical load L_c and critical depth D_c of scratch. This example clearly suggests that the scratch technology is a powerful tool to examine the adhesion strength and durability of thin solid films.

6 Other Measuring Techniques

6.1 Fluorometry Technique

6.1.1 Introduction

The solid-liquid two-phase flow is widely applied in modern industry, such as chemical-mechanical polish (CMP), chemical engineering, medical engineering, bioengineering, and so on [80,81]. Many research works have been made focusing on the heat transfer or transportation of particles in the micro scale [82–88]. In many applications, e.g., in CMP process of computer chips and computer hard disk, the size of solid particles in the two-phase flow becomes down to tens of nanometres from the micrometer scale, and a study on two-phase flow containing nano-particles is a new area apart from the classic hydrodynamics and traditional two-phase flow research. In such an area, the forces between particles and liquid are in micro or even to nano-Newton scale, which is far away from that in the traditional solid-liquid two-phase flow.

For most existing measuring methods, the actual motion of individual nano-particles in two-phase flow cannot be observed easily. Conventional particle image velocimetry (PIV) apparatus can measure the particles in micro scale

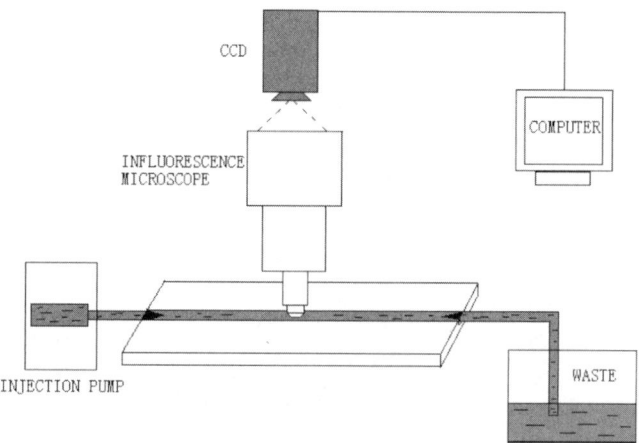

Fig. 35—Sketch map of the measuring system.

[89]. Confocal laser scanning microscope can see nano-particles, but it requires a long scanning time and cannot observe the real-time motion of particles [90]. Most of the scanning electron microscopes (SEM) cannot observe the fast motion of nano-particles in liquid. Therefore, development of accurate measuring methods for monitoring solid-liquid two-phase flow with the particles in the nanoscale become an important demand and it is crucial for the understanding of behaviors of microflow.

6.1.2 Experiment

A system, including a fluorescence microscope, a high sensitive CCD called Evolution QE cooled CCD camera, a microchannel, a precision injection pump, and other assistant equipment, has been installed for the experiment of observing the motion of spherical nano-particles in liquid as shown in Fig. 35 [80]. The system can capture and save high resolution images with 1,360 by 1,036 pixels for a very high sensitivity to the fluorescence intensity. In order to avoid splashing on the lens of the microscope, the liquid in a channel is covered by a glass sheet. The injection flowing rate is from 0.05 mL/min to 10 mL/min.

These spherical nano-particles about 55 nm in diameter have a fluorescent material of ruthenium pyridine inside, and the shell of silicon dioxide, as shown in Fig. 36. The excitation wavelength of the ruthenium pyridine is 480 nm and the emission wavelength is 592 nm [81]. In order to get a clear image of nano-particles, the mass concentration of the fluorescent particles should be limited to a very low level.

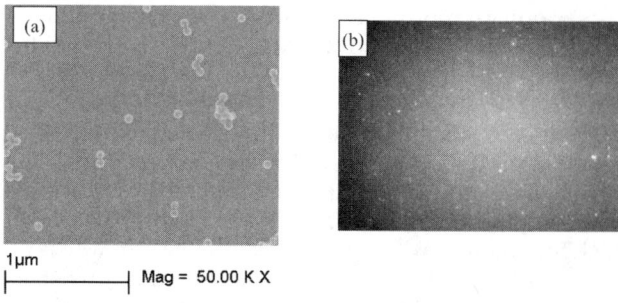

Fig. 36—Photograph of nano-particles, (a) SEM photograph, (b) photograph under fluorescence microscope.

	Mass Concentration of Fluorescent Particles	Mass Concentration of Silicon Dioxide Particles
Sample 1	0.005 %	1.2 %
Sample 2	0.005 %	12 %

TABLE 1—Mass concentrations of nano-particles in the liquid.

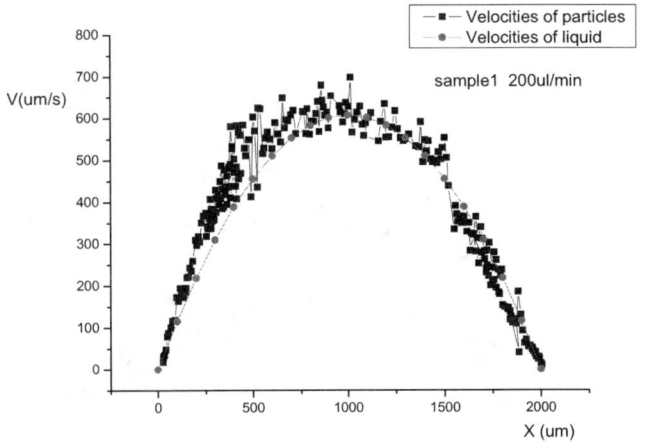

Fig. 38—Velocity distribution comparison of particles and liquid.

Therefore, the samples are prepared by mixing fluorescent particles solution and silicon dioxide nano-particles solution as shown in Table 1.

6.1.3 Characteristics of Two-Phase Flow Containing Nano-particles—A Study by the Present Authors

From the image sequences, information on the velocities of nano-particles can be extracted. The statistical effect of Brownian motion on the flowing speed of the mixed liquid is found small enough to be ignored as shown in Fig. 37 where most of the particles trajectories in the liquid are straight lines and parallel with the wall basically. Therefore, Brownian diffusive motion is ignorable.

The distribution of particles velocities in the liquid under a small flowing rate is much closer to that calculated from Eq (31) [85] when the width and depth of channels are big enough (about 2 mm) to ignore the effect of the surface force of the solid wall as shown in Fig. 38.

$$V = \frac{1}{2\mu_0}\frac{dp}{dx}X(2H - X) \qquad (31)$$

where μ_0 is the viscosity of water, H is the half-width of the channel, X is the shortest distance from the calculated point to the solid wall. In the given condition of experiments, dp/dx is a constant. The velocities of liquid should also change along the direction of the depth in the channel; however, the CCD is only able to capture the velocity distribution for the particles near the upper surface of the flow because the work distance of the microscope is only 0.2 mm.

Therefore, the velocities of liquid are consistent with the velocities of particles, that is, the motion of nano-particles can reflect the flow of liquid verily in the given condition. Figures 39 and 40 show the comparison of the two liquid samples with different mass concentration of the nano-particles at different flow rates. Generally, particle velocity increases synchronously with the liquid flow rate, but the velocity becomes dispersive when it exceeds 300 μL/min. The more the particles were added in the liquid, the more dispersive of the velocities of the particles were observed. Several possible causes can result in this phenomenon. One possible reason is that when the velocity of flow becomes large enough, the bigger particles in the liquid cannot follow the flow as the smaller particles do, or bigger particles will move slower than the liquid around them, so the velocities of particles will distribute dispersedly. Another possible reason is that when the velocity of flow increases the time for particles to traverse, the view field of the microscope will decrease. As a result, the number of data points in the trace of a particle

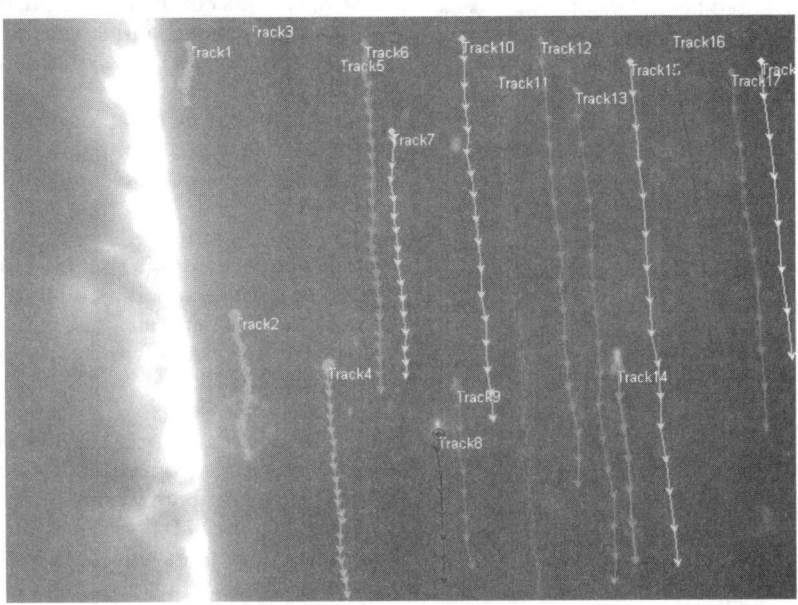

Fig. 37—Tracking particles image of Sample 1 in 200 μL/min.

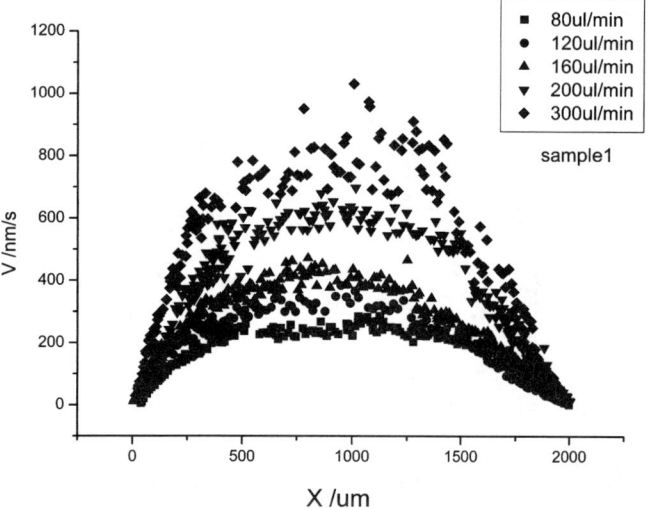

Fig. 39—Different velocities of flow for Sample 1.

will be reduced, and the shortage of original data will lead to a poor statistical description for particle movement and the influence of its Brownian motion on the calculation of the velocity will appear. So the distributions of velocities will be dispersive.

Therefore, fluorescent nano-particles can be used as the tracer in studying micro-fluids. The visualizing approach to nano-flow is not very mature and more experiments should be practiced. Because of its wide connection to modern technologies, the research of the solid-liquid two-phase flow will attract more and more attention.

6.2 Scanning Acoustic Microscopy (SAM)

Scanning acoustic microscopy (SAM) is a relatively new technique which broke through in the mid-seventies and was commercialized recently. The SAM uses sound to create visual images of variations in the mechanical properties of samples. The ability of acoustic waves to penetrate optically opaque materials makes it possible to provide surface or subsurface structural images nondestructively, which might

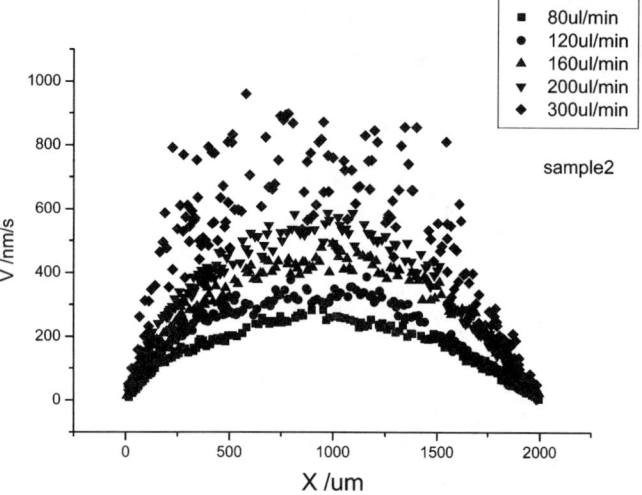

Fig. 40—Different velocities of flow for Sample 2.

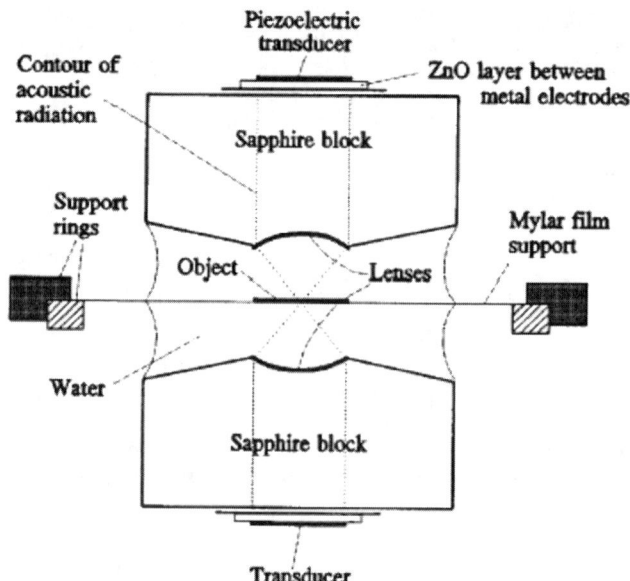

Fig. 41—Lens geometry of the transmission acoustic microscope.

not be observable by other techniques. Up to now, SAM has already been used in a wide range of applications, such as biology, integrated circuits, metrology, and semiconductor devices.

The concept of acoustic microscopy was first put forward by Sokolov in 1949 [91]. Because an acoustic wave interacts directly with the elastic bodies of the material through which it propagates, the wavelength of sound at a high frequency can be very short, from which it should be possible to build an acoustic microscope with a resolution comparable to that of the optical microscope. However, until the early 1970s, this was not achieved because techniques for producing high-frequency sound waves were not readily available. In 1974, the scanning acoustic microscope was developed at Stanford University by Lemon and Quate [92]. Now, various commercial acoustic microscope instruments are available for scientific and industrial applications.

There are two types of scanning acoustic microscopes. If the illumination and the reception of the acoustic waves are performed by two identical lenses arranged confocally, the SAM is called a transmission SAM. The lens geometry used for transmission imaging is shown schematically in Fig. 41 [93].

This system consists of a symmetrical pair of lens elements connected by a small volume of liquid. Each lens consists of a single spherical interface between the liquid and a lens rod. The lens element is formed by polishing a small concave spherical surface in the end of a sapphire rod. At the opposite end of the rod, a thin film piezoelectric transducer is centered on the axis of the lens surface.

In the reflection SAM, the second type, the transmission arrangement is conceptually folded over, so that the same lens is used for both transmitting and receiving the acoustic signal (see Fig. 42) [94]. The transmission version can use a simple continuous wave, but in the reflection mode, pulsed signals should be used in order to separate the reflected signal from the transmitted signal. In the transmission mode, the ultrasonic beam passes through the object placed be-

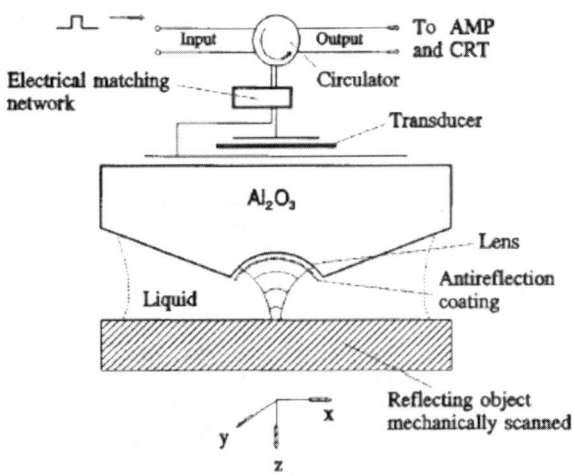

Fig. 42—Schematic of a reflection SAM.

tween the objective and collector. The transmission acoustic microscope is especially suitable for investigation of samples with acoustic impedances and attenuation comparable to those of water, i.e., for biological specimens. The reflection mode is more promising for the investigation of samples of high acoustic impedance and attenuation, i.e., for solid materials. The operation of a transmission SAM requires the lenses to be set up so that they are accurately confocal, which becomes difficult at high frequencies and shorter wavelengths. In the reflection SAM, the need to adjust the lenses to be confocal is obviated. Specimens to be imaged in the transmission version must be thin in order to enable acoustic waves to propagate through. For biological specimens, this often presents few problems, but for many solids, such a thin specimen is quite difficult to prepare. Therefore, although the first scanning acoustic microscope operated in the transmission mode, there is an increasing shift from transmission microscopy to reflection microscopy. Most of the recent development and application of the SAM has been associated with the reflection mode.

As shown in Fig. 42, a sapphire rod cut along the crystallographic c axis of the sapphire is used to be an acoustic lens in a reflection scanning acoustic microscope. In the center of one face of the rod, a concave spherical surface is ground. This surface provides the focusing action and, to optimize transmission of the acoustic waves, it is coated with a quarter-wavelength-thick matching layer. On the other face of the rod a piezoelectric transducer, usually a thin film of rf-sputtered ZnO, is deposited. In operation, a coupling fluid (usually water) is necessary between the lens and the specimen. When the transducer is energized with a short rf pulse (approximately 30 ns in duration), plane acoustic waves are generated, which travel through the rod and are focused on the axis of the lens by refraction at the spherical interface between the lens and the liquid. The object to be imaged is placed at the focus of this lens. The acoustic waves are partially reflected at the interface and the echoes thus produced traverse the system in reverse order and are converted back into an electrical pulse by the transducer, which acts in this case like a light-sensitive receptor and coherent detector. The strength of this pulse is proportional to the acoustic reflectivity of the object at the point being investigated. By mechanically scanning the object in a raster fashion, one can build up an acoustic image of the desired field of view and display it on a TV monitor.

The resolution of an acoustic lens is determined by diffraction limitations, and is $R = 0.51\lambda_w/\text{N.A}$ [95], where λ_w is the wavelength of sound in liquid, and N.A is the numerical aperture of the acoustic lens. For smaller (high-frequency) lenses, N.A can be about 1, and this would give a resolution of $0.5 \lambda_w$. Thus a well designed lens can obtain a diameter of the focal spot approaching an acoustic wavelength (about 0.4 μm at 2.0 GHz in water). In this case, the acoustic microscope can achieve a resolution comparable to that of the optical microscope.

As the resolution is proportional to the wavelength in the liquid λ_w, the way to improve the resolution is therefore to make the wavelength smaller. The wavelength depends on the velocity of sound in the liquid, v_w, and the frequency f such that $\lambda_w = v_w/f$. That is, if the frequency could be indefinitely increased, unlimited resolution could be achieved. Unfortunately, the application of higher frequencies is restricted due to the attenuation of the coupling medium and the available radius of curvature of the lens. Acoustic waves need a medium to support their propagation (in acoustics there is no analogy to a vacuum in optics). Between the acoustic lens and the specimen, the medium must be a liquid. Most liquids at or near room temperature exhibit linear viscosity, which causes the attenuation of acoustic waves propagating through them to be proportional to the frequency squared. To increase the frequency, it is necessary to reduce the liquid path length between the lens and the specimen, which means that the focal length of the lens and in turn its radius of curvature must be small. Grinding lenses with a very small radius is a quite difficult art. Moreover, even if it is possible to grind lens radii as small as required (about 15 microns), for a lens operating in a pulsed mode with higher frequencies, there are major problems with high-speed switches to obtain sufficiently narrow pulses to prevent the specimen from being swamped by the lens echo. Usually, the highest frequency for a microscope with water coupling at 60°C is about 2 GHz. The attenuation of acoustic waves in water decreases with increasing temperature. By raising the temperature of the water, it is possible to reduce the attenuation. Using this feature, and by stretching the existing technology to its limits, the reflection SAM has been operated in water at 3.5 GHz, with a corresponding wavelength of 0.425 μm [96]. The utilization of the nonlinear properties of the coupling liquid has been proposed to enhance resolution. The generation of harmonics makes possible an improvement in the resolution of the microscope by at least a factor of 1.4 [97]. Finding a liquid that has a lower velocity, a lower absorption coefficient, or preferably both, further improves the resolution. One possibility is to use cryogenic fluids such as super-fluid helium. In liquid helium at 0.1 K, sound velocity is equal to 238 m/s, and attenuation is so small that it becomes negligible. In this type of cryogenic SAM, operating with 8 GHz frequency, the resolution of micrographs obtained was better than 0.025 μm [98]. At this level, cryogenic acoustic microscopy as a research tool may offer an alternative to electron microscopy.

The ability to image below the surface of solids is another attractive property of the acoustic microscope. Many

TABLE 2—Practical resolution limits and penetration depth in a SAM [100].

Operation Frequency	Resolution Limit	Penetration Depth
20 MHz	100 μm	4 mm
200 MHz	8 μm	300 μm
1,000 MHz	1.5 μm	25 μm
2,000 MHz	0.7 μm	10 μm

materials that are opaque to light are transparent to acoustic waves. This property of acoustic waves has long been exploited in ultrasonic nondestructive testing. Indeed, due to this ability, acoustic microscopy provides valuable insights regarding material structures and subsurface imaging, which cannot be obtained by any other way. The penetration ability of a SAM can be estimated as equal to the penetration of excited surface waves, which is about the same magnitude as the wavelength of the surface waves. The practical penetration depth depends on the elastic parameters of the object, the signal-to-noise ratio, and the operating frequency of the acoustic microscope [99]. A higher acoustic mismatch between the object and liquid will lower the penetration depth, and a higher signal-to-noise ratio will improve it. Some changes in the parameters of the acoustic lens system, such as optimizing the lens opening angle, can maximize penetration for a given material. On the other hand, increasing the operating frequency improves the resolution, but reduces the penetration due to the increase in attenuation with frequency. Trade-offs between resolution and penetration depth must be made for acoustic microscope instruments. At the moment, the most promising frequency range for subsurface imaging analysis would appear to be 10–150 MHz, where penetration up to a few mm is easily attained. Table 2 gives practical resolution limits and penetration depths for copper and brass with a SAM [100].

The significance of the scanning acoustic microscope does not lie in its resolution alone. There is a stronger interest, image contrast. In acoustic microscopy the near-surface of the specimen is examined, and therefore the acoustic image contains information about the way that acoustic waves interact with the properties of the specimen. Image contrast observed in acoustic microscopy can be related to the elastic properties at the surface as well as below the surface of the sample. Thus fringes are seen as an interference effect associated with waves that can be excited in the surface of a specimen. Interpretation of the contrast of these fringes is not a simple matter. One cannot simply say that a brighter area corresponds to a higher (or lower) density, or to a greater (or smaller) elastic modulus. Moreover, the contrast varies very sensitively with the distance between the lens and the surface of the specimen. This behavior is best visualized as a $V(z)$ curve. The $V(z)$ effect is a "source of contrast" and is used to record quantitative information on the elastic properties of a specimen with microscopic precision. By mechanically scanning the object plane, one can obtain the scanning image of a specimen at the surface or subsurface of an object. Rather than scan in the plane, one keeps the lens and object at a fixed (x,y) position and translates the lens towards the object in the z direction; one then observes a series of oscillations in the transducer video output as a function of z. This dependence of the variation of the signal output, V, on the defocus z is known as the acoustic material signature or simply $V(z)$ curve. Historically, the study of this effect was pioneered experimentally by Weglein and co-workers [101] and theoretically by a number of authors [102,103]. This effect gives the acoustic microscope an important edge over the optical microscope.

The scanning acoustic microscope is a powerful new tool for the study of the physical properties of materials and has been successfully used for imaging interior structures and for nondestructive evaluation in materials science and biology.

The acoustic microscopy's primary application to date has been for failure analysis in the multibillion-dollar microelectronics industry. The technique is especially sensitive to variations in the elastic properties of semiconductor materials, such as air gaps. SAM enables nondestructive internal inspection of plastic integrated-circuit (IC) packages, and, more recently, it has provided a tool for characterizing packaging processes such as die attachment and encapsulation. Even as ICs continue to shrink, their die size becomes larger because of added functionality; in fact, devices measuring as much as 1 cm across are now common. And as die sizes increase, cracks and delaminations become more likely at the various interfaces.

By examining the dispersion properties of surface acoustic waves, the layer thickness and mechanical properties of layered solids can be obtained using the SAM. It can be used to analyze the wear damage progress [104], and detect the defects of thermally sprayed coatings [105].

By measuring $V(z)$, which includes examining the reflectance function of solid material, measuring the phase velocity and attenuation of leaky surface acoustic waves at the liquid-specimen boundary, the SAM can be used in determining the elastic constants of the material.

By examining the dispersion properties of surface acoustic waves, the layer thickness and mechanical properties of layered solids can be obtained using the SAM.

Many biological materials have a wider range of values for their elastic properties—which vary as much as two orders of magnitude—than for their optical properties, whose variation is only 0.5 %. Thus, optical microscopes have a limited contrast capability. Specimens must be prepared with appropriate stains designed to bring out particular features of the sample, such as specific pathologies or biochemical processes. Acoustic microscopy, however, provides a sensitive tool for imaging soft-tissue structures without the need for staining or elaborate sample preparation. Acoustic microscopy could provide an immediate assessment of pathology long before conventional methods. Acoustic microscopy of cells or tissue in culture enables scientists to examine living structures without killing them, as happens using optical means when tissue requires the use of light at extremely high frequencies to obtain adequate resolution, which in turn damages or destroys the cells.

6.3 Laser Doppler Vibrometer (LDV)

Laser Doppler Vibrometry (LDV) is a sensitive laser optical technique well suited for noncontact dynamic response measurements of microscopic structures. Up to now, this technology has integrated the micro-scanning function for

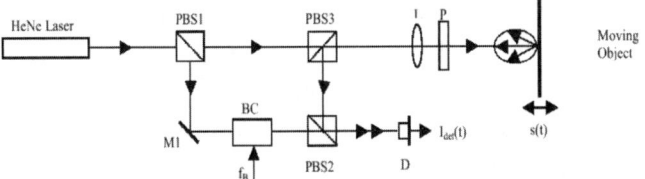

Fig. 43—Schematic of a modified Mach-Zehnder Interferometer.

automated scan measurement and display of deflection shapes with sub-nanometer resolution.

Laser Doppler Technique was first reported by Yeh and Cummins in 1964 [106]. Now it has been a valuable technique for noncontact vibration measurements on large, macroscopic structures. The earliest designs dating over 20 years ago were used for measuring resonance and mode shapes on large objects such as automotive components. Use of LDV for MEMS research began in the early 1990s at universities such as Stanford and Cornell. The first systems used conventional single-point LDV systems with relatively coarse, manual beam-positioning capability for the micron scale needed on MEMS structures. Microscope-based scan measurements were first conducted in 1996 with the group at Cornell University (McDonald et al.) using a scanning vibrometer system pointed through a microscope for a scan measurement on MEMS [107]. The advantages of the microscope-based technique for LDV measurements were clearly understood given the micron level laser spot sizes and capability to position the spot accurately anywhere on the visible microscope image. The first prototype microscope adapter for the LDV was developed from this collaboration and made commercially available in 1998. Soon following, the microscope-based scanning vibrometer was developed and released in the year 2000 [108]. Laser Doppler Vibrometry (LDV) is now an established tool for characterization of numerous MEMS technologies such as micro-optics, accelerometers, actuators, gyros, oscillators, and fluidic pumps.

Laser-Doppler vibrometry relies on the fact that light back-scattered from a moving target contains information about its velocity and displacement. Displacement of the surface modulates the phase of the light wave while instantaneous velocity shifts the optical frequency. As the optical frequency of the laser is far too high to demodulate directly (about 5×10^{14} Hz), interferometric techniques are employed to reveal the measurement quantities. In an interferometer, the received light wave is mixed with a reference beam so that the two signals heterodyne onto the surface of a photodetector. The basic arrangement of a modified Mach-Zehnder interferometer, which is used as an optical sensor in the majority of all laser Doppler vibrometers, is depicted in Fig. 43.

In the interferometer, coherent light emitted from the HeNe laser is split into measurement and reference beams by polarizing beam splitter PBS1. While the reference beam is directed via mirror M1, Bragg cell BC and PBS2 directly to the photodetector D, the measurement beam is directed to the vibrating target via PBS3, focusing lens L and quarter-wave plate P. The polarized back-scattered portion (collected by lens L) is directed to the detector via PBS3 and PBS2. The Bragg cell BC is an acousto-optical component, which pre-shifts the optical frequency of the reference beam by the frequency f_0 of an electric control signal.

The resulting intensity on the surface of the photodetector is determined by relative phase and frequency of the heterodyning light waves. Obviously, the phase difference between reference and measurement beam depends on the optical length difference between the reference and measurement paths, which changes with the target displacement $s(t)$. In case of a stationary target, the output current of the detector $i_{\text{det}}(t)$ is given by

$$i_{\text{det}}(t) = I_{DC} + \hat{i}\cos(2\pi f_0 t + \varphi_0) \quad (32)$$

where I_{DC} = DC component, $\hat{i}$ = AC amplitude, f_0 = Bragg cell drive frequency, and φ_0 = offset phase angle, defined by the initial object position.

The second term of Eq (32) represents a high-frequency signal at frequency f_0, which is a characteristic of the so-called heterodyne interferometer. This signal can carry both direction-sensitive frequency and phase modulation information resulting from target motion. In case of a moving target, displacement $s(t)$ results in a phase modulation, i.e., φ_0 becomes superimposed by a time-dependent portion $\varphi_m(t)$:

$$\varphi_m(t) = \frac{4\pi s(t)}{\lambda} \quad (\lambda = \text{laser wavelength}) \quad (33)$$

A phase modulation can also be expressed as frequency modulation. The corresponding frequency deviation is the time derivative of the modulated phase angle $\varphi_m(t)$. According to the basic relationships $d\varphi/dt = 2\pi f$ and $ds/dt = v$, object velocity $v(t)$ results in a frequency deviation $\Delta f(t)$ with respect to the carrier frequency f_0, commonly known as the Doppler frequency shift

$$\Delta f(t) = \frac{2v(t)}{\lambda} \quad (34)$$

As long as in the presence of negative velocity values, the absolute value of Δf does not exceed the carrier frequency f_0, i.e., $f_0 > |\Delta f|$, the resulting frequency of the detector output signal correctly preserves the directional information (sign) of the velocity vector. In the case of a vibrating object where $v(t) = v \sin 2\pi f_{vib} t$, the bandwidth of the modulated heterodyne signal is theoretically infinite, but practically estimated to be

$$BW_{het} = 2(\Delta f + f_{vib}) \quad (35)$$

Consequently, f_0 has to be at least $\Delta f_{\text{peak}} + f_{vib}$. With f_0 chosen at 40 MHz, this condition is maintained up to a peak velocity of about 10 m/s and a maximum vibration frequency of 2 MHz.

Equations (33) and (34) demonstrate that the motion quantities s (displacement) and v (velocity) are encoded in phase and frequency modulation of the detector output signal, purely referenced to the laser wavelength λ. To be able to recover the time histories $s(t)$ and $v(t)$ from the modulated detector signal, adequate phase and frequency demodulation techniques, or both, are utilized in the signal decoder blocks of a laser vibrometer.

By scanning the laser beam in the x and y directions, the LDV technique can be extended to full area measurement and display. This technique scans an area on a point-by-

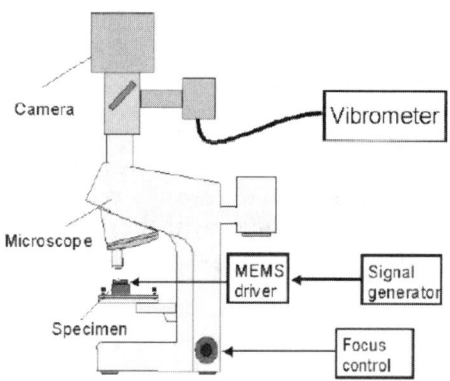

Fig. 44—Polytec Micro Scanning Vibrometer.

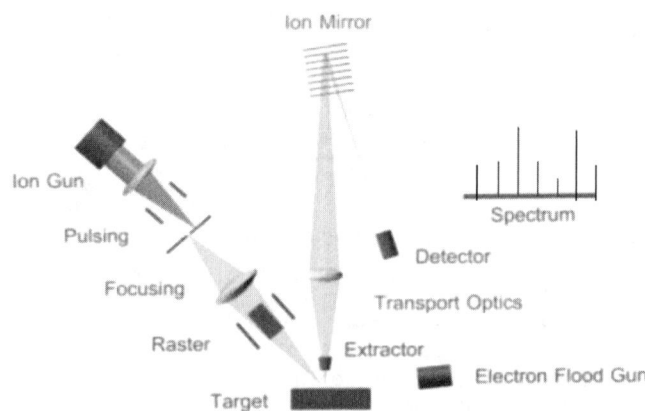

Fig. 45—The schematic of a Time-of-Flight Secondary Ion Mass Spectrometer.

point basis to measure the velocity field of the structure. From these data the operating deflection shape at any given frequency or time sample can be determined. For moving the laser beam across smaller micron scale structures, a mirror system with piezo-electrical actuation is commonly used. In conjunction with a proper relay lens, the system allows scanning of small flat fields with parallel scanning beams. Such a scanning system has been designed for operation through microscopes. The device is attached to the C-mount of a compound microscope and can measure the response of micron-sized mechanical devices. The setup for this is illustrated in Fig. 44 [108].

The advantages of microscope-based LDV technology are realized, including real-time analog output, high resolution (picometer), small spot size (μm), wide frequency range (MHz), wide dynamic range (160 dB), and high accuracy. With the scanning ability to automatically acquire, analyze, and reconstruct complex vibration modes, the Micro Scanning Laser Vibrometer is an ideal tool for displaying out-of-plane deflections.

6.4 Time-of-Flight Secondary Ion Mass Spectrometry (TOF-SIMS)

Secondary Ion Mass Spectrometry (SIMS) is a surface sensitive instrument for mass analysis, based upon the emission of positively and negatively charged atomic and molecular surface species. Mass analysis of these secondary ions by time-of-flight (TOF) instrumentation guarantees parallel mass registration, high mass range, high mass resolution at high transmission, and precise mass determination. In combination with a fine focused primary ion beam, this results in an extreme sensitivity (<1 ppm of one ML) combined with high depth (1 nm) and lateral (<50 nm) resolution. Today TOF-SIMS is a mature and well established analytical technique. Its extreme sensitivity in combination with its unique sensitivity for even large and involatile molecular species makes the technique indispensable in many high technology areas, ranging from microelectronic, chemistry and material sciences, to nanotechnology and life sciences.

TOF-SIMS was pioneered by Professor Benninghoven and his group in the early 1980s [109], originally developed in static mode and applied for the chemical analysis (elemental as well as molecular) of the uppermost monolayer of solid surfaces [110]. By the introduction and further development of the dual beam technique, the TOF-SIMS can be operated in the static mode as well as in high performance depth profiling modes [111]. Continuous improvement in TOF-SIMS instrumentation over the past two decades resulted in impressive sensitivities for elements as well as for molecular surface species and in excellent depth as well as lateral resolutions. Today more than 200 TOF-SIMS instruments are operated worldwide, particularly in industry, but also in many university and government laboratories.

As mentioned previously, SIMS is based upon the emission of positively and negatively charged surface species (atoms, clusters, molecules, fragments), which are sputtered from any solid surface under the bombardment of energetic (0.1–10 keV range) primary ions [112]. Nearly all of these secondary ions originate from the uppermost monolayer of the bombarded sample. Their mass analysis supplies detailed information on the chemical composition of the sample surface. TOF mass spectrometry is based on the fact that ions with the same energy but different masses travel with different velocities. Figure 45 presents a schematic layout for a Time-of-Flight Secondary Ion Mass Spectrometer (TOF-SIMS).

The mass separated, pulsed, and focused primary ions with the energy of 1–25 keV, typically liquid metal ions such as Ga^+, Cs^+, and O^-, are used to bombard the sample surface, causing the secondary elemental or cluster ions to emit from the surface. The secondary ions are then electrostatically accelerated into a field-free drift region with a nominal kinetic energy of:

$$E_k = eV_0 = mv^2/2 \qquad (36)$$

where V_0 is the accelerating voltage, m the mass of ion, v the flight velocity of ion, e its charge. It is obviously seen from the above formula that the ion with lower mass has higher flight velocity than one with higher mass, thus they will reach the secondary-ion detector earlier. As a result, the mass separation is achieved according to the flight time t from the sample to the detector. The flight time t is expressed by:

$$t = L/(2eV_0/m)^{1/2} \qquad (37)$$

where L is the effective length of the mass spectrometer. A variety of mass ions is recorded by the detector with the time sequence to give the SIMS spectrum.

TOF-SIMS is operated in three different modes, for (large area) surface mass spectrometry, for surface imaging, and for depth profiling. For surface spectrometry and imaging, static SIMS [111] has to be applied. These operation modes are characterized by a very low primary ion dose density, guaranteeing that only a small fraction of the uppermost monolayer of the sample is removed during the whole process of analysis. In surface spectrometry, the primary ion beam is rastered over a larger area, and all secondary ion spectra are summed up. In the imaging mode for each pixel, the complete spectrum is storaged. Images for any mass and any selected surface area can be reconstructed. Depth profiling requires fast surface erosion, i.e., high primary ion dose densities (dynamic SIMS). For depth profiling, TOF-SIMS instruments are operated with two separate primary ion beams (dual beam technique): a high current rastered sputter beam generates a flat crater, meanwhile a fine focused analysis beam is used for analyzing or mapping continuously the center area of the actual crater bottom [113].

State-of-the-art TOF-SIMS instruments feature surface sensitivities well below one ppm of a mono layer, mass resolutions well above 10,000, mass accuracies in the ppm range, and lateral and depth resolutions below 100 nm and 1 nm, respectively. They can be applied to a wide variety of materials, all kinds of sample geometries, and to both conductors and insulators without requiring any sample preparation or pretreatment. TOF-SIMS combines high lateral and depth resolution with the extreme sensitivity and variety of information supplied by mass spectrometry (all elements, isotopes, molecules). This combination makes TOF-SIMS a unique technique for surface and thin film analysis, supplying information which is inaccessible by any other surface analytical technique, for example EDX, AES, or XPS.

TOF-SIMS can be applied to identify a variety of molecular fragments, originating from various molecular surface contaminations. It also can be used to determine metal trace concentrations at the surface. The use of an additional high current sputter ion source allows the fast erosion of the sample. By continuously probing the surface composition at the actual crater bottom by the analytical primary ion beam, multi element depth profiles in well defined surface areas can be determined. TOF-SIMS has become an indispensable analytical technique in modern microelectronics, in particular for elemental and molecular surface mapping and for multielement shallow depth profiling.

The successful application of TOF-SIMS in chemistry and material sciences is based upon its capability to identify, with high sensitivity, any sort of elements and molecular surface species. The high element and isotope sensitivity, even for the smallest atom of hydrogen, has opened manifold applications in many different fields of material sciences, including the detection of hydrogen, oxygen, carbon, nitrogen, and other elements in a wide variety of materials such as metals, alloys, oxide, glasses, etc. In these applications, surface spectrometry and mapping as well as depth profiling and 3-D analysis have to be applied.

Today the successful application of TOF-SIMS in chemistry and material sciences covers all kinds of organic and inorganic materials—functionalized surfaces as well as monomolecular or multilayer coatings with lateral structures down to the 50 nm range. TOF-SIMS is here applied in the areas where surfaces and their chemical composition play an important role, not only for research and development but most successfully also for routine product control and failure analysis.

In nanotechnology, dimensions of interest are shrinking from the μm to the nm range. For many microelectronic devices, such as laterally structured surfaces, particles, sensors, their physical as well as their chemical properties are decisively determined by their chemical composition. Its knowledge is mandatory for understanding their behavior, as well as for their successful and reliable technical application. This presents a challenge for TOF-SIMS, because of its demand for the unique combination of spatial resolution and sensitivity.

TOF-SIMS has important potentials in many areas of life science, in fundamental and applied research as well as in product development and control. This holds for the characterization of biological cells and tissues, of sensor and microplate arrays, of drug delivery systems, of implants, etc. In all these areas, relevant surfaces feature a very complex composition and structure, requiring the parallel detect ion of many different molecular species as well as metal and other elements, with high sensitivity and spatial resolution requirements, which are exactly met by TOF-SIMS.

References

[1] Tabor, D., and Winterton, R. H., "The Direct Measurement of Normal and Retarded van der Waals Forces," *Proc. R. Soc. London, Ser. A*, Vol. 312, No. 1511, 1969, pp. 435–450.

[2] Binning, G., Rohrer, H., Gerber, C., and Weibel, E., "Surface Studies by Scanning Tunneling Microscopy," *Phys. Rev. Lett.*, Vol. 49, No. 1, 1982, pp. 57–61.

[3] Goher, R., and Cammeron, A., "The Mapping of Elastohydrodynamic Contact," *ASLE Trans.*, Vol. 10, No. 2, 1967, pp. 215–225.

[4] Luo, J. B., "Study on the Measurement and Experiments of Thin Film Lubrication," Ph.D. thesis, Tsinghua University, Beijing, China, 1994.

[5] Luo, J. B., Wen, S. Z., and Huang, P., "Thin Film Lubrication, Part I: The Transition Between EHL and Thin Film Lubrication," *Wear*, Vol. 94, No. 1–2, 1996, pp. 107–115.

[6] Kong, F. R., "Experimental Technique and Characteristics of Mixed Lubrication," Ph.D. thesis, Tsinghua University, Beijing, 1993.

[7] Zou, Q., Kong, F. R., and Wen, S. Z., "A New Method for Measuring the Film Thickness of Mixed Lubrication in Line Contacts," *Tribo. Trans.*, Vol. 38, No. 4, 1995, pp. 869–874.

[8] Zhou, K., Leng, Y. S., and Hu, Y. Z., "Surface Force Apparatus and Its Application to Research of the Surface Contact," *Chinese Science Bulletin*, Vol. 44, No. 11, 1999, pp. 992–995.

[9] Lu, X. C., Wen, S. Z., and Meng, Y. G., et al., "A Friction Force Microscope Employing Laser Beam Deflection for Force Detection," *Chinese Science Bulletin*, Vol. 41, No. 22, 1996, pp. 1873–1876.

[10] Johnston, G. J., Wayte, R., and Spikes, H. A., "The Measurement and Study of Very Thin Lubricant Films in Concentrated Contact," *STLE, Trib. Trans.*, Vol. 34, No. 2, 1991, pp. 187–194.

[11] Westlake, F. J., "An Interferometric Study of Ultra-Thin Fluid Films," Ph.D. thesis, Univ. of London, 1970.

[12] Guangteng, G., and Spikes, H. A., "Properties of Ultra-Thin

Lubricating Films Using Wedged Spacer Layer Optical Interferometry," *Proceedings, 14th Leeds-Lyon Symposium on Tribology, Interface Dynamics*, Leeds, 1988, pp. 275–279.

[13] Ratoi, M., Anghel, V., Bovington, C., and Spikes, H. A., "Mechanisms of Oiliness Additives," *Tribol. Int.*, Vol. 33, No. 3-4, 2000, pp. 241–247.

[14] Wedeven, L., "Optical Measurements in Elastohydrodynamic Rolling Contacts," Ph.D. thesis, Univ. of London, 1970.

[15] Kauffman, A. M., "A Simple Immersion Method to Determine the Refractive Index of Thin Silica Films," *Thin Solid Films*, Vol. 1, No. 1, 1967, pp. 131–136.

[16] Born, M., and Wolf, E., *Principle of Optics*, 5th ed., Pergamon Press, 1975, pp. 257–270.

[17] Luo, J. B., Shen, M. W., and Wen, S. Z., "Tribological Properties of Nanoliquid Film Under an External Electric Field," *J. Appl. Phys.*, Vol. 96, No. 11, 2004, pp. 6733–6738.

[18] Luo, J. B., Huang, P., and Wen, S. Z., "Characteristics of Liquid Lubricant Films at the Nano-Scale," *ASME Trans., J. of Tribology*, Vol. 121, No. 4, 1999, pp. 872–878.

[19] Hartl, M., Krupka, I., Poliscuk, R., and Liska, M., "An Automatic System for Real-Time Evaluation of EHD Film Thickness and Shape Based on the Colorimetric Interferometry," *STLE Tribo. Trans.*, Vol. 42, No. 2, 1999, pp. 303–309.

[20] Hartl, M., Krupka, I., and Liska, M., "Differential Colorimetry: Tool for Evaluation of Chromatic Interference Patterns," *Optical Engineering*, Vol. 36, No. 9, 1997, pp. 2384–2391.

[21] Hartl, M., Krupka, I., Poliscuk, R., Molimard, J., Querry, M., and Vergne, P., "Thin Film Colorimetric Interferometry," *STLE Tribo. Trans.*, Vol. 44, No. 2, 2001, pp. 270–276.

[22] Goher, R., and Cammeron, A., "The Mapping of Elastohydrodynamic Contact," *ASLE Trans.*, Vol. 10, No. 2, 1967, pp. 215–225.

[23] Tabor, D., and Winterton, R. H., "The Direct Measurement of Normal and Retarded van der Waals Forces," *Proc. R. Soc. London, Ser. A*, Vol. 312, No. 1511, 1969, pp. 435–450.

[24] Israelachvili, J. N., and Tabor, D., "The Measurement of van der Waals Dispersion Forces in the Range 1.5 to 130 nm," *Proc. R. Soc. London, Ser. A*, Vol. 331, 1972, pp. 19–38.

[25] Israelachvili, J. N., and Adams, G. E., "Measurement of Forces Between Two Mica Surfaces in Aqueous Electrolyte Solutions in the Range 0–100 nm," *J. Chem. Faraday Trans.*, Vol. I74, 1978, pp. 975–1001.

[26] Israelachvili, J. N., "Direct Measurements of Forces Between Surfaces in Liquids at the Molecular Level," *Proc. Natl. Acad. Sci. U.S.A.*, Vol. 84, 1987, pp. 4722–4736.

[27] Israelachvili, J. N., and McGuiggan, P. M., "Adhesion and Short-Range Forces Between Surfaces. Part I: New Apparatus for Surface Force Measurement," *J. Mater. Res.*, Vol. 5, No. 10, 1990, pp. 2223–2231.

[28] Klein, J., "Forces Between Mica Surfaces Bearing Adsorbed Macromolecules in Liquid Media," *J. Chem. Soc., Faraday Trans.*, Vol. I79, 1983, pp. 99–118.

[29] Parker, J. L., Christenson, H. K., and Ninham, B. W., "A Device for Measuring the Force and Separation Between Two Surfaces Down to Molecular Separation," *Rev. Sci. Instrum.*, Vol. 60, No. 10, 1989, pp. 3135–3138.

[30] Prieve, D. C., Luo, F., and Lanni, F., "Brownian Motion of a Hydrosol Particle in a Colloidal Force Field," *Faraday Discuss. Chem. Soc.*, Vol. 83, 1987, pp. 297–308.

[31] Tonock, A., Georges, J. M., and Loubet, J. L., "Measurement of Intermolecular Forces and the Rheology of Dodecane Between Alumina Surfaces," *J. Colloid Interface Sci.*, Vol. 126, No. 1, 1988, pp. 150–163.

[32] Tolansky, S., *Multiple Beam Interferometry of Surfaces and Films*, Oxford University Press, 1948.

[33] Israelachvili, J. N., Homola, A. M., and McGuiggan, P. M., "Dynamic Properties of Molecularly Thin Liquid Films," *Science*, Vol. 240, 1988, pp. 189–191.

[34] Homola, A. M., Israelachvili, J. N., McGuiggan, P. M., and Gee, M., "Fundamental Experimental Studies in Tribology: The Transition from 'Interfacial' Friction of Undamaged Molecularly Smooth Surfaces to 'Normal' Friction with Wear," *Wear*, Vol. 136, No. 1, 1990, pp. 65–83.

[35] Alsten, J. V., and Granick, S., "Friction Measured with a Surface Forces Apparatus," *Trib. Trans.*, Vol. 32, 1988, pp. 246–250.

[36] Peachey, J., Alsten, J., and Granick, S., "Design of an Apparatus to Measure the Shear Response of Ultrathin Liquid Films," *Rev. Sci. Instrum.*, Vol. 62, No. 2, 1991, pp. 463–473.

[37] Homola, J., "Thin Films Study by Means of Optically Excited Surface Plasmons, Nano'94—International Conference on Nanometrology Scanning Probe Microscopy and Related Techniques," Proceedings, Nano'94, Brno, Czech Rep., 1994, pp. 84–87.

[38] Israelachvili, J. N., "Adhesion, Friction and Lubrication of Molecularly Smooth Surfaces," *Fundamentals of Friction*, I. L. Singer and H. M. Pollock, Eds., Kluwer Academic Publishers, 1991, pp. 351–385.

[39] Chen, Y. L., Gee, M. L., and Helm, C. A., et al., "Effects of Humidity on the Structure and Adhesion of Amphiphilic Monolayers on Mica," *J. Phys. Chem.*, Vol. 93, 1989, pp. 7057–7059.

[40] Berman, A., and Israelachvili, J. N., "Surface Forces and Microrheology of Molecularly Thin Liquid Films," *Handbook of Micro/Nanotribology*, 2nd ed., B. Bhushan, Ed., CRC Press, Boca Raton, FL, 1999.

[41] Granick, S., "Motion and Relations of Confined Liquids," *Science*, Vol. 253, 1991, pp. 1374–1379.

[42] Giaever, I., "Energy Gap in Superconductors Measured by Electron Tunneling," *Phys. Rev. Lett.*, Vol. 5, 1960, pp. 147–148.

[43] Binning, G., Rohrer, H., Gerber, C., and Weibel, E., "Surface Studies by Scanning Tunneling Microscopy," *Phys. Rev. Lett.*, Vol. 9, 1982, pp. 57–61.

[44] Woedtke, S., "Inst. F. Exp. U. Ang. Phys." der CAU Kiel, Ph.D. thesis, 2002.

[45] Bhushan, B., Fuchs, H., and Kawata, S., *Applied Scanning Probe Methods V*, Springer-Verlag, Berlin, Heidelberg, 2007.

[46] Güntherodt, H. J., and Wiesendanger, R., *Scanning Tunneling Microscopy I*, Springer-Verlag, Berlin, 1994.

[47] Vasileios, K., "Physical Properties of Grafted Polymer Monolayers Studied by Scanning Force Microscopy: Morphology, Friction, Elasticity," Dissertations, University of Groningen, Denmark, 1997.

[48] Binning, G., Quate, C. F., and Gerber, C., "Atomic Force Microscopy," *Phys. Rev. Lett.*, Vol. 56, 1986, pp. 930–933.

[49] Meyer, G., and Amer, N. M., "Novel Optical Approach to Atomic Force Microscopy," *Appl. Phys. Lett.*, Vol. 53, 1988, pp. 1045–1047.

[50] Bhushan, B., *Handbook of Micro/Nano Tribology*, 2nd ed., CRC Press, Boca Raton, FL, 1999.

[51] Martin, Y., Williams, C. C., and Wickramasinghe, H. K., "Atomic Force Microscope-Force Mapping and Profiling on Sub 100-A Scale," *J. Appl. Phys.*, Vol. 61, 1987(a), pp. 4723–4729.

[52] McClelland, G. M., Erlandsson, R., and Chiang, S., "Atomic Force Microscopy: General Principles and a New Implementation," *Review of Progress in Quantitative Nondestructive Evaluation*, Vol. 6B, 1987, pp. 1307–1314.

[53] Maivald, P., Butt, H. J., Prater, S. A., and Drake, C. B., et al., "Using Force Modulation to Image Surface Elasticities with the Atomic Force Microscope," *Nanotechnology*, Vol. 2, 1991, pp. 103–106.

[54] Radmacher, M., Tillman, R. W., Fritz, M., and Gaub, H. E., "From Molecules to Cells: Imaging Soft Samples with the Atomic Force Microscope," *Science*, Vol. 257, 1992, pp. 1900–1905.

[55] Mate, C. M., McClelland, G. M., Erlandsson, R., and Chiang, S., "Atomic Scale Friction of a Tungsten Tip on a Graphite Surface," *Phys. Rev. Lett.*, Vol. 59, 1987, pp. 1942–1945.

[56] Lu, X. C., Wen, S. Z., and Dai, C. C., et al., "A Friction Force Microscopy Study of Au Film, CD-Disk and LB Film," *Chinese Science Bulletin*, Vol. 41, No. 11, 1996, pp. 900–903.

[57] Doerner, M. F., and Nix, W. D., "A Method for Interpreting the Data from Depth-Sensing Indentation Instruments," *J. Mater. Res.*, Vol. 1, No. 4, 1986, pp. 601–609.

[58] Pharr, G. M., and Oliver, W. C., "Measurement of Thin Film Mechanical Properties Using Nanoindentation," *MRS Bull.*, Vol. 17, 1992, pp. 28–33.

[59] Oliver, W. C., and Pharr, G. M., "An Improved Technique for Determining Hardness and Elastic Modulus Using Load and Displacement Sensing Indentation Experiments," *J. Mater. Res.*, Vol. 7, 1992, pp. 1564–1583.

[60] Loubet, J. L., Georges, J. M., Marchesini, O., and Meille, G., "Vickers Indentation Curves of Magnesium Oxide (Mgo)," *J. Tribol.*, Vol. 106, No. 1, 1984, pp. 43–48.

[61] Oliver, W. C., "Progress in the Development of a Mechanical Properties Microprobe," *MRS Bull.*, Vol. 11, No. 5, 1986, pp. 15–19.

[62] Nix, W. D., "Mechanical Properties of Thin Films," *Metall. Trans.*, Vol. 20A, 1989, pp. 2217–2245.

[63] Pethica, J. B., Hutchings, R., and Oliver, W. C., "Hardness Measurement at Penetration as Small as 20 nm," *Philos. Mag. A*, Vol. 48, 1983, pp. 593–606.

[64] Nastasi, M., Hirvonen, J. P., Jervis, T. R., Pharr, G. M., and Oliver, W. C., "Surface Mechanical Properties of C Implanted Ni," *J. Mater. Res.*, Vol. 3, 1988, pp. 226–232.

[65] Pharr, G. M., "Measurement of Mechanical Properties by Ultra-Low Load Indentation," *Mater. Sci. Eng., A*, Vol. 253, 1998, pp. 151–159.

[66] Bhushan, B., and Li, X. D., "Nanomechanical Characterisation of Solid Surfaces and Thin Films," *Int. Mater. Rev.*, Vol. 48, No. 3, 2003, pp. 125–163.

[67] Pharr, G. M., Oliver, W. C., and Brotzen, F. R., "On the Generality of the Relationship Among Contact Stiffness, Contact Area, and Elastic Modulus During Indentation," *J. Mater. Res.*, Vol. 7, No. 3, 1992, pp. 613–617.

[68] Li, X. D., and Bhushan, B. A., "Review of Nanoindentation Continuous Stiffness Measurement Technique and Its Applications," *Mater. Charact.*, Vol. 48, 2002, pp. 11–36.

[69] Qi, J., Luo, J. B., Wang, K. L., and Wen, S. Z., "Mechanical and Tribological Properties of Diamond-Like Carbon Films Deposited by Electron Cyclotron Resonance Microwave Plasma Chemical Vapor Deposition," *Tribology Letters*, Vol. 14, No. 2, 2003, pp. 105–109.

[70] Bhattacharya, A. K., and Nix, W. D., "Analysis of Elastic and Plastic Deformation Associated with Indentation Testing of Thin Films on Substrates," *Int. J. Solids Struct.*, Vol. 24, 1988, pp. 1287–1298.

[71] Bhushan, B., and Li, X. D., "Nanomechanical Characterisation of Solid Surfaces and Thin Films," *Int. Mater. Rev.*, Vol. 48, No. 3, 2003, pp. 125–164.

[72] Harding, D. S., Olive, W. C., and Pharr, G. M., "Cracking During Nanoin: Denation and Its Use in the Measurement of Fracture Toughness," *Materials Research Society Symposium Proceedings*, Vol. 356, 1995, pp. 663–668.

[73] Pharr, G. M., Harding, D. S., and Oliver, W. C., *Mechanical Properties and Deformation Behavior of Materials Having Ultra-Fine Microstructures*, M. Nastasi, D. M. Parkin, and H. Gleiter, Eds., Kluwer Academic Publishers, Netherlands, 1993.

[74] Lawn, B. R., Evans, A. G., and Marshall, D. B., "Elastic-Plastic Indentation Damage in Ceramics: The Median-Radial Crack System," *J. Am. Ceram. Soc.*, Vol. 63, 1980, pp. 574–581.

[75] Lankford, J., and Davidson, D. L., "The Crack-Initiation Threshold in Ceramic Materials Subject to Elastic/Plastic Indentation," *J. Mater. Sci.*, Vol. 14, 1979, pp. 1662–1668.

[76] Heavens, O. S., "Some Factors Influencing Adhesion of Films Produced by Vacuum Evaporation," *J. Phys. Rad.*, Vol. 11, 1950, pp. 355–359.

[77] Benjamin, P., and Weaver, C., "Measurement of Adhesion of Thin Films," *Proc. R. Soc. London, Ser. A*, Vol. 254, 1960, pp. 163–176.

[78] Burnett, P. J., and Rickerby, D. S., "The Relationship Between Hardness and Scratch Adhesion," *Thin Solid Films*, Vol. 154, 1987, pp. 403–416.

[79] Bull, S. J., and Rickerby, D. S., "New Development in the Modelling of the Hardness and Scratch Adhesion of Thin Films," *Surf. Coat. Technol.*, Vol. 42, 1990, pp. 149–164.

[80] Serizawa, A., Feng, Z. P., and Kawara, Z., "Two-Phase Flow in Microchannels," *Exp. Therm. Fluid Sci.*, Vol. 26, No. 6–7, 2002, pp. 703–714.

[81] Zhang, L., "Experimental Research on the Properties of SiO_2-Hydrosol Flows with Fluorescence Technique," Beijing, Ph.D thesis, Tsinghua University, 2004.

[82] Du, H. S., "Study on Elastohydrodynamic Lubrication with Fluorescent Technique," Beijing, Ph.D thesis, Tsinghua University, 1999.

[83] Santiago, J. G., Wereley, S. T., and Meinhart, C. D., et al., "A Particle Image Velocimetry System for Microfluidics," *Exp. Fluids*, Vol. 25, No. 4, 1998, pp. 316–319.

[84] Kim, M. J., Beskok, A., and Kihm, K. D., "Electro-Osmosis-Driven Micro-Channel Flows: A Comparative Study of Microscopic Particle Image Velocimetry Measurements and Numerical Simulations," *Exp. Fluids*, Vol. 33, No. 1, 2002, pp. 170–180.

[85] Sinton, D., Canseco, C. E., and Ren, L. Q., et al., "Direct and Indirect Electro-Osmotic Flow Velocity Measurements in Micro-Channels," *J. Colloid Interface Sci.*, Vol. 254, No. 1, 2002, pp. 184–189.

[86] Devasenathipathy, S., Santiago, J. G., and Takehara, K., "Particle Tracking Techniques for Electrokinetic Micro-channel Flows," *Anal. Chem.*, Vol. 74, No. 15, 2002, pp. 3704–3713.

[87] Duan, J. H., Wang, K. M., and Tan, W. H., et al., "A Study of a Novel Organic Fluorescent Core-shell Nanoparticle," *Chemical Journal of Chinese Universities*, Vol. 24, No. 2, 2003, pp. 255–259.

[88] Wen, S. M., *Theories of Micro-fluidic Boundary Layers and Its Application*, Beijing, Metallurgical Industry Press, 2002.

[89] Sun, H. Q., Kang, H. G., and Li, G. W., "Planar Measurement of the Fluid: PIV," *Instrument Technique and Sensor*, Vol. 6, 2002, pp. 43–45.

[90] Zhang, X., and Xu, W. Q., "Development of Confocal Microscopy and Its Application," *Modern Scientific Instruments*, Vol. 2, 2001, pp. 21–23.

[91] Sokolov, S., *The Ultrasonic Microscope*, Doklady Akademia Nauk SSSR (in Russian), Vol. 64, 1949, pp. 333–336.

[92] Lemons, R. A., and Quate, C. F., "Acoustic Microscope—Scanning Version," *Appl. Phys. Lett.*, Vol. 24, 1974, pp. 163–165.

[93] Lemons, R. A., and Quate, C. F., "Acoustic Microscopy," *Physical Acoustics XIV*, W. P. Mason and R. N. Thurston, Eds., Academic Press, New York, 1979, pp. 2–92.

[94] Yu, Z. L., "Scanning Acoustic Microscopy and Its Applications to Material Characterization," *Rev. Mod. Phys.*, Vol. 67, No. 4, 1995, pp. 863–891.

[95] Kino, G. S., *Acoustic Waves: Devices, Imaging and Analog Signal Processing*, Prentice-Hall, Englewood Cliffs, NJ, 1987.

[96] Rugar, D., "Cryogenic Acoustic Microscopy," Ph.D. thesis, Stanford University, 1981.

[97] Rugar, D., "Resolution Beyond the Diffraction Limit in the Acoustic Microscope: A Nonlinear Effect," *J. Appl. Phys.*, Vol. 56, 1984, pp. 1338–1346.

[98] Hadimioglu, B., and Foster, J. S., "Advances in Superfluid Helium Acoustic Microscopy," *J. Appl. Phys.*, Vol. 56, 1984, pp. 1976–1980.

[99] Atalar, A., "Penetration Depth of the Scanning Acoustic Microscope," *IEEE Trans. Sonics Ultrason.*, Vol. 32, No. 2, 1985, pp. 164–167.

[100] Block, H., Heygster, G., and Boseck, S., "Determination of the OTF of a Reflection Scanning Acoustic Microscope by a Hair Crack in Glass at Different Ultrasonic Frequencies," *Optik*, Vol. 82, 1989, pp. 147–154.

[101] Weglein, R. D., "A Model for Predicting Acoustic Materials Signatures," *Appl. Phys. Lett.*, Vol. 34, No. 3, 1979, pp. 179–181.

[102] Atalar, A., "An Angular Spectrum Approach to Contrast in Reflection Acoustic Microscopy," *J. Appl. Phys.*, Vol. 49, 1978, pp. 5130–5139.

[103] Wickramasinghe, H. K., "Contrast and Imaging Performance in the Scanning Acoustic Microscope," *J. Appl. Phys.*, Vol. 50, 1979, pp. 664–672.

[104] Blau, P. J., and Simpson, W. A., "Applications of Scanning Acoustic Microscopy in Analyzing Wear and Single-Point Abrasion Damage," *Wear*, Vol. 181–183, 1995, pp. 405–412.

[105] Yang, Y. Y., Jing, Y. S., and Luo, S. Y., "SAM Study on Plasma Sprayed Ceramic Coatings," *Surface and Coatings Technology*, Vol. 91, 1987, pp. 95–100.

[106] Yeh, Y., and Cummins, H. Z., "Localized Fluid Flow Measurements with an He–Ne Laser Spectrometer," *Appl. Phys. Lett.*, Vol. 4, 1964, pp. 176–178.

[107] Lewin, A. D., "Non-Contact Surface Vibration Analysis Using a Monomode Fiber Optic Interferometer," *J. Phys. E: Sci. Instrum.*, Vol. 18, 1985, pp. 604–608.

[108] Lawton, R., "MEMS Characterization Using Scanning Laser Vibrometer," *Reliability, Testing, and Characterization of MEMS/MOEMS II*, R. Ramesham and D. M. Tanner, Eds., *Proceedings of SPIE*, Vol. 4980, 2003, pp. 51–62.

[109] Benninghoven, A., "Surface Investigation of Solids by the Static Mmethod of Secondary Ion Mass Spectroscopy (SIMS)," *Surf. Sci.*, Vol. 35, 1973, pp. 427–457.

[110] Benninghoven, A., "Developments in Secondary Ion Mass Spectroscopy and Applications to Surface Studies," *Surf. Sci.*, Vol. 53, 1975, pp. 596–625.

[111] Benninghoven, A., "Static SIMS Applications—From Silicon Single Crystal Oxidation to DNA Sequencing," *J. Vac. Sci. Technol. A*, Vol. 3, 1985, pp. 451–460.

[112] Benninghoven, A., "Surface Analysis by Secondary Ion Mass Spectroscopy (SIMS)," *Surf. Sci.*, Vol. 299/300, 1994, pp. 246–260.

[113] Benninghoven, A., and Cha, L. Z., "TOF-SIMS—A Powerful Tool for Practical Surface, Interface and Thin Film Analysis," *Vacuum of China*, Vol. 5, 2002, pp. 1–10.

[114] Hartl, M., Křupka, I., and Liška, M., "Experimental Study of Boundary Layers Formation by Thin Film Colorimetric Interferometry," *Sci. China, Ser. A: Math., Phys., Astron. Technol. Sci.*, Vol. 44(supp), 2001, pp. 412–417.

3

Thin Film Lubrication—Experimental Study

Jianbin Luo[1] and Shizhu Wen[1]

1 Introduction

1.1 Advancements in Thin Film Lubrication

OIL FILM WITH A THICKNESS IN THE NANOSCALE has been well studied from the beginning of the 1990s [1–3]. Thin film lubrication (TFL), as the lubrication regime between elastohydrodynamic lubrication (EHL) and boundary lubrication, has been proposed from 1996 [3,4]. The lubrication phenomena in such a regime are different from those in elastohydrodynamic lubrication (EHL) in which the film thickness is strongly related to the speed, viscosity of lubricant, etc., and also are different from that in boundary lubrication in which the film thickness is mainly determined by molecular dimension and characteristics of the lubricant molecules.

In lubrication history, research has been mainly focused for a long period on two fields—fluid lubrication and boundary lubrication. In boundary lubrication (BL), lubrication models proposed by Bowdeon and Tabor [5], Adamson [6], Kingsbury [7], Cameron [8], and Homola and Israelachvili [9] indicated the research progressed in the principle of boundary lubrication and the comprehension about the failure of lubricant film. In fluid lubrication, elastohydrodynamic lubrication proposed by Grubin in 1949 has been greatly developed by Dowson and Higginson [10], Hamrock and Dowson [11], Archard and Cowking [12], Cheng and Sternlicht [13], Yang and Wen [14], and so on. The width of the chasm between fluid lubrication and boundary lubrication has been greatly reduced by these works. The research on micro-EHL and mixed lubrication has been trying to complete the whole lubrication theory system. Nevertheless, the transition from EHL to boundary lubrication is also an unsolved problem in the system of lubrication theory. Thin film lubrication [3,4] bridges the EHL and boundary lubrication [15].

Thin film lubrication (TFL) investigated by Johnston et al. [1], Wen [2], Luo et al. [3,4,16–19], Tichy [20–22], Matsuoka and Kato [23], Hartal et al. [24], Gao and Spikes [25] et al. has become a new research area of lubrication in the 1990s. However, some significant progress can retrospect to 60 years ago. In the 1940s, it had been proven by using the X-ray diffraction pattern that a fatty acid could form a polymolecular film on a mercury surface and the degree of molecular order increased from outside towards the metal surface [26]. Allen and Drauglis [27] in 1996 proposed an "ordered liquid" model to explain the experimental results of Fuks on thin liquid film. However, they thought the thickness of ordered liquid is more than 1 μm, which is much larger than that shown in Refs. [4,17,18]. The surface force apparatus (SFA) developed by Israelachvili and Tabor [28] to measure the van der Waals force and later becoming a more advanced one [29] has been well used in the tribological test of thin liquid layer in molecular order. Using SFA, Alsten et al. [30], Granick [31], and Luengo et al. [32] observed that the adsorptive force between two solid surfaces was strongly related to the distance between the two solid surfaces and the temperature of the lubricant. In 1989, Luo and Yan [33] proposed a fuzzy friction region model to describe the transition from EHL to boundary lubrication. In their model, the transition region was considered as a process in which the characteristics of lubricant changed with the variation of quantitative parameters, e.g., the film thickness. Johnston et al. [1] found that EHL phenomenon did not exist with films less than 15 nm thick. Tichy [20–22] proposed the models of thin lubricant film according to the improved EHL theory. Luo and Wen [3,4,18,34,35] have got the relationship between the transition thickness from EHL to TFL and the viscosity of lubricant, and proposed a physical model of TFL, and a lubrication map of different lubrication regimes.

In the later 1990s, some other advancements have been made in experiments made by Gao and Spikes [36] and Hartl et al. [37] and in theory by Thompson et al. [38] and Hu et al. [39]. However, there are many unsolved problems in the area of nano-lubrication [40].

1.2 Definition of Thin Film Lubrication and Boundary Lubrication

The definition of different lubrication regimes is a historic problem [41]. In boundary lubrication, molecules will be absorbed on a solid surface of a tribo-pair and form a monomolecular absorbed layer as described by Hardy [42] as shown in Fig. 1(a). If the film thickness of lubricants in the contact region is from a few nanometres to tens of nanometres, different layers will be formed as shown in Fig. 1(b) proposed by Luo et al. [3,4]. The layer close to the surfaces is the adsorbed film that is a monomolecular layer. The layer in the

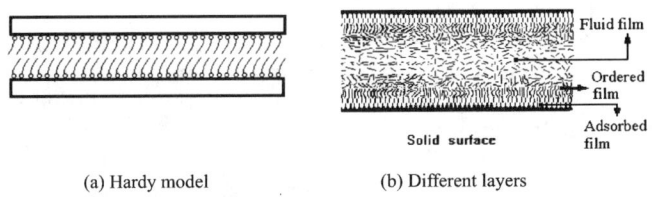

(a) Hardy model (b) Different layers

Fig. 1—Physical model of lubrication [2].

---

[1] State Key Laboratory of Tribology, Tsinghua University, Beijing, China.

center of the gap or apart from the solid surface is the dynamic fluid film which is formed by hydrodynamic effects and its thickness mainly depends on fluid factors, e.g., speed, lubricant viscosity, and pressure. The layer between the adsorbed and the dynamic fluid layers is the ordered liquid film, which is formed during the tribological process. By the solid surface force, the shearing stress, and the interaction of rod-shaped molecules with one another, the lubricant molecules form an ordered region between the adsorbed and dynamic layers.

The molecules in the adsorbed layer are more ordered and solid-like than that in the fluid state because the adhesion force between the lubricant molecules and the solid surface are mostly much stronger than that among lubricant molecules. Therefore, the ordered liquid layer is different from both the adsorbed layer and the dynamic fluid layer because it is closer to the solid surfaces than the fluid layer, but not in contact with the solid surface as the adsorbed layer. The molecular order degree of such a layer should become smaller along the direction apart from the solid surface. The tribological properties of such a layer are also different from that of boundary lubrication because they are still related to the lubricant viscosity and speed, not like that of the boundary lubrication, which does not depend on them. In another side, such a layer is also different from the hydrodynamic fluid layer because characteristics of such a layer are related to the surface force, molecular polar characters, molecular sizes, the distance from the surface, and so on. Therefore, the lubrication in such a layer is called thin film lubrication (TFL) [3,4,20,24]. In the point contact region, the lubrication regime should be divided into four types, i.e., hydrodynamic lubrication (HDL), elastohydrodynamic lubrication (EHL), thin film lubrication (TFL), and boundary lubrication (BL). For very smooth surfaces of tribo-pairs where surface roughness in the contact region is very small compared to the film thickness, the transition between the lubrication regimes will take place according to the variation of the film thickness as shown in Fig. 1(b). When the film is very thick, the fluid layer plays the leading role and the lubrication properties obey the EHL or DHL rules. As the film becomes thinner, the proportion of the ordered layer thickness to the total film thickness will become larger. When the proportion is large enough, the ordered liquid layer will play the chief role and the lubrication regime changes into TFL. After the ordered layer is crushed, the monomolecular adsorbed layer will act in a leading role and the lubrication belongs to boundary lubrication.

1.3 Lubrication Map

For lubrication engineers, it is very important to know which lubrication regime plays the leading role in the contact region. Two methods are usually used to distinguish different lubrication regimes. One is using the Streibeck curve [6]. However, it is very difficult for engineers to make sure what a lubrication regime is in the contact region just based on the friction coefficient of the lubricant without the whole Streibeck curve of its working condition because the friction coefficient of different kinds of lubricants are far away from each other. Another method is using the ratio of the film thickness h to the combined surface roughness R_a to judge the lubrication regime [43], i.e., the lubrication regime is

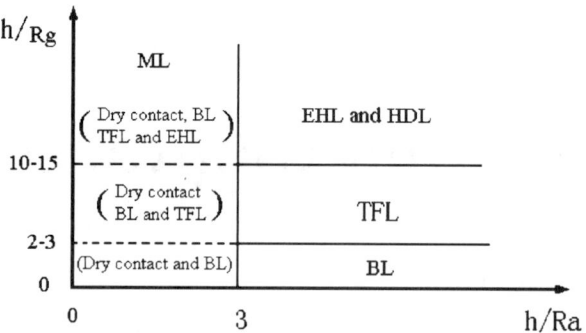

Fig. 2—Lubrication map [35] Ra: combined surface roughness, h: film thickness, Rg: effective molecular radius.

EHL or HDL if the ratio h/R_a is larger than 3, mixed lubrication if the ratio ranges from 1 to 3, and boundary lubrication when the ratio is smaller than 1. This method is very useful for engineers, but it is not applicable for the ultra-flat solid surfaces. For example, the surface roughness R_a of some smooth surfaces, e.g., silicon wafer, is usually one nanometre or smaller, and the thickness of only two molecular layers is 5 nm, e.g., $C_{15}H_{31}COOH$, so the ratio h/R_a is larger than 5, but the lubrication regime is a typical boundary lubrication. Therefore, the distinguishing method for different lubrication regimes should not only include the ratio h/R_a, but also the number of the molecular layers, or the ratio of film thickness to lubricant molecular size. Thus, a method using the ratio h/R_a and h/R_g together to distinguish the lubrication regimes as shown in Fig. 2 has been proposed by Luo et al. [34] in which h is film thickness, R_a is the combined surfaces roughness, and R_g is the effective radius of lubricant molecules. The ratio h/R_g indicates the number of molecular layers in the gap of tribo-pair. When the ratio h/R_a is larger than 3, three lubrication regimes can be formed separately according to the ratio h/R_g. The regime is boundary lubrication when the ratio h/R_g is smaller than about 3, is TFL when the ratio h/R_g is larger than 3 and lower than 10 to 15, which is related to the surface force of the solid surfaces and molecular polarity, and is EHL or HDL when the ratio h/R_g is higher than 15. The mixed lubrication will be found when the ratio h/R_a is smaller than 3. Different types of mixed lubrication with different values of the ratio h/R_g are shown in Fig. 2.

The relationship of film thickness and its influence factors in different regions with different lubrication regimes is shown in Fig. 3 when the film thickness is larger than three times of the combined surface roughness in the contact region [34]. In the EHL region, the film thickness changes with different parameters according to the EHL rules. As the speed decreases or pressure increases, these films with different viscosity or molecular length will meet different critical points, and then go into the TFL region. The pressure has a slight influence on the film thickness in such a region. These critical points are related to the viscosity of lubricant, the solid surface energy, chemical characteristics of lubricant molecules, etc. When the speed decreases and the pressure increases further, these films will meet the failure points where the liquid film will collapse. If the polar additives have been added into the lubricants, the TFL regime will be maintained at a smaller speed shown as the dotted lines in Fig. 3.

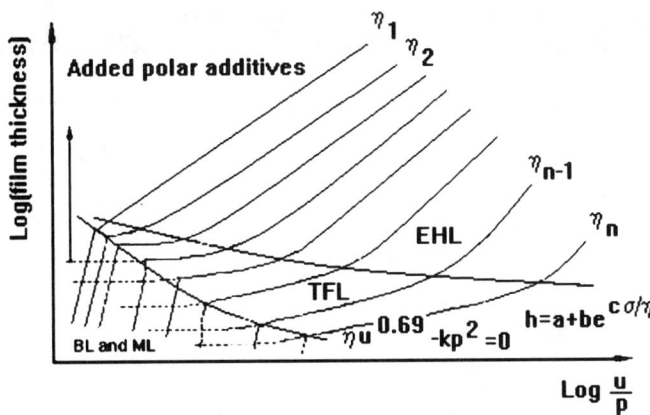

Fig. 3—Film characteristics in different regimes [35]. Viscosity: $\eta_1 > \eta_2 > \ldots > \eta_{n-1} > \eta_n$, $h/R_a > 3$.

If the speed is smaller than that of the failure point, the film thickness will suddenly decrease to a few molecular layers, or it is in boundary lubrication [34].

2 Properties of Thin Film Lubrication

2.1 Relationship between Film Thickness and Influence Factors

In elastohydrodynamic lubrication, the formula of film thickness was given by Hamrock and Dowson [44] as:

$$h = 2.69 U^{0.67} G^{0.53} W^{-0.067}(1 - 0.61 e^{-0.73k}) \quad (1)$$

where U is the speed parameter, W is the load parameter, G is the materials parameter, and k is the ellipticity of the contact bodies. In TFL, more factors should be considered to estimate the film thickness because it is not only related to load w, initial viscosity η_0, reduced modulus of elasticity E', and radius R, but also related to molecular interactions. The relationship between the film thickness and the speed in TFL can be written as:

$$h = k_1 u^\phi \quad (2)$$

where ϕ is a speed index and k_1 is a coefficient. The speed index ϕ can be obtained by Eq (3) from two measured values of the film thickness at two speeds in the experiment. If the two points of speed are close enough and other condition parameters do not change, k_1 is constant, and then ϕ, which is a key factor to distinguish EHL and TFL is:

$$\phi = \frac{\log\left(\dfrac{h_2}{h_1}\right)}{\log\left(\dfrac{u_1}{u_2}\right)} \quad (3)$$

2.1.1 Effect of Rolling Speed

The film thickness varies with the rolling speed as shown in Fig. 4 in which Curve (a) is from the measured data and Curve (b) is the measured value of thickness minus the static film thickness, that is, the thickness of fluid film. The data of Curve (c) are calculated from the Hamrock-Dowson formula [44]. In the higher speed region (above 5 mm/s) of Fig. 4, the film becomes thinner as speed decreases and the speed index ϕ is about 0.69 (Fig. 4, Curve b), which is very close to that in Eq (1). When the film thickness is less than 15 nm, the speed

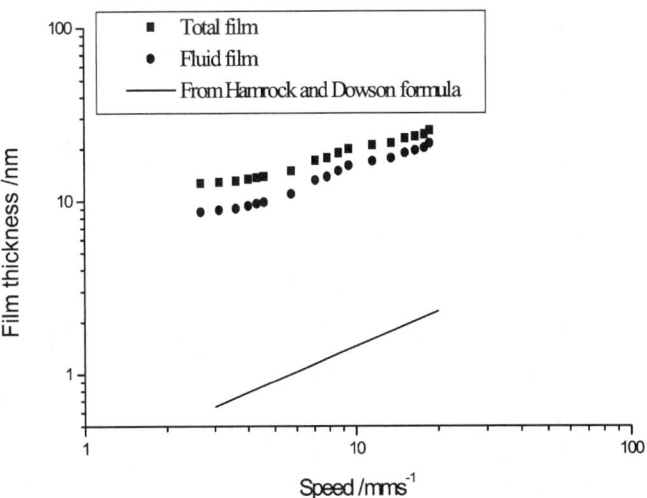

Fig. 4—Film thickness in the central contact region [18]. The ball is 23.5 mm in diameter and the lubricant is mineral oil CN13604 with no additives. Temperature is 25°C and load 4 N. The film thickness in Curve b is the data of the total thickness (Curve a) minus the static film thickness. The data of Curve c is calculated from Hamrock-Dowson formula (1981).

index is only about half of that in EHL and the relation between the film thickness and the speed does not obey Eq (1) anymore. This point, a critical point where the slope of the film-speed curve changes significantly, is generally believed as the transition point from EHL to TFL [3,4,18]. While for Curve (a) in Fig. 4, even if the relationship between the film thickness and the speed in the logarithmic coordinates is linear in the higher speed region, the speed index is only 0.45, which is smaller than that in EHL. When the film thickness is below the critical point, the index ϕ quickly decreases to 0.18. Therefore, the total film can be divided into three kinds of layers, i.e., a static film, which is a layer adsorbed on the metal surfaces, a solid-like layer, which is more ordered than fluid films, and a dynamic fluid layer. These layers play important roles in TFL.

Since the dynamic layer is very thick compared to total film thickness in EHL, the biggest part of the speed index is contributed by that of dynamic fluid film, and the difference between the total and the dynamic film thickness is so small that it can be ignored. However, in the TFL region, the dynamic film only takes a smaller part of total film and the difference between the thickness of total film and that of the dynamic one becomes significant. Therefore, if only the dynamic film is used to consider the influence of the speed, the index ϕ in the linear region will rise to 0.69 (Fig. 4(b)) that is very close to the EHL one. While for the total film including static and dynamic film, the index ϕ is about 0.45 (Fig. 4(c)). In other words, using the total film, not the dynamic one, is one of the main reasons why the index of the speed in the linear region is smaller than that in EHL. However, there is also a transition point in Curve (b), which does not contain the static film. Thus, another kind of film that is similar to static film in characteristics may exist. Such film is called ordered liquid film or solid-like film [4,18].

Hartle et al. have also observed the transition point in their experiment as shown in Fig. 5 [45]. The film thickness of octamethylcyclotetrasiloxane (OMCTS) exhibits a devia-

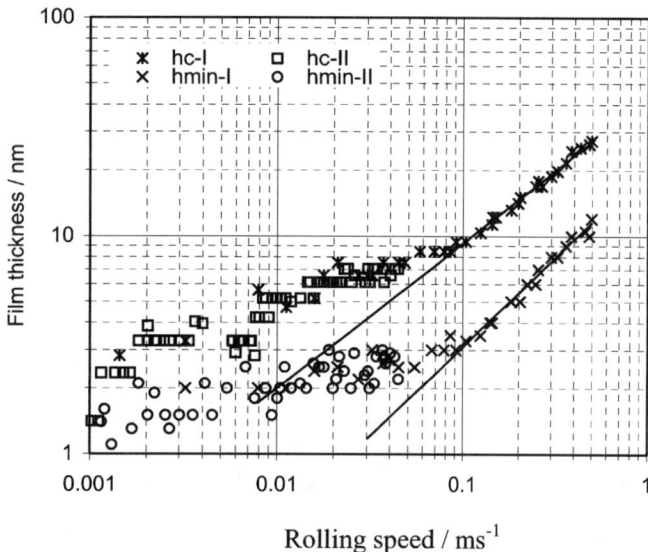

Fig. 5—Variation of film thickness with speed for octamethylcyclotetrasiloxane [56].

tion from linearity on a log (film thickness) versus log (rolling speed) in the thin film region, which is thicker than that predicted from EHD theory in the low speed region. This is explained as follows.

Because EHD film thickness is determined by the viscosity of the fluid in the contact inlet [46], it is obvious that the viscosity of OMCTS remains at the bulk value down to approximately 0.1 m/s. However, below this speed the discretization of both central and minimum film thicknesses can be observed. The central film thickness begins to deviate from the theory at about 10 nm and the interval of the discretization is approximately 2 nm. If the molecular diameter of OMCTS that is about 1 nm is taken into account, it corresponds to approximately two molecular layers.

2.1.2 Effect of Lubricant Viscosity

As shown in Fig. 6 [19], for the lubricant with higher viscosity (kinetic viscosity from 320 mm²/s to 1,530 mm²/s), the film is thick enough so that a clear EHL phenomenon can be observed, i.e., the relationship between the film thickness and speed is in liner style in the logarithmic coordinates.

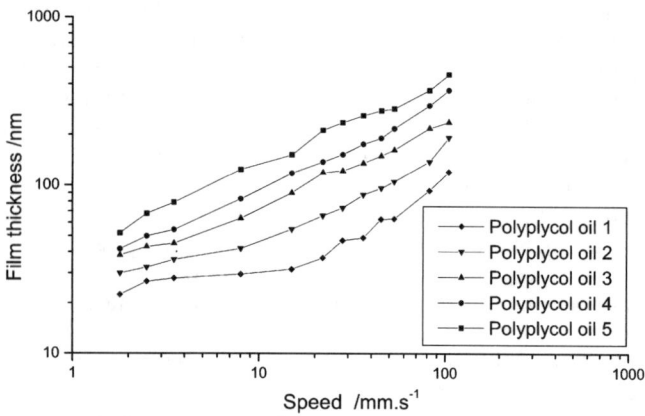

Fig. 6—Film thickness with viscosity [19]. Pressure: 0.184 GPa, Temperature: 28°C.

TABLE 1—Lubricants in the experiment.

Oil Sample	Viscosity (mm²/s)/(20°C)	Index of Refraction
Polyglycol oil 0	100	1.444
Polyglycol oil 1	47	1.443
Polyglycol oil 2	145	1.454
Polyglycol oil 3	329	1.456
Polyglycol oil 4	674	1.456
Polyglycol oil 5	1,530	1.457
Mineral oil 13604	20.6	1.482
White oil No. 1	61	1.472
White oil No. 2	96	1.483
Paraffin liquid	25.6	1.471
Decane	7.3	1.461

However, for the lubricants with lower viscosity, e.g., Polyglycol oil 1 and 2 with the kinetic viscosity of 47 mm²/s to 145 mm²/s in Table 1, the transition from EHL to TFL can be seen at the speed of 8 mm/s and 23 mm/s, i.e., the relationship between film thickness and speed becomes much weaker than that in EHL. The transition regime can be explained when the film reduces to several times the thickness of the molecular size, the effect of solid surface forces on the action of molecules becomes so strong that the lubricant molecules become more ordered or solid like. The thickness of such a film is related to the lubricant viscosity or molecular size.

2.1.3 Isoviscosity of Lubricant in TFL

The isoviscosity of the lubricant close to the solid surface in TFL is much different from that of bulk fluid, which has been discussed by Gao and Spikes [47] and Shen et al. [48]. For the homogeneous fluids, Eq (4) predicts the elastohydrodynamic central film thickness by a power relationship of about 0.67 with velocity and viscosity, i.e.,

$$h = ku^{0.67}\eta^{0.67} \quad (4)$$

where h is central film thickness, u average rolling speed, and η the dynamic viscosity of lubricant under atmospheric pressure. The constant k is related to the geometry and modulus of the tribo-pair and the pressure-viscosity coefficient of the lubricant. Assuming that the pressure viscosity coefficient of any boundary layer is approximately equal to the bulk fluid, the relation between isoviscosity and initial viscosity can be obtained by rearranging Eq (4).

$$\frac{\bar{\eta}}{\eta_0} \approx \left(\frac{u_0}{u}\right)\left(\frac{h}{h_0}\right)^{1/0.67} \quad (5)$$

where $\bar{\eta}$ is effective viscosity, u is the speed at which the film thickness is h, η_0 is the viscosity of the bulk fluid, and h_0 is the film thickness based on bulk fluid, which is formed at speed u_0.

The relationship of isoviscosity calculated by Eq (5) and a distance apart from the solid surface is shown in Fig. 7. For different kinds of solid materials with different surface energy, the isoviscosity becomes very large as the film thickness becomes thinner. It increases about several to more than ten times that of bulk fluid when it is close to the solid surface. In the thick film region, the isoviscosity remains a constant, which is approximately equal to the dynamic viscosity of bulk liquid. Therefore, the isoviscosity of lubricant smoothly

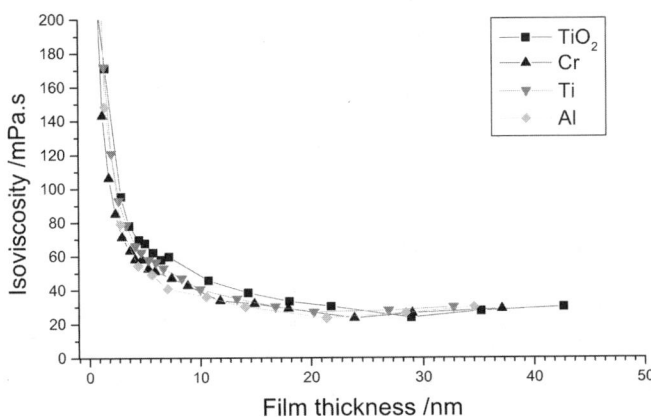

Fig. 7—Isoviscosity of lubricant versus film thickness [59].

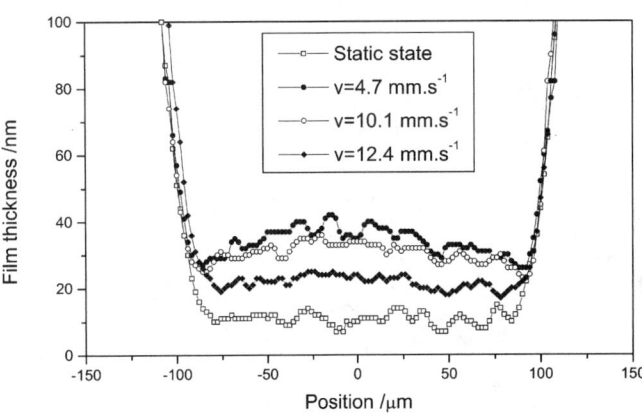

Fig. 9—Film thickness in the central cross section [2]. Lubricant: mineral oil with viscosity of 36 mPa·s at 20°C, Temperature: 25°C, Diameter of ball: 20 mm, Load: 6.05 N.

increases or the film becomes more solid like with the decrease of film thickness in TFL. When the film decreases down to very thin or the two solid surfaces are very close to each other, the isoviscosity of the lubricant becomes very much larger and a solid layer is formed.

2.1.4 Interference Fringes and Shapes of the Film Thickness Curves

The shape of interference fringes of oil film in the contact region at different speeds are shown in Fig. 8. In the static state, fringes are regular circles as in Fig. 8(a). When the ball starts rolling, an outlet effect appears as shown in Fig. 8(b), which will become much stronger as the speed increases.

The shapes of film thickness along A—A in Fig. 8 are given in Fig. 9. With the decrease of speed, the curve of film thickness in the central region becomes flat. The thickness of the film in a cross section of the central region is about 24 nm for the mineral oil with a viscosity of 36 mPa·s (20°C). However, for the lubricant with a viscosity of 17.4 mPa·s as in Fig. 10, the curve is quite crooked when the average thickness is about 24 nm, and the curve becomes flat at a thickness of about 14 nm. These indicate that the thinner the film is, the flatter the film in the central region will be. The thickness at which the shape of the film curve becomes flat is related to the critical film thickness where EHL transfers to TFL. The thicker the critical film is, the thicker will be the average film at which the film curve turns flat.

The film shape for the elliptoid contact region with an ellipticity parameter $k=2.9$ is shown in Fig. 11 [45], which gives all characteristic features of medium loaded point EHD contacts, i.e., nearly uniform film thickness in the central region with a horseshoe-shaped restriction. With decreasing value of k, the side lobes are developed and the locations of the film thickness minima move towards the sides. The side lobes provide an effective seal to axial flow resulting in a central plateau whose extent depends on the ellipticity of the contact. Smaller values of k imply a larger region of nearly uniform film thickness. Figure 12 shows the variation of film thickness transverse to the direction of rolling for all three ellipticities and a speed of 0.012 ms^{-1}. The figure shows the influence of the ellipticity on the lubricant compressibility.

2.2 Transition from EHL to Thin Film Lubrication

The capacity of the ordered film for supporting loads is between that of the static film and that of the dynamic fluid film. The orientation property of the ordered layer gradually becomes weak with the distance apart from the metal surface. The transition occurs as the ordered film appears more important between the two solid surfaces. The thickness of the ordered film is related to the initial viscosity or molecular size of the lubricant, as shown in Fig. 13, so that we can generally write the critical film thickness as follows:

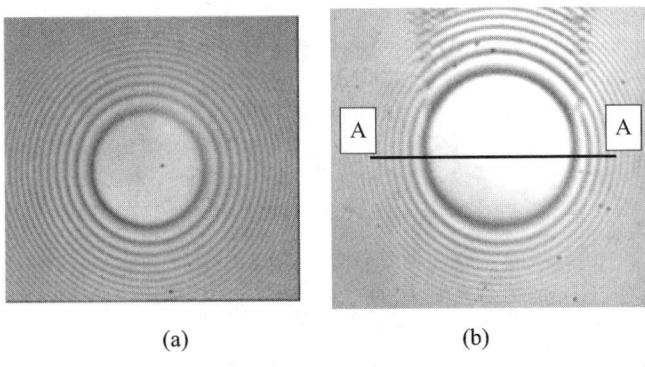

(a)   (b)

Fig. 8—Interference fringes.

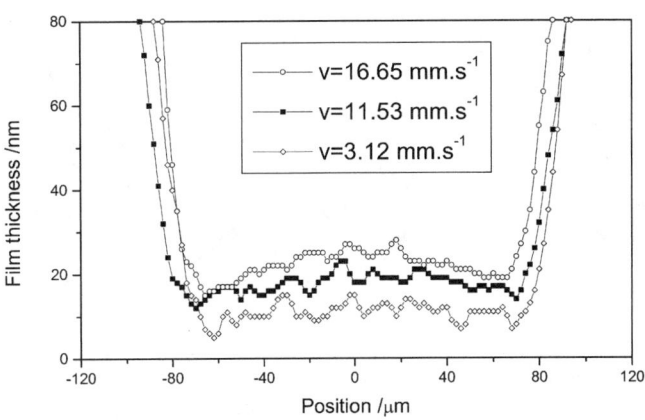

Fig. 10—Film thickness in the central cross section [2]. Lubricant: mineral oil with viscosity of 17.4 mPa·s at 20°C, Temperature: 25°C, Diameter of ball: 20 mm, Load: 4 N.

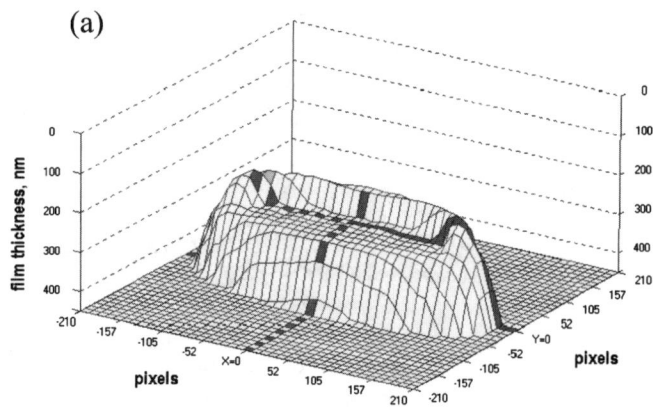

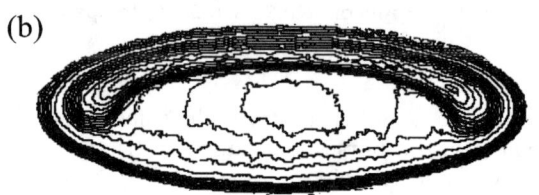

Fig. 11—Isometric views and associated contour line plots of film shape for $k=2.9$ and rolling speed of 0.021 and 0.042 ms^{-1} (Ref. [56]).

$$h_{ct} = a + be^{c/\eta_0} \qquad (6)$$

where h_{ct} is the critical film thickness; η_0 is the initial viscosity of lubricant; a, b, and c are coefficients that are related to the molecular structure and polarity of lubricants, the solid surface energy. These coefficients can be determined from the experimental data. From Eq (6) it can be seen that the thickness bounder between EHL and TFL is not a definite value, but the one varying with viscosity or molecular size, molecular polarity, and solid surface tension. The following equation can be used for common mineral oils without polar additives:

$$h_{ct} = 9 + 17.5e^{-8.3/\eta_0} \qquad (7)$$

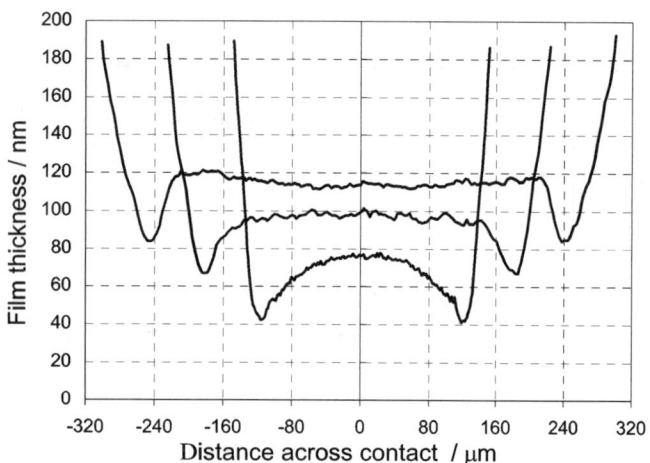

Fig. 12—Variation of film thickness transverse to the direction of rolling for $k=1$, 1.9, and 2.9 and rolling speed of 0.011 ms^{-1} (Ref. [56]).

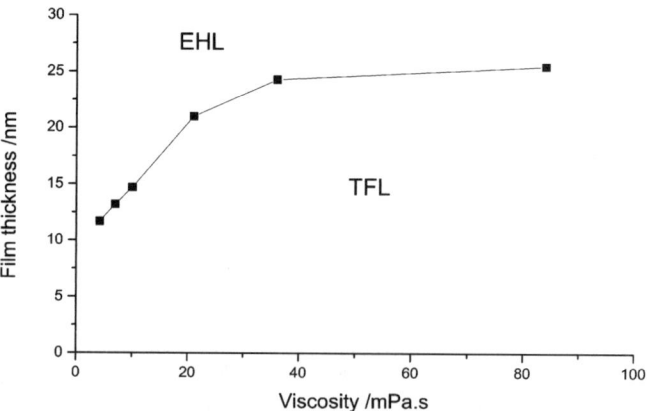

Fig. 13—Critical transition film thickness versus viscosity [2]. Load: 4 N, Ball: ϕ20 mm, Lubricant: mineral oils.

Equation (7) indicates that the critical film thickness changes very slowly with an increase in initial viscosity when it is close to 26 nm.

2.3 Effects of Solid Surface Energy

The influence of the solid surface tension or surface energy on the behavior of lubricants in TFL is shown in Fig. 14 [27]. A chromium coating, which is close to steel in the surface energy and an aluminum coating with a surface energy much smaller than that of steel are chosen to be formed on the surfaces of the steel (E52100) balls by ion beam assisted deposition (IBAD) as shown in Fig. 15. The surface roughness of these balls in the contact region is around 5 nm. As shown in Fig. 14, the film thickness decreases as the rolling speed decreases. In the higher speed region, the film thickness changes significantly with the rolling speed. When the film thickness is less than a critical thickness (27 nm in curve a, 25 nm in curve b, and 18 nm in curve c, Fig. 14), the film thickness decreases slightly with speed. From the thickness difference among the three curves in Fig. 14, it can be seen that the films with the substrate of Fe and Cr are very close to each other in thickness, while they are much thicker than the film with the substrate of Al, which has lower surface energy than the former. Thus, the larger the solid surface energy (or tension), the thicker the total film and the critical film will be.

In addition, the film thickness enhanced with time is

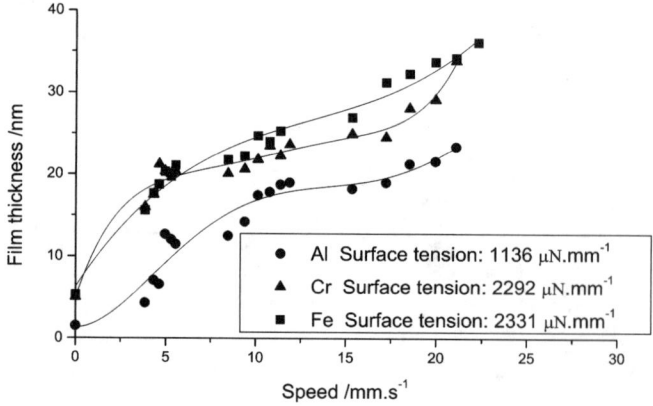

Fig. 14—Film thickness with different substrate [17]. Lubricant: mineral oil, Load: 4 N, Temperature: 18.5°C, Ball: ϕ23.5 mm.

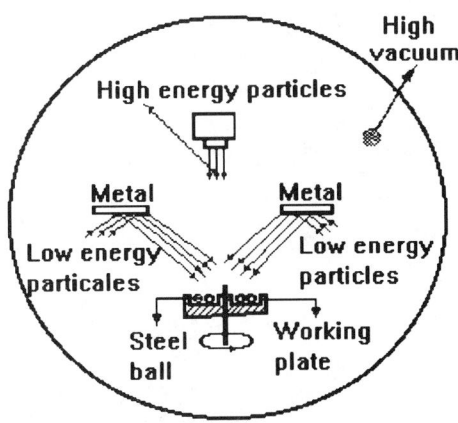

Fig. 15—Coating system [17].

also affected by the surface energy of the substrate. As shown in Fig. 16, the film in the contact region becomes thicker with time during the running process; however, when the ball stops running, the film thickness rapidly turns thinner and no longer changes with time. Before running, the static film thickness is 5 nm for the surface of steel and about 2 nm for Al. When the ball starts rolling, the film for the solid surface of steel immediately increases to 15 nm and that for Al to 5 nm. The film thickness with the steel surface increases to 30 nm within 60 minutes but for that of the Al surface it is about 15 nm. The static film after rolling for Fe is about 13 nm thick, which is much larger than that with Al (7 nm). Therefore, the higher the surface energy, the larger will be the thickness of the static film and total film.

2.4 Friction Force of TFL

In order to investigate the friction properties of lubricant film in TFL, an apparatus with a floating device was developed by Shen et al. as shown in Fig. 17 [48]. The steel ball is fixed so that it does not roll in the experiment and a pure sliding has been kept. The measuring system of micro-friction force is composed of a straining force sensor with a resolution of 5 μN, a dynamic electric resistance strain gage, an AD data-collecting card, and a computer.

The lower surface of the glass disk is coated with Al, Cr,

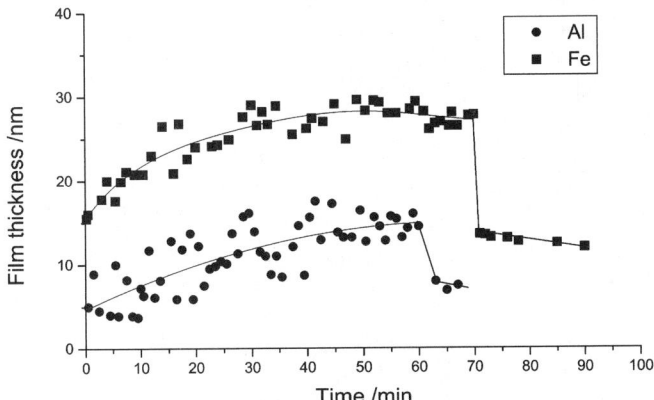

Fig. 16—Film thickness with different substrates and time [17]. Lubricant: mineral oil, Load: 7 N, Temperature: 18.5°C, Ball: ϕ23.5 mm, Speed: 4.49 mm·s^{-1}.

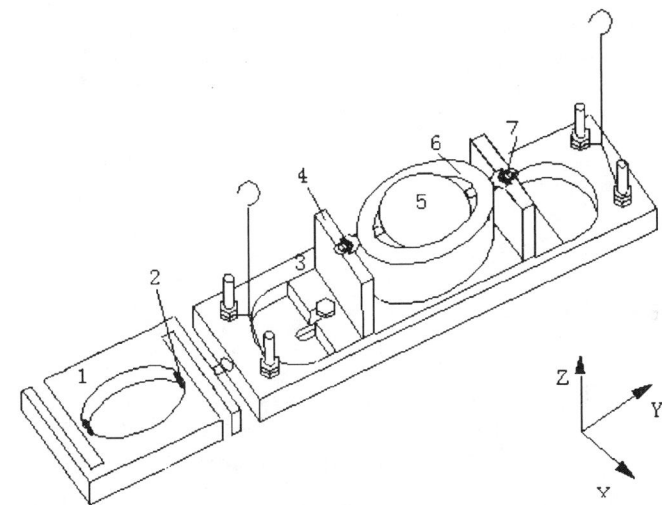

Fig. 17—Floating device for friction measurement [48]. 1—Carrier of strain gage; 2—Strain gage; 3—Beam; 4—Plank; 5—Steel ball; 6—Oil cup; 7—Mandril.

Ti, and TiO$_2$ layer separately by vacuum braising plating at 160°C. These films should be plated up to a certain thickness in order to get clear interference rings. The steel ball (E52100) is 25.4 mm in diameter. The free surface roughness of the glass disk and the steel ball are about 7 nm and Yang's modulus is 54.88 GPa and 205.8 GPa, respectively. When the lubricant film was in nanometre scale, the roughness in the contact area was much smaller than that of the free surface because of the elastic deformation [17]. Table 2 lists the surface energy γ of metal coatings at 25°C [49]. Table 3 shows the characteristics of test liquids.

Different substrates make little difference in the friction coefficients in the higher sliding speed region as shown in Fig. 18 [48]. The friction coefficient increases with the decrease of sliding speed. The friction coefficient with the substrate of TiO$_2$ is the highest, those with Cr and Ti are close to each other, and that with Al is the lowest. When the sliding speed decreases to a certain value, the friction coefficient increases sharply. It indicates that the viscosity increases with the decrease of the sliding speed in TFL. While at a very slow speed, the lubrication regime changed into the boundary regime or mixed lubrication and the friction coefficient changes from 0.15 to 0.3, which is approximately equal to that in the dry rubbing state. Furthermore, the friction is the force required to break interfacial bonds of the two solid sur-

TABLE 2—Surface energy of metal substrates.

Substrate	Al	Ti	Cr	Fe
γ(mN·m^{-1})	1,136	2,082	2,292	2,331

TABLE 3—Characteristics of the lubricants.

Lubricant	Viscosity (mPas) (22°C)	Surface Tension (mN·m^{-1})(24°C)	Refractive Index
Paraffin liquid	30	29.4	1.4612
Base oil 13604	17.4	29.57	1.4711
13604+10 % EP	15.4	30.5	1.4675
EP	4.9	27.5	...

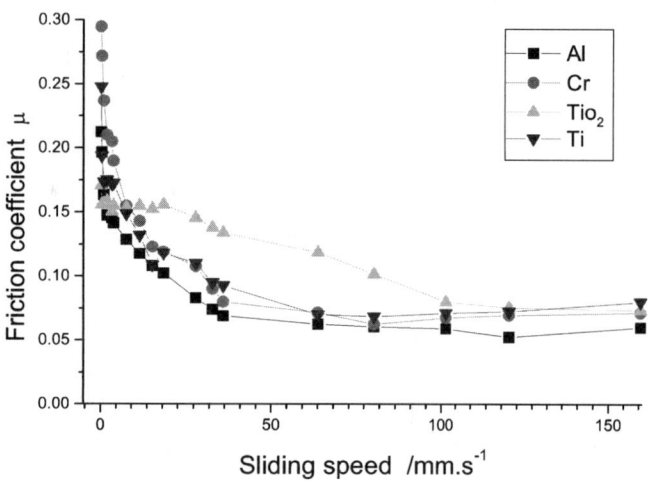

Fig. 18—Friction coefficient with different substrates [48]. Lubrication: paraffin liquid; Load: 2 N.

faces during sliding, from which it follows that the surface with the lower surface energy will exhibit a lower adhesion force, and therefore a lower friction coefficient. So the friction coefficient is closely related to the characteristics of the substrate in TFL, boundary lubrication, or dry contact. The effect rules of solid surface energy on friction are the same as that on film thickness. Higher energy leads to high friction coefficient.

Figure 19 shows that the friction coefficients with different materials of solid surfaces decrease slightly at first and then increase as the load increases. It also can be seen that the higher the surface energy is, the larger the friction coefficient is. A slight decrease in the friction coefficient with the increase of load is in accordance with the usual rule. But, when the load further increases, the fluid film becomes thin and converts to a more viscous or solid-like layer [4,39,40], which inevitably gives a great friction coefficient.

2.5 Effect of Slide Ratio on Thin Film Lubrication

The relation between the slide ratio $[s = 2(u_1 - u_2)/u_1 + u_2]$ and the film thickness is shown from Figs. 20–22, which indicate that the film thickness hardly changes with the slide

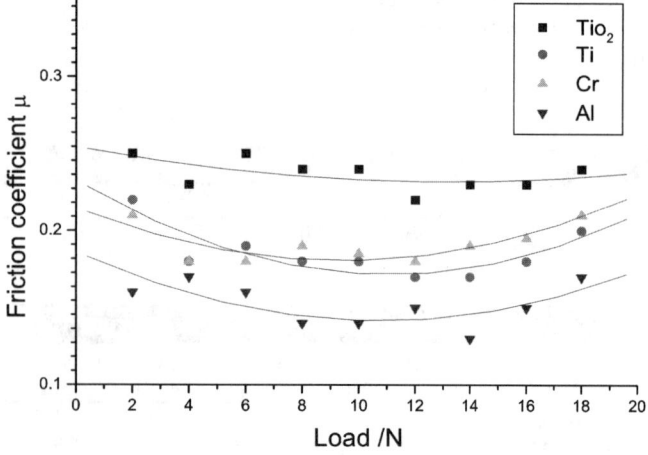

Fig. 19—Friction coefficient and load on different substrates [48]. Velocity: 15.6 mm/s; Lubricant: paraffin liquid.

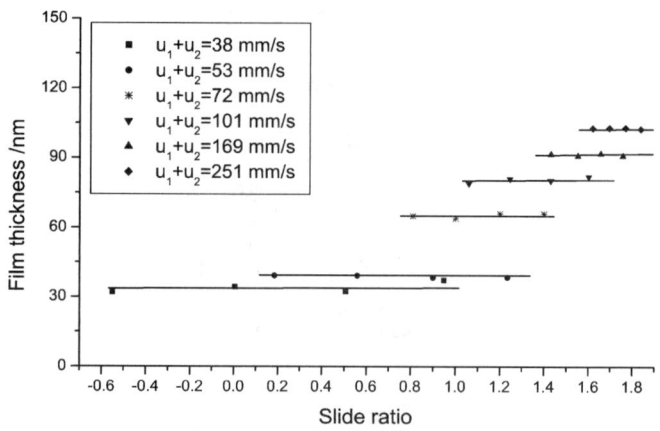

Fig. 20—Film thickness with slide ratio. Lubricant: White oil No. 1+5 % nonionic acid, Temperature: 20°C, Relative humid: 76 %, Ball diameter: 25.4 mm, Load: 2 N.

ratio, but is closely related to the average speed of the two solid surfaces $(u_1 + u_2)$ for white oil No. 1 (with 5 % nonionic acid in volume as an additive) under load of 2 N [34]. When the load increases to 4 N, the thickness also does not change with the slide ratio S in the range from −0.9 to 1.9 except it

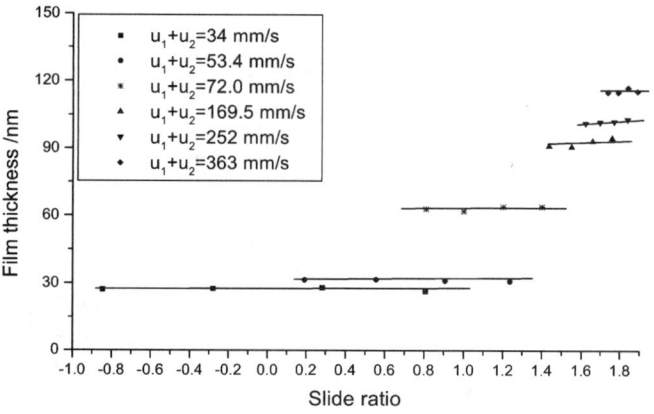

Fig. 21—Film thickness with slide ratio. Lubricant: White oil No. 1+5 % nonionic acid, Temperature: 20°C, Relative humid: 76 %; Ball diameter: 25.4 mm, Load: 4 N.

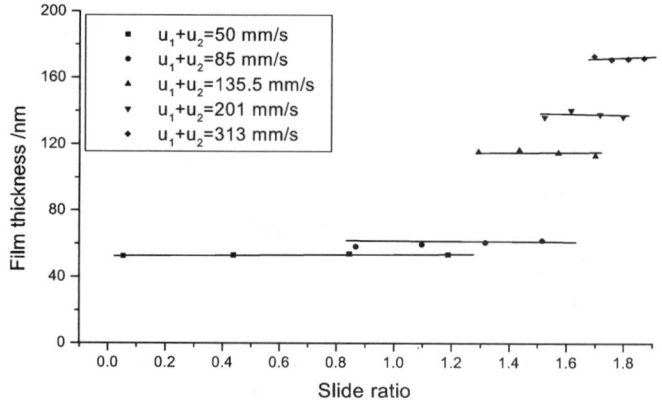

Fig. 22—Film thickness with slide ratio. Lubricant: White oil No. 2+5 % nonionic acid, Temperature: 20°C, Relative humid: 76 % Ball diameter: 25.4 mm, Load: 4 N.

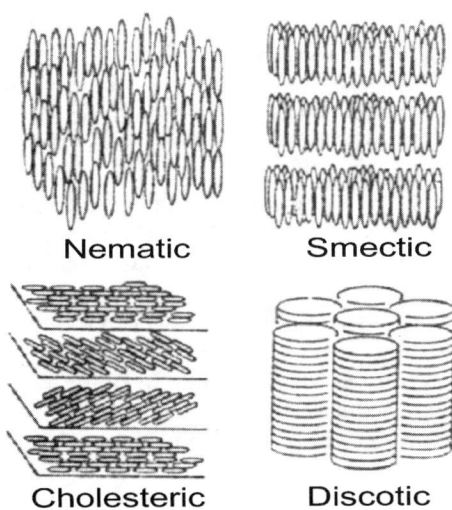

Fig. 23—The ideal models of LCs [55].

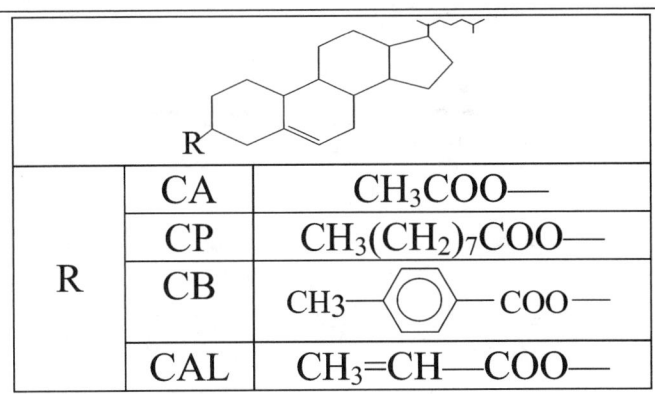

has a slight decrease as shown in Fig. 21. The similar phenomenon can be seen for white oil No. 2 with a higher viscosity in Fig. 22. Therefore, when the film thickness is beyond about 30 nm, the film is related to the total speed of the two solid surfaces but not the slide ratio. Such phenomenon is very similar to the EHL theory. However, if a pure sliding condition happens, the film will be easy to collapse at such speed range and wear can be observed.

2.6 Effect of Electric Field

The electric voltage also has an influence on the film thickness in TFL. Shen et al. [50] used hexadecane with the addition of cholesteryl LCs in chemically pure as the lubricant to check the variation of its film thickness by applying an external DC voltage. With the technique of ROII [3,4,51], the effects of LCs' polarity and concentration on film thickness and the effects of lubricant molecules on a film-forming mechanism were investigated by them.

2.6.1 Styles of Liquid Crystal

Liquid crystals (LCs) are organic liquids with long-range ordered structures. They have anisotropic optical and physical behaviors and are similar to crystal in electric field. They can be characterized by the long-range order of their molecular orientation. According to the shape and molecular direction, LCs can be sorted as four types: nematic LC, smectic LC, cholesteric LC, and discotic LC, and their ideal models are shown in Fig. 23 [52,55].

The sample LCs used in their experiments were cholesteryl esters with low molecular weight: cholesteryl acetate (CA), cholesteryl pelargonate (CP), cholesteryl benzoate (CB), cholesteryl acrylate (CAL) [50]. In their molecular structures listed in Table 4, four tightly conjoining rings are composed of 17 carbon atoms (three rings consist of six carbon atoms, respectively, the fourth of five carbon atoms). Such rings form a comparatively linear and quasi-rigid rod-like structure.

LCs exhibit many good lubricating properties as lubricant or additives and have attracted much attention of tribologists [53–55]. When the external electrical field is strong enough, LC molecules rearrange in the direction where the lubricant viscosity is the maximum, and an electroviscous effect occurs [53]. Kimura [54] studied the active control of friction coefficient by applying a dc voltage of 30 V to the lubricants of nematic LCs and observed a 25 % reduction of friction coefficient. Once the voltage was turned off, the reduction disappeared. Using nematic LCs and smectic LCs as lubricants, Mori [55] reported that the tribological properties depended closely on their structures and their friction coefficient decreased to about 0.04. However, many other results showed that LCs used as additives exhibited better properties than as pure lubricants due to the higher stability and adaptation to temperature [56].

2.6.2 Experiment

ROII technique [4,51] is shown in Figs. 5 and 7 in Chapter 2. In order to apply a dc voltage on the oil film, the lower surface of the glass disk is coated with a thin semi-reflecting layer of Cr, on top of which an insulative layer of SiO_2 about 300 nm is coated as shown in Fig. 24. The lubricant film is formed between the surface of the glass disk and the surface of super-polished steel ball (E52100, 25.4 mm in diameter). The main shaft which keeps in contact with the Cr layer and the steel ball with a welded shaft were used as two electrodes. A pure rolling and a temperature of 30 ± 0.5°C were kept in the test to ensure the LCs were in the liquid crystalline phase. The employed external dc voltage varies from 0 to 1,000 mV, corresponding to the electrical field intensity from 0 to 35.2 V/mm. Table 5 shows the characteristics of tested lubricants.

2.6.3 Film Thickness of LCs

The relationship between film thickness of hexadecane with the addition of cholesteryl LCs and rolling speed under different pressures is shown in Fig. 25 [50], where the straight line is the theoretic film thickness calculated from the Hamrock-Dowson formula based on the bulk viscosity under the pressure of 0.174 GPa. It can be seen that for all lubricants, when speed is high, it is in the EHL regime and a speed index ϕ about 0.67 is produced. When the rolling speed decreases and the film thickness falls to about 30 nm, the static adsorption film and ordered fluid film cannot be negligible, and the gradient reduces to less than 0.67 and the transition from EHL to TFL occurs. For pure hexadecane, due to the weak interaction between hexadecane molecules and metal surfaces, the static and ordered films are very thin. EHL

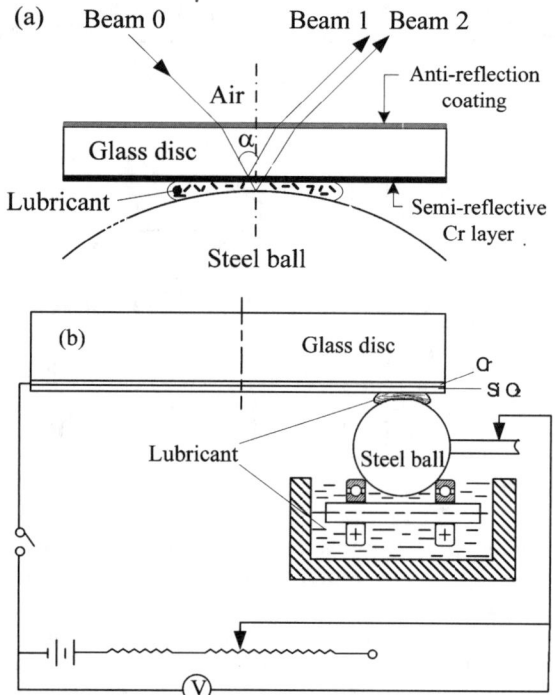

Fig. 24—Measuring instrument: (a) interference lights of the measuring system; (b) the scheme of the experiment system. For applying an external voltage to the oil film, one end of direct current (dc) power keeps contacting with the Cr layer on the glass disk, and the other end contacts a shaft welded on the steel ball through an electric brush. The SiO_2 layer with a thickness of 300 nm has been coated on the surface of the Cr film.

theory can maintain down to a very low speed. However, for mixture oils, metal surfaces prefer to adsorb polar LC molecules and form a stronger interaction and surface force field. This leads to a thicker ordered film. The transition from EHL to TFL occurs at a high speed, i.e., the system enters into TFL when speed decreases to 101 mm/s for CB, 98.6 mm/s for CAL, 89.1 mm/s for CA, and 80 mm/s for CP, and the film thickness correspondingly reduces to 23 nm for CB, 21 nm for CAL, 18 nm for CA, and 15 nm for CP. Moreover, their practical film thickness is larger than that expected from EHL theory. A lower speed produces a greater difference between them, and the former is three to five times as large as the latter. Therefore, the addition of LC is a benefit in the formation of the ordered film.

It also can be seen that the polarity of LC has great effects on the film thickness in TFL. For the tested cholesteryl LCs, they have the same cholesteryl structures but different R groups. The R group in CB is a benzene ring and has the highest polarity. CAL includes C=C bond and its polarity is next to CB. CA and CP belong to the alkyl chain compound. Their polarity is less than the former two. Moreover, less polarity results from more carbon number in the alkyl chain. Thus, at the speed of less than 80 mm/s (for 0.174 GPa), the sequence of film thickness of the hexadecane with 2 wt. % LCs is CB>CAL>CA>CP (according to LC additives).

The effect of the amount of CA in hexadecane on the film thickness is shown in Fig. 26. It is clear that a higher concentration of LC results in thicker film in BL or TFL. While in EHL, for hexadecane with different LCs, the film thickness is almost independent of the concentration of LC. It is due to more LC molecules being adsorbed by the metal surfaces to form a thicker static film and ordered film in the low speed region under the higher concentration of LCs. However, the fluid washing effect becomes stronger and the thickness of the fluid film becomes larger with increasing speed, so that LC molecules are difficult to be adsorbed and the thickness is too small to be compared to the total film. The fluid film plays a main role and the bulk viscosity becomes the key factor to determine the film thickness. When the concentration is less than 5 % as shown in Table 5, there is only a little enhancement in the viscosities of hexadecane with LCs.

In order to investigate the relationship between the voltage and the film thickness, Shen and Luo [50,51] applied a certain external voltage on the lubricant film. Before each measurement, the solid surfaces were sufficiently cleared. The profile of film thickness of hexadecane with LC versus the external dc voltage is shown in Fig. 27. When the voltage is zero, the film thickness is from 14 nm to 18 nm for the four kinds of tested lubricants. As the voltage rises and the electrical field becomes stronger, the LC molecules rearrange in the direction of maximum viscosity, causing an enhancement in film thickness. When the voltage increases further to 500 mV, the film thickness reaches a maximum value of about 30 nm and the intensity of the electrical field is 17 kV/mm or so. Then the film thickness hardly changes with increasing voltage.

It was explained that because a higher voltage or a stronger electric field makes LC molecules tending to rearrange in the direction of maximum viscosity, it gives rise to a greater electricoviscous effect to form a thicker film. However, when the electric field becomes strong to some value (depending on the type of LCs), LC molecules rearrange to the largest ex-

TABLE 5—Characteristics of tested lubricants.			
LC Additive	Concentration	Surface Tension $(mN \cdot m^{-1})$ 25°C	Apparent Dynamic Viscosity (mPas) 22°C
...	...	27.1 (base oil)	2.297 (base oil)
CA	2 %	25.42	2.3
CA	3 %	26.96	2.579
CA	5 %	27.25	2.988
CP	2 %	25.49	2.583
CB	2 %	25.59	2.356
CAL	2 %	25.18	2.684
CAL	3 %	25.83	2.384
CAL	5 %	26.22	3.068

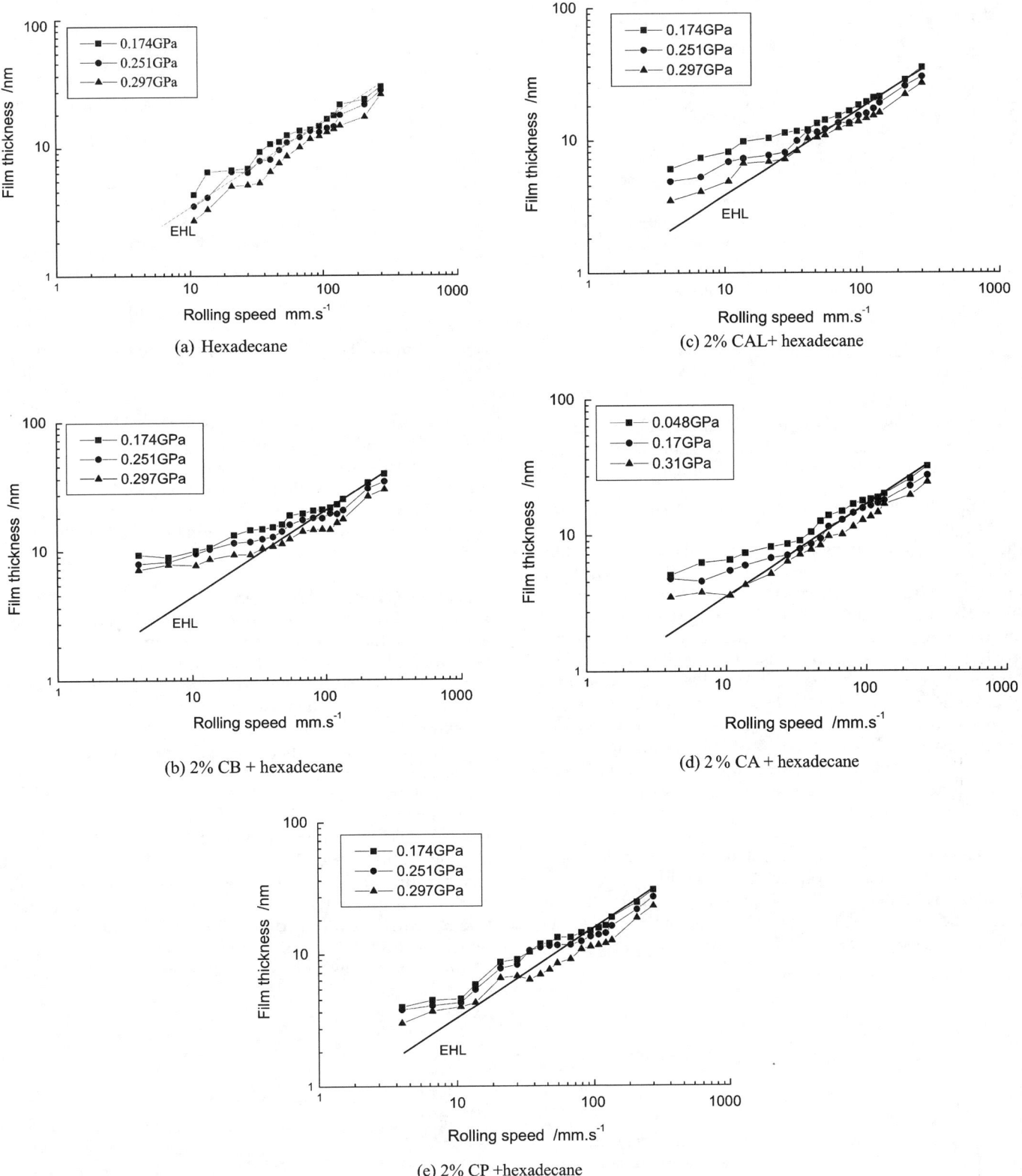

Fig. 25—The film thickness with rolling speed [50]. (a) Hexadecane, (b) 2 % CB+hexadecane, (c) 2 % CAL+hexadecane, (d) 2 % CA+hexadecane, (e) 2 % CP+hexadecane.

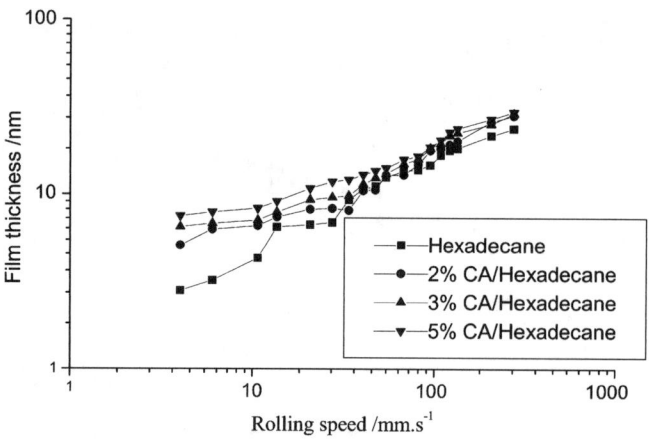

Fig. 26—Influence of LC concentration on thickness under 0.174 GPa [50].

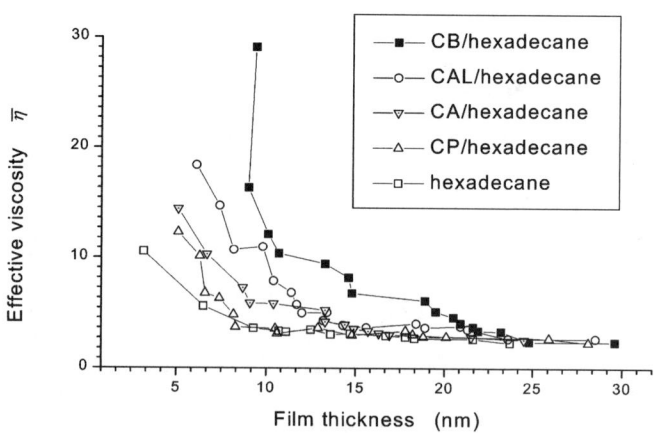

Fig. 28—Isoviscosity with film thickness without voltage [50]. Concentration: 2 wt. %, Load: 0.174 GPa.

tent. The viscosity and film thickness also reach their maximum, respectively, so that the film thickness hardly changes anymore with a further increase of the voltage [50,51].

2.6.4 Isoviscosity of LCs Film Close to Solid Surface

The relationship between the isoviscosity $\bar{\eta}$ and the film thickness is shown in Fig. 28 in which $\bar{\eta}$ is calculated according to Eq (5) from the experimental data as shown in Fig. 25. When the film is thicker than 25 nm, the isoviscosities of hexadecane with or without LC remain a constant that is approximately equal to the bulk viscosity. As the film becomes thinner, their isoviscosities increase at different extents for different additives. When the film thickness is about 7 nm, the isoviscosity of pure hexadecane is about two times its bulk viscosity, about three times for CP, four times for CA, six times for CAL, and more than ten times for CB. Thus, it can be concluded that the addition of a polar compound into base oil is a benefit to the formation of thicker solid-like layer.

The isoviscosity is not only related to the film thickness but also to the external voltage. As shown in Fig. 29 where the relationship between isoviscosity and external voltage is from experimental data (Fig. 27), when the voltage increases, the electricoviscous effect occurs and the isoviscosity rises. When the voltage rises to 500 mV, the electricoviscous effect and viscosity enhance to their largest extent. Then, it hardly changes further with increasing external voltage. A high concentration or a strong polarity of LCs gives rise to a greater isoviscosity, therefore, a stronger molecular polarity is a benefit to the formation of a thicker film in the nano scale.

2.7 Time Effect of Thin Film

In the TFL regime, a phenomenon that the film thickness varies with the running time under some conditions has been found by Luo et al. [3,18]. Factors such as load, rolling speed, and viscosity of lubricant have an influence on the relationship between the film thickness and the rolling time. Shown in Fig. 30, the film thickness of paraffin liquid in the central contact region increases with the running time. The enhancement of the film thickness produced in the running process with the running time is defined as the thickness of a time dependent film. Before running, the thickness of the static film is 5 nm under the load of 4 N (Curve A and C, Fig. 30) and 4.2 nm under 7 N (Curve B, Fig. 30). When the ball starts rolling at the speed of 4.49 mm/s (Curve A), the film thickness immediately increases to 16 nm and then turns thicker with the running time. Within the 80 minutes of running, the film thickness increases from 16 nm to about 22 nm. However, as long as the ball stops rolling, the film col-

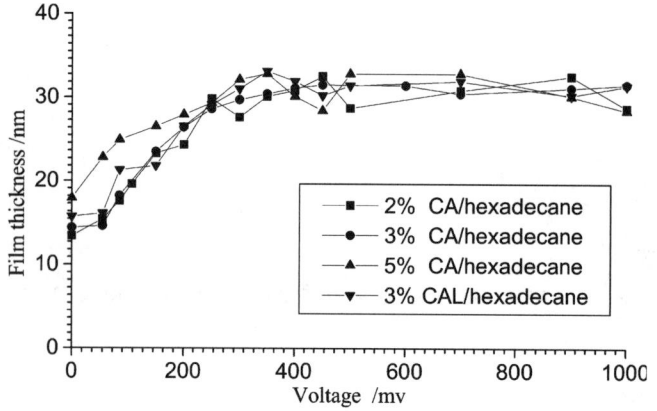

Fig. 27—Influence of dc voltage on thickness [50]. Load: 0.174 GPa, Rolling speed: 68 mm·s⁻¹.

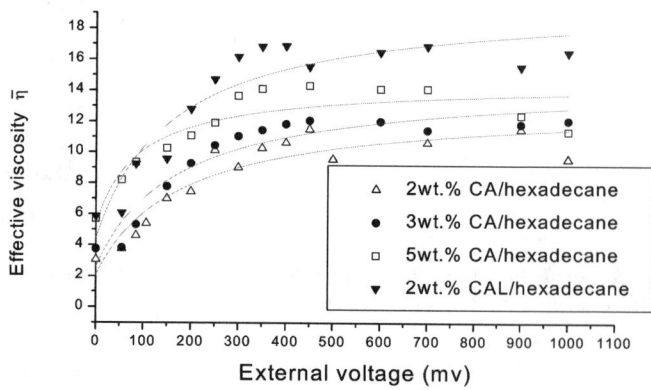

Fig. 29—Isoviscosity under external voltage. Load: 0.174 GPa, Rolling speed: 68 mm·s⁻¹.

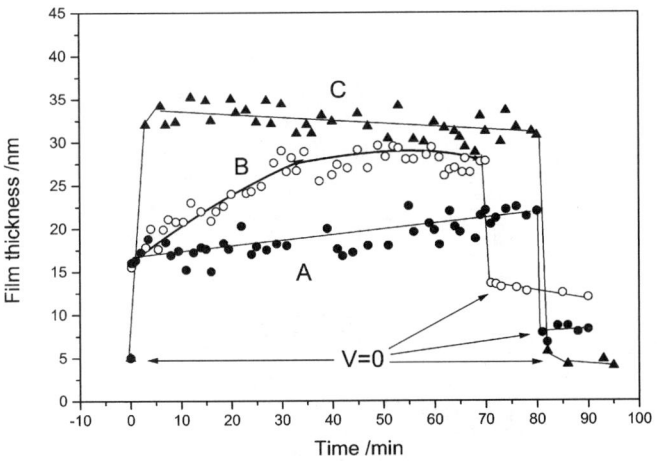

Fig. 30—Film thickness versus time [18]. Lubricant: Paraffin liquid, temperature: 18°C, Ball: φ23.5 mm A: speed v=4.49 mm/s, load w=4 N; B: speed v=4.49 mm/s, load w=7 N; C: speed v =17.28 mm/s, load w=6 N.

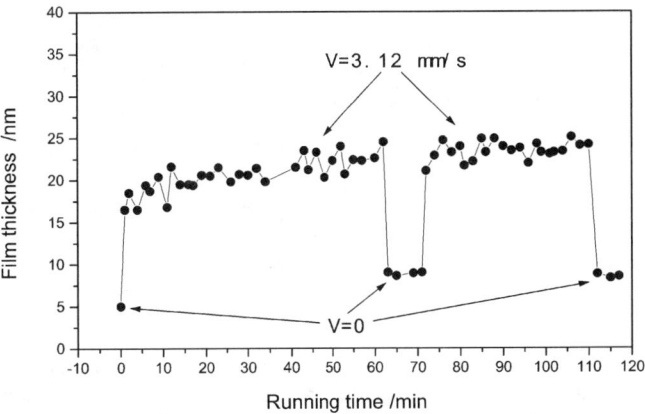

Fig. 32—Film thickness with time [18]. Lubricant: paraffin liquid; temperature: 18°C, Ball: φ23.5 mm, Load: 4 N.

lapses suddenly and the thickness decreases down to about 8 nm, which is 3 nm thicker than the static film before rolling. When the load is raised from 4 N to 7 N as shown in Curve B of Fig. 30, the film thickness increases intensively with the time and keeps this increasing tendency for about 40 minutes, and then hardly changes. The increment of the total film thickness within 70 minutes of running is about 13 nm. When the system stops running, the film drops down to about 10 nm thick. However, when the speed increases to 17.28 mm/s under the load of 6 N, the film thickness is up to about 33 nm, which is usually in the EHL region and it hardly varies with the running time.

The time dependent phenomenon also can be seen in Fig. 31. The static film of decane added with 3 % palimitic acid (Curve A in Fig. 31) starts at 5 nm and then subsides by about 2.5 nm for nearly 15 minutes, and it returns to 5 nm and keeps constant for more than two hours. But when the system starts pure rolling at a speed of 3.12 mm/s as shown in Fig. 31 (Curve B), the film keeps the increasing tendency for about 110 minutes, and the thickness changes from 10 nm up to 45 nm, and then it hardly changes with time. However, when the higher viscosity of lubricant is used at a higher speed, as shown in Fig. 31 (Curve C), the film thickness is 45 nm at the beginning and is reduced slightly with time.

The time dependent film is also related to the running history as shown in Fig. 32. Before the system starts running, the static film is 5 nm thick. After 60 minutes of running at a speed of 3.12 mm/s, the system stops running for 10 minutes, and the film drops down to 9 nm. When the system restarts running, the film rapidly recovers. This phenomenon indicates the films on the surface of substrates will adhere to the surface for a period of time when the rolling action has stopped, and the film thickness can pick up its value quickly once the rolling is restarted.

Figure 33 shows the shape of the film curve in the cross section of the central region. Even though the average film thickness (about 30 nm) of Curve (a) in the central contact region is nearly the same as that of Curve (b), the shapes of the two thickness curves are quite different from each other. Curve (a) is at the beginning of the running at a higher speed. It is crooked in the range from −60 μm to 60 μm; that is, the

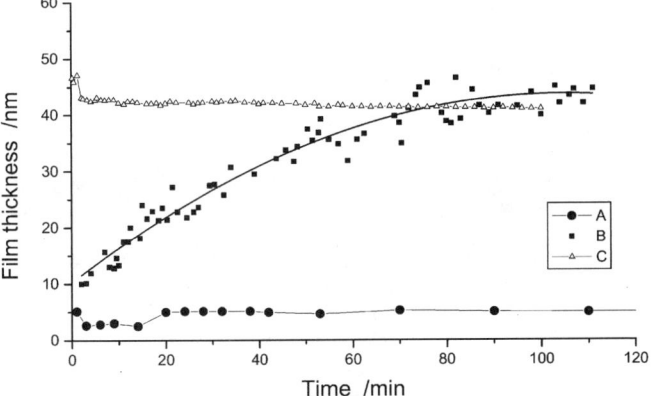

Fig. 31—Film thickness with time [18]. A: Decane +3 % palimitic, Load: 4 N, T=30°C, v=0 mm/s, Ball: φ25.4 mm. B: Decane +3 % palimitic, Load: 4 N, T=30°C, v=3.12 mm/s, Ball: φ25.4 mm. C: White oil No. 1, Load: 20 N, T=20°C, v=54.5 mm/s, Ball: φ25.4 mm.

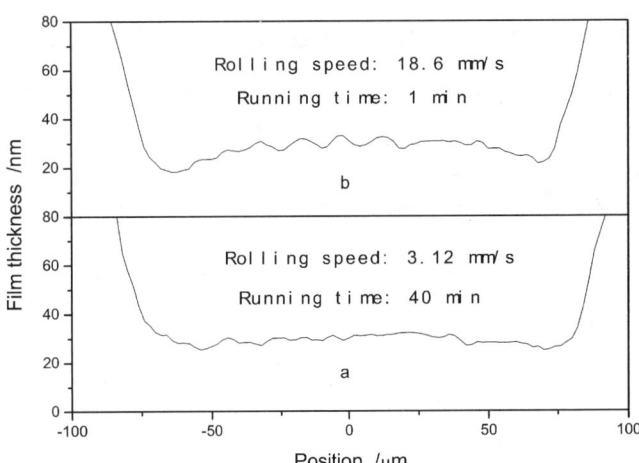

Fig. 33—Film in cross section of Hertz region [18]. Lubricant: Paraffin liquid, Load: Temperature: 30°C, Ball: φ23.5 mm.

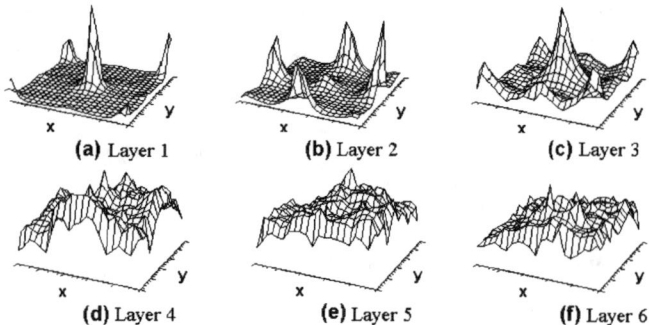

Fig. 34—Spatial distribution in six layers nearest a solid wall under a constant pressure $P^*_{zz}=3.0$, Temperature: 140 K, $\varepsilon_{wf}/\varepsilon=3.5$, where ε_{wf} is the ratio of characteristics energy of wall-fluid interaction, ε is characteristics energy of molecular interaction [58].

film thickness near the edge of the contact region is much thinner than that in the central. The phenomenon is due to the hydrodynamic effect and the side leakage of pressure [57]. However, the shape of Curve (b) after 40 minutes of running under the condition of lower speed is quite flat in the range from -60 μm to 60 μm. The mechanism of time effect was explained as follows [18].

The orienting force, the absorption potential of the solid surfaces, as well as the shear stress will convert lubricant molecules near the solid-liquid interface into the ordered state or solid-like in the running process. Such film will move with the ball together. Therefore, the film on the surface of the ball will be getting thicker with the running time, which is similar to a snow ball getting larger when it is rolling on a snowy surface. Similar to this idea, the isoviscosity should increase also near the solid surface, and thus the total film becomes thicker with time. When the ordered film or solid-like film becomes thick enough, the number of the molecules turned to the ordered state will balance that of the molecules washed away by fluid, and therefore, the film thickness will hardly change further with time. That the thickness of the ordered layer increases with pressure has also been proven by Hu et al. [58] by using molecular dynamic simulation. Their results showed that the ordered structure originated from the wall-fluid interface and grew toward the middle of the film as the system pressure increases as shown in Fig. 34.

2.8 Properties of Thin Film Lubrication with Nano-Particles

In the 1990s, much progress in preparation and application of spherical or surface modified nano-particles has been made, especially of nanometre-sized spherical particles, such as ultra-fine diamond powders (UDP) and C60. UDPs have been reported to be successfully synthesized by the detonation of explosives with a negative oxygen balance [59]. Since they are formed at very high velocity, high pressure and high temperature, UDPs have many unusual physical and chemical properties, such as spherical shape, nanometre sizes, large specific surface area, high surface activity, high defect density, and so on. The behavior of a basic oil added with UDP nano-particles are investigated by Shen et al.[60] using the technique of ROII [3,4] and a high-precision force measuring system. UDPs were prepared by the detonation of explosives [59]. Their images of transmission electron microscopy (TEM) and electron diffraction are shown in Fig. 35. In the ED image, the three rings counted clearly correspond to (111), (220), and (311) crystal surfaces, respectively, suggesting UDP particles have a complete crystal face and the structure of cubic diamond. Diameters of UFD range from 1 to 15 nm, and the average size is 5 nm. Table 6 lists some physical parameters of UDP.

A dispersant of polyoxyethylene nonylphenol ether by 1 wt. % and UDP with different weight ratios were added into base oils such as polyethylene glycol (PEG) or polyester

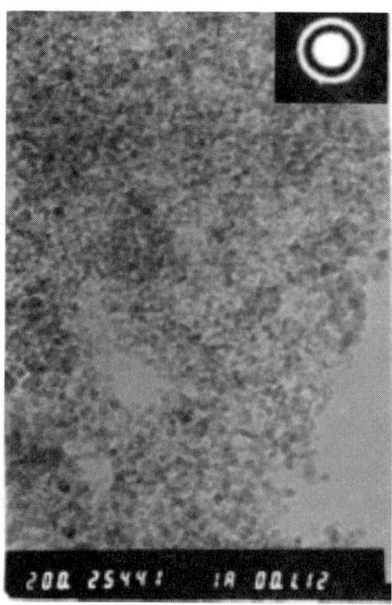

Fig. 35—The TEM and ED image of UDP (200,000×).

TABLE 6—Main parameters of synthesized UDP by detonation of explosives.	
Technical Parameters	**Description**
Crystal structure	Single crystal and polycrystal, single crystal appearing as cubiccrystal structure, and lattice constant is 0.3559–0.3570 nm.
Mineralogical properties	1b style diamond
Size of crystal	1–50 nm (distribution); 5–6 nm (average)
Specific surface area	200–420 m^2/g
Specific gravity	>3 g/cm^3
Initial oxidation temperature	700 K (in air)
Productivity of diamond	Up to 10% of the detonator weight
Chemical component	86–90 % C, 0.1–1 % H, 2.5–3.7 % N, 8.9–11.4 % O

TABLE 7—Parameters of lubricants.

Lubricants	η_0/mPas	n	Lubricant	η_0/mPas	n
PEG	98 (28°C)	1.458	PE	102.9 (23°C)	1.471
...	...	...	PE +0.05 wt % UDP	104.1	1.472
PEG +0.1 wt % UDP	110.6	1.459	PE +0.1 wt % UDP	112.3	1.482
PEG +0.3 wt % UDP	123.5	1.460	PE +0.3 wt % UDP	125.6	1.475
PEG +0.5 wt % UDP	134.9	1.462	PE +0.5 wt % UDP	136.8	1.476

Note: PEG is Polyethylene glycol and PE is Polyester, n is refractive index.

(PE). Then, the mixtures were placed in an ultrasonic cleaner for 15 minutes to obtain a uniform mixture oil. The viscosity and the optical refractive index of these base oils and mixture oils were given in Table 7.

In order to explore the effect of UDP on tribological properties of lubricant oils, especially on friction force, a glass disk driven by a motor and a steel ball fixed on a mount lubricated by PE+0.5 % UDP or pure PE rub each other for 30 minutes under a load of 4 N at a speed of 2 mm/s. Then, the SEM images of the wear scar, which is about 0.3 mm in diameter on the steel ball were observed (shown in Fig. 36). The surface of the steel ball without rubbing (Fig. 36(a)) is very smooth and that lubricated by PE+0.5 % UDP (Fig. 36(b)) exhibits some smooth micro-grooves on it due to hard UDP particles' plowing. However, the rubbing surface lubricated by pure PE (Fig. 36(c)) is much rougher than the former and many pits or spalls can be observed due to contact fatigue and adhesive fatigue [60,61].

The logarithmic profiles of film thickness versus rolling speed for the mixture oils with different concentrations of UDP are shown in Fig. 37. Under the pressure of 174 MPa shown in Fig. 37(a), the film of pure PE liquid obeys the EHL theory in the higher speed region. When the film thickness decreases to about 26 nm with reducing speed, it deviates from EHL theory and is larger than that calculated from EHL theory. Due to the strong adsorption between the metal surface and the functional group (-COOR) in PE, the film maintains a certain thickness and load-carrying capability even at a very low speed. When UDP are added into PE, the variation trend of the film thickness with rolling speed is similar to that of pure PE liquid. However, the addition of UDP causes a greater deviation of the film thickness than that of pure PE in the low speed region because of the enhancement in viscosity resulting from the inlet viscosity effect and UDP nano-particle adsorption effect [61,62]. In addition, the more UDPs added, the thicker the films produced.

The cases under the pressure of 297 MPa as shown in Fig. 37(b) are similar to that under 174 MPa, except that the films are a little thinner than that under the latter pressure. Furthermore, when speed is less than 1 mm/s, the film thickness of pure PE under a higher pressure decreases by a larger slope with the rolling speed, and there is little difference in the film thickness of UDP-containing PEs under different loads.

The relationship between the friction coefficient and the sliding distance at different sliding speeds is shown in Fig. 38. At the beginning, the friction coefficient of PEG +0.3 wt. % UDP is almost the same for different sliding speeds, then decreases with the sliding distance, and at last trends to a constant. At the sliding speed of 4 mm/s, the friction coefficient reduces to the lowest stable value of 0.055, that at 0.4 mm/s falls off to 0.095, and that at 0.2 mm/s to 0.125. Therefore, with the same sliding distance, higher sliding speed produces a lower friction coefficient. It indicates that except for the abrasive wear effect of hard UDP particles, another new friction force-decreasing mechanism takes effect. With the increase of speed, the film becomes thicker and then the collision probability between asperities of two surfaces and that between UDP particles and surface asperities will be lessened. Therefore, the friction force at a higher speed is smaller than that at a lower speed.

The concentration of UDP also affects the friction coefficient as shown in Fig. 39. It is discovered that the friction coefficient of pure PEG also decreases gradually and reaches a somewhat reduced value due to a time effect of the film thickness [16,18]. At the speed of 2 mm/s and pressure of 174 GPa, the friction coefficient of pure PEG is the highest. That for PEG+0.5 % UDP ranks second. Those for PEG +0.1 % UDP and PEG+0.3 %UDP are almost the same and have the lowest friction coefficient among all tested oils. Therefore, there is a good concentration extent of UDP in the basic oil. If the concentration is out of such extent, the effect

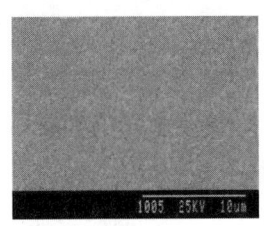

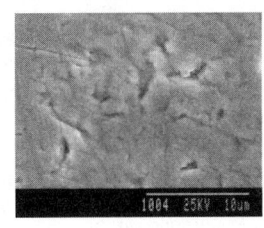

(a) Without rubbing   (b) PE+0.5%UDP   (c) PE

Fig. 36—The SEM picture of scar on steel ball [60]. Load: 4 N, Rubbing time: 30 min, (a) Without rubbing, (b) PE+0.5 % UDP, (c) PE.

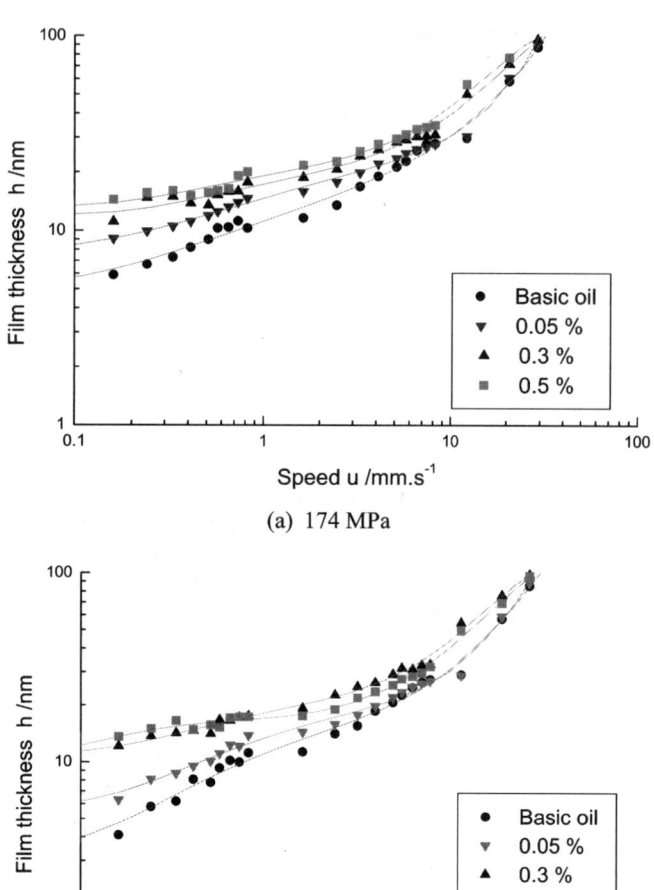

Fig. 37—Film thickness for different rolling speed [60]. Base oil: PE; Temperature: 20°C, Pressure: (a) 174 MPa, and (b) 297 MPa.

of UDP on friction becomes less. The reason that the friction coefficient decreases greatly with the sliding distance when UDP is added into the basic oil is proposed in Ref. [60]. UDP particles have the probability to run in the central region as

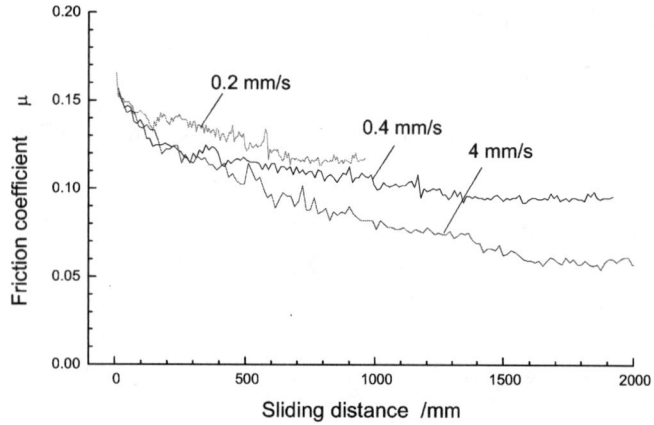

Fig. 38—Friction coefficient at different sliding speed [60]. Load: 2 N, Concentration of UDP: 0.3 %, Base oil: PEG.

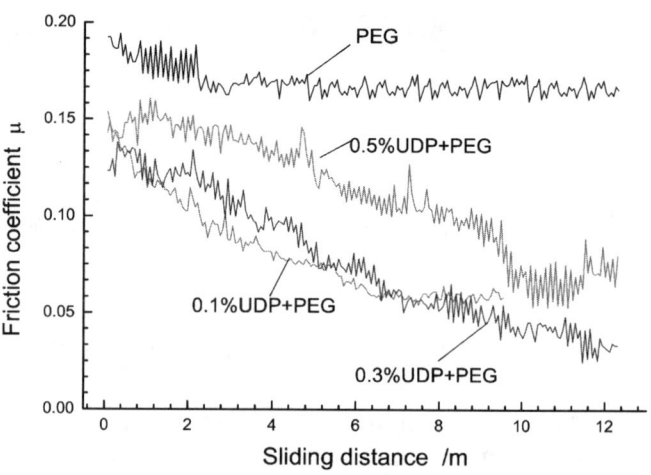

Fig. 39—Friction coefficient for different concentrations [60]. Load: 2 N, Base oil: PEG, Sliding speed: 2 mm·s^{-1}.

the tiny ball bearing when the micro-grooves have been formed as shown in Fig. 36(b). At the beginning, the hard UDP particles plow the surfaces under the load, and the friction force is mainly composed of mechanical plowing forces. With the increase of the sliding distance, the micro-grooves are formed and become smoother, and then more spherical UDP particles are brought into the contact area rolling as tiny ball bearings so that cracks and pits in the solid surface are avoided. However, their supposition is very difficult to be proven because it is difficult to observe the particles' rolling in the running process directly in the experiment. In addition, a greater amount of UDP results in an enhancement in viscosity and then friction.

Figure 40 shows the relationship between the friction coefficient and the sliding speed for PEGs with different concentrations of UDP. It is clear that the friction coefficient of all lubricants is about 0.22 at the speed of about 0.1 mm/s. With increasing sliding speed, it drops sharply in the speed range from 0.1 mm/s to 15 mm/s, and then maintains about 0.03 when the speed is more than 15 mm/s.

It can be concluded that the spherical diamond nano-

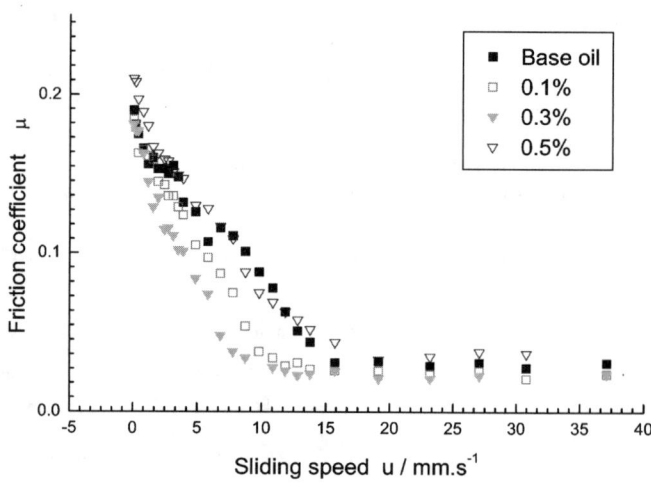

Fig. 40—Friction coefficient versus sliding speed [60]. Base oil: PEG, Temperature: 20°C, Maximum Hertz pressure 219 MPa.

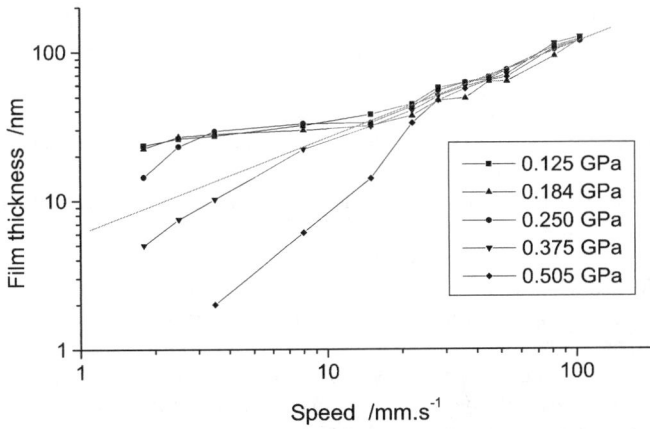

Fig. 41—Film thickness with pressure [19]. Lubricant: Polyglycol oil 1, Temperature: 28°C.

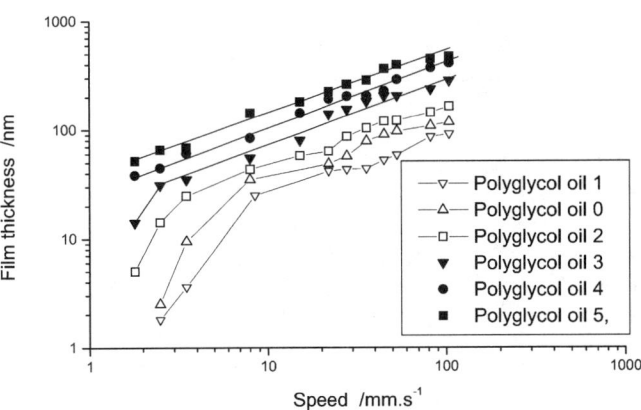

Fig. 42—Film thickness versus viscosity [19]. Pressure: 0.505 GPa, Temperature: 28°C.

particles in base oil exhibit a viscosity-increasing effect making a thicker film in the pure rolling state in the TFL regime, and a good antifriction property. The friction coefficient of UDP-containing oils decreases with sliding distance. Higher sliding speed leads to a larger decreasing slope of the friction coefficient with the sliding distance and a smaller stable friction coefficient. There is an optimal concentration range of UDP from 0.1 % to 0.3 % causing the lowest friction force.

3 The Failure of Lubricant Film

The failure of the fluid film at nano-scale and the relation between the failure point and pressure, velocity, and viscosity have been investigated [19]. The lubricants used in the experiments are given in Table 1.

The variation of the film thickness of Polyglycol oil No. 1 with the velocity appears in different forms under different pressures as shown in Fig. 41. Under the pressures lower than 0.250 GPa, an EHL phenomenon is observed in the higher speed range (>20 mm/s). In such a case, the film thickness increases linearly with the velocity in logarithmic coordinates. However, when the film reduces to a critical thickness (about 30 nm), the further decrease in velocity only can lead to little drops in thickness and this point is usually known as the transition from EHL to TFL because the EHL rules are no longer applicable [4,18]. When the speed decreases to sufficiently low, another point (at the speed about 2.5 mm/s) has been reached at which the film thickness rapidly drops from 25 nm to 12 nm. Because the surface roughness R_a is about 2 nm in the contact region and much smaller than that of the free surface due to the elastic deformation of the ball [16], so the film thickness is three times thicker than the surface roughness and a full lubricant film has been formed in the contact region. Thus, the sudden drop in the film thickness indicates that the hydrodynamic effect becomes too small to bring liquid molecules into the contact region and the fluid film cannot support the applied pressure, and thus, the failure of the fluid film has taken place. When the pressure changes from 0.125 GPa to 0.505 GPa, the failure point occurs at different velocities for different pressures as shown in Fig. 41. The failure tends to happen at higher velocity values under higher applied pressures; in other words, a larger hydrodynamic effect is required to sustain the enhanced pressure if

the fluid film wants to be guaranteed in the contact region.

Figure 42 shows the failure of different viscosities of lubricants. Under the pressure of 0.505 GPa, there is also no film failure occurring when the viscosity is above 674 mm^2/s (Polyglycol oil 4 and 5). However, when the velocity is reduced to about 2 mm/s for the lubricant with the viscosity of 329 mm^2/s (Polyglycol oil 3), the fluid film collapsed. For the lower viscosity lubricants (Polyglycol oil 0, 1, and 2), the failure happens at higher velocities. A similar phenomenon also can be seen in Fig. 43 [53]. The hexadecane ($C_{16}H_{34}$) film did not fail under the pressure of 0.185 GPa in the speed ranging from 10 mm/s to 500 mm/s. However, for the lubricants of shorter carbon chains ($C_{12}H_{26}$, $C_{10}H_{22}$, C_8H_{18}), the failure points can be clearly seen, and the shorter the carbon chain is, the higher the speed can be found at which the failure point appears. Therefore, the lower the lubricant viscosity is, the higher the velocity is required to maintain a successful fluid film in the contact region.

The failure of lubricant film is also related to characteristics of lubricants as shown in Fig. 44. Hexadecanethiol ($C_{16}H_{34}S$) film changes from EHL to TFL at the speed of 120 mm/s and no failure point appears in the speed range

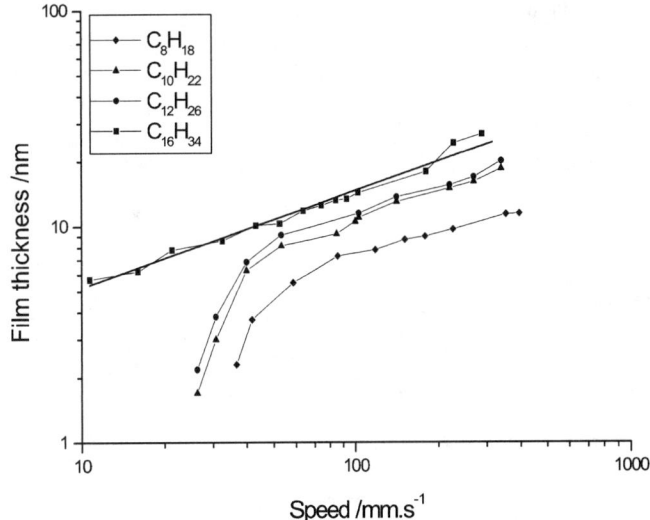

Fig. 43—Film thickness with different lubricant [42]. Pressure: 0.185 GPa, Temperature: 25°C.

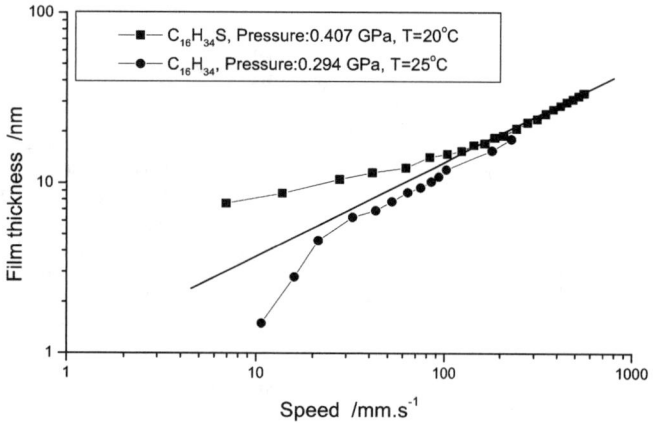

Fig. 44—Film thickness with speed [35].

from 7 mm/s to 553 mm/s under a heavy pressure of 0.407 GPa. However, failure point of hexadecane film appears at the speed of 21 mm/s under the pressure of 0.294 GPa. This indicates that the polar molecules, e.g., hexadecanethiol, form a TFL film, which is difficult to collapse.

In order to set up the relationship among the pressure, speed, and viscosity of lubricants without polar additives, Luo et al. [19] define the fluid factor L as:

$$L = \eta_k u^{0.69} \quad (8)$$

where η_k is the kinetic viscosity of lubricant in mm²/s; u is the velocity in mm/s. The velocity index 0.69 is determined from the experimental data according to the rule that the same failure fluid factor L_f with different failure velocities and failure viscosity should meet the same pressure where the fluid film failure has taken place. The curve in Fig. 45 is called the failure curve where the lubricants are of alkyl without any polar additives. If the fluid factor L ($\eta_k u^{0.69}$) in the point contact region of a pure rolling system is in the region above the curve in Fig. 45, the fluid film will be formed and the lubrication is in the TFL or EHL regime. Whereas if it is in the region below the curve, the fluid film will fail and the lubrication regime becomes boundary lubrication or mixed lubrication. Therefore, when the film thickness is thicker than three times of surface roughness R_a, the relation between failure fluid factor L_f and failure pressure P_f at which the failure of the fluid film takes place is [19]:

$$\eta_k u^{0.69} - kP_f^2 = 0 \quad (9)$$

The unit of P_f here is MPa; η_k is the kinetic viscosity of lubricant in mm²/s; u is the velocity in mm/s, for the oil without polar additives, k is 23.5×10^4. If the tribo-pairs need to be lubricated with the fluid film in the TFL and EHL regime, the lubricant and the rolling speed should be chosen according to the pressure applied as Eq (9) so as to make that the liquid factor L is larger than the failure fluid factor L_f. Otherwise, the liquid film cannot be maintained under the pressure added.

The mechanism of the film failure is due to the limiting shear stress of the fluid film in the nano-scale [34,53]. According to Newtonian fluid theory:

$$\tau = \eta \frac{\partial u}{\partial z} \quad (10)$$

when the relative speed u and the lubricant viscosity η are constant, the shear stress τ will increase as the film thickness decreases. If the thickness is close to zero, shear rate $\partial u / \partial z$ and the stress τ will become infinitively large according to Eq (10). However, the maximum shear stress, τ_{max}, cannot be larger than the maximum inter-force between lubricant molecules, and therefore, a critical thickness must exist. If film thickness is approaching the critical value, τ is close to the limiting shear stress [63], and it will not increase anymore when the shear rate increases further. When there is an increase in pressure and the film is thinner than the critical thickness, the decrease of the film thickness just causes the increase of shear rate, but not the increase of shear stress. Therefore, the liquid film cannot support the extra pressure, and it will collapse.

4 Thin Film Lubrication of Ionic Liquids

Recently, room temperature ionic liquids (RT-ILs) have attracted much attention for their excellent properties, e.g., wide temperature range of liquid phase, ultra-low vapor pressure, chemical stability, potential as "green" solvents, and high heat capacities [64,65]. These properties make them good candidates for the use in many fields, such as thermal storage [66], electrochemical applications, homogeneous catalysis [67], dye sensitized solar cells [68], and lubricants [69,70].

It was reported that RT-ILs could act as a universal lubricant for different sliding pairs such as steel/steel, steel/aluminum, steel/copper, steel/silica, and steel/silicon nitride. They exhibit favorable friction properties with reduced wear and high load-bearing capacity [70]. A group at TU Ilmenau recently studied the electronic structure of typical RT ILs (namely [emim][Tf$_2$N]) deposited on a polycrystalline Au surface and studied with different techniques such as MIES, UPS, XPS, and HREELS. After preparation, the sample was investigated by XPS using monochromatic Al Kα radiation. As seen in a survey spectrum (Fig. 46), a closed [emim][Tf$_2$N] film was successfully prepared. The expected elements F, N, O, S, and C are detected and no evidence of impurities could be found. On another hand, tribology measurements of the above ionic liquid have also been performed in ambient to investigate its interfacial behavior when confined to nanometres or micrometres thickness. With a trace amount of

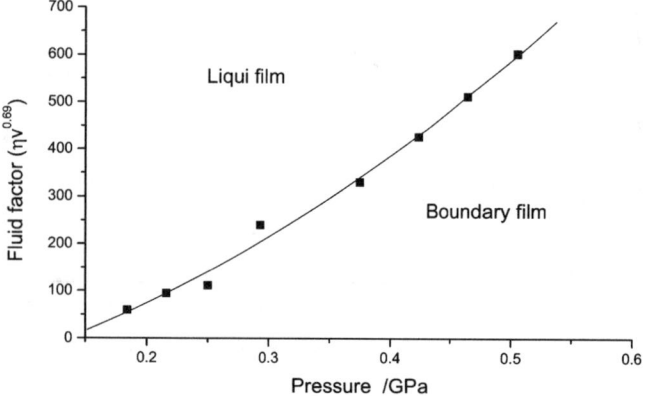

Fig. 45—Failure of liquid film [19].

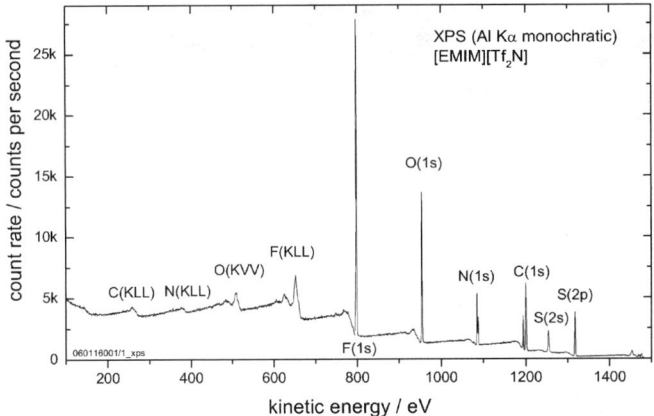

Fig. 46—XPS spectrum of [emim][Tf$_2$N] thin film deposited on a polycrystalline Au substrate.

ionic liquid ([emim][Tf$_2$N]) confined between the surfaces of a nano-scale smooth steel sphere and a glass plate, its interfacial adhesive force has been investigated by Zhang et al. [71]. For a critical confined volume of pitolitres to nanolitres, and a critical clearance of tens to hundreds of nanometres, confinement-induced spontaneous spreading and abrupt shrinking were observed, accompanied by the presence of a stable interfacial adhesive force of remarkable magnitude. This spreading/shrinking-induced adhesive force was proven to be fundamentally different from the common meniscus capillary force and was considered to stem from confinement-induced solidification according to our further investigations. Additional, load-carrying and shearing behavior of ILs thin film confined at a sphere-disk contact area was investigated and compared to conventional lubricant Pao (poly-α-olefin) oil. As seen in Fig. 47, ILs thin film confined at a sphere-disk contact area (Si$_3$N$_4$ ball and glass substrate) can sustain a pressure about several hundred MPa

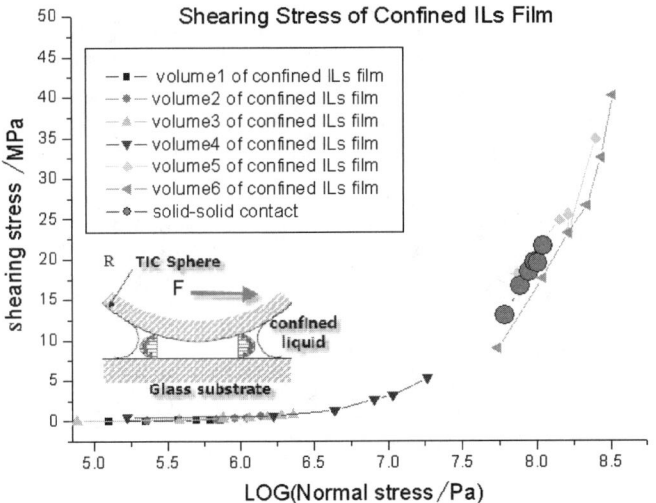

Fig. 47—Load-carrying and shearing behavior of confined ILs thin film. Liquid volume is decreased from volume 1 to volume 6 corresponding to a decreasing thickness of ILs films. The confined thin film of ILs exists at the contact area under a normal load of hundreds of MPa and undertakes shearing stress like a solid-solid contact.

without breakage, and meanwhile undertakes a significant shearing stress as a solid-solid contact. These results indicate that ionic liquids can work as lubricants under severe operating conditions and act as an effective anti-wear boundary film, which is attributed to its strong intermolecular force and its inherent polarity structure.

RT-ILs are also selected as lubricant additives. Usage of ionic liquids as boundary lubricant additives for water has resulted in dramatically reduced running-in periods for silicon nitride materials [70]. When ILs were mixed into a neat mineral oil, the mixture has proven to produce lower wear on aluminum flat than either the oil or the ionic liquid alone [72], which indicates that a small amount of ILs in the mineral oil may function as an anti-wear additive.

However, investigations up to now have mainly concentrated themselves on ambient environments even though it is known that ionic liquids have a very low vapor pressure, making them suitable for vacuum applications such as in space mechanisms, the disk drive industry, and microelectromechanical systems (MEMS). Due to the ultra-low vapor pressure of most ionic liquids, they have been expected to be good lubricants in vacuum. Further experimental works are required to evaluate lubrication behavior of ionic liquids under ultra-high vacuum conditions and in inert atmospheres.

5 Gas Bubble in Liquid Film under External Electric Field

In the past decade, effects of an EEF on the properties of lubrication and wear have attracted significant attention. Many experimental results indicate that the friction coefficient changes with the intensity of the EEF on tribo-pairs. These phenomena are thought to be that the EEF can enhance the electrochemical reaction between lubricants and the surfaces of tribo-pairs, change the tropism of polar lubricant molecules, or help the formation of ordered lubricant molecular layers [51,73–77]. An instrument for measuring lubricant film thickness with a technique of the relative optical interference intensity (ROII) has been developed by Luo et al. [4,48,51,78] to capture such real-time interference fringes and to study the phenomenon when an EEF is applied, which is helpful to the understanding of the mechanism of thin film lubrication under the action of the EEF.

An interesting phenomenon that some gas microbubbles emerged in the thin liquid film in a nanogap under an EEF was observed by Luo's group [78–80]. They have investigated the influence factors on and the mechanism of the emergence of these micro-bubbles.

The schematic experimental set and the measuring principle are shown in Fig. 24. On the upper surface of the glass disk, there is an anti-reflective coating (Fig. 24(a)). A liquid film is formed between the lower surface of the glass disk coated with a semi-reflective chromium layer and the surface of a highly polished steel ball (E52100) with a diameter of 22.23 mm. The appearance of the liquid film is exhibited by the interference pattern [4]. In order to investigate the effects of an EEF on the films of liquids with different characteristics, four kinds of pure chemical liquids were used in the experiment listed in Table 8 [78,80], and the real-time interference patterns of the films were observed under an external voltage increasing gradually from 0 V to the voltage at which the gas micro-bubbles appeared. The load applied to

TABLE 8—Physical parameters of experimental liquid samples (23°C and normal atmosphere pressure).

	Glycerin	Glycol	Hexadecane	Liquid Paraffin
Molecular formula	$(CH_2OH)_2CHOH$	$(CH_2OH)_2$	C_nH_{2n+2}, $n=16$	C_nH_{2n+2}, $n=14\sim18$
Viscosity (mPa·s)	1,069	19.9	3.591	30
Boiling point (°C)	290.0	197.8	286.8	>340
Water content (%)	0.175	2.940	1.400	0.374
Relative permittivity	42.5	38.66	2.046	2.2~4.7
Electric conductivity ($\Omega^{-1}\cdot m^{-1}$)	6.40×10^{-6}	3.00×10^{-5}	2.04×10^{-10}	2.00×10^{-10}
Polarity	Polar	Polar	Nonpolar	Nonpolar

the contact point was from 4 N to 20 N, and the corresponding Hertz pressure inside the central region was from 0.15 GPa to 0.26 GPa, respectively. Their experiments were performed at a room temperature of 23 ± 1°C.

The interference pattern of the static film of glycerin without applying any EEF is shown as Fig. 48(a). The central white region is a constructive fringe or Hertz contact region, where the film thickness is 10 nm and its radius is 117 μm under the load of 4 N. When the external voltage was lower than 6 V (the corresponding field intensity was 19 kV/cm), the interference pattern remained unchanged. As soon as the voltage was raised to 6 V, there were two places around the edge of the central region as shown in Fig. 48(b), where the bubbles about a few micrometres in size emerged one after another, and then moved off the central region roughly in the radial direction as shown in Fig. 48(b)–48(f). This process was also observed by a video camera. The value (6 V) is defined as the critical voltage of the micro-bubbles emerging in liquid glycerin. When the time reached 0.2 second under such a voltage, the number of the places where the micro-bubbles emerged increased to 7, and 9 at 0.5 second, and more than 40 at 1.0 second (Fig. 48(b)–48(d)). Then it became stable as time increased further (Fig. 48(f)).

Luo and He [78,80] also observed a similar phenomenon in glycol under the EEF, and the critical voltage of the micro-bubbles emerging became 5 V and the film thickness in the central region was 6 nm (the corresponding field intensity was about 17 kV/cm). For hexadecane and liquid paraffin, the critical voltages became 9 V and 7 V, and the film thicknesses in the central region were 4 nm and 7 nm with the corresponding field intensities of 571 kV/cm and 302 kV/cm, respectively. However, the number of the places where the micro-bubbles emerged in hexadecane and liquid paraffin under their critical voltages decreased to 0 in 100 ~300 seconds. Moreover, the sizes of the micro-bubbles in glycerin and glycol were bigger than those in hexadecane and liquid paraffin as shown in Fig. 49. From Table 8, it can be seen that the number of the micro-bubbles that emerge in polar liquids is much bigger than that in nonpolar ones, and also the size of the micro-bubbles in the former are larger than that in the latter ones.

The interference pattern without an EEF is shown in Fig. 50(a). As the external voltage was raised, the number of the places where micro-bubbles emerged increased (Fig.

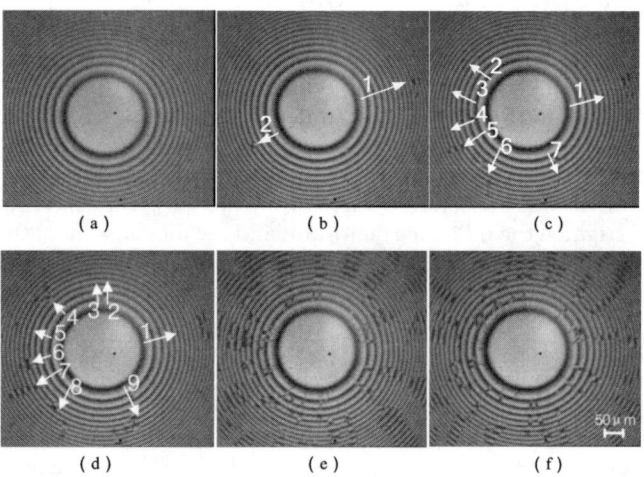

Fig. 48—Interference patterns of the film of glycerin under a load of 4 N. The film thickness in the central region was 10 nm, 105 nm in the first dark ring, and about 315 nm in the second dark ring. (a) External voltage was 0 V, which means no EEF was applied to the liquid film. (b)–(f) The external voltage was kept constant (6 V) for a time of "t," and (b) t=0 s, (c) t=0.5 s, (d) t=1.0 s, (e) t=5.0 s, (f) t=100 s.

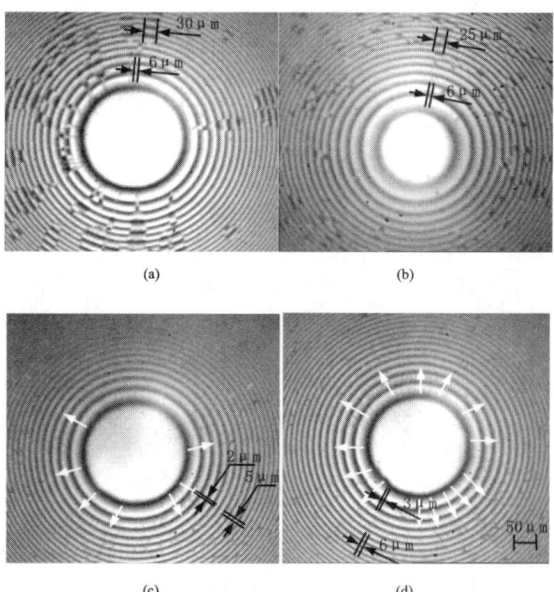

Fig. 49—Interference patterns of four pure chemical liquid films. All of the patterns were caught when the respective voltages were kept for 0.5 s. Load is 4 N and the scale bar is 50 μm. (a) glycerin, 6 V; (b) glycol, 5 V; (c) hexadecane, 9 V; (d) liquid paraffin, 7 V. The black arrows help to show the micro-bubbles' moving routes.

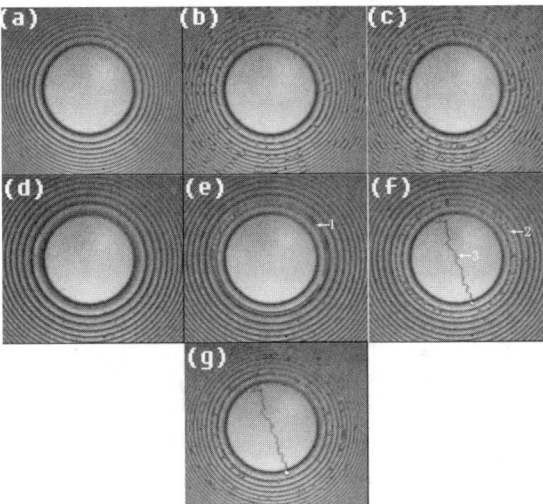

Fig. 50—Interference patterns of the lubricant film. Liquid: glycerin. Load: 20 N. All of the patterns were caught when each external voltage had been kept for five minutes, and each pattern was taken under the corresponding external voltages as follows: (a) 0 V, (b) 10 V, (c) 20 V, (d) 40 V, (e) 60 V, (f) 80 V, (g) removal of the dc field.

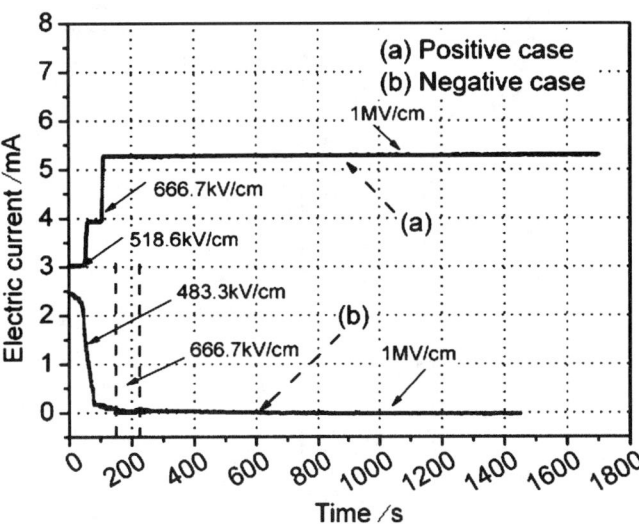

Fig. 51—Dependence of electric current on time for different EEF polarities. Liquid: glycerin. Load: 4 N. Line (a) and (b) represent the positive and negative cases, respectively. Positive EEF intensities of 518.6 kV/cm during the initial period of 60 s then 668.7 kV/cm during the second period of 60 s, finally 1 MV/cm during the remaining period were applied (Line (a)); Negative EEF intensities of 483.3 kV/cm during the initial period of 150 s then 668.7 kV/cm, finally 1 MV/cm during the remaining period were applied (Line (b)).

50(b)–50(f)) and these micro-bubbles moved more quickly (as being observed in video). When the voltage was greater than or equal to 40 V, the micro-bubble became too many to be distinguished from one another (Fig. 50(d)–50(f)). As the voltage reached 60 V or higher, because the speed of micro-bubble emergence was greater than that of the micro-bubble movement, many micro-bubbles could not be displaced in time, and the accumulation of the micro-bubbles near the edge of the central region was formed as shown in Arrow 1 in Fig. 50(e) and Arrow 2 in Fig. 50(f). As soon as the external voltage reached 80 V, the electric field intensity inside the central region became strong enough to cause the breakdown of the SiO_2 layer, as indicated by the black curve in the central region (Arrow 3 in Fig. 50(f)). When the EEF was removed, the micro-bubbles' emerging speed and moving velocity decreased rapidly over time (Fig. 50(g)), and all the micro-bubbles disappeared after ten minutes.

The electric current in the circuit during the process of the emergence of the micro-bubbles was detected by Xie and Luo [81]. Based on this observation, the EEF polarity effect on the emergence of these micro-bubbles in the glycerin film was examined and the results are shown in Figs. 51 and 52. The SiO_2 layer was removed in order to avoid the qualitative nature in analyzing the micro-bubble emergence. The variation curves of the electric current in Fig. 51 were recorded by a sampling resistance with an oscilloscope. Line (a) represents the positive polarity case with the EEF intensity increasing from 518.6 kV/cm where micro-bubbles began to emerge in the edge area as shown in Fig. 52(a) to 666.7 kV/cm, and then to 1 MV/cm where the electric current hardly changed even to 1,700 s. A very slight reduction in the micro-bubble intensity could be seen by comparing Fig. 52(b) (t = 200 s) with Fig. 52(c) (t = 1,600 s). However, in the case of the negative polarity, there was a decreasing trend of the electric current in the initial period (Line (b) in Fig. 51), and the amount of the micro-bubbles also decreased nearly at the same pace as the electric current, which indicated that the emergence of the micro-bubbles was related to the electric current. Two snapshots of the negative polarity at the time instants of 10 s and 100 s are shown in Fig. 52(d) and 52(e). When the electric current dropped to a very low level, these micro-bubbles began to disappear. When the EEF intensity was raised to 666.7 kV/cm and 1 MV/cm, neither the electric current nor the number of the micro-bubbles (Fig. 52(f)) was enhanced.

The film thickness in the contact region was measured in order to find its correlation with the micro-bubble emerging. The dependence of the average film thickness of glycerin on time at different positive EEF intensities is shown in Fig. 53. At a lower intensity (518.6 kV/cm) where micro-bubbles began to emerge, the film thickness did not change too much over time. When it increased to 666.7 kV/cm, a general slight decrease in the film thickness could be seen, and correspondingly the amount and intensity of the micro-bubble emerging increased as compared with that in the case of the lower EEF intensity of 518.6 kV/cm. When the EEF intensity increased further to 1 MV/cm, a more rapid reduction trend of the film thickness could be observed, which dropped by nearly 30 % at the initial period of about 100 s. Also, a more intensive micro-bubble emergence can be observed in the experiment. Then the film thickness tended to level off, at least a slow variation in the remaining 500 s. For the hexadecane film, as a representative nonpolar liquid film, the thickness hardly changed at all.

The chemical composition of the glycerin liquid after the EEF test was measured with a Raman microscopy as shown in Fig. 54. Curve (a) is a typical Raman spectrum of glycerin without any EEF applied, and Curve (b) is the Raman spectrum of the glycerin after the positive EEF intensity

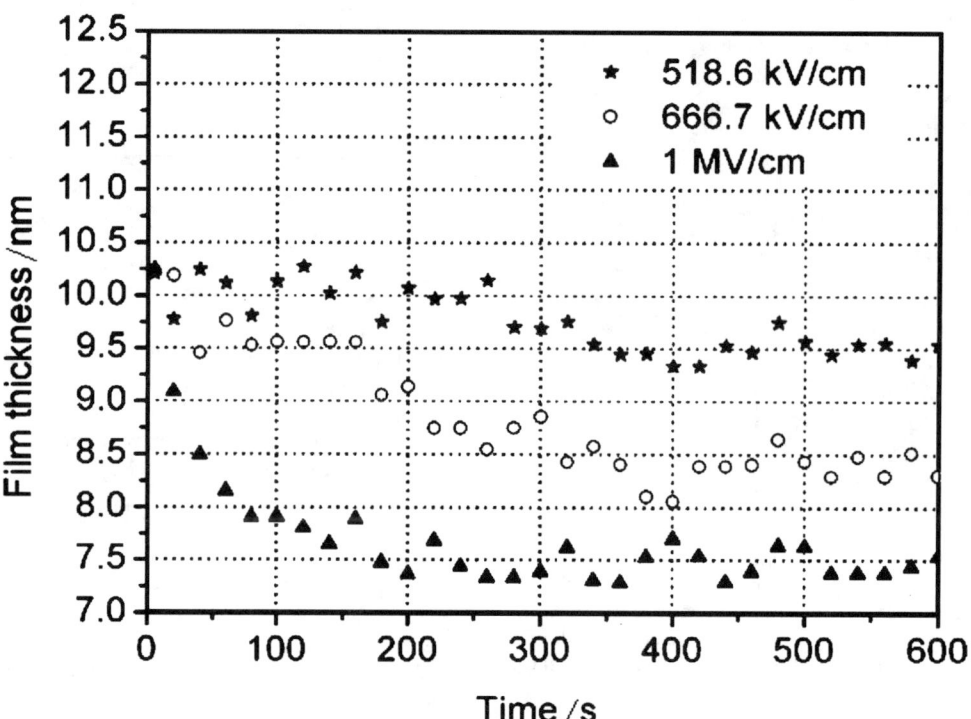

Fig. 52—Interference patterns of glycerin film at different EEF polarities. (a) positive EEF intensity=518.6 kV/cm, t=10 s; (b) positive EEF intensity=1 MV/cm, t=200 s; (c) positive EEF intensity=1 MV/cm, t=1,600 s; (d) negative EEF intensity=483.3 kV/cm, t=10 s; (e) negative EEF intensity=483.3 kV/cm, t=100 s; (f) negative EEF intensity=1 MV/cm, t=1,400 s.

of 1 MV/cm was applied for 60 minutes. Clearly, no difference could be observed in the chemical composition between the glycerin liquids before and after the EEF was applied. Furthermore, the micro-bubbles emerging near the contact region tended to merge together to form bigger ones away from the contact region. After the experiment, a lot of micro-bubbles could be seen under atmospheric circumstance. Nevertheless, most of the micro-bubbles disappeared in a short time period; only a few micro-bubbles could exist for several hours.

The overheating effect is believed to play a dominant role in explaining the physical mechanism of the micro-bubble formation, and three main reasons could be underpinned: First, it was observed that a small amount of the glycerin injected into the gap would eventually disappear after a few hours at the positive EEF intensity of 1 MV/cm. Second, no remarkable difference in the chemical composition between the glycerin before and after the EEF was applied could be found in the experiment, which indicated physical effects might predominate. Besides, a rough estimation of the temperature rise in the contact region due to the electro-thermal effect will be conducted as follows.

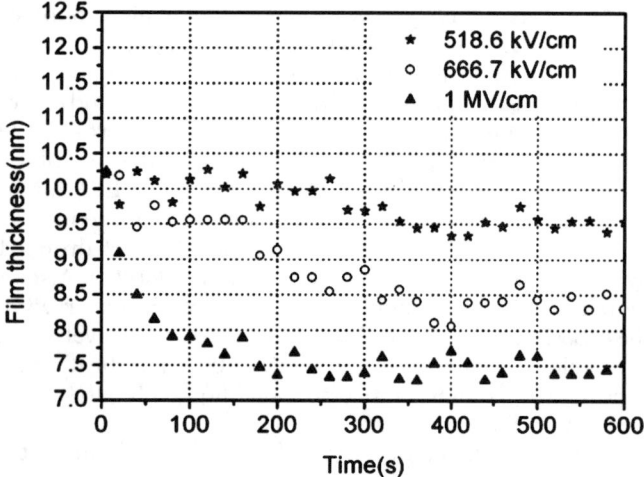

Fig. 53—Dependence of the average thickness of the glycerin film in the contact region over time at positive EEF intensities of 518.6 kV/cm (★), 666.7 kV/cm (○), and 1 MV/cm (▲).

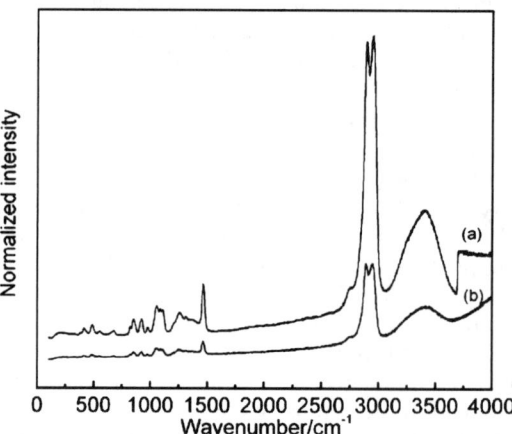

Fig. 54—Raman spectra of the tested glycerin liquid. (a) before EEF applied; (b) after a positive EEF intensity of 1 MV/cm applied for 60 minutes.

The contact region is simplified as a cylindrical heated volume with only lateral heat exchange, and it is an optimized case that the heat flow towards the interfaces is neglected. The heat flow Q at a steady state can be given by [82]

$$Q = k\frac{dT}{dx} \quad (11)$$

where k is the thermal conductivity of the tested liquid, dT/dx is the temperature gradient around the heated region. If r is the contact radius, and the film thickness in the contact region is defined as h, the heating power P can be written as:

$$P = 2\pi h r k \frac{dT}{dx} \quad (12)$$

On the other hand, the power generating in the liquid film due to the electro-thermal effect is given by [83]

$$P = \pi r^2 h \sigma E^2 = \pi r^2 h \sigma \frac{U^2}{h^2} \quad (13)$$

where σ is the electric conductivity. By a comparison of Eqs (12) and (13), and considering $dx \approx r$, it can be obtained

$$\Delta T = dT \approx \frac{\sigma U^2 r^2}{2 k h^2} \quad (14)$$

The temperature rise in the contact region can be evaluated based on Eq (14), and sufficient heat would gasify the liquid film when the boiling point is reached. Moreover, it is to be noted that a relatively high contact pressure would cause an increase in the boiling point, and it is known that the pressure p_r in the contact region is nonuniform with a distribution, which can be written as

$$p_r = p_m \left(1 - \frac{x^2}{r^2}\right)^{1/2} \quad (15)$$

where x is the distance away from the central contact point, and the maximum Hertz contact pressure $p_m = 3F/2\pi r^2$, F is the applied load. Thus, the boiling point of the liquid film will increase due to the elevated contact pressure, and the boiling point T_b at a given pressure p_r can be determined by the Clausius-Clapeyron equation [84]

$$\ln\left(\frac{p_r}{p_0}\right) = \frac{\Delta H}{R}\left(\frac{1}{T_0} - \frac{1}{T_b}\right) \quad (16)$$

where p_0 and T_0 are the atmospheric pressure and the corresponding boiling point, respectively. $R (= 8.3145 \text{ J mol}^{-1} \text{ K}^{-1})$ and ΔH is the gas constant and enthalpy of vaporization (62,490 J/mol for glycerin, 51,500 J/mol for hexadecane), respectively. Substituting Eq (15) into Eq (16), the boiling points T_b at different locations along the radial direction in the contact region become

$$T_b = 1 \left/ \left[\frac{1}{T_0} - \frac{R\left(\ln(3F) - \ln(2\pi r^2) + \frac{1}{2}\ln\left(1 - \frac{x^2}{r^2}\right) - \ln p_0\right)}{\Delta H}\right]\right. \quad (17)$$

Equation (17) is plotted in Fig. 55 for the glycerin film (solid line) and the hexadecane film (dotted line) in the contact region ($x \leq r$). Clearly, a rapid decrease in the values of the boiling point in the edge region can be found. The calcu-

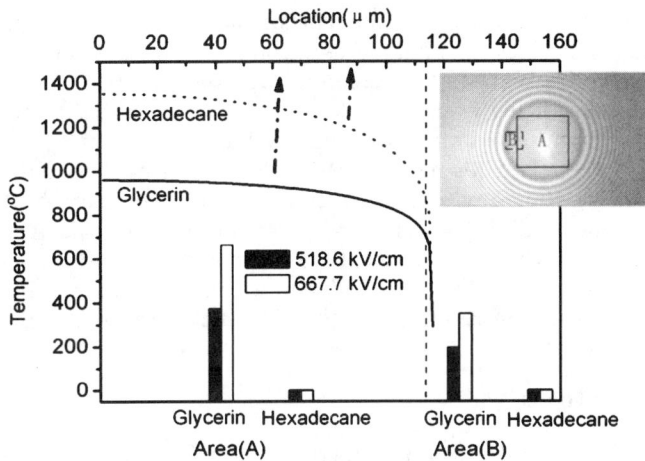

Fig. 55—Temperature calculation for different liquid films (glycerin and hexadecane) at different locations in the contact region: Area A is the central area in the inset photo and area B is the edge area. The filled histogram represents the positive EEF intensity of 518.6 kV/cm, and the empty one of 667.7 V/cm. The solid (glycerin) and dotted (hexadecane) lines are variation curves of the boiling point along the radial direction in the contact region.

lated temperature values for different liquid films, different EEF intensities, and different locations in the contact region based on Eq (14) are also shown in Fig. 55. Taking into consideration that the pressure is nonuniform and the pressure-viscosity of glycerin is not easy to be determined, two film thicknesses of area A and B (the inset photo of Fig. 55), were chosen to approximate the film thickness variation. Concretely, one represents the average film thickness of the central contact area A (9.6 nm for the glycerin film, 4 nm for the hexadecane film), and the other represents the thickness of the edge area B (13.7 nm for the glycerin film, 5.8 nm for the hexadecane film). The calculated temperature value of the central region in the glycerin film is nearly two times higher than the edge area at the threshold EEF of 518.6 kV/cm, e.g., the former temperature value is around 400°C and the latter one is 200°C. However, the value of 200°C in the edge area is closer to the corresponding boiling point, suggesting the micro-bubble is easier to emerge there. In contrast, the temperature rise is insignificant for the hexadecane film and too low to initiate nucleate boiling.

The reduction trend of film thickness in the central region is thought to be ascribed to the transfer of the lubricant from the contact region. Micro-bubbles are considered to generate in the edge zone, where a vacancy is left by the micro-bubbles. Then the vacancy can be partly replenished by the transfer of the lubricant molecules from the central contact region due to the difference of the contact pressure. Besides, a prominent variation of EEF along the radial direction outside the contact region is present due to the change in film thickness. It was proposed that most polar materials moved toward the place with the greatest field intensity due to the polarization effect in the nonuniform EEF, and such a phenomenon was called "dielectrophoresis" [85]. Thus, it is likely that the glycerin molecules can move from the outer region to the vicinity of the contact region. In this instance, the transfer of the glycerin molecules from inside the contact region and the effect of "dielectrophoresis" can collectively contribute to the replenishment for the vacancy left by the micro-bubbles. In other words, a competing mechanism between the micro-bubble emerging and the material replenishment of the glycerin molecules can be proposed to account for the variation in the film thickness.

The mechanism of the EEF polarity dependence of the micro-bubble emerging is believed to be that the electrolysis of water molecules absorbed plays an important role. The deposited Cr layer is susceptible to be oxidized, and cracks tend to form and propagate due to the interfacial stress between the oxidized film and the glass disk, resulting in the damage of the electrode.

6 Summary

In thin film lubrication (TFL), the film thickness is related to the surface energy of tribo-pairs, the index of effective viscosity of lubricants, molecular characteristics, rolling time, speed, load, dc voltage applied, and so on. TFL is a lubrication regime between boundary lubrication and elastohydrodynamic lubrication. The critical film thickness, at which the transition from EHL to TFL takes place, is related to the lubricant viscosity η_0 under the atmospheric pressure, the solid surface energy, molecular polarity of lubricants, etc. The typical kind of the lubricant film response to the TFL regime is recognized as an ordered liquid layer or a solid-like layer in which the viscosity becomes stronger as it is close to the solid surface.

References

[1] Johnston, G. J., Wayte, R., and Spikes, H. A., "The Measurement and Study of Very Thin Lubricant Films in Concentrate Contact," *STLE Tribol. Trans.*, Vol. 34, 1991, pp. 187–194.

[2] Wen, S. Z., "On Thin Film Lubrication," *Proceedings of 1st International Symposium on Tribology*, International Academic Publisher, Beijing, China, 1993, pp. 30–37.

[3] Luo, J. B., "Study on the Measurement and Experiments of Thin Film Lubrication," Ph.D. thesis (directed by S.Z. Wen), Tsinghua University, Beijing, China, 1994, pp. 10–50.

[4] Luo, J. B., Wen, S. Z., and Huang, P., "Thin Film Lubrication, Part I: The Transition between EHL and Thin Film Lubrication," *Wear*, Vol. 194, 1996, pp. 107–115.

[5] Bowden, F. P., and Tabor, D., *The Friction and Lubrication of Solid*, Oxford University Press, 1954, pp. 233–250.

[6] Adamson, A. W., *The Physical Chemistry of Surfaces*, Interscience, 3rd ed., New York, 1976, pp. 447–448.

[7] Kingsbury, E. P., "Some Aspects of the Thermal of a Boundary Lubrication," *J. Appl. Phys.*, Vol. 29, 1958, pp. 888–891.

[8] Cammera, A., "A Theory of Boundary Lubrication," *ASLE Trans.*, Vol. 2, 1959, pp. 195–198.

[9] Homola, A. M., and Israelachvili, J. N., "Fundamental Studies in Tribology: The Transition from Interfacial Friction of Undamaged Molecularly Smooth Surfaces to 'Normal' Friction with Wear," *Proceedings of the 5th International Congress on Tribology*, Finland, 1989, pp. 28–49.

[10] Dowson, D., and Higginson, G. R., "A Numerical Solution to the Elastohydrodynamic Problem," *J. Mech. Eng. Sci.*, Vol. 1, 1959, pp. 6–15.

[11] Hamrock, B. J., and Dowson, D., "Isothermal Elastohydrodynamic Lubrication of Point Contact: Part I—Theoretical Formulation," *ASME J. Lubr. Technol.*, Vol. 98, 1976, pp. 375–383.

[12] Archard, J. F., and Cowking, E. W., "Elastohydrodynamic Lubrication at Point Contacts," *Proceedings, Institute of Mechanical Engineering*, Vol. 180 (Part 3B), 1966, pp. 1965–1966.

[13] Cheng, H. S., and Sternlicht, B., "A Numerical Solution for the Pressure, Temperature and Film Thickness between Two Infinitely Long, Lubricated Rolling and Sliding Cylinders, under Heavy Loads," *ASME Trans. J. of Basic Engineering*, Vol. 87, 1965, pp. 695–707.

[14] Yang, P. R., and Wen, S. Z., "A Generalized Reynolds Equation Based on Non-Newtonian Thermal Elastohydrodynamic Lubrication," *Trans. ASME, J. Tribol.*, Vol. 112, 1990, pp. 631–639.

[15] Hu, Y. Z., and Granick, S., "Microscopic Study of Thin Film Lubrication and Its Contributions to Macroscopic Tribology," *Tribology Letter*, Vol. 5, 1998, pp. 81–88.

[16] Luo, J. B., and Wen, S. Z., "Study on the Mechanism and Characteristics of Thin Film Lubrication at Nanometer Scale," *Sci. China, Ser. A: Math., Phys., Astron. Technol. Sci.,*, Vol. 35, 1996, pp. 1312–1322.

[17] Luo, J. B., Wen, S. Z., and Li, K. Y., "The Effect of Substrate Energy on the Film Thickness at Nanometer Scale," *Lubr. Sci.*, Vol. 10, 1998, pp. 23–29.

[18] Luo, J. B., Huang, P., and Wen, S. Z., "Characteristics of Liquid Lubricant Films at the Nano-Scale," *Trans. ASME, J. Tribol.*, Vol. 121, 1999, pp. 872–878.

[19] Luo, J. B., Qian, L. M., Lui, S., and Wen, S. Z., "The Failure of Liquid Film at Nano-Scale," *STLE Tribol. Trans.*, Vol. 42, 1999, pp. 912–916.

[20] Tichy, J. A., "Modeling of Thin Film Lubrication," *STLE Tribol. Trans.*, Vol. 38, 1995, pp. 108–118.

[21] Tichy, J. A., "A Surface Layer Model for Thin Film Lubrication," *STLE Tribol. Trans.*, Vol. 38, 1995, pp. 577–582.

[22] Tichy, J. A., "A Porous Media Model of Thin Film Lubrication," *Trans. ASME, J. Tribol.*, Vol. 117, 1995, pp. 16–21.

[23] Matsuoka, H., and Kato, T., "An Ultral-thin Liquid Film Lubrication Theory—Calculation Method of Solvation Pressure and Its Application to the EHL Problem," *Trans. ASME, J. Tribol.*, Vol. 119, 1997, pp. 217–226.

[24] Hartl, M., Krupka, I., Poliscuk, R., Molimard, J., Querry, M., and Vergne, P., "Thin Film Colorimetric Interferometry," *STLE Tribol. Trans.*, Vol. 44, 2001, pp. 270–276.

[25] Gao, G. T., and Spikes, H. A., "Boundary Film Formation by Lubricant Base Fluid," *STLE Tribol. Trans.*, Vol. 39, 1996, pp. 448–454.

[26] Iliuc, I., *Tribology of Thin Layers*, Elsevier Scientific Publishing Company, 1980, pp. 58–59 (translated from the Romanian).

[27] Allen, C. M., and Drauglis, E., "Boundary Layer Lubrication: Monolayer or Multilayer," *Wear*, Vol. 14, 1969, pp. 363–384.

[28] Israelachvili, J. N., and Tabor, D., "The Measurement of van der Waals Dispersion Force in the Range of 1.5 nm to 130 nm," *Proc. R. Soc. London, Ser. A*, Vol. 331, 1972, pp. 19–38.

[29] Israelachvili, J., *Intermolecular and Surface Force*, 2nd ed., Academic Press, San Diego, CA, 1992.

[30] Alsten, J. V., and Granick, S., "Friction Measured with a Surface Forces Apparatus," *Tribol. Trans.*, Vol. 32, 1989, pp. 246–250.

[31] Granick, S., "Motions and Relaxation of Confined Liquid," *Science*, Vol. 253, 1991, pp. 1374–1379.

[32] Luengo, G., Schmitt, F., and Hill, R., "Thin Film Rheology and Tribology of Confined Polymer Melts: Contrasts with Bulk Properties," *Macromolecules*, Vol. 30, 1997, pp. 2482–2494.

[33] Luo, J. B., and Yan, C. N., "Fuzzy View Point in Lubricating Theory," *Lubrication Engineering*, Vol. 4, 1989, pp. 1–4 (in Chinese).

[34] Luo, J. B., Lu, X. C., and Wen, S. Z., "Developments and Unsolved Problems in Nano-Lubrication," *Progress in Natural Science*, Vol. 11, 2001, pp. 173–183.

[35] Liu, S. H., Ma, L. R., Zhang, C. H. et al., "Effect of Surface Hydrophilicity on the Confined Water Film," *Appl. Phys. Lett.*, Vol. 91, 2007, pp. 253110.

[36] Gao, G., and Spikes, H. A., "The Control of Friction by Molecular Fractionation of Base Fluid Mixtures at Metal Surface," *STLE Tribol. Trans.*, Vol. 40, 1997, pp. 461–469.

[37] Hartl, M., Krupka, I., Poliscuk, R., and Liska, M., "An Automatic System for Real-Time Evaluation of EHD Film Thickness and Shape Based on the Colorimetric Interferometry," *STLE Tribol. Trans.*, Vol. 42, 1999, pp. 303–309.

[38] Thompson, P. A., Grest, G. S., and Robbin, M. O., "Phase Transitions and Universal in Confined Films," *Phys. Rev. Lett.*, Vol. 68, 1992, pp. 3448–3451.

[39] Hu, Y. Z., and Granick, S., "Microscopic Study of Thin Film Lubrication and Its Contributions to Macroscopis," *Tribology Letters*, Vol. 5, 1998, pp. 81–88.

[40] Luo, J. B., and Wen, S. Z., "Progress and Problems in Nano-Tribology," *Chinese Science Bulletin*, Vol. 43, 1998, pp. 369–378.

[41] Jost, H. P., "Tribology: The First 25 Years and Beyond—Achievements, Shortcomings and Future Tasks," *"Tribology 2000," 8th International Colloquium, Technische Akademic Esslingen*, 1992, pp. 14–16.

[42] Hardy, W. B., and Doubleday, I., "Boundary Lubrication—The Paraffin Series," *Proc. R. Soc. London, Ser. A*, Vol. 100, 1922, pp. 550–574.

[43] Bhushan, B., *Introduction to Tribology*, John Wiley & Sons, Inc., New York, 2005.

[44] Hamrock, B. J., and Dowson, D., "Isothermal Elastohydrodynamic Lubrication of Point Contacts, Part III—Full Flooded Results," *ASME J. Lubr. Technol.*, Vol. 99, 1977, pp. 264–276.

[45] Hartl, M., Křupka, I., and Liška, M., "Experimental Study of Boundary Layers Formation by Thin Film Colorimetric Interferometry," *Sci. China, Ser. A: Math., Phys., Astron. Technol. Sci.*, Vol. 44 (supp), 2001, pp. 412–417.

[46] Wedeven, L. D., Evans, D., and Cameron, A., "Optical Analysis of Ball Bearing Starvation," *ASME J. Lubr. Technol.*, Vol. 93, 1971, pp. 349–363.

[47] Gao, G. T., and Spikes, H., "Fractionation of Liquid Lubricants at Solid Surfaces," *Wear*, Vol. 200, 1996, pp. 336–345.

[48] Shen, M. W., Luo, J. B., and Wen, S. Z., "Effects of Surface Physicochemical Properties on the Tribological Properties of Liquid Paraffin Film in the Nanoscale," *Surf. Interface Anal.*, Vol. 32, 2001, pp. 286–288.

[49] Yao, Y. B., Xie, T., and Gao, Y. M., *Handbook of Physicochemistry*, Science and Technology Press of Shanghai, Shanghai, China, 1985.

[50] Shen, M. W., Luo, J. B., Wen, S. Z., and Yao, J. B., "Nano-Tribological Properties and Mechanisms of the Liquid Crystal as an Additive," *Chinese Science Bulletin*, Vol. 46, 2001, pp. 1227–1232.

[51] Luo, J. B., Shen, M. W., and Wen, S. Z., "Tribological Properties of Nanoliquid Film under an External Electric Field," *J. Appl. Phys.*, Vol. 96, 2004, pp. 6733–6738.

[52] Demus, D., Goodby, J., and Gray, G. W., *Handbook of Liquid Crystals*, New York: Wiley-VCH, 1998; Biresaw, G., "Tribology & the Liquid-Crystalline State," *ACS Symposium Series 441*, Lavoisier, 1990.

[53] Morishita, S., Nakano, K., and Kimura, Y., "Electroviscous Effect of Nematic Liquid Crystal," *Tribol. Int.*, Vol. 26, 1993, pp. 399–403.

[54] Kimura, Y., Nakano, K., and Kato, T., "Control of Friction Coefficient by Applying Electric Fields across Liquid Crystal Boundary Films," *Wear*, Vol. 175, 1994, pp. 143–149.

[55] Mori, S., and Iwata, H., "Relationship between Tribological Performance of Liquid Crystals and Their Molecular Structure," *Tribol. Int.*, Vol. 29, 1996, pp. 35–39.

[56] Bermudez, M. D., Gines, M. N., Vilches, C., and Jose, F., "Tribological Properties of Liquid Crystals as Lubricant Additives," *Wear*, Vol. 212, 1997, pp. 188–194.

[57] Wen, S. Z., and Yang, P. R., "Elastohydrodynamic Lubrication," Tsinghua University Press, Beijing, China, 1992, pp. 261–288.

[58] Hu, Y. Z., Wang, H., Guo, Y., and Zheng, L. Q., "Simulation of Lubricant Rheology in Thin Film Lubrication, Part I: Simulation of Poiseuille Flow," *Wear*, Vol. 196, 1996, pp. 243–248.

[59] Chen, P. W., Yun, S. R., and Huang, F. L., "The Properties and Application of Ultrafine Diamond Synthesized by Detonation," *Ultrahard Materials and Engineering*, Vol. 3, 1997, pp. 1–5.

[60] Shen, M. W., Luo, J. B., and Wen, S. Z., "The Tribological Properties of Oils Added with Diamond Nano-Particles," *STLE Tribol. Trans.*, Vol. 44, 2001, pp. 494–498.

[61] Liu, J. J., Cheng, Y. Q., and Chen, Y., "The Generation of Wear Debris of Different Morphology in the Running-In Proccess of Iron and Steels," *Wear*, Vol. 154, 1982, pp. 259–267.

[62] Ryason, P. R., Chan, I. Y., and Gilmore, J. T., "Polishing Wear by Soot," *Wear*, Vol. 137, 1990, pp. 15–24.

[63] Bair, S., Khonsari, M., and Winer, W. O., "High-Pressure Rheology of Lubricants and Limitations of the Reynolds Equation," *Tribol. Int.*, Vol. 31, 1998, pp. 573–586.

[64] Yoshimura, D., Yokoyama, T., Nishi, T., Ishii, H., Ozawa, R., Hamaguchi, H., and Seki, K., "Electronic Structure of Ionic Liquids at the Surface Studied by UV Photoemission," *J. Electron Spectrosc. Relat. Phenom.*, Vol. 144–147, 2005, pp. 319–322.

[65] Wasserscheid, P., and Welton, T., *Ionic Liquids in Synthesis*, Wiley-VCH, Weinheim, 2003.

[66] Crosthwaite, J. M., Muldoon, M. J., Dixon, J. K., Anderson, J. L., and Brennecke, J. F., "Phase Transition and Decomposition Temperatures, Heat Capacities and Viscosities of Pyridinium Ionic Liquids," *Journal Chem. Thermodynamics*, Vol. 37, No. (6), 2005, pp. 559–568.

[67] Welton, T., "Room-Temperature Ionic Liquids. Solvents for Synthesis and Catalysis," *Chemical Review*, Vol. 99, No. (8), 1999, pp. 2071–2083.

[68] Pinilla, C., Del Popolo, M. G., Lynden-Bell, R. M., and Kohanoff, J., "Structure and Dynamics of a Confined Ionic Liquid. Topics of Relevance to Dye-Sensitized Solar Cells," *J. Phys. Chem. B*, Vol. 109, No. (38), 2005, pp. 17922–17927.

[69] Ye, C. F., Liu, W. M., Chen, Y. X. et al., "Room-Temperature Ionic Liquids: A Novel Versatile Lubricant," *Chem. Commun. (Cambridge)*, Vol. 21, 2001, pp. 2244–2245.

[70] Phillips, B. S., and Zabinski, J. S., "Ionic Liquid Lubrication Effects on Ceramics in a Water Environment," *Tribology Letters*, Vol. 17, 2004, pp. 533–541.

[71] Zhang, X. H., Zhang, X. J., Liu, Y. H., Schaefer, J. A., and Wen, S. Z., "Impact of Confined Liquid Thin Film Upon Bioadhesive Force between Insects and Smooth Solid Surface," *Acta Phys. Sin.*, Vol. 56, 2007, pp. 4722–4727.

[72] Qu, J., Truhan, J., Dai, S. et al., "Ionic Liquids with Ammonium Cations as Lubricants Or Additives," *Tribol. Lett.*, Vol. 22, 2006, pp. 207–214.

[73] Morishita, S., Matsumura, Y., and Shiraishi, T., "Control of Film Thickness of a Sliding Bearing Using Liquid Crystal as Lubricant," *Journal of Japanese Society of Tribologists*, Vol. 47, 2002, pp. 846–851 (in Japanese).

[74] Lavielle, L., "Electric Field Effect on the Friction of a Polythylene-Terpolymer Film on a Steel Substrate," *Wear*, Vol. 176, 1994, pp. 89–93.

[75] Morishita, S., Nakano, K., and Kimura, Y., "Electroviscous Effect of Nematic Liquid Crystals," *Tribol. Int.*, Vol. 26, 1993, pp. 399–403.

[76] Wasan, D. T., and Nikolov, A. D., "Spreading of Nanofluids on Solids," *Nature*, Vol. 423, 2003, pp. 156–159.

[77] Chang, Q. Y., Meng, Y. G., and Wen, S. Z., "Influence of Interfacial Potential on the Tribological Behavior of Brass/Silicon Dioxide Rubbing Couple," *Applied Surface Science*, Vol. 202, 2002, pp. 120–125.

[78] Luo, J. B., He, Y., Zhong, M., and Jin, Z. M., "Gas Bubble Phenomenon in Nanoscale Liquid Film under External Electric Field," *Appl. Phys. Lett.*, Vol. 88, 2006, pp. 25116–25119.

[79] Shen, M. W., "Study on Film-Forming Mechanisms and Tribological Properties of Lubricating Film in the Nanoscale," Tsinghua University Ph.D. thesis (directed by S. Z. Wen and J. B. Luo) Beijing, China, 2000.

[80] He, Y., Luo, J. B., and Xie, G. X., "Characteristics of Thin Liquid Film under an External Electric Field," *Tribol. Int.*, Vol. 40, No. 10-12, 2007, pp. 1718–1723.

[81] Xie, G. X., Luo, J. B., Liu, S. H., Zhang, C. H., Lu, X. C., and Guo, D., "Effect of External Electric Field on Liquid Film Confined within Nanogap," *J. Appl. Phys.* Vol., 103, 2008, p. 094306

[82] Gidon, S., Lemonnier, O., Rolland, B., Bichet, O., and Dressler, C., "Electrical Probe Storage Using Joule Heating in Phase Change Media," *Appl. Phys. Lett.*, Vol. 85, 2004, pp. 6392–6394.

[83] Mustafa, M. M., and Wright, C. D., "An Analytical Model for Nanoscale Electrothermal Probe Recording on Phase-Change Media," *J. Appl. Phys.*, Vol. 99, 2006, pp. 03430101–03430112.

[84] Landau, L., and Lifshitz, E., *Statistical Physics*, Pergamon Press, New York, 1964, pp. 322–323.

[85] Moghaddam, S., and Ohadi, M. M., "Effect of Electrode Geometry on Performance of an EHD Thin-Film Evaporator," *Journal of Microelectromechical Systems*, Vol. 15, 2005, pp. 978–986.

4

Thin Film Lubrication—Theoretical Modeling

Chaohui Zhang[1]

1 Introduction

THIN FILM LUBRICATION (TFL) DEALS WITH THE lubrication region wherein the film gap is of the order of nano metres or molecular scale, i.e., the clearance is usually between several nanometres to several tens of nanometres [1–3]. To date, it is clear that TFL is distinctive enough to be qualified as a separate lubrication regime (for instance, Refs. [4,5]), leaving many mysteries to be unveiled.

In literature, some researchers regarded that the continuum mechanic ceases to be valid to describe the lubrication behavior when clearance decreases down to such a limit. Reasons cited for the inadequacy of continuum methods applied to the lubrication confined between two solid walls in relative motion are that the problem is so complex that any theoretical approach is doomed to failure, and that the film is so thin, being inherently of molecular scale, that modeling the material as a continuum ceases to be valid. Due to the molecular orientation, the lubricant has an underlying microstructure. They turned to molecular dynamic simulation for help, from which macroscopic flow equations are drawn. This is also validated through molecular dynamic simulation by Hu et al. [6,7] and Mark et al. [8]. To date, experimental research had "got a little too far forward on its skis;" however, theoretical approaches have not had such rosy prospects as the experimental ones have. Theoretical modeling of the lubrication features associated with TFL is then urgently necessary.

As revealed through experimental works, however, the flow of lubricants in TFL provides a hint that the macroscopic properties, such as the viscosity and the elastic modulus remain to be a measurement of the fluid characteristics. In addition, the transition from EHL to TFL is inherently *progressive*, wherein no abrupt transform in lubrication states are found. Thus, the continuum theory is validated to some extent. Furthermore, one can arrives at a continuum viewpoint, but in a different way from conventional fluid mechanics, by considering the material to be a continuum one in an ensemble averaged, rather than a spatial averaged, sense.

Size effects can be seen in TFL and are regarded as the deviation of TFL from EHL [9], i.e., the film thickness versus velocity, viscosity, pressure, et al. relations are no longer linear ones in log-log coordinates. Figure 1 is a schematic view of the typical test curve from experimental data, wherein the axes are logarithm coordinates. Three regions can be seen from the figure: thick film region (Section 1), thin film region (Section 2), and failure region (Section 3). It can be seen that

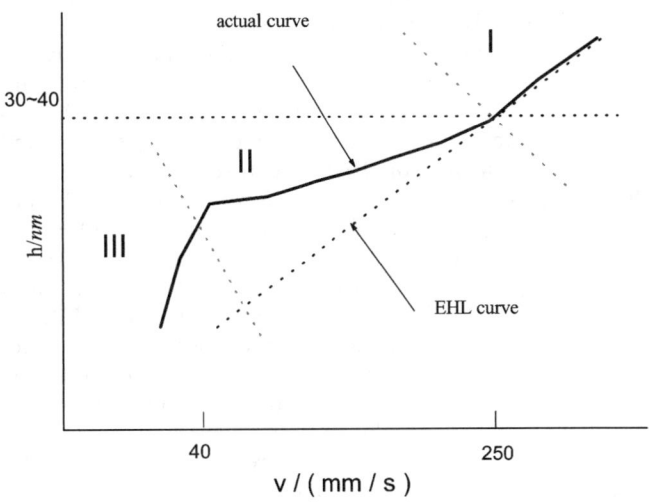

Fig. 1—Schematic view of various regimes of film thickness.

the film thickness varies linearly with velocity (or viscosity and pressure) in the thick film region. This is the EHL regime, where film thickness variation complies with EHL theory predictions. While in the thin film region, with the decreasing of film thickness, the film thickness curve levels off progressively, differing from that of EHL predictions. The thinner the film is, the more the difference can be seen. With further decreasing film thickness, either due to the shear limit of lubricant being achieved, or due to the solid surface ceasing to sustain an available absorption, the lubricant can no longer be mobile and bear no loads. This heralds a failure region.

The failure of TFL only means a loss of mobility here, but monolayers can stay on solid surfaces to separate the solid surfaces in relative motion, and subsequently sustain a feasible boundary lubrication state [10]. Because the film thickness of TFL is of the nano scale or molecular order, from a mechanical point of view, TFL is the last one of the lubrication regimes where the Reynolds equation can be applied.

Boundary lubrication (BL) can also evolve into TFL in a bottom-up way [11]. Compared to BL, TFL has a thicker film. In the vicinity of solid walls, the liquid molecules take the states of that of the boundary film, the ordered one, and the disordered one, from the wall surfaces to the center of the gap. From a mechanical point of view, the existence of an ordered film makes the lubricant film differ strongly from the boundary film, which can form a glassy state or solid-like

---

[1] *School of Mechanical, Electronic, and Control Engineering, Beijing Jiaotong University, Beijing, China.*

state [12]. On the other hand, a lubricant of TFL can only sustain some viscosity, expressing a trait of mobility to some extent.

For an engineering theory, the predictability is the "holy grail" other than to gain knowledge. Driven by this ambition, some theoretical models will be discussed in this chapter. These models, nevertheless, are still in limbo. From the film thickness point of view, the governing equation of the continuum fluids, i.e., the Reynolds equation [13] will also be applicable to TFL, by adding some new insights to it.

As noted before, thin film lubrication (TFL) is a transition lubrication state between the elastohydrodynamic lubrication (EHL) and the boundary lubrication (BL). It is widely accepted that in addition to piezo-viscous effect and solid elastic deformation, EHL is featured with viscous fluid films and it is based upon a continuum mechanism. Boundary lubrication, however, featured with adsorption films, is either due to physisorption or chemisorption, and it is based on surface physical/chemical properties [14]. It will be of great importance to bridge the gap between EHL and BL regarding the work mechanism and study methods, by considering TFL as a specific lubrication state. In TFL modeling, the microstructure of the fluids and the surface effects are two major factors to be taken into consideration.

This chapter will focus on theoretical modeling on TFL properties. First, the concept of ensemble average that validates the use of the Reynolds equations will be introduced. Second, the velocity field analysis for the films of the ordered structure will be given, which can be regarded as a pure science, i.e., in the intention to obtain fundamental knowledge. The following sections will describe the studies of more practical facets, including the simulation with fluid theories taking into consideration the microstructures and a viscosity modification method. Last, results from the studies with different considerations will be provided.

2 Spatial Average and Ensemble Average

This section provides an alternative measurement for a material parameter: the one in the ensemble averaged sense to pave the way for usage of continuum theory from a hope that useful engineering predictions can be made. More details can be found in Ref. [15]. In fact, macroscopic flow equations developed from molecular dynamics simulations agree well with the continuum mechanics prediction (for instance, Ref. [16]).

Typically, the arguments considered for a continuum depend on molecules being very small relative to the problem scale (i.e., the film thickness), as shown in Fig. 2(a), which implies a spatial averaging. One must choose a small region of space (the point), which contains many particles, but is still much smaller than the problem scale. If certain ratios remain constant as the region of space is reduced in size, i.e., if a limit exists, a smoothly varying continuum spatial averaged property (e.g., density) can be defined:

$$\rho_{\text{space}}(\vec{x},t) = \lim\left(\frac{\Delta m}{\Delta V}\right)_{\Delta V_2 \to \Delta V_1} \quad (1)$$

$$\Delta V_{\text{molecule}} \Delta V_1 < \Delta V_2 \Delta V_{\text{problem}} \quad (2)$$

Similarly, smooth time averaged quantities are defined as:

Spatial average versus ensemble average

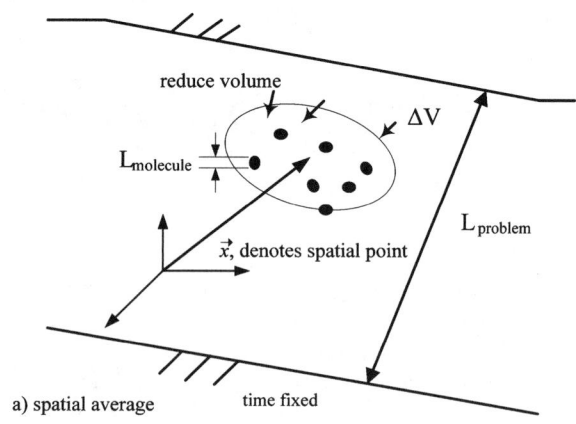

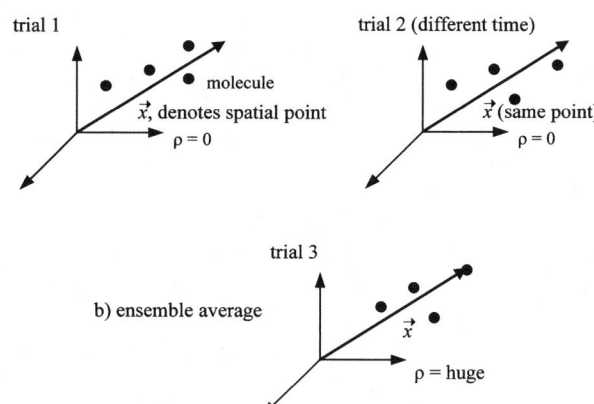

Fig. 2—Ensemble average versus spatial average.

$$\rho_{\text{time}}(\vec{x},t) = \lim\left(\frac{\Delta m}{\Delta t}\right)_{\Delta t_2 \to \Delta t_1} \quad (3)$$

$$\Delta t \Delta t_1 < \Delta t_2 \Delta t_{\text{problem}} \quad (4)$$

Instead of spatial averaging, the ensemble averaging is introduced to determine the value of the expected parameter as follows. The first example is for a large number of trials T at a precise point in space (not for a region), as shown in Fig. 2(b). Although the process is considered to be steady in a global sense, the i_{th} trial may represent a different instant of time in which the molecules are arranged differently. Average properties are determined at each point in space for the T trials. For most of the trials, the density at the given point is zero—there is no molecule at the point. For a few trials, the density at the point is huge from a global perspective, i.e., the density of the molecule itself. Thus, for the ensemble-average density, one obtains:

$$\rho_{\text{ensemble}}(\vec{x}) = \frac{1}{T}\sum_{i=1}^{T}\rho_i(\vec{x}) \quad (5)$$

The most useful of such properties would appear to be an ensemble-averaged momentum and density (their ratio being velocity) or elastic modulus, weighting the likelihood

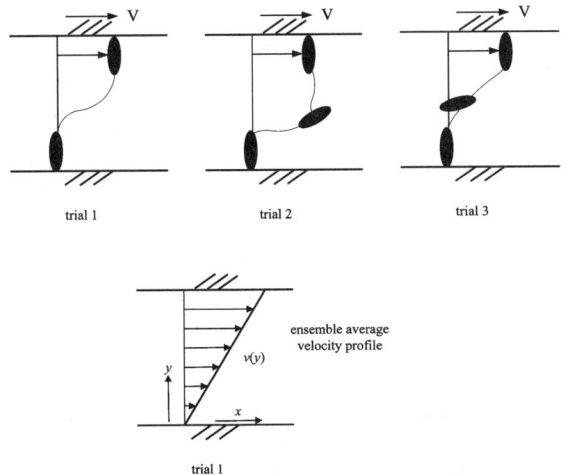

Fig. 3—Ensemble average velocity profile.

of finding a molecule at a given point, and the momentum it is most likely to have. In the ensemble-averaged case, no restriction on the length scale of the film is needed.

The case of velocity profiles for one sliding and one stationary surface is shown in Fig. 3. The first trial might be only two mass particles spanning the film, one sticking to each surface. The second trial might be three mass particles spanning the film, one sticking to each surface and the third at an intermediate velocity, etc. A smooth profile results from the average of many trials at each spatial point spanning the gap.

In tribological research, the global friction of a given surface rubbing against another is, in fact, the ensemble average of hundreds of micro contacts. Therefore, the ensemble average approach directly calculates the engineering quantities of interest and it can be applied to the modeling of the TFL process.

3 Velocity Field of Lubricants with Ordered Molecules

3.1 General Flow Properties for Nematic-like Ordered Fluids

TFL is inherently a lubrication regime in which the ordered layers play a dominant role in determining the lubricating performances. Surface guiding effects and anchoring potential of the solid walls impose the molecules into an ordered state. Their behaviors expressed during the friction process depend also on the elasticity and viscous properties of the lubricants' components. Similar to the nematics, the ordering orientation of the molecules can be described by a vector, namely, the *director*. But they are physically different. It must be noted that the nematic is inherently ordered, i.e., the *director* exists in the conventional state and the parameters such as the Leslie coefficients and elastic modulus are prescribed [17–20], but to TFL, the alignments of molecules and their orientation are imposed by external sources [21], so that the parameters are strongly dependent on the properties of the surface force field of the walls and the mutual matching conditions between the solid walls and the liquid molecules.

A nematic flows very much like a conventional organic liquid with molecules of similar size. In the vicinity of the walls, since the molecules must adjust to a prescribed boundary condition, their orientation changes progressively; this takes place in a certain "transition layer" of thickness. While in TFL, the molecules of lubricants are rearranged due to the effects of the shear inducing forces and surface adsorption potentials so that they hold an ordering orientation to some extent. This makes it possible to analyze the ordered flow in TFL with the help of nematic theories.

From a continuous mechanical point of view, an ordered flow is characterized by the existence of a *director* n_j. The *director* is a unit vector, subject to

$$n_j n_j = 1 \quad (6)$$

Incompressible inertialess flow of these substances is described by the following equations of continuity:

$$0 = v_{j,j} \quad (7)$$

momentum:

$$0 = \sigma_{ij,j} - \left(\frac{\partial W}{\partial n_{k,j}} n_{k,i}\right)_{,j} \quad (8)$$

and the cross product of n_j and the angular momentum equation:

$$0 = \varepsilon_{ijk}\left(n_j \frac{\partial W}{\partial n_{k,p}}\right)_{,p} + \varepsilon_{ijk}\tau_{kj} \quad (9)$$

Following usual conventions, repeated indices indicate summation and $f_{,j}$ denotes $\partial f/\partial x_j$. The permutation symbol ε_{ijk} is used to present the vector cross product in indicial notation. Due to the anisotropic nature, traction and body couples can exist, and thus the angular momentum equation must be considered. For purely viscous fluids this equation says simply that the deviatoric stresses are symmetric.

Elastic forces and couples are calculated from the elastic energy W:

$$2W = K_1(n_{j,j})^2 + K_2(\varepsilon_{pkj}n_k n_{j,p})^2 + K_3(n_p n_k n_{j,p} n_{j,k}) \quad (10)$$

where K_1, K_2, and K_3 are material property parameters which govern the effects of splay, twist, and bend of a static element of homogeneous nematic material, respectively. Elastic stress τ^e_{ij}, elastic couple stress s^e_{ij}, and elastic body couple g^e_i, are determined from the elastic energy by:

$$\tau^e_{ij} = -\frac{\partial W}{\partial n_{k,j}} n_{k,i} \quad s^e_{ij} = \frac{\partial W}{\partial n_{i,j}} \quad g^e_i = \frac{\partial W}{\partial n_i} \quad (11)$$

Viscous stress is computed from the following expressions:

$$\sigma_{ij} = -p\delta_{ij} + \tau_{ij} \quad (12)$$

$$\tau_{ij} = \alpha_1 n_k n_p A_{kp} n_i n_j + \alpha_2 N_i n_j + \alpha_3 N_j n_i + \alpha_4 A_{ij} + \alpha_5 A_{ik} n_k n_j + \alpha_6 A_{jk} n_k n_i \quad (13)$$

where δ_{ij} is the Kronecker delta, σ_{ij} the total dynamic stress, τ_{ij} the deviatory stress, and p the pressure. The material property parameters $\alpha_1,\ldots,\alpha_6$ are Leslie [14] coefficients, each with units of viscosity. It must be noted that these coefficients are functions of temperature in the nematic case, namely, if the temperature remains unchanged, they can be treated as constants. In TFL, however, they are not only determined by the properties of the lubricant but also the properties of the solid walls. The rate of strain tensor is A_{ij} and N_i represents the rotation rate of the director relative to the fluid itself.

3.2 TFL Lubrication Flow Formulation

Now consider the case of two-dimensional flow subjected to the lubrication geometry assumptions that result from analyzing the order of magnitude for the velocities in thin film flow:

$$v_x \gg v_y, \quad v_z = 0; \quad v_{i,x} \ll v_{i,y}, \quad v_{i,z} = 0 \tag{14}$$

where the x and y directions are along and across the film, respectively. Here we also assume that the *director* is one-dimensional, hence the two components can be represented by a single variable $\theta(x,y)$, in which

$$n_x = \cos(\theta), \quad n_y = \sin(\theta), \quad n_z = 0 \tag{15}$$

And it also subjects to $\theta_{,y} \gg \theta_{,x}$.

Here, for simplicity, we allow the one-constant approximation and then set $K_1 = K_3 = K$.

Taking dimensionless variables and parameters as

$$\gamma_1^* = \frac{\gamma_1}{\gamma_2} \quad \alpha_i^* = \frac{\alpha_i}{\gamma_2} \quad x^* = \frac{x}{B} \quad y^* = \frac{y}{h} \quad v^* = \frac{v}{U} \quad \eta^*(\theta) = \frac{\eta(\theta)}{\gamma_2}$$

$$\sigma^* = \frac{\sigma h^2}{\gamma_2 U B} \quad \lambda_{ve} = \frac{\gamma_2 U h}{2K}$$

where U is a reference velocity, B the film breadth, h the film thickness, η the equivalent viscosity, the governing equations then become

$$\frac{d\sigma^*}{dx^*} = -\frac{d}{dy^*}\left(\eta^*(\theta)\frac{dv^*}{dy^*}\right) \tag{16}$$

$$\eta^*(\theta) = \frac{1}{2}(\alpha_4^* + 2\alpha_1^* \sin^2\theta \cos^2\theta + (\alpha_5^* - \alpha_2^*)\sin^2\theta + (\alpha_6^* + \alpha_3^*)\cos^2\theta) \tag{17}$$

$$0 = \frac{d^2\theta}{dy^{*2}} + \lambda_{ve}\frac{dv^*}{dy^*}(\gamma_1^* - \cos(2\theta)) \tag{18}$$

and

$$0 = \frac{d\int_0^1 v^* dy^*}{dx^*}. \tag{19}$$

With the boundaries:

$$y^* = 0: \quad v^* = 1 \quad \theta = \theta_1$$

$$y^* = 1: \quad v^* = 0 \quad \theta = \theta_2$$

$$x^* = 0, \quad x^* = 1: \quad \sigma^* = 0.$$

Parameter λ_{ve} is termed as the viscosity-to-elasticity ratio here, which labels the relative ratio of viscosity to elasticity. In the typical case of EHL, this value is strong enough to assure it pertinent to omit the elastic effect. But in the typical TFL case, its order of magnitude is close to one and the elasticity must be taken into account.

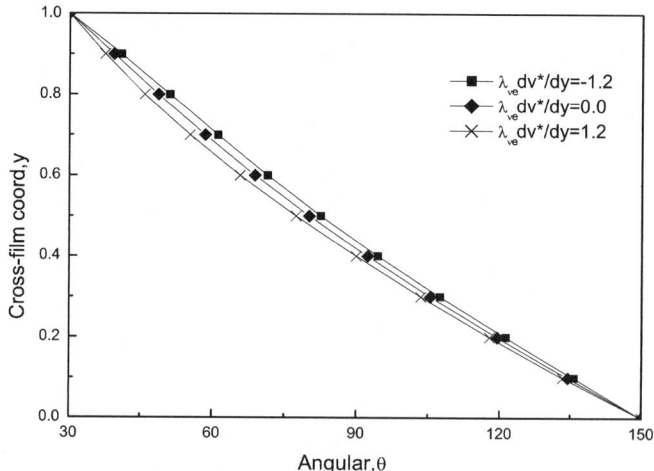

Fig. 4—Director angular distribution profile.

3.3 Results

On one hand, if elasticity dominates the lubricating field, namely, $\lambda_{ve} \to 0$, then $d^2\theta/dy^{*2} = 0$ and θ varies linearly with respect to coordinate y^*. On the other hand, if viscosity dominates the lubricating field, namely, $\lambda_{ve} \to \infty$, then $\theta = 1/2 \arccos(\gamma_1^*)$ holds, which means θ does not vary in the transverse direction, thereby the effective viscosity does not vary either. This is in consistency with conventional nematic lubricants. But it must be pointed out that it may not comprise with the boundary conditions. For lack of better information, the same set of values of Leslie coefficients used in Tichy [19] is accepted here; the parameter values are: $\gamma_2 = 0.02$Pa·s, $\alpha_1^* = -0.0826$, $\alpha_2^* = 0.9848$, $\alpha_3^* = 0.0152$, $\alpha_4^* = -1.0572$, $\alpha_5^* = -0.5883$, $\alpha_6^* = 0.4371$, and $\gamma_1^* = -0.9895$. The elastic constant takes the value of $K = 1.0E - 11N$. The *director* angles at boundaries are: $\theta_1 = \pi/6$, $\theta_2 = \pi - \theta_1$.

3.3.1 Angular Distributions and Effective Viscosity Variation

Figure 4 shows the *director* profile at various $\lambda_{ve}dv^*/dy^*$ values. Figure 5 shows the corresponding effective viscosity values. Only one set of Leslie coefficients are presented, so no general conclusions can be drawn from the data on the rela-

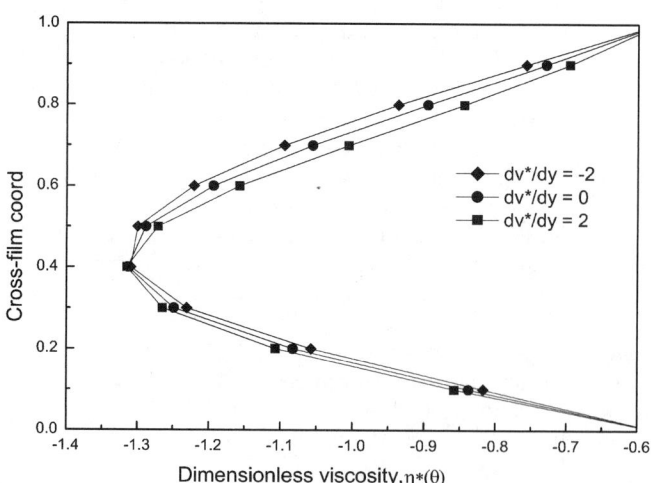

Fig. 5—Effective dimensionless viscosity profile at $\lambda_{ve} = 0.6$.

tion between the *director*, consequently the effective viscosity, and the product of viscosity-to-elasticity ratio and dimensionless velocity gradients.

3.3.2 Velocity Profile

Velocity profiles are demonstrated from Figs. 6 and 7. Figure 6 shows the effects of the viscosity-to-elasticity ratio at negative and positive normal stress gradients. Figure 7 shows the effects of various normal stress gradients on velocity profiles at the value of $\lambda_{ve}=0.6$. No general conclusions can be drawn from these data for the scarcity of practically available information; one only can say that the viscosity-to-elasticity ratio does play an important role in determining the velocity profiles and consequently the lubricating features.

Even though the load carrying capacity and friction behavior are apparently of interest to the authors, they are not presented here because no general law can be constructed if there are no practically reliable parameters presented. This remains to be done in our further research.

3.4 Conclusions

A preliminary theory accounting for the molecule ordering in TFL is presented. The alignment in molecules can be described by a unit vector, the *director*, and then the lubricating features can be analyzed by taking advantage of nematic theory. In TFL, Leslie coefficients are determined by combining the property of solid walls with that of the lubricants in TFL. The specific methods for determining these coefficients are for further research.

Discrepancy of TFL from EHL results from the elasticity of lubricants. The viscosity-to-elasticity ratio is introduced to account for the effects of the elasticity. If the viscosity plays a dominant role, it keeps constant, but if the elasticity prevails, the angle of the *director* varies linearly; consequently the viscosity varies.

The angular momentum conservation equation couples the viscous and the elastic effects. The angular profiles of the *director* and the effective viscosity data are computed for one set of material parameters based on published data in literature. The velocity profiles are also attained from the same dataset. The results show that the alignment of molecules has a strong influence on the lubrication properties.

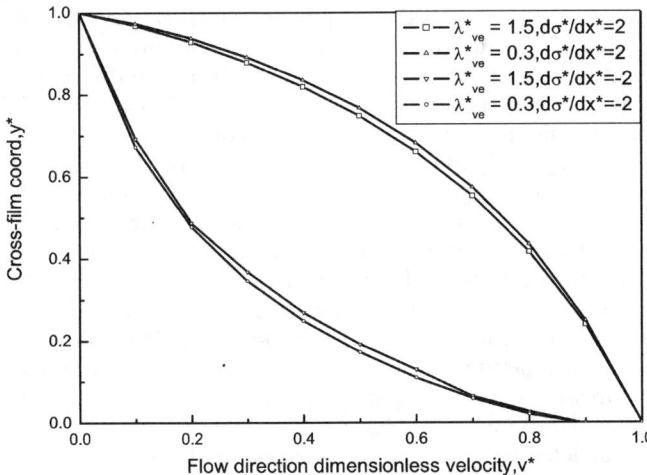

Fig. 6—Velocity profile at various viscosity-to-elasticity ratios.

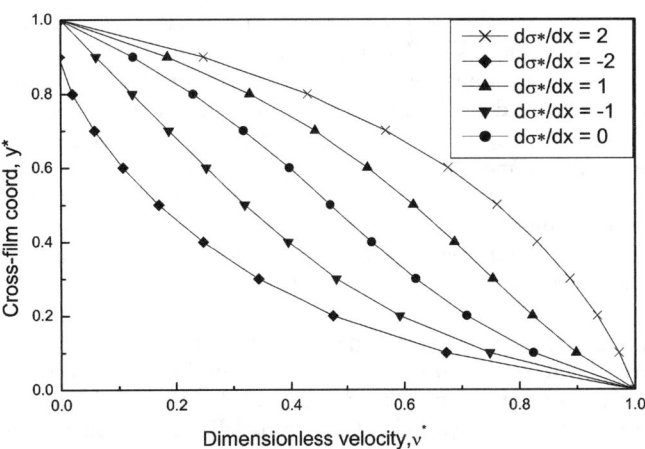

Fig. 7—Velocity profile at different normal stress gradients while $\lambda_{ve}=0.6$.

No general conclusions on the relations among these parameters can be drawn for the scarcity of better information. Experiments should be designed to measure the tribological properties of TFL for engineering applications.

The velocity analysis is of great important for a lubrication theory, which will lay the foundation for further processes, to obtain the flux, the pressure distribution, the load and the friction, etc. As shown before, however, the present model requires a complex procedure to achieve the results. Thus, it can be regarded as a more "purely scientific one," i.e., there is a long way to the success of predictive ability. For a practical purpose, from an engineering point of view, some simplifications should be conducted in an attempt to get the parameters of interest.

4 Simulations via Micropolar Theory

From experimental results, the variation of film thickness with rolling velocity is continuous, which validates a continuum mechanism, to some extent in TFL. Because TFL is described as a state in which the film thickness is at the molecular scale of the lubricants, i.e., of nanometre size, common lubricants may exhibit microstructure in thin films. A possible way to use continuum theory is to consider the effect of a spinning molecular confined by the solid-liquid interface. The micropolar theory will account for this behavior.

When the length scale approaches molecular dimensions, the "inner spinning" of molecules will contribute to the lubrication performance. It should be borne in mind that it is not considered in the conventional theory of lubrication. The continuum fluid theories with microstructure were studied in the early 1960s by Stokes [22]. Two concepts were introduced: couple stress and microstructure. The notion of couple stress stems from the assumption that the mechanical interaction between two parts of one body is composed of a force distribution and a moment distribution. And the microstructure is a kinematic one. The velocity field is no longer sufficient to determine the kinematic parameters; the spin tensor and vorticity will appear. One simplified model of polar fluids is the micropolar theory, which assumes that the fluid particles are rigid and randomly ordered in viscous media. Thus, the viscous action, the effect of couple stress, and the direct coupling of the microstructure to the velocity field will simultaneously affect the motion.

Size-dependent effects will be seen in fluids with microstructure, and the thinner the gap is, the greater the effect will be. During lubrication when the film approaches molecular dimensions, the microstructure will play a dominant role. In the present paper, micropolar theory is incorporated into the theory accounting for very thin film EHL. The effects of coupling number and characteristic length are analyzed for specific conditions.

4.1 General Properties of Micropolar Fluids

The discussion in this section is a simplified description of the theory for micropolar fluids, and detailed explanations are provided by Brulin [23]. The generalized Reynolds equation that takes into account micropolar effects is given by Bessonov [24]. Micropolar fluids are a subclass of microfluids, a viscous media whose behavior and properties are affected by the local motion of particles in its micro-volume. It allows particle micro-motion to take place, but prevents the deformation of microelements. The stress tensor and couple stress tensor of a micropolar fluid, a polar and isotropic fluid, are given by:

$$T_{ij} = (-p + \lambda v_{k,k})\delta_{ij} + \eta(v_{i,j} + v_{j,i}) + \eta_r(v_{j,i} - v_{i,j}) - 2\eta_r \varepsilon_{mij}\omega_m \quad (20)$$

$$C_{ij} = c_1 \omega_{k,k} \delta_{ij} + c_2(\omega_{i,j} + \omega_{j,i}) + c_3(\omega_{j,i} - \omega_{i,j}) \quad (21)$$

The conservation laws of the hydrodynamics of isotropic polar fluids (conservation of mass, momentum, angular momentum, and energy, respectively) are written as follows:

$$\frac{d\rho}{dt} = -\rho \nabla v \quad (22)$$

$$\rho \frac{dv}{dt} = -\nabla p + (\lambda + \eta - \eta_r)\nabla \text{ div } v + (\eta + \eta_r)\Delta v + 2\eta_r \text{ rot } \omega + \rho f' \quad (23)$$

$$\rho I \frac{d\omega}{dt} = 2\eta_r(\text{rot } v - 2\omega) + (c_1 + c_2 - c_3)\nabla \text{ div } \omega + (c_3 + c_2)\Delta \omega + \rho g \quad (24)$$

$$\rho \frac{dU}{dt} = -p \text{ div } v + \rho \Phi' - \nabla q \quad (25)$$

where the dissipation function of mechanical energy per unit mass is

$$\rho \Phi' = \lambda (\text{div } v)^2 + 2\eta D:D + 4\eta_r \left(\frac{1}{2}\text{rot } v - \omega\right)^2 + c_1(\text{div } v)^2 + (c_2 + c_3)\nabla \omega : \nabla \omega + (c_2 - c_3)\nabla \omega : (\nabla \omega)^T \quad (26)$$

and

$$D = \frac{1}{2}(v_{i,j} + v_{j,i}) \quad (27)$$

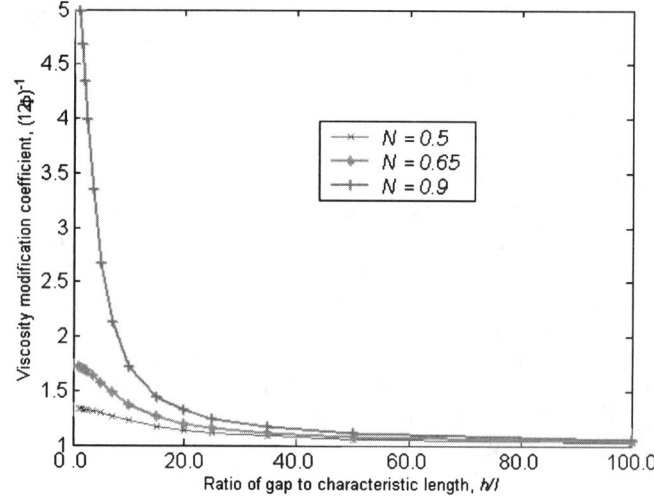

Fig. 8—Modification effects on the viscosity with micro-polarity.

4.2 Governing Equations

The point contact thin film lubrication problem with micropolar fluids requires the simultaneous solution of several governing equations as described below.

4.2.1 Reynolds Equation for a Micropolar Model

Under the usual assumptions made for lubricating films, and through a detailed order-of-magnitude analysis, Singh et al. derived a generalized Reynolds equation for micropolar fluids [25]. Based on the lubrication equation, having incorporated a cavitation algorithm, Lin studied the performances of finite journal bearing with micropolar fluids [26]. Here we adopt the Reynolds equation as [27]

$$\frac{\partial}{\partial x}\left(\frac{h^3}{\eta}\Phi(N,l,h)\frac{\partial p}{\partial x}\right) + \frac{\partial}{\partial y}\left(\frac{h^3}{\eta}\Phi(N,l,h)\frac{\partial p}{\partial y}\right) = \frac{\partial}{\partial x}\left(\frac{Uh}{2}\right) \quad (28)$$

and

$$\Phi(N,l,h) = \frac{1}{12} + \frac{l^2}{h^2} - \frac{Nl}{2h}\coth\left(\frac{Nh}{2l}\right) \quad (29)$$

where N is a dimensionless parameter called the coupling number, which labels the coupling of the linear and angular momentum equations. For $N=0$, these equations are independent of each other. The parameter l is called the characteristic length as it characterizes the micropolar fluid and the film thickness and has a dimension of length.

Being compared to conventional Reynolds equations, $\Phi^{-1}/12$ can be regarded as a modification coefficient of the micropolar effects on viscosity, and its effects are shown in Fig. 8. This shows that the microstructure and microrotation will add an increase in lubricant viscosity. When the ratio h/l increases, the viscosity enhancement decreases; further increasing the ratio, the modification approaches unit. Because l is related to the molecular size, and h is the film gap, this means that if the problem scale is much larger than the molecular dimension, microrotation and the microstructure of particles will contribute insignificantly to the macroscopic properties. The larger N is, the more the increase is, as also evidenced by Fig. 8.

4.2.2 Viscosity-Pressure Relation Equation

The lubricant viscosity is assumed to be dependent on pressure, following the Roelands relation given below [28]

$$\eta = \eta_0 \exp\left((\ln(\eta_0 + 9.67))\left(\left(1 + \frac{p}{1.96 \times 10^8}\right)^Z - 1\right)\right) \quad (30)$$

4.2.3 Film Thickness Equation

For a point contact problem, the local lubricant film thickness can be expressed as:

$$h(x,y) = h_0 + \frac{x^2}{2R_x} + \frac{y^2}{2R_y} + h_e(x,y) \quad (31)$$

and

$$h_e(x,y) = \frac{2}{\pi E} \int\int \frac{p(x',y')}{\sqrt{(x-x')^2 + (y-y')^2}} dx'dy' \quad (32)$$

4.2.4 Load Balance Equation

The applied load must be balanced by the integral of the pressure over the entire solution domain, i.e.,

$$w = \int\int p(x,y) dx dy \quad (33)$$

This equation is used to determine the reference film thickness in the film thickness equation.

4.2.5 Pressure Boundary Condition

$$p(x,y) = 0 \quad (x,y) \in \Gamma \quad (34)$$

where Γ is the contact area boundary, at which the atmospheric pressure is used.

4.3 Simulation Results

The system of equations, including the Reynolds, film thickness, load balance, viscosity-pressure, and viscosity modification equations, is simultaneously solved with the help of a multilevel technique described by Venner and Lubrecht with modifications to take account of the variation in viscosity. The invariant parameters used in the computation are: $Z = 0.67$, $E = 226$ GPa, $\eta_0 = 0.04$ Pa·s, $p_h = 0.46$ GPa, $R = 12.7$ mm, Mmoes = 465, Lmoes = 3.6.

4.3.1 Effects on Minimum Film Thickness

Figure 9 shows the minimum lubricant film thickness as a function of velocity, with or without micro-polarity. The film thickness curves with micro-polarity are larger than those with the nonpolar molecules. This means that the microstructure and microrotation will have an influence on the film thickness. The simple exponential relation between film thickness and velocity, which holds in EHL, is no longer valid for thin film lubrication if the microstructure and the microrotation are taken into account. However, if the minimum film thickness is sufficiently large, as the velocity increases, the discrepancy between results with and without the consideration of the polar effect is very small. With an increase in both the characteristic length l and coupling number N, the minimum film thickness becomes much larger than that of the nonpolar case. This reveals a size-dependent effect which accords well with experimental re-

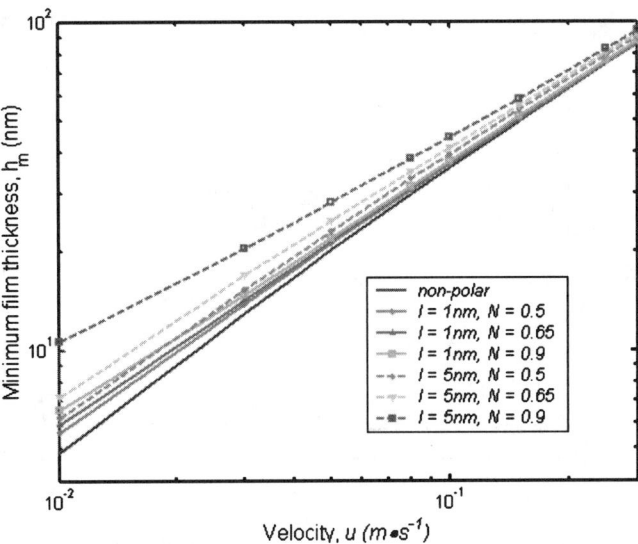

Fig. 9—Relation of minimum film thickness to velocity.

sults. With a large value of N, the effect of the microstructure becomes significant, while with a small value of N, the influence of the substructure is much less pronounced. In the limiting case, as N approaches zero, the micro-polarity disappears and the lubricant behaves as a nonpolar fluid. In addition, as shown in Fig. 9, the effect of microstructure becomes more and more obvious with the increasing values of the characteristic length l.

4.3.2 Pressure Profile and Film Shape

The pressure profile and film shape with or without micro-polarity are shown in Figs. 10–17. The polarity does not alter the positions of the second pressure spike and the minimum thickness, and it has a minor influence on the pressure profile and the film shape. In the case of the pressure profile, the micro-polarity affects the pressure distribution in the vicinity of the second pressure spike. It should be noted that, in Figs. 10 and 12, the second pressure spikes are not clear enough due to low velocities. With an increase in character-

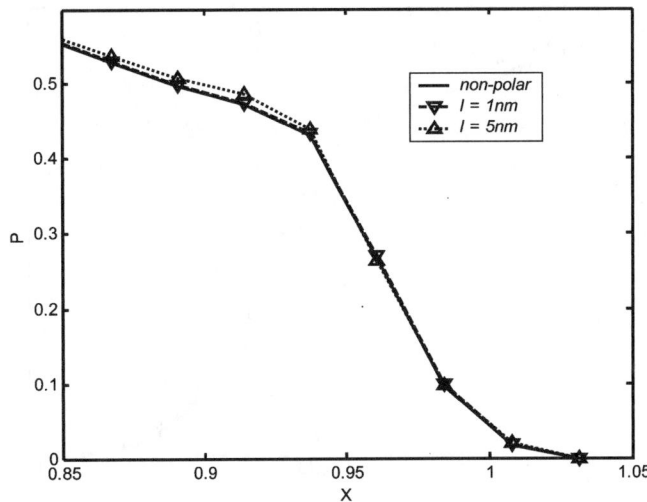

Fig. 10—Pressure profile near the second pressure spike for $v = 10$ mm/s and $N = 0.5$.

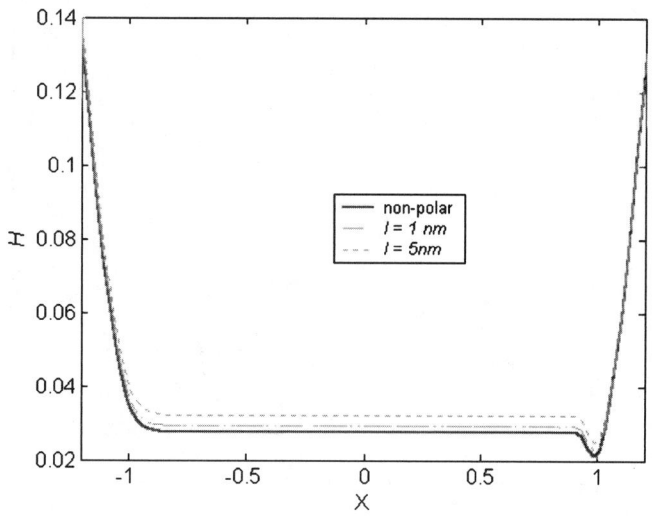

Fig. 11—Film profile for $v=10$ mm/s and $N=0.5$.

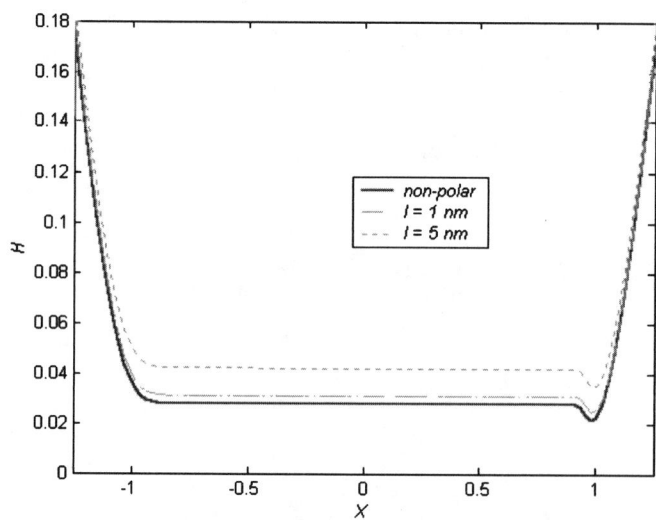

Fig. 13—Film shape for $v=10$ mm/s and $N=0.9$.

istic length l, the deviation of the pressure profile from that of the nonpolar case becomes larger. There is no large discrepancy in the film shape between the polar and nonpolar cases, such as the position of the minimum film thickness; it is obvious that the film thickness with micro-polarity is larger than that of the nonpolar case, and it is more evident for a larger value of characteristic length l. As to the effect of the coupling number N, a similar phenomena, but less obvious, can be seen.

4.3.3 Equivalent Viscosity Comparison with Experiment

In the literature, Hamrock and Dowson's formula on EHL theory is widely accepted, which can be written as [29]

$$h_m = k \eta_0^{0.67} \quad (35)$$

where k is a coefficient. Shen et al. introduced an effective viscosity to account for the effect of thin film close to solid surface [30]. Here, assuming that this idea can also be extended to very thin film EHL, we introduce the effective viscosity as:

$$\eta_{\text{effective}} = \left(\frac{h}{k}\right)^{1/0.67} \quad (36)$$

then the ratio of very thin film EHL with polarity over nonpolarity can be obtained.

$$\bar{\eta} = \frac{\eta_{\text{effective}}}{\eta_0} \quad (37)$$

For a given film thickness, we can get the effective viscosity from Eq (17), then the ratio can be obtained from Eq (18). The relation between the ratios versus the film thickness is plotted in Fig. 18. Interestingly, this figure is very close to the experimental results (see, for example, Fig. 7 in Chapter 5). In the thick film regime, the ratio approaches unity. In the thin film regime, however, it increases with the diminishing film thickness. Therefore, microrotation and

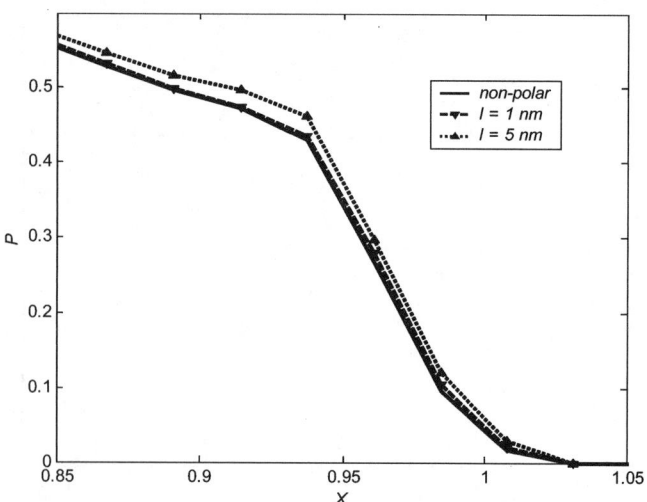

Fig. 12—Pressure profile near the second pressure spike for $v=10$ mm/s and $N=0.9$.

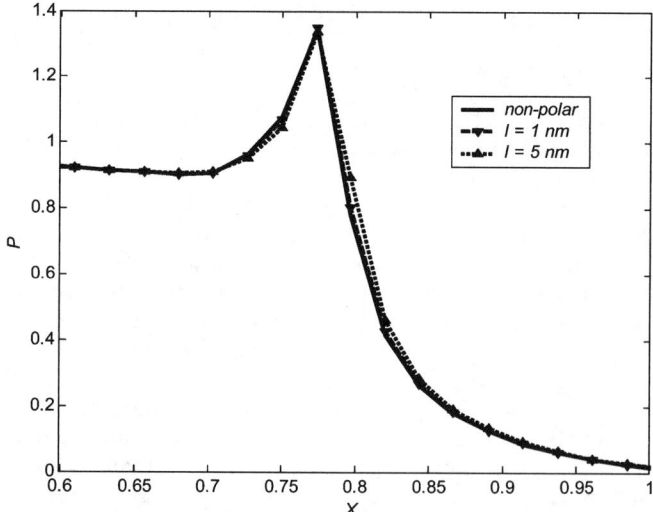

Fig. 14—Pressure profile near the second pressure spike for $v=300$ mm/s and $N=0.5$.

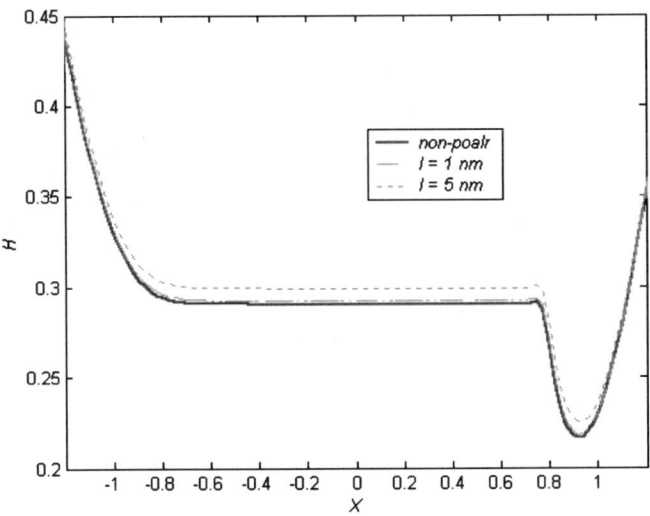

Fig. 15—Film shape for $v=300$ mm/s and $N=0.5$.

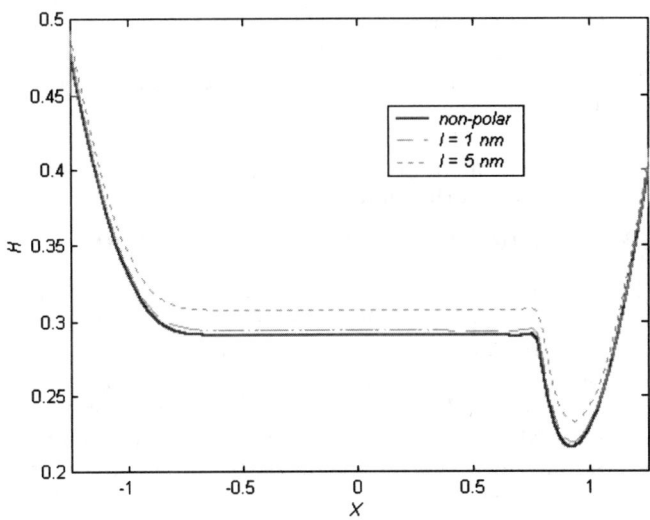

Fig. 17—Film shape for $v=300$ mm/s and $N=0.9$.

microstructure give rise to an increase in effective viscosity. Similarly, the size-dependent effect also makes its appearance. For the given data, the abrupt increase in effective viscosity appears in the film thickness ranging from 10 nm to 30 nm.

4.3.4 Discussion and Analysis

In micropolar theory, the fluid particles are rigid and randomly distributed in a viscous medium. If the micro-polarity is taken into account, the motion is affected by the viscous action, the effect of couple stress, and the direct coupling of the microstructure to the velocity field. On the contrary, in the nonpolar case, the motion is only affected by the viscous action; the other two factors are ignored. They are accounted for in the present model by introducing a characteristic length and a coupling number, which reveals particular features in thin film lubrication. The microstructure and the microrotation effects are integrated Eq (29) and may be regarded as having an influence on the viscosity of lubricants.

One of the apparent results of introducing couple stress is the size-dependent effect. If the problem scale approaches molecular dimension, this effect is obvious and can be characterized by the characteristic length l. The size effect is a distinctive property while the film thickness of EHL is down to the nanometre scale, where the exponent index of the film thickness to the velocity does not remain constant, i.e., the film thickness, if plotted as a function of velocity in logarithmic scale, will not follow the straight line proposed by Hamrock and Dowson. This bridges the gap between the lubrication theory and the experimental results.

The effective viscosity is also affected by the microrotation of the rigid particles. If the gap is much larger than the molecular dimensions, the boundary walls will have little influence on the microrotation motion. This means that if the gap between the solid walls is sufficiently large, the micropolarity can be reasonably taken out of consideration without losing precision. The microrotation in thin film lubrication will result in viscosity-enhancements and consequently higher film thicknesses, which contribute to a better performance of lubrication.

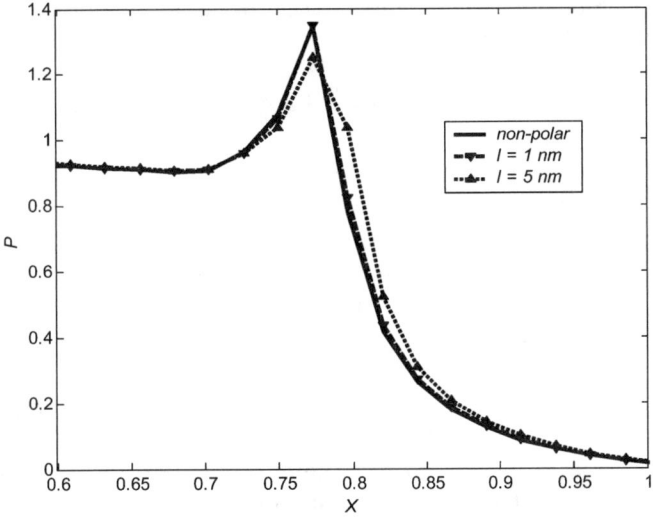

Fig. 16—Pressure profile near the second pressure spike for $v=300$ mm/s and $N=0.9$.

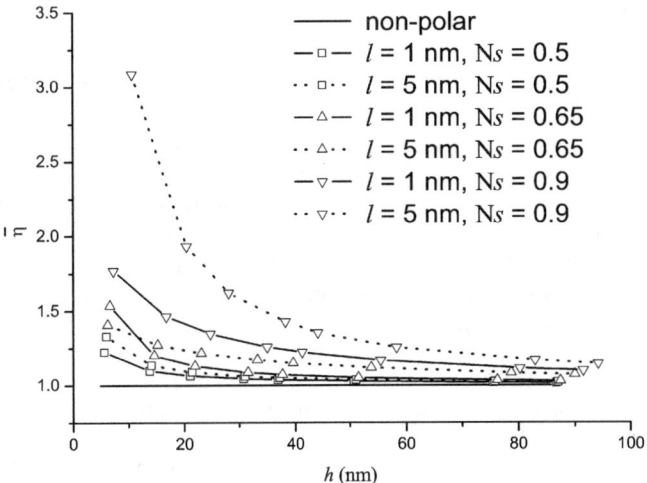

Fig. 18—Effective viscosity versus minimum film thickness.

The effective viscosity predicted with micropolar theory is in very close agreement with that found by experimental results in a previous work. This does not adequately assure that it is the only possible way to explain the traits of thin film lubrication, but it shows the roles the microstructure and microrotation of the particles will play in the lubrication process in the very thin film EHL situation.

5 Rheology and Viscosity Modification

Thin film lubrication is essentially a transition lubrication regime between elastohydrodynamic lubrication and boundary lubrication regimes. Papers devoted to the investigations of this lubrication regime are not enough for engineering needs. In this section, a function to describe the viscosity distribution is proposed in order to attain predictive results and to describe the characteristics of TFL in the viewpoint of engineering.

5.1 Fluid Model and Viscosity Modification Equation

The materials used to be employed for making tribo-pairs are metals or metal oxides that belong to a high-energy surface so that it will exert influence on the arrangement of lubricant molecules nearby the solid surfaces. This contributes mainly to the viscosity variety of lubrication oil near to the solid surface.

As we already know, the viscosity of oil near to the solid wall is limited to a finite magnitude (whose value approaches the value of the relevant solid). Thus, the variation of viscosity along the direction normal to the wall plane can be described by a function as follows [31]

$$\eta = \eta' \cdot \varphi(\bar{z}) \tag{38}$$

where η is the effective viscosity or "true viscosity" of lubrication oil with considerations of the influence of solid surface and ordered property, etc.; η' is the bulk viscosity of lubrication oil, which varies in accord with conventional theories; $\varphi(\bar{z})$ is the modification function of viscosity; $\bar{z}$ is the dimensionless displacement to solid surface, normalized by the effective diameter of molecule Dg, or ($\bar{z} = z/Dg$). Function $\varphi(\bar{z})$ is used to describe the variation of viscosity along the direction normal to the solid surfaces, which can be presented as:

$$\varphi(\bar{z}) = 1 + \frac{b}{1 + \bar{z}^a} \tag{39}$$

where b ($\gg 1$), a ($>= 2$) are constants to be determined. The constant b is mainly determined by coupling conditions. If z approaches zero, it simply indicates how many times the oil viscosity is increased due to the existence of solid walls and it roughly is equal to the viscosity while oil is in solid-like form. a is a constant mainly determined by the molecular structure and properties related to the physical and chemical characteristics of the solid walls. It describes the effect of the composition, the structure, and the ordered degree of the lubricant molecules arrangement and the interactions between lubricant oil molecules and solid walls.

Here, according to the experimental result, we roughly choose $a = 4$, $b = 1,000$, and $Dg = 1.8$ nm in a computing case. For these data, the effective viscosity is shown in Fig. 19.

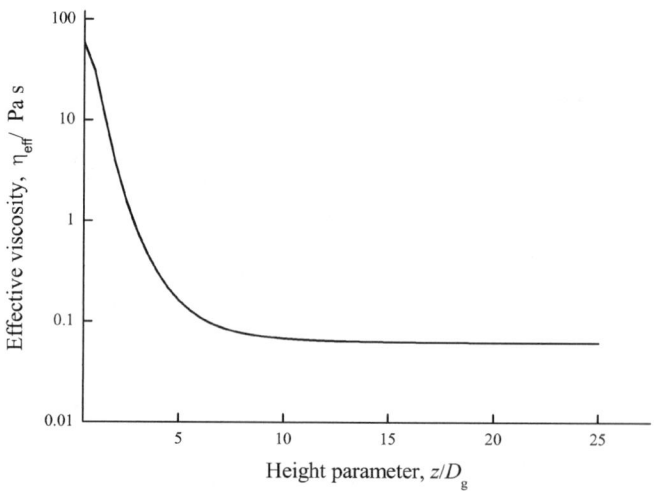

Fig. 19—Modified effective viscosity.

5.2 Governing Equations

Considering an example of isothermal, incompressible body in elastic contact with the presumption that there are adequate molecular layers on the minimum film thickness spot, we will get the governing equations as follows

5.2.1 Viscosity Modification Equation

$$\eta = \varphi(\bar{z}) \cdot \eta' \tag{40}$$

where $\varphi(\bar{z}) = 1 + (b/1 + \bar{z}^a)$, $\bar{z} = (z/Dg)$.

5.2.2 Viscosity-Pressure Relation

$$\eta' = \eta_0 \cdot e^{\alpha p} \tag{41}$$

where η_0 is the viscosity of lubricant in atmosphere pressure, α is the pressure-viscosity index, η' is the bulk viscosity in the conventional sense (with the absence of the influence of viscosity modification from the presented model, and its variation does not depend on the thickness coordinate in the considered case), and p is the pressure.

5.2.3 Modified Reynolds Equation

The momentum equations and the continuity equation governing the motion are:

$$\frac{\partial}{\partial z}\left(\eta \frac{\partial u}{\partial z}\right) = \frac{\partial p}{\partial x} \tag{42}$$

$$\frac{\partial}{\partial z}\left(\eta \frac{\partial v}{\partial z}\right) = \frac{\partial p}{\partial y} \tag{43}$$

$$0 = \frac{\partial p}{\partial z} \tag{44}$$

$$\frac{\partial u}{\partial x} + \frac{\partial v}{\partial y} + \frac{\partial w}{\partial z} = 0 \tag{45}$$

where u, v, w are velocity components in the x, y, z directions, respectively, and η is defined by Eq (40).

Taking consideration of the boundary conditions,

$$z = 0:$$

$$u = U, v = 0$$

$$z = h:$$

$$u = 0, v = 0$$

we can get the velocity components as

$$u = \frac{1}{\eta'}\frac{\partial p}{\partial x}\left[\int_0^z \frac{zdz}{\phi} - \left(\int_0^h \frac{zdz}{\phi} \Big/ \int_0^h \frac{dz}{\phi}\right)\int_0^z \frac{dz}{\phi}\right] + \left[U \Big/ \int_0^h \frac{dz}{\phi}\right]\int_0^z \frac{dz}{\phi} \quad (46)$$

$$v = \frac{1}{\eta'}\frac{\partial p}{\partial y}\left[\int_0^z \frac{zdz}{\phi} - \left(\int_0^h \frac{zdz}{\phi} \Big/ \int_0^h \frac{dz}{\phi}\right)\int_0^z \frac{dz}{\phi}\right] \quad (47)$$

Consequently, the Reynolds equation takes the form

$$\frac{\partial}{\partial x}\left(F_2\frac{\partial P}{\partial x}\right) + \frac{\partial}{\partial y}\left(F_2\frac{\partial P}{\partial y}\right) = U\frac{\partial}{\partial x}\left(h - \frac{F_1}{F_0}\right) \quad (48)$$

where

$$F_0 = \frac{1}{\eta'}\int_0^h \frac{dz}{\varphi(\bar{z})}$$

$$F_1 = \frac{1}{\eta'}\int_0^h \frac{zdz}{\varphi(\bar{z})}$$

$$F_2 = \frac{1}{\eta'}\int_0^h \frac{z^2 dz}{\varphi(\bar{z})} - \frac{F_1^2}{F_0}$$

$$\bar{z} = \frac{z}{Dg}$$

5.2.4 Film Thickness Equation

$$h(x,y) = h_0 + \frac{x^2 + y^2}{2 \cdot R} + \frac{2}{\pi \cdot E'}\int\int_\Omega \frac{P(x',y')dx'dy'}{\sqrt{(x-x')^2 + (y-y')^2}} \quad (49)$$

where h_0 is the initial central thickness of film, R is the radius of curvature, E' is the equivalent Young's elasticity modulus, and Ω is the load distribution regime.

5.3 Computation Results and Discussion

To compare with experimental results, the parameters corresponding to real conditions were used in our computation cases. The lubricant used in the experiment is polyglycol oil. The diameter of the steel ball is 25.4 mm, elastic modulus of the balls is 2.058×10^{11} Pa, and the elastic modulus of the glass disk is 5.488×10^{10} Pa. The circumstance temperature is $28 \pm 1\,°C$. The oil viscosity-pressure index is taken as 1.5×10^{-8} Pa^{-1}.

5.3.1 Film Shape and Pressure Distribution

Figure 20 shows the film shape and the pressure distribution in the moving direction, taken from the section at the central line of the contact regime. The atmosphere viscosity of the lubricant is 0.062 Pa·s. Figure 20 tells us that the film shape and the pressure distribution are both in the forms similar to

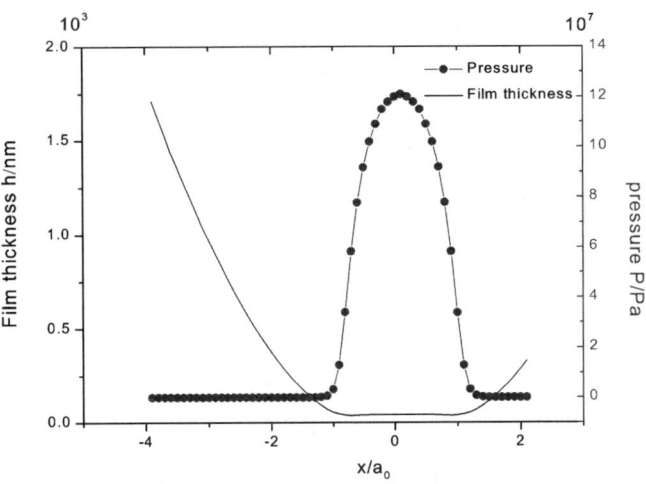

Fig. 20—Oil shape and pressure distribution computed from the ordered model.

the Hertz contact state (in steady and stillness case). It is expected from the fact that the main factor imposing the greatest influence on lubrication properties is velocity, which is consistent with the EHL theory of film profiles versus the velocity relation of the lubricant. Figure 21 shows the film shape at the central line section. While the film is as thin as 15 nm (resulting from low velocity), the centerline along the film direction is flat. However, while the film is thicker (which results from the high velocity), it shows a typical EHL earlobe shape.

5.3.2 Film Thickness Versus Velocity Relation

Figure 22 shows variation of the film thickness with velocities. The three curves in the figure are results from the EHL solution, experimental data, and TFL solution, respectively. The maximum Hertzian contact pressure is 0.125 GPa and the atmosphere viscosity of oil is 0.062 Pa·s. While the velocity is higher than 100 mm·s^{-1}, i.e., the film is thicker than 50 nm, all the results from EHL, TFL, and experimental data are very close to each other, which indicate that when in the EHL lubrication regime, bulk viscosity plays the main role and the results of three types are close to each other. When

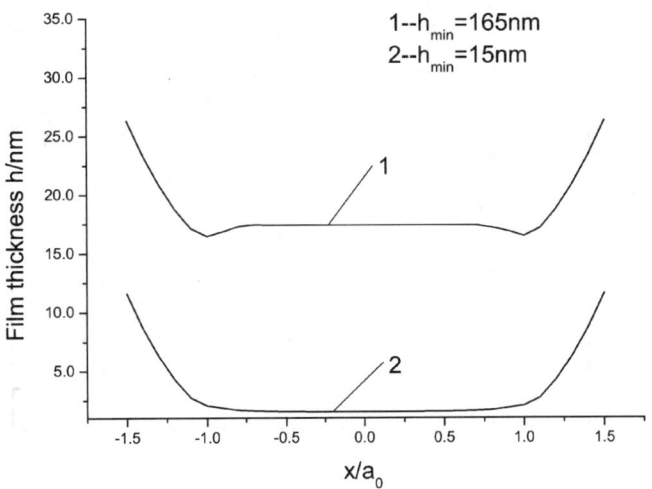

Fig. 21—Oil shape at different conditions.

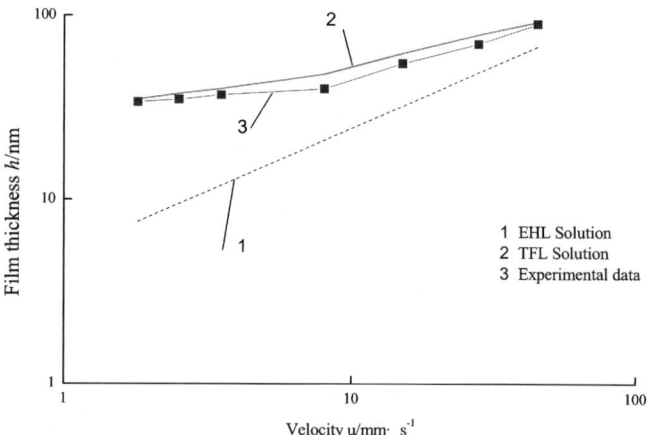

Fig. 22—Film thickness at various velocities.

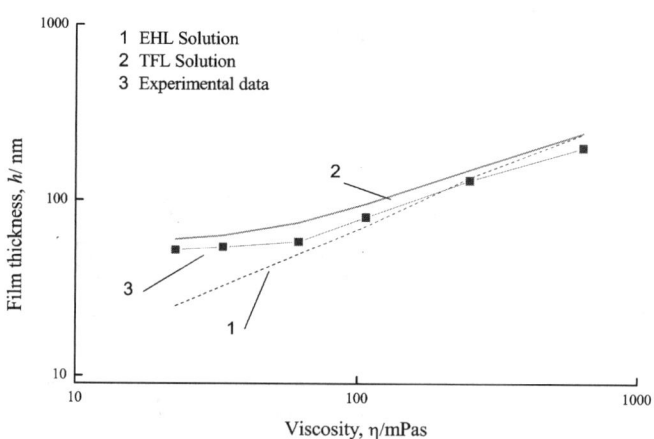

Fig. 24—Film thickness at various loads exerted.

the velocity decreases to a low value, the EHL results are much lower than that of experimental data. Thus, it cannot be regarded as in the EHL lubrication regime. In EHL, the film thickness varies with the velocity in an exponential relation. However, it is not the case in thin film lubrication, as shown in Fig. 22. The thinner the film is, the less the correlation between velocity and bulk viscosity will be.

The computed film thickness distribution of the cross line at the contact center is shown in Fig. 23. The abscissa axis is the dimensionless coordinate in the flow direction whereas the ordinate axis is the dimensionless film thickness. It is very clear that the film thickness distribution is similar to that of EHL predictions.

5.3.3 Film Thickness Versus Inlet Viscosity Relation

Figure 24 shows the film thickness varying with lubricant atmosphere viscosity at the velocity of 28 mm·s^{-1}. As mentioned above, while the viscosity is relatively high as η_0 = 0.6371 Pa·s, results of EHL solution, TFL solution, and experimental data are close to each other, from which we can deduce the lubrication is in the EHL regime. When the film thickness decreases to about 15 nm, the EHL result is much less than that of the experimental data, which can no longer be attributed to the EHL lubrication regime. In this case, the result coming from the model proposed in this chapter is very close to the experimental data. In addition, the film thickness varies with the bulk viscosity of lubricant in an exponential way. However, it is not true in thin film lubrication. The decreased pattern was demonstrated by the TFL solution and experiments.

5.3.4 Film Thickness Versus Load Relation

Figure 25 shows the film thickness varying with the load. The film thickness obtained from TFL solution is higher than that of EHL because the effective viscosity is prominently larger than that in bulk. The film thickness decreases slightly with load in both cases. However, the exponential coefficient of TFL is larger than that of EHL (in absolute value sense).

6 Other Approaches Related to TFL Theories

6.1 Solvation Pressure and Its Application to Very Thin EHL Problem

The dimension of machine elements has been reduced continuously in recent years with the advance of micromachining technology, and the separation between surfaces in rela-

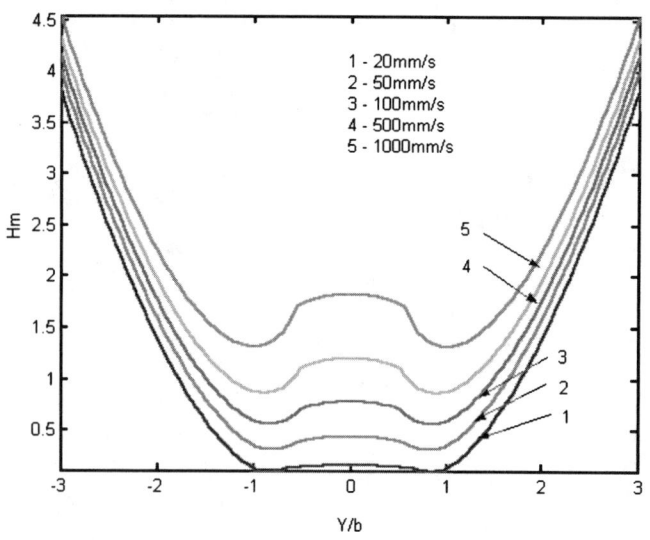

Fig. 23—Film thickness at various atmosphere viscosities.

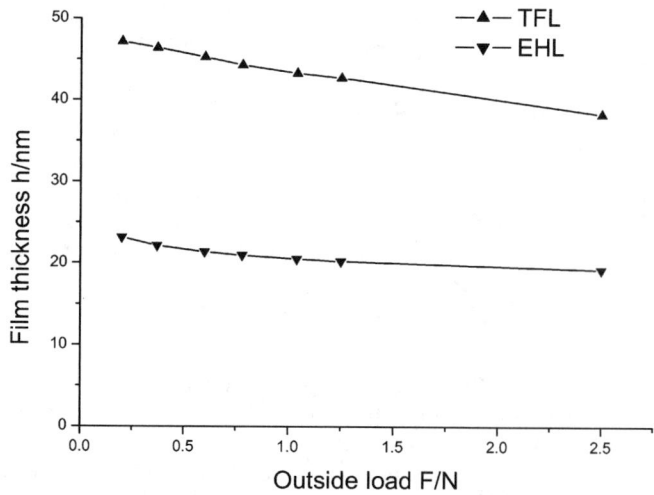

Fig. 25—Cross line film thickness of TFL (simulation results).

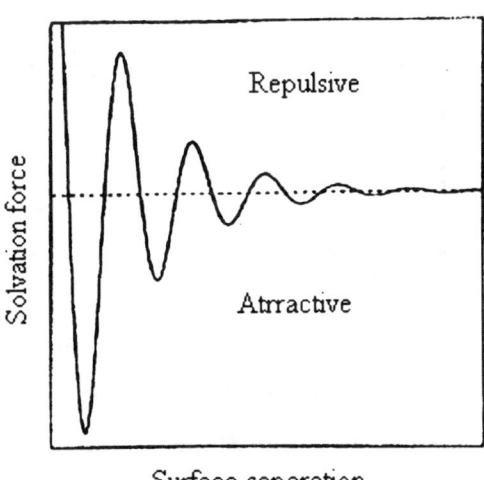

Fig. 26—Solvation force.

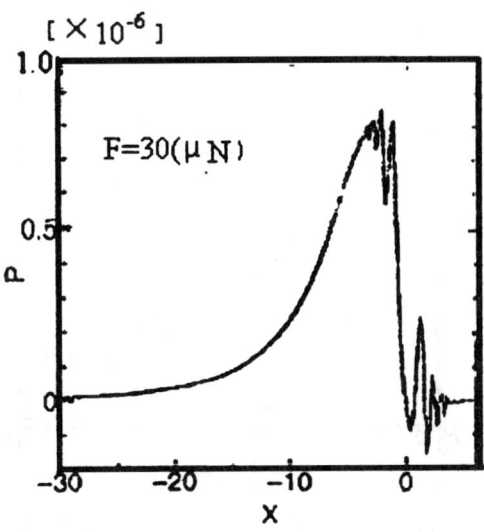

Fig. 27—Pressure distribution considering solvation force.

tive motion becomes smaller and smaller. For example, Matsuoka observed discretization of the film thickness due to solvent force when it is less than about 10 nm, based on the result of the film thickness measurements using the apparatus for ultra-thin fluid lubrication [32]. Among surface effects, solvent force (usually refers to structural force) is a unique force that acts between two solid surfaces when the two surfaces approach each other to a very small distance in a liquid, and is schematically shown in Fig. 26. It is known that the solvation force has the following characteristics [33]:

1. An oscillatory force which decays exponentially with surface separation;
2. The period of the oscillation with surface separation is roughly the same as the diameter of an intervening liquid molecule.

Subsequently, a lubrication theory considering the solvation force is deduced accordingly in such ultra-thin film lubrication.

Pressure between solid walls separated by ultra-thin liquid film is then expressed as follows:

$$P = P_{visc} + P_{solv} + P_{vdw} \quad (50)$$

where P_{visc}, P_{solv}, P_{vdw} are viscous force, solvation force, and van der Waals force, respectively.

Figure 27 shows an example of calculated pressure distribution for OMCTS [32]. It is found that the oscillatory solvation force lies on the conventional hydrodynamic viscous pressure. If the film thickness is smaller than that in this example, the viscous pressure becomes negligible and the solvation pressure dominates. According to the Hamrock-Dowson diagram [29], the sliding condition of this example falls into the Rigid-Isoviscous region where elastic surface deformation is negligible. The surface deformation due to the solvation pressure was also small and the noted EHL film shape, namely, the flattened center and the occurrence of minimum film thickness at the trailing edge, was not obtained under this sliding condition. Figure 28 shows the relation between the fluid force, F, and the film thickness, h, which is obtained by summarizing raw data such as those in Fig. 26. Black dots represent experimental results, the thin solid line is the conventional theoretical line in the R-I regime, the thin dashed line and thin dot-dashed line are theoretical minimum and central film in the E-I regime, respectively, and the thick solid line shows results of the EHL calculation by using solvation pressure model for OMCTS. The solvation pressure is calculated assuming $\sigma = 1$ nm. It is seen in Fig. 28 that the film thickness begins to deviate from the conventional EHL theory when the film thickness is less than about 7–8 nm and changes stepwise. The present model gives good agreement with experiments in general, though it overestimates the fluid force slightly when the film thickness is about 1–2 nm. An advantage of the calculation is that laborious and difficult experiments are not required in order to obtain unknown parameters.

Figure 29 shows a comparison of experimental results with calculation results obtained for cyclohexane. The discretization of film thickness is again observed as in the OMCTS results, and the interval is 0.5–0.6 nm, which is roughly the same as the molecular diameter of cyclohexane.

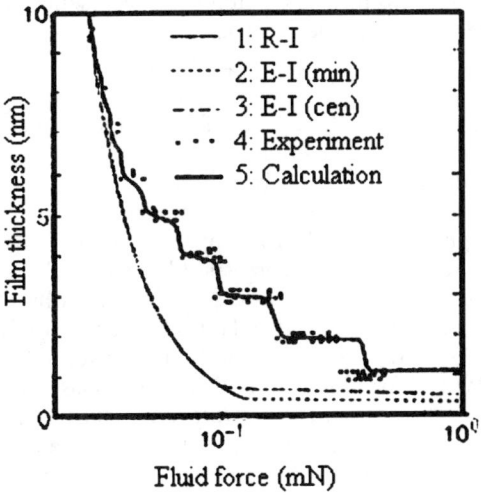

Fig. 28—EHL calculation result for OMCTS.

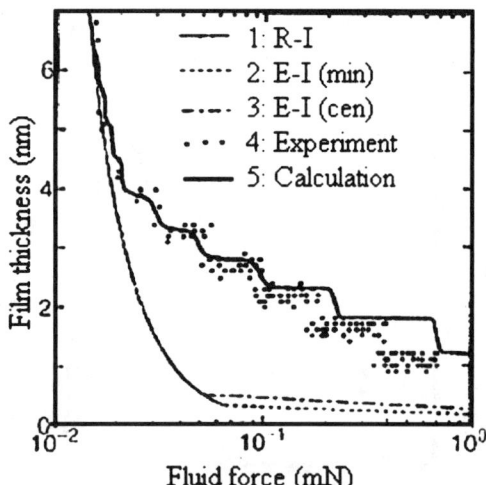

Fig. 29—EHL calculation result for cyclohexane.

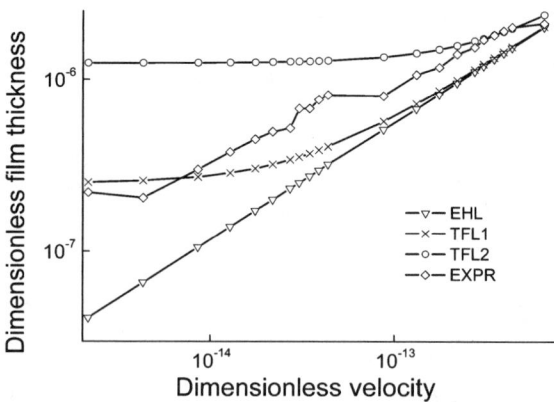

Fig. 30—Dimensionless minimum film thickness versus dimensionless velocity.

The present calculation agrees well with the experimental data also in the case of cyclohexane.

It is considered that the calculation described in this section agrees well with the experiments for the liquids that have a strong solvation force such as OMCTS and cyclohexane. It may be more difficult to apply this theory to the liquids that have a molecular shape far from spherical and exhibit weak solvation force.

6.2 Couple Stress and Its Application to Two-Phase Flows

The couple stress method can be used for modeling a special case of micro-polar fluids, i.e., the two-phase flow, wherein the constitutive equation is given by [22,34–38]

$$\tau_{ij} = (-p + \lambda d_{rr})\delta_{ij} + 2\mu d_{ij} - \frac{1}{2}e_{ijk}(m_{tk,t} + \rho l'_k) \quad (51)$$

$$m_{ij} = \frac{1}{3}m_{rr}\delta_{ij} + 4\eta k_{ij} + 4\eta' k_{ji} \quad (52)$$

If we ignore inertia force and follow the conventional assumptions in liquid lubrication, motion equation (here we consider the incompressible fluids in the absence of volume force and volume momentum) is:

$$0 = -\nabla p + \mu \nabla^2 v - \eta \nabla^4 v \quad (53)$$

$l = \sqrt{\eta/\mu}$ is named as the characteristic length which contributes to the couple stress effect.

Simulations for the two-phase flow containing nanoparticles were carried out and comparisons were taken to the EHL predictions [39]. To compare with the experimental results, the parameters corresponding to real conditions were used in our computation examples. The lubricant used in the experiment is polyglycol oil. The diameter of the steel ball is 25.4 mm, the elastic modulus of the balls is 2.058×10^{11} Pa, and the elastic modulus of the glass disk is 5.488×10^{10} Pa. The ambient temperature is $(20 \pm 1)°C$. The oil's viscosity-pressure index is 1.5×10^{-8} Pa^{-1}. The average diameter of nano-particles as additives is 5 nm. The concentration of additive particles is 0.3 %. In all the following figures, EHL, TFL1, TFL2 correspond to $l = 0$ nm, 1 nm, and 5 nm, respectively, and EXPR corresponds to the experimental data. Figure 30 shows the relation of dimensionless central film thickness with respect to dimensionless velocity. While the couple stress effect exists ($l > 0$), the film thickness is larger than that of EHL. The TFL solution with characteristic length 1 nm conforms to the experiment data well. The larger the characteristic length is, the greater the enhancement effect is. The influence of couple stress on film thickness at very high velocity is insignificant. The simple exponential relation between film thickness and velocity, which holds in EHL, is invalid in TFL where couple stress exists and plays an important role. The smaller the film thickness or the larger the characteristic length is, the more obvious the effect is. This reveals a size-dependent effect that conforms well to the experimental results. Therefore we can draw the following conclusions: in the case of thick film thickness, the volume ratio of ordered molecules is small, and the lubrication is characterized by conventional fluids and can thus be regarded as EHL. With the decrease of the film thickness, couple stress does have an increasing effect, which leads to an increased volume ratio of the ordered fluids molecules, and subsequently the lubrication transfers gradually into the TFL regime.

TFL is an important sub-discipline of nano tribology. TFL in an ultra-thin clearance exists extensively in micro/nano components, integrated circuit (IC), micro-electromechanical system (MEMS), computer hard disks, etc. The impressive developments of these techniques present a challenge to develop a theory of TFL with an ordered structure at nano scale. In TFL modeling, two factors to be addressed are the microstructure of the fluids and the surface effects due to the very small clearance between two solid walls in relative motion [40].

The mechanism of TFL can be summarized to the ordered film lubrication, concluded from previous researches. It should be noted that the property and mechanism of TFL are not fully understood to date, and there remains a wide blank area to be covered between experiments and theoretical predictions and it leaves a far cry to achieve predictive ability. Some key problems associated with theoretical modeling which need to be addressed are:

1. The relation of the effective viscosity and the property of the lubricant molecules;

2. Influence factors relating to transition between the liquid phase and the solid-like phase;
3. Further perfection of the mathematical model based on the ordered film;
4. The property of electro-rheology, magneto-rheology, intelligibility, and controllability of ordered liquids, and so on.

7 Conclusions

TFL is essentially a transition lubrication regime between EHL and boundary lubrication. A new postulation based on the ordered model and ensemble average (rather than bulk average) was put forward to describe viscosity in the nanoscale gap. In TFL, EHL theories cannot be applied because of the large discrepancies between theoretical outcomes and experimental data. The effective viscosity model can be applied efficiently to such a condition. In thin film lubrication, the relation between film thickness and velocity or viscosity accords no longer with an exponential one. The studies presented in this chapter show that it is feasible to use a modified continuous scheme for describing lubrication characteristics in TFL.

References

[1] Zhang, C. H., "Research on Thin Film Lubrication: State of the Art," *Tribol. Int.*, Vol. 38, No. 4, 2005, pp. 443–448.

[2] Luo, J. B., Wen, S. Z., and Huang, P., "Thin Film Lubrication. Part I. Study on the Transition Between EHL and Thin Film Lubrication Using Relative Optical Interference Intensity Technique," *Wear*, Vol. 194, 1996, pp. 107–115.

[3] Luo, J. B., Huang, P., Wen, S. Z., and Li, K. Y., "Characteristics of Fluid Lubricant Films at the Nano-Scale," *Trans. ASME*, Vol. 121, No. 4, 1999, pp. 872–878.

[4] Luo, J. B., and Wen, S. Z., "Study on the Mechanism and Characteristics of Thin Film Lubrication at Nanometer Scale," *Sci. China, Ser. A: Math., Phys., Astron. Technol. Sci.*, Vol. 35, No. 12, 1996(b), pp. 1312–1322.

[5] Tichy, J. A., "Ultra Thin Film Structured Tribology," Proc. 1st Int. Symp. Tribol. 19–23 Oct, 1993, Beijing, pp. 48–57.

[6] Hu, Y. Z., Wang, H., Guo, Y., and Zheng, L. Q., "Simulation of Lubricant Rheology in Thin Film Lubrication, Part I: Simulation of Poiseuille Flow," *Wear*, Vol. 196, 1996, pp. 243–248.

[7] Hu, Y. Z., Wang, H., Guo, Y., Shen, Z. J., and Zheng, L. Q., "Simulation of Lubricant Rheology in Thin Film Lubrication, Part II: Simulation of Couette Flow," *Wear*, Vol. 196, 1996, pp. 249–253.

[8] Mark, O. R., Thompson, P. A., and Grest, G. S., "Simulations of Nanometer-Thick Lubricating Films," *MRS Bull.*, 1993.

[9] Zhang, C. H., "Numerical Analysis on Tribological Performances of Lubricating Film in the Nano Scale," Ph.D. thesis, Beijing: Tsinghua University, 2002 (in Chinese).

[10] Dorinson, A., and Ludema, K. C., *Mechanics and Chemistry in Lubrication*, Elsevier, Berlin, 1985.

[11] Glovnea, R. P., and Spikes, H. A., "Elastohydrodynamic Film Formation at the Start-Up of the Motion," Proc Instn Mech Engrs (part J), Vol. 215, 2001, pp. 125–138.

[12] Homola, A. M., Israelachvili, J. N., Gee, M. L., et al., "Measurements of and Relation Between the Adhesion and Friction of Two Surfaces Separated by Molecularly Thin Liquid Films," *Trans. ASME, J. Tribol.*, Vol. 111, No. 3, 1989, pp. 675–682.

[13] Oscar, P., "The Reynolds Centennial: A Brief History of the Theory of Hydrodynamic Lubrication," *Trans. ASME, J. Tribol.*, Vol. 109, No. 1, 1987, pp. 2–20.

[14] Luo, J. B., Lu, X. C., and Wen, S. Z., "Progress and Problems of Thin Film Lubrication," *Progress in Natural Science*, 2000, pp. 1057–1065.

[15] Tichy, J. A., "Modeling of Thin Film Lubrication," *Tribol. Trans.*, Vol. 38, No. 1, 1995, pp. 108–118.

[16] Bashir, Y. M., and Goddard, J. D., "A Novel Method for the Quasi-Static Mechanics of Granular Assemblages," *J. Rheol.*, Vol. 35, No. 5, 1991, pp. 849–884.

[17] De Gennes, P. G., and Prost, J., *The Physics of Liquid Crystals*, 2nd ed., Oxford Science Publication, New York, 1993.

[18] Leslie, F. M., "Theory of Flow in Nematic Liquid Crystals," *The Breadth and Depth of Continuum Mechanics—A Collection of Papers Dedicated To J. L. Ericksen*, C. M. Dafermos, D. D. Joseph, and F. M. Leslie, Eds., Springer-Verlag, Berlin, 1986.

[19] Tichy, J. A., "Lubrication Theory for Nematic Liquid Crystals," *Tribol. Trans.*, Vol. 33, No. 3, 1990, pp. 367–370.

[20] Ericksen, J. L., and Kinderlehrer, D., *Theory and Applications of Liquid Crystals*, Springer-Verlag, 1987.

[21] Zhang, C. H., Wen, S. Z., and Luo, J. B., "Lubrication Theory for Thin Film Lubrication Accounting for the Ordered Film Model," *Int. J. Nonlinear Sci. Numer. Simul.*, Vol. 3, No. 3–4, 2002, pp. 481–485.

[22] Stokes, V. K., *Theories of Fluids with Microstructure—An Introduction*, Springer-Verlag, Berlin, 1984.

[23] Brulin, O., and Hsieh, R. K. T., "Mechanics of Micropolar Media," *World Scientific*, Singapore, 1982.

[24] Bessonov, N. M., "A New Generalization of the Reynolds Equation for a Micropolar Fluid and Its Application to Bearing Theory," *Tribol. Int.*, Vol. 27, No. 2, 1994, pp. 105–108.

[25] Singh, C., and Sinha, P., "The Three-Dimensional Reynolds Equation for Micropolar-Fluid Lubricated Bearings," *Wear*, Vol. 76, No. 2, 1982, pp. 199–209.

[26] Lin, T. R., "Analysis of Film Rupture and Re-formation Boundaries in a Finite Journal Bearing with Micropolar Fluids," *Wear*, Vol. 161, 1993, pp. 145–153.

[27] Zhang, C. H., Luo, J. B., and Wen, S. Z., "Exploring Micropolar Effects in Thin Film Lubrication," *Sci. China, Ser. G*, Vol. 47 (supp), 2004, pp. 65–71.

[28] Roelands, C. J. A., Vlugter, J. C., and Waterman, H. I., "The Viscosity Temperature Pressure Relationship of Lubrication Oils and Its Correlation with Chemical Constitution," *ASME J. Basic Eng.*, 1963, pp. 601–610.

[29] Hamrock, B. J., *Fundamentals of Fluid Film Lubrication*, McGraw-Hill, New York, 1994.

[30] Shen, M. W., Luo, J. B., Wen, S. Z., and Yao, J. B., "Investigation of the Liquid Crystal Additive's Influence on Film Formation in Nano Scale," *Lubr. Eng.*, Vol. 58, No. 3, 2002, pp. 18–23.

[31] Zhang, C. H., Wen, S. Z., and Luo, J. B., "A New Postulation of Viscosity and Its Application in Computation of Film Thickness in TFL," *Trans. ASME, J. Tribol.*, Vol. 124, No. 4, 2002, pp. 811–814.

[32] Matsuoka, H., and Kato, T., "An Ultrathin Liquid Film Lubrication Theory-Calculation Method of Solvation Pressure and Its Application to the EHL Problem," *Trans. ASME, J. Tribol.*, Vol. 119, 1997, pp. 217–226.

[33] Jang, S., and Tichy, J. A., "Rheological Models for Thin Film

EHL Contacts," *Trans. ASME, J. Tribol.*, Vol. 117, No. 1, 1995, pp. 22–28.

[34] Naduvinamani, N. B., Hiremath, P. S., and Gurubasavaraj, G., "Squeeze Film Lubrication of a Short Porous Journal Bearing with Couple Stress Fluids," *Tribol. Int.*, Vol. 34, 2001, pp. 739–747.

[35] Das, N. C., "Elastohydrodynamic Lubrication Theory of Line Contacts: Couple Stress Fluid Model," *Tribol. Trans.*, Vol. 40, No. 2, 1997, pp. 353–359.

[36] Lin, J. R., "Squeeze Film Characteristics Between a Sphere and a Flat Plate: Couple Stress Fluid Model," *Computers & Structures*, Vol. 75, 2000, pp. 73–80.

[37] Lin, J. R., Yang, C. B., and Lu, R. F., "Effects of Couple Stresses in the Cyclic Squeeze Films of Finite Partial Journal Bearings," *Tribol. Int.*, Vol. 34, 2001, pp. 119–125.

[38] Wang, X. L., Zhu, K. Q., and Wen, S. Z., "Thermohydrodynamic Analysis of Journal Bearings Lubricated with Couple Stress Fluids," *Tribol. Int.*, Vol. 34, 2001, pp. 335–343.

[39] Zhang, C. H., Wen, S. Z., and Luo, J. B., "On Characteristics of Lubrication at Nano-Scale in Two-Phase Fluid System," *Sci. China, Ser. B: Chem.*, Vol. 45, No. 2, 2002, pp. 166–172.

[40] Zhang, C. H., Luo, J. B., and Huang, Z. Q., "Analysis on Mechanism of Thin Film Lubrication," *Chinese Science Bulletin*, Vol. 50, No. 18, 2005, pp. 2645–2649.

5

Molecule Films and Boundary Lubrication

Yuanzhong Hu[1]

1 Introduction

WHEN TWO SURFACES IN RELATIVE MOTION ARE completely separated by a liquid film that carries the applied load, a preferred state with low friction and without wear is achieved. This state of hydrodynamic or elastohydrodynamic lubrication (EHL), however, is not always the case in tribology practices. An increase in load, decrease in velocity or changes of surface roughness can lead to disappearance or discontinuity of hydrodynamic films and transition of the lubrication condition. Boundary lubrication refers to a lubrication regime where hydrodynamic lubrication is no longer effective, and the physics and chemistry of the interfacial substance play a dominant role in protecting the surfaces from direct contact [1]. The transition to boundary lubrication is a progressive process that goes across a region known as mixed lubrication where hydrodynamic films gradually disappear as the film thickness decreases continuously and the "asperity contacts" carry an increasing portion of the applied loads. In this sense, boundary lubrication can be regarded as a lower limit of the mixed lubrication region. The Stribeck curve shown in Fig. 1 provides an illustrative description for the transition from hydrodynamic lubrication to mixed and boundary lubrication.

Due to the absence of a hydrodynamic effect, boundary film thickness is expected to be independent of speed of surface movement, as can be observed in the left part of the Stribeck curve. This is a significant criterion that distinguishes boundary lubrication from EHL and mixed lubrication, and provides an opportunity for measuring boundary film thickness using an interferometer, for example. More details will be presented in Section 2.

Boundary lubrication is an extremely complex process in which numerous mechanisms, including rheology transition, adsorption, tribochemical reactions, selective transfer, etc., simultaneously participate in the play. In a sense, the term of boundary lubrication may take different meanings to different investigators. Physicists may regard boundary lubrication as a process where adsorbed monolayer plays a primary role in reducing friction; rheology scientists are investigating rheological transitions in confined lubricant films and their response to shearing, and chemists are more interested in tribochemical processes and formation of reaction layers that protect surfaces under severe conditions of wear. The essential function of boundary lubrication, however, is to reduce or eliminate immediate contacts between two solids, via adsorbed or reactive surface layers.

As a general rule, the functions of boundary lubrication can be classified into two groups, the friction modification and wear resistance, which are connected with each other. The intention of this book is to describe tribological phenomena occurring in micro and nano scale systems, such as MEMS and nano-devices, where a large-scale surface destruction or material transfer is not a major concern as much as in conventional machines. For this reason, the emphasis in this chapter is given to the discussions on molecule packing on surfaces, the shear strength of boundary layers, and the use of ordered molecule monolayers as boundary lubricant. For information on tribochemical processes, formation of chemical reacting films, and functions of additives that are designed to prevent wear, readers are referred to the tribology textbooks and monograph.

Section 2 describes mechanisms of boundary lubrication, more specifically, the formation of surface films, including adsorbed layers and chemical reaction films. The main concern is given to the discussion on molecule arrangement and orientation in boundary films. Section 3 introduces the idea that regards boundary lubrication as a limiting state of hydrodynamic lubrication, which leads to a technique, developed in the State Key Laboratory of Tribology (SKLT), Tsinghua University, and other research groups around the world, for determining the thickness of boundary layers by measuring EHL film thickness as a function of speed. Based on results from experiments and the author's own molecular dynamics simulations, the role of thin film rheology in boundary lubrication has been discussed. Section 4 gives an introduction to the highly ordered molecular

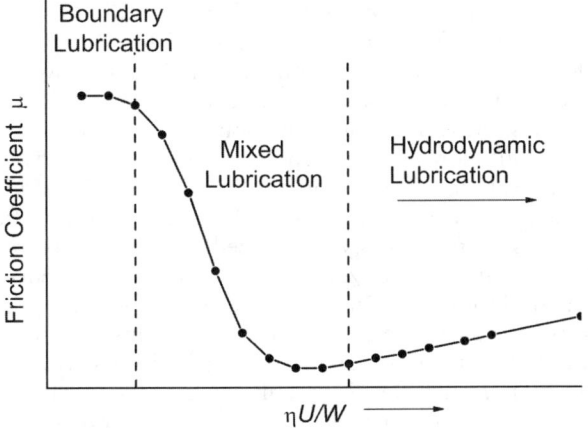

Fig. 1—Schematic Stribeck curve—an illustration of transition in lubrication conditions.

[1] State Key Laboratory of Tribology, Tsinghua University, Beijing, China.

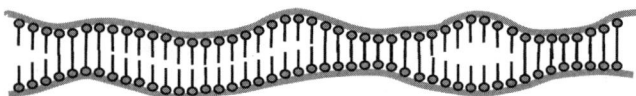

Fig. 2—Model of boundary lubrication proposed by Hardy, suggesting molecule adsorption in perfect order.

films, e.g., Langmuir-Blodgett films and self-assembled monolayers, and their performances in tribology applications, as have been revealed in experimental studies and computer simulations performed at SKLT. A brief discussion on boundary friction is presented in Section 5, where the issues to be addressed include the inconsistency between the current lubrication models, the difficulty in determining the shear strength of boundary films, and possible solutions proposed by the present author.

2 Mechanisms of Boundary Lubrication

2.1 Formation of Adsorbed Monolayers

The mechanism of boundary lubrication is a subject that has been studied intensively for nearly a century, but is still an on-going investigation, in which the mystery has not been fully revealed. Formation of monolayers through physical or chemical adsorption is a widely accepted mechanism of boundary lubrication under light duty conditions. An early model proposed by Hardy [2] is sketched in Fig. 2, which shows that a monolayer of lubricant molecules is formed on each solid surface through physical adsorption. It is suggested that the long-chain molecules arrange themselves in an ordered structure with head groups packed on the solids and molecule chains stretching normal to the surfaces. As a result, the sliding takes place between tail groups of the adsorbed molecules, something like rubbing between two brushes, so that friction has been reduced.

The ordered structure and molecule orientation in the monolayers, as suggested by the Hardy model, have been studied by various means. Electron diffraction techniques, for example, including both reflection and transmission, have been employed to examine the molecular orientation of adsorbed monolayers or surface films. The observations from these studies can be summarized as follows [3].

1. The long-chain molecules in the monolayers are fairly well oriented relative to the substrates. Unlike the bulk crystals, however, the ordering found in the monolayers is short range in nature, extending over regions of a few hundreds angstroms.
2. The short-range order gradually disappears as the temperature rises, and the structure becomes almost completely disordered near the bulk melting point of the monolayer materials.
3. A correlation between the packing in the monolayer and its lubricating properties is detected, which suggests that monolayers in higher degree order may exhibit better boundary lubricating properties.

The expectation of the structural dependence of lubrication motivated great numbers of investigations that intended to prepare and use highly ordered organic films, such as the Langmuir-Blodgett (L-B) films and Self-Assembled Monolayers (SAMs), as solid lubricant, which will be discussed more specifically in Section 4.

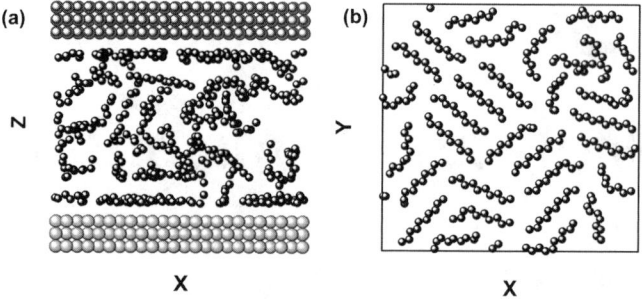

Fig. 3—Molecular configurations for a five-layer decane film at equilibrium: (a) cross section in the x-z direction; (b) projection in the x-y plane of the first layer adsorbed on the lower surface.

Another concern associated with molecule architecture is the chain length effect on lubricating properties. In the experiments when a stainless steel sphere slid on a glass surface covered with a fatty acid monolayer, the measured kinetic friction was found decreasing as the chain length of lubricant molecule increased, and reached a minimum for a chain length of about 14 carbon atoms. Similar chain-length dependence was reported [4] on the basis of AFM measured friction for self-assembled monolayers of alkane molecules on mica surfaces. The poor lubricating performance of short chain molecules may be interpreted in terms of the Hardy model that only the molecules longer than the minimum chain length can maintain themselves in an ordered arrangement, via inter-chain van der Waals forces. Likewise, the model may also help to understand why the branched molecules are a less effective boundary lubricant because the side chains simply impair close packing of the molecules [5].

However, the assumption of molecule orientation normal to the surface is not convincing enough for this author, and it does not consist well with the results of the molecular dynamics simulations for the alkane confined between solid walls. An example in Fig. 3 shows that the chain molecules near the wall are found mostly lying parallel, instead of normal, to the wall [6]. This means that the attractions between lubricant molecules and solid wall may readily exceed the inter-molecule forces that are supposed to hold the molecules in the normal direction. Results in Fig. 3 were obtained from simulations for liquid alkane with nonpolar molecules, but similar phenomenon was observed in computer simulations for the functional lubricant PFPE (perfluoropolyether) adsorbed on a solid substrate [7], confirming that molecules near a solid wall lie parallel to the surface.

It has been gradually recognized that the Hardy model presents a highly idealized picture for the real scenario of boundary lubrication. In fact, lubricant molecules confined within boundary films would organize themselves into ordering structures in much more complex forms. The structure depends on molecular architecture of the lubricant. In general, spherical molecules can rearrange more quickly in thin films while the lubricants composed of chain molecules, especially branched chain molecules, are more difficult to form the ordering structure so that such lubricants tend to give a relatively smooth sliding. More details will be further discussed in the next section.

Despite the uncertainness in molecule orientation, the formation of adsorbed monolayers is confirmed convincingly by the fact that the presence of a tiny portion of surfactant additives in base oil would lead to a great reduction in friction [5]. In this sense, the coverage ratio and shear strength of the adsorbed monolayers are perhaps the more crucial issues to be considered. It was confirmed that metallic contact might be reduced by a factor of 100,000 in the presence of an effective lubricant [5].

The shear strength of the monolayers, a factor that determines friction, remains to be the most mysterious part in the studies of boundary lubrication. Typical values of the shear strength, measured on surface force apparatus (SFA) for lubricant films confined between two solid walls, were reported ranging from 10^6 to 10^7 N/m^2 [8], depending on various factors, such as the terminal groups of adsorbed molecules, the number of layers, the molecule orientation and architecture, the pressure to which the film was subjected, and so on. Yet it has not been fully understood why the frictional behavior diverges so dramatically sometimes when one made only a small change in lubricant composition. This is a subject to be further discussed when we describe the origin of friction in Chapter 9.

When the surface is subjected to wear, the weakly bonded monolayers would be worn off in a few passes of sliding, but active molecules in the lubricant would adsorb successively on the surface to repair the worn part of the film. The wear resistance depends on the balance of the rate of film removal and generation.

A key limitation to the application of the monolayer as a boundary lubricant is the sensitivity to temperature. Molecules are desorbed from the surfaces as the temperature rises, and the coverage ratio would drop to zero above 200°C at which the monolayer becomes completely ineffective. Boundary lubrication functions at high temperature regions rely mainly on chemical reactive films.

In summary, impressive progress has been achieved to understand the role of adsorbed layers in boundary lubrication, but the effect of molecular orientation on tribology performance and the shear strength of adsorbed monolayers are the issues that remain to be clarified in a long future study of boundary lubrication.

2.2 Formation of Surface Films Via Chemical Reactions

It is unlikely in real tribological events that adsorbed monolayers work solely to provide lubrication. Instead, adsorption and chemical reactions may occur simultaneously in most cases of boundary lubrication. For example, fatty acid is usually regarded as a friction modifier due to good adsorptivity, meanwhile its molecules can react with metal or a metal oxide surface to form metallic soap which provides protection to the surface at the temperature that is higher than its own melting point.

Chemical reactions in boundary lubrication are different from static reactions even if the reactive substances involved are the same. The temperature to activate a chemical reaction on rubbing surfaces is usually lower than that required in the static chemical process. Some believe this is because of the naked surfaces and structural defects created by the friction/wear process, which are chemically more active. Kajdas proposed a new concept that accumulations of stress and strain in friction contacts could cause emission of low-energy electrons (exoelectrons) from metal surfaces, which ionized the molecules of a lubricant. This results in the formation of anions and anion radicals interacting with metal surfaces positively charged by exoelectron emission [9,10]. This process, known as the Negative Ion Radical Mechanism (NIRAM), is believed to play a significant role in boundary lubrication. Clearly, boundary lubrication involves a new kind of chemistry connected with the chemical reactions activated and accelerated by friction and wear, which is well known to most tribology scientists as "Tribochemistry."

The recognition of the function of tribochemical reactions motivates applications of a wide variety of lubricant additives containing sulfur or phosphor, in order to produce metallic sulfides or phosphides, which act as protective surface films. Among the increasing number of lubricant additives, zinc dialkyldithiophosphate (ZDDP) has received the most attention in the literature as highly effective antiwear (AW) and extreme-pressure (EP) additives. Intensive investigations have been carried out, by use of almost every tool in the field of surface analysis, to explore the mechanisms of ZDDP, from which a general picture starts to emerge. In the case of steel components lubricated by alkyl-ZDDP, for example, it is found that surface films are formed through a complex process of chemical reactions and thermal-oxidation decomposition. The reactive surface films, with a film thickness ranging from 40 nm to a few hundreds nanometres, are reportedly composed of a mixture of short-chain phosphates and long-chain polyphosphate. This chapter will not engage in a detailed description of formation, structure, and function of the ZDDP originated surface films, and readers are referred to the review paper by Gellman and Spencer [5] for further information.

The tribochemical processes in lubricated sliding are not limited to the reactions between the lubricant and solid surface. Other agents like oxygen and water molecules from the environment can also come into play. It is well known that oxidation of metal surfaces provides effective surface layers that prevent surfaces from direct metal contact and scuffing. In addition, organic molecules in the lubricant react with oxygen, which results in degradation of the lubricant, and the oxidation may be catalyzed due to the presence of the active metals in tribological systems [11].

It has been recognized that tribopolymerization plays an important role in the formation of boundary lubrication films. The polymerization takes place under certain conditions by use of monomers to form high molecular weight polymer films that reduce friction and wear. The process can be activated by temperature generated in friction or by exoelectron emission; more importantly, the protective films are formed "in situ" and can be effective on metal or ceramic surfaces [12]. Studies also revealed that the products that are so-called "friction polymers" are composed of particles of metal or metal oxides intertwined with high molecular weight organometallic compounds, which form surface films with different colors and appearances [13].

In summary, for metal surfaces in boundary lubrication, complex tribochemical reactions occur along with the physical/chemical adsorptions, which lead to the formation of surface films, consisting of reaction products, oxide layer, the mixture of particles and organometallic polymer, and perhaps a viscous layer. The surface films operate as a sacrificial layer, friction modifying layer, or a wear resistant and load bearing layer that provide protection to the surfaces in sliding.

The investigations on boundary lubrication used to focus on the friction elements made of metallic materials, and of steel in particular. This is, of course, due to the fact that a great majority of machines are built from metal and steel, but it is also because the hydrocarbon-based oils have been proven to be an extraordinarily good lubricant for metal surfaces. Unfortunately, the conventional oils are not so effective to lubricate the components made of other materials, like ceramics, rubbers, silicon, etc., so that the study on new types of lubricants suitable for such materials has attracted great attention in recent years.

Finally, there is another category of lubricants, including the laminated materials, highly ordered organic monolayers, and various thin solid films, which provides effective lubrication via their properties of low shear strength or high wear resistance. Lubrication via ordered molecular films and other solid lubricants, which have been considered by some investigators as a sub-discipline of boundary lubrication, will be discussed more specifically in Section 4.

3 Properties of Boundary Films as Confined Liquid

3.1 Boundary Lubrication as a Limiting State of Hydrodynamic Lubrication

In hydrodynamic lubrication, two surfaces in relative motion are separated by a fluid film, and it is the lubricant rheology that determines the performance of lubrication. As the film thickness decreases to the dimension of lubricant molecules, a transition from hydrodynamic lubrication to boundary lubrication would take place. The transition of the lubrication regime, as a result of reduction of film thickness, is accompanied by the change in lubricant rheology. Consequently, the mechanical properties of the fluid film will undergo a continuous transition that will finally transform the lubricant into a boundary film. In this sense, boundary lubrication may be regarded as a limiting state of hydrodynamic lubrication or EHL. In fact, the transition has to be considered as a process that starts from hydrodynamic lubrication, goes through some intermediate states known as "thin film lubrication" and "mixed lubrication," and ends up at boundary lubrication. The topics on thin film lubrication and mixed lubrication are discussed in Chapter 3 and Chapter 7, respectively, and here we focus on the properties of the film at the final stage of transition. The discussions are limited to the films confined between perfectly smooth surfaces since the roughness effects are not of concern in this chapter.

At the thin film limit, the hydrodynamic pressure will approach a distribution that is consistent with the pressure between the two solid surfaces in dry static contact, while the shear stress experienced by the fluid film will reach a limiting value that is equal to the shear strength of a boundary film.

The present author has performed computer simulations to examine the transition of pressure distributions and shear response from a hydrodynamic to boundary lubrication. Figure 4(a) shows an example of a smooth elastic sphere in contact with a rigid plane, the EHL pressure calculated at a very low rolling speed coincides perfectly with the

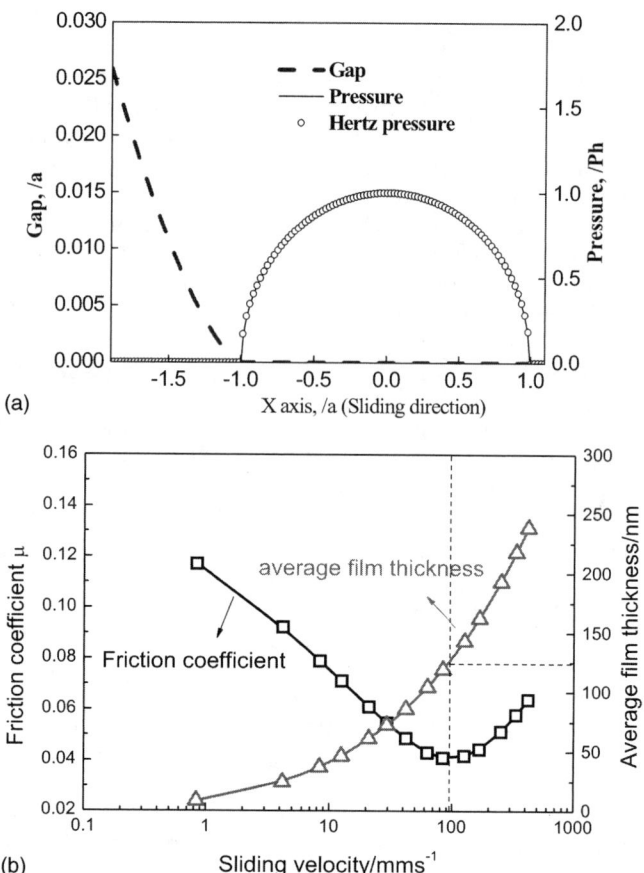

Fig. 4—Illustration of the transition from hydrodynamic to boundary lubrication: (a) a comparison of pressure of thin EHL film with Hertzian distribution; (b) a schematic stress-velocity map showing the dependence of shear stress of lubricating films on sliding velocity.

classical Hertz pressure distribution [14]. Friction coefficients measured between two smooth surfaces ($Ra = 0.002$ μm) in a circular point contact have been plotted in Fig. 4(b) as a function of sliding velocity, and the film thickness, obtained from numerical simulations under the same conditions, is also displayed in the figure. The film thickness decreases monotonously with the velocity but the friction varies in a more complicated way. On the right-hand side of the curve, friction or traction declines with decreasing velocity and reaches the minimum near 70 mm/s when the corresponding film thickness is about 100 nm. As the velocity further decreases, however, the friction starts to rise even though the magnitude of film thickness suggests that there is no direct contact between the two surfaces in sliding. The friction finally becomes independent of velocity at the left end of the curve, giving an example for a transition from fluid traction to boundary friction. It is interesting to note that before reaching the end of transition, friction rises with diminishing film thickness, indicating a progressive transition and the existence of an intermediate state where the lubricant gradually converts itself from fluid into a solid-like state so that the nature of friction changes correspondingly. This is the subject to be discussed more specifically in Section 4.

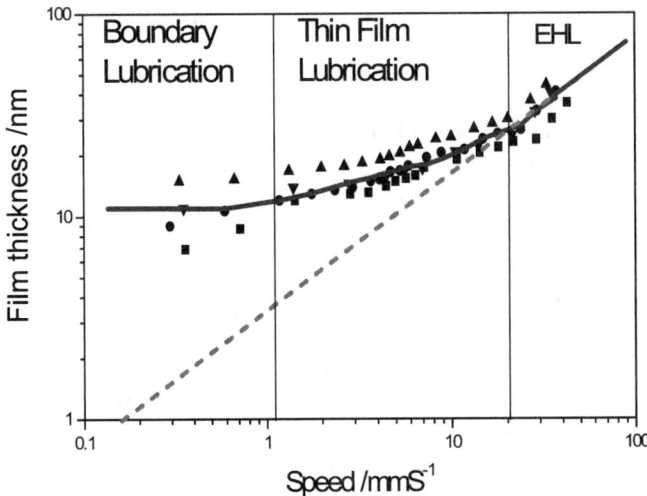

Fig. 5—Film thickness measured by Luo et al. using the interferometer versus the rolling speed.

3.2 Measurement of Boundary Film Thickness

The idea of regarding the boundary lubrication as a limiting state of EHL due to reduction of film thickness motivates the attempts to detect boundary film thickness by means of an ultra-thin film interferometer, which used to be employed for measuring EHL film thickness but recently with an improved resolution that enables us to examine thin films of a few nanometres thick [15,16]. A typical result of the measurements was obtained in the State Key Laboratory of Tribology (SKLT), Tsinghua University, and reported by one of the authors of this book, Professor Luo [17], as shown in Fig. 5 where the film thickness between a steel ball and glass disk is plotted in logarithmic scales as a function of the rolling speed. In the EHL region where rolling speed is relatively high, the film thickness drops with decreasing speed, following a power law predicted by Hamrock and Dowson [18], as marked in the figure by the dashed line. When the film thickness goes below a certain threshold, the pattern of the velocity dependence changes considerably such that the h-V curve turns to a much gentler slope, indicating the film becomes less and less dependent on the surface speed. As the speed further decreases, the film thickness seems to stabilize around a limiting value, as symbolized by the solid line in Fig. 5. The stabilized value has been regarded as an indication of boundary film thickness by some investigators since the velocity-independence is a remarkable identification of boundary lubrication.

Similar results were reported by other investigators, [19,20], but attention was paid to investigating the effect of lubricant additives on the boundary film thickness. It is speculated that there should be no adsorbed layers formed on rubbing surfaces if purified and nonpolar lubricants are applied. The interferometer measurements show that in the case of using base oils, the relation between the film thickness and rolling speed follows the EHL power law pretty well down to 1 nm (Fig. 6(a)), or sometimes the film thickness may deviate from the Hamrock and Dowson's line and turn down quickly (Fig. 6(b)). If there is a small percentage of additives in the lubricant, on the other hand, the deviation from the power law occurs in a different way that the h-V curve levels off and stabilizes at the film thickness around a few nanometres, as shown in Fig. 6(c), which is attributed to the formation of adsorbed boundary layers. The results in Fig. 6(a) and 6(b) imply that the thickness of the boundary layer formed from base oils may be too small to be detected by means of the interferometry technique. However, one has to be cautious to state that the EHL prediction holds even in molecularly thin films since the coincidence between the experimental data and the Hamrock and Dowson's line could be a result of a collective play by several factors. Theoretically, it is expected that there is an increase in film thickness in comparison to the classical EHL prediction, regardless of the lubricant polarity. This is because of the viscosity enhancement of confined liquid, as will be discussed in next sections.

3.3 Thin Film Rheology—Experiments

Liquid lubricant confined in molecularly thin films would experience dramatic changes in its physical properties, such as increased viscosity, slow relaxation, and solidification. Progress in studies of thin film rheology has greatly improved our understanding of boundary lubrication, which is the subject to be discussed in this section and in the next.

The rheological transition in thin liquid films has been studied by means of various experimental methods throughout recent decades. The effective viscosity of confined dodecane and hexadecane, measured by Granick on SFA [21] and by one author of this book on interferometer [22], respectively, are plotted in Fig. 7(a) and 7(b) as a function of film thickness. In both experiments the effective viscosities are found to be much higher than the bulk values if film thickness goes below 3 ~ 5 nm and grows rapidly with diminishing film thickness. This can be explained in terms of the confinement from solid walls which restrict the molecule movements within the film and greatly reduce the lubricant flow and its fluidity. The enhanced viscosity would lead to a film thickness larger than the predicted value by the EHL theory, which was exactly what we observed in Fig. 5 and Fig. 6(c).

It is also observed in SFA experiments [23] that as shear rate increases, the effective viscosity declines in a power law, $\eta_{\mathrm{eff}} \propto \dot{\gamma}^n$ where $n = -2/3 \sim -1$, which is the phenomenon well known in the field of rheology as the shear thinning, a typical behavior of non-Newtonian fluid. The present author also measured the effective viscosity as a function of shear rates, as shown in Fig. 8, from which the shear thinning, i.e., the decline of the viscosity with increasing shear rate can be clearly seen. The measurements were made on SFA at different levels of film thickness, ranging from 3.6 to 122 nm. Another important message delivered by Fig. 8 is that as film thickness increases, the shear thinning gradually disappears, i.e., $n \to 0$; meanwhile, the effective viscosity decreases with the growing film thickness and finally approaches to the bulk value, which means that the Newtonian behavior resumes again.

The shear thinning is also found for the bulk fluid if it is sheared at a very high rate, but in thin films the shear thinning occurs at much lower shear rates. In other words, a Newtonian lubricant in the bulk would exhibit the non-Newtonian shear response—the shear thinning, when it is confined in molecularly thin films. The observations of the

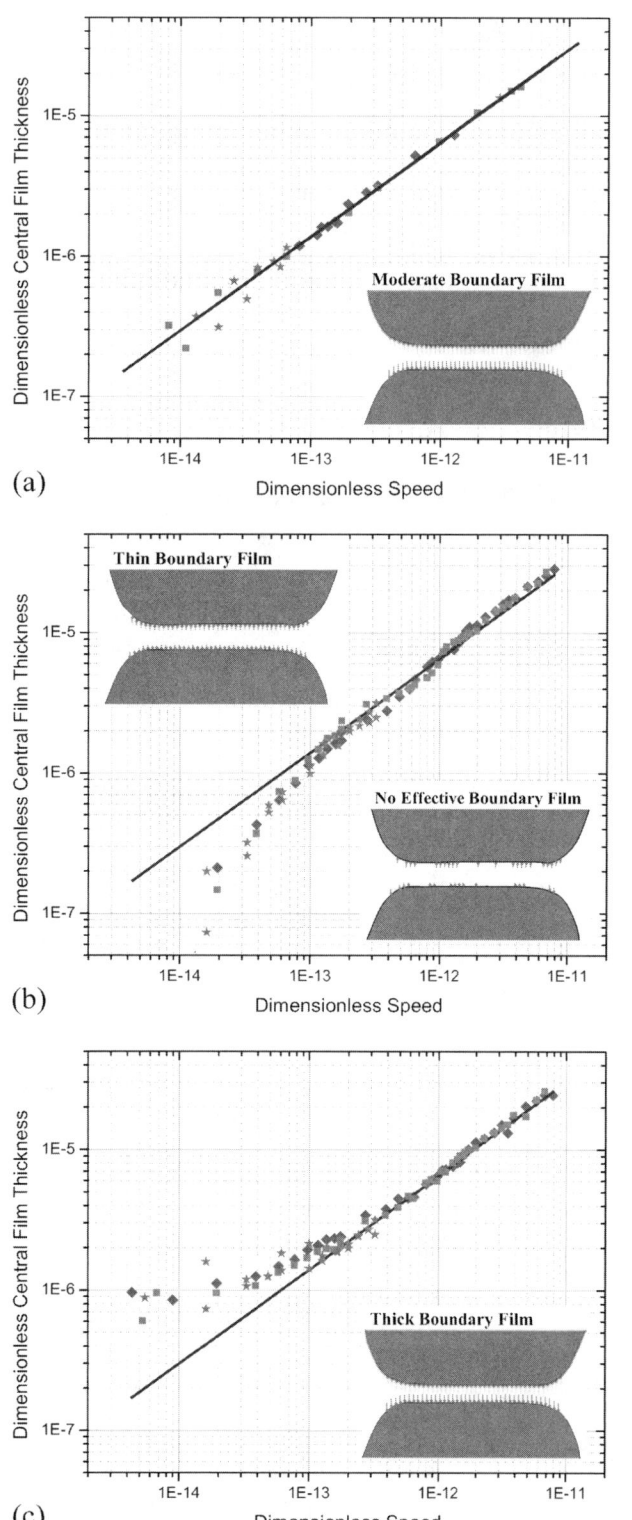

Fig. 6—Detect boundary films by means of the interferometry technique: (a) h-v curve follows the Hamrock and Dowson line all the way down to a couple of nanometres when there are few molecules adsorbed on surfaces; (b) the curve exhibits a downward turn if the adsorbed layer completely disappears and direct contacts occur; (c) the curve levels off due to the existence of thick adsorbed layers.

dynamic shear response of confined liquid imply that the relaxation process in thin films is much slower and the relaxation time for the confined molecules can increase by several orders [23].

Another remarkable feature of thin film rheology to be discussed here is the "quantized" property of molecularly thin films. It has been reported [8,24] that measured normal forces between two mica surfaces across molecularly thin films exhibit oscillations between attraction and repulsion with an amplitude in exponential growth and a periodicity approximately equal to the dimension of the confined molecules. Thus, the normal force is "quantized," depending on the thickness of the confined films. The quantized property in normal force results from an ordering structure of the confined liquid, known as the "layering," that molecules are packed in thin films layer by layer, as revealed by computer simulations (see Fig. 12 in Section 3.4). The quantized property appears also in friction measurements. Friction forces between smooth mica surfaces separated by three layers of the liquid octamethylcyclotetrasiloxane (OMCTS), for example, were measured as a function of time [24]. Results show that friction increased to higher values in a quantized way when the number of layers falls from $n = 3$ to $n = 2$ and then to $n = 1$.

The next topic involves another important process occurring in thin films: the phase transition and solidifications, and we start the discussion by examining the viscosity increase shown in Fig. 7. A sharp increase in the effective viscosity, observed in Fig. 7(a) at the film thickness ~26 Å, indicates that at this distance the confined liquid may undergo a transition to the solid phase. The solidification has been further confirmed by a rapid increase in the elasticity module measured for the films less than a few nanometres thick [25], which suggests that the confined film exhibits a strong elasticity and behaves more like a solid. A more substantial evidence for the solidification comes from the observation when a tangential force is applied on a solid body lubricated by thin films, it will not start to move until the stress reaches a critical value, as we frequently see in static friction, which means that a boundary film may deform in responding to an applied shear stress but the lubricant may not flow. In other words, the film under shear is broken like a solid, rather than a liquid in which the flow can be initiated by any small shear stress. The phenomenon of solidification is usually attributed to the strong adsorption but the real mechanism remains to be further clarified.

A solidified boundary layer tends to resist the shearing until it yields under a critical shear stress, which gives rise to a static friction and initiates the motion. The yield stress of OMCTS, measured on SFA at film thickness 9 Å (corresponding to one adsorbed layer), was reported by Homola et al. [8] to be around 4×10^6 N/m² (~4 MPa). It is important to note that the critical shear stress depends on the number of adsorbed layers and the time of contact. For the OMCTS film consisting of two adsorbed layers, the SFA measurement gave a yield stress of 7×10^5 N/m², five times less than that in one layer film. The direct measurements of the yield stress of thin films are generally in agreement with predictions of the cobblestone model, first proposed by Tabor and developed by several investigators [24], which estimates that the shear strength for a typical adsorbed hydrocarbon layer

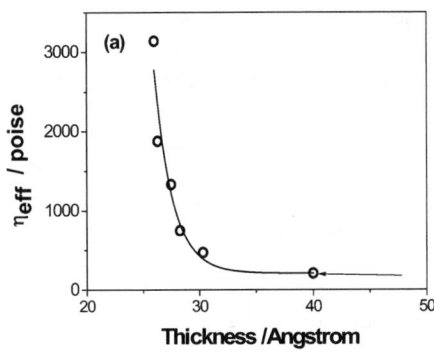
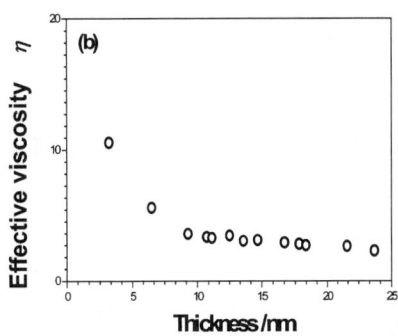

Fig. 7—Viscosity enhancement of confined liquid: (a) effective viscosity of dodecane measured on SFA [18]; (b) effective viscosity of hexadecane measured on interferometer [19].

on a van der Waals surface is in the order of $\sim 10^7$ N/m^2, and it is proportional inversely to the surface separation or film thickness.

There is evidence to suggest that the yield stress of thin films grows with the time of experiments, over a remarkably long duration—minutes to hours, depending on the liquid involved. Figure 9 gives the critical shear stress of OMCTS, measured by Alsten and Granick [26], as a function of experiment time. The yield stress on the first measurement was ~3.5 MPa, comparable to the result presented in Ref. [8], but this value nearly tripled over a 10-min interval and then became stabilized as the time went on. This observation provides a possible explanation for the phenomenon that static friction increases with contact time.

The solidified layer yields and returns to the liquid phase if the shear stress excesses the critical value, which initiates the sliding. When the stress is relaxed as a result of slip, the solid phase resumes again. The periodic transition between the solid and liquid states has been interpreted in the literature as a major cause of the stick-slip motion in lubricated sliding. Understanding the stick-slip and static friction in terms of solid-liquid transitions in thin films makes a re-

markable contribution to the study of boundary lubrication.

In summary, as the film thickness decreases to molecule dimension, the confinement of walls would induce dramatic changes in rheological properties of molecularly thin films, including the viscosity enhancement, non-linear response to the shear, formation of ordering structures, and liquid-solid transition. It is now generally recognized that the boundary friction depends not only on the composition of the lubricating films but also on the molecule packing within the confined space, though the actual physical picture requires further explorations.

3.4 Thin Film Rheology—from Author's Simulations

It has been generally recognized in recent years that Molecular Dynamics (MD) simulations can be a powerful approach to explore thin film rheology and boundary lubrication. In MD simulations, the solids in relative motion and the liquid lubricants confined between the solid surfaces are considered consisting of atoms and molecules which interact with each other via some sort of potential functions, so that the molecule movements can be calculated by solving Newton's equation of motion, which will lead to a prediction of the material properties of concern. There are a great many papers on MD simulations published over the past decade. This section, however, is dedicated to presenting the results ob-

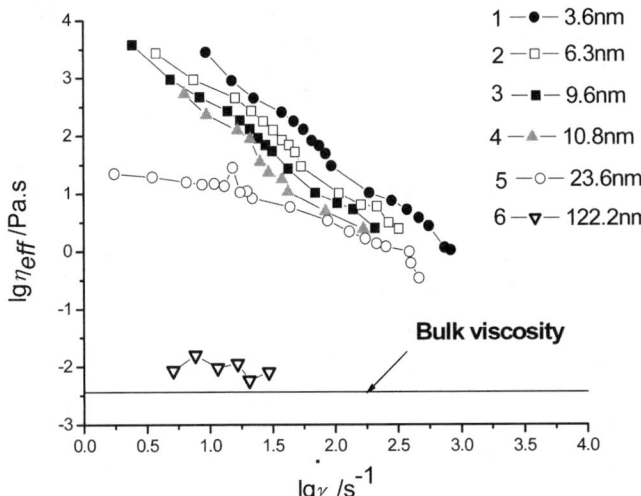

Fig. 8—Effective viscosity of confined hexadecane measured on SFA as a function of shear rate and film thickness, from which it is seen that the shear thinning gradually disappears as the film thickness increases and the viscosity finally has approached the bulk values at h = 122 nm.

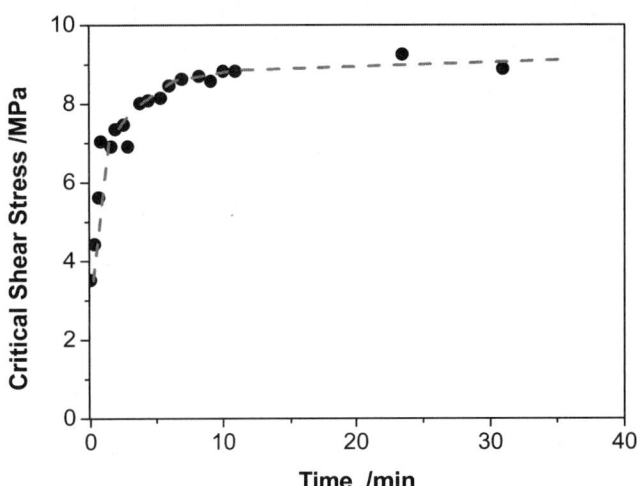

Fig. 9—Shear stress of confined OMCTS versus experiment time [23].

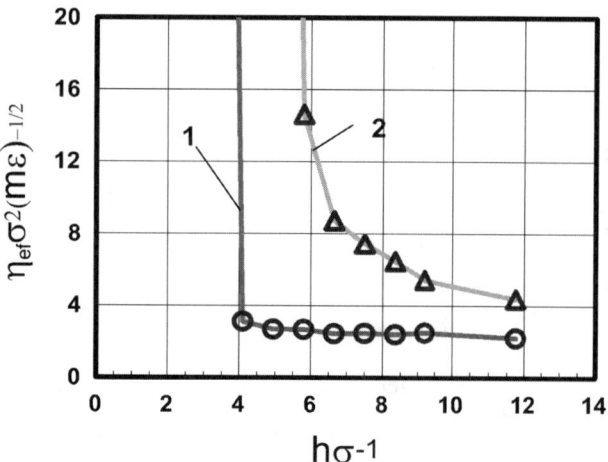

Fig. 10—Effective viscosity versus film thickness: results from simulations of confined argon, in which Curves 1 and 2 correspond to the cases with weak and strong liquid/wall interactions.

tained from author's own MD simulations, with emphasis on the confinement induced phenomena such as the enhanced viscosity, phase transition, molecule ordering, and interfacial slip.

The system used in the simulations usually consists of solid walls and lubricant molecules, but the specific arrangement of the system depends on the problem under investigation. In early studies, hard spherical molecules, interacting with each other through the Lennard-Jones (L-J) potential, were adopted to model the lubricant [27], but recently we tend to take more realistic models for describing the lubricant molecules. The alkane molecules with flexible linear chains [28,29] and bead-spring chains [7,30] are the examples for the most commonly used molecular architectures. The inter- and intra-molecular potentials, as well as the interactions between the lubricant molecule and solid wall, have to be properly defined in order to get reliable results. Readers who intend to learn more about the specific techniques of the simulations are referred to Refs. [27–29].

First, we present the results of the effective viscosity, which were obtained from our simulations of liquid argon confined between solid walls [27], and plotted against the film thickness in Fig. 10 where Curves 1 and 2 correspond to the cases with weak and strong wall-lubricant interactions, respectively. The viscosity begins to rise at relatively large film thickness, about ten diameters of argon atom, it grows with the diminishing film thickness and goes to divergent when the film thickness decreases to a critical value. It is also apparent from the simulations that the strong wall-lubricant interactions lead to a higher effective viscosity.

Results in Fig. 11, also from the simulations of liquid argon, are presented to illustrate a process of phase transition in the film. The effective viscosity and density of the confined liquid argon were calculated at a fixed film thickness and at a given normal pressures, P_{zz}, applied on the system. After each calculation, the normal pressure was increased by a small step while the film thickness remains constant, and computations were repeated under the new setup. In such a way one may be able to plot the viscosity and density against the normal pressure, as shown in Fig. 11(a), which reveals the growth of viscosity and density with the increasing pressure. A remarkable feature found from the results is that there is a stepwise increase in the density curve when the normal pressure grows up to a certain value, accompanied by a sharp increase in the viscosity. The pressure at this point is defined as the critical pressure beyond which the confined liquid transits into a solid state. Through similar procedures the critical pressures at different values of film thickness can be obtained, as presented in Fig. 11(b). The investigation confirms the occurrence of liquid-solid phase transition, but more importantly, it reveals that the transition pressure declines with diminishing film thickness so that the solidification in thin films could take place under relatively low pressure. This is the major message carried out from the MD simulations.

MD simulations also provide an opportunity to detect the structure of molecularly thin films. The most commonly known ordering structure induced by the confinement, the *layering*, has been revealed that the molecules are packed layer by layer within the film and the atoms would concentrate on several discrete positions. This has been confirmed in the simulations of liquid decane [29]. The density profile of unite atoms obtained from the simulations is given in Fig. 12 where two sharp density peaks appear at the locations near the walls, as a result of adsorption, while in the middle of the film smaller but obvious peaks can be observed on the density profile. The distance between the layers is largely identical to the thickness of the linear chain of decane molecules, which manifests the layered packing of molecules.

The confinement may also induce another sort of structure known as the in-plane order, i.e., an ordering structure formed in the plane parallel to the wall. This can be demon-

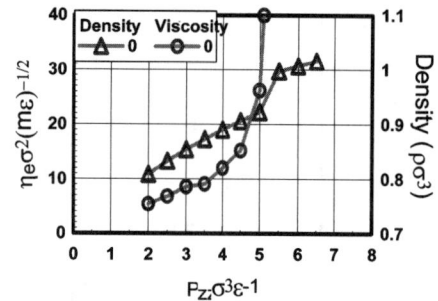
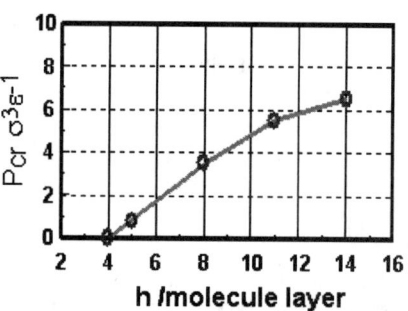

Fig. 11—Demonstration of phase transition in thin films: (a) viscosity and density versus system pressure for a five-layer argon film; (b) critical transition pressure as a function of the number of argon atoms.

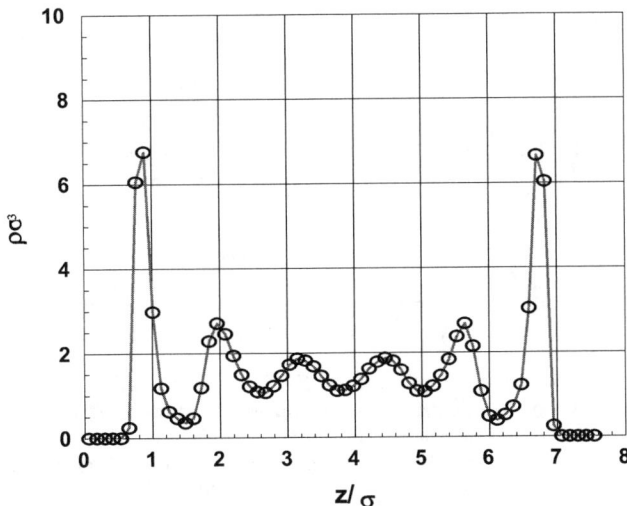

Fig. 12—Density profile from simulations of liquid decane, the peaks in density and fluctuations across the film reveal the layering structure in thin films.

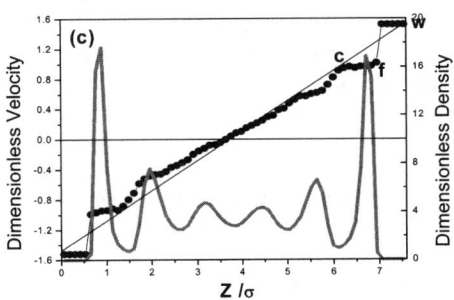

Fig. 14—Interfacial slip revealed by the velocity profile from simulations of confine liquid decane: the step in the profile at location f indicating a velocity discontinuity between the wall and the molecules adjacent to the wall [26].

strated by the results from the simulations of liquid argon [27]. The film was divided into several thin layers, by imaginary planes parallel to the wall, so that each layer would contain a certain number of atoms. The probability for an atom to appear at a specific planar location within each layer could be calculated in the process of simulations, which gave rise to the probability distributions shown in Fig. 13. The charts in the figure, from (a) through (f), correspond to the six layers positioned at different distances from the solid wall, and the squares are aligned with the wall lattice. Chart (a), for example, describes the particle distribution in the first layer adjacent to the wall, where the peaks indicate that atoms are most likely to be found at the center and four corners of the square. In the second layer, the peaks appear at the middle of four sides of the square. As the layers are more distant from the wall, the distributions tend to be more and more irregular, indicating an increasing disorder in the atomic packing. When reaching the middle of the film, i.e., the sixth layer, the in-plane order disappears completely, as depicted in Fig. 13(f). The ordering structure may not be observed if lubricants consist of chain molecules or the solid walls are composed of amorphous materials.

A basic assumption in classical fluid mechanics is that the fluid adjacent to the wall moves with the same velocity as that of the wall, but in practice slip may take place at the solid-liquid interface due to high shear rate. MD simulations are proven to be a useful means to study the phenomenon of interfacial slip, especially when the liquid is confined in thin films. In our simulations of liquid decane [29], the average velocities of unite atoms are calculated at different positions across the film, and the results are plotted in Fig. 14. A remarkable jump in the velocity profile can be clearly identified at the position near the wall (point f), indicating a velocity discontinuity between the wall and the lubricant molecules near the wall, which is a manifest of interfacial slip. It is equally interesting to note a smaller jump appearing at point c, corresponding precisely to the interface of two molecule layers, which can be identified by comparing the velocity to the density profile in the same figure, providing an evidence for the inter-layer slip.

As we discussed previously that if the shear rate increases continuously, the film is subjected to a growing shear stress, which will eventually reach a critical value that results in the slip either at the wall-lubricant interface or within the films. This provides additional information about

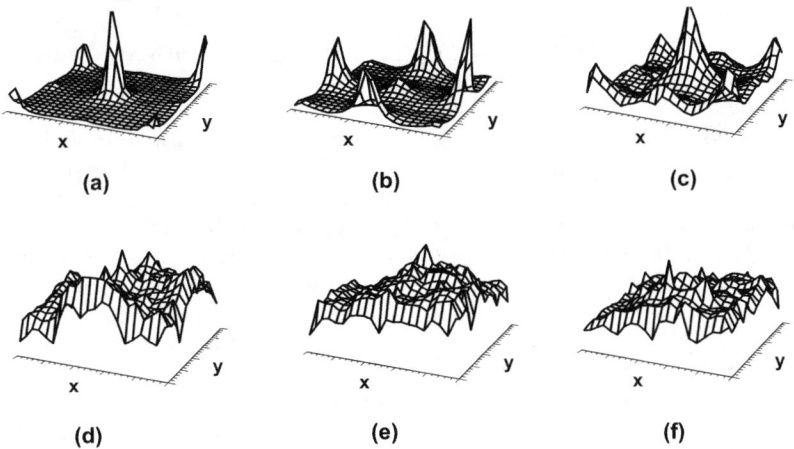

Fig. 13—In-plane structure illustrated by results from simulations of liquid argon: charts from (a) through (f) show the probability of particle distribution in different layers across the film, from the place adjacent to the wall stretching to the middle of the film.

the performance of thin films under shear that sliding may be initiated by two possible causes, the yield of solidified lubricant, or the slip occurred at the wall-fluid interface. However, it requires further investigation to understand which mechanism would prevail under given conditions. The coexistence of different slip mechanisms reveals the complex nature of boundary lubrication that the shear strength of boundary layers should not be regarded simply as a material property of the lubricating substance, but also a question of how the interfacial structure responds to the shear.

4 Ordered Molecular Films

The understanding that boundary lubrication works through the formation of adsorbed layers and surface films encourages the attempt to use artificially prepared surface films for providing lubricating functions: friction reduction and material protection. In this sense any surface film capable of reducing friction and protecting a surface from wear can be regarded as some sort of lubricant. Ordered molecular films are to be discussed in this section as a solid lubricant, with emphasis on the interrelation between molecule structure of the films and their tribological performance.

Recent achievements in techniques of preparing highly ordered molecule films, such as Langmuir-Blodgett films (L-B films) and Self-Assembled Monolayers (SAMs), provide a new category of lubricant. Tribological application of L-B films and SAMs for reducing friction or adhesion has been a topic that receives great attention and intensive investigations. The enthusiasm comes from the speculation that the surfaces with ordered molecule arrangements, as proposed by Hardy and illustrated in Fig. 2, would lead to a state of low friction while sliding. It is more important to realize that the ability to tailor molecular composition and structure, e.g., chain length, orientation, species of end groups, packing density, etc., would provide an opportunity to explore the interrelation between molecular architecture and tribological functions, which would lead to fundamental progress in achieving the long term target of preparing surface films with expected properties via molecular design.

4.1 Preparation of L-B Films and SAMs

The terminology of L-B films originates from the names of two scientists who invented the technique of film preparation, which transfers the monolayer or multilayers from the water-air interface onto a solid substrate. The key of the L-B technique is to use the *amphiphile* molecule insoluble in water, with one end hydrophilic and the other hydrophobic. When a drop of a dilute solution containing the amphiphilic molecules is spread on the water-air interface, the hydrophilic end of the amphiphile is preferentially immersed in the water and the hydrophobic end remains in the air. After the evaporation of solvent, the solution leaves a monolayer of amphiphilic molecules in the form of two-dimensional gas due to relatively large spacing between the molecules (see Fig. 15(a)). At this stage, a barrier moves and compresses the molecules on the water-air interface, and as a result the intermolecular distance decreases and the surface pressure increases. As the compression from the barrier proceeds, two successive phase transitions of the monolayer can be observed. First a transition from the "gas" to the "liquid" state,

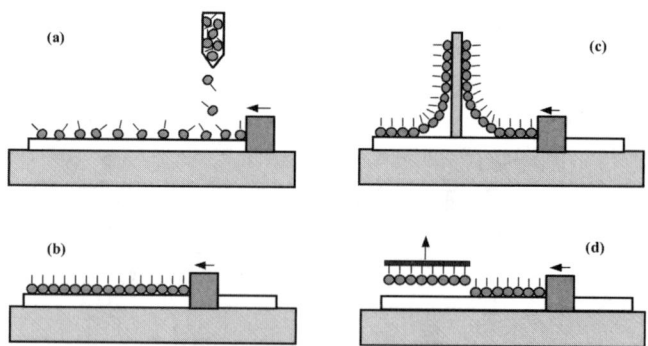

Fig. 15—Sketch of preparation of L-B films: (a) spread amphiphilic molecules on water surface, (b) compress the molecules using the barrier to get close packed and ordered molecular film, (c) transfer the film onto a substrate through the vertical immerse/retreat process, (d) transfer the film by horizontal lifting.

which occurs at the area per molecule ~24 Å^2 for a monolayer of stearic acid, and the second transition from the "liquid" to the "solid" state takes place when the area per molecule is about 20 Å^2 (for a single alkyl chain, e.g., stearic acid) [31]. In this condensed phase the molecules are closely packed and uniformly oriented (Fig. 15(b)).

To transfer the monolayer formed on the water-air interface onto a solid substrate there are two different methods, namely the vertical deposition and the horizontal lifting. The vertical deposition was first developed by Blodgett and Langmuir. They showed that a monolayer of amphiphiles at the water-air interface could be deposited to a solid substrate by immersing a plate vertically into the water, through the monolayer, and then withdrawing from it. For a hydrophilic substrate, the monolayer is pulled out of the water-air interface and transferred on the plate during the process of retraction, as schematically shown in Fig. 15(c). It is recognized that there is a critical velocity U_m, above which the transfer process does not work. This gives a limitation to the efficiency of the deposition, especially for the viscous layers.

The second way of preparing L-B monolayer structures, the horizontal lifting method, was introduced by Langmuir and Schaefer. In this method, a compressed monolayer first is formed at the water-air interface, and a flat substrate is then placed horizontally on the monolayer film. When the substrate is lifted and separated from the water surface, the monolayer is transferred onto the substrate, as depicted in Fig. 15(d).

Multiayer L-B films can be prepared in both methods, by repeated deposition of monolayer on the substrate, with the molecular direction changing alternatively after each deposition (Y-type), or keeping the same molecule direction in all monolayers (X-type or Z-type).

Self-assembled monolayers are formed spontaneously by the immersion of an appropriate substrate into a solution of active surfactant in an organic solvent. After the substrate is immersed for a time from minutes to hours, it is rinsed with ligroin, methanol, distilled water, and dried in a steam of nitrogen. An apparent effect of the monolayer coating is the drastic change in wettability of the surface so that the measurement of the contact angle can be considered as an effective way to detect the formation of the SAMs.

Molecules that constitute the self-assembled monolay-

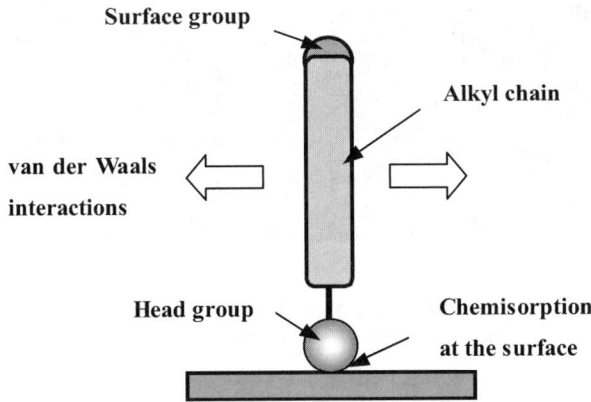

Fig. 16—A sketch of a molecule in self-assembled monolayers.

ers can be divided into three parts, as schematically drawn in Fig. 16. The head group is chemically adsorbed to the substrate via covalent or ionic bond. The alkyl chain, the second part, is closely packed to the nearest chains through van der Waals interactions. The third part is the terminal group that may exhibit an ordered or disordered feature, depending on the temperature.

The SAMs are considered as ordered, closely packed molecule assemblies in comparison to the polymer glass or liquid paraffin. If compared to inorganic crystals, however, they are highly disordered systems because of the number of defects. The defects are created as a result of the adsorption process, which increases packing density and order in the assembly, and does not match properly with the kinetics of monolayer-to-surface bond formation that prevent molecules from free packing. Tribological performance of both L-B films and SAMs is closely related to the defects created within the monolayers and associated energy dissipation. This will be further discussed in the following.

Readers who want more systematic information on ordered molecular films are referred to the book by Ulman [31].

4.2 Experimental Studies on Tribological Properties of Molecular Films

The most commonly used amphiphiles to build L-B films for tribological applications are the straight chain hydrocarbon compounds with simple functional groups such as the fatty acids, including stearic acids, arachidic acids, and behenic acids [32], but other amphiphilic molecules, e.g., 2,4-heneicosanedione and 2-docosylamina-5-nitropyridine, are also applied in some cases. There are two major systems of self-assembled monolayers, namely the alkylsilance derivatives (e.g., OTS, octadecyltrichlorosilane) on hydroxylated surfaces and the alkanethiols on metal substrates, which have been investigated extensively to examine their properties as solid lubricants and protective surface films [31].

Phenomenal studies were made to observe the frictional behavior of L-B films and SAMs and its dependence on applied load and sliding velocity, which has been summarized in a review article by Zhang [33]. It has been confirmed that in comparison to the bare surface of the substrates, the friction on molecular films is significantly reduced, with friction coefficients in a range of 0.05–0.1. Friction forces are found

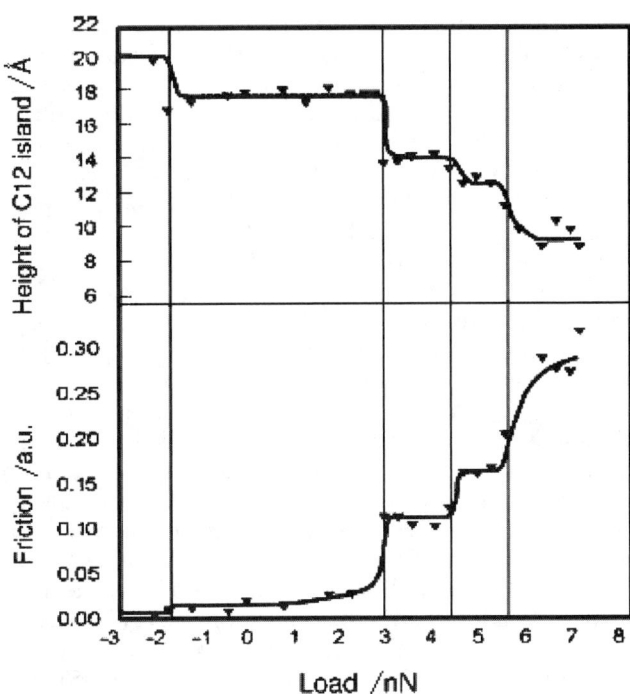

Fig. 17—Quantized changes in both film height and friction force, measured on an island of C_{12} thiols on Au(111) as a function of applied load (taken from Ref. [31]).

generally increasing with the growing normal load, following Amonton's Law, but the friction coefficient declines slightly as the load increases. However, Salmeron [34] and other investigators addressed a significant phenomenon in friction-load relations that under an increasing applied load, the molecule tilts occur only to discrete angles, due to the ratchet-like bounding between adjacent molecules and the zig-zag nature of the C-C skeleton in the chain. The discrete changes in tilt angles result in a discrete decrease in the film height and a stepwise increase in friction, which has been confirmed in experiments under increasing load, as illustrated by the results in Fig. 17. The quantized changes in film heights and friction forces have been observed, too, in confined liquid films as we discussed in Section 3.3, which should be considered as a fundamental feature of highly ordered and closely packed molecular films.

The dependence of friction on sliding velocity is more complicated. Apparent stick-slip motions between SAM covered mica surfaces were observed at the low velocity region, which would disappear when the sliding velocity excesses a certain threshold [35]. In AFM experiments when the tip scanned over the monolayers at low speeds, friction force was reported to increase with the logarithm of the velocity, which is similar to that observed when the tip scans on smooth substrates. This is interpreted in terms of thermal activation that results in depinning of interfacial atoms in case that the potential barrier becomes small [36].

Perhaps, the most interesting observation on the molecular films is the dependence of friction on chain length. It has been reported from independent experiments conducted on different sorts of L-B films and SAMs that the friction decreases with an increasing number of carbon atoms, but once the chain length exceeds a certain limit the friction becomes stabilized [4,37]. The dependence is in fact a manifest of the effects from the packing density and defects created in the films. The higher friction on the short chain films is attributed to the poor packing of molecules, which results in more energy dissipating modes, including chain bending and tilting, rotations, formation of gauche defects, etc. Salmeron [34] presented a systematic discussion on the formation of defects in molecular films, such as terminal gauche, internal gauche, and rigid chain tilts. He concluded that longer molecular chains would result in higher packing density due to stronger van der Waals inter-chain interactions and thus produced less defects.

Another issue of great concern relates to the effects of terminal groups. In AFM experiments, Frisbie et al. studied adhesion and friction between self-assembled monolayers with different functional groups [38]. Results show that the monolayers with the same hydrophilic terminal groups, COOH/COOH, exhibit the largest adhesion and friction in comparison to the other combinations, manifesting a direct correlation between adhesion and friction. This is due to the hydrogen bond formation that raises the energy barrier for the sliding. As a general rule, the films with terminal groups of low free energy are expected to show weak adhesion and low friction, but in reality the situations are more complicated. Attempt was made by Kim et al. to change the terminal CH_3 group of the alkane chain by CF_3 groups [39], which led to a noticeable increase in friction, contrary to the expectation that fluorinated films would exhibit lower friction. The increase in friction has been interpreted by some investigators [34] in terms of the incompatibility between the relatively small lattice space of CF_3-terminated films (~4.9 Å) and larger CF_3 ends, (~5.8 Å), which results in numerous point defects that provide additional channels for energy dissipation.

One limitation to the applications of molecular films is the relatively short durability under sliding wear. Nanosize species, e.g., C_{60}, MoS_2, or metallic particles, are introduced as additives during the preparation of L-B films in the expectation of improving tribological properties. It is reported by Zhang et al. that fatty acid L-B films-C_{60} composite shows better wear resistance [40]. A possible mechanism is suggested that C_{60} particles are dispersed in the form of nano scale islands which carry the applied load. L-B films containing metal nano-particles have received much attention recently and impressive improvements in tribological performance have been reported [41]. Another option for strengthening the molecule films is the polymerization of alkane chains, by means of UV illumination, for instance, to enhance the stability of the monolayers. As a result, the molecular weight of the lubricant greatly increases, which is expected to yield a lower friction. In addition, the polymerization prevents inter-chain slip during sliding so that the stepwise change in friction mentioned in Ref. [34] disappears.

Tribology performances and applications of ordered molecular films have been a long-standing research subject in SKLT, the workplace for the authors of this book. Hu and Luo [42] prepared SAMs of fluoroalkylsilane (FAS) and polyfluorealkylmethacrylate (PFAM) on the magnetic head of computer hard disk drivers. Experiment results show that the molecular films greatly improve the performance of the

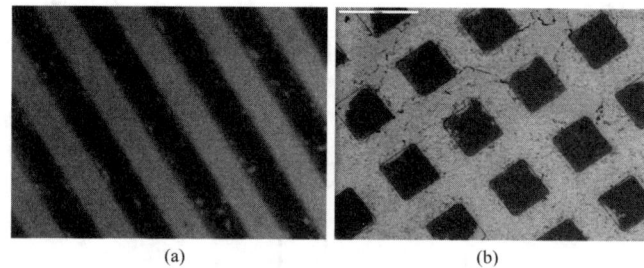

Fig. 18—Patterned SAMs composed of OTS (light) and APTES (dark) arranged in striped (a) and dotted patterns (b).

system in contact start-stop (CSS) tests. Meanwhile, the films also enhance corrosion resistance, and provide an effective protection to the Head/Disk system from the damage caused by static electrical charge or adhesion. Moreover, attempts have been made to apply the molecular films to the Micro Electrical and Mechanical Systems (MEMS) as an effective lubricant for preventing the components from stiction and wear during manufacturing and operation.

Incorporation of optical lithography with the preparation of monolayers may give rise to the patterned films, i.e., two or more types of SAMs of different natures distribute in a designed pattern. Figure 18 shows the photos of patterned SAMs prepared by Peng et al. [43]. The films are composed of two kinds of monolayers, octadecyltrichlorosilane (OTS) and (3-Aminopropyl)-triethoxysilane (APTES), distributed in striped (Fig. 18(a)) and dotted (Fig. 18(b)) patterns. The development creates a new technology, known as the *patterned lubrication*, which provides an effective way to control the tribological performance of a surface by arranging several types of lubricants of different natures into a certain geometric array. As a result, the overall property of the surface can be modified through adjusting the size and distribution of patterned monolayers.

4.3 MD Simulations of Molecular Films Performed at SKLT

Study on mechanisms of ordered molecular films as a model lubricant is of particular importance in the field of micro and nano-tribology, for it would help to understand how molecules contribute to the creation of friction and wear. The understanding has been much improved in recent years through MD simulations, performed by investigators around the world, to detect interactions between the molecular films in relative motion, and to reveal the process and specific mode of energy dissipation.

It has been found from MD simulations that friction of SAMs on diamond decreases with the increasing chain length of hydrocarbon molecules, but it remains relatively constant when the number of carbon atoms in the molecule chain exceeds a certain threshold [44], which confirmed the experimental observations. In simulations of sliding friction of L-B films, Glosli and McClelland [45] identified two different mechanisms of energy dissipation, namely, the *viscous mechanism*, similar to that in viscous liquid under shear, and the *plucking mechanism* related to the system instability that transfers the mechanical energy into heat, similar to that proposed in the Tomlinson model (see Chapter 9). On the basis of a series work of simulations performed in the similar

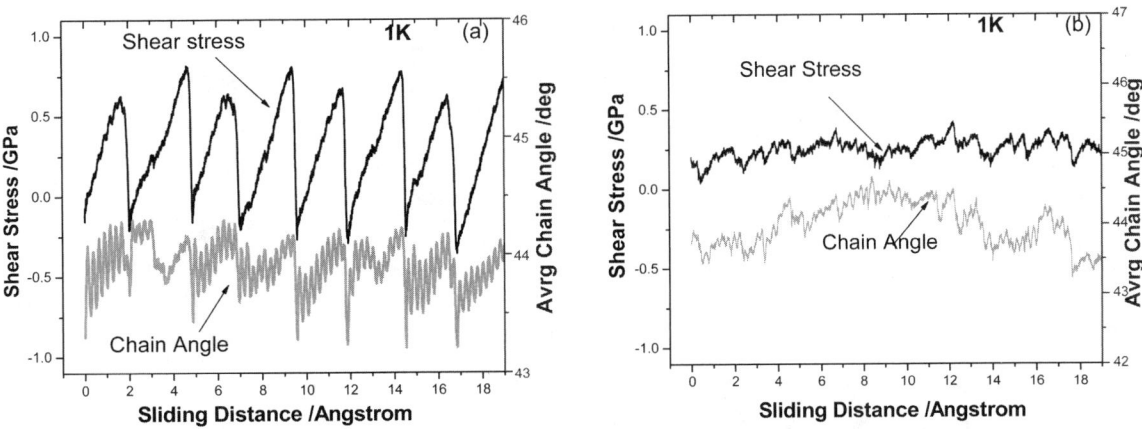

Fig. 19—Shear stress and chain angle as a function of sliding distance, from simulations of alkanethiolates on Au(111) at temperature 1 K: (a) results from commensurate sliding show a stick-slip motion with a period of 2.5 Å, (b) in incommensurate case both shear stress and chain angle exhibit random fluctuations with a much smaller average friction [45].

systems of ordered molecular films, Ohzono et al. suggested that there were two dissipation modes associated with *soft* and *hard* springs in the molecular system, corresponding to the van der Waals interactions and covalent bonds, respectively, and the amplitude of friction was mainly determined by the dissipation process in the soft springs [46].

The author's research team at SKLT also performed MD simulations to study sliding friction between two self-assembled monolayers of alkanethiolates chemisorbed on Au(111) substrates, with special interest in tracing the movements of individual molecules and revealing the effect of structural commensurability on frictional behavior [47]. The friction force and chain tilt angle are presented in Fig. 19 as a function of sliding distance. For the sliding between commensurate films, results in Fig. 19(a) demonstrate the molecule plucking and apparent stick-slip motions with a period of 2.5 Å, while in incommensurate cases, Fig. 19(b) shows that both friction and chain angles fluctuated in random patterns, which gave rise to a much smaller average value of friction, about one-fifth of that observed in the commensurate sliding. This investigation shows the significant impact of the commensurability on frictional behavior. However, there is a discrepancy between the finite friction observed in Fig. 19(b) and the prediction based on the Tomlinson-type models that friction would disappear for incommensurate systems in quasi-static sliding, so that the mechanisms of energy dissipation in such systems require further investigations.

To examine the molecule motions, the movements of two tail groups were monitored during the simulations. One tail group, TC, belongs to a molecule that moves along with the upper monolayer, and TA is a tail group selected from the molecules on the stationary lower monolayer. The relative distances between TC and TA have been calculated and displayed in Fig. 20(a) and Fig. 20 for the commensurate and incommensurate cases, respectively. The abscissa in Fig. 20 represents the traveling distance of the moving tail group, TC, while the ordinates in the top and bottom panels of the figure stand for the x and y components of the distance increments, $\Delta x = x_{TC} - x_{TA}$ and $\Delta y = y_{TC} - y_{TA}$. The stick-slip nature of molecule motion is visible in Fig. 20(a) that the tail groups, TC and TA, change their relative positions only at the moments of slip, i.e., at the end of each period, which means TC suddenly jumps from one equilibrium site (three-fold hol-

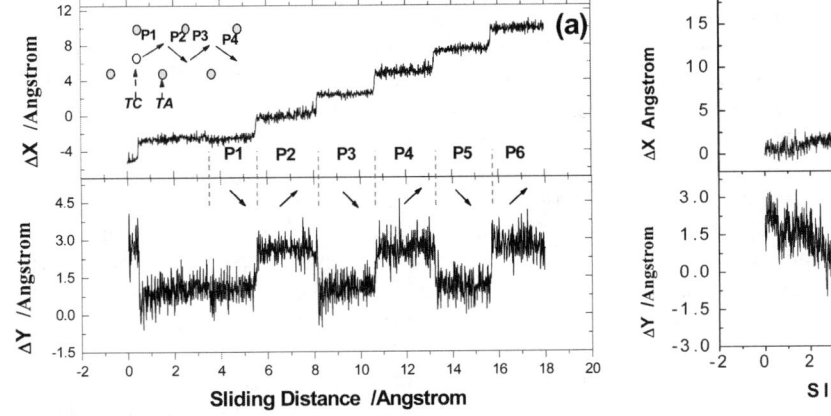
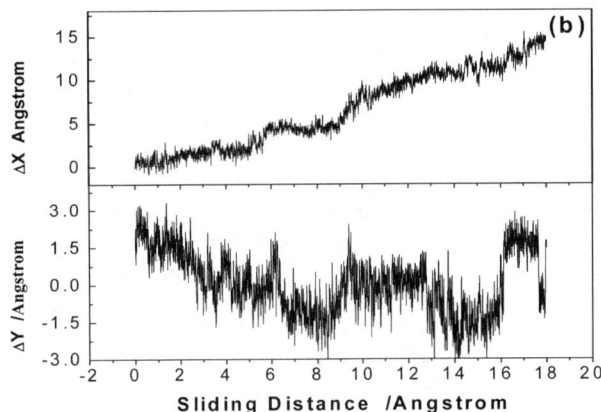

Fig. 20—Distance between two terminal groups, TC and TA, in relative sliding, results from simulations of (a) commensurate and (b) incommensurate SAMs in which the top and bottom panels show the x and y components of the increments plotted as a function of sliding distance.

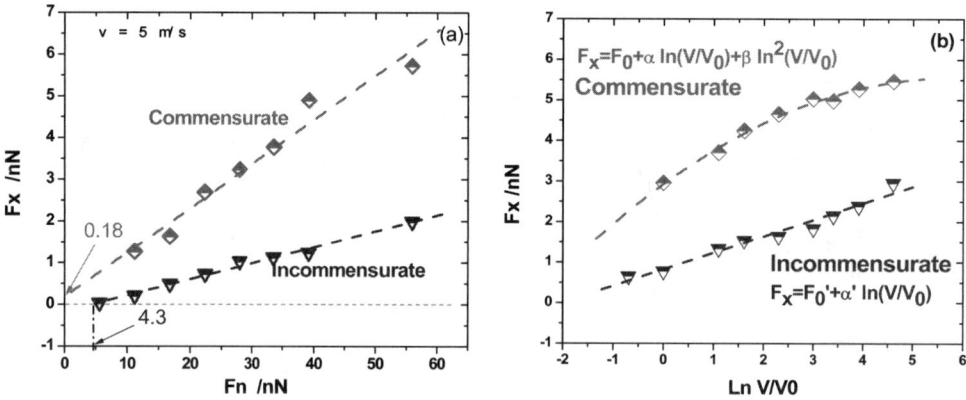

Fig. 21—Comparisons of friction forces from simulations of commensurate and incommensurate SAMs in relative sliding: (a) friction exhibits a linear dependence on applied normal load, (b) friction increases logarithmically with the sliding velocity.

low) to the next when slip occurs. As a result, Δx increases step by step while Δy oscillates up and down periodically, suggesting that TC dances over TA in a zigzag way, as sketched in the upper-left corner of Fig. 20(a). The increments of Δx and Δy match exactly with the distances between the nearest neighboring equilibrium sites, i.e., $\Delta x_e = 0.5a$ and $\Delta y_e = \sqrt{3}a/6$ where $a = 5$ Å is the lattice constant of the monolayers. For the incommensurate case shown in Fig. 20(b), on the other hand, Δx exhibits a continuous growth while Δy fluctuates randomly without a periodicity.

Figure 20 provides important information on the activities of individual molecules, considering the fact that the performance of tail groups is a manifestation of molecule motion. The periodic jumps observed in Fig. 20(a) indicate that the alkane chains are plucked by the opposite monolayer at the moment of slip. Looking at Fig. 20(b), however, it is difficult to tell whether or not the plucking mechanism also involves in an incommensurate sliding.

Next, friction forces on sliding SAMs were examined as a function of normal loads and sliding speeds, which gave a further comparison between commensurate and incommensurate friction. Figure 21(a) shows friction forces obtained under various normal loads but at a fixed sliding velocity of 5 m/s and a temperature of 300 K. It can be seen clearly that under the same conditions friction forces in commensurate sliding are always higher than those of incommensurate cases. For the SAMs in commensurate sliding, all alkane chains are deformed in a similar way during the time of stick, accumulating more mechanical energies that are then released at the moment of slip. Another impressive feature is that the friction in both cases grows linearly with increasing load. Considering that the contact area, i.e., the area of the simulation cell, is fixed, the growth of friction implies that the shear stress at the interface increases with normal load. This is in contrast to the assumption that surface films used to show a constant limiting shear stress while the friction-load dependence attributes to the real contact area that increases with applied load. The result for the load-dependent shear stress, however, consists with the prediction of the cobblestone model [24] which relates the shear stress to surface film thickness, and then to the normal load. In present simulations, the normal load is applied to press the two monolayers closer, shortening the distance between the interacting tail groups.

There are two additional details that may draw attention. In commensurate sliding, a finite value of friction, 0.18 nN, remains at the zero normal load, which is the friction usually attributed to adhesion. In incommensurate sliding, on the other hand, the friction vanishes below the load of 4.3 nN, suggesting a possible state of zero friction.

The frictions at a fixed normal load (39 nN) but various sliding speeds were also simulated for both commensurate and incommensurate cases, and the results are given in Fig. 21(b) as a function of $\ln(v)$. Frictions in commensurate sliding are again found to be much higher than those in the incommensurate case under the same conditions. A remarkable feature in Fig. 21(b) is the nonlinear growth of friction with increasing velocity, and the dependences takes logarithmic forms:

$$F_x = F_0 + \alpha \ln(v/v_0) + \beta \ln^2(v/v_0), \text{ in commensurate sliding, and}$$

$$F_x = F_0' + \alpha' \ln(v/v_0), \text{ for incommensurate cases.}$$

The logarithmic dependence of friction on velocity has been also found in experiments for various systems, though at much lower sliding velocities. The mechanism for the phenomenon has been well interpreted recently in terms of a modified Tomlinson model, taking into account the effects of thermal activation [48]. For the self-assembled monolayers in relative sliding, interfacial molecules may transit from one metastable state to another. The probability for the transition to take place is found related to the temperature T, energy barrier ΔE, and sliding velocity ϖ. According to this theory, the jump is more difficult to occur for the monolayers in faster sliding. The delay means more energies to be stored before a jump and thus leads to a higher friction at the moments of transition. The simulation results for commensurate SAMs show a remarkably different velocity dependence containing a term of $\ln^2(v/v_0)$. The mechanism for such velocity dependence remains to be explored further.

4.4 Other Types of Solid Lubricants

One of the major themes of boundary lubrication is to transfer the shear stress at the interface of direct solid contact to somewhere inside the lubricating layer, to achieve low friction and high wear resistance. In this sense, materials with low shear strength, such as liquid films, soft metals, and lamella solids, can be employed as candidate lubricants.

Molybdenum disulfide (MoS_2), graphite, hexagonal boron nitride, and boric acid are examples of lamella materials commonly applied as solid lubricants. The self-lubricating nature of the materials results from the lamella crystalline structure that can shear easily to provide low friction. Some of these materials used to be added to oils and greases in powder forms to enhance their lubricity. Attention has been shifted in recent years to the production and use of nanosize particles of MoS_2, WS_2, and graphite to be dispersed in liquid lubricants, which yields substantial decreases in friction and wear.

Soft metals such as In, Ag, Sn, Pb, and Au can lead to reasonably low friction coefficients, when used as solid lubricants, due to their low shear strength. The metals were generally applied as thin films prepared by the vacuum deposition process. Especially, in applications to the high temperature conditions where liquid lubricants fail due to the evaporation, the thin films of soft metals can provide effective protection to the surfaces in sliding.

Thin films of hard materials, e.g., diamond, diamond-like carbon (DLC), and carbon nitride, deposited on surfaces are considered as solid lubricants for they can provide both high wear resistance and low friction. In particular, DLC films prepared by chemical vapor deposition with introducing a specific ratio of hydrogen, exhibit very low friction coefficients, less than 0.01, when measured in inert gas or vacuum environments [49]. This causes a great interest in understanding the role of hydrogen in the DLC films and how it gives rise to the superlow friction. Several models have been proposed, but it is obvious that more investigations are required to achieve an agreement and full understanding of the mechanism of superlubricity. DLC films have been used extensively in magnetic storage systems of computer hard disk drives, to prevent the magnetic materials from direct contact, scratch, and corrosion. As the areal density of storage increases by 100 % each year, accompanied by the continuous reduction of head/disk spacing, a crucial task at present is to prepare ultra-thin films, less than 3 nm thick according to the current standard.

Solid lubricants are mostly applied to the sliding surfaces in the forms of thin films. Some of these films are prepared by the advanced vacuum deposition process, such as ion beam-assisted deposition, magnetron sputtering, ion plating and pulsed laser deposition, while others are produced by molecular grafting, sulfurizing, and vapor phase deposition. The advances of surface engineering also provide opportunities to develop a new strategy of boundary lubrication. Examples include the use of a combined system of solid/liquid lubrication so that if the liquid lubricant fails, the solid lubricant can carry the load and prevent direct metal contact, and the modification of properties of a tribological system via nanostructured coatings such as multilayer coatings and patterned molecular films. Readers are referred to Chapter 8 and the recent review paper written by Erdemir [49] for more discussions on this subject.

5 Discussions on Boundary Friction

On concluding this chapter, it is worthwhile to address a few issues that associate in particular with the boundary friction, i.e., the friction originated from the shear of surface films. The emphasis is given to the discussion of the shear strength and two parallel models that describe the boundary friction in terms of different mechanisms.

(1) Adsorbed Layer Versus Solidified Liquid

In most textbooks, surface films in boundary lubrication are usually regarded as products of adsorption or chemical reactions. From the point of view of transition of lubrication regimes, on the other hand, boundary lubrication can be considered as a limit state of hydrodynamic lubrication or EHL when the hydrodynamic film becomes molecularly thin, which implies that the boundary film is simply a layer of confined lubricant. Confusions and disputes exist in current investigations, as to how to describe and model the boundary lubrication. The thickness of adsorbed layer of liquid alkane, for instance, is assumed to be very small (1 nm or less), according to some investigators [21] who suggested that the adsorption would not be able to take place unless the lubricant contains polar molecules, but studies of thin film rheology of confined liquid indicate that solidified boundary films of a few nanometres thick are found even for the purified alkane [23]. The difference also leads to two parallel models for describing boundary friction, in which the friction may be viewed either as a result of the slip between two adsorbed and reacted layers, or originating from the yield of confined lubricants under shear.

(2) Shear Strength of Boundary Films

In both models, the magnitude of friction can be determined in terms of the shear strength of boundary films, i.e., the critical shear stress when slip occurs, regardless of how the films are formed, by adsorption, reaction, or solidification. The existence of a critical shear stress, τ_c, independent of applied load has been long expected, for example in Bowden and Tabor's friction model, so that the Amonoton's law for the linear friction-load dependence can be interpreted as a result of an increase in real contact areas. The first measurement was obtained on SFA, indicating that typical values of the shear strength for boundary films of OMCTS ranges from 10^6 to 10^7 N/m^2, depending on the number of layers of lubricant molecules [8], which provided the experimental evidence for the assumption in friction models. It is found latterly that the shear strength, τ_c, contains a secondary component in direct proportion to the normal pressure, but the dominating part of the shear strength remains to be independent of the applied load, which forms a basic theme in modeling friction. The prediction of the shear strength and understanding its correlations with the compositions and structures of lubricant molecules will be a mission of high priority in future studies of boundary friction.

(3) Effects of Molecular Structure and Arrangement

When boundary friction is considered as a result of sliding between the adsorbed layers, studies on the shear response of surface films are mostly concentrated on the influential factors related to the lubricant molecule itself, such as the chain length, molecular architecture, and the nature of terminal groups. Impressive progress has been made in this respect, as presented in Sections 2 and 4, but the mystery as to why a slight change in molecular structure or chemical composition could result in widely diverged frictional behavior remains largely unsettled.

In the studies that attribute the boundary friction to confined liquid, on the other hand, the interests are mostly in understanding the role of the spatial arrangement of lubricant molecules, e.g., the molecular ordering and transitions among solid, liquid, and amorphous states. It has been proposed in the models of confined liquid, for example, that a periodic phase transition of lubricant between frozen and melting states, which can be detected in the process of sliding, is responsible for the occurrence of the stick-slip motions, but this model is unable to explain how the chemical natures of lubricant molecules would change the performance of boundary lubrication.

(4) A Further Discussion

The previous discussions on the shear strength of boundary films are based on the model that friction is associated with real contact and shear of adhesive conjunctions, which has been widely accepted over past 50 years and considered as one of the most successful models in studies of friction, in the sense that it is logic and simple yet capable of explaining many complex phenomena observed in the real world. It has to be realized, however, that the word of *shear strength* is borrowed from the vocabulary of engineers for describing the material properties, rather than a fundamental scientific concept for revealing the origin of friction. If one considers the slip between two solids in contact as a result of molecular depinning at the interface, the shear of a contact conjunction, with or without lubricating films, is therefore a result of pinning-depinning transition. In this sense, friction should be modeled at a more fundamental level in terms of molecular interactions and energy dissipation, as will be the topic of Chapter 9. This would also solve the contradiction between two parallel models, adsorption/reaction versus rheology transition, and lead to a unified theory of friction.

Finally, it has to be mentioned again that wear resistance has been a major subject of investigations in boundary lubrication, but this chapter is contributed mainly to the discussions of the mechanism of boundary friction. Readers who are interested in the wear process in microscopic scale are referred to Chapter 10 for further information.

6 Summary

Unlike traditional textbooks of tribology, in this book we regard boundary lubrication as a limit state of hydrodynamic lubrication when film thickness is down to molecular dimension and independent of the velocity of relative motion. The discussions are based on the existing results, some from literatures but mostly from the authors' own work. The topics are mainly focused on the mechanical properties of boundary films, including rheology transitions, molecular ordering, and shear responses. Ordered molecule films, such as L-B films and SAM, are discussed, with emphasis on the frictional performance, energy dissipation and the effects from structural features. Boundary films can be modeled either as a confined substance, or an adsorbed/reacted layer on the surface, but both contribute to improving our knowledge of boundary lubrication.

References

[1] Halling, J., *Principles of Tribology*, Macmillan, London, 1975.

[2] Dowson, D., *History of Tribology*, Longmans, London, 1979.

[3] Bowden, F. P. and Tabor, D., *Friction and Lubrication of Solids*, Oxford University Press, 1964.

[4] Xiao, X. D., Hu, J., Chayuch, D. H., and Salmeron, M., "Chain Length Dependence of the Frictional Properties of Alkylsilane Molecules Self-Assembled on Mica Studied by Atomic Force Microscopy," *Langmuir*, Vol. 12, 1996, pp. 235–240.

[5] Gellman, A. J. and Spencer, N. D., "Surface Chemistry in Tribology," Proc Instn Mech Engrs Part J: J Engineering Tribology, Vol. 216, 2002, pp. 443–461.

[6] Wang, H. and Hu, Y. Z., "Molecular Dynamics Study on Interfacial Slip Phenomenon of Ultra-thin Lubricating Films," presented at *ITC2000*, Nagasaki, Japan, Oct. 2000.

[7] Li, X., Hu, Y. Z., and Wang, H., "A Molecular Dynamics Study on Lubricant Perfluoropolyether in Hard Disk Driver," *Acta Phys. Sin.*, Vol. 54, 2005, pp. 3787–3792.

[8] Homola, A. M., Israelachvili, J. N., Gee, M. L., and McGuigan, P. M., "Measurement of and Relation Between the Adhesion and Friction of Two Surfaces Separated by Molecularly Thin Liquid Film," *J. Tribol.*, Vol. 111, 1989, pp. 675–682.

[9] Buyanovsky, I. A., Ignatieva, Z. V., and Zaslavsky, R. N., "Tribochemistry of Boundary Lubrication Processes," *Surface Modification and Mechanisms*, G. E. Totten and H. Liang, Eds., Marcel Dekker, New York and Basel, 2004, pp. 353–404.

[10] Kajdas, C., "Tribochemistry," *Surface Modification and Mechanisms*, G. E. Totten and H. Liang Eds., Marcel Dekker, New York and Basel, 2004, pp. 99–104. "Importance of the Triboemission Process for Tribochemical Reaction," *Tribol. Int.*, Vol. 38, 2005, pp. 337–353.

[11] Hucknell, D. J., *Selective Oxidation of Hydrocarbons*, Academic Press, New York, 1974.

[12] Furey, M. J. and Kajdas, C., "Tribopolymerization as a Mechanism of Boundary Lubrication," *Surface Modification and Mechanisms*, G. E. Totten and H. Liang, Eds., Marcel Dekker, New York and Basel, 2004, pp. 165–201.

[13] Hsu, S. M., "Boundary Lubrication: Current Understanding," *Tribology Letters*, Vol. 3, 1997, pp. 1–11.

[14] Hu, Y. Z., Wang, H., and Zhu, D., "A Computer Model of Mixed Lubrication in Point Contacts," *Tribol. Int.*, Vol. 34, No. (1), 2001, pp. 65–73.

[15] Spikes, H. A., "Direct Observation of Boundary Layers," *Langmuir*, Vol. 12, 1996, pp. 4567–4573.

[16] Luo, J. B., Wen, S. Z., and Huang, P., "Thin Film Lubrication, Part I: The Transition Between EHL and Thin Film Lubrication," *Wear*, Vol. 194, 1996, pp. 107–115.

[17] Luo, J. B., Wen, S. Z., Sheng, X. Y., et al., "Substrate Surface Energy Effects on Liquid Lubricant Film at Nanometer Scale," *Lubr. Sci.*, Vol. 10, No. 11, 1998, pp. 23–35.

[18] Hamrock, B. J. and Dowson, D., "Isothermal Elastohydrodynamic Lubrication of Point Contacts-1: Theoretical Formulation," *J. Lubr. Technol.*, Vol. 98, No. 2, 1976, pp. 223–229.

[19] Rattoi, M., Anghel, V., Bovington, C. H., and Spikes, H. A., "Mechanism of Oilness Additives," *Tribol. Int.*, Vol. 33, 2000, pp. 241–247.

[20] Zhu, D. and Hartl, M., "A Comparative Study on the EHL Film Thickness Results from Both Numerical Solutions and Experi-

mental Measurements," presented at *STLE Annual Meeting*, May 2004, Toronto, Canada.

[21] Granick, S., "Motions and Relaxations of Confined Liquids," *Science*, Vol. 253, 1991, pp. 1374–1379.

[22] Shen, M. W., Luo, J. B., Wen, S. Z., and Yao, Z. B., "Investigation of the Liquid Crystal Additive's Influence on Film Formation in Nano Scale," *Lubr. Eng.*, Vol. 58, No. 3, 2002, pp. 18–23.

[23] Hu, H. W., Carson, G. A., and Granick, S., "Relaxation Time of Confined Liquid Under Shear," *Phys. Rev. Lett.*, Vol. 66, 1991, pp. 2758–2761.

[24] Berman, A. and Israelachivili, J. N., "Surface Force and Micro-Rheology of Molecularly Thin Liquid Films," *Hanbook of Micro/Nanotribology*, 2nd ed., B. Buushan, Ed., CRC Press, Boca Raton, FL.

[25] Hu, H. W. and Granick, S., "Viscoelastic Dynamics of Confined Polymer Melts," *Science*, Vol. 258, 1992, pp. 1339–1342.

[26] Alsten, J. V. and Granick, S., "The Origin of Static Friction in Ultrathin Liquid Films," *Langmuir*, Vol. 6, 1990, pp. 876–880.

[27] Hu, Y. Z., Wang, H., and Guo, Y., "Simulation of Lubricant Rheology in Thin Film Lubrication, Part I: Simulation of Poiseuille Flow," *Wear*, Vol. 196, 1996, pp. 243–248.

[28] Wang, H., Hu, Y. Z., and Zou, K., "Nano-tribology Through Molecular Dynamics Simulations," *Sci. China, Ser. A: Math., Phys., Astron.*, 44, No. 8, 2001, pp. 1049–1055.

[29] Wang, H., Hu, Y. Z., and Guo, Y., "Molecular Dynamics Study of The Interfacial Slip Phenomenon in Ultrathin Lubricating Film," *Lubr. Sci.*, 16, No. 3, 2004, pp. 303–314.

[30] Li, X., Hu, Y. Z., and Wang, H., "Modeling of Lubricant Spreading on a Solid Substrate," *J. Appl. Phys.*, Vol. 99, No. 2, 2006, art. No. 024905.

[31] Ulman, A., *An Introduction to Ultrathin Organic Films, from Langmuir-Blodgett to Self-Assembly*, Academic Press, Boston, 1991.

[32] Zhang, P. Y., Xue, Q. J., Du, Z. L., and Zhang, Z. J., "The Tribological Behavior of L-B Films of Fatty Acids and Nanoparticles," *Wear*, Vol. 242, 2000, pp. 147–151.

[33] Zhang, S. W. and Lan, H. Q., "Development in Tribological Research on Ultrathin Films," *Tribol. Int.*, Vol. 35, 2002, pp. 321–327.

[34] Salmeron, M., "Generation of Defects in Model Lubricant Monolayers and Their Contribution to Energy Dissipation in Friction," *Tribology Letters*, Vol. 10, No. 1–2, 2001, pp. 69–79.

[35] Berman, A. D., Ducker, W. A., and Israelachivili, J. N., "Experimental and Theoretical Investigations of Stick-Slip Friction Mechnisms," *Physics of Friction*, B. N. J. Persson and E. Tosatti, Eds., Kluwer Academic Publishers, 1996, pp. 51–67.

[36] Gnecco, E., Bennewitz, R., Gyalog, T., et al., "Velocity Dependence of Atomic Friction," *Phys. Rev. Lett.*, Vol. 84, No. 6, 2000, pp. 1172–1175.

[37] Dominguez, D. D., Mowery, R. L., and Turner, N. H., "Friction and Durabilities of Well-Ordered, Close Packed Carboxylic Acid Monolayers Deposited on Glass and Steel Surfaces by the Langmuir-Blodgett Technique," *Tribol. Trans.*, 37, No. 1, 1994, pp. 59–66.

[38] Frisbie, C. D., Lawrence, F. R., et al., "Functional Group Imaging by Chemical Force Microscopy," *Science*, Vol. 265, 1994, pp. 2071–2074.

[39] Kim, H. I., Graupe, M., Oloba, O., et al., "Molecularly Specific Studies of the Frictional Properties of Monolayer Films: A Systematic Comparison of CF_3-, $(CH_3)_2CH-$, and CH_3-Terminated Films," *Langmuir*, Vol. 15, 1999, pp. 3179–3185.

[40] Zhang, P. Y., Lu, J. J., Xue, Q. J., and Liu, W. M., "Microfrictional Behavior of C_{60} Particles in Different C_{60} L-B Films Studied by AFM/FFM," *Langmuir*, Vol. 17, No. 7, 2001 pp. 2143–2145.

[41] Sarkar, J., Pal, P., and Talapatra, G. B., "Self-Assembly of Silver Nano-particles on Stearic Acid Langmuir-Blodgett Film: Evidence of Fractal Growth," *Chem. Phys. Lett.*, Vol. 401, No. 4–6, 2005, pp. 400–404.

[42] Hu, X. L., Luo, J. B., and Wen, S. Z., "Research on Formation and Tribological Properties of Polyfluoroalkylmethacrylate Film on the Magnetic Head Surface," *Chinese Science Bulletin*, Vol. 50, 2005, pp. 2385–2390.

[43] Peng, Y. T., Hu, Y. Z., and Wang, H., "Patterned Deposition of Multi-walled Carbon Nanotubes on Self-assembled Monolayers," *Chinese Science Bulletin*, Vol. 51, No. 2, 2006, pp. 147–150.

[44] Tutein, A. B., Stuart, S. J., and Harrison, J. A., "Role of Defects in Compression and Friction of Anchored Hydrocarbon Chains on Diamond," *Langmuir*, Vol. 16, 2000, pp. 291–296.

[45] Glosli, J. N. and McClelland, G., "Molecular Dynamics Study of Sliding Friction of Ordered Organic Monolayers," *Phys. Rev. Lett.*, Vol. 70, 1993, pp. 1960–1963.

[46] Ohzono, T., Glosli, J. N., and Fujihira, M., "Simulations of Wearless Friction at a Sliding Interface Between Ordered Organic Monolayers," *Jpn. J. Appl. Phys.*, Vol. 37, No. 12A, 1998, pp. 6535–6543.

[47] Hu, Y. Z., Zhang, T., and Wang, H., "Molecular Dynamics Simulations on Atomic Friction Between Self-Assembled Monolayers: Commensurate and Incommensurate Sliding," *Comp. Mater. Sci.*, Vol. 38, 2006, pp. 98–104.

[48] Sang, Y., Dube, M., and Grant, M., "Thermal Effects on Atomic Friction," *Phys. Rev. Lett.*, Vol. 87, 2001, art. No. 174301.

[49] Erdemir, A., Eryilmaz, O. L., and Fenske, G. R., "Synthesis of Diamond-like Carbon Films with Superlow Friction and Wear Properties," *J. Vac. Sci. Technol. A*, Vol. 18, 2000, pp. 1987–1992.

6
Gas Lubrication in Nano-Gap

Meng Yonggang[1]

1 History of Gas Lubrication

IN COMPARISON WITH OTHER LUBRICATION TECHnologies, gas lubrication has the advantages of extremely low friction, no lubricant leakage, capabilities of withstanding high or low temperatures, and severe radiation. In 1897, Kingsbury [1] successfully developed the first air journal bearing. In the 1930s, air bearings were applied to gyroscope systems for navigation of missiles [2]. Air bearings were also successfully used in the man-made satellite launched by NASA in 1959. Today, gas lubrication has become a technology widely applied to precision instruments, the aerospace industry, and magnetic recording systems.

Since the first hard disk drive, IBM350RAMAC, was invented in 1956 as a major external data storage element for computers, hard disk technology has continuously rapidly developed for five decades. When a hard disk drive works, a read/write magnetic head, also referred to as a slider, floats above a rotating platen medium disk with a small spacing, which was 20 μm at the time of IBM350RAMAC. The position of the slider relative to the rotating disk in the vertical direction is determined by the balance of the force of a flexible suspension spring with the resultant of the air pressure generated due to the hydrodynamic effect within the head-disk clearance. With a continuous increase of the areal recording density of hard disks, the head-disk physical gap, also called flying height, has been demanded to reduce substantially, from 20 μm in the first hard disk drive down to 10 nm in the latest products at present, keeping the head-disk interface be noncontacted. It is expected that the flying height may drop down to 3 nm approximately when the areal density of recording approaches to 1,000 Gb/in.^{2} in the near future [3]. Because air pressure within the head-disk working area is a decisive factor to ensure a steady and proper flying height, gas lubrication theory has been a foundation of the engineering design of a head-disk system. Meanwhile, study on ultra-thin film gas lubrication problems has become one of the most attractive subjects in the field of tribology during the past three decades.

Historically, gas lubrication theory was developed from the classical liquid lubrication equation—Reynolds equation [4]. The first gas lubrication equation was derived by Harrison [5] in 1913, taking the compressibility of gases into account. Because the classical gas lubrication equation is based on the Navier-Stokes equation, it does not incorporate some gas flow characteristics rooted in the rarefaction effects of dilute gases. As early as 1959, Brunner's experiment [6] showed that the classical gas lubrication equation was not applicable to the situation where gas film thickness was less than 10 μm. One of the gas flow characteristics is the discontinuities in velocity and temperature on walls. The velocity discontinuity is referred to as velocity slip, which was recognized and predicted by Maxwell [7] in 1867. According to Maxwell theory, velocity slip is related to Knudsen number, Kn, which is a dimensionless characteristic parameter defined as the ratio of the mean free path of gas molecules to a characteristic length in a problem. By accounting for the effect of velocity slip into gas flow, Burgdorfer [8] derived a modified Reynolds equation for thin film gas lubrication in the small Knudsen number conditions in 1959. His theory is commonly known as the first-order slip-flow model because he used the linear model proposed by Maxwell to deal with flow slippage. Later, two other modified Reynolds equations based on second-order slip-flow and 1.5-order slip-flow models were respectively proposed for higher Knudsen number conditions [9,10]. However, the applicability of these slip-flow models at ultra-thin gas film conditions is questionable from the viewpoints of physical background and experiment verification. To deal with the ultra-thin film gas lubrication problem for arbitrary Knudsen numbers, Fukui and Kaneko [11] treated it in a different approach in their derivation. Instead of the fundamental momentum equilibrium equations of continuum mechanics, the linearized Boltzmann equation of the kinetic theory was solved to obtain the flow rates of fundamental Couette flow and Poiseuille flow. By applying the mass flow conservation law, they subsequently reached a generalized Reynolds equation, also called the F-K model. The F-K model is a refinement of Gans's work [12], and is widely accepted for analyzing the ultra-thin film gas lubrication problem in magnetic storage systems. Because the model is originated from the kinetic theory of dilute gases, it is also referred to as the molecular gas film lubrication (MGL) theory.

Since the middle of the 1990s, another computation method, direct simulation Monte Carlo (DSMC), has been employed in analysis of ultra-thin film gas lubrication problems [13–15]. DSMC is a particle-based simulation scheme suitable to treat rarefied gas flow problems. It was introduced by Bird [16] in the 1970s. It has been proven that a DSMC solution is an equivalent solution of the Boltzmann equation, and the method has been effectively used to solve gas flow problems in aerospace engineering. However, a disadvantageous feature of DSMC is heavy time consumption in computing, compared with the approach by solving the slip-flow or F-K models. This limits its application to two- or three-dimensional gas flow problems in microscale. In the

[1] *State Key Laboratory of Tribology, Tsinghua University, Beijing, China.*

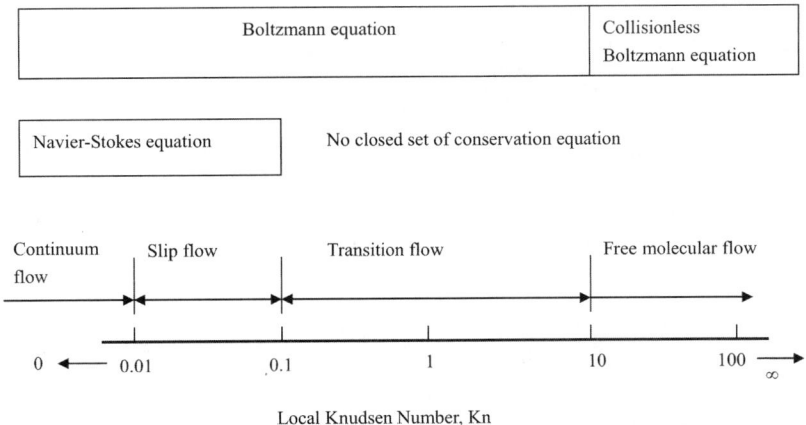

Fig. 1—Gas flow regimes and mathematical models.

study of the ultra-thin film gas lubrication, the DSMC is often taken as a reference solution for checking the accuracy of various Reynolds-type models.

Rapid progress in microsystem technology in recent years has extended gas lubrication practices to the areas at microscale. The motivation to apply gas lubrication theory to microsystems comes from the following needs. The first is to study the slip flow and rarefaction effects which are inherent characteristics of some microsystems such as micro-gas turbines, micro-gas pumps, and microsensors working in nonvacuum environments. The performance of these devices is closely related to the gas flow behavior in microchannels. The second is the attempt to solve the existing severe friction and wear problems in micromotors and microrotation machinery by actively utilizing preferable gas bearing designs. Nano gas lubrication has become an important fraction in today's micro/nano science and technology.

2 Theory of Thin Film Gas Lubrication

2.1 Gas Flow Regimes and Models

There are two levels, discrete particle level and continuum level, for describing and modeling of the macroscopic behaviors of dilute and condensed matters. The physics laws concerning the conservation of mass, momentum, and energy in motion, are common to both levels. For simple dilute gases, the Boltzmann equation, as shown below, provides the governing equation of gas dynamics on the discrete particle level

$$\frac{\partial (nf)}{\partial t} + \vec{c} \cdot \frac{\partial (nf)}{\partial \vec{r}} + \vec{F} \cdot \frac{\partial (nf)}{\partial \vec{c}} = \int_{-\infty}^{\infty} \int_{0}^{4\pi} n^2 (f^* f_1^* - f f_1) c_r \sigma d\Omega d\vec{c}_1 \quad (1)$$

where $\vec{c}$ is the velocity of a particular particle with components u, v, and w in the direction of the Cartesian axes x, y, and z; f is the velocity distribution function; n is the particle number in a unit volume of physical space; t is time; $\vec{r}$ is the position vector of the particle in physical space; and $\vec{F}$ is the external force per unit mass.

The second term on the left-hand side of Eq (1) expresses the convection of gas molecules across the face of $d\vec{r}$ in physical space by the molecular velocity $\vec{c}$. The third term on the left-hand side of Eq (1) represents the convection of molecules across the surface of $d\vec{c}$ in velocity space due to the external force $\vec{F}$. The right-hand side of Eq (1) means the scattering of gas molecules into and out of $d\vec{c}d\vec{r}$ in phase space as a result of intermolecular collisions. The superscript * identifies the post-collision variables, and the subscript 1 denotes the variables in a different class. σ is the cross section of a collision. For a gas mixture consisting of several species or a gas composed of polyatomic molecules with internal degrees of freedom, the Boltzmann equation can be extended and generalized.

On the continuum level of gas flow, the Navier-Stokes equation forms the basic mathematical model, in which dependent variables are macroscopic properties such as the velocity, density, pressure, and temperature in spatial and time spaces instead of nf in the multi-dimensional phase space formed by the combination of physical space and velocity space in the microscopic model. As long as there are a sufficient number of gas molecules within the smallest significant volume of a flow, the macroscopic properties are equivalent to the average values of the appropriate molecular quantities at any location in a flow, and the Navier-Stokes equation is valid. However, when gradients of the macroscopic properties become so steep that their scale length is of the same order as the mean free path of gas molecules, λ, the Navier-Stokes model fails because conservation equations do not form a closed set in such situations.

The applicability of the two different models of gas flow is generally judged from the gas flow regimes according to the magnitude of the local Knudsen number, Kn, defined as

$$Kn = \frac{\lambda}{L} \quad (2)$$

where L is a characteristic length in a gas flow problem.

Gas flow is usually classified into four regimes as below:
(i)   Continuum flow for $Kn \leq 0.01$
(ii) Slip flow for $0.01 < Kn < 0.1$
(iii) Transient flow for $0.1 \leq Kn < 10$
(iv) Molecular flow for $Kn \geq 10$

Figure 1 shows the four gas flow regimes and the applicable models. The Boltzmann equation is valid for the whole range of Kn, from 0 to infinity. A simplified Boltzmann equation or the collisionless Boltzmann equation, where the right-hand side reduces to zero, is suitable when Kn is very large. The Navier-Stokes equation is valid in the continuum flow regime, and it is still applicable in the slip flow regime if a proper velocity slip model is incorporated. However, significant errors would be induced if it is extended to the transient flow regime.

In most of the gas lubrication problems in nano-gaps, gas flow usually locates in the slip flow or the transient flow regime, depending on working conditions and local geometry. Therefore, both of the macroscopic and microscopic models are introduced to analyze the gas lubrication problems.

2.2 Modified Reynolds Equations for Ultra-Thin Film Gas Lubrication

In spite of the different assumptions adopted in derivation, all of the modified Reynolds equations presented in the literature up to now has the same form as below

$$\frac{\partial}{\partial X}\left(PQH^3 \frac{\partial P}{\partial X}\right) + \left(\frac{L}{B}\right)^2 \frac{\partial}{\partial Y}\left(PQH^3 \frac{\partial P}{\partial Y}\right) = \Lambda \frac{\partial(PH)}{\partial X} + \Psi \frac{\partial(PH)}{\partial \bar{\tau}} \quad (3)$$

where

X, Y = dimensionless coordinates; $x/L, y/L$;
x, y = coordinates in Cartesian axes;
L = length of lubrication region;
B = width of lubrication region;
$\bar{\tau}$ = dimensionless time;
H = dimensionless film thickness; h/h_0;
h = film thickness;
h_0 = characteristic film thickness;
P = dimensionless pressure; p/p_0;
p = pressure;
p_0 = ambient pressure;
Λ = bearing number; $6\mu UL/p_0 h_0^2$;
μ = viscosity;
U = boundary velocity in the x direction;
ψ = squeeze number; $12\mu\omega_0 L^2/p_0 h_0^2$;
ω_0 = characteristic angular velocity;
Q = relative flow rate; $Q_p(D)/Q_{con}$;
Q_p = flow rate coefficient of Poiseuille flow; $(-q_p\sqrt{2RT_0})/[h^2(dp/dx)]$;
q_p = mass flow rate of Poiseuille flow;
Q_{con} = flow rate of the continuum Poiseuille flow; $D/6$;
D = inverse Knudsen number; $\sqrt{\pi}/2Kn$ or $D_0 PH$;
D_0 = characteristic inverse Knudsen number; $p_0 h_0/\mu\sqrt{2RT_0}$;
T_0 = characteristic temperature;
R = gas constant.

However, different models have different expressions of q_p, or $Q_p(D)$, and hence different relative flow rates Q. Here four well known models are summarized as below.

(i) First-order slip model:

$$Q_p = \frac{D}{6} + \frac{\sqrt{\pi}a}{2} \quad (4)$$

(ii) Second-order slip model:

$$Q_p = \frac{D}{6} + \frac{\sqrt{\pi}}{2} + \frac{\pi}{4D} \quad (5)$$

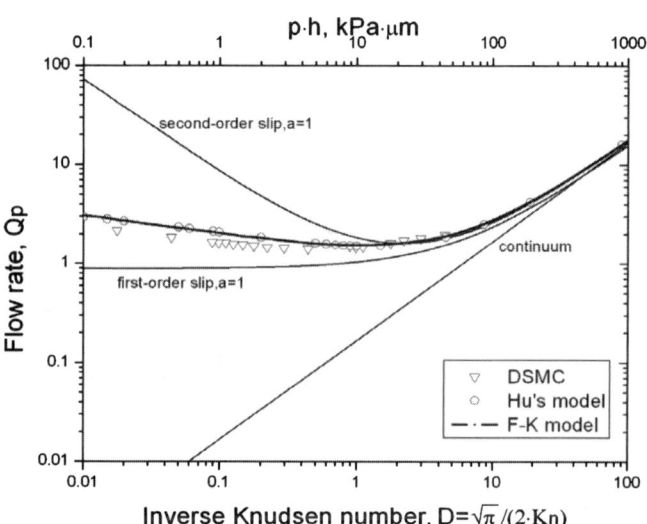

Fig. 2—Comparison of flow rates between different models.

(iii) Fukui-Kaneko model:

$$Q_p = \frac{D}{6} + 1.0162 + \frac{1.0653}{D} - \frac{2.1354}{D^2}, \quad 5 \leq D \quad (6)$$

$$Q_p = 0.13852D + 1.25087 + \frac{0.01653}{D} - \frac{0.00969}{D^2}, \quad 0.15 \leq D < 5 \quad (7)$$

$$Q_p = -2.22919D + 2.10673 + \frac{0.01653}{D} - \frac{0.0000694}{D^2}, \quad 0.01 \leq D < 0.15 \quad (8)$$

(iv) Free molecular flow:

$$Q_p = -\frac{1}{\sqrt{\pi}}\log D, \quad D \to 0 \quad (9)$$

In Eq (4), a is the surface correcting coefficient defined by $a = c(2 - \alpha/\alpha)$, where α is the accommodation coefficient at the boundary surface ($0 < \alpha < 1$ in general), and $c \approx 1$.

Among these models, the Fukui-Kaneko model is regarded as the most accurate one because it is derived from the linearized Boltzmann equation. However, the flow rate coefficient, Q_p, in the Fukui-Kaneko model, is not a unified expression, as defined in Eqs (6)–(8). This would cause some inconveniences in a practical numerical solution of the modified Reynolds equation. Recently, Huang and Hu [17] proposed a representation of Q_p, as shown in Eq (10), for the whole range of D from 0.01 to 100, by a data-fitting method, to replace the segmented expressions of the Fukui-Kaneko model

$$Q_p = \frac{D}{6} + 1.0162 + 0.40134\ln(1 + 1.2477/D) \quad (10)$$

In Fig. 2, the flow rate coefficients, Q_p, calculated from these different models are plotted in the range of inverse Knudsen number D from 0.01 to 100. We can see that for

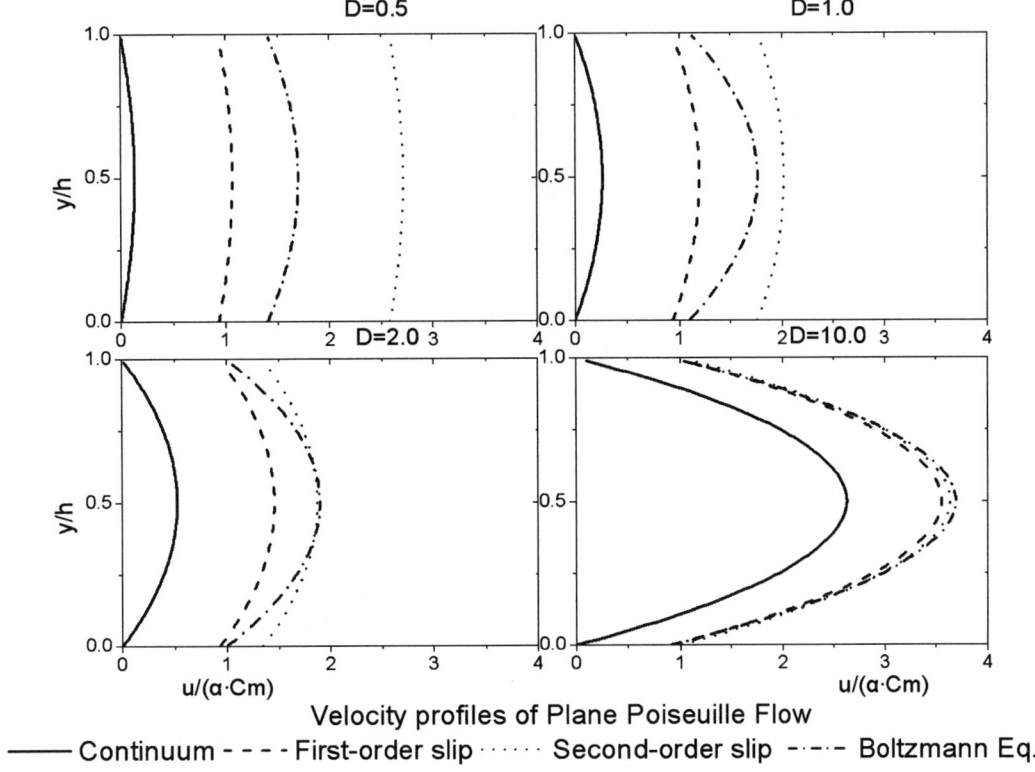

Velocity profiles of Plane Poiseuille Flow
—— Continuum - - - - First-order slip ······ Second-order slip -·-·- Boltzmann Eq.

Fig. 3—Comparison of velocity profiles of plane Poiseuille flow between different models.

small inverse Knudsen numbers ($D < 1$), or high Knudsen numbers ($Kn > 0.886$), the first-order slip model underestimates the flow rate coefficient, while the second-order slip model overestimates it, comparing with the Fukui-Kaneko model. In the range of $1 < D < 10$, the differences of the flow rate coefficients between these models become small, but all of them is much greater than the value calculated from the continuum model. When $D > 10$, the flow rate coefficients of both slip models approach to that of the continuum model as well as that of the Fukui-Kaneko model. It is also shown that Eq (10) is a good approximation of the power series functions of Eqs (6)–(8). In the figure, the flow rate coefficients calculated with the DSMC are also plotted. The DSMC results are slightly smaller than those of the Fukui-Kaneko model, but very close to the free molecular flow model, Eq (9), for the lower inverse Knudsen numbers.

The flow rate coefficient can be considered as a correction factor of mass flow rate in the plane Poiseuille condition. The larger the flow rate coefficient is, the greater the mass flow rate at a given pressure gradient condition. This can be understood from the velocity profiles, the integration of which over the gas flow spacing gives the mass flow rate. Figure 3 shows the velocity profiles based on the different models at several typical inverse Knudsen numbers. The flow velocity is expressed as below.

(i) Continuum model:

$$U(Y) = 2D/(5\pi) \cdot (1 - Y^2), \quad Y \in [0,1] \quad (11)$$

(ii) First-order slip model:

$$U(Y) = 2D/(5\pi) \cdot (1 + 2\sqrt{\pi}/D - Y^2), \quad Y \in [0,1] \quad (12)$$

(iii) Second-order slip model:

$$U(Y) = 2D/(5\pi) \cdot (1 + 2\sqrt{\pi}/D + \pi/D^2 - Y^2), \quad Y \in [0,1] \quad (13)$$

(iv) Fukui-Kaneko model:

$$U(Y) = \frac{1}{D}(A_{0P} \cdot Y^2 + B_{0P} - 1), \quad Y \in [0,1] \quad (14)$$

where

$$A_{0P} = \frac{c_1 c_{22} - c_2 c_{12}}{c_{11} c_{22} - c_{12}^2}, \quad B_{0P} = \frac{c_2 c_{11} - c_1 c_{12}}{c_{11} c_{22} - c_{12}^2}$$

$$c_{11} = \pi^{-1/2}\left(8 - \frac{\pi^{1/2}D^3}{12} + \frac{D^4}{16} - 2D(4 + D^2)T_0(D)\right.$$
$$\left. - \left(16 + 8D^2 + \frac{D^4}{8}\right)T_1(D) - D(16 + D^2)T_2(D)\right)$$

$$c_{12} = \pi^{-1/2}\left(2 - \frac{\pi^{1/2}D}{2} + \frac{D^2}{4} - 2DT_0(D)\right.$$
$$\left. - \left(4 + \frac{D^2}{2}\right)T_1(D) - 2DT_2(D)\right)$$

$$c_{22} = \pi^{-1/2}(1 - 2T_1(D))$$

$$c_1 = \tfrac{1}{12}D^3$$

$$c_2 = D$$

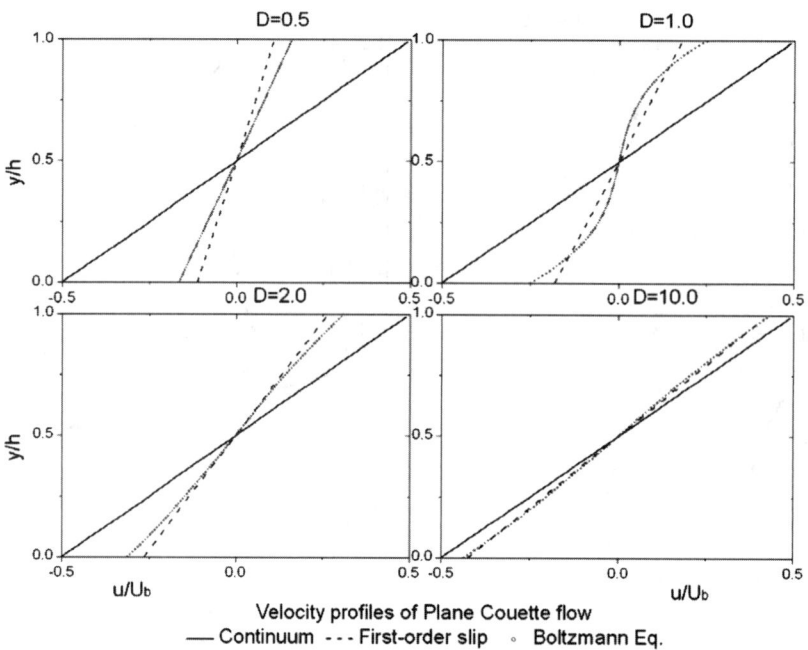

Fig. 4—Comparison of velocity profiles of plane Couette flow between different models.

$$T_n(t) = \int_0^\infty t^n e^{-t^2 - D/t} dt$$

Y is the dimensionless coordinate, y/h; and U is the dimensionless velocity, $u/(\alpha C_m)$.

We can see that the velocity profiles of these models become more and more different from each other with the decrease in the inverse Knudsen number. The difference between the velocity profiles of the first-order and second-order slip models is in the magnitude of the slip velocity, but the shapes of the velocity profiles are almost same. The second-order slip model predicts a larger and larger slip velocity than the first-order slip model when the inverse Knudsen number decreases. The Fukui-Kaneko model gives a medium slip velocity and velocity profile.

It should be pointed out that the flow rate in the case of the Couette flow is independent of the inverse Knudsen number, and is the same as the prediction of the continuum model, although the velocity profiles predicted by the different flow models are different as shown in Fig. 4. The flow velocity in the case of the plane Couette flow is given as follows:

(i) Continuum model:

$$U(Y) = Y, \quad Y \in [0,1] \quad (14')$$

(ii) First-order and second-order slip models:

$$U(Y) = Y/(1.0 + \sqrt{\pi}/D), \quad Y \in [0,1] \quad (15)$$

(iii) Fukui-Kaneko model:

$$U(Y) = A_{0C} \cdot Y + B_{0C} \cdot Y^3, \quad Y \in [0,1] \quad (16)$$

where

$$A_{0C} = \frac{c_1 c_{22} - c_2 c_{12}}{c_{11} c_{22} - c_{12}^2}, \quad B_{0C} = \frac{c_2 c_{11} - c_1 c_{12}}{c_{11} c_{22} - c_{12}^2}$$

$$c_{11} = \pi^{-1/2}\left(\tfrac{1}{4}D^2 - 1 + DT_0(D) + \left(\tfrac{1}{2}D^2 + 2\right)T_1(D) + 2DT_2(D)\right)$$

$$c_{12} = \pi^{-1/2}\left(\frac{1}{16}D^4 - \frac{\pi^{1/2}}{8}D^3 + \frac{3}{4}D^2 - 12 + D\left(\frac{9}{4}D^2 + 12\right)T_0(D)\right.$$
$$+ \left(\frac{1}{8}D^4 + \frac{21}{2}D^2 + 24\right)T_1(D)$$
$$\left. + D(D^2 + 24)T_2(D)\right)$$

$$c_{22} = \pi^{-1/2}\left(\frac{1}{64}D^6 - \frac{3\pi^{1/2}}{80}D^5 + \frac{3}{16}D^4 - 216 + D\left(\frac{21}{16}D^4 + 54D^2\right.\right.$$
$$\left. + 216\right)T_0(D) + \left(\frac{1}{32}D^6 + \frac{69}{8}D^4 + 216D^2 + 432\right)T_1(D)$$
$$\left. + D\left(\frac{3}{8}D^4 + 36D^2 + 432\right)T_2(D)\right)$$

$$c_1 = \pi^{-1/2}\left(\frac{D}{2} - \frac{\pi^{1/2}}{2} + DT_1(D) + 2T_2(D)\right)$$

$$c_2 = \pi^{-1/2}\left(\frac{1}{8}D^3 - \frac{3\pi^{1/2}}{8}D^2 + 3D - \frac{9\pi^{1/2}}{2} + 3D^2 T_0(D) + \left(\frac{3}{2}D^2\right.\right.$$
$$\left. + 18\right)T_2(D) + D\left(\frac{1}{4}D^2 + 12\right)T_1(D)\right)$$

$$T_n(t) = \int_0^\infty t^n e^{-t^2 - (D/t)} dt$$

Y is the dimensionless coordinate, y/h; and U is the dimensionless velocity, u/U_b, where U_b is the relative speed between the upper and the lower plates.

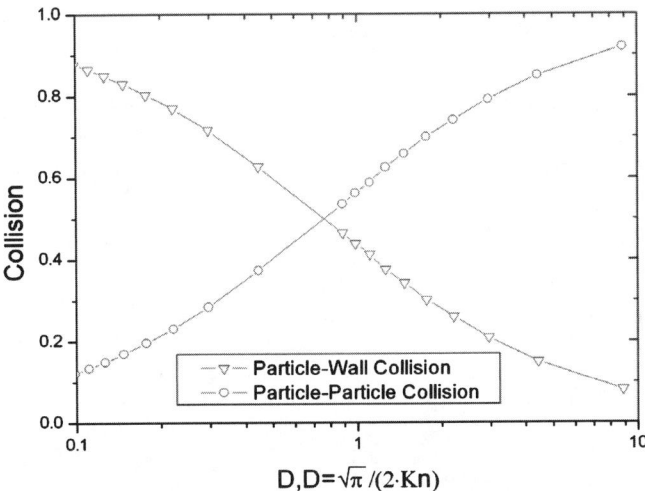

Fig. 5—Change of the rates of particle-particle and particle-wall collisions with the inverse Knudsen number.

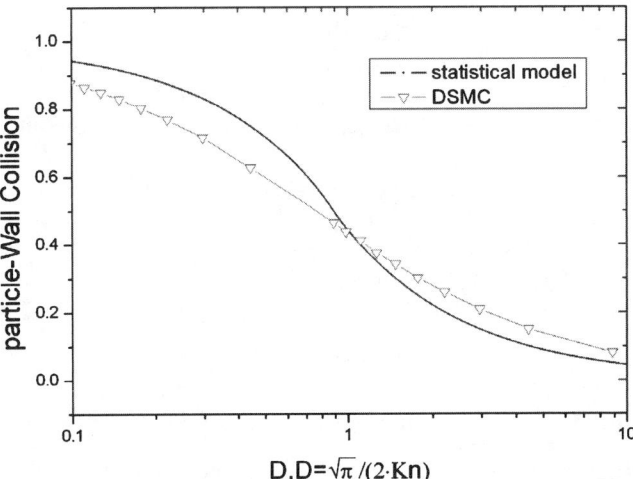

Fig. 6—Change of the particle-wall collision ratio with the inverse Knudsen number.

2.3 Effect of Gap Width on Gas Flow

As described above, the magnitude of Knudsen number, Kn, or inverse Knudsen number, D, is of great significance for gas lubrication. From the definition of Kn in Eq (2), the local Knudsen number depends on the local mean free path of gas molecules, λ, and the local characteristic length, L, which is usually taken as the local gap width, h, in analysis of gas lubrication problems. From basic kinetic theory we know that the mean free path represents the average travel distance of a particle between two successive collisions, and if the gas is assumed to be consisted of hard sphere particles, the mean free path can be expressed as

$$\lambda_h = \frac{1}{\sqrt{2}\,\pi d^2 n} \qquad (17)$$

where d is the diameter of the particles and n is the number of the particles in a unit volume, or the number density. Hence, λ_h is independent on the gap width. In derivation of Eq (17), infinite space was presumed, i.e., no bounds existing to confine the gas particles. In a practical problem, especially in ultra-thin film gas lubrication, however, the existence of bounding walls should be considered in calculation of the mean free path.

When bounding walls exist, the particles confined within them not only collide with each other, but also collide with the walls. With the decrease of wall spacing, the frequency of particle-particle collisions will decrease, while the particle-wall collision frequency will increase. This can be demonstrated by calculation of collisions of particles in two parallel plates with the DSMC method. In Fig. 5 the result of such a simulation is shown. In the simulation [18], 2,000 representative nitrogen gas molecules with 50 cells were employed. Other parameters used here were viscosity $\mu = 1.656 \times 10^{-5}$ Pa·s, molecular mass $m = 4.65 \times 10^{-26}$ kg, and the ambient temperature $T_{ref} = 273$ K. Instead of the hard-sphere (HS) model, the variable hard-sphere (VHS) model was adopted in the simulation, which gives a better prediction of the viscosity-temperature dependence than the HS model. For the VHS model, the mean free path becomes:

$$\lambda_v = 2.591\left(\frac{m}{2\pi kT}\right)^{1/2}\frac{\mu}{\rho} \qquad (18)$$

Because the total collision number, C_T, is the sum of the number of particle-particle collisions, C_p, and the number of particle-wall collisions, C_w, we can define the particle-wall collision ratio, M_w, as below.

$$M_w = \frac{C_w}{C_p + C_w} \qquad (19)$$

By adopting the basic assumption that the probability density of the direction distribution of particle velocity vector is uniform in the whole space, i.e., $\Theta(\theta,\beta) = \sin\theta/4\pi$, the probability of collision between the wall and a particle located δ away from the wall (see Fig. 7) can be expressed as the following equation:

$$P_w(\delta) = \begin{cases} \dfrac{1}{2}\left(1 - \dfrac{\delta}{\lambda}\right) & \delta \leq \lambda \\ 0 & \delta > \lambda \end{cases} \qquad (20)$$

Because the distance δ is distributed randomly, the particle-wall collision ratio can be given by the expectance of P_w over the wall spacing, h, as Eq (21).

$$M_w = \begin{cases} 1 - \dfrac{h}{2\lambda} = 1 - \dfrac{1}{2Kn} & \langle h \leq \lambda, Kn \geq 1 \rangle \\ \dfrac{\lambda}{h} = Kn & \langle h > \lambda, Kn < 1 \rangle \end{cases} \qquad (21)$$

Figure 6 shows the change of M_w with the Knudsen number Kn. When $Kn \ll 1$, M_w is near zero. In the range of $Kn < 2$, M_w increases rapidly with Kn. When $Kn > 2$, M_w approaches to 1 gradually. In the same figure, the calculated M_w with the DSMC method is also plotted. The DSMC results are similar in tendency with that of the statistical expression (21), although the former is somewhat lower than the latter in the range of $Kn > 1$ probably because the direction distribution of particle velocity vector in DSMC did not accurately follow the uniform assumption.

Considering the fact that the particle-wall collision rate

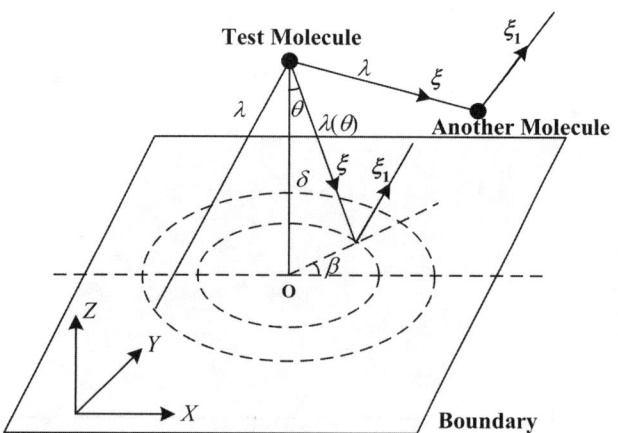

Fig. 7—Collisions between the gas molecules and boundary [19].

gets greater when the gap width is near or less than the conventional particle mean free path, λ_h, Peng et al. [19] made an extension of the concept of the mean free path. Their definition of the particle mean free path, denoted as λ_m hereafter, is the travel distance of a particle between two successive collisions including the collisions between two particles and those between the particles and the walls. They introduced a so-called nanoscale effect function, $N_p = \lambda_m/\lambda_h$, to relate the λ_m with the conventional mean free path λ_h. N_p can be derived from the statistic average of the distance between successive collisions as follows.

As shown in Fig. 7, let δ denote the distance between the wall and a test particle, θ be the angle from the particle velocity vector $\vec{v}$ to the Z axis, and β be the angle from the part of the $\vec{v}$ in XY plane to the Z axis, then we can write the mean free path of the test particle as

$$\lambda(\theta,\beta) = \lambda(\theta) = \begin{cases} \delta/\cos\theta & 0 \leq \theta \leq \arccos(\delta/\lambda_h) \\ \lambda_h & \arccos(\delta/\lambda_h) \leq \theta \leq \pi \end{cases} \quad (22)$$

where, both θ and β are random variables distributed uniformly in the whole space.

It is clear that if $\delta > \lambda_h$, particle-particle collision happens most probably and $\lambda(\theta,\beta) = \lambda_h$. If $\delta < \lambda_h$, however, particle-wall collision happens within the travel distance of $\delta/\cos\theta$. Then, the local free path of the test particle, λ^*, is the average of $\lambda(\theta,\beta)$ over the whole ranges of θ and β. That is,

$$\lambda^* = E[\lambda(\theta,\beta)] = \iint_\Sigma \lambda(\theta,\beta)\Theta(\theta,\beta)d\theta d\beta$$

$$= \int_0^\pi \lambda(\theta)d\theta \int_0^{2\pi} \Theta(\theta,\beta)d\beta$$

$$= \int_0^{\arccos(\delta/\lambda_h)} \frac{\delta}{\cos\theta} \cdot \frac{\sin\theta}{2} d\theta + \int_{\arccos(\delta/\lambda_h)}^\pi \lambda_h \cdot \frac{\sin\theta}{2} d\theta$$

$$= \frac{\lambda_h}{2}\left(1 + \frac{\delta}{\lambda_h} - \frac{\delta}{\lambda_h}\ln\left(\frac{\delta}{\lambda_h}\right)\right), \quad \delta \leq \lambda_h$$

$$\lambda^* = \lambda_h, \quad \delta > \lambda_h \quad (23)$$

Assuming that the gas molecules within the gap are in chaotic state, then λ_m can be written as

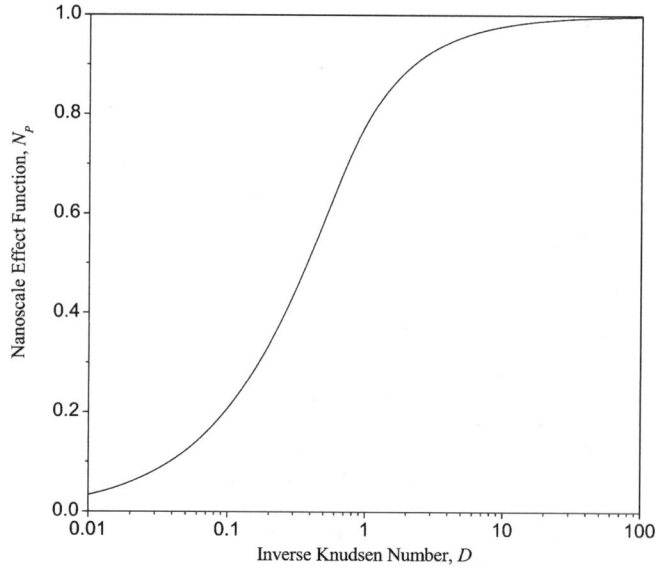

Fig. 8—Dependence of the nanoscale effect function on the inverse Knudsen number.

$$\lambda_m = \int_0^h \lambda^*(\delta) \cdot \frac{1}{h} d\delta = \begin{cases} \lambda_h \cdot \left(1 - \frac{1}{4}\frac{\lambda_h}{h}\right), & h \geq \lambda_h \\ \lambda_h \cdot \left(\frac{3}{4}\frac{h}{\lambda_h} - \frac{h}{2\lambda_h}\ln\left(\frac{h}{\lambda_h}\right)\right), & h < \lambda_h \end{cases} \quad (24)$$

or

$$N_p = \begin{cases} 1 - \frac{\sqrt{\pi}}{8} \cdot \frac{1}{D}, & D \geq \frac{\sqrt{\pi}}{2} \\ \frac{3}{2\sqrt{\pi}} \cdot D - \frac{1}{\sqrt{\pi}} \cdot D \cdot \ln\left(\frac{2D}{\sqrt{\pi}}\right), & D < \frac{\sqrt{\pi}}{2} \end{cases} \quad (25)$$

where

$$D = \frac{\sqrt{\pi}}{2} \cdot \frac{1}{Kn} \quad \text{and} \quad Kn = \frac{\lambda_h}{h}$$

Figure 8 shows the change of N_p with the inverse Knudsen number D. When $D > 1$, the nanoscale effect is weak, and the λ_m is slightly less than λ_h. When $D < 1$, however, the difference between λ_m and λ_h becomes significant. It is worth to note that although the nanoscale effect function is a two-segment function, the curve is very smooth at the joint point of $D = \sqrt{\pi}/2$. Figure 9 presents the comparison of the nanoscale effect functions calculated from the Eq (25) with those by the DSMC method. They are in reasonable agreement.

Based on the nanoscale effect function, Peng et al. further derived the modified Reynolds equations of the first-order slip, the second-order slip and the Fukui-Kaneko models [19]. The flow rate coefficients become the following expressions.

(i) New first-order slip model:

$$Q_P = \frac{D}{6 \cdot N_P(D)} + \frac{a\sqrt{\pi}}{2} \quad (26)$$

(ii) New second-order slip model:

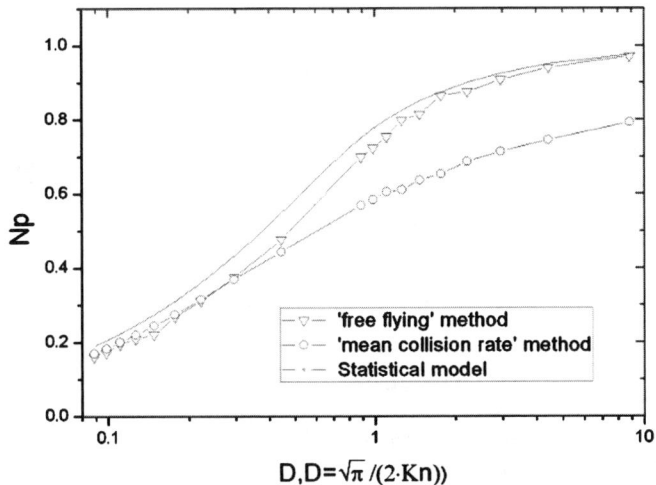

Fig. 9—Comparison of the nanoscale effect function between three calculation methods.

$$Q_P = \frac{D}{6N_P} + \frac{\sqrt{\pi}}{2} + \frac{\pi}{4}\frac{N_P}{D} \quad (27)$$

(iii) New Fukui-Kaneko model:

$$Q_p = -\frac{N_P}{D} + \frac{N_P^2}{D^2} \cdot \frac{c_{22}c_1^2 - 2c_{12}c_1c_2 + c_{11}c_2^2}{c_{11}c_{22} - c_{12}^2} \quad (28)$$

where the coefficients, $c_1, c_2, c_{11}, c_{22}, c_{12}$, are given by

$$c_{11} = \pi^{-1/2}\left[8 - \frac{\pi^{1/2}D^3}{12N_P^3} + \frac{D^4}{16N_P^4} - \frac{2D}{N_P}\left(4 + \frac{D^2}{N_P^2}\right)\mathbf{T}_0\left(\frac{D}{N_P}\right)\right.$$
$$\left. - \left(16 + \frac{8D^2}{N_P^2} + \frac{D^4}{8N_P^4}\right)\mathbf{T}_1\left(\frac{D}{N_P}\right)\right.$$
$$\left. - \frac{D}{N_P}\left(16 + \frac{D^2}{N_P^2}\right)\mathbf{T}_2\left(\frac{D}{N_P}\right)\right] \quad (29)$$

$$c_{12} = \pi^{-1/2}\left[2 - \frac{\pi^{1/2}D}{2N_P} + \frac{D^2}{4N_P^2} - \frac{2D}{N_P}\mathbf{T}_0\left(\frac{D}{N_P}\right)\right.$$
$$\left. - \left(4 + \frac{D^2}{2N_P^2}\right)\mathbf{T}_1\left(\frac{D}{N_P}\right) - \frac{2D}{N_P}\mathbf{T}_2\left(\frac{D}{N_P}\right)\right] \quad (30)$$

$$c_{22} = \pi^{-1/2}\left[1 - 2\mathbf{T}_1\left(\frac{D}{N_P}\right)\right] \quad (31)$$

$$c_1 = \frac{1}{12}\frac{D^3}{N_P^3} \quad (32)$$

$$c_2 = \frac{D}{N_P} \quad (33)$$

and $T_n(t) = \int_0^\infty t^n e^{-t^2 - D/t} dt$.

The correction of mean free path, λ_h, by the nanoscale effect function results in a smaller mean free path, or a smaller Knudsen number in other word. As a matter of fact, a similar effect is able to be achieved even if we use the conventional definition of mean free path, $\lambda_h = 1/\sqrt{2}\pi d^2 n$, and the Chapmann-Enskog viscosity equation, $\mu = (5/16)$ $\times(\sqrt{\pi m k T}/\pi d^2)$, or $\lambda_h = (\sqrt{\pi}\mu\sqrt{2RT}/2p)$, as long as substituting the gap-dependent viscosity rather than the bulk viscosity. Because the effective viscosity decreases as the Knudsen number enters the slip flow and transition flow ranges, and thus the mean free path becomes smaller as discussed by Morris [20] on the dependence of slip length on the Knudsen number.

3 Application of Gas Lubrication Theory

3.1 Solution Scheme of Modified Reynolds Equation

The basic thin film gas lubrication equation, or modified Reynolds equation as described in Section 2.2, is a two-dimensional nonlinear partial differential equation. To solve the equation for a specific bearing configuration and working condition, various computational methods developed for solving partial differential equations in the past are available, such as the finite difference method, finite element method, finite volume method, etc. With the conventional numerical methods, however, numerical instability or divergence is often confronted when solving the thin film gas lubrication equation due to the following reasons. (a) Bearing number, Λ, is extremely high, reaching hundreds of thousand or even millions when the gap width is in nanometer scale and sliding speed is high. (b) Local pressure gradient is usually very large because there exists abrupt geometrical changes such as steps or grooves on bearing surfaces. (c) The Knudsen number differs remarkably from place to place, partly because of the severe nonuniformity of bearing spacing distribution and partly because of the even more nonuniform distribution of gas pressure over the bearing surface. Therefore, a lot of effort has been made to overcome the computation difficulty.

In 1980, White and Nigam [21] implemented a factored implicit scheme for the numerical solution of the gas lubrication equation at very low spacing. They introduced a new variable $Z = Ph$ to replace the original pressure P and solved the equation for a gas film thickness of $h = 127$ nm. With White and Nigam's scheme, Miu and Bogy [22] analyzed the gas lubrication problem of the Taper-Flat type slider in 1986. In 2003, Wu and Bogy [23] introduced a multi-grid scheme to solve the slider air bearing problem. In their approach, two types of meshes, with unstructured triangles, were used. They obtained solutions with the minimum flying height down to 8 nm. Lu et al. [24] also used a multilevel method to solve the air bearing problem. In their calculations, they implemented a relative truncation error estimation as a grid adaptation criterion. The pressure distribution was obtained for a minimum flying height of 15 nm. Wu and Chen [25] used a finite element method with an operator decomposition technique to obtain the pressure distribution with a flying height of 25 nm. Recently, Huang et al. [26] proposed a new simple scheme for the numerical solution of Reynolds equation at ultra-thin gas film condition. Numerical experiments have proven that Huang's scheme is effective when the bearing number is very large. The numerical scheme is a modification of the conventional finite difference method, and the key points are explained as below.

For steady state lubrication, the general modified Reynolds equation can be written as

$$\frac{\partial}{\partial X}\left(PQH^3\frac{\partial P}{\partial X}\right) + \left(\frac{L}{B}\right)^2 \frac{\partial}{\partial Y}\left(PQH^3\frac{\partial P}{\partial Y}\right) = \Lambda\frac{\partial(PH)}{\partial X} \quad (34)$$

The terms on the left-hand side represents the flux due to Poiseuille flow, or diffusion part, while the term on the right-hand side represents the convective transport of Couette flow. For problems with very large bearing numbers, the convective term dominates in most of the bearing domain. The left-hand side terms become significant only in the regions next to the boundary and the regions where bearing gap has a sudden change. By taking this feature into account, the convective term is discretized by using an upwind finite difference scheme, and the diffusion terms on the left-hand side are expressed by a centered difference scheme. i.e.,

$$2\Lambda(P_{i,j}H_{i,j} - P_{i-1,j}H_{i-1,j})/\Delta X_i = [q_{i+1/2,j}(P_{i+1,j}^2 - P_{i,j}^2) - q_{i-1/2,j}(P_{i,j}^2 - P_{i-1,j}^2)]/\Delta X_i^2 + A^2[q_{i,j+1/2}(P_{i,j}^2 - P_{i,j+1}^2) - q_{i,j-1/2}(P_{i,j}^2 - P_{i,j-1}^2)]/\Delta Y_j^2 \quad (35)$$

where $q = QH^3$ is evaluated at the midway of two neighboring grid points, and $A = L/B$. The dimensionless pressure P is set as 1.0, or $p = p_0$, on the boundary of the gas bearing.

It should be pointed out that if there exists a step between two neighboring grid points, or a discontinuity of H in other words, $q_{i+1/2,j}$, $q_{i-1/2,j}$, $q_{i,j+1/2}$, and $q_{i,j-1/2}$ should be evaluated by using the harmonic average instead of the simple average. For example,

$$q_{i+1/2,j} = \frac{2q_{i+1,j}q_{i,j}}{q_{i+1,j} + q_{i,j}} \quad (36)$$

The harmonic average is effective to overcome troubles due to jumps of q from grid point (i,j) to $(i+1,j)$.

The discretized Eq (35) can be solved iteratively. A simple point-by-point iterative scheme is

$$\hat{P}_{i,j} = \frac{[2\bar{P}_{i-1,j}\Lambda H_{i-1,j}/\Delta X_i + (\tilde{q}_{i+1/2,j}\tilde{P}_{i+1,j}^2 + \tilde{q}_{i-1/2,j}\bar{P}_{i-1,j}^2)/\Delta X_i^2 + A^2(\tilde{q}_{i,j+1/2}\tilde{P}_{i,j+1}^2 + \tilde{q}_{i,j-1/2}\bar{P}_{i,j-1}^2)/\Delta Y_j^2]}{[2\bar{P}_{i-1,j}\Lambda H_{i,j}/\Delta X_i + \tilde{P}_{i,j}(\tilde{q}_{i+1/2,j} + \tilde{q}_{i-1/2,j})/\Delta X_i^2 + A^2\tilde{P}_{i,j}(\tilde{q}_{i,j+1/2} + \tilde{q}_{i,j-1/2})/\Delta Y_j^2]} \quad (37)$$

$$\bar{P}_{i,j} = \beta\tilde{P}_{i,j} + (1-\beta)\hat{P}_{i,j} \quad (38)$$

where the variables with $\sim$, $\wedge$ or — are the variables before, during or after the current iteration, respectively; β is the relaxation factor, $0 < \beta \leq 1$. During iteration, the relaxation factor is often set as a small value, saying 0.1, at the start, and then is gradually increased to 1.0 for accelerating the calculation.

The convergence of calculation can be indicated by evaluating the residual of Eq (35) at each node point during iteration. As a criterion of convergence, the maximum residual should be less than a preassigned value, say 10^{-3}.

$$E_{i,j} = 2\Lambda(P_{i,j}H_{i,j} - P_{i-1,j}H_{i-1,j})/\Delta X_i - [q_{i+1/2,j}(P_{i+1,j}^2 - P_{i,j}^2) - q_{i-1/2,j}(P_{i,j}^2 - P_{i-1,j}^2)]/\Delta X_i^2 - A^2[q_{i,j+1/2}(P_{i,j+1}^2 - P_{i,j}^2) - q_{i,j-1/2}(P_{i,j}^2 - P_{i,j-1}^2)]/\Delta Y_j^2 \quad (39)$$

3.2 Numerical Solution of Head-Disk Gas Lubrication

An important area in which gas lubrication theory plays the major role is the gas-lubricated sliding bearing in magnetic recording systems. The flying head slider (just head or slider for abbreviation) is used to carry a magnetic transducer above a rotating magnetic medium disk (just disk for abbreviation) by virtue of a thin air film. One of the features of the head-disk gas lubrication is that the minimum film thickness is extremely small, about 10 nm in today's technology, under high speed working conditions, and is expected to get down to 5 nm in the near future. Another feature is that the size of a slider becomes smaller and smaller while the air bearing surface of a slider gets more and more sophisticated. Figure 10 illustratively shows the evolutions of head sliders and their air bearing surfaces during the past 30 years. In the early development stage before 1990, taper-flat two-rail or tri-rail design was employed to generate air pressure, and the weight of a slider, which is a very important parameter for determining the load capacity of an air bearing, was in the order of several milligrams or higher. In recent years, slider weight has reduced to sub-milligram, and air bearing surface design has become so sophisticated that the cavity-and-step surface structure looks like a labyrinth. The ultra-thin gas film and the complicated surface structure make the solution of head-disk gas lubrication be a challenging problem in the viewpoints of both the mathematical modeling and numerical implementation.

Let us take a look at the working condition of a slider first. As shown in Fig. 11, the slider is flexibly supported by a suspension over a smooth magnetic disk, which rotates at a fixed speed. The diameter of the disk is normalized as 3.5 in., 2.5 in., or less, and the rotation speed is generally 5,400 r/min, 7,500 r/min, or higher. On the disk there are a number of recording circular tracks, on which magnetic information is read or written by a magnetic head, with a pitch of a few micrometres. When carrying out a read/write task, the slider is moved to a position in the recording area, the (r, θ) coordinates of which are controlled by a voice coil motor and a piezo-actuator for secondary fine positioning. The real distance, or flying height as denoted in the figure, between the slider and the disk in the direction perpendicular to the disk plane during working depends on the air film thickness, which is crucial for a reliable read/write action. Because the constraint from the connecting suspension is flexible, the slider has three freedoms of movement, floating

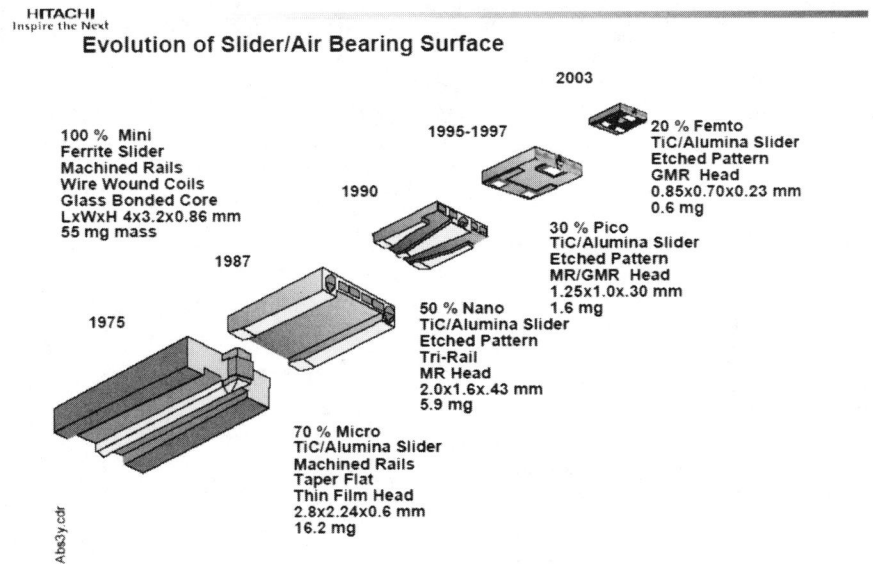

Fig. 10—Evolution of head slider and air bearing surface, cited from www.hitachigst.com/hdd/

in the vertical direction, rolling around the X axis, and pitching around the Y axis (see Fig. 11). The attitude of the slider relative to the disk is determined by the balance between the weight of the slider, the suspension force, the air pressure, and possible surface force from the disk in some cases.

The main task of head-disk gas lubrication analysis is to calculate the air pressure distributions for various known slider surface structures and given working conditions. In the following calculation examples, the modified Reynolds equation derived by Fukui and Kaneko was used, and the flow rate coefficient Q_p was calculated by using the Eq (10).

The first example is for a two-rail slider, the shape and dimensions of which are shown in Fig. 12. The input parameters are listed in Table 1. The direction of gas flow is along the rail direction X. The roll angle was set as zero. The calculated pressure distribution is plotted in Fig. 13. We can see that the air pressure quickly rises at the end of the wedge of the front taper, then gently increases to the pitch angle of the slider, and reaches the maximum near the end of the rails. At the tail, the pressure steeply drops to the ambient value.

The second example is for a Ω type slider as shown in Fig. 14. The dimensions of the slider and working conditions are listed in Table 2. The direction of gas flow is also along the X axis, and the roll angle is zero. Figure 15 shows the calculated dimensionless pressure distribution in the whole range for four different pitch angles. When the pitch angle is zero, pressure at the location of the front bump is higher than the pressure spikes at the three tail peaks, pressure in the center recession zone is negative, and the resultant lifting force is positive and high. When the pitch angle increases to 10 μrad, the pressure spikes at the three tail peaks rise, while pressure in the front bump region gets down, and the resultant lifting force goes down too. In the case of 300 μrad pitch angle, gas pressure in the front bump region drops to a low level, while the pressure spike at the central tail peak gets very high, and the floating force becomes negative, i.e., attractive to the disk due to the large negative (lower than the ambient) pressure in the center recession zone. In the case of pitch angle as large as 3,000 μrad, pressure in the whole region almost disappears, except at the central tail peak where a large pressure spike remains. Figure 16 shows the variation of the floating force with the change of pitch angle from 0.1 μrad to 0.1 rad. We can see that the floating force

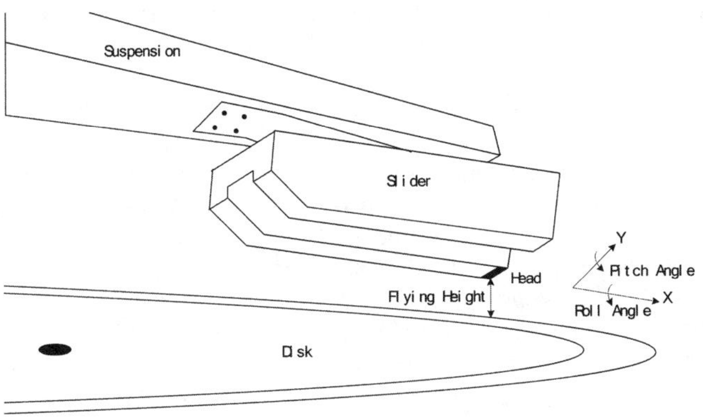

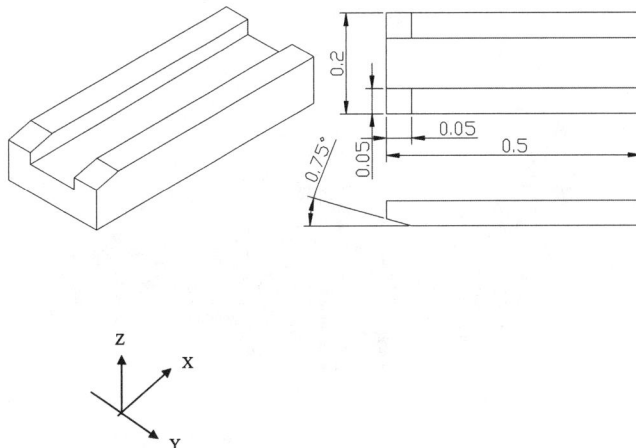

Fig. 11—Schematic drawing of slider position and movement.

Fig. 12—Shape and dimensions of a two-rail slider.

TABLE 1—Parameters of a two-rail slider.						
Parameter	Values	Unit	Parameter	Values	Unit
Minimum film thickness h_0	10×10^{-9}	m	Length of slider L	0.5×10^{-3}	m
Velocity of the disk	40	m/s	Width of head B	0.202×10^{-3}	m
Pitch angle γ	12.9	°/100	Width of slider B_1	0.0515×10^{-3}	m
Gas viscosity η	1.83×10^{-5}	Pa·s	Wedge angle α	0.75	°/10
Bearing number Λ	219600	...	Ambient pressure p_0	1×10^5	Pa

slightly decreases with the increase of pitch angle in the range of 0.1–1 μrad, and then sharply drops to a negative value in the range of 1–300 μrad, rises again with the pitch angle in the range of 300–3,000 μrad, and goes down again in the range of 3,000–100,000 μrad.

It is worthy to note that a reasonable calculation accuracy of pressure can be attained with a medium fine grid, provided that the solution scheme described in the preceding section is utilized. This can be demonstrated with a one-dimensional gas lubrication problem as shown in Fig. 17. For comparison, two grids, one with 481 nodes and the other with 161 nodes, were tested. Table 3 lists the input parameters of calculations. Figure 18 presents the results of gas pressure with the two grids. It is clear that the pressure distributions in the two cases are almost identical, except the little differences in the regions where the spacing changes abruptly. The relative errors in terms of the load, maximum pressure, and minimum pressure are all less than 0.1 % as listed in Table 4.

In order to analyze the dynamic behavior of a magnetic head, time-dependent pressure should be calculated. This can be attained by solving the modified Reynolds equation including a squeezing term, i.e.,

$$\Lambda \frac{\partial}{\partial X}(PH) + \frac{\Lambda_1}{2}\frac{\partial}{\partial T}(PH) = \frac{\partial}{\partial X}\left(PH^3 Q \frac{\partial P}{\partial X}\right) + \frac{\partial}{\partial Y}\left(PH^3 Q \frac{\partial P}{\partial Y}\right) A^2 \quad (40)$$

where $\Lambda_1 = 24\eta L^2/h_0^2 p_0 t_1$ and $\Lambda = 6\eta UL/p_0 h_0^2$, with the similar discrete procedure.

The integration of the gas pressure gives the load carrying capacity, F, of a bearing,

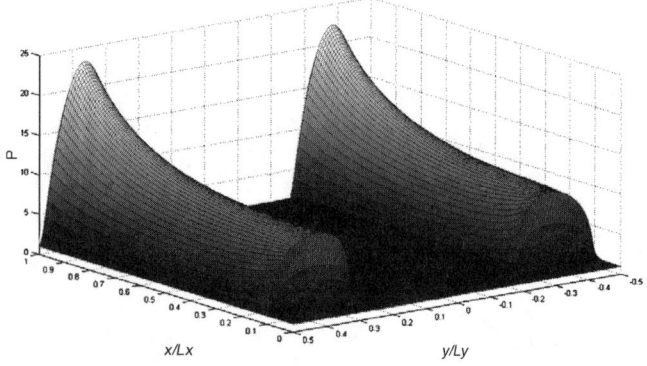

Fig. 13—Dimensionless pressure distribution of the gas film within the two-rail slider. Lx: slider length; Ly: slider width.

$$F = \iint_S p\,dxdy \quad (41)$$

And the dynamic moments around the X and Y axes can be calculated from the following equations, respectively.

$$M_\beta = \iint_S yp\,dxdy$$

$$M_\alpha = \iint_S xp\,dxdy \quad (42)$$

In the above integrations, S represents the air bearing surface. F, M_α and M_β are closely related to the flying height h_0, the pitch angle α_P and the rolling angle α_R. Figure 20(a) shows the dependence of F on the flying height h_0, and Fig. 20(b) shows the variation of M_α with the pitch angle α_P for the slider shown in Fig. 19 [26]. The input parameters for calculations are listed in Table 5. From Fig. 20 we can find that F increases almost linearly with the decrease in h_0, meaning that the air cushion resists the approaching of the slider to the disk. The relationship between M_α and the pitch angle α_P, however, is not monotonous. When $\alpha_P < 0.007$ rad, M_α increases rapidly with the decrease of α_P, showing a restoring tendency. When α_P exceeds 0.007 rad, M_α increases slightly with α_P. In the latter case, the slider has a tendency of losing balance.

3.3 Discussions on Rarefaction and Surface Force Effects on Head-Disk Gas Lubrication

It has been addressed that the rarefaction effect is a unique characteristic of ultra-thin film gas lubrication. Let us discuss the problem that how much rarefaction affects head-disk gas lubrication. Take a practical slider with the bearing surface structure shown in Fig. 21 as an example. For the sake of simplicity, we first take the cross section along the center line ($Y=0$) for analysis. Assuming the flying height $h_{CT0} = 13$ nm (flying height at a selected point within the center tail), the pitch angle $\alpha_p = 1.68 \times 10^{-4}$ rad and the rolling angle $\alpha_R = 0.0$ rad, then the profile of the head cross section can be shown in Fig. 22 with a few characteristic lines, the base line indicating the disk top surface, 10 nm spacing line and 63.5 nm spacing line which means the mean free path of air molecules, λ_0, at the ambient pressure. It should be noted that the scales of the X and Y axes in the figure are intentionally different, to show an enlarged pitch angle. From the figure, we can see that the local spacing or gas film thickness is different from point to point along the cross section, and that

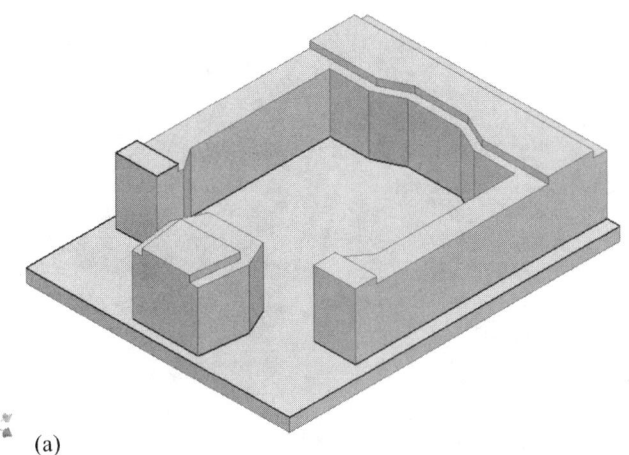

(a)

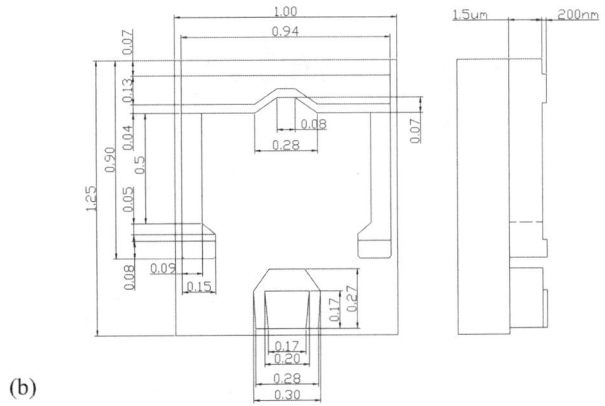

(b)

Fig. 14—Shape and dimensions of a Ω type slider. (a) a three-dimensional view of a Ω type slider; (b) dimensions of the Ω type slider.

$$\lambda_h = \frac{\sqrt{\pi}\mu\sqrt{2RT}}{2p} \qquad (43)$$

we can calculate the local Kn or D after solving the modified Reynolds equation for a given slider surface structure and working condition. And then we can account for the area percentage for different ranges of Kn or D. Figure 24 shows the results of area percentages of four inverse Knudsen number ranges for the slider shown in Fig. 21. At a relative low speed, $v = 9.57557$ m/s, 66.79 % of the whole bearing surface area is in the regime of high inverse Knudsen number ($D > 5$), or low Knudsen number ($Kn < 0.177$), the remaining area is in the regime of medium inverse Knudsen number ($0.15 < D < 5$), or medium Knudsen number ($0.177 < Kn < 5.908$). At a higher speed, $v = 40$ m/s, the area percentage of the medium inverse Knudsen number range ($0.15 < D < 5$) increases to 58.11 %, while that of high D range ($D > 5$) decreases to 41.89 %. We can conclude that both the slip flow and transition flow regions are observed in the slider investigated, and with the increase in sliding speed, the transition flow region extends.

Table 5 gives the comparisons of the calculated load carrying capacity F, the pitch moment M (moment around Y axis), the maximum and minimum dimensionless pressures P_{max}, P_{min} at the speed of 9.57557 m/s, with and without consideration of the rarefaction effect. All of the absolute values of F, M, P_{max}, and P_{min} become lower when the rarefaction effect is accounted. About 20 % reductions in F and M are observed in the case.

Apart from the rarefaction effect, direct interaction between the slider and disk surfaces appears when head-disk spacing gets down to several nanometres. There are several sources of surface interaction forces, including the van der Waals force, electrical force, capillary force, and chemical bond force, etc. In Chapter 9, a brief description of surface forces will be provided. Detailed discussions on surface forces can be found in Ref. [27]. Among the various types of surface forces, the van der Waals force is considered as an essential one, and has been investigated in relation to the head-disk gas lubrication. In 2001, Zhang and Nakajima [28] extended the basic van der Waals force model to an inclined plane slider bearing and analyzed the effects of the minimum spacing, the pitch angle, and the overlayer thickness on the load carrying capacity of the slider. Wu and Bogy [29] incorporated Zhang's approach to numerically analyze a Ω type slider of sub-5 nm flying attitude, and found that the surface force would be important when flying height is less than 0.5 nm. Thornton and Bogy's experiment [30] showed that the surface forces can induce instability of the flying head when the flying height drops to a few nanometres. To show the effect of the van der Waals force on head-disk gas lubrication, analysis of a two-rail slider is given here with the

film thickness in most of the line portions is much greater than the mean free path λ_0.

If we take A_0 as the whole contact area of the magnetic head, and calculate the area A_h of the segments in which film thickness is less than h, then we can get a plot of A_h/A_0 versus h as shown in Fig. 23. For $h = \lambda_0$, the ratio of A_h to A_0 is no more than 6.34 %. For $h = 20$ nm, the ratio of A_h to A_0 is only 2.37 %. Consequently, the portions for actual low flying height, say $h < 10$ nm, have a very small percentage of the whole bearing surface.

To evaluate the intensity of the rarefaction effect, the film thickness h itself is not enough. We have to know the Knudsen number Kn or the inverse Knudsen number D. From the definition of Kn, Eq (2), and the relationship between λ_h and pressure p,

TABLE 2—Parameters of multi-rail slider.

Parameter	Values	Unit	Parameter	Values	Unit
Minimum film thickness h_0	6×10^{-9}	m	Length of head	1.25×10^{-3}	m
Velocity of the disk	25	m/s	Width of head	1.1×10^{-3}	m
Roll angle γ	0.0	rad	Gas viscosity	1.83×10^{-5}	Pa·s
Slider mass	1.6	mg	Ambient pressure	1×10^5	Pa

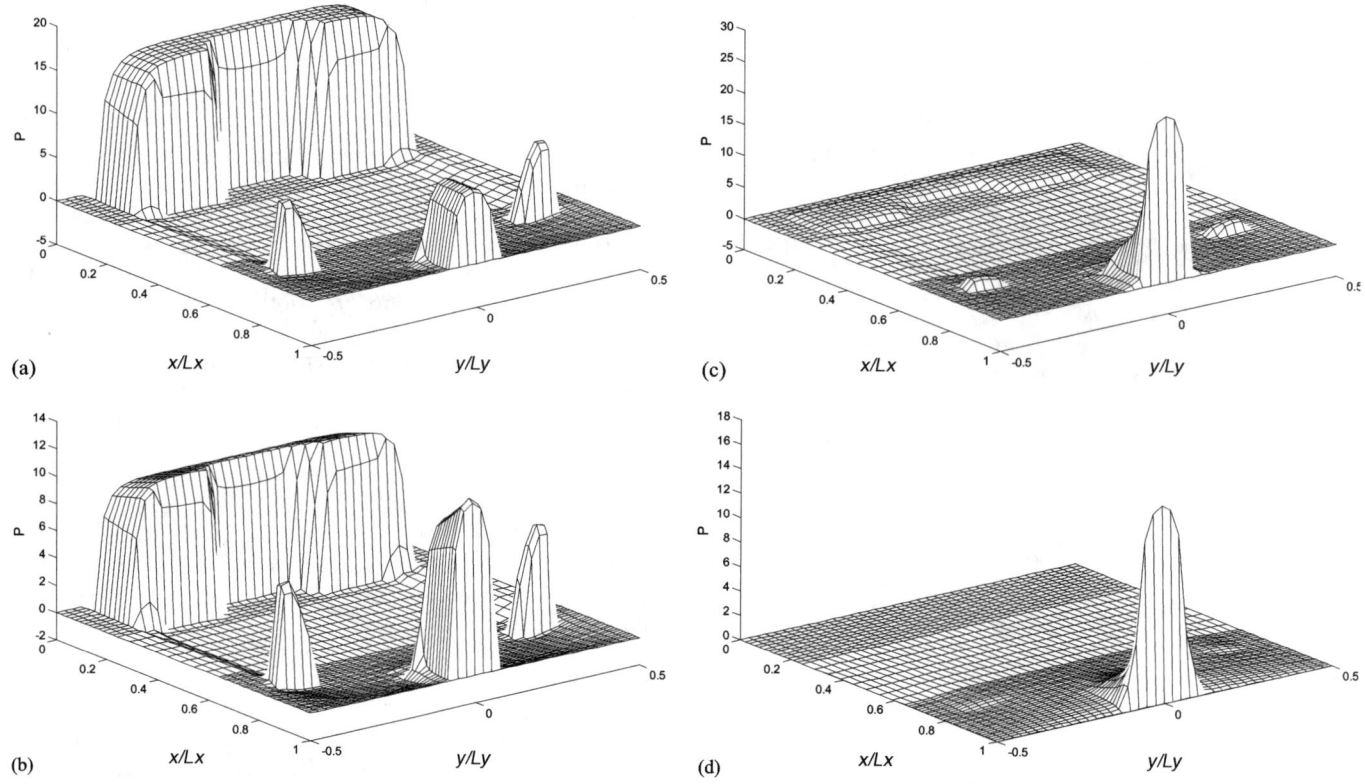

Fig. 15—Dimensionless pressure distributions of the gas film in the Ω type slider under several different pitch angle conditions. (a) Pitch angle $\theta=0$ μrad, calculated floating force F=17.83 g; (b) Pitch angle $\theta=10$ μrad, calculated floating force F=9.76 g; (c) Pitch angle $\theta=300$ μrad, calculated floating force F=−0.75 g; (d) Pitch angle $\theta=3,000$ μrad, calculated floating force F=3.64 g.

Zhang's approach, and the results are compared with those of the inclined plane slider.

The van der Waals force between two parallel plates on unit area is given as follows:

$$F(h) = -\frac{A}{6\pi h^3} \quad (44)$$

where A is the Hamaker constant, and h is the spacing between the plates. The minus sign means that the force is attractive.

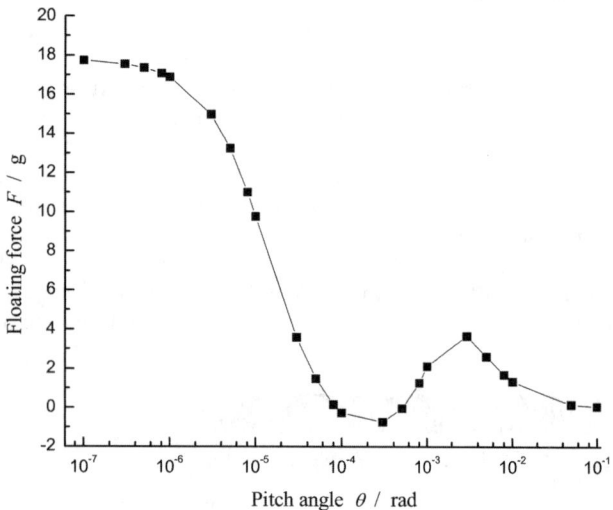

Fig. 16—Influence of pitch angle on the floating force of a Ω type slider.

If there exists a film 2 with thickness δ on one of the plates (see Fig. 25), then van der Waals force on unit area becomes the expression (45) according to Israelachvili's model.

$$F(h) = -\frac{1}{6\pi}\left(\frac{A_{231}}{h^3} + \frac{A_{312}}{(h+\delta)^3}\right) \quad (45)$$

The Hamaker constant A_{231} can be expressed by

$$A_{231} = A_{132} = \frac{3}{4}\kappa T\left(\frac{\varepsilon_1 - \varepsilon_3}{\varepsilon_1 + \varepsilon_3}\right)\left(\frac{\varepsilon_2 - \varepsilon_3}{\varepsilon_2 + \varepsilon_3}\right) + \frac{3\bar{h}v_e}{8\sqrt{2}}$$
$$\times \frac{(n_1^2 - n_3^2)(n_2^2 - n_3^2)}{(n_1^2 + n_3^2)^{1/2}(n_2^2 + n_3^2)^{1/2}\{(n_1^2 + n_3^2)^{1/2} + (n_2^2 + n_3^2)^{1/2}\}}$$
$$(46)$$

where ε_i is the dielectric constant of medium i, n_i is the re-

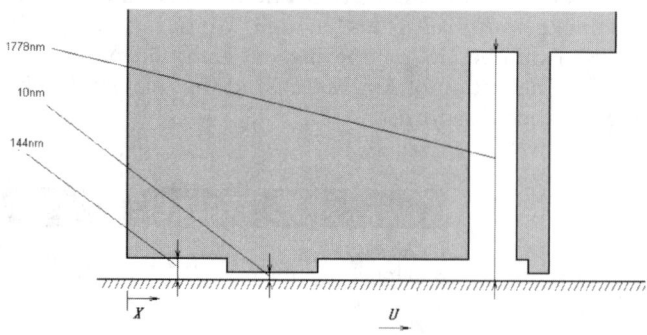

Fig. 17—A cross section of a Ω type slider with zero pitch angle.

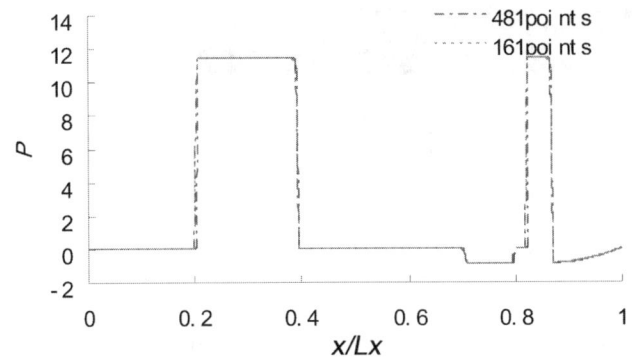

Fig. 18—Dimensionless pressure distributions with two different mesh sizes.

fractive index of medium i, κ the Boltzmann constant, T the absolute temperature, $\bar{h}$ the Plank constant, and $v_e = 3 \times 10^{15}$ s^{-1}.

In general, the first term on the right-hand side of Equation (46) is much less than the second term so as to be neglected,

$$A_{231} \approx \frac{3\bar{h}v_e}{8\sqrt{2}} \times \frac{(n_1^2 - n_3^2)(n_2^2 - n_3^2)}{(n_1^2 + n_3^2)^{1/2}(n_2^2 + n_3^2)^{1/2}\{(n_1^2 + n_3^2)^{1/2} + (n_2^2 + n_3^2)^{1/2}\}} \quad (47)$$

And the Hamaker constant A_{312} is given by

$$A_{312} \approx \frac{3\bar{h}v_e}{8\sqrt{2}} \times \frac{(n_1^2 - n_3^2)(n_1^2 - n_2^2)}{(n_1^2 + n_3^2)^{1/2}(n_1^2 + n_2^2)^{1/2}\{(n_1^2 + n_3^2)^{1/2} + (n_1^2 + n_2^2)^{1/2}\}} \quad (48)$$

The total van der Waals force between a slider and a disk is the integration of the van der Waals pressure over the slider surface Ω as below.

$$W_v = \iint_\Omega F(h)dxdy \quad (49)$$

For an inclined plane slider, the above equation yields,

$$W_p = \frac{-B}{12\pi\alpha_p}\left(A_{231}\left[\frac{1}{h_0^2} - \frac{1}{(h_0 + \alpha_p L)^2}\right] + A_{312}\left[\frac{1}{(h_0 + \delta)^2} - \frac{1}{(h_0 + \alpha_p L + \delta)^2}\right]\right) \quad (50)$$

where L represents the slider length, B the slider width, α the pitch angle, and h_0 the spacing at the exit, as shown in Fig. 26.

In the circumstance of head-disk interface, Medium 1 can be regarded as the DLC films coated on the head and disk substrates, Medium 2 is the PFPE film on the DLC film of disk, and Medium 3 is the air film. Thus, $n_1 = 1.9$, $n_2 = 1.3$, and $n_3 = 1.0$. By substituting these values into Eqs (47) and (48), we have $A_{231} = 7.12 \times 10^{-20}$ J and $A_{312} = 1.20 \times 10^{-19}$ J. For the two-rail slider shown in Fig. 12, we can also derive an expression,

$$W_{tr} = \frac{-(B - B')}{12\pi\alpha_p}\left(A_{231}\left[\frac{1}{h_0^2} - \frac{1}{h_1^2}\right] + A_{312}\left[\frac{1}{(h_0 + \delta)^2} - \frac{1}{(h_1 + \delta)^2}\right]\right) + \frac{-(B - B')}{12\pi(\alpha_p + \beta)}\left(A_{231}\left[\frac{1}{h_1^2} - \frac{1}{h_2^2}\right] + A_{312}\left[\frac{1}{(h_1 + \delta)^2} - \frac{1}{(h_2 + \delta)^2}\right]\right) \quad (51)$$

where h_2, h_0, and h_1 are the gas film thicknesses at the inlet, outlet, and the end of the taper portion, respectively, α_p is the pitch angle, β is the wedge angle of the front taper, and $(B-B')/2$ is the width of each rail.

In the derivation, the van der Waals force acting on the recession of the slider is neglected because the depth from the rail to the recession is in the order of micrometre, much greater than the film thickness.

Figure 27 shows the total van der Waals forces for the simple inclined plane and two-rail sliders at different outlet film thicknesses. For comparison, the same dimensions of the slider length $L = 0.5$ mm, the slider width $B = 0.2$ mm, and

TABLE 3—Parameters for the one-dimensional problem.

Parameter	Values	Unit	Parameter	Values	Unit
Minimum film thickness h_0	10×10^{-9}	m	Length of head	1.25×10^{-3}	m
Velocity of the disk	40	m/s	Gas viscosity	1.83×10^{-5}	Pa·s
Roll angle γ	0.0	rad	Ambient pressure	1×10^5	Pa

TABLE 4—Comparison of the results with different node points.

	Values		
Parameter (dimensionless)	161 Node Points	481 Node Points	Relative Error[a]
Load W	2.46623	2.46741	0.0478234 %
Maximum pressure P_{max}	11.4000	11.4000	0.000000 %
Minimum pressure P_{min}	−0.930101	−0.929473	−0.0675652 %

[a] Where, error = $(a_{481} - a_{161})/a_{481}$.

TABLE 5—Comparison of calculation results with and without rarefaction effect.

	W (N)	M (Nm)	P_{max} (−)	P_{min} (−)	h_{max} (nm)	h_{min} (nm)
Without rarefaction effect ($Q=1$)	0.187	-7.90×10^{-5}	87.4	−0.817	2015	5.22
With rarefaction effect ($Q>1$)	0.147	-6.35×10^{-5}	73.9	−0.781		
Relative difference	22 %	20 %	15 %	4 %		

the pitch angle $\alpha_p = 12.9''$ were used. The width of the recession of the two-rail slider was taken as 0.1 mm as shown in Fig. 12. It can be seen that the van der Waals force is smaller in the case of the two-rail slider than that in the case of the plane slider, because of the difference of the effective bearing areas.

Let us compare the magnitude of the van der Waals pressure with that of air bearing pressure. In Section 3.2, we have shown the numerical solution of gas pressure distribution for a two-rail slider and a Ω type slider. By simply summing up the contributions of gas pressure and van der Waals pressure and integrating over the whole slider surface area, the total dimensionless load carrying capacity becomes

$$W_T = \iint_\Omega P_T dX dY = W_P + W_v \qquad (52)$$

where W_p and W_v denote the load carrying capacities due to the gas pressure and the van der Waals pressure, respectively. Note that the van der Waals pressure is always attractive, while the gas pressure may be repulsive or attractive, depending on the air bearing surface design and working conditions of the slider. Figure 28 shows the changes of the load carrying capacities of an inclined plane slider and a two-rail slider with the minimum gas film thickness when

van der Waals force was taken into account. In calculation, the pitch angle was set as $\alpha_p = 10$ μrad, and the bearing number as 219,600. In the case of inclined plane slider, the load carrying capacity reaches its maximum at $h_0 \approx 4$ nm if no PFPE film exists on disk, i.e., $d = 0$ nm. When the spacing gets lower than 2.5 nm, the load capacity becomes negative. Increasing the thickness of PFPE film leads to a shift of the

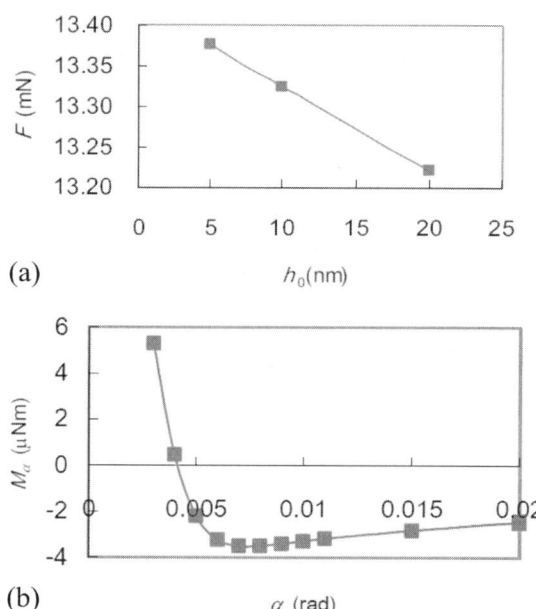

Fig. 20—Effects of slider attitude on the floating force and moment. (a) influence of flying height on the floating force; (b) Influence of pitch angle on the pitch moment.

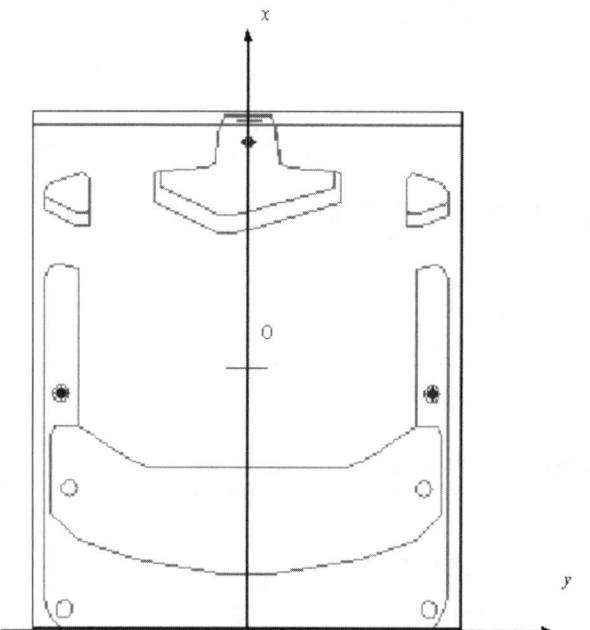

Fig. 19—A slider surface structure.

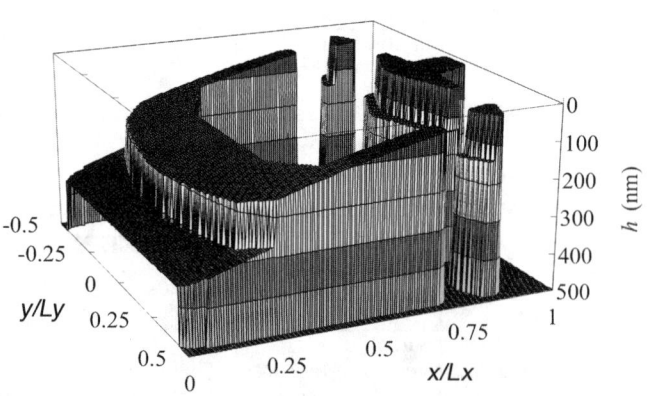

Fig. 21—Surface structure of a Ω type slider.

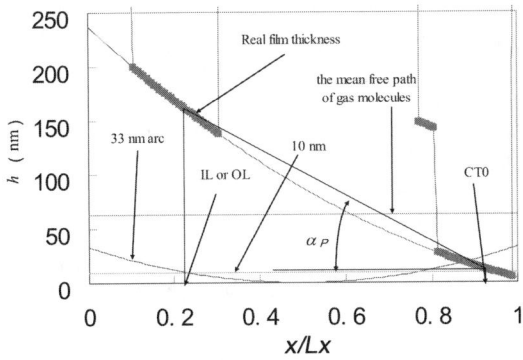

Fig. 22—An enlarged cross section view of the head-disk clearance, h is the height; X is the dimensionless coordinate.

curve towards left, meaning that load capacity can be enhanced. A similar effect can be seen in the case of a two-rail slider, where the peak load capacity appears at a lower spacing $h_0 \approx 1$ nm for $d=0$ nm because van der Waals force is relative weak as shown in Fig. 27. The van der Waals force is also strongly affected by the pitch angle when the minimum spacing is fixed. Figure 29 shows the load capacity of a Ω type slider (see Fig. 14) when van der Waals effect is accounted. The van der Waals effect is negligible when the pitch angle is greater than 4 μrad. However, in the cases of pitch angle smaller than 1 μrad, the van der Waals effect is remarkable.

3.4 Topology Design of Flying Head Air Bearing Surface

In order to meet the strict performance requirements, such as ultra low head-disk spacing, high stiffness of air bearing, and constant flying height over the entire recording area, etc., the air bearing surface of the flying head slider should be optimized in structure. On the other hand, the processing techniques, such as photolithography and plasma etching, available for the head manufacturing, have provided the feasibility for producing complicated air bearing geometries. In addition to a correct gas lubrication theory and a general and accurate solution scheme, automatic and efficient strategies for the flying head structure design are strongly demanded in recent years.

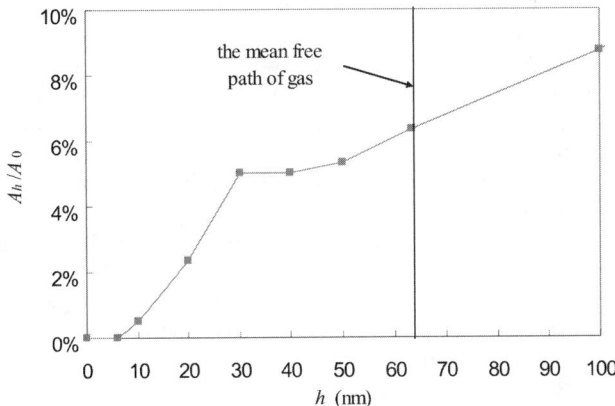

Fig. 23—Area ratio A_h/A_0 versus h, h is the flying height of the magnetic head; A_h is the area where the flying height is below h; and A_0 is the total area of the head.

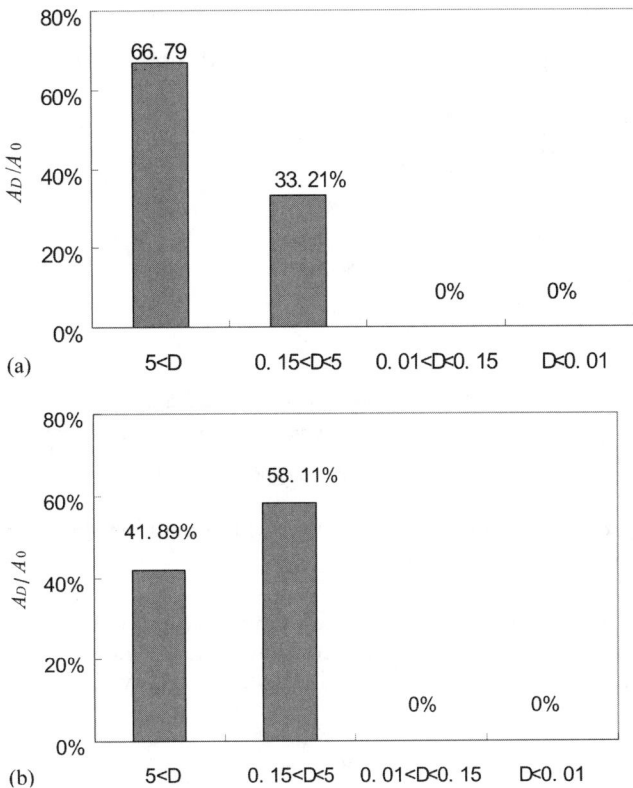

Fig. 24—Area percentage in different inverse Knudsen number ranges, D is inverse Knudsen number; A_D is the area satisfied with the inequalities in the horizontal abscissa. (a) sliding speed v =9.57557 m/s; (b) sliding speed v=40 m/s.

There are two approaches for structure optimization design, parametric optimization and topology optimization. The parametric optimization aims at attaining the best set of geometric or structural parameters defining the configura-

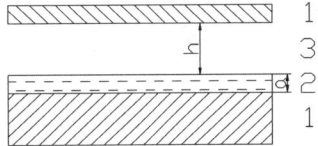

Fig. 25—Surface force between two parallel plates.

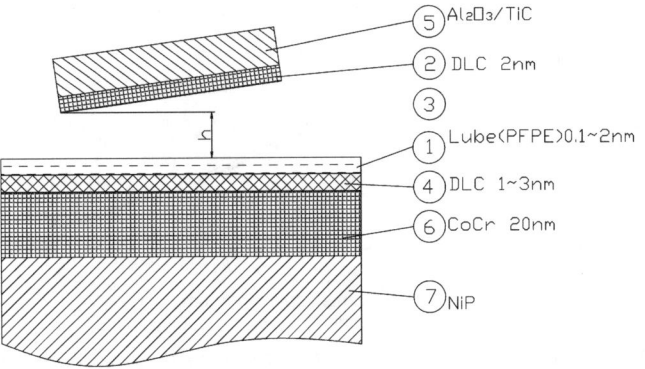

Fig. 26—Film structures on disk and head.

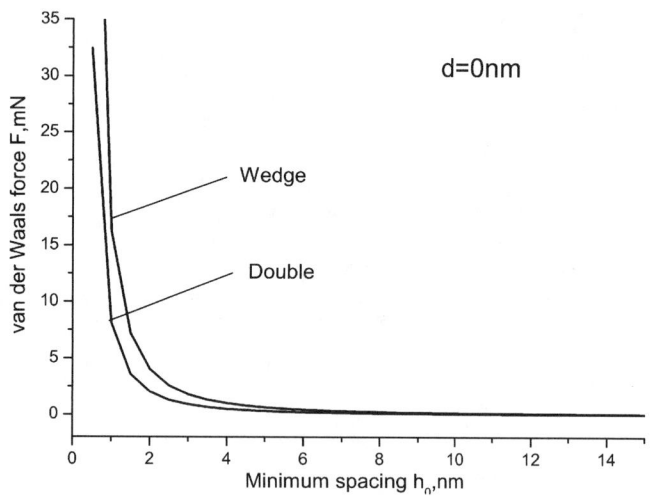

Fig. 27—Comparison of van der Waals force between simple inclined and two-rail sliders.

tion of the object to be optimized. Optimizations of taper-flat sliders, transverse pressure contour sliders, and negative pressure sliders have been studied by using various paramet-

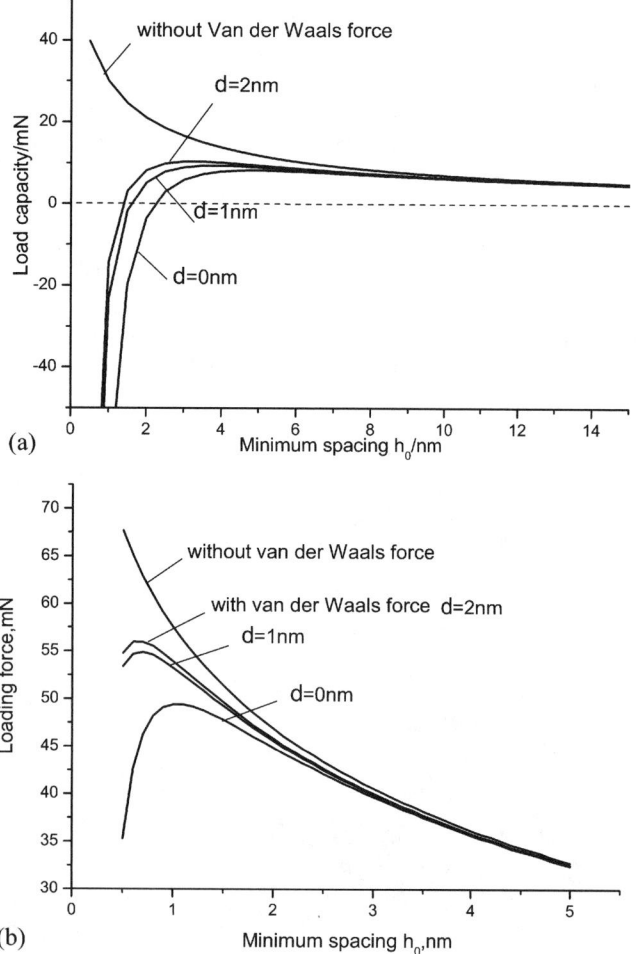

Fig. 28—Effect of van der Waals force on loading capacities of simple inclined and two-rail sliders, d is the thickness of PFPE lubrication film on disk. (a) simple inclined slider; (b) two-rail slider.

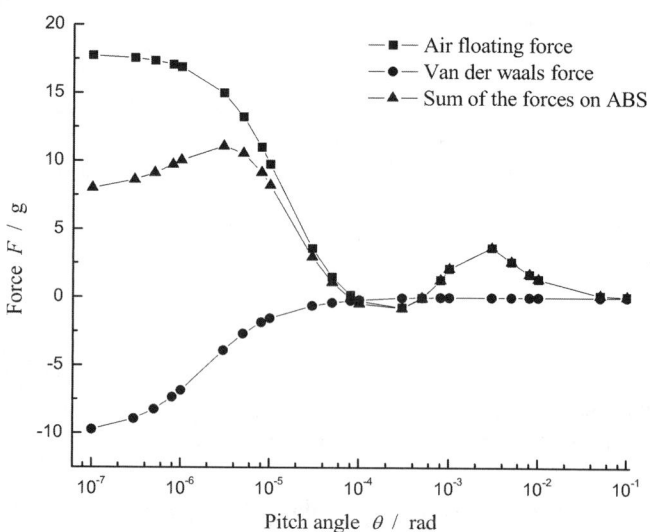

Fig. 29—Effect of van der Waals force on loading capacity of a Ω type slider under different pitch angles. Input parameter was set as: minimum film thickness $h_0=6$ nm, roll angle $\Phi=0$, sliding speed $u=25$ m/s, length of the slider $L=1.25$ mm, width of the slider $B=1.1$ mm, mass of the slider $M=1.6$ mg.

ric optimization techniques. The results of these parametric optimizations depend on the initial topology design which is either chosen intuitively or inspired by already existing designs. An alternative approach is the topology optimization, the goal of which is to find the best feasible and admissible topological structure in the initial stage. Recently, the genetic algorithm has been employed to search for the optimal configuration of the flying head air bearing surface (ABS), and the effectiveness of the ABS topology optimization method has been demonstrated [31].

The main steps of the ABS topology design with the genetic algorithm are as follows:

1. Coding the air bearing surface. The initial flying head structure can be assumed as a rectangular block with a flat surface and a tapered front as shown in Fig. 30. The role of the front taper is to produce a lifting force in virtue of the convergent wedge effect. The design domain is the two-dimensional top surface layer shown in Fig. 30, which can be discretized as a two-dimensional matrix composed of equal-sized elements with different thickness δ_1, δ_2, δ_3, etc. If we assign digital 2 to the elements which should be removed to form a cavity with the depth of δ_2, digital 1 to the elements which should be removed to form a cavity with the depth of δ_1, and digital 0 to those of solid elements, then an admissible design domain can be represented by a two-dimensional matrix composed of 2, 1, and 0. In this way, the topology of the air bearing surface is coded with a two-dimensional matrix of three integers, 0, 1, and 2. With more integers the resolution of the depth of elements can be improved, but the cost of computation would increase. This matrix is referred to as an ABS code.

2. Defining a cost function and constraint conditions. For optimizing air bearing surface design, both static and dynamic performance of a slider should be considered. The static factors include flying height, pitch angle, and

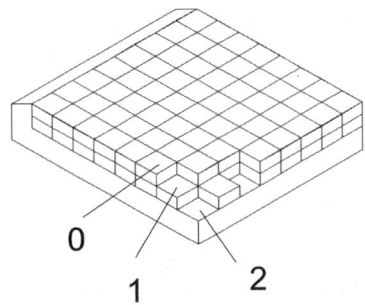

Fig. 30—Schematic of the coded slider.

roll angle over the entire recording band, and the dynamic factors are the stiffness and damping coefficients in the three directions of freedom. The cost function can be defined with a linear combination of the deviations of these static and dynamic parameters from the target values. Meanwhile, constraint conditions such as the force balance and moment balance requirements should be satisfied.

3. Evaluating population. If a slider surface is discretized with $m \times n = k$ elements, then there are a total of 3^k possible slider ABS codes. A group of ABS codes is said to be a population. The initial population can be generated randomly, and then be evolved with a genetic algorithm. When an ABS code is given, the static and dynamic parameters of the coded slider can be calculated with the simulation scheme described in Section 3.2, and then the cost function can be determined. The fitness value of each individual ABS code is evaluated based on its cost function. Once all individuals in the population have been evaluated, a new population can be formed in the following two steps.

4. Performing selection. According to their fitness values, the ABS codes in the current population are replicated into the next generation. The greater the fitness value of an ABS code is, the higher the probability of the ABS code is selected in the next generation. In this way, low-fitness individuals may be eliminated from the population.

5. Performing crossover and mutation. In addition to replication, crossover and mutation are also two effective ways to form a new population. Crossover manipulation is the combination of two ABS codes to form two new ABS codes. Mutation changes one or two elements, saying 0 to 1, or 1 to 0, of a selected ABS code. The crossover and mutation are performed probabilistically. The replication, crossover, and mutation processes are repeated until the termination criterion is reached.

Figure 31 shows an example of the optimized topology design with the above scheme. In the optimization, the cost function was set to be $f(x) = abs(h(x) - h_0)$, where x represents a possible ABS code, $h(x)$ is the flying height corresponding to the ABS code, and h_0 is the target flying height of 20 nm. A total of ten generations were evolved, and the optimal design was found at the ninth generation. The pressure distribution with the optimal design is shown in Fig. 32.

A practical ABS is very complicated and needs a fine grid for describing its geometry. This would lead to an extremely long computation time for finding the optimal design. Today,

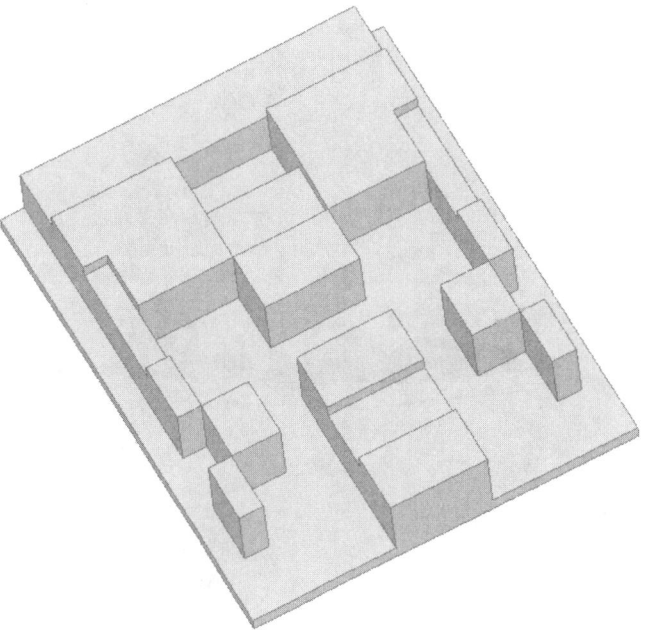

Fig. 31—Configuration of the optimal topology of the coded slider.

the topology optimization of ABS is in the conception stage, and a lot of work is required to reduce the computation time to a reasonable amount.

3.5 Gas Flows in Microsystems

Gas flows are encountered in many microsystems like micro-motors, micro-turbines, micro-sensors, and microfluidic systems in the presence of air or gas environment. Since the ratio of surface area to volume increases in such microsystems, surface forces become dominant over the body forces, and gas flows have great affects on the performance and reliability of many microdevices.

There are three basic categories of gas flow problems in microsystems according to the main driving type. The first category is the squeeze flow, which is encountered in many MEMS devices, such as micro-accelerometers, microphones, micro-gyroscopes, where parallel plates are often employed for the purposes of capacitive sensing and actuation. To increase the sensitivity of capacitive sensing and actuation, the gap between a vibrating plate and a fixed plate gets smaller and smaller with the progress in the microfabrication processes. The presence of gas in the small gap leads to energy dissipation if the gas is squeezed between the two plates due to the transverse motion of the movable plate. This energy dissipation is known as the squeeze film damping, which is dominant over other types of damping mechanisms for determining the quality factor of a dynamic microdevice. The MGL equation has been used for evaluating the squeeze film damping in some MEMS devices, taking the rarefaction effect into account. To incorporate the surface roughness effect on gas flow, Li et al. [32] derived the so-called modified molecular gas film lubrication equation (MMGL), which includes the coupled effects of surface roughness and gas rarefaction, and the MMGL theory has been applied to the analysis of squeeze damping of gas between two parallel plates [33]. It has been found from the cal-

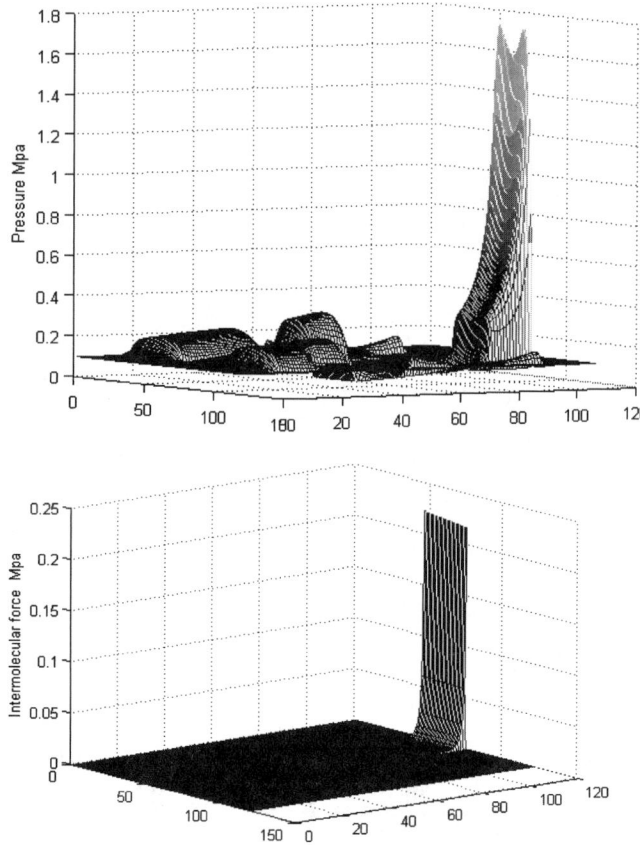

Fig. 32—Pressure distribution of the optimal coded slider.

culation that rarefaction generally lowers the damping force without affecting the spring force. However, when the surface roughness is large and the amplitude of vibration is greater than 5 % of the nominal gap, both the spring force and the damping force increase considerably with the frequency of vibration.

The second category is the shear-driven gas flows, where there exists a relative sliding velocity between the two solid surfaces of a gap, just like the case of the head-disk interface in hard disk drives. The shear-driven gas flow is encountered in micro-motors, micro-turbines, and micro-combdrives, etc. In recent years, self-acting air bearing design has been proposed to overcome the existing severe friction and wear on the rotor-stator interfaces in micromotors [34]. Chen et al. [35] analyzed the load carrying capacity of a step-type bushing based on the MMGL equation and concluded that the air pressure generated beneath the step bushing can support the micromotor, and thus decreases the possibility of contact and wear. Kim et al. [36] investigated the hydrodynamic performance of a microbearing fabricated with deep X-ray lithography and electroplating through the solution of the MGL equation, and proposed improved bearing designs for better stability.

The last category is the pressure-driven gas flows, which are typical in micro gas fluidic and micro heat transfer systems. Because the channel diameter or width in micro gas fluidic systems is in the scale of sub-micrometer or less, ultra-thin gas lubrication theory plays an important role in the design of devices. In this field, the DSMC method is a powerful tool suitable for detailed analysis.

4 Summary

Gas lubrication is an important part of tribology. This chapter focuses on the theory and applications of ultra-thin film gas lubrication. The development of gas lubrication in the past century is reviewed at first. Continuous and discrete models and their application ranges in solving gas flow problems are introduced. After describing the modified Reynolds equations for ultra-thin film gas lubrication, effect of gap width on gas flow has been discussed with statistic analysis and the Direct Simulation Monte Carlo method. Numerical solution procedure of ultra-thin film gas lubrication problems in high density hard disk drives is described and demonstrated with a few examples. The effects of rarefaction and surface force on head-disk gas lubrication are emphasized and analyzed. In addition, the concept of topology optimization of air bearing surfaces is given. At last, the applications of gas lubrication theory in microsystems are briefly introduced.

Acknowledgments

The author is greatly in debt to Professor Huang Ping, who contributed valuable materials including Figs. 17–24 and Tables 3–5, and made a careful check of the manuscript. Thanks are due to Bai Shaoxian, Zhang Yanrui, Tang Weiwei, Lin Jing, and Yue Zhaoyang, who work with the author and contributed a lot to the chapter. The author would also like to thank the Ministry of Science and Technology, China, for the financial support with grant No. 2003CB716205 and the National Natural Science Foundation of China for the financial support with grant No. 50525515 and No. 50721004.

References

[1] Kingsbury, A., "Experiments with an Air Lubricated Journal," *J. Am. Soc. Nav. Eng.*, Vol. 9, 1897, pp. 267–292.

[2] Gross, W. A., *Gas Film Lubrication*, John Wiley and Sons, Inc., , New York, 1962, pp. 4–7.

[3] Wood, R., "The Feasibility of Magnetic Recording at 1 Terabit per Square Inch," *IEEE Trans. Magn.*, Vol. 36, No. 1, 2000, pp. 36–42.

[4] Reynolds, O., "On the Theory of Lubrication and Its application to Mr. Beauchamp Tower'S Experiments Including an Experimental Determination of the Viscosity of Olive Oil," *Philos. Trans. R. Soc. London*, Vol. 177, 1886, pp. 157–234.

[5] Harrison, W. J., "The Hydrodynamical Theory of Lubrication with Special Reference to Air as a Lubricant," *Trans. Cambridge Philos. Soc.*, Vol. 22, 1913, pp. 39–54.

[6] Brunner, R. K. and Harker, J. M., "A Gas Film Lubrication Study Part III: Experimental Investigation of Pivoted Slider Bearings," *IBM Journal of Research and Development*, Vol. 8, 1959, pp. 260–274.

[7] Maxwell, J., "On Stress in Rarefied Gases Arising from Inequalities of Temperature," *Philos. Trans. R. Soc. London*, Vol. 170, 1879, pp. 231–261.

[8] Burgdorfer, A., "The Influence of the Molecular Mean Free Path on the Performance of Hydrodynamic Gas Lubricated Bearings," *ASME J. Basic Eng.*, Vol. 81, 1959, pp. 94–100.

[9] Hsia, Y. T. and Domoto, G. A., "An Experimental Investigation of Molecular Rarefaction Effects in Gas Lubricated Bearings

at Ultra-low Clearances," *ASME J. Lubr. Technol.*, Vol. 105, 1983, pp. 120–130.

[10] Mitsuya, Y., "Modified Reynolds Equation for Ultra-thin Film Gas Lubrication Using 1.5-order Slip-Flow Model and Considering Surface Accommodation Coefficient," *ASME J. Tribol.*, Vol. 115, 1993, pp. 289–294.

[11] Fukui, S. and Kaneko, R., "Analysis of Ultrathin Gas Film Lubrication Based on Linearized Boltzmann Equation: First Report—Derivation of a Generalized Lubrication Equation Including Thermal Creep Flow," *ASME J. Tribol.*, Vol. 111, 1988, pp. 253–262.

[12] Gans, R. F., "Lubrication Theory at Arbitrary Knudsen Number," *ASME J. Tribol.*, Vol. 107, 1985, pp. 431–433.

[13] Alexander, F. J., Garcia, A. L., and Alder, B. J., "Direct Simulation Monte Carlo for Thin-film Bearings," *Phys. Fluids*, Vol. 6, No. 12, 1994, pp. 3854–3860.

[14] Eddie, Y. N. and Liu, N., "Stress-Density Ratio Slip-corrected Reynolds Equation for Ultra-thin Film Gas Bearing Lubrication," *Phys. Fluids*, Vol. 14, No. 4, 2002, pp. 1450–1457.

[15] Fukui, Y., "DSMC/MGL Comparisons of Stress on Slider Air Bearing with Nanometer Spacings," *IEEE Trans. Magn.*, Vol. 38, No. 5, 2002, pp. 2153–2155.

[16] Bird, G. A., *Molecular Gas Dynamics*, Oxford University Press, Bristol, 1976, pp. 67–72.

[17] Huang, P., Niu, R. J., and Hu, H. H., "A New Numerical Method to Solve Modified Reynolds Equation for Magnetic Head/Disk Working in Ultra Thin Gas Films," *Sci. China, Ser. E: Technol. Sci.* Vol. 51, No. 4, 2008. pp. 337–480.

[18] Zhang, Y. R. and Meng, Y. G., "Direct Simulation Monte Carlo Method on Nanoscale Effect of Gas Mean Free Path," *Lubr. Eng.*, No. 7, 2006, pp. 5–7/11 (in Chinese).

[19] Peng, Y. Q., Lu, X. C., and Luo, J. B., "Nanoscale Effect on Ultrathin Gas Film Lubrication in Hard Disk Drive," *ASME J. Tribol.*, Vol. 126, 2004, pp. 347–352.

[20] Morris, D. L., Hannon, L., and Garcia, A. L., "Slip Length in a Dilute Gas," *Phys. Rev. A*, Vol. 46, No. 8, 1992, pp. 5279–5281.

[21] White, J. W. and Nigam, A., "A Factored Implicit Scheme for the Numerical Solution of the Reynolds Equation at Very Low Spacing," *ASME J. Lubr. Technol.*, Vol. 102, No. 1, 1980, pp. 80–85.

[22] Miu, D. and Bogy, D. B., "Dynamics of Gas Lubricated Slider Bearing in Magnetic Recording Disk Files: Part II—Numerical Simulation," *ASME J. Tribol.*, Vol. 108, 1986, pp. 589–593.

[23] Wu, L. and Bogy, D. B., "Numerical Simulation of the Slider Air Bearing Problem of Hard Disk Drives by Two Multi-Dimensional Unstructured Triangular Meshes," *J. Comput. Phys.*, Vol. 172, 2001, pp. 640–657.

[24] Lu, C. J., Chiou, S. S., and Wang, T. K., "Adaptive Multilevel Method for the Air Bearing Problem in Hard Disk Drives," *Tribol. Int.*, Vol. 37, 2004, pp. 473–480.

[25] Wu, J. K. and Chen, H. X., "Operator Splitting Method to Calculate Pressure of Ultra-Thin Gas Film of Magnetic Head/Disk," *Tribology*, Vol. 23, No. 5, 2003, pp. 402–405 (in Chinese).

[26] Huang, P., Wang, H. Z., Xu, L. G., Meng, Y. G., and Wen, S. Z., "Numerical Analysis of the Lubrication Performances for Ultrathin Gas Film Lubrication of Magnetic Head/Disk with a New Finite Difference Method," *Proceedings of IMECE05*, Paper No. IMECE2005-80707, 2005.

[27] Israelachvili, J. N., *Intermolecular and Surface Forces*, 2nd ed., Academic Press, Dublin, 1992.

[28] Zhang, B. and Nakajima, A., "Possibility of Surface Force Effect in Slider Air Bearings of 100 Gbit/in.^{2} Hard Disks," *Tribol. Int.*, Vol. 36, 2003, pp. 291–296.

[29] Wu, L. and Bogy, D. B., "Effect of the Intermolecular Forces on the Flying Attitude of Sub-5 nm Flying Height Air Bearing Sliders in Hard Disk Drives," *ASME J. Tribol.*, Vol. 124, No. (3), 2002, pp. 562–567.

[30] Thornton, B. H. and Bogy, D. B., "A Parametric Study of Head-Disk Interface Instability Due to Intermolecular Forces," *IEEE Trans. Magn.*, Vol. 40, No. (1), 2004, pp. 337–343.

[31] Lu, C. and Wang, T., "New Designs of HDD Air-Lubricated Sliders Via Topology Optimization," *ASME J. Tribol.*, Vol. 126, No. (1), 2004, pp. 171–176.

[32] Li, W. L. and Weng, C. I., "Modified Average Reynolds Equation for Ultra-Thin Film Gas Lubrication Considering Roughness Orientations at Arbitrary Knudsen Numbers," *Wear*, Vol. 209, 1997, pp. 292–230.

[33] Pandey, A. K. and Pratap, R., "Coupled Nonlinear Effects of Surface Roughness and Rarefaction on Squeeze Film Damping in MEMS Structures," *J. Micromech. Microeng.*, Vol. 14, 2004, pp. 1430–1437.

[34] Fukui, S. and Kaneko, R., "Estimation of Gas Film Lubrication Effects Beneath Sliding Bushings of Micromotors Using a Molecular Gas Film Lubrication Equation," *Wear*, Vol. 168, 1993, pp. 175–179.

[35] Chen, M. D., Lin, J. W., Lee, S. C., Chang, K. M., and Li, W. L., "Application of Modified Molecular Gas Film Lubrication Equation to the Analysis of Micromotor Bushings," *Tribol. Int.*, Vol. 37, 2004, pp. 507–513.

[36] Kim, D. J., Lee, S. H., Bryant, M. D., and Lin, F. F., "Hydrodynamic Performance of Gas Microbearings," *ASME J. Tribol.*, Vol. 126, 2004, pp. 711–718.

7

Mixed Lubrication at Micro-scale

Wen-zhong Wang,[1] Yuanzhong Hu,[2] and Jianbin Luo[2]

1 Introduction

TECHNOLOGY DEVELOPMENT DEMANDS MORE EFficient machines capable of operating in more critical conditions, e.g., under heavier loads and higher operating temperatures or using lower-viscosity oils, which result in thinner lubricant films. As a result, machine components operate in a regime of mixed lubrication where hydrodynamic lubrication and asperity contact act simultaneously, and lubrication performances are dominated by surface roughness. Great efforts were made over the past years to understand the role of surface roughness in mixed lubrication, and it is a first but substantial step in exploring the microscopic mechanisms of tribology.

The classical theory of hydrodynamic lubrication and the Reynolds equation were published in 1886, but the complete solutions for the problems that combine the effects of lubrication and solid deformation, known as the elastohydrodynamic lubrication (EHL), were not available until the 1960s. Only after that were the attentions shifted to understanding the role of surface roughness, which led to the concept of mixed lubrication (ML) that emerged in the 1970s, also termed as the partial EHL at that time [1]. The idea is simply to consider that the hydrodynamic film, which separates two surfaces in relative motion, is penetrated by surface roughness when the film thickness becomes smaller than the asperity height. As a result, the lubrication domain is divided into two subregions, the hydrodynamic lubrication areas and the asperity contact areas, and the applied normal load is shared by hydrodynamic pressure and asperity contacts. Impressive progress has been achieved in studies of mixed lubrication, but due to the highly random and irregular nature of surface roughness and the consequent difficulties in measurements and numerical solutions, mixed lubrication remains a still poorly understood regime.

The existence of asperity contacts in mixed lubrication causes great many local events and significant consequences. For example, the parameters describing lubrication and contact conditions, such as film thickness, pressure, subsurface stress, and surface temperature, fluctuate violently and frequently over time and space domain. It is expected that these local events would have significant effects on the service life of machine elements, but experimental measurements are difficult because of the highly random and time-dependent nature of the signals. Only a few successes were reported so far in experimental studies of mixed lubrication, mostly limited to the artificially manufactured surface roughness. Numerical simulations are thus considered to be a powerful means for exploring mixed lubrication, especially for extracting the local information of lubrication. Considerable efforts have been devoted during past 20–30 years to developing numerical models of mixed lubrication. This chapter is contributed to the description of approaches in modeling mixed lubrication, namely the statistic or average model and deterministic model, and the applications of the model to the studying mixed lubrication and the transition of the lubrication regime.

The present authors have been working on the deterministic mixed lubrication (DML) model for more than ten years, and a large part of this chapter is for presenting the results from the authors' own studies. The contents are arranged as follows. Section 2 gives a brief review for the statistic approach of mixed lubrication with the major concern on the average flow model. The DML (deterministic mixed lubrication) model proposed by the authors and corresponding numerical techniques for evaluating surface deformation and temperature are presented in Section 3. Section 4 contributes to the model validations, by comparing the results of the present numerical model with those obtained from other sources of numerical analyses and experiments. In Section 5, the performances of mixed lubrication, including the surface roughness effects, the deformation mode of asperity, and the transition of lubrication regimes, are discussed based on the results from the simulations performed by the present authors. Finally, conclusion remarks are presented in Section 6.

2 Statistic Approach of Mixed Lubrication

The statistic models consider surface roughness as a stochastic process, and concern the averaged or statistic behavior of lubrication and contact. For instance, the average flow model, proposed by Patir and Cheng [2], combined with the Greenwood and Williamsons statistic model of asperity contact [3] has been one of widely accepted models for mixed lubrication in early times.

2.1 Stochastic Models for Rough Surface Lubrication

When the film thickness is of the order of roughness heights, the effects of roughness become significant which have to be taken into account in a profound model of mixed lubrication. The difficulty is that the stochastic nature of surface roughness results in randomly fluctuating solutions that the numerical techniques in the 1970s are unable to handle. As

---
[1] School of Mechanical and Vehicular Engineering, Beijing Institute of Technology, Beijing, China.
[2] State Key Laboratory of Tribology, Tsinghua University, Beijing, China.

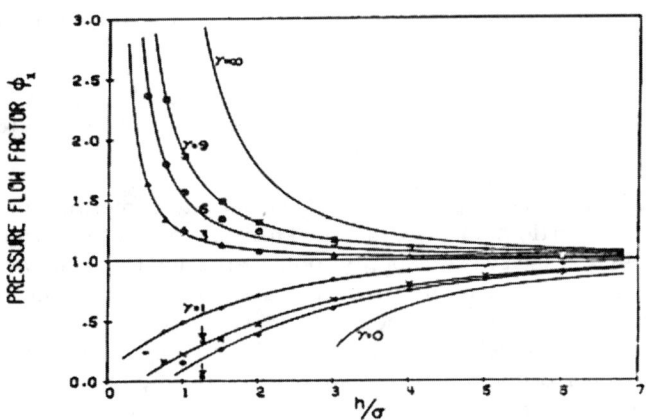 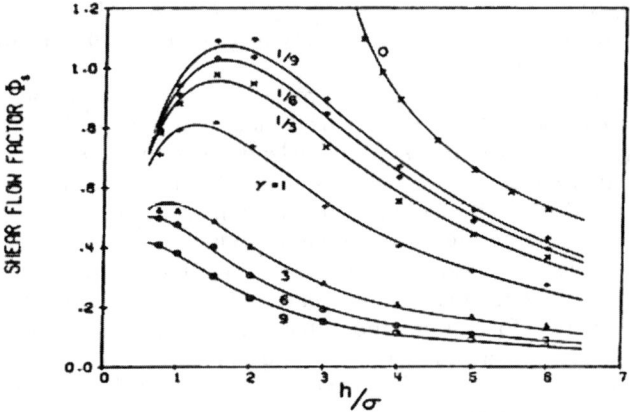

Fig. 1—Flow factor for Gaussian surfaces from Ref. [9].

an alternative, the statistical approach was adopted to deal with the problem in order to reveal the averaged properties of lubrication.

Early work on rough surface lubrication was mainly based on the idea that regards surface roughness as a stochastic process so that a statistical form of the Reynolds equation can be developed, which would give rise to the solutions of mean pressure distributions. The stochastic concepts were first introduced to the application of a slider bearing with transverse roughness by Tzeng and Saibel [4]. Christensen and Tonder [5–8] extended the work by developing a stochastic Reynolds equation for predicting the mean pressure on the surfaces having transverse or longitudinal ridges, and used this equation to analyze the hydrodynamic lubrication of slider and journal bearings. However, a limitation of the stochastic Reynolds equation is that it can only be applied to two specific types of roughness pattern, transverse or longitudinal texture. The solutions on three-dimensional roughness were not available until the average flow model was presented.

2.2 The Average Flow Model

Patir and Cheng [2] developed an average flow model able to predict the mean pressure distributions on general rough surfaces without limitation to the roughness pattern. They introduced pressure flow factors and shear flow factor into the Reynolds equation to describe the effect of roughness on lubricant flow. The average Reynolds equation for incompressible lubricant can be written in the form of

$$\frac{\partial}{\partial x}\left(\frac{\phi_x h^3}{12\eta}\frac{\partial \bar{p}}{\partial x}\right) + \frac{\partial}{\partial y}\left(\frac{\phi_y h^3}{12\eta}\frac{\partial \bar{p}}{\partial y}\right)$$
$$= \frac{U_1 + U_2}{2}\frac{\partial \bar{h}_T}{\partial x} + \frac{U_1 - U_2}{2}\sigma\frac{\partial \phi_s}{\partial x} + \frac{\partial \bar{h}_T}{\partial t} \quad (1)$$

where $\bar{p}$ is the mean pressure, η is lubricant viscosity, h and $\bar{h}_T$ stand respectively for the nominal and mean film thickness, ϕ_x and ϕ_y denote the pressure flow factors, ϕ_s is the shear flow factor, and σ gives the root mean square (rms) value of the roughness.

The pressure flow factors describe how the surface roughness would affect the lubricant flow driven by the pressure gradient. It is defined as the ratio of pressure flow in an element rectangular bearing with a given roughness structure to that in a similar element bearing but with a smooth surface, as expressed as,

$$\phi_x = \frac{1}{L_x}\int_0^{L_x}\left(\frac{h_T^3}{12\eta}\frac{\partial p}{\partial x}\right)dy \bigg/ \frac{h^3}{12\eta}\frac{\partial \bar{p}}{\partial x} \quad (2)$$

where $h_T = h + \delta_1 + \delta_2$, the nominal film thickness h is assumed to be constant within the element bearing.

The shear flow factor represents additional flow transport introduced by the roughness in relative sliding. It is defined as the mean flow in an element bearing where the two surfaces are in relative sliding, but the mean pressure gradient would be zero.

$$\phi_s = \frac{2}{u_s\sigma}\mathbf{E}\left(-\frac{h_T^3}{12\eta}\frac{\partial p}{\partial x}\right) \quad (3)$$

where u_s is the sliding velocity, $u_s = u_1 - u_2$.

The average flow model can be applied to lubrication analyses for the rough surfaces with different patterns, such as transversal, longitudinal, and isotropic roughness. For a certain type of roughness, the flow factors are determined by solving the Reynolds equation in the element bearing. The computations are carried out repeatedly for the same problem with different but statistically identical roughness, and the statistical average gives the mean flow factors. Results presented in Fig. 1 [9] reveal that the flow factors are dominated by two parameters, namely the film thickness ratio h/σ and a pattern parameter defined as $\gamma = \lambda_x/\lambda_y$, where λ_x and λ_y denote the correlation lengths in the x and y directions. The value of γ is considered as a measure for the orientation feature of the roughness. For large film thickness ratios, the flow factors all approach to the values corresponding to the smooth surface, i.e., $\phi_x = \phi_y = 1$ and $\phi_s = 0$. For the longitudinal surfaces roughness ($\gamma > 1$), the pressure flow factors ϕ_x are always greater than 1, corresponding to an increase in pressure flow, while for the transverse roughness ($\gamma < 1$), there is $\phi_x < 1$ on all film thickness ratios, indicating a reduction of the flow caused by the transversal roughness. It is important to note that in the mixed lubrication regime $h/\sigma < 3$, the flow factors become very sensitive to the orientation of the surface roughness. The shear flow factors grow with decreasing film ratio, h/σ for all roughness patterns, revealing that the sliding roughness would always induce additional lubricant flow, but the effect is more profound for the transversal pattern.

However, the flow factors presented by Patir and Cheng were calculated for the rough surfaces without considering asperity deformation, which can lead to significant errors in predicting lubrication behavior at very low film thickness ratio h/σ, as will be shown in later sections of this chapter. Today, the increasing power of the computer and development of efficient algorithm enables one to investigate the effects of surface roughness in greater details. More variables, such as asperity deformation and inter-asperity cavitation, have been incorporated recently into the calculations of flow factors that will lead to more accurate estimations of roughness effects on lubrication.

2.3 Statistic ML Models Combining Lubrication and Asperity Contact

In addition to the lubrication analysis of rough surfaces, asperity contact is another issue that has to be incorporated into the model of mixed lubrication. Greenwood and Williamson [3] presented a statistic model in 1966 for estimating the mean contact pressure between two solid rough surfaces. They assumed that spherical asperities of the same curvature distributed on surfaces in a density of η; the asperity height followed a distribution function, $\phi(z)$, and the inter-asperity interactions during the contact were ignored. When two surfaces with such asperities are brought into contact, the real contact area A_c and the load carrying capacity F can be estimated through a statistic approach as described as follows:

$$A_c = \pi \eta A R \sigma^* F_1(\xi) \quad (4)$$

$$F = \tfrac{4}{3} \eta A E^* R^{1/2} \sigma^{*3/2} F_{3/2}(\xi) \quad (5)$$

where A is the area of nominal contact, R stands for the radium of the spherical asperities, E^* denotes an equivalent Yang's modulus, σ^* gives the rms of asperity height, and ξ defines a dimensionless gap between the surfaces, $\xi = d/\sigma^*$. The function $F_n(\xi)$ in Eqs (4) and (5) is an integration for statistically describing the involvement of asperities in contact, which is defined as

$$F_n(\xi) = \int_\xi^\infty (s-\xi)^n \phi^*(s) ds \quad (6)$$

where $\phi^*(\xi)$ is a normalized distribution function.

The average pressure, p_c, due to asperity contact can be obtained from Eq (5) by defining $p_c = F/A$, which gives

$$p_c = \tfrac{4}{3} \eta E^* R^{1/2} \sigma^{*3/2} F_{3/2}(\xi) \quad (7)$$

As the statistical models for rough surface lubrication and contact are established, a mixed lubrication model can be thus constructed in the following procedure.

As a result of asperity contact, the nominal contact zone is split into a number of discrete areas that can be cataloged either to the lubrication region or asperity contact area (Fig. 2). The mean hydrodynamic pressure in the lubrication regions, p_l, can be calculated by the average flow model, while contact pressure is estimated via Eq (7). Consequently, the film thickness is determined through numerical iterations to ensure that the following equation of load balance is to be satisfied.

$$W = \iint_\Omega (p_l + p_c) \cdot d\Omega \quad (8)$$

where W is the applied normal load and Ω denotes the computation domain. After film thickness and pressure distribution are solved, the asperity contact area, load sharing ratio between lubrication and contact, and friction force contributed by the two components can be estimated.

The model has been applied successfully to predicting the performances of bearings, gears, seals, and engines [10–12]. A fundamental limitation of the statistic models is their inability to provide detailed information about local pressure distribution, film thickness fluctuation, and asperity deformation, which are crucial for understanding the mechanisms of lubrication, friction, and surface failure. As an alternative, researchers paid a great interest to the deterministic ML model.

3 A DML Model Proposed by the Present Authors

3.1 An Historical Review

The word "deterministic" means that the model employs a specific surface geometry or prescribed roughness data as an input of the numerical procedure for solving the governing equations. The method was originally adopted in micro-EHL to predict local film thickness and pressure distributions over individual asperities, and it can be used to solve the mixed lubrication problems when properly combined with the solutions of asperity contacts.

The DML (deterministic mixed lubrication) models have been developing over the past years, starting with line contact problems, like the partial EHL model presented by Chang [13], and then being extended to the point contact problems that require more computational efforts. The strategy was straightforward at the beginning, as proposed by Hua et al. [14], for example, that each point in the computation domain had to be identified and labeled that it belongs to either a lubrication region or asperity contact area, and the method for calculating the pressure at this point depends on the nature of the location. For example, the pressure in the lubrication region was calculated in terms of the Reynolds equation and asperity contact pressure was obtained from the Hertz theory of elastic contacts. In 1999, Jiang et al. [15] made a successful application of the strategy that solved hydrodynamic and asperity contact pressure simultaneously. In their solutions, the identification of actual contact regions were carried out through a trial-and-error procedure, which required laborious computational work, and the convergent solutions presented were limited only to relatively large film thickness ratios.

All the models of mixed lubrication developed previously were based on a traditional idea, as schematically shown in Fig. 2, that the nominal contact zone, Ω, has to be divided into two different types of areas: the lubricated area, Ω_l, where two surfaces are separated by a lubricant film and the asperity contact area, Ω_c, where two surfaces are assumed to be in direct contact. The present authors and Dr. Zhu [16,17] proposed a different strategy for modeling

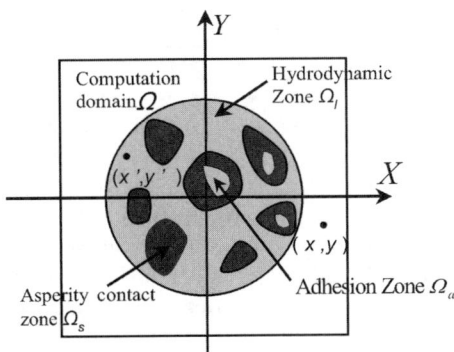

Fig. 2—Schematic contour plot of point-contact mixed lubrication zone.

mixed lubrication by assuming that there were always thin lubricating films existing between contacting asperities and the pressure over the thin films could be predicted by the Reynolds equation. Consequently, the pressure distributions on the entire domain can be calculated from a unified Reynolds equation system, without having to identify the hydrodynamic and asperity contact areas. The DML model proposed by the authors allows one to study the entire regime of lubrication, from full EHL to boundary lubrication without much convergence problem even though a significant amount of asperity contacts may be involved in the problems analyzed.

3.2 Governing Equations in Isothermal DML Model

3.2.1 Pressure Control Equation

Instead of dividing the computation domain into lubrication regions and asperity contact areas, the mixed lubrication model proposed by the present authors assumes that the pressure distribution over the entire domain follows the Reynolds equation:

$$\frac{\partial}{\partial x}\left(\frac{\rho h^3}{12\eta^*}\frac{\partial p_l}{\partial x}\right) + \frac{\partial}{\partial y}\left(\frac{\rho h^3}{12\eta^*}\frac{\partial p_l}{\partial y}\right) = u_e\frac{\partial(\rho h)}{\partial x} + \frac{\partial(\rho h)}{\partial t} \quad (9)$$

with the following boundary conditions to be satisfied:

$$p(x_0,y) = p(x_e,y) = p(x,y_0) = p(x,y_e) = 0 \quad (10)$$

In this model, there will be no asperity contacts in the traditional sense, but as the film thickness between the interacting asperities decreases below a certain level, the right-hand terms in Eq (9), which represent the lubricant flow caused by pressure gradient, become so insignificant that the pressure can be predicted by a reduced Reynolds equation [16,17]:

$$u_e\frac{\partial(\rho h)}{\partial x} + \frac{\partial(\rho h)}{\partial t} = 0 \quad \text{when } h \to 0 \quad (11)$$

The asperity contact areas in conventional models of mixed lubrication correspond here to the areas where Eq (11) is applied. More discussions about this reduced equation are left to the next section.

3.2.2 Gap Equation

The gap between two mating surfaces is evaluated via a geometric equation,

$$h(x,y,t) = h_0(t) + h_s(x,y) + \delta(x,y) + v(x,y,t) \quad (12)$$

where $h_0(t)$ is the distance between surfaces at $x=0$ without accounting for deformation, $h_s(x,y)$ describes the geometry of mating surfaces, $\delta(x,y)$ is the sum of roughness heights of surfaces, $v(x,y,t)$ is the surface deformation calculated through the formula given by [18]

$$v(x,y,t) = \frac{2}{\pi E'}\iint_{\Omega_l+\Omega_c}\frac{p_l(x,y,t)+p_c(x,y,t)}{\sqrt{(x-\xi)^2+(y-\zeta)^2}}d\xi d\zeta \quad (13)$$

3.2.3 Load Balance Equation:

$$W(t) = \iint_\Omega [p_l(x,y,t)+p_c(x,y,t)]dxdy \quad (14)$$

3.2.4 Lubricant Properties:

An effective viscosity η^* has been introduced in the Reynolds equation to describe the non-Newtonian lubricant properties. Ignoring the variation of viscosity across the film thickness, one may evaluate the effective viscosity via the following rheological model that considers a possible shear-thinning effect [19],

$$\frac{1}{\eta^*} = \frac{1}{\eta}\frac{\tau_0}{\tau_1}\sinh\left(\frac{\tau_1}{\tau_0}\right) \quad (15)$$

where τ_0 is a reference shear stress whose value has been given as $\tau_0 = 18.0$ MPa for a typical mineral oil. τ_1 denotes the shear stress acting on the lower surface, and η is the limiting viscosity at low shear rate, which is determined in terms of the Roelands viscosity relation:

$$\eta = \eta_0 \exp\left\{(\ln \eta_0 + 9.67)\left[(1+5.1\times 10^{-9}p_l)^z \times \left(\frac{T-138}{T_0-138}\right)^{-s_0} - 1\right]\right\} \quad (16)$$

where η_0 is the viscosity at ambient pressure, T_0 is ambience temperature, $s_0 = \beta(T_0-138)/(\ln \eta_0+9.67)$, $\beta=0.04$ and $z=\alpha/5.1\times 10^{-9}(\ln \eta_0+9.67)$.

The density of the lubricant is given by,

$$\frac{\rho}{\rho_0} = 1 + \frac{Ap_l}{1+Bp_l} + D(T-T_0) \quad (17)$$

where ρ_0 is the ambience density. For a mineral oil, constants A, B, and D are $A=0.6\times 10^{-9}$ m^2/N, $B=1.7\times 10^{-9}$ m^2/N, and $D=-0.0007$ K^{-1}. In Eqs (16) and (17), T refers to the temperature in the middle layer of lubricant films.

Note that the viscosity and density are the functions of pressure and temperature, through which the interactions between pressure and temperature are presented.

3.3 Thermal Analysis of Mixed Lubrication

3.3.1 Development of Thermal DML Model

The models mentioned above were developed under the assumption of isothermal conditions. However, frictional heating is an important factor that has to be taken into account in order to establish a more comprehensive model of mixed lubrication.

The deterministic thermal analysis of mixed lubrication

was pioneered by Lai and Cheng [20] in 1984, and further developed by Qiu and Cheng [21] more recently. Using the pressure distribution obtained from isothermal solutions, they calculated transient temperature distribution in point-contact mixed lubrication, through a numerical procedure of the moving-point-heat-source integration [22]. Zhai and Chang [23,24] developed another transient thermal model for the mixed lubrication in line contacts, by solving energy equations for both solid and fluid subject to thermal interfacial conditions. Progress in this subject can be found in Refs. [25,26], but for the classical approach of thermal analysis of lubrication, the difficulty is to handle the energy equation when film thickness becomes molecularly thin, so great efforts are required to achieve a satisfactory thermal model of mixed lubrication.

A common feature in the models reviewed above was to calculate pressure and temperature distributions in a sequential procedure so that the interactions between temperature and other variables were ignored. It is therefore desirable to develop a numerical model that couples the solutions of pressure and temperature. The absence of such a model is mainly due to the excessive work required by the coupling computations and the difficulties in handling the numerical convergence problem. Wang et al. [27] combined the isothermal model proposed by Hu and Zhu [16,17] with the method proposed by Lai et al. for thermal analysis and presented a transient thermal mixed lubrication model. Pressure and temperature distributions are solved iteratively in a iterative loop so that the interactions between pressure and temperature can be examined.

3.3.2 Determining the Surface Temperature

A typical method for thermal analysis is to solve the energy equation in hydrodynamic films and the heat conduction equation in solids, simultaneously, along with the other governing equations. To apply this method to mixed lubrication, however, one has to deal with several problems. In addition to the great computational work required, the discontinuity of the hydrodynamic films due to asperity contacts presents a major difficulty to the application. As an alternative, the method of moving point heat source integration has been introduced to conduct thermal analysis in mixed lubrication.

If the heat flux from friction or viscous shear is properly estimated, the surface temperature, which is of interest in most engineering problems, can be determined through integrating an analytical solution of temperature rise caused by a moving point heat source, without having to solve the energy equation. For two solid bodies with velocity u_1 and u_2 in dry contacts, the temperature rises at the surfaces can be predicted by the formula presented in Ref. [22],

$$\Delta T_1(x,y,t) = \int_0^t \iint_\Omega \frac{[1-f(x',y',t')]q(x',y',t')}{4\rho_1 c_1 [\pi\alpha_{s1}(t-t')]^{3/2}}$$
$$\times e^{\{[(x-x')-u_1(t-t')]^2+(y-y')^2\}/[4\alpha_{s1}(t-t')]} dx'dy'dt' \quad (18)$$

$$\Delta T_2(x,y,t) = \int_0^t \iint_\Omega \frac{f(x',y',t')q(x',y',t')}{4\rho_2 c_2 [\pi\alpha_{s2}(t-t')]^{3/2}}$$
$$\times e^{\{[(x-x')-u_2(t-t')]^2+(y-y')^2\}/[4\alpha_{s2}(t-t')]} dx'dy'dt' \quad (19)$$

where ΔT_1 and ΔT_2 are the temperature rises at the two surfaces, $q(x',y',t')$ denotes a point heat source at the location (x',y') and time t', and $f(x',y',t')$ is the heat partition coefficient that describes how the heat fluxes are assigned between the two bodies.

Equations (18) and (19) are valid too, in hydrodynamic regions for calculating surface temperature, if we assume that viscous heating concentrates on the middle layer of lubricating films and temperature varies linearly across the film [20]. The Fourier law of heat conduction gives rise to the following expressions:

$$K_f \frac{T(x,y,t) - T_1(x,y,t)}{h(x,y)} = (1 - f(x,y,t)) \cdot q(x,y,t) \quad (20)$$

$$K_f \frac{T(x,y,t) - T_2(x,y,t)}{h(x,y)} = f(x,y,t) \cdot q(x,y,t) \quad (21)$$

where T is the lubricant temperature in the middle layer, T_1 and T_2 are the temperatures at the solid surfaces, estimated by the sum of bulk temperatures of solid bodies T_{bi} and temperature rises ΔT_i, and K_f denotes the thermal conductivity of fluid.

The heat flux partition coefficient $f(x,y,t)$ is determined from the following equation obtained by eliminating T from Eq (20) and (21).

$$[T_{b2} + \Delta T_2(x,y,t)] - [T_{b1} + \Delta T_1(x,y,t)] = K_f \cdot [1 - 2f(x,y,t)] \cdot h(x,y) \cdot q(x,y,t) \quad (22)$$

If contacts occur at position (x,y), namely $h(x,y)=0$, Eq (22) is reduced to

$$[T_{b2} + \Delta T_2(x,y,t)] - [T_{b1} + \Delta T_1(x,y,t)] = 0 \quad (23)$$

which means that the temperatures at two contacting surfaces are the same.

In summary, the temperature rises ΔT_i and the heat partition coefficient $f(x,y,t)$ can be determined through solving Eqs (18) and (19) simultaneously with (22) or (23), if the heat flux $q(x,y,t)$ has been given in advance. In the simplest way, an estimation of heat generation can be made by calculating frictional work in terms of experiential friction coefficients.

$$q = \mu p V_s \quad (24)$$

The friction coefficient, μ, is assigned with different values in different friction conditions. V_s is the relative sliding velocity of two contacting bodies.

3.4 Two Approaches in DML Models for Solving Pressure Distributions

As described previously, two different strategies have been developed in the DML model to solve the pressure distributions for hydrodynamic lubrication and asperity contacts, simultaneously. The advantage and disadvantage of the two methods deserve a further discussion.

3.4.1 Method Based on Separated Solutions from Two Regions

In this method, the pressure distributions in the lubricated zone and asperity contacts zone are solved separately. The hydrodynamic pressure in lubricated regions is determined by solving the coupling Eqs (9), (12), and (13), while contact pressure is calculated from contact mechanics. The method is sometimes referred to as "conventional" because it agrees with the common sense for the mixed lubrication that nominal contact zone is divided into hydrodynamic and contact regions where different mechanics apply. A typical example was presented by Jiang et al. [15] for solving a point-contact problem in mixed lubrication, as described in Section 3.1. They reported that convergent solutions were achieved for a typical case, with an average gap of 1.16 (normalized by the rms of the roughness) and an asperity contact area ratio of 12%.

A major shortage of the method is that the border positions and boundary pressure distributions between the hydrodynamic and contact regions have to be calculated at every step of computation. It is a difficult and laborious procedure because the asperity contacts may produce many contact regions with irregular and time-dependent contours, which complicates the algorithm implementation, increases the computational work, and perhaps spoils the convergence of the solutions.

3.4.2 Method Based on Unified Reynolds Equations

In developing the DML model, the present authors proposed a new approach, which uses the Reynolds equation consistently in both hydrodynamic and contact regions. This strategy is based on the idea that there will be lubricating films remaining between asperities even though they are in "contact," which means the films become molecularly thin, or fluid films transit to boundary layers. The idea is supported by the experience that the solution of the Reynolds equation under the limit of $h=0$ would give the same result as that from the contact mechanics. It is well known that the physical interpretation of Reynolds equation (9) is a balance of fluid flow. The left-hand side of the equation represents the lubricant flow due to the hydrodynamic pressure gradient, while the two terms on the right-hand side stand for the lubricant flow caused by surface motion in both tangential and normal directions. As the film thickness approaches to zero, the pressure flow gradually decreases and ultimately vanishes, thus the Reynolds equation is reduced to the form in Eq (11). During the process of decrease of film thickness, the pressure for smooth contacts, obtained from the Reynolds equation, will gradually approach to the Hertzian distribution. In other words, the pressure obtained under the condition of ultra-thin film is the same as that of dry contact analysis under the same operating conditions. This means that the reduced Reynolds equation can be applied for solving the contact pressure.

The approach has proven to be advantageous in comparison with the "conventional" method. Because the reduced equation is a special form of the Reynolds equation, a full numerical solution over the entire computation domain, including both the hydrodynamic and the contact areas, thus obtained through a unified algorithm for solving one equation system. In this way, both hydrodynamic and contact pressure can be obtained through the same iteration loop and the pressure continuity at the borders will be satisfied automatically. From the point of view of computation practice, this approach has been proven to be successful to reach the convergent solutions under severe operating conditions. More than 100 cases have been analyzed so far, and no convergence problem has been encountered even for those with λ ratios as low as 0.02–0.03 at rolling speeds as low as 0.001 mm/s.

The success in application of the method provides an example that transition from full-film to boundary lubrication can be simulated via a unified mathematical approach, and boundary lubrication can be regarded as a limiting case of hydrodynamic lubrication.

3.5 Numerical Methods for Evaluating Surface Deformation

The calculation of normal deformation of the surfaces in contact plays a key role in numerical solutions of EHL, and consumes a great part of computer time. The question is to evaluate the deformation caused by a distributed load on elastic half-space. There are accurate closed-form solutions for the normal deformation caused by the forces in several special forms, such as a point load, a uniformly distributed pressure, or a Hertzian pressure [18]. However, for a more general form of pressure distributions, a closed-form solution does not exist so that people have to rely on the numerical techniques to calculate the deformations. In a conventional way, the normal deformation can be numerically evaluated through constructing a matrix of influence coefficients, and conducting a direct summation based on the linear elasticity theory. To obtain influence coefficients, the pressure distribution has to be interpolated on each subdomain, according to the nodal pressure values. The interpolation functions may take different forms, such as the constant [28,29], piecewise biquadratic polynomial [30], or a bilinear polynomial [31]. For a line contact problem, the direct summation demands as many as N^2 times of multiplication where N is the grid number. The number of multiplications is even greater for the three-dimensional problems. Such a large amount of computation work greatly limits the efficiency of numerical procedure. Ren et al. [32] developed a Moving Grid Method (MGM), which has an advantage over the conventional matrix method in that the memory size required to store a flexibility matrix of a dimension N is only $O(N)$ instead of $O(N^2)$.

During the past three decades, many efforts have been made to calculate the surface deformation more efficiently and with less computer storage. A new method, known as the Multi-level Multi-integration (MLMI), which was orders of magnitude faster than the conventional methods, was developed by Lubrecht and Ioannides [33,34]. The method has been proven to be very efficient in saving CPU times though it costs a complicated procedure in programming.

Noticing the fact that the formula for determining surface deformation takes the form of convolution, the fast Fourier transform (FFT) technique has been applied in recent years to the calculations of deformation [35,36]. The FFT-based approach would give exact results if the convolution functions, i.e., pressure and surface topography take periodic form. For the concentrated contact problems, however, the application of FFT will create so-called periodicity errors. Previous studies in this field tried to reduce the periodicity errors, mostly by extending the computation domain and padding pressure with zero [35,36], but at the cost of computation efficiency. Recently, Liu et al. [37] proposed an improved FFT-based method (DC-FFT) resulting from the discrete convolution theorem. The present authors have successfully implemented the method into the DML model [38]. In the following we are going to give a brief description for the improved FFT-based approach, and present a comparison of efficiency and accuracy between several currently used methods. This may shed a light on the uncertainty as for which method would perform better in calculation of deformation.

Consider a distributed pressure acting on an elastic half-space, and let the pressure distribution and the normal surface deformation be denoted by $p(x)$ and $v(x)$ for line contacts, or by $p(x,y)$ and $v(x,y)$ for point contacts, respectively. According to the theory of contact mechanics [18], the normal surface deformation $v(x)$ or $v(x,y)$ caused by a distributed pressure may be written in the forms of

$$v(x) = -\frac{4}{\pi E^*} \int \ln|x - s| p(s) ds \quad (25)$$

for line loading, and

$$v(x,y) = \frac{2}{\pi E^*} \iint \frac{p(\xi, \eta)}{\sqrt{(x-\xi)^2 + (y-\eta)^2}} d\xi d\eta \quad (26)$$

for point contacts.

In numerical analysis, both functions of normal surface deformation and pressure distribution have to be discretized in a space domain over N grid points for a line load, or $N_x^* N_y$ grid points for two-dimensional distributed load. As an example, the deformation for line loading can be rewritten in discrete form as follows:

$$v(x_i) = -\frac{4}{\pi E^*} \sum_{j=0}^{N-1} K(x_i - x_j) p(x_j) \quad (27)$$

where $K(x_i - x_j)$, or denoted as K_j^i, is known as the influence coefficient that refers to the normal deformation at point x_i due to a unit load acting on a position x_j. It can be seen from Eq (27) that the calculation of normal surface deformation includes two steps:
1. Determine the influence coefficients K_j^i.
2. Calculate the multi-summation.

For simplicity, the following discussion will be limited to the line contact problems only, and it can be easily extended to the point contact problems.

3.5.1 Determination of Influence Coefficients

The influence coefficient (IC) K_j^i has been interpreted physically as the deformation at point x_i due to unit point load acting on x_j. For the distributed load, K_j^i can be obtained through calculating the deformation at x_i, caused by the pressure distributed over a small area around the position x_j:

$$K_j^i = \int_{x_j-\Delta x/2}^{x_j+\Delta x/2} h(x_i - \xi) n(\xi) d\xi \quad (28)$$

where Δx denotes the dimension of a element in discrete grid, $n(x)$ is an interpolation function to approximate the pressure distribution within the element, and $h(x)$ is known as the response or Green's function, which can be written in the form of $\ln(x)$ for the line contact problems. For calculation of normal deformation on a half-space of homogeneous materials, K_j^i depends only on the distance between point x_i and x_j. When x_i is fixed on the coordinate origin, the distance between x_i and x_j is $(x_j - x_i) = x_j$, so that Eq (28) may be rewritten as

$$K_j = \int_{x_j-\Delta x/2}^{x_j+\Delta x/2} h(\xi) n(\xi) d\xi \quad (29)$$

where influence coefficients K_j denote the deformation at the origin due to a load acting on x_j.

Different interpolation functions have been used for determining the influence coefficients, as summarized briefly in the following.

(1) Green's Function Based Scheme
When $n(x)$ in Eq (29) is taken as a unit constant, and the Green's function $h(x)$ is assumed to be invariant within a element, approximated by $h(x_j)$, influence coefficient K_j can be simply determined as:

$$K_j = \Delta x \times h(x_j) \quad (30)$$

It is noticed that the Green's function $h(x)$ has one singularity when $x_j = 0$, but it can be eliminated by an integration over the element around the point $x_j = 0$. A computation experiment shows that Eq (30) may result in a significant numerical error when coarser grids are employed.

(2) Constant Function Based Scheme
If $n(x)$ in Eq (29) is taken as a constant function on the element surrounding point x_j, the influence coefficients are given by performing the following integration

$$K_j = \int_{x_j-\Delta x/2}^{x_j+\Delta x/2} h(\xi) d\xi \quad (31)$$

which leads to an analytical solution

$$K_j = x_m \ln\left(\frac{2x_m}{\Delta x}\right)^2 - x_p \ln\left(\frac{2x_p}{\Delta x}\right)^2 \quad (32)$$

with $x_m = x_j + \Delta x/2$, $x_p = x_j - \Delta x/2$

For two-dimensional problems, readers can refer to Ref. [39].

(3) Linear Interpolation Based Scheme
If a linear interpolation function $n(x)$ is applied to approximate the pressure distribution within the element, the influence coefficients can be obtained by performing the integration in Eq (29), which results in

$$K_j = x_j \frac{x_{j-1} + x_{j+1}}{\Delta x} F(x_j) - \frac{x_{j-1}^2}{\Delta x} F(x_{j-1}) - \frac{x_{j+1}^2}{\Delta x} F(x_{j+1}) \quad (33)$$

with $F(x) = 2 \ln|x|$, and $F(0) = 0$.

For two-dimensional problems, if a bilinear interpolation function is employed, the influence coefficients can be computed likewise in analytical form [31].

One may adopt higher order polynomials, such as bi-quadratic polynomial [30], but the methods are conceptually the same. Theoretically, the numerical accuracy corresponding to higher order interpolations is expected to be improved, but computation practices show this is not always true. As a matter of fact, when the grid becomes very fine, there is little difference between the results from different orders of interpolation.

3.5.2 Calculation of Multi-Summation

Having the influence coefficients obtained, the normal surface deformations can be obtained from the multi-summation as described in Eq (27). The computation may be implemented using different numerical approaches, including direct summation (DS), MLMI, and DC-FFT based methods, which will be briefly described in this section.

(1) Direct Summation (DS)
A direct computation of Eq (27) may reach accuracy up to the level of discrete error, but this needs N^2 multiplications plus $(N-1)^2$ additions. For two-dimensional problem, it needs $N^2 \times M^2$ multiplications and $(N-1)^2 \times (M-1)^2$ additions. The computational work will be enormous for very large grid numbers, so a main concern is how to get the results within a reasonable CPU time. At present, MLMI and discrete convolution and FFT based method (DC-FFT) are two preferential candidates that can meet the demands for accuracy and efficiency.

(2) DC-FFT Method
Equation (27) is in fact a discrete linear convolution whose calculation can be speeded up by applying the discrete Fourier transform (DFT), but periodic errors may result from two sources:

1. Green's function $h(x)$ was truncated at the boundary of computation domain. In reality, however, function $h(x)$ may possess non-zero values beyond the computation domain. The truncation of $h(x)$ will therefore result in errors in the convolution. The error is expected to be minimal at the center of computation domain but increases at the positions close to the border of the domain.
2. When the DFT was employed to calculate the convolution, it means all signals and output were converted into periodic functions. This, more or less but inevitably, will bring about the periodic errors.

To use the DFT properly for evaluating normal surface deformation, the linear convolution appearing in Eq (27) has to be transformed to the circular convolution. This requires a pretreatment for the influence coefficient $\{K_j\}$ and pressure $\{p_j\}$ so that the convolution theorem for circle convolution can be applied. The pretreatment can be performed in two steps:

(a) Extend the dimension of $\{p_j\}$ by adding an appropriate number of zeros to the array;
(b) Extend the dimension of $\{K_j\}$ to the same length as $\{p_j\}$ and then circulate or wrap-around the series $\{K_j\}$.

In summary, when the DC-FFT algorithm is applied, the calculation of normal surface deformation within a region $[x_0, x_e]$ can be implemented in the following procedure:

1. Discretize the influence coefficient $K(x)$ into a series $\{K_j\}_{2N}$ of size $2N$ in a region whose sides are twice as that of the computation domain.
2. Reorder the indexes of $\{K_j\}_{2N}$. The rule is to copy the components $K_1, \ldots, K_N$ of $\{K_j\}_{2N}$ into a new series $\{KN_j\}_{2N}$ as $KN_{N+1}, \ldots, KN_{2N}$, and copy $K_{N+1}, \ldots, K_{2N}$ into the series $\{KN_j\}_{2N}$ as $KN_1, \ldots, KN_N$.
3. Discretize the pressure function $p(x)$ into a series $\{p_i\}_N$ of size N in the computation domain, and then extend the length of the series $\{p_i\}_N$ into a new series $\{pN_i\}_{2N}$ through zero padding.
4. Apply FFT to the series $\{pN_i\}_{2N}$ and $\{KN_j\}_{2N}$, and the results are denoted as $\{\hat{p}N_i\}_{2N}$ and $\{\hat{K}N_j\}_{2N}$.
5. Compute the component-wise product of $\{\hat{p}N_i\}_{2N}$ and $\{\hat{K}N_i\}_{2N}$, resulting in $\{\hat{T}_i\}_{2N}$.
6. Perform the inverse FFT to $\{\hat{T}_i\}_{2N}$ which gives a new series $\{T_i\}_{2N}$. The desired deformation $\{v_i\}_N$ in the region $[x_0, x_e]$ is obtained by setting $v_i = T_i, i = 1, \ldots, N$.

The total complexity of the numerical analysis is $O(N \ln(N))$. Readers who want more details are referred to Liu [37].

(3) Multi-Level Multi-Integration Method (MLMI)
Another preferable method for evaluating the deformation is the multi-level multi-integration method. An outline of the MLMI algorithm is presented as below.

In Eq (25), the integration kernel function $K(x) = \ln|x|$ is smooth everywhere except the singularity point ($x = 0$). In a numerical analysis, the integration has to be evaluated in discrete form over a grid with the mesh size h

$$v_i^h = h \sum_j K_{i,j}^{hh} p_j^h \qquad (34)$$

where v_i^h and p_j^h denote the discrete value of function $v(x)$ and $p(x)$ on the grids with mesh size h, respectively, and the value of $K_{i,j}^{hh}$, which approximates the kernel function $K(x)$, used to be predetermined by analytical approaches [31].

If a coarse grid with a mesh size H ($H = 2h$, for example) is employed to evaluate the same integration (25), its discrete form can be written as

$$v_I^H = H \sum_J K_{I,J}^{HH} p_J^H \qquad (35)$$

The idea of MLMI method is to calculate the integration v_J^H first on the coarse grid (H), and then to evaluate the v_j^h on the fine grid (h) through an interpolation,

$$v^h = I_H^h v^H \qquad (36)$$

where I_H^h denotes an interpolation operator from the coarse grid (H) to the fine grid (h), whose specific form depends on the choice of the interpolation function. If this process has been applied to a system with several levels of grid, the reduction in computation times can be very significant.

The problem is, however, the evaluation (36) may create a considerable error due to the singularity of the kernel function. In these cases, a correction process has to be applied. Here a two-level grid system with $H = 2h$ is used as an example to illustrate the approach of correction.

1. Calculate the integration (25) on the coarse grid (H) to get v_J^H.
2. For the points $i = 2I$ on the fine grid, the value of v_i^h can be evaluated based on v_J^H, but with the following correction in the neighborhood of $i = j$ if the kernel $K(x)$ is supposed to be singular at $x = 0$,

TABLE 1—Time for calculating multi-summation on different grids (units: s).

Methods/Mesh	64×64	128×128	256×256	512×512
DS	0.1	1.64	117.66	2604.81
MLMI	0.09	0.38	1.58	6.5
DC-FFT	0.025	0.1	0.44	2.06

$$v_i^h \approx v_I^H + h \sum_{\|i-j\|\leq m} C_{i,j}^{hh} p_j^h \quad \text{if } i = 2I \quad (37)$$

3. Calculate the interpolation $[I_H^h v_I^H]_i$ for the points where $i = 2I + 1$, and evaluate v_j^h using the correction,

$$v_i^h = [I_H^h v_I^H]_i + h \sum_{\|i-j\|\leq m} D_{i,j}^{hh} p_j^h \quad \text{if } i = 2I + 1 \quad (38)$$

The correction terms, $C_{i,j}^{hh}$ and $D_{i,j}^{hh}$ in Eqs (37) and (38), have to be precomputed. For more details, refer to Brandt and Lubrecht [33].

The above algorithm can be easily extended to two-dimensional cases by applying the same procedure alternately in the x dimension and the y dimension.

(4) Comparison and Discussion

Comparisons of the accuracy and efficiency for three numerical procedures, the direct summation, DC-FFT-based method and MLMI, are made in this section. The three methods were applied to calculating normal surface deformations at different levels of grids, under the load of a uniform pressure on a rectangle area $2a \times 2b$, or a Hertzian pressure on a circle area in radius a. The calculations were performed on the same personal computer, the computational domain was set as $-1.5a \leq x \leq 1.5a$ and $-1.5a \leq y \leq 1.5a$, and covered by a uniform mesh with the node number ranging from 64 × 64, 128 × 128, 256 × 256 to 512 × 512 for all computational cases.

The CPU times required for the computations on different grids are given in Table 1 and displayed in Fig. 3 as well. It can be seen that the DC-FFT based method is the fastest among the three methods. If one multiplication is defined as a unit operation and the total number of nodes is N, the computation in the DS method will take N^2 operations, while in theory both computations in MLMI and FFT require $N \ln(N)$ operations. In practice, however, the MLMI method performs much slower than the FFT probably due to the complicated programming in MLMI which causes additional operations.

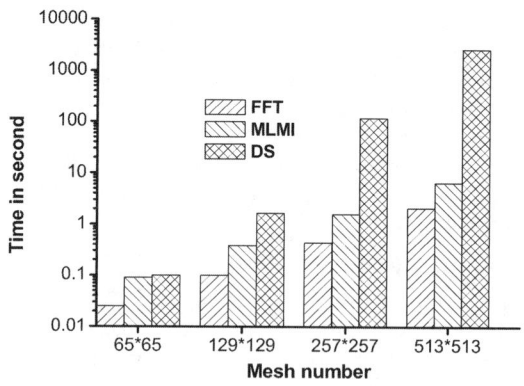

Fig. 3—A graphic illustration for the data in Table 1.

Concerning the numerical accuracy, the closed form solutions of normal surface deformation have been compared to the numerical results calculated through the three methods of DS, DC-FFT, and MLMI. The influence coefficients used in the numerical analyses were obtained from three different schemes: Green' function, piecewise constant function, and bilinear interpolation. The relative errors, as defined in Eq (39), are given in Table 2 while Fig. 4 provides an illustration of the data.

$$\varepsilon = \frac{\sum_i \sum_j |v_{i,j} - v_{i,j}^c|}{\sum_i \sum_j |v_{i,j}^c|} \quad (39)$$

where $v_{i,j}$ and $v_{i,j}^c$ are the deformations obtained through numerical and closed-form solutions, respectively.

The first conclusion drawn from the comparison is that the numerical accuracy is dominated largely by the discrete errors. When the same influence coefficients were employed,

TABLE 2—Relative errors for DS, FFT-based method and MLMI over different grids (%) (Green, Constant and Bilinear stand, respectively, for the schemes based on Green's function, constant function, and linear interpolation in determining the influence coefficients).

Mesh Number	Methods for IC	Uniform Pressure on Rectangular			Hertzian Pressure on Circle		
		DS	MLMI	DC-FFT	DS	MLMI	DC-FFT
64×64	Green	1.2779	1.2779	1.2779	0.17757	0.17757	0.17757
	Constant	1.4504	1.4518	1.4504	0.034789	0.036657	0.03479
	Bilinear	1.4461	1.4889	1.4461	0.051467	0.055390	0.051468
128×128	Green	0.80975	0.80975	0.80975	0.088167	0.088168	0.088167
	Constant	0.72369	0.71965	0.72369	0.014601	0.018734	0.014601
	Bilinear	0.72476	0.71667	0.72476	0.014877	0.022698	0.014878
256×256	Green	0.31889	0.31889	0.31889	0.043488	0.043488	0.043488
	Constant	0.36285	0.36912	0.36289	0.0028181	0.0091057	0.0028184
	Bilinear	0.36262	0.37509	0.36262	0.0034759	0.015171	0.0034759
512×512	Green	0.20334	0.20333	0.20334	0.021720	0.02172	0.021721
	Constant	0.18135	0.17387	0.18135	0.0011437	0.0087182	0.0011439
	Bilinear	0.18141	0.16645	0.18142	0.0010846	0.016247	0.0010833

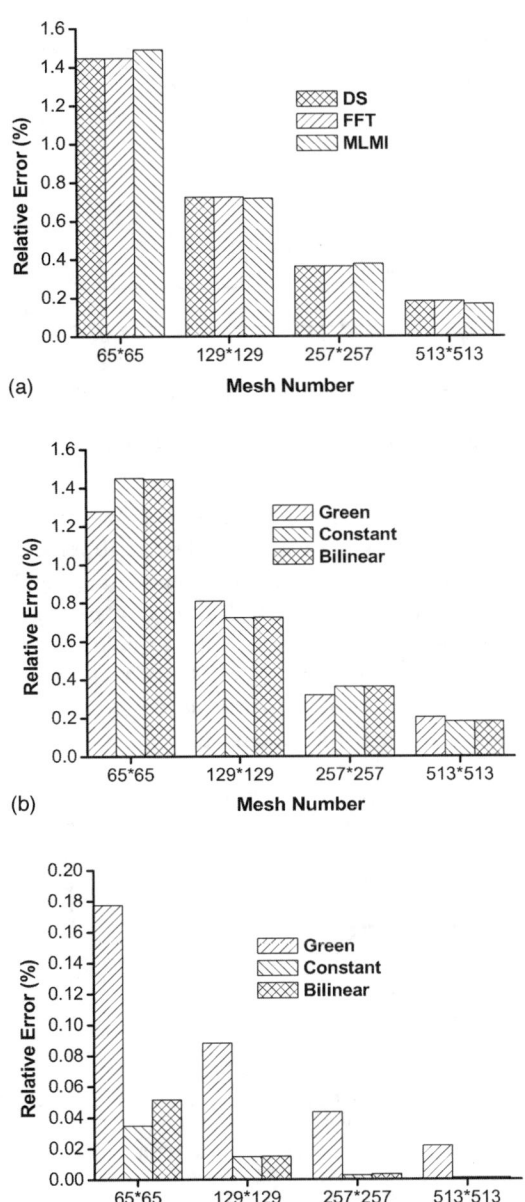

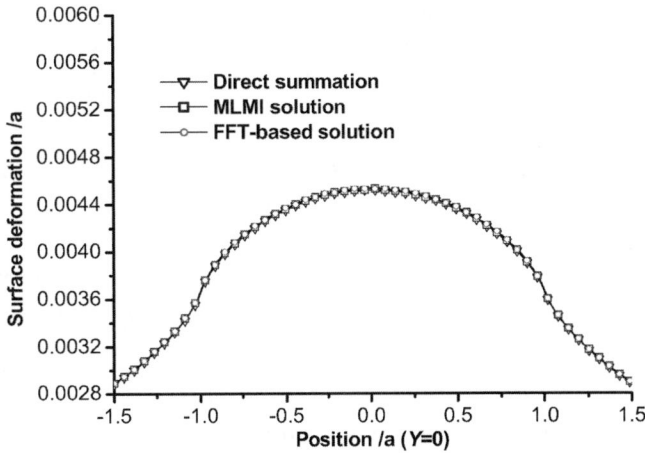

Fig. 5—A comparison of deformation profiles at $Y=0$ on 256×256 grid for a uniform pressure on a rectangle area $2a \times 2b$, obtained from the three methods.

Fig. 4—Comparison of relative error for different schemes. (a) A comparison of relative errors for a uniform pressure on a rectangle area $2a \times 2b$, in which the multi-summation is calculated via DS, FFT, and MLMI, and IC is determined through bilinear interpolation based scheme. (b) A comparison of relative errors for a uniform pressure on a rectangle area $2a \times 2b$, in which the multi-summation is calculated via DS and IC is determined through the Green, constant, and bilinear-based schemes. (c) A comparison of relative errors for a Hertzian pressure on a circular region in radius a, in which the multi-summation is calculated via DS, and IC is determined through the Green, constant, and bilinear-based schemes.

the DS, MLMI, and FFT-based methods produced similar errors, indicating that fast computation methods did not introduce additional errors. The second conclusion is that the errors decrease roughly by a factor of two if the calculations are performed on a finer grid with half mesh size. In addition, high order interpolation functions may or may not give higher accuracy depending on the form of pressure distribution. For uniform pressure, Green' function method or piecewise constant function may perform better, but for the Hertzian distribution, linear or higher order interpolation is recommended. Figure 5 shows a comparison of the profiles of normal surface deformation obtained from the three methods on 256×256 grid.

A major message carried in this section is that the FFT-based method is faster and easier to implement but with similar accuracy in comparison to the MLMI, so it is expected to be a powerful numerical approach suitable for the deformation computations, especially when a larger number of grid nodes are involved.

4 Validation of the DML Model

The DML model, proposed by the present authors, exhibits a good computational efficiency and robustness in different operating conditions, but a careful examination and complete validation are essential for the model to be accepted in engineering applications. The validation presented in this section is made by comparing the results from this model with those published in the literatures, or obtained by other investigators from their simulations and experiments.

4.1 Comparison of Results in Full-Film EHL Regime

Considerable numbers of the numerical solutions of full-film EHL for different surface geometries, such as the smooth surfaces, surfaces with single asperity, and sinusoidal waviness, were published over the past years. They provide good reference data for the purpose of model validation.

4.1.1 Smooth Surface

Figure 6 shows the comparison between the present solutions and those from Venner et al. [40] and Holmes [41] for smooth EHL contacts in full-film regime. The central and minimum film thicknesses are listed in Table 3. From Fig. 6 and Table 3, it can be seen that as long as in full-film EHL regime the solutions from the present model are in good agreement with those by Venner et al. and Holmes.

4.1.2 Single Transversal Ridge

Figures 7 and 8 compare results of the present model in two specific cases with the corresponding results from Venner et

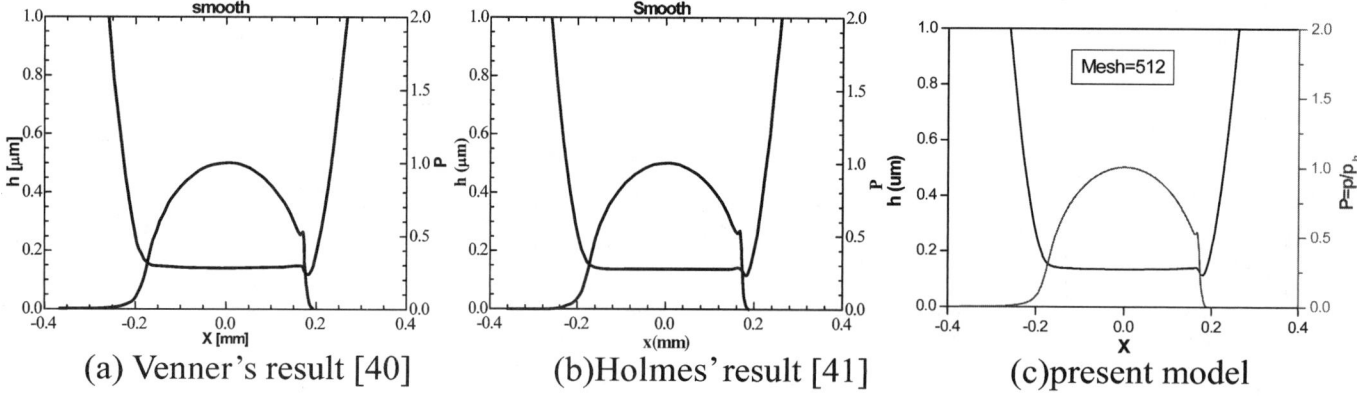

(a) Venner's result [40]   (b) Holmes' result [41]   (c) present model

Fig. 6—Result comparison for smooth surface in point contact EHL. (a) Venner's result [40]; (b) Holmes' result [41]; and (c) present model.

TABLE 3—Comparison of film thickness for smooth surface in point contact EHL.

	H_{cen} (nm)	H_{min} (nm)
Venner and Lubrecht [40]	143	63
Holmes [41]	136	61
Present method	135	61

al. [40], and Holmes [41], obtained in the same conditions. In the first case, a stationary transversal ridge superposed on a smooth ball is in contact with a moving plane (Fig. 7) while in the second, a transversal ridge carried by the smooth ball moves over the plane (Fig. 8). The film thickness and pressure profiles, shown in Figs. 7 and 8, are taken along the x direction at the position of $y = 0$ and plotted in the same scale. Again, good agreement between our results and those ob-

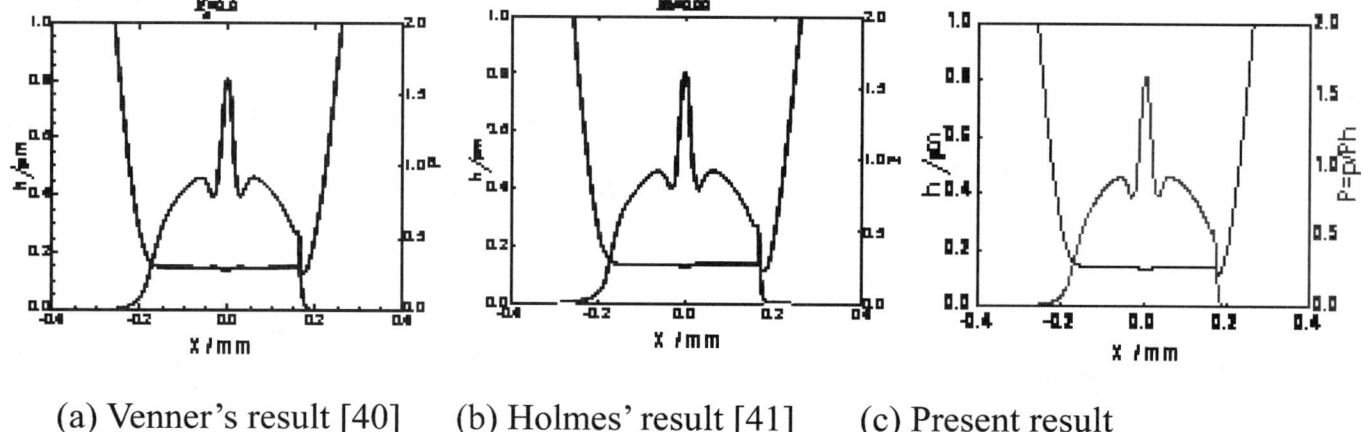

(a) Venner's result [40]   (b) Holmes' result [41]   (c) Present result

Fig. 7—Comparison of film and pressure profiles with Venner and Holmes' results (for a stationary transversal ridge). (a) Venner's result [40]; (b) Holmes' result [41]; and (c) present result.

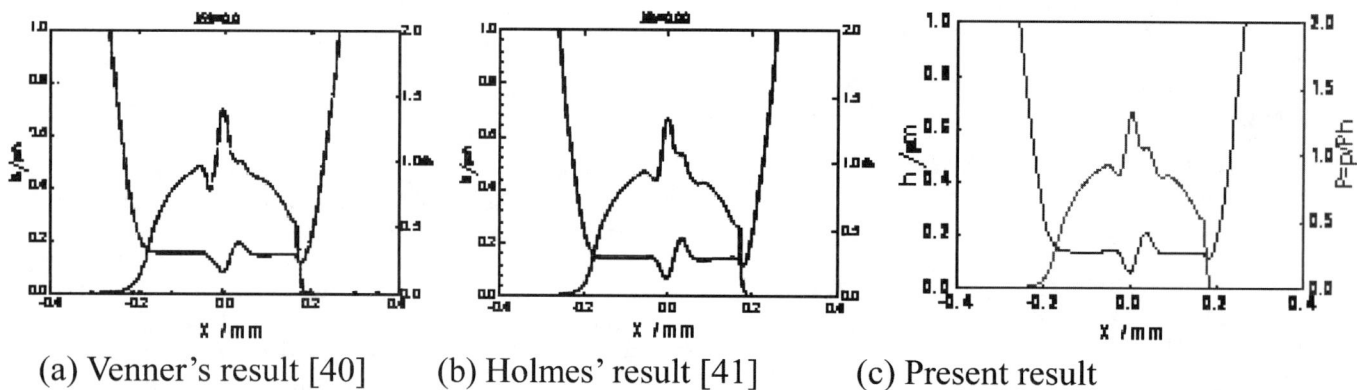

(a) Venner's result [40]   (b) Holmes' result [41]   (c) Present result

Fig. 8—Comparison of film and pressure profiles with Venner and Holmes' results (for a moving transversal ridge, slip=0.0). (a) Venner's result [40]; (b) Holmes' result [41]; and (c) present result.

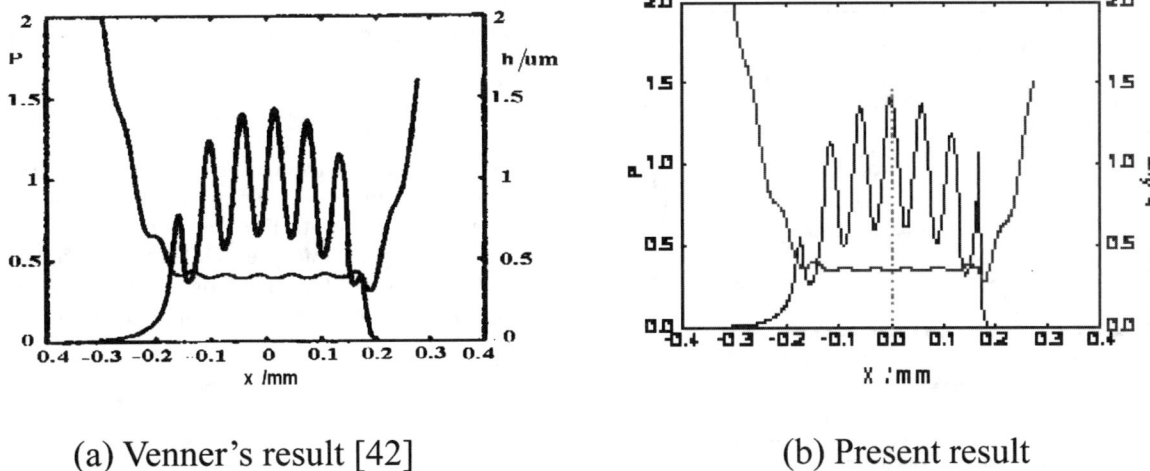

(a) Venner's result [42]   (b) Present result

Fig. 9—Comparison of film thickness and pressure profiles with Venner's results (for transversal waviness, stationary). (a) Venner's result [42] and (b) present result.

tained by other investigators can be observed.

4.1.3 Sinusoidal Surface

Figures 9 and 10 compare the film thickness and pressure profiles from the present model with the results calculated by Venner et al. [42] obtained under the same conditions. The comparison in Fig. 9 is made for the cases of transverse stationary sinusoidal waviness while Fig. 10 is for longitudinal waviness. Again, the results exhibit a good similarity. The good agreements in solutions from different researchers confirm that the numerical schemes implemented in the present model are reliable, at least for the analyses in full-film EHL regime.

4.2 Comparison to the Solutions of Static Dry Contact

Due to lack of deterministic solutions of mixed lubrication in the literatures, it is difficult to conduct a direct result comparison in the mixed lubrication regime. Considering the fact that at the ultra-low speed mixed lubrication would transit to the boundary lubrication or lubricated contact with no effective hydrodynamic films, comparisons of the mixed lubrication solutions at ultra-low speeds with those of static solid contact would help to evaluate the effectiveness of the present mixed-lubrication model. This is based on a reasonable expectation that the pressure predicted by the Reynolds equation over an area with an extremely thin film should approach to the results from the corresponding solid contact, calculated under the same load and same surface geometry.

The comparisons presented in the following are made for several types of surface geometry: the single asperity, sinusoidal waviness, and machined engineering surfaces. All simulations are performed for the steady cases, in which a stationary surface with specified asperities contacts against a moving plane. The solid contact pressure is calculated by means of the conjugate gradient method and the FFT technique [43].

4.2.1 Single Asperity

The first case concerns a single asperity attached to a stationary smooth upper surface in contact with a moving plane. Its

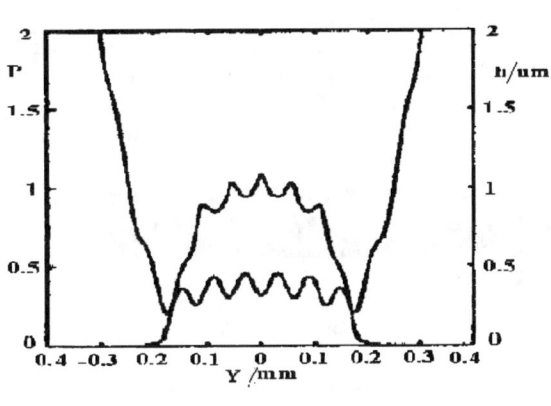

(a) Venner's result [42]

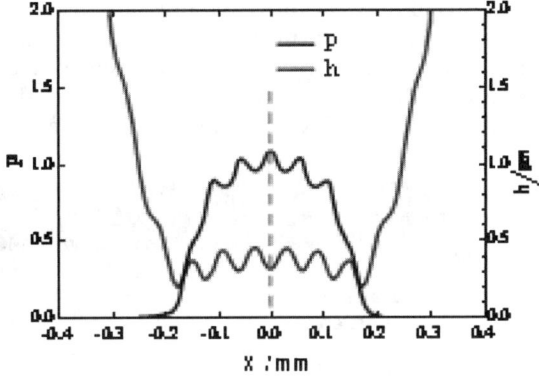

(b) Present result

Fig. 10—Comparison of film thickness and pressure profiles with Venner's results (for longitudinal waviness, stationary). (a) Venner's result [42] and (b) present result.

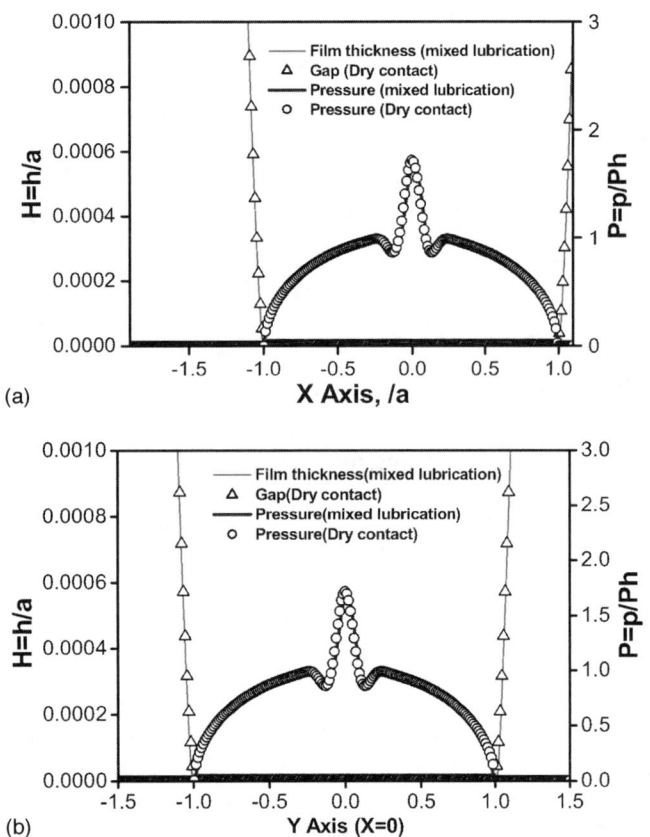

Fig. 11—Pressure and film thickness profiles along x and y directions, in comparison with the corresponding results of dry contact. (a) Pressure and film thickness profiles along the x direction at y = 0. (b) Pressure and film thickness profiles along the y direction at x = 0.

geometric shape is defined in Eq (40), with the height A given as 0.4 μm and width ω as $0.4a$

$$\delta(x,y) = A \times 10^{-10(x^2+y^2)/\omega^2} \cos\left(2\pi\frac{\sqrt{x^2+y^2}}{\omega}\right) \quad (40)$$

When the plane moves at 0.1 mm/s and under a normal load of 38.29 N, the numerical mixed lubrication model yields the solutions of film thickness and pressure distribution shown in Fig. 11 where the profiles are taken along the central lines in the x and y directions, respectively. Corresponding results from dry contact analyses are given in the same figure for comparison. Excellent agreements observed from these plots verify the accuracy of the mixed lubrication model in predicting pressure at the low-speed limit.

4.2.2 Sinusoidal Waviness

Comparisons in this part are made on sinusoidal surfaces with transverse, longitudinal, and isotropic patterns. Geometric shapes for the three types of waviness are specified by Eqs (41)–(43). Amplitude of waviness, A, is set at 0.4 μm for transverse and longitudinal waviness but adjusted to 0.2 mm for the isotropic one, while the wavelength is $\omega_x = \omega_y = 0.5a$ in all cases. The operation conditions used in the simulations of mixed lubrication are summarized in Table 4.

$$\delta(x,y) = A\cos\left(2\pi\frac{x}{\omega_x}\right) \quad (41)$$

$$\delta(x,y) = A\cos\left(2\pi\frac{y}{\omega_y}\right) \quad (42)$$

$$\delta(x,y) = A\cos\left(2\pi\frac{x}{\omega_x}\right)\cos\left(2\pi\frac{y}{\omega_y}\right) \quad (43)$$

The film thickness and pressure distributions for the three cases are given in Figs. 12–14 where the profiles are taken along the central lines in the x and y directions. Excellent agreements are observed in all cases for solutions obtained from the two sources, which convinces once more that the pressure predicted by the mixed-lubrication model at low-speed converge to dry contact solutions. The contour plot of film thickness and 3-D display of pressure distribution are given in Figs. 14(c) and 14(d), respectively. Note that the sinusoidal waves in the contact zone have not been completely flattened under the given load.

4.2.3 Measured Engineering Surfaces

In this section, two typical engineering surfaces, a ground surface and a shaved surface, are employed for the comparison. The original roughness data were measured with an optical profilometer, but the roughness amplitude has been rescaled for the convenience of computation. In simulations, a relatively small load of 50 N is applied to guarantee that asperities will not be completely flattened while other operation conditions are listed in Table 4.

The first example given in Fig. 15 compares the solutions for the ground surface with transverse texture and r.m.s. set as 0.1 μm. It can be seen from Fig. 15(a) that surface roughness causes significant fluctuations in film thick-

TABLE 4—Working conditions for mixed lubrication of ultra-low speed cases.

Surface Geometry		Amplitude (μm)	Wavelength ($\times a$)	Load (N)	Sliding speed (mm/s)
Single asperity		0.4	0.4	38.29	0.1
Sinusoidal waviness	Transversal	0.4	0.5	800	0.00001
	Longitudinal	0.4	0.5	800	0.001
	Isotropic	0.2	0.5	80	0.001
Measured surfaces	Transverse	0.1		50	0.000001
	Isotropic	0.08		50	0.0001

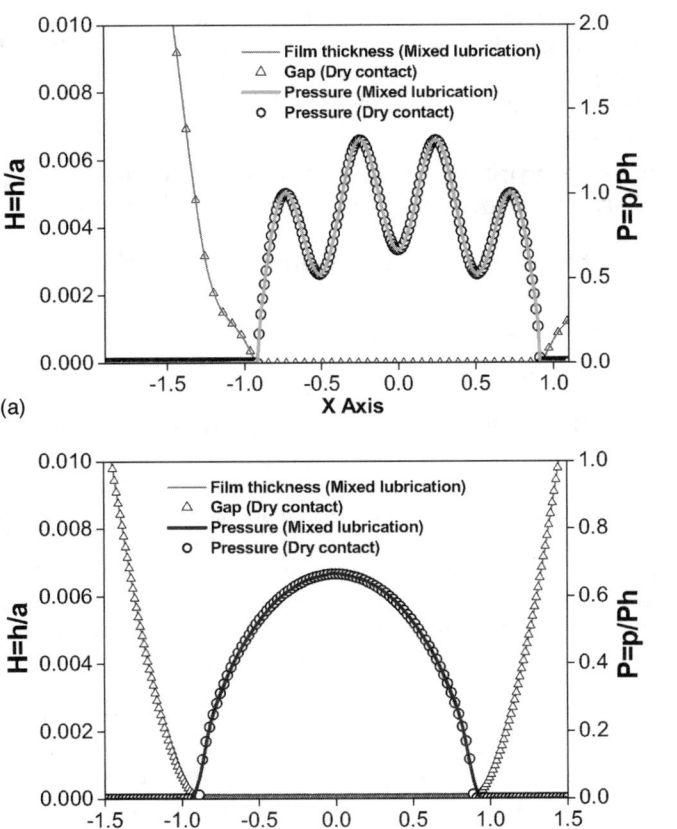

Fig. 12—Comparison between low-speed mixed lubrication and dry contact: transverse waviness. (Profiles are taken alone the central lines in *x* and *y* directions.) (a) Pressure and film thickness profiles along the *x* direction at *y*=0. (b) Pressure and film thickness profiles along the *y* direction at *x*=0.

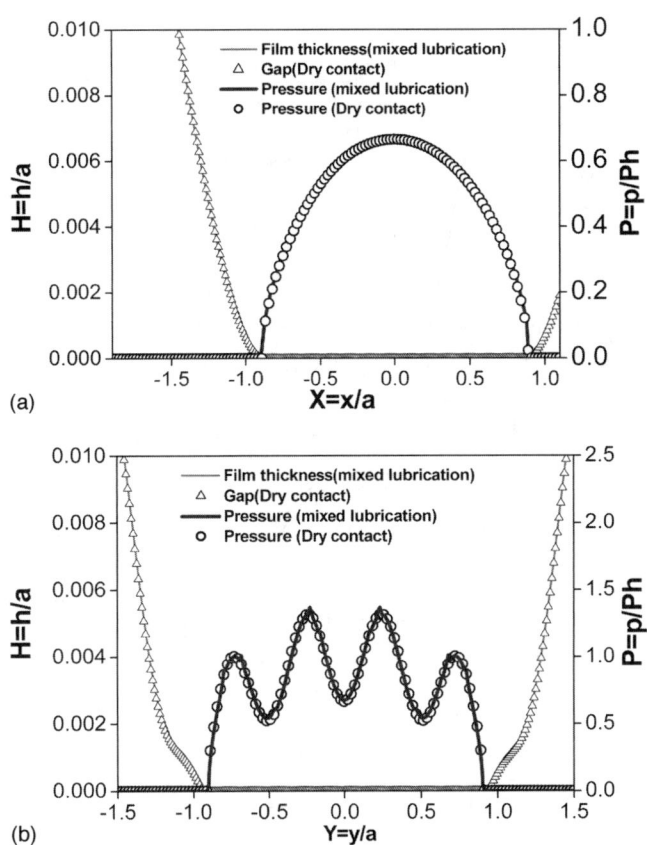

Fig. 13—Comparison between low-speed mixed lubrication and dry contact: longitudinal waviness. (Profiles are taken along the central lines in *x* direction at *y*=0, and in the *y* direction at *x*=0.) (a) Pressure and film thickness profiles along the *x* direction at *y* =0. (b) Pressure and film thickness profiles along the *y* direction at *x*=0.

ness and pressure distributions. The pressure profiles from low-speed mixed lubrication and dry contact generally well agree with each other. In the places where the small valley remains, however, the differences between the two solutions are visible, which is illustrated in the enlarged view in Fig. 15(*b*). This means that if asperities are not completely flattened, a small quantity of lubricant may be entrapped in the valleys and carry a small proportion of the applied load, which is estimated around 3 % in this case. The results for the engineering surface with isotropic pattern (rms = 0.08 μm) are given in Fig. 16, which again illustrate a good agreement between the solutions of mixed lubrication and dry contact. In this case, there is only 1 % of applied load carried by the lubricant confined in the small valleys.

4.3 Comparison to Experiment Results

Due to technical difficulties in measurement, experimental results for the local film thickness and pressure distributions caused by asperities are quite limited that only two experiment cases recently have been reported which are employed here as the comparison reference.

4.3.1 Comparison with Felix's Experimental Results at Full-Film EHL

Felix et al. [44] recently measured the film thickness profiles on a patterned surface with artificially manufactured asperi-

ties as shown in Fig. 17(*a*). The experiments were conducted under full-film EHL conditions for three sliding-to-rolling ratios, −1.0, 0, and 1.0. Figures 17(*b*)–17(*d*) compare the measured results of film thickness with those predicted by simulations of the mixed lubrication. The simulations were performed under the same surface geometry and at the same operating conditions, exactly as described in the experiments, and proper formulations have to be adapted to describe viscosity variations and non-Newtonian effects. It can be seen in Fig. 17 that simulated film thickness profiles accurately match those from the experiments. A significant conclusion from the comparisons is that the numerical simulation cannot only predict the amplitude of film thickness but also the precise film shape over the asperities. This gives further credit to the deterministic approach and present numerical model. The slight difference may result from the shape difference between digitalized asperities used in simulations and that in experiments.

4.3.2 Comparison with Choo's Experimental Results in Thin Films

The model validation in mixed lubrication should be made under the conditions when asperity contacts coexist with lubrication. Choo et al. [45] measured film thickness on the surface distributed with artificial asperities. The experi-

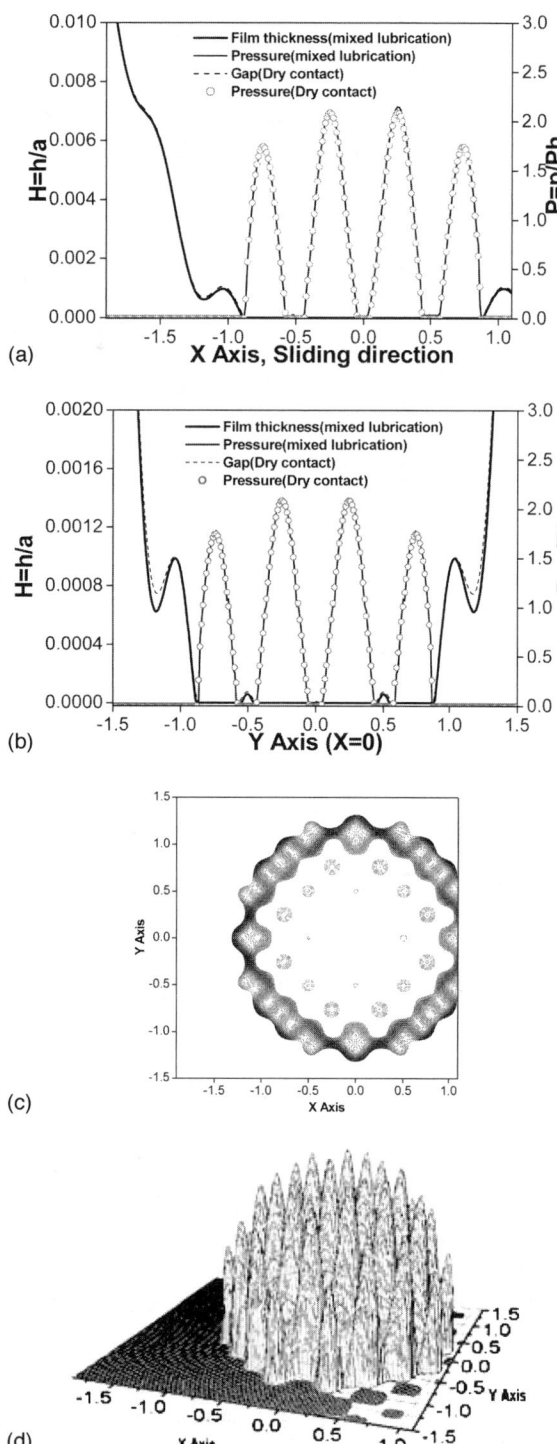

Fig. 14—Comparison between low-speed mixed lubrication and dry contact: isotropic waviness. (Profiles are taken along the central lines in x direction at y=0 and in y direction at x=0.) (a) Pressure and film thickness profiles along the x direction at y=0. (b) Pressure and film thickness profiles along the y direction at x=0. (c) The contour plot of film thickness. (d) 3-D plot of pressure distribution.

ments were conducted under thin film conditions, which provide reference data for the comparisons in the cases when the inter-asperity film thickness is in nanometre scale.

The numerical simulations were performed at the same conditions as in Choo's experiments. It can be found from Fig. 18 that good agreements between the simulation and experiment results are achieved even at the film thickness less than 100 nm.

5 Performance of Mixed Lubrication—Numerical and Experimental Studies

5.1 Characterization of Roughness and Effects on Lubrication

5.1.1 Generation of Rough Surfaces

The first step to investigate the roughness effects on lubrication is to get digitized roughness data to be used as the input in numerical solutions. There are several ways to define a rough surface: to superpose asperities in regular shapes, such as spherical or sinusoidal, onto a smooth surface, to use the measured data of a real rough surface, obtained, for instance, from an optical instrument, or to generate a rough surface through computer simulation, which presents a property similar to that of a engineering surface. This section gives a brief description of the methods proposed by the present authors to generate 3-D random surfaces with specified statistic features.

If a surface with random roughness is described mathematically as a stochastic process, the statistical properties of the surface are mainly defined by two functions, namely the height distribution and auto-correlation function (ACF) [46]. Most engineering surfaces exhibit a Gaussian distribution, but they may become non-Gaussian as a result of surface finishing or wear. For non-Gaussian rough surfaces, two additional parameters, the skewness (S_k) and kurtosis (K), are needed to characterize the surface. The skewness describes the symmetry of the height distribution and the kurtosis defines the distribution range. On a Gaussian surface, for example, the two parameters take the values of $S_k = 0$ and $K = 3$.

Autocorrelation function provides the information as for how the random heights are correlated to each other. The ACF in two-dimensional form is defined as

$$R(x,y) = \iint_\Omega z(\xi,\psi) \cdot z(\xi+x, \psi+y) d\xi d\psi \quad (44)$$

where $z(\xi,\psi)$ is the height coordinate of the roughness. Usually, the autocorrelation function is found to be in an exponential form,

$$R(x,y) = \sigma^2 e^{-2.3[(x/\beta_x)^2 + (y/\beta_y)^2]^{1/2}} \quad (45)$$

where β_x and β_y define the correlation length in the x and y directions. The correlation length defines a distance at which the ACF has decreased to 0.1 of the original value. The Fourier transform of ACF is defined as the spectral density function, which is an important concept to the surface generation.

Consequently, the surface generation becomes a question of producing a matrix of random data that obey the defined height distribution and a prescribed ACF. This can be carried out efficiently through the procedure of a 2-D digital filter, proposed by Hu and Tonder [47], as summarized as follows.

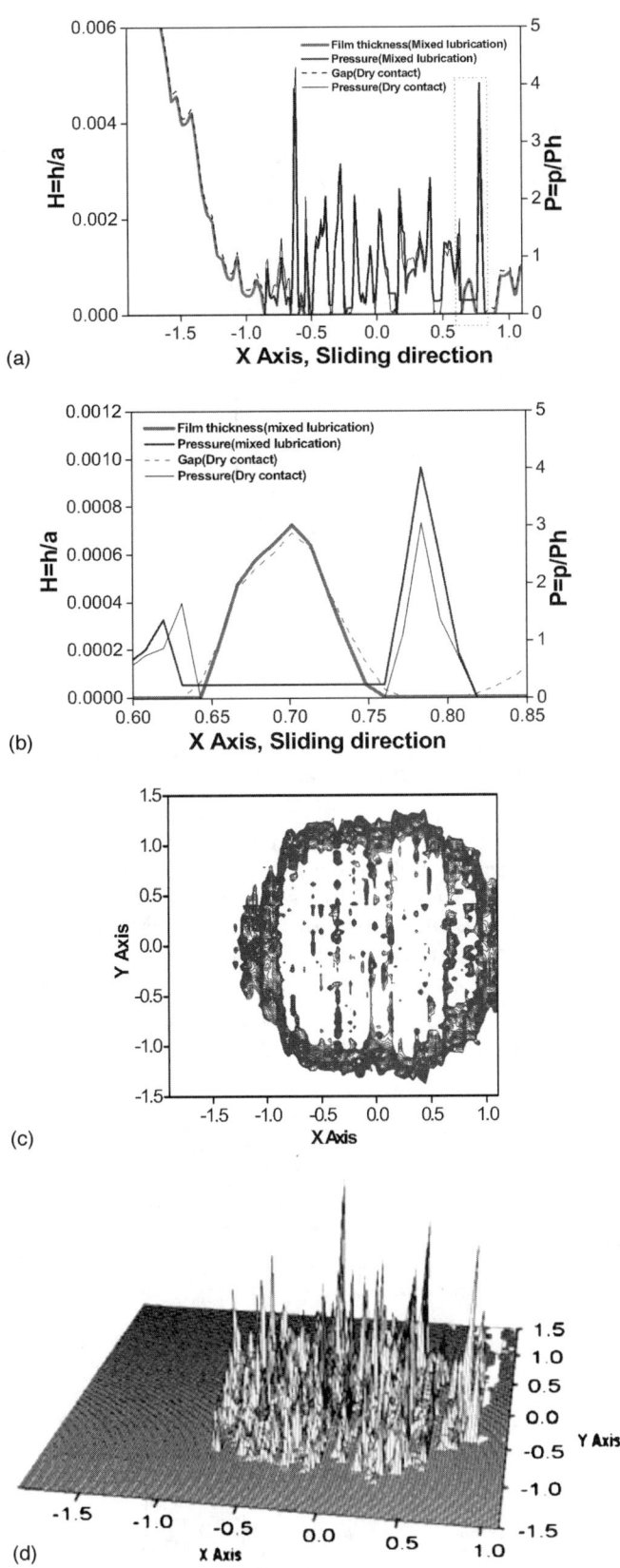

Fig. 15—Comparison between low-speed mixed lubrication and dry contact: a ground surface with transverse texture. (a) Profiles of pressure and film thickness along the x direction at y=0. (b) An enlarged view for the marked zone in 15(a). (c) The contour plot of film thickness. (d) 3-D plot of pressure distribution.

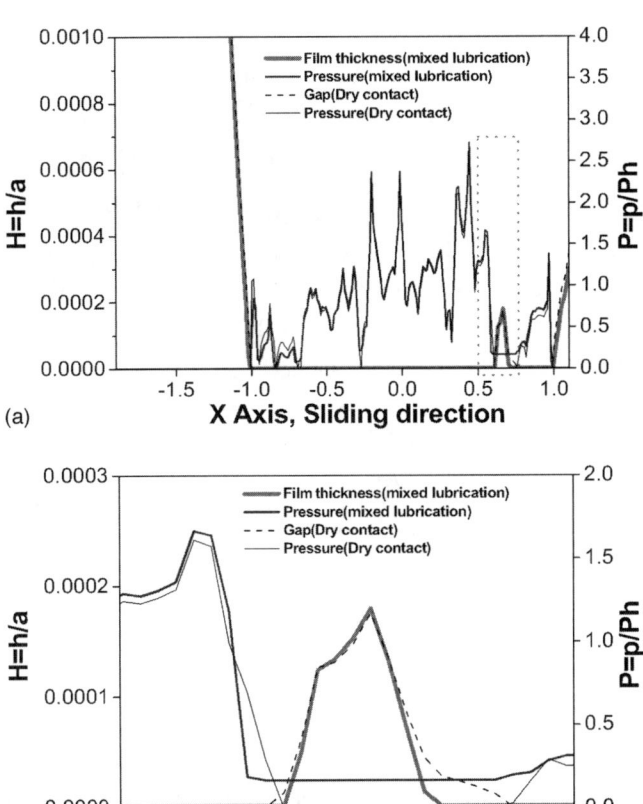

Fig. 16—Comparison between low-speed mixed lubrication and dry contact: an engineering surface with isotropic roughness. (a) Profiles of pressure and film thickness along the x direction at y=0. (b) An enlarged view for the marked zone in Fig. 16(a).

Suppose $\eta(i,j)$ is a sequence of random numbers produced by computer generation, which takes a Gaussian distribution of a known standard deviation σ, the target roughness heights $z(i,j)$ that obeys a given ACF can be written as

$$z(i,j) = \sum_{k=0}^{n-1} \sum_{l=0}^{m-1} h(k,l)\eta(i+k,j+l) \quad (46)$$

where $h(i,j)$ is the coefficient, called the FIR filter, which modifies the correlation function of the system. According to signal process theory, $z(i,j)$ will possess the same height distribution as the input sequence $\eta(i,j)$. Equation (46) can be transformed to the frequency domain as

$$\hat{z}(i,j) = \sum_{i=0}^{n-1} \sum_{j=0}^{m-1} H(i,j)\hat{\eta}(i,j) \quad (47)$$

where $\hat{z}(i,j)$, $H(i,j)$, and $\hat{\eta}(i,j)$ are the Fourier transforms of $z(i,j)$, $h(i,j)$, and $\eta(i,j)$, respectively. $H(i,j)$ is also called the transfer function of system, and can be determined by the following relation [48]

$$|H(i,j)|^2 = \frac{S_z(i,j)}{S_\eta(i,j)} \quad (48)$$

where $S_\eta(i,j)$ denotes the spectral density of input sequence $\eta(i,j)$, which gives a constant value for a random sequence in

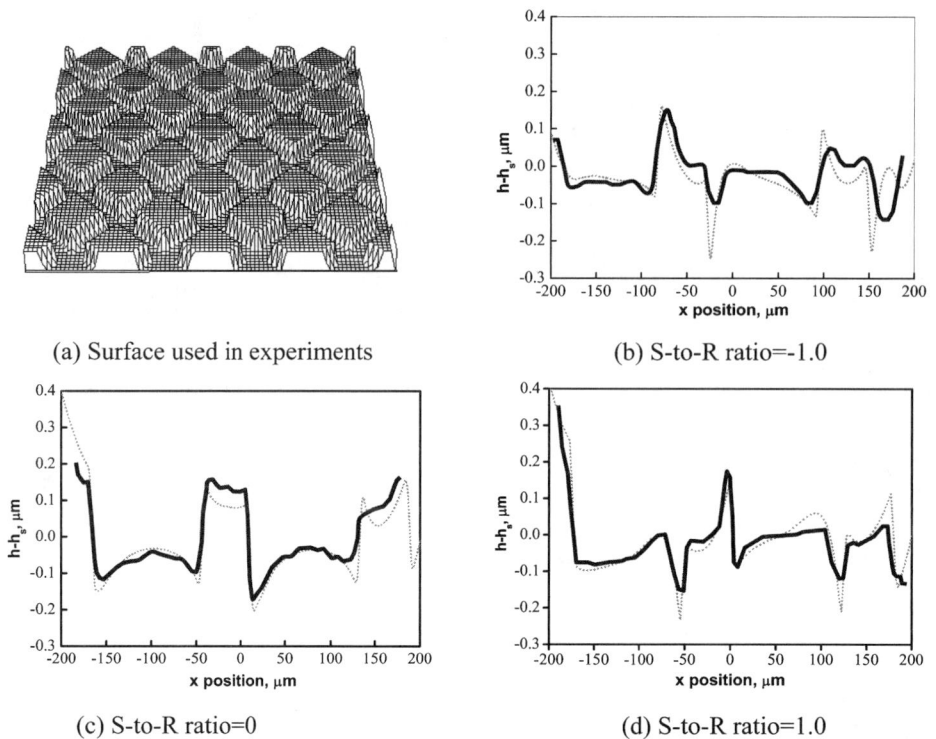

Fig. 17—Comparison between numerical results and experiments by Felix [44], thick line is Felix's experimental results and thin line is present results. (a) Surface used in experiments; (b) S-to-R ratio=−1.0; (c) S-to-R ratio=0; and (d) S-to-R ratio=1.0.

white noise type. $S_z(i,j)$ is the spectral density of output sequence $z(i,j)$, i.e., the Fourier transform of the expected ACF, $R(i,j)$. Thus, the filter coefficients $h(i,j)$ can be obtained by applying inverse Fourier transforms to $H(i,j)$.

When a non-Gaussian rough surface is expected, the Gaussian input sequence $\eta(i,j)$ should be transformed first to another sequence $\eta'(i,j)$ with appropriate skewness (SK') and kurtosis (K') using the Johnson translator system of distribution, then let $\eta'(i,j)$ pass through the filter to obtain the output sequence $z(i,j)$ which possesses the specified autocor-

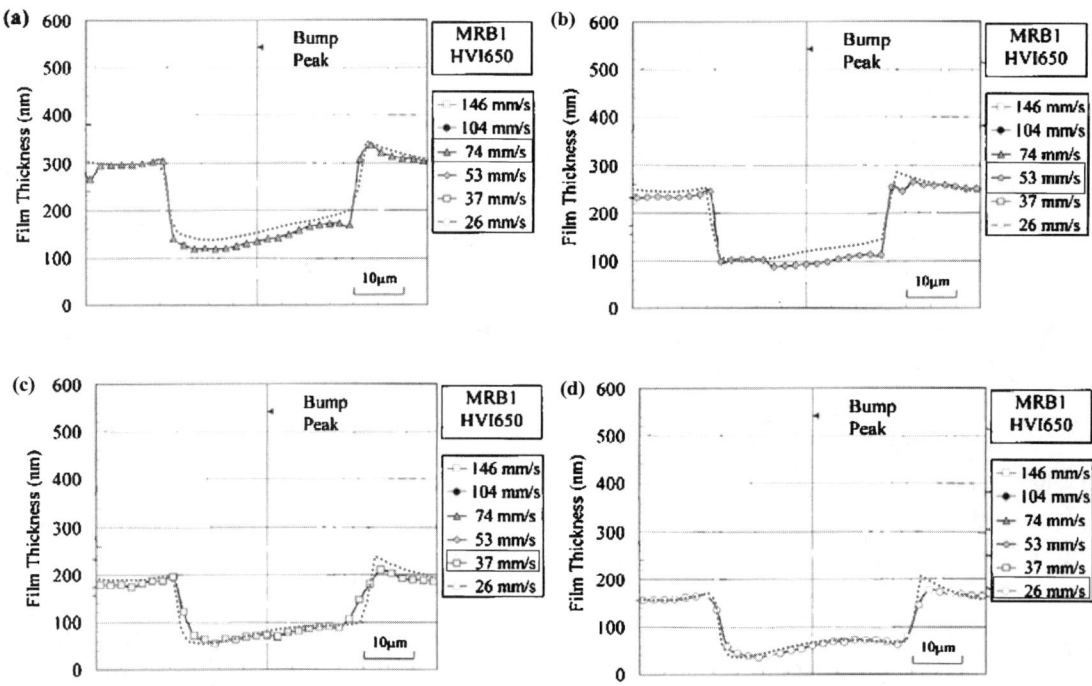

Fig. 18—Comparison between numerical results and experiments by Choo et al. [45]. Line and symbol for experiment results and thin line for present results. (a) 74 mm/s; (b) 53 mm/s; (c) 37 mm/s; and (d) 26 mm/s.

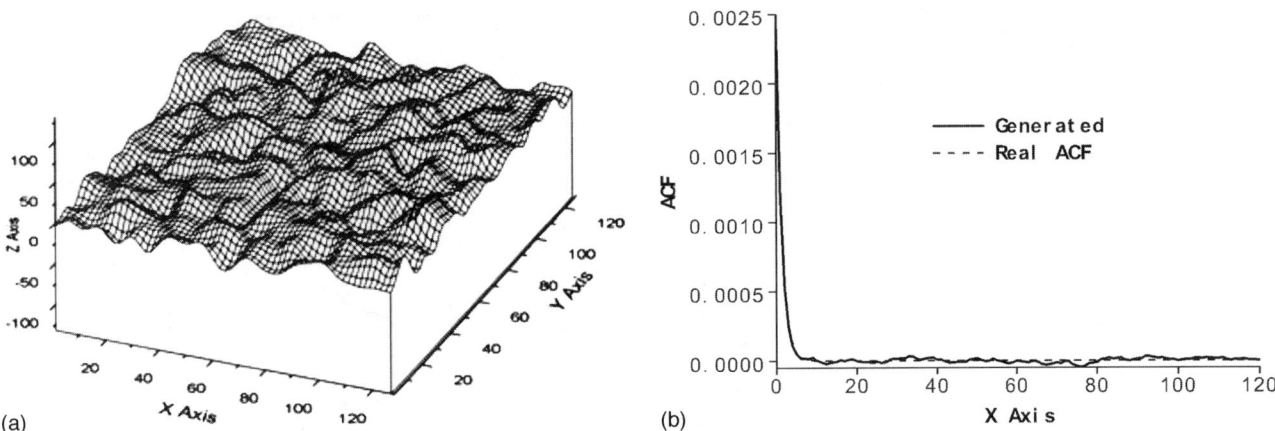

Fig. 19—A generated rough surface with Gaussian distribution and exponential ACF. (a) 3-D display of generated Gaussian surface and (b) comparison of ACF.

relation function, skewness, and kurtosis. The skewness (SK') and kurtosis (K') for the transformed input sequence $\eta'(i,j)$ can be obtained by the following formulations

$$SK_z = SK' \sum_{i=1}^{q} \theta_i^3 \bigg/ \left(\sum_{i=1}^{q} \theta_i^2 \right)^{3/2} \quad (49)$$

$$K_z = \left[K' \sum_{i=1}^{q} \theta_i^4 + 6 \sum_{i=1}^{q-1} \sum_{j=i+1}^{q} \theta_i^2 \theta_j^2 \right] \bigg/ \left(\sum_{i=1}^{q} \theta_i^2 \right)^2 \quad (50)$$

where SK_z and K_z are the prescribed skewness and kurtosis of the surface to be generated.

Figure 19(a) shows a generated Gaussian rough surface generated through this approach. The comparison of ACF of the generated surface to its specified values given in Fig. 19(b) shows a good agreement.

Figure 20 shows a 3-D view of a generated non-Gaussian rough surface with an exponential autocorrelation and desired skewness and kurtosis of −1.75 and 5.0, respectively. The surface shows an outlook of a typical worn surface due to the negative skewness. The real values of SK and K were calculated as −1.7827 and 5.1104, a good agreement between specified and real values.

5.1.2 Effects of Height Distribution on Mixed Lubrication

To examine the effects of height distribution on mixed lubrication, rough surfaces with the same exponential autocorrelation function but different combinations of skewness and kurtosis have been generated, following the procedure described in the previous section. Simulations were performed for the point contact problem with geometric parameters of $R_x = R_y = 19.05$ mm. In all the cases analyzed, the applied load is fixed at $w = 800$ N, rolling speed $u = 0.625$ m/s, the slide-to-roll ratio $S = 2.0$. The material properties are $E' = 219.78$ GPa, $\eta_0 = 0.096$ Pa·s, and $\alpha = 1.82 \times 10^{-8}$ m^2/N.

The results from the simulations for rough surface with r.m.s fixed at 0.1 μm are given in Fig. 21 where the lubrication parameters, such as the area ratio (real contact area divided by nominal contact area), load ratio (asperity shared load divided by total load), maximum pressure, maximum surface temperature, and average film thickness, are plotted as a function of kurtosis. It can be seen that the area and load ratios rise quickly with the increasing kurtosis, and so does the maximum pressure. Meanwhile the maximum temperature and average film thickness are relatively insensitive to the kurtosis. The phenomena can be understood that the increase of kurtosis makes rough surface "peakier" so that contact will initiate at higher peaks as a smooth plane normally approaches the rough surface, resulting in an increase in real contact area while the average film thickness remains largely unchanged as shown in Fig. 21(e). Since the change in the kurtosis would modify the local film thickness and pressure distribution, the effects of kurtosis on lubrication are different from those from dry contact studied in Ref. [49] where the maximum pressure was reported decreasing with the increase of kurtosis.

Figure 22 shows the area ratio and average film thickness as a function of skewness. Similar trends to those shown in Fig. 21 are observed. When a smooth surface normally approaches a rough surface, a surface with positive skewness will be much more engaged in contact than the one with negative skewness. Since the average film thickness remains almost the constant, as demonstrated in Fig. 22(b), the real contact area will increase with skewness.

In summary, the height distribution of surface roughness, characterized by the skewness and kurtosis, may present a significant influence on certain performances of mixed lubrication, such as the real contact area, the load carried by asperities, and pressure distribution, while the average film thickness and surface temperature are relatively unaffected.

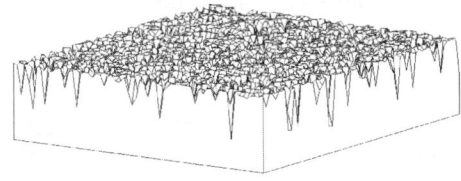

Fig. 20—Generated non-Gaussian rough surface ($Sk = -1.75$, $K = 5.0$).

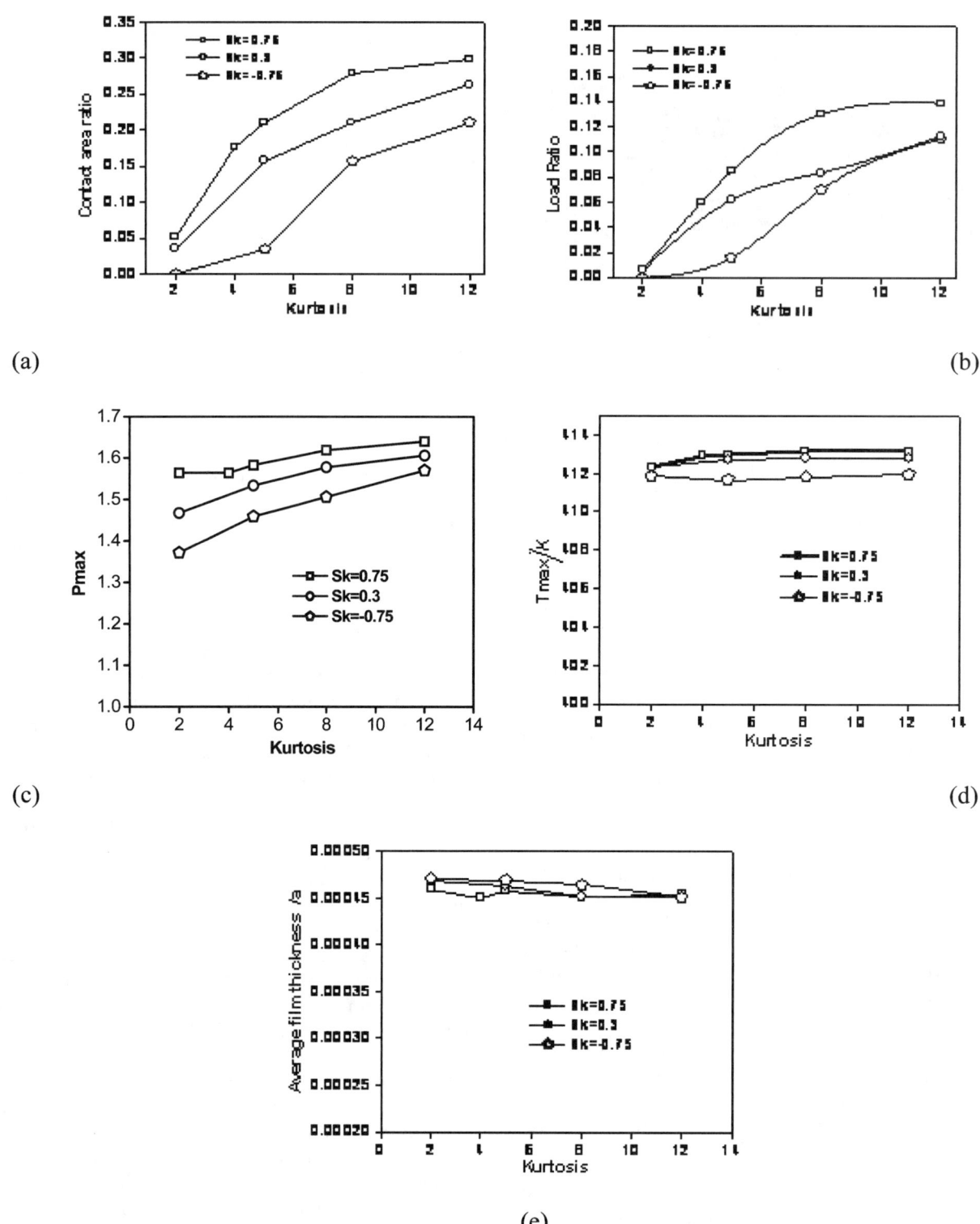

Fig. 21—The changes of lubrication properties with the kurtosis, simulated at $\sigma=0.1$ μm and different values of skewness. (a) Contact area ratio versus skewness. (b) Load ratio versus skewness. (c) Maximum pressure versus kurtosis. (d) Maximum temperature versus kurtosis. (e) Average film thickness versus kurtosis.

5.2 Asperity Deformations

Both numerical analyses and experiment studies show that the roughness experiences significant deformation in contact zone, which will certainly make a strong impact on the properties of mixed lubrication, but the behavior of the asperity deformation and its role in lubrication is worth a further discussion.

It has been generally recognized that asperities are flattened by contact or hydrodynamic pressure, and the extent of the deformation depends on the load, speed, and slide-to-

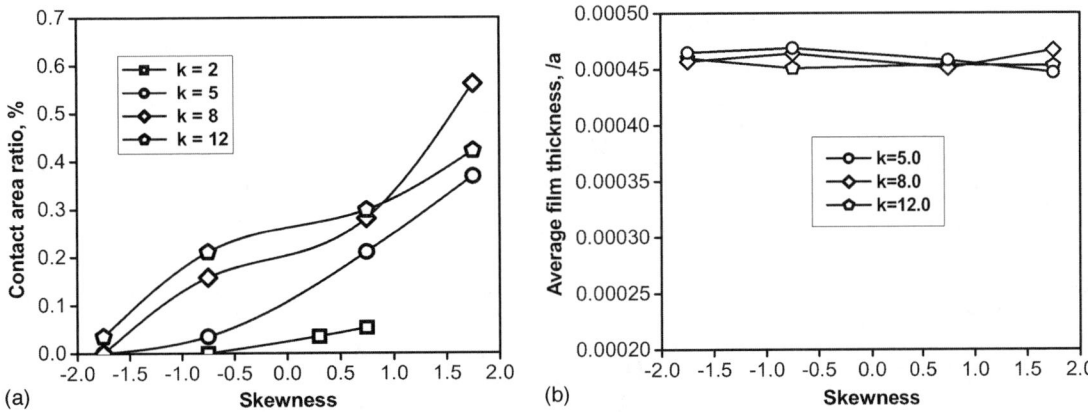

Fig. 22—The change of contact area ratio and average film thickness with skewness, simulated at $\sigma=0.1$ μm and for different kurtosis. (a) Area ratio versus skewness and (b) average film thickness versus skewness.

roll ratio. Theoretical and experimental efforts have been made in order to understand the interplays among the contact, lubrication, and asperity deformation. A widely adopted strategy is to examine the lubrication performance of asperities in a well-defined patterns and shapes, such as single or uniformly distributed sphere, or sinusoidal waviness.

In the experiments based on optical interferometer, Kaneta et al. [50] measured the film thickness variation and asperity deformation as an artificial circular bump or latticed asperities pass through the contact zone. They found that in the case of the circular bump, slide-to-roll ratio plays a very important role for the effects of surface roughness on EHL film, as displayed in Fig. 23. When slide-roll ratio is positive which means the fluid in the EHL conjunction goes faster than asperities, there is a deformed pocket downstream the bump in addition to the flattened bump itself while the upstream film is little interfered (Fig. 23(a)). On the other hand, when slide-roll ratio is negative the deformed pocket is found at upstream while the downstream film is not affected, and the bump is flattened, too. For a transversely oriented ridge in pure sliding condition, when the ridge is located at the entrance of the contact, there is a reduction of film thickness in the whole downstream area of the bump. When the ridge goes to the center of contact area, the ridge is flattened but the overall film shape is almost the same as that of the smooth surface.

In the experiments of pure rolling cases, Guangteng [51], Ehret [52], Felix [53] reported that the surface feature of a defect deformed prior to entering the high-pressure region of the contact. After passing through the inlet zone, the deformed feature retained almost unchanged as it went throughout the high-pressure region of the contact. They also found a locally enhanced region of lubricant film formed in front of the defect when entering the inlet zone, and traveled at the same speed as the feature during its passage through the high pressure conjunction. As the entraining speed increases, the normal deformation of the defect decreases due to a general growth in film thickness.

When sliding is involved, the behavior of asperity deformation depends strongly on the slide-to-roll ratio, mainly due to the interactions between the fluid and asperities traveling in different speeds [53]. For a positive slide-roll ratio when a ridge moves faster than the lubricant, the ridge becomes flattened as it progresses through the high-pressure region. A constriction underneath the ridge is formed at its trail edge when it is at the inlet of the contact. The negative slide-roll ratio, however, introduces very different deformation behavior. A constriction underneath the ridge is also formed but at the leading edge. The ridge is initially deformed upon its entry to the contact and, it gradually undergoes further deformation as it proceeds through the high-pressure region.

In Felix's experiments conducted on the surfaces prepared with arrayed defects and in lubricated contact [54], the dependence of deformation on the slide-to-roll ratio was also observed, but in a more complex mode, as presented in Fig. 24. In pure rolling, the defects deform and entrap lubricant underneath during their passage through the inlet zone. For the same reason as discussed above in single defect case, the

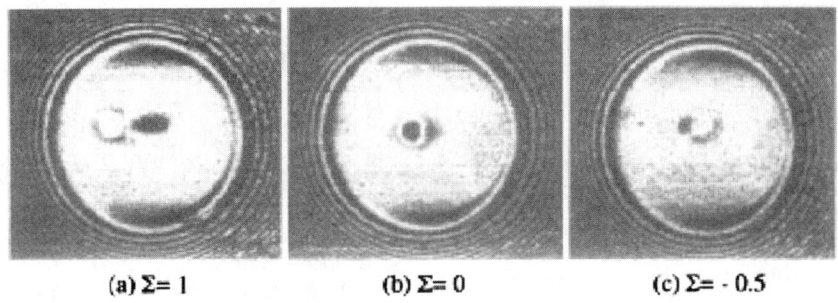

(a) $\Sigma = 1$ (b) $\Sigma = 0$ (c) $\Sigma = -0.5$

Fig. 23—Interferograms with a circular bump passing through the contact area with $u = 9 \times 10^{-12}$ (fluid flows from left to right) [50].

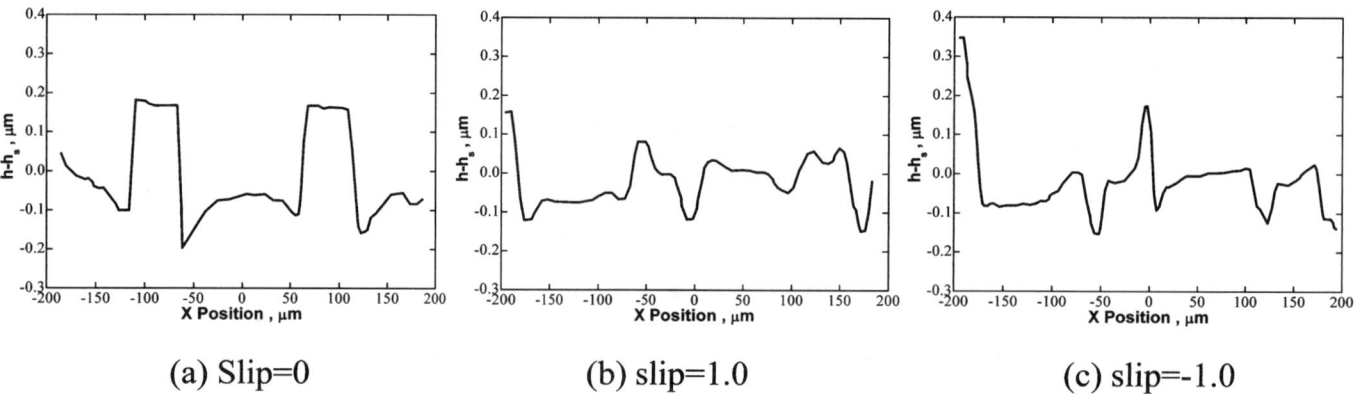

(a) Slip=0   (b) slip=1.0   (c) slip=-1.0

Fig. 24—Measured film thickness perturbation for square features passing through the contact area under different sliding condition, taken from Felix [54]. (a) Slip=0; (b) Slip=1.0; and (c) Slip=−1.0.

introduction of sliding leads to more deformations than in the pure rolling case, to the extent that the defects are almost completely flattened. The defects behave similarly to the pure rolling case when entering the inlet zone, but they sustain additional deformations in the high-pressure region as they meet the highly viscous lubricant inside the contact region, which is either moving faster or slower than the defect depending on the sliding conditions.

In parallel with experimental studies, numerical and theoretical studies are also conducted to explain the experimental observations. A two-wave mechanism for the roughness effect on lubrication, for example, was proposed by Greenwood et al. [55,56]. First they expected that the original surface roughness would tend to be flattened under contact in the center of the conjunction, except for the case of pure rolling. Second, the model assumed that the roughness, as entering the conjunction, would produce a clearance ripple that traveled through the contact at the entrainment velocity, thus the film thickness variation would consist of two components: an attenuated form of the original roughness and a traveling ripple generated at the inlet, as schematically shown in Fig. 25. The wavelength and amplitude of the ripple are found to be affected by the speed difference between lubricant and asperities, which leads to a film shape depending on the slide-to-roll ratios. More discussions about the mechanism can be found in Refs. [55,56].

Numerical simulations by Hooke and Venner et al. [57,58] for small amplitude wavy surfaces show that the amplitude attenuation of waviness in an EHL contact depends on the operating conditions and the orientation of the roughness pattern. They found that the attenuation appeared to be governed by a single nondimensional parameter: the ratio of the wavelength to the length of the inlet pressure sweep at the entrance to the conjunction, it can be evaluated by

$$\frac{\lambda}{L} = \frac{1}{\bar{L}}\left(\frac{128}{27}\right)^{0.25} q \frac{\lambda}{a} P^{1.5} S^{-2} \quad (51)$$

where λ is wavelength of roughness, L denotes the length of the inlet pressure sweep, $P = \alpha p_H$ and $S = GU^{0.25}$ define the Greenwood load and speed parameter, respectively, a is semiwidth of the Hertzian contact in the direction of entrainment, $\bar{L}$ remains constant for all contacts and all operating conditions within the elastic piezoviscous regime, and q is a coefficient relating the Hertzian contact width and pressure, $a/R_x = 4q^2 P_H/E'$.

From Fig. 26 [59], it can be found that for short wavelengths the roughness remains unchanged when passing through the contact; for long wavelengths it is almost completely flattened; for a fixed wavelength, the amplitude reduction is largest when the roughness lies in the direction of entrainment.

For the irregular surface roughness, however, the deformation in EHL contact is much more complex as shown in experiments. Full numerical simulations are necessary to explore the behavior of asperity deformations in more realistic conditions, and then incorporate the information into the design to optimize the lubrication conditions of engineering surfaces. An alternative proposed by Venner et al. [60] is to combine the "master" curves via a FFT-based approach. Using FFTs, the roughness is decomposed into a set of sinusoidal waves, whose amplitude is modified according to the "master" curves before applying the inverse Fourier transformation to obtain the deformed roughness. However, the

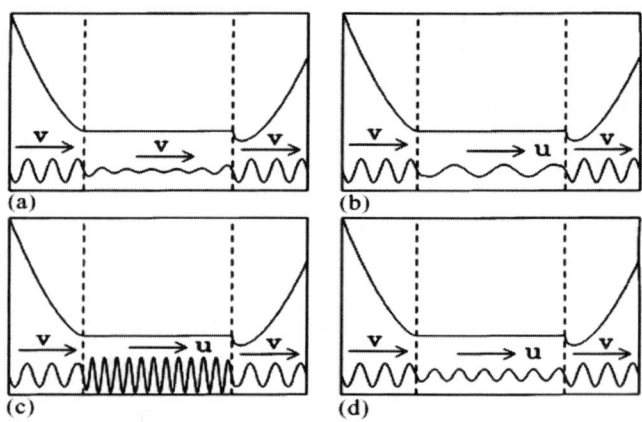

Fig. 25—Idealized behavior of sinusoidal surface roughness under a heavily loaded contact, taken from Hooke [56]. (a) Attenuation of the original roughness due to relative siding of surfaces. (b) Replacement of the original roughness by longer wavelength clearance variation; S=v/u=0.5. (c) Replacement of the original roughness by shorter wavelength clearance variation; S=v/u=2. (d) Modification of the original roughness at the inlet in pure rolling contacts; S=v/u=1.

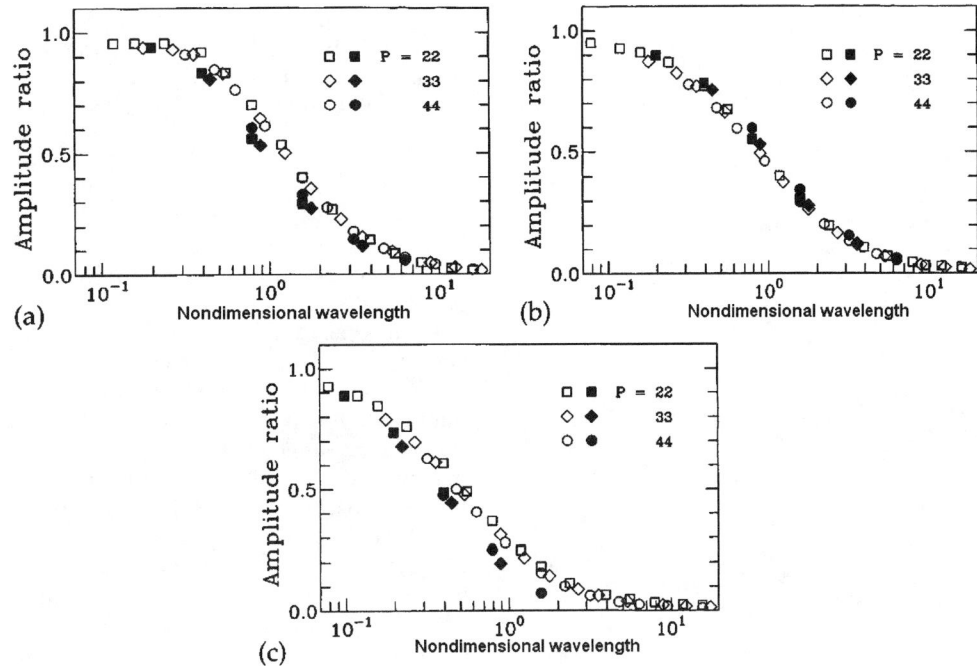

Fig. 26—Amplitude ratio for line contacts (□, ◇, ○) and point contacts (■, ◆, ●), taken from Ref. [59]. (a) Transversal roughness; (b) isotropic roughness; and (c) longitudinal roughness.

accuracy of this FFT-based approach needs to be further improved.

5.3 Transition of Lubrication Regime—A Study Based on the DML Model

The transition of lubrication regimes, from full-film EHL to boundary lubrication, is a subject of great importance for theoretical understanding of lubrication failure, and for tribological design in engineering practices. The DML model proposed by the present authors provides a powerful tool to detect the tribological events occurring during the process of transition.

The process of transition from hydrodynamic to boundary lubrication can be described qualitatively by plotting the measured friction coefficients against film thickness, which depends on the operational conditions, such as load, sliding velocity and lubricant viscosity. A typical diagram known as the "Stribeck Curve" is schematically shown in Fig. 27, in which the friction coefficients are given as a function of λ,

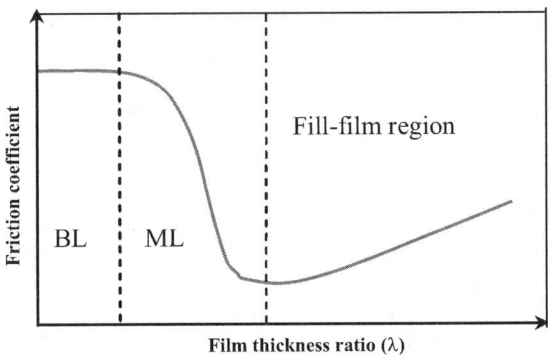

Fig. 27—A schematic Stribeck curve.

defined as $\lambda = h/\sigma$, the ratio of film thickness to the rms of surface roughness. The value of λ has been widely accepted as the criterion of lubrication regime, and $\lambda = 3$ is usually regarded as an indication for the occurrence of asperity contact since the roughness on most engineering surfaces exhibits a Gaussian distribution that more than 99 % of asperities are in the range of $(-3\sigma, 3\sigma)$.

In the range of $\lambda > 3$, friction declines linearly with the decreasing film thickness. This is a regime of full-film hydrodynamic lubrication where two surfaces are separated by liquid films, as sketched in Figs. 28(a) and 28(b), and the friction is caused by shearing lubricant which follows the viscous law. As the λ value decreases and approaches 3, roughness effects prevail and asperities start to interfere with each other, as illustrated by Figs. 28(c) and 28(d). If the asperity contact occurs occasionally and carries only a small fraction of load, friction would remain in the decline track but the linear relation can no longer sustain. At a certain point, $\lambda \sim 3$, friction reaches a minimum and then goes up as a result of increase in asperity contact areas where the friction coefficients are higher than viscous shear stress. As film thickness further decreases to the range where asperities carry a most part of applied load [see Fig. 28(e)], the friction becomes stabilized and independent of film thickness. This is the regime of boundary lubrication [61].

The Stribeck curve gives a general description for the transition of lubrication regime, but the quantitative information, such as the variations of real contact areas, the percentage of the load carried by contact, and changes in friction behavior, are not available due to lack of numerical tools for prediction. The deterministic ML model provides an opportunity to explore the entire process of transition from full-film EHL to boundary lubrication, as demonstrated by the examples presented in this section.

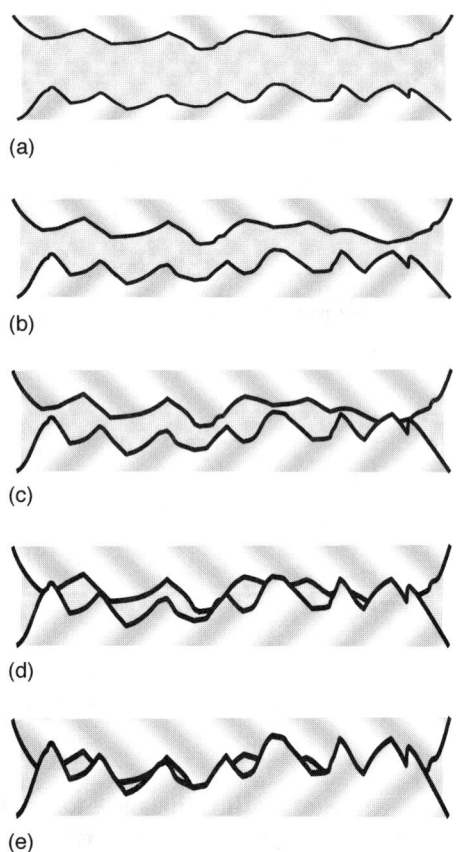

Fig. 28—Different stages in transition of lubrication regimes. (a) Full-film lubrication with film thickness much larger than roughness ($h/\sigma > 1$). (b) Surfaces are separated but roughness effect becomes significant ($5 > h/\sigma > 3$). (c) Asperities interfere with each other but hydrodynamic films carry the most load ($h/\sigma \approx 3$). (d) Typical mixed lubrication with load shared by lubrication and asperity ($h/\sigma < 3$). (e) Boundary lubrication when asperities carry the most part of load ($h/\sigma < 0.5$).

5.3.1 Smooth Surface

Take into consideration a lubricated system composed of a smooth steel ball and a rigid plane. As the entraining speed decreases under a fixed load, the changes of film thickness, pressure distribution, and contact area ratio with speed are presented in Figs. 29 and 30.

It can be seen from Fig. 29, that at high speed, the two surfaces are fully separated by a fluid film, which exhibits typical EHL characteristics, such as horseshoe shape for film thickness distribution, and the second pressure spike at the outlet, etc. As the speed decreases, the film thickness drops quickly but the surfaces are still separated by lubricant film in the range of $v > 0.01$ m/s. In this process, there is no significant change in the pressure distribution, except that the amplitude of the second pressure spike gradually disappears. Continuous reduction of entraining speed would finally lead to films that are molecularly thin when the two surfaces close to a distance of molecular dimension, which is defined in this chapter as "contact." The contact first appears at two sides of nominal Hertz contact zone, and spreads toward the center of Hertz contact zone, and finally covers the whole Hertz contact zone. Figure 30 shows that the average film thickness decreases with the declining entraining speed, following a power law proposed by Hamrock and Dowson [62] until the contact occurs. After that the average film thickness deviates from the Hamrock-Dowson formula and drops quickly, accompanied by a rapid growth of contact area.

5.3.2 Rough Surface

In mixed lubrication, the involvement of surface roughness will induce significant fluctuations in film thickness and pressure distribution, but the general trend of the changes in film thickness and pressure distribution is similar to that of smooth surface, except that "contact" occurs at a much higher speed in comparison with the smooth case. This implies that for rough surfaces, mixed lubrication extends over a very wide range of operation conditions, and the interacting surfaces are not completely separated by hydrodynamic films in most cases unless the speed is very high.

A steel ball with transversely oriented roughness in contact with a rigid plane is analyzed in this part of work to show its performance in mixed lubrication. The three-dimensional roughness was measured using an optical profilometer, which gives average roughness height $R_a = 455$ nm. A 3-D view of the measured surface is shown in Fig. 31.

The digitized roughness data were recorded and used as the input of numerical solutions of mixed lubrication. The simulations were performed for simply sliding, i.e., a smooth surface moving over a stationary roughness. The results are presented in Fig. 32 for a wide range of λ ratios when the sliding speed changes from 10 m/s down to 0.001 m/s, in order to show the transition of lubrication regime from the full-film to the boundary lubrication.

It can been observed in Fig. 32(a) that when the sliding speed is 10 m/s, the lubricant film is thick enough to separate two mating surfaces. Figure 32(b) shows the second case in which the speed decreases to 1 m/s and the film thickness is greatly reduced but the system remains in the full-film lubrication region. The roughness causes significant fluctuations in pressure distributions, especially at low λ values, meanwhile the tops of the asperities are flattened due to elastic deformation. As the sliding speed further decreases, more and more asperities are engaged in contact where the films are molecularly thin, and the contact areas marked by the shadow in Fig. 32 grow with the declining speed. When the rolling speed becomes very low, below 0.1 mm/s, for example, the hydrodynamic function is insignificant, and the load is mainly supported by the asperity contacts, which is practically in a regime of boundary lubrication. Further reduction of speed does not make any significant contribution to the film thickness, pressure distribution, and contact areas, as demonstrated by Figs. 32(d)–32(f). The average film thickness and asperity contact ratio obtained from the numerical simulations are plotted in Fig. 33 as a function of rolling speed. The average film thickness h_a drops quickly as the speed decreases, and exhibits a linear decline in log scale for the speed range from 10 m/s to 0.25 m/s, indicating that the Hamrock-Dowson formula holds in rough surface EHL. At the same time, contact area rapidly increases, from 0 to 33 %, which means that hydrodynamic effect still prevails in this stage. When the speed is further reduced from 0.25 m/s down to 0.001 m/s, the film thickness curve levels off and the contact area ratio approaches slowly to a limit value. Since

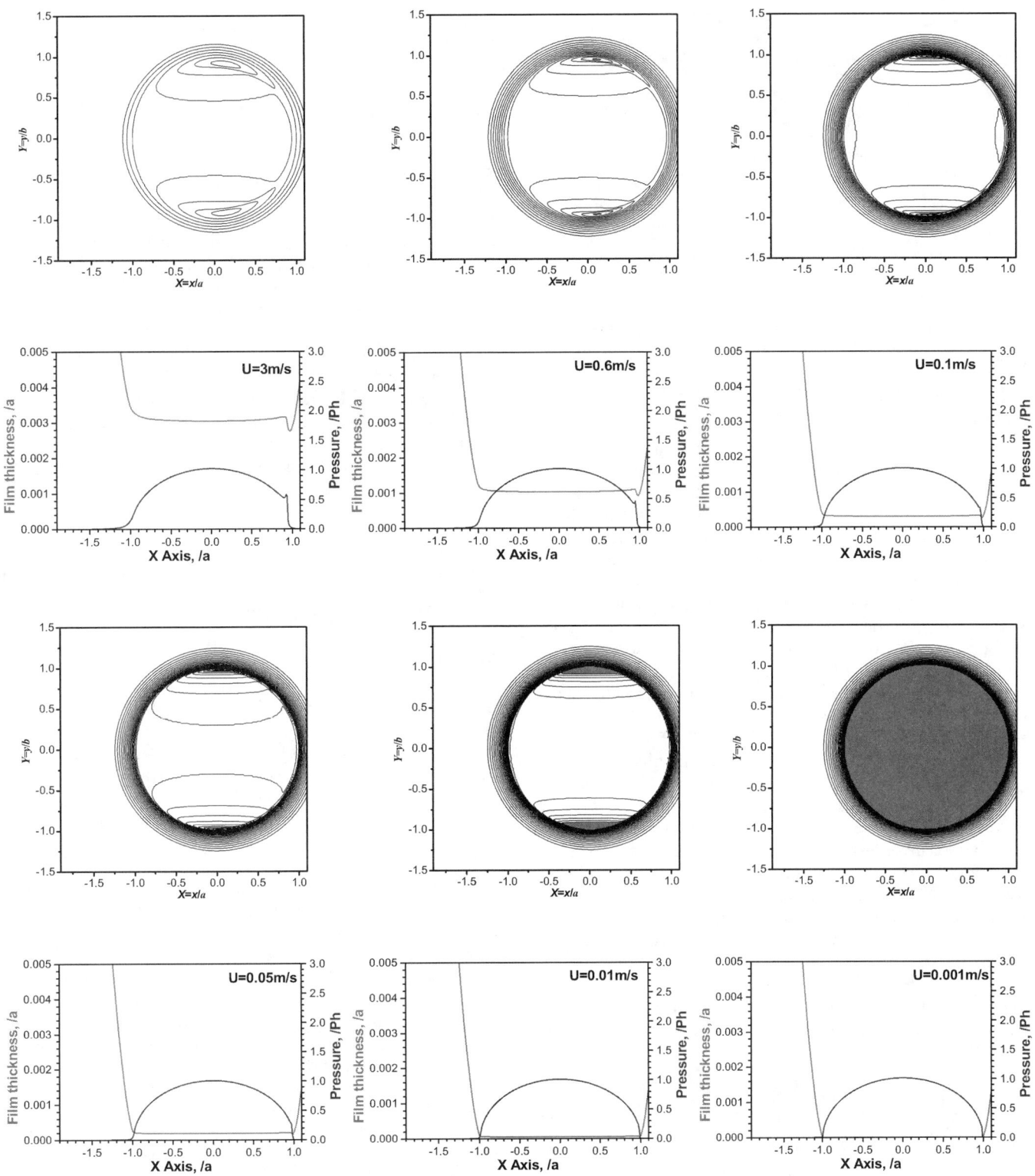

Fig. 29—Film thickness and pressure distributions under different entraining speeds for smooth EHL contact, p_H=2.24 GPa, a=0.413 mm, η_0=0.1756 Pa·s, α=18.8 GPa^{-1}.

there will be a small amount of lubricant trapped in the "pockets" formed by remaining roughness, the contact area ratios can hardly reach 100 % even at extremely low speeds.

It has been demonstrated through this example that the numerical simulations are capable of revealing the entire process of transition of lubrication regimes from full-film EHL, to mixed lubrication, and eventually to boundary lubrication.

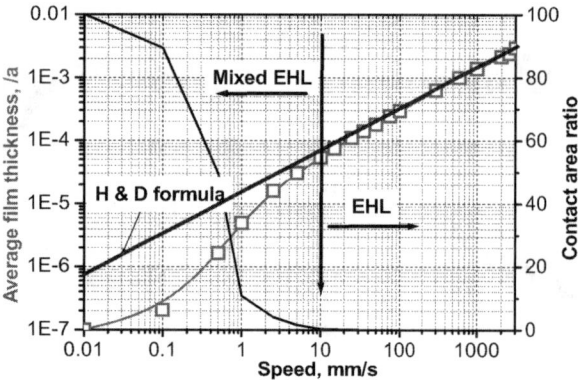

Fig. 30—Average film thickness and contact area ratio versus entraining speeds for a smooth EHL contact in the same conditions as Fig. 31.

5.3.3 Discussion

In the study of the transition of lubrication regime, the variation of film thickness versus the entrainment speed has been an issue that caused many discussions recently. Theoretically, the dependence of film thickness on speed is a manifest of hydrodynamic effect. Hence, the dependence would become weaker with the decreasing speed and vanishing hydrodynamic functions, and the film thickness would eventually approach a limit value that is independent of the speed. It is suggested that the limit film thickness can be considered as a measure of boundary layer created by adsorption or chemical reactions (see Chapter 5). The new technique based on optical interference provides an efficient means to measure the film thickness in nanometer scale and to observe the speed dependence in mixed lubrication [63–65]. The experiment results confirm the expectation that the film thickness follows the Hamrock and Dowson formula (H&D Line) at high speeds, corresponding to the full-film lubrication region, but the h-v curve gradually levels off as the speed decreases and converge to a constant value. In some experimental conditions, however, the film thickness was reported to follow the H&D line all the way down to 1 nm [63], or the film thickness shows a downward turn from the H&D line [64,66,67]. The observations cause many discussions but the mechanisms for the phenomena have not been fully understood.

The downward turn is also found in simulations of lubrication transition in smooth contacts, as can be seen in Fig. 30, while level-off phenomenon can be observed from the simulation result given in Fig. 33 for the rough surface in mixed lubrication. It is unclear at the present whether the downward turn shown in Fig. 30 is a manifest of real physical process, or a false phenomenon due to numerical inaccuracy.

5.4 Measurement of Contact Ratio—An Experimental Study in SKLT

In the previous section, numerical simulations predicted the change of contact ratio as a function of film thickness. Professor J. B. Luo in the State Key Laboratory of Tribology (SKLT) [68] has conducted an experimental study for measuring the relationships between the contact ratio and influential factors, which are presented in the following.

Contact ratio, α, is defined as the real contact area divided by the nominal contact zone, where the real contact area is referred to as the sum of all areas where film thickness is below a certain criterion in molecular scale. The contact area was measured by the technique of Relative Optical Interference Intensity (ROII) with a resolution of 0.5 nm in the vertical direction and 1 μm in the horizontal direction [69].

5.4.1 Effect of the Applied Load on Contact Ratio

For static contacts under different loads, the film thickness shapes in a selected region are shown in Fig. 34 where the deformations of a specific valley under different loads haven been displayed. From Fig. 34(a), we can see that under the load of 1.2 N only a small part of the observed region is in contact state while the rest of the parts are separated by lubricant film, and the contact ratio is only 0.08. When the load rises to 3.6 N, as shown in Fig. 34(b), the contact ratio grows to 0.24. With further increase of loads, we can see from (c) and (d) of Fig. 34 that the contact area grows constantly, and finally the contact ratio arrives at 0.78 under the load of 20.3 N corresponding to the maximum Hertz pressure of 0.375 GPa. The height of asperities becomes smaller as a result of asperity deformation. Figure 34(e) gives the contact ratio curve as a function of load.

The variations of contact ratio under various loads and velocities are shown in Fig. 35. For the surfaces with combined roughness R_a of 17 nm in the static contact or at very low speed, the contact ratio is close to 1, and it decreases with the increasing speed. The contact ratio also increases with loads if speed is fixed. Under the lower load (0.185 GPa), the contact ratio becomes zero at the speed of 6.3 mm/s, corresponding to the full-film EHL. With the increase of load, the critical speed for the transition to zero contact ratio and full-film EHL becomes larger, for example, 7.1 mm/s under 0.233 GPa, 11.2 mm/s under 0.293 GPa, and 12.7 mm/s under 0.370 GPa. Similarly it can be seen in Fig. 35 that the critical speed for reaching full contact, i.e., $\alpha = 1$, also increases with pressure. This is due to the factor that the increase of load results in larger deformation of asperities and thinner oil film; therefore, a larger contact area and contact ratio can be achieved at higher speed.

5.4.2 The Relation Between Contact Ratio and Film Thickness

Figure 36 shows the contact ratios measured for different types of lubricant, at the speed ranging from 0.13 to 62

Fig. 31—Transversely ground surface.

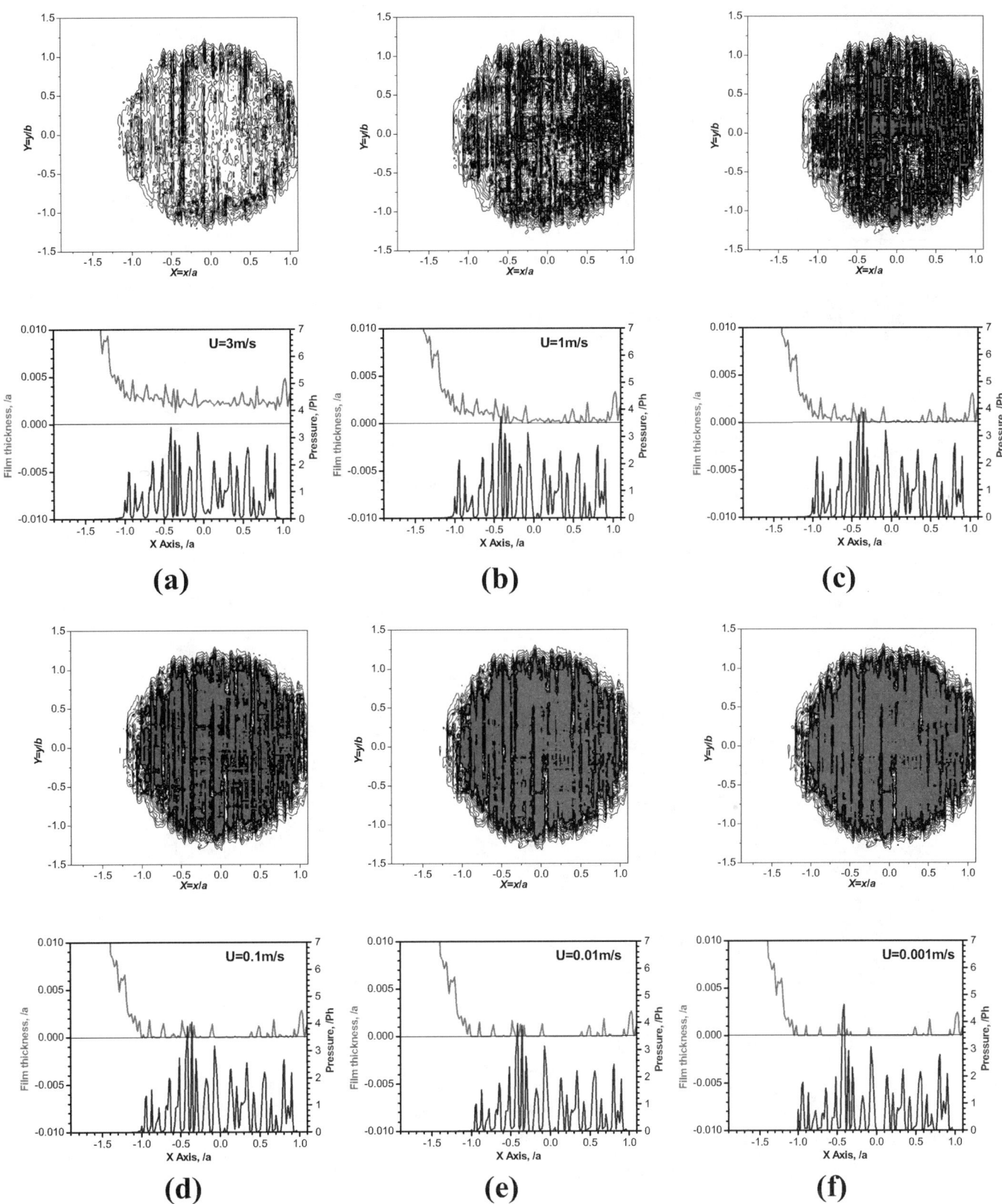

Fig. 32—Film thickness and pressure distributions under different entraining speeds for a ground surface, p_H=2.24 GPa, a=0.413 mm, η_0=0.1756 Pa·s, α=18.8 GPa^{-1}, R_a=455 nm.

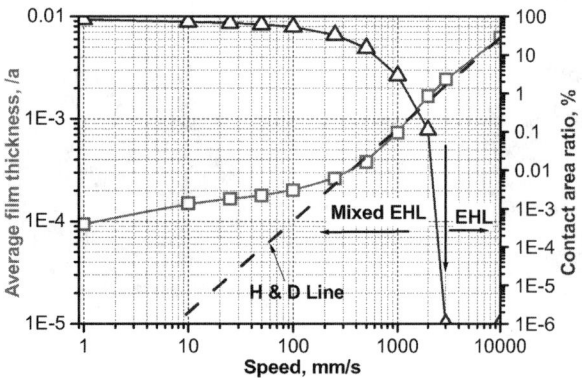

Fig. 33—The changes of average film thickness and contact area ratio with entraining speed for a ground surface. p_H=2.24 GPa, a=0.413 mm, η_0=0.1756 Pa·s, α=18.8 GPa^{-1}, R_a=455 nm.

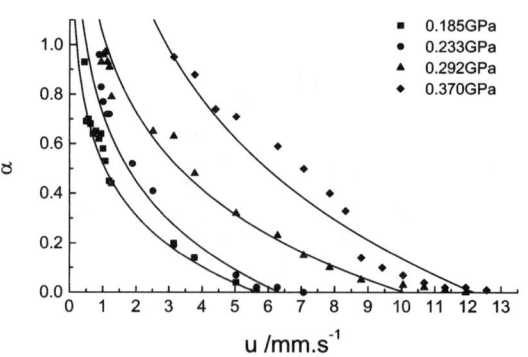

Fig. 35—Effect of pressure on the contact ratio α, combined surface roughness: 17 nm, lubricant: 13604.

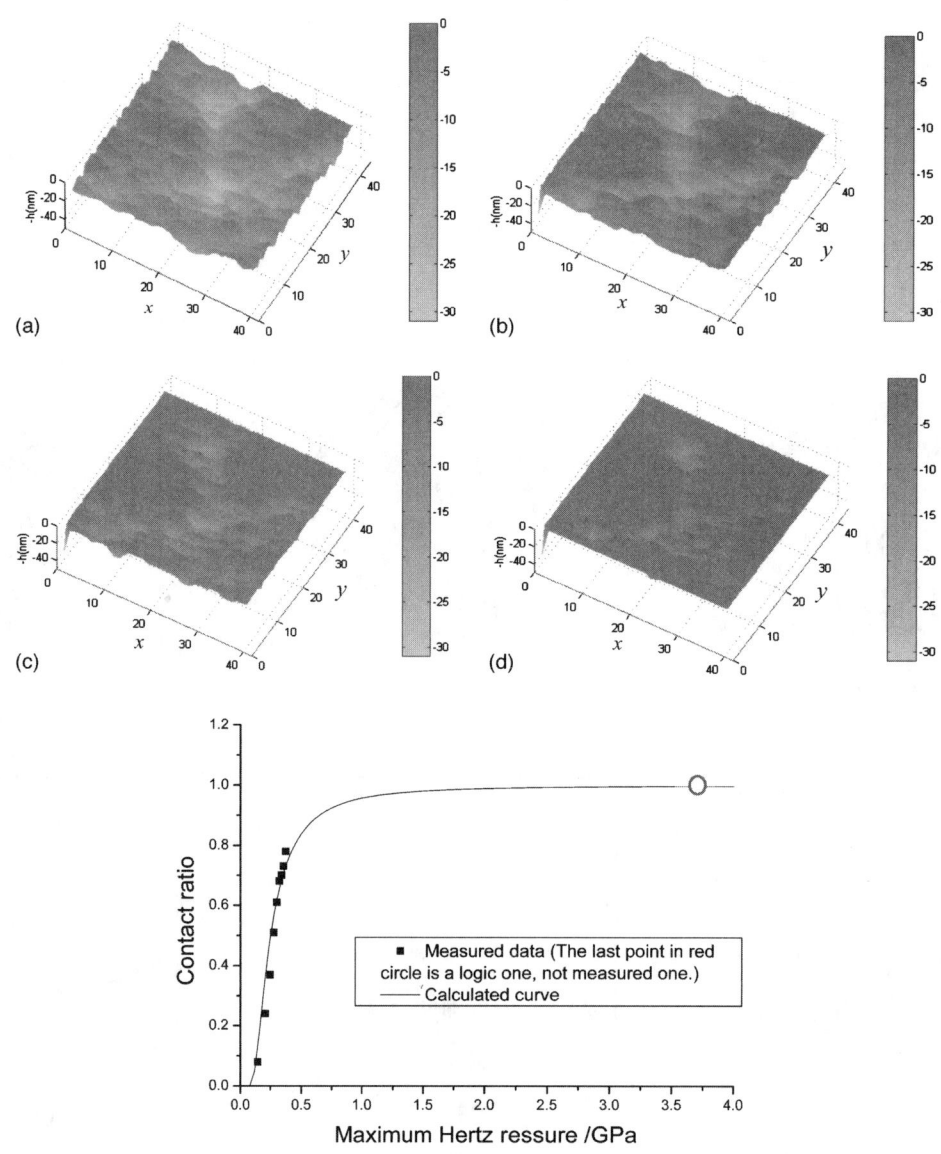

(e) Contact ratio with pressure in static state

Fig. 34—Contact under different loads for ball with rms 31.7 nm at temperature 27°C. (a) Load: 1.2 N, α=0.08; (b) Load: 3.6 N, α=0.24; (c) Load: 8.4 N, α=0.51; (d) Load: 20.3 N, α=0.78; and (e) Contact ratio with pressure in static state.

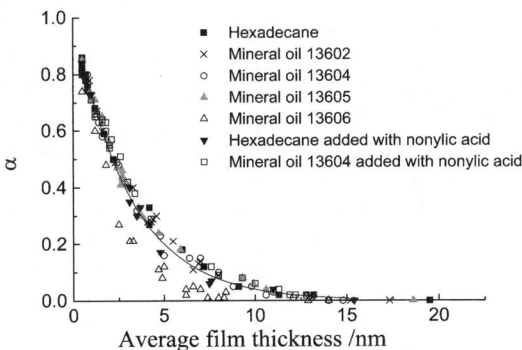

Fig. 36—Contact ratio α versus average film thickness, combined surface roughness: 17 nm, maximum Hertz pressure: 0.293 GPa.

mm/s, and under the pressure of 0.293 GPa. The results indicate that the contact ratio α increases with the decrease of average film thickness, and the trend holds for all kinds of tested lubricants. When the average film thickness rises from zero to about 15 nm the contact ratio drops from 0.86 to 0, as shown in Fig. 36, indicating a change of the lubrication regime from mixed lubrication to full-film EHL. Above the film thickness of 15 nm, the contact ratio remains zero and EHL is fully developed. It has to be noticed that the critical film thickness about 15 nm for the transition to full-film state is close to the combined roughness of free surface (17 nm), which is different from the general idea that a full fluid film lubrication will not realize until the ratio of film thickness to combined surface roughness R_a is larger than 3. This phenomenon indicates that the combined surface roughness in the contact region will be much smaller than its original value. In addition, it also can be seen from Fig. 36 that if the lubricant with high viscosity or larger molecules (Mineral oil 13606) is used, a smaller contact ratio can be detected at the same film thickness.

5.4.3 Effect of Lubricant Viscosity

The contact ratios for lubricants with different viscosities are plotted in Fig. 37 as a function of speed. The decreasing rates vary significantly due to the difference in viscosities. For the maximum viscosity of tested oil 13606 (100 mm²/s), the decreasing rate of contact ratio with speed is highest, and the contact ratio becomes zero at the speed lower than 2 mm/s. For the lower viscosity of lubricants such as oil

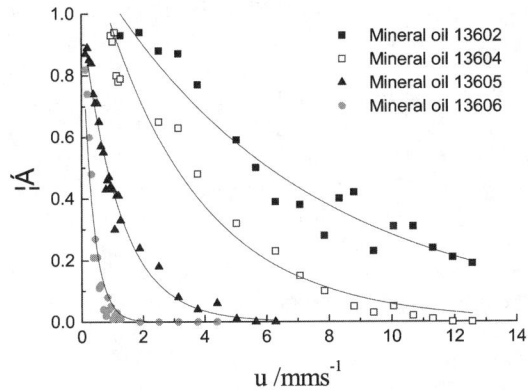

Fig. 37—Effect of viscosity on contact ratio, combined surface roughness: 17 nm, maximum Hertz pressure: 0.292 GPa.

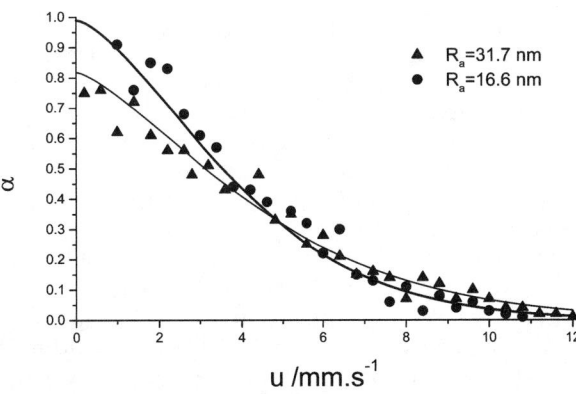

Fig. 38—Relation between contact ratio α and surface roughness, lubricant: 13604, maximum Hertz pressure: 0.292 GPa.

13605 (50 mm²/s) and oil 13604 (20 mm²/s), the critical speeds where α becomes zero are 5.5 mm/s and 12.7 mm/s, respectively. For the lowest viscosity of tested oil 13602 (5 mm²/s), contacts of asperities always exist even though the speed is greater than 13 mm/s at which the contact ratio remains about 0.2.

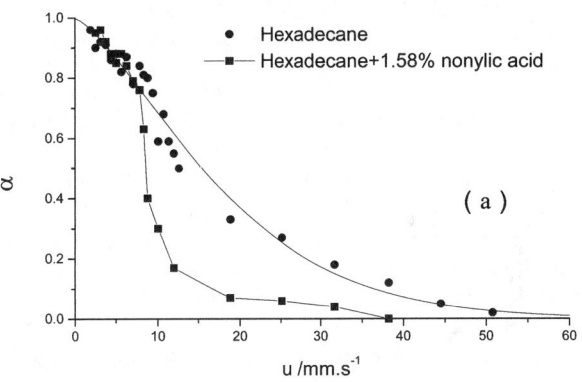

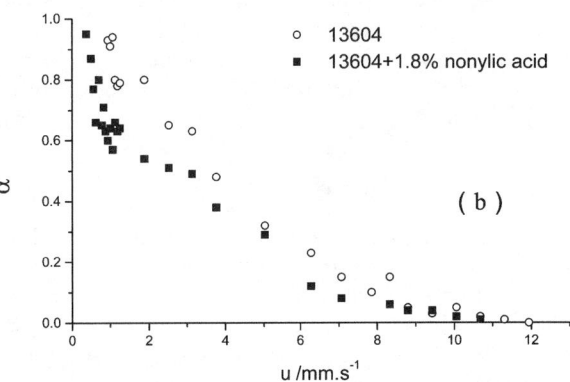

Fig. 39—Effect of polar additives on contact ratio, combined surface roughness: 17 nm, maximum Hertz pressure: 0.292 GPa. (a) Hexadecane with polar additive. (b) Mineral oil with polar additive.

5.4.4 Effect of Surface Roughness

Figure 38 shows the contact ratios from two experiments in which the R_a values of the combined roughness of the glass disk and ball are 17 nm and 32 nm, respectively.

In the higher speed range from 6 mm/s to over 12 mm/s, the contact ratio is greater for the surfaces with larger combined roughness. When the speed decreases below 6 mm/s, however, the situation reverses that the contact ratio of the surfaces with larger combined roughness becomes lower than that with the small roughness. For the rough ball, the contact initiates at 12 mm/s while for the smooth ball it starts at 11 mm/s. At the low speed limit when the speed decreases to 0.13 mm/s, the contact ratio for the rough ball is close to 0.75 while for the smooth ball it is close to 1. Therefore, the asperity contact on rough surfaces initiate at higher speed in comparison to the smooth surfaces, but its varying rate with speed is smaller. This is due to a higher peak that prevents the contact from further growth and slows the increase in contact ratio, but for the smooth surfaces, once the contact occurs, asperities will be engaged in contact at a much greater rate.

5.4.5 Effect of Polar Additives

When the polar additive nonylic acid was added into hexadecane liquid, the contact ratio becomes much smaller than that of pure hexadecane, which is shown in Fig. 39. For hexadecane liquid, the critical speed to reach zero contact ratio is 50 mm/s, which is much higher than that of mineral oil 13604 because of its much lower viscosity. However, when nonylic acid was added into the hexadecane liquid, the critical speed decreased from more than 50 mm/s to 38 mm/s. The same phenomenon can be seen in Fig. 39(b) which shows the comparison of oil 13604 and that added with 1.8 %wt. nonylic acid. The addition of polar additive reduces the contact ratio, too, but its effect is not as strong as that in hexadecane liquid because the oil 13604 has a much larger viscosity. Therefore, it can be concluded that the addition of polar additives will reduce the contact ratio because the polar additives are easy to form a thick boundary layer, which can separate asperities of the two rubbing surfaces.

5.4.6 Discussions

According to the experimental data presented above, the dependence of the contact ratio α on the test conditions and material properties, such as viscosity, speed, surface roughness, pressure, and elastic module can be described by an exponential function as:

$$\alpha = e^{f(\eta, u, R'_a, P_m, E')} \qquad (52)$$

where $f(\eta, u, R'_a, P_m, E')$ is a function related to the lubricant viscosity η, rolling speed u, combined surface roughness R'_a, maximum Hertz pressure P_m, and the combined elastic module of surfaces E'. For the mineral oil without any polar additives, Equation (52) can be written into a specific form by fitting the experimental data shown above:

$$\alpha = e^{-[(0.0114 \eta^2 u^2 R'^{-0.1}_a + R'_a)(E')^2]/(3.62 \times 10^7 P_m^2)} \qquad (53)$$

where η is in mm^2/s, u in mm/s, R'_a in nm, P_m in GPa, and E' in GPa. The formula can be used to predict the contact between a steel ball and a hard disk.

6 Summary

1. Understanding the role of surface roughness in mixed lubrication is a first step toward the microscopic study of tribology. It has been an effort for more than 30 years, starting from statistic models, but it is the deterministic approach that provides a powerful means to explore the tribological events occurring at the micrometre scale.
2. This chapter describes a DML model proposed by the authors, based on the expectation that the Reynolds equation at the ultra-thin film limit would yield the same solutions as those from the elastic contact analysis. A unified equation system is therefore applied to the entire domain, which gives rise to a stable and robust numerical procedure, capable of predicting the tribological performance of the system through the entire process of transition from full-film to boundary lubrication.
3. Numerical accuracy and efficiency are critical issues for the DML model to be accepted in engineering applications. FFT-based algorithm has been introduced to speed up the computations, and attempts have been made to validate the model by comparing the results from the present DML model with those from other sources of numerical analyses and experiments.
4. The model has been applied successfully to examine the performances of mixed lubrication, such as the effects of the roughness height distribution, the behavior of asperity deformation under different conditions of contact and sliding, and the transition of lubrication regime from hydrodynamic to boundary lubrication.
5. The transition of the lubrication regime can be described efficiently in two different ways: to observe the changes in friction as a function of film thickness (the Stribeck curve), or to plot the film thickness, h, against the surface speed (the h-v curve). In mixed lubrication, the friction deviates from the linear viscous law, grows up quickly with the decreasing h, and becomes stabilized in the regime of boundary lubrication. The film thickness follows the Harmrock and Dowson formula in the full-film EHL region, but in mixed lubrication it turns away from the H & D line, and eventually reaches a limit value that characterizes the thickness of boundary layers.

References

[1] Tallian, T. E., "The Theory of Partial Elastohydrodynamic Lubrication," *Wear*, Vol. 19, No. 1, 1972, pp. 91–108.

[2] Patir, N. and Cheng, H. S., "Average Flow Model for Determining Effects of Three-Dimensional Roughness on Partial Hydrodynamic Lubrication," *ASME J. Lubr. Technol.*, Vol. 100, No. 1, 1978, pp. 12–17.

[3] Greenwood, J. A. and Williamson, J. B. P., "Contact of Nominally Flat Surface," *Proc. R. Soc. London, Ser. A*, Vol. 295, 1966, pp. 300–319.

[4] Tzeng, S. T. and Saibel, E., "Surface Roughness Effect on Slider Bearing Lubrication," *ASLE Trans.*, Vol. 10, 1967, p. 334.

[5] Christensen, H., "Stochastic Models for Hydrodynamic Lubrication of Rough Surfaces," *Proc. Inst. Mech. Eng., Part J: J. Eng. Tribol.*, Vol. 184, 1969–70, p. 1013.

[6] Christensen, H. and Tonder, K., "The Hydrodynamic Lubrica-

[6] tion of Rough Bearing Surfaces of Finite Width," *Journal of Lubrication Technology, Trans. ASME Series F*, Vol. 93, 1971, p. 324.
[7] Christensen, H. and Tonder, K., "The Hydrodynamic Lubrication of Rough Journal Bearings," *Journal of Lubrication Technology, Trans. ASME Series F*, Vol. 95, 1973, p. 324.
[8] Tonder, K. and Christensen, H., "Waviness and Roughness in Hydrodynamic Lubrication," *Proc. Inst. Mech. Eng., Part J: J. Eng. Tribol.*, Vol. 186, 1972, p. 807.
[9] Patir, N., "Effects of Surface Roughness on Partial Film Lubrication Using an Average Flow Model Based on Numerical Simulation." Ph.D. Thesis, Northwestern University, 1978.
[10] Patir, N. and Cheng, H. S., "Application of Average Flow Model to Lubrication Between Rough Sliding Surfaces," *ASME J. Lubr. Technol.*, Vol. 101, No. 2, 1979, pp. 220–230.
[11] Shi, F. H. and Salant, R. F., "Numerical Study of a Rotary Lip Seal with a Quasi-Random Sealing Surface," *ASME J. Tribol.*, Vol. 123, No. 3, 2001, pp. 517–524.
[12] Zhu, D. and Cheng, H. S., "Effect of Surface Roughness on the Point Contact EHL," *ASME J. Tribol.*, Vol. 110, No. 1, 1988, pp. 32–37.
[13] Chang, L., "Deterministic Model for Line-Contact Partial Elasto-Hydrodynamic Lubrication," *Tribol. Int.*, Vol. 28, No. 2, 1995, pp. 75–84.
[14] Hua, D. Y., Qiu, L., and Cheng, H. S., "Modeling of Lubrication in Micro Contact," *Tribol. Lett.*, Vol. 3, 1997, pp. 81–86.
[15] Jiang, X. and Hua, D. Y., "A Mixed Elastohydrodynamic Lubrication Model With Asperity Contact," *ASME J. Tribol.*, Vol. 121, 1999, pp. 481–491.
[16] Hu, Y. Z. and Zhu, D., "A Full Numerical Solution to the Mixed Lubrication in Point Contacts," *ASME J. Tribol.*, Vol. 122, 2000, pp. 1–9.
[17] Zhu, D. and Hu, Y. Z., "A Computer Program Package for the Prediction of EHL and Mixed Lubrication Characteristics, Friction, Subsurface Stresses and Flash Temperatures Based on Measured 3-D Surface Roughness," *Tribol. Trans.*, Vol. 44, No. 3, 2001, pp. 383–390.
[18] Johnson, K. L., *Contact Mechanics*, Cambridge University Press, 1996.
[19] Yang, P. and Wen, S., "A Generalized Reynolds Equation for Non-Newtonian Thermal Elastohydrodynamic Lubrication," *ASME J. Tribol.*, Vol. 112, 1990, pp. 631–636.
[20] Lai, W. T. and Cheng, H. S., "Temperature Analysis in Lubricated Simple Sliding Rough Contacts," *ASLE Trans.*, Vol. 128, No. 3, 1984, pp. 303–312.
[21] Qiu, L. and Cheng, H. S., "Temperature Rise Simulation of Three-Dimensional Rough Surface in Mixed Lubricated Contact," *ASME J. Tribol.*, Vol. 120, No. 2, 1998, pp. 310–318.
[22] Carslaw, H. S. and Jaeger, J. C., *Conduction of Heat in Solids*, 2nd ed., Oxford at the Clarendon Press, London, 1958.
[23] Zhai, X. and Chang, L., "A Transient Thermal Model for Mixed-Film Contacts," *Tribol. Trans.*, Vol. 43, No. 3, 2000, pp. 427–434.
[24] Zhai, X. and Chang, L., "Some Insights into Asperity Temperatures in Mixed-Film Lubrication," *Tribol. Int.*, Vol. 34, 2001, pp. 381–387.
[25] Zhao, J., Farshid, S., and Hoeprich, M. H., "Analysis of EHL Circular Contact Start Up: Part 1—Mixed Contact Model with Pressure and Film Thickness Results," *ASME J. Tribol.*, Vol. 123, 2001, pp. 67–74.
[26] Zhao, J., Farshid, S., and Hoeprich, M. H., "Analysis of EHL Circular Contact Start Up: Part 2—Surface Temperature Rise Model and Results," *ASME J. Tribol.*, Vol. 123, 2001, pp. 75–82.
[27] Wang, W. Z., Liu, Y. C., Wang, H., and Hu, Y. Z., "A Computer Thermal Model of Mixed Lubrication in Point Contacts," *ASME J. Tribol.*, Vol. 126, No. 1, 2004, pp. 162–170.
[28] Dowson, D. and Hamroc, B. J., "Numerical Evaluation of the Surface Deformation of Elastic Solids Subjected to a Hertz Contact Stress," *ASLE Trans.*, Vol. 19, 1976, pp. 279–286.
[29] Chang, L. "An Efficient and Accurate Formulation of the Surface Deformation Matrix in Elastohydrodynamic Point Contacts," *ASME J. Tribol.*, Vol. 111, 1989, pp. 642–647.
[30] Biswas, S. and Snidle, R. W., "Calculation of Surface Deformation in Point Contact EHD," *ASME J. Lubr. Technol.*, Vol. 99, 1977, pp. 313–317.
[31] Ai, X., "Numerical Analyses of Elastohydrodynamically Lubricated Line and Point Contacts with Rough Surfaces by Using Semi-System and Multi-Grid Methods," Ph.D. Thesis, Northwestern University, 1993.
[32] Ren, N. and Lee, S., "Contact Simulation of Three-Dimensional Rough Surfaces Using Moving Grid Method," *ASME J. Tribol.*, Vol. 115, 1993, pp. 597–601.
[33] Brandt, A. and Lubrecht, A. A., "Multilevel Matrix Multiplication and Fast Solution of Integral Equations," *J. Comput. Phys.*, Vol. 90, No. 2, 1990, pp. 348–370.
[34] Lubrecht, A. A. and Ioannides, E. A., "Fast Solution of the Dry Contact Problem and Associated Surface Stress Field, Using Multilevel Techniques," *ASME J. Tribol.*, Vol. 113, 1991, pp. 128–133.
[35] Ju, Y. and Farris, T. N., "Spectral Analysis of Two-Dimensional Contact Problems," *ASME J. Tribol.*, Vol. 118, 1996, pp. 320–328.
[36] Nogi, T. and Kato, T., "Influence of a Hard Surface Layer on the Limit of Elastic Contact. Part 1: Analysis Using a Real Surface Model," *ASME J. Tribol.*, Vol. 119, 1997, pp. 493–500.
[37] Liu, S., Wang, Q., and Liu, G., "A Versatile Method of Discrete Convolution and FFT (DC-FFT) for Contact Analyses," *Wear*, Vol. 243, 2000, pp. 101–111.
[38] Wang, W. Z., Wang, H., Liu, Y. C., Hu, Y. Z., and Zhu, D., "A Comparative Study of the Methods for Calculation of Surface Elastic Deformation," *Proc. Inst. Mech. Eng., Part J: J. Eng. Tribol.*, Vol. 217, 2003, pp. 145–153.
[39] Venner, C. H. and Lubrecht, A. A., *Multilevel Methods in Lubrication*, Elsevier, Amsterdam, 2000.
[40] Venner, C. H. and Lubrecht, A. A., "Numerical Simulation of a Transverse Ridge in a Circular EHL Contact Under Rolling/Sliding Source," *ASME J. Tribol.*, Vol. 116, No. 4, 1994, pp. 751–761.
[41] Holmes, M. J. A., "Transient Analysis of the Point Contact Elastohydrodynamic Lubrication Problem Using Coupled Solution Methods," Ph.D. Thesis, Cardiff University, 2002.
[42] Venner, C. H. and Lubrecht, A. A., "Numerical Analysis of the Influence of Waviness on the Film Thickness of a Circular EHL Contact," *ASME J. Tribol.*, Vol. 118, No. 4, 1996, pp. 153–161.
[43] Hu, Y. Z., Barber, G. C., and Zhu, D., "Numerical Analysis for the Elastic Contact of Real Rough Surfaces," *Tribol. Trans.*, Vol. 42, No. 3, 1999, pp. 443–452.
[44] Felix-Quinonez, A., Ehret, P., and Summers, J. L., "On Three-Dimensional Flat-Top Defects Passing Through an EHL Point

[45] Choo, J. W., Glovnea, R. P., Olver, A. V., and Spikes, H. A., "The Effects of Three-Dimensional Model Surface Roughness Features on Lubricant Film Thickness in EHL Contacts," *ASME J. Tribol.*, Vol. 125, 2003, pp. 533–542. Contact: A Comparison of Modeling with Experiments," *ASME J. Tribol.*, Vol. 127, No. 1, 2005, pp. 51–60.

[46] Thomas, T. R., *Rough Surfaces*, Longman, New York, 1982.

[47] Hu, Y. Z. and Tonder, K., "Simulation of 3-D Random Surface by 2-D Digital Filter and Fourier Analysis," *Int. J. Mach. Tools Manuf.*, Vol. 32, 1992, pp. 82–90.

[48] Oppenheim, A. V. and Schaferm, R. W., *Digital Signal Processing*, Englewood Cliffs, NJ: Prentice-Hall, 1975.

[49] Chilamakuri, S. K. and Bhushan, B., "Contact Analysis of Non-Gaussian Random Surfaces," *Proc. Inst. Mech. Eng., Part J: J. Eng. Tribol.*, Vol. 212, No. J1, 1998, pp. 19–32.

[50] Kaneta, M. and Nishikawa, H., "Experimental Study on Micro-Elastohydrodynamic Lubrication," *Proc. Inst. Mech. Eng., Part J: J. Eng. Tribol.*, Vol. 213, 1999, pp. 371–381.

[51] Guangteng, G., Cann, P. M., Olver, A. V., and Spikes, H. A., "Lubricant Film Thickness in Rough Surface, Mixed Elastohydrodynamic Contact," *ASME J. Tribol.*, Vol. 122, 2000, pp. 65–76.

[52] Ehret, P., Felix-Quinonez, A., Lord, J., Jolkin, A., Larsson, R., and Marklund, O., "Experimental Analysis of Micro-Elastohydrodynamic Lubrication Conditions," *Proceedings of the International Tribology Conference*, Nagasaki Japan, 2000, pp. 621–624.

[53] Felix, Q. A., Ehret, P., and Summers, J. L., "New Experimental Results of a Single Ridge Passing Through an EHL Conjunction," *ASME J. Tribol.*, Vol. 125, 2003, pp. 252–259.

[54] Felix, Q. A., Ehret, P., and Summers, J. L., "On Three-Dimensional Flat-Top Defects Passing Through an EHL Point Contact: A Comparison of Modeling with Experiments," *ASME J. Tribol.*, Vol. 127, 2005, pp. 51–60.

[55] Greenwood, J. A. and Morales-Espejel, G. E., "The Behavior of Transverse Roughness in EHL Contacts," *Proc. Inst. Mech. Eng., Part J: J. Eng. Tribol.*, Vol. 208, 1994, pp. 121–132.

[56] Hooke, C. J., "Surface Roughness Modification in Elastohydrodynamic Line Contacts Operating in the Elastic Piezoviscous Regime," *Proc. Inst. Mech. Eng., Part J: J. Eng. Tribol.*, Vol. 212, 1998, pp. 145–162.

[57] Hooke, C. J., "Surface Roughness Modification in EHL Line Contacts—The Effect of Roughness Wavelength, Orientation and Operating Conditions," *Proceedings of the 25th Leeds-Lyon Symposium on Tribology*, Lyon, 1998, pp. 193–202.

[58] Venner, C. H. and Lubrecht, A. A., "Amplitude Reduction of Non-Isotropic Harmonic Patterns in Circular EHL Contacts, Under Pure Rolling," *Proceedings of the 25th Leeds-Lyon Symposium on Tribology*, Lyon, 1998, pp. 151–162.

[59] Hooke, C. J. and Venner, C. H., "Surface Roughness Attenuation in Line and Point Contacts," *Proc. Inst. Mech. Eng., Part J: J. Eng. Tribol.*, Vol. 214, 2000, pp. 439–444.

[60] Morales-Espejel, G. E., Venner, C. H., and Greenwood, J. A., "Kinematics of Transverse Real Roughness in Elastohydrodynamically Lubricated Line Contacts Using Fourier Analysis," *Proc. Inst. Mech. Eng., Part J: J. Eng. Tribol.*, Vol. 214, No. J6, 2000, pp. 523–534.

[61] Spikes, H. A., "Mixed Lubrication—An Overview," *Lubr. Sci.*, Vol. 9, No. 3, 1997, pp. 221–252.

[62] Hamrock, B. J. and Dowson, D., "Isothermal Elastohydrodynamic Lubrication of Point Contacts—1: Theoretical Formulation," *ASME J. Lubr. Technol.*, Vol. 98, No. 2, 1976, pp. 223–229.

[63] Luo, J. B., Qian, L. M., Liu, S., Wen, S. Z., and Li, L. K. Y., "The Failure of Fluid Film at Nano-scale," *Tribol. Trans.*, Vol. 42, No. 4, 1999, pp. 912–916.

[64] Spikes, H. A., "Thin Films in Elastohydrodynamic Lubrication: The Contribution of Experiment," *Proc. Inst. Mech. Eng., Part J: J. Eng. Tribol.*, Vol. 213, 1999, pp. 335–352.

[65] Hartl, M., Krupka, I., Poliscuk, P., and Liska, M., "Thin Film Colorimetric Interferometry," *Tribol. Trans.*, Vol. 44, No. 2, 2001, pp. 270–276.

[66] Zhu, D., Hartl, M., and Krupka, I., "A Comparative Study on the EHL Film Thickness Results From Numerical Solutions and Experimental Measurements," presented at *2004 STLE Annual Meeting*, Toronto, Canada.

[67] Johnston, G. J., Wayte, R., and Spikes, H. A., "The Measurement and Study of Very Thin Lubrication Films in Concentrated Contacts," *Tribol. Trans.*, Vol. 34, 1991, pp. 187–194.

[68] Luo, J. B. and Liu, S., "The Investigation of Contact Ratio in Mixed Lubrication," *Tribol. Int.*, Vol. 39, No. 5, 2006, pp. 409–416.

[69] Luo, J. B., "Study on the Experimental Technique and Properties of Thin Film Lubrication," Ph.D. Thesis, Beijing: Tsinghua University, 1994.

8

Thin Solid Coatings

Chenhui Zhang[1] and Tianmin Shao[1]

1 Introduction

MOSTLY, THE PERFORMANCE OF A MECHANICAL system depends on, to a large extent, surface properties and interfacial behavior of/between its individual components. Especially, in the application of micro/nano tribology, surface behaviors of materials often play a key role in the actualization of designed functions. Techniques of thin solid coatings are extensively used in many fields, which provide desired surface properties by producing thin layers on different substrates. The materials of such layers differ from that of the substrates and their thicknesses are generally in the scale of nanometre to micrometre. Figure 1 gives a typical system of thin solid coating/substrate. By selecting coating materials and deposition methods, desired properties such as low friction, wear resistance, corrosion resistance, etc., could be attained. As one of the most attractive hot spots in material science and technology, research on thin solid coatings has been extensively carried out and the emphasis is generally put on seeking proper coating materials and corresponding preparation techniques. Nowadays, numerous coating materials such as metals, polymers, alloys, ceramics, and their hybrids are available for various applications. As for techniques of coating production, physical vapor deposition (PVD), chemical vapor deposition (CVD), and their derived techniques are extensively used.

In the past few decades, some new thin coating materials appeared and attracted great attention. The most attractive coating should be called a diamond-like carbon coating (DLC). With both sp^2 and sp^3 structures, DLC coatings exhibited many excellent properties like low friction coefficient, high hardness, good bio-consistence, and so on, which make it a useful coating material for a wide range of applications.

Carbon nitrides are another kind of coating material, on which much attention was paid in the past 20 years. Carbon nitride is practically a material first worked out by calculation. Being claimed possibly to have a bulk modulus greater than diamond, this exciting "calculated material" has attracted great attention since its first prediction in 1989 [1] and considerable research has been directed toward its synthesis as the form of surface coatings.

Despite the fact that numerous coating materials are available, frequently, single layer structure of a coating does not meet the requirement. Composite coatings with multilayer structure or hybrid structure could sometimes provide peculiar properties.

In this chapter, a brief introduction will be given mainly to the DLC, CNx, multilayer films and nano-composite coatings. Detailed and comprehensive introduction of the conventional thin solid coating technique is not the objective of this chapter. Readers are referred to relevant publications to attain the knowledge in this area.

2 Diamond-like Carbon (DLC) Coatings

Diamond-like carbon (DLC) coating is the name for the amorphous carbon coatings which have similar mechanical, optical, electrical, and chemical properties to natural diamond. They have an amorphous structure, and consist of a mixture of sp^3 and sp^2 carbon structures with sp^2-bonded graphite-like clusters embedding in an amorphous sp^3-bonded carbon matrix. Some DLC coatings also contain significant quantities of hydrogen to form C-H alloy. By the sp^3 fraction and whether containing hydrogen or not in the coatings, DLCs can be divided into four types: amorphous carbon (a-C), amorphous C-H alloy (a-C:H), tetrahedral amorphous carbon (ta-C, sp^3 fraction >70 %), and tetrahedral amorphous C-H alloy (ta-C:H, sp^3 fraction >70 %).

2.1 Deposition Methods of DLC Coatings

DLC coating was first prepared by Aisenberg and Chabot using ion beam deposition in 1971 [2]. At present, PVD, such as ion beam deposition, sputtering deposition, cathodic vacuum arc deposition, pulsed laser deposition, and CVD, like plasma enhanced chemical vapor deposition are the most popular methods to be selected to fabricate DLC coatings.

In ion beam deposition, hydrocarbon gas such as methane or ethyene is ionized into plasma by an ion source such as the Kaufman source [3]. The hydrocarbon ions are then extracted from the ion source and accelerated to form an ion beam. The ions and the unionized molecules condense on the substrate surface to form DLC coating. However, in this method, ionized ratio of precursor gases could hardly exceed 10 %. In order to obtain a better quality of DLC coatings,

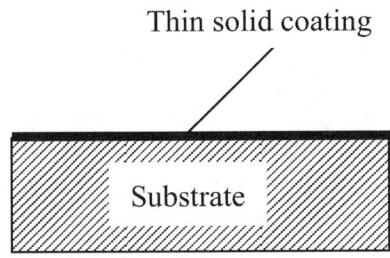

Fig. 1—A typical system of thin solid coating/substrate.

[1] State Key Laboratory of Tribology, Tsinghua University, Beijing, China.

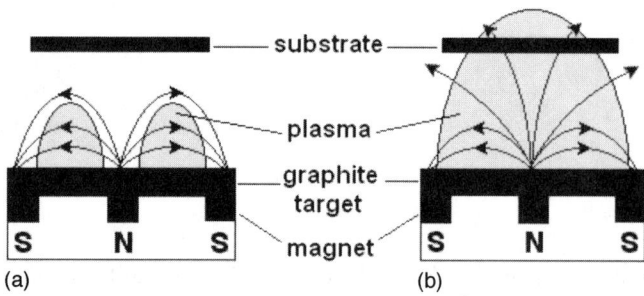

Fig. 2—Schemetics of (a) magnetron sputtering and (b) unbalanced magnetron sputtering.

relative higher ion energy of 100~1,000 eV is usually required.

Sputtering method is widely used in industrial application for DLC deposition. During the sputtering process, a dc or rf power is applied on a graphite cathode in Ar atmosphere to generate Ar⁺ plasma. Carbon atoms and ions are sputtered from the graphite cathode by the plasma and are deposited onto the substrate surface to form DLC coating. Normally, a magnet is placed behind the cathode to increase the sputtering yield of graphite by complicating the electrons path, which is called magnetron sputtering as shown in Fig. 2(a). If the magnetic field is configured to pass across to the substrate (unbalance magnetron, as shown in Fig. 2(b)) and a negative bias is applied on the substrate, part of the Ar⁺ ions can be attracted to bombard the growing coating, which is helpful to increase sp^3 fraction in the coating. In order to improve the ionization rate of Ar, a method named closed field unbalance magnetron sputtering was invented, in which at least two unbalanced magnetrons are used to form a closed magnetic field and the electrons are restricted in the closed magnetic field [4]. Hydrogen contained DLC coatings can be synthesized by sputtering graphite in a mixture of Ar gas and reactive gas such as hydrogen or methane. Another sputtering technique, named ion beam assistance deposition (IBAD) [5], employs an Ar⁺ ions beam to sputtering a graphite target to create a flux of carbon particles. Simultaneously, a second Ar⁺ beam is used to bombard the growing film to promote the formation of sp^3 bonding. The drawback of the sputtering method is the low ions fraction in the sputtered species, and the DLC coating with high sp^3 fraction cannot be produced by using this method.

Cathodic vacuum arc is an effective method to deposit DLC coatings with a high fraction of sp^3 and high hardness. Arc is initiated by touching the graphite cathode with a trigger, and carbon plasma with mean energy of 10–30 eV is generated [6,7]. However, besides the plasma, neutral particulates with the size varying from submicron to several micrometres are also produced. In order to remove the particles, a toroidal magnetic filter duct always follows the arc source to filter the neutral particulates, so-called filtered cathodic vacuum arc (FCVA), as shown in Fig. 3 [8]. The plasma can go along with the magnetic line and passes through the bent magnetic field that is parallel to the duct axis. The neutral particulates cannot follow the bent field and hit on the walls of the duct, which prevents the particulates from reaching the substrate. For metal cathode, one bend is enough to remove the particulates. However, in the case of graphite cathode, part of particulates can reach the filter exit by multi-bouncing on the wall [9]. Adding another bend to form double bend or "S-bend" filter will remove more than 99 % particulates from the flux. At the filter exit, the ions fraction is nearly 100 % of the flux, which is advantageous for forming sp^3 bonding. Additional energy can be provided to the carbon ions by applying a negative bias on the substrate. The properties of ta-C coating are dominated by the C⁺ ion energy. Normally, the coating shows high sp^3 fraction (>70 %), high hardness (>50 GPa) and large Young's modulus (about 300 GPa), and small roughness (Ra≈0.1 nm) at the ion energy of 100 eV [10–13]. A disadvantage accompanying the high sp^3 fraction and high hardness is the large biaxial compressive stress in ta-C coatings [10], which will result in the coating peeling off from the substrate even as the coating thickness is only about 100 nm. Annealing ta-C coating at the temperature of 500–600 °C can relax the compressive stress to nearly zero without reducing the sp^3 fraction and the hardness of the coating. Thick ta-C coating with the thickness larger than 1 μm can be achieved by combining deposition and annealing periodically [14].

Vacuum arc can also be initiated by pulsed laser, as shown in Fig. 4 [15,16]. The target material is first vaporized into plasma by pulsed laser, and vacuum arc is subsequently initiated and more plasma is generated under the function of the electric field between the anode and the cathode (target). Since the arc is controllable, the surface quality might be evi-

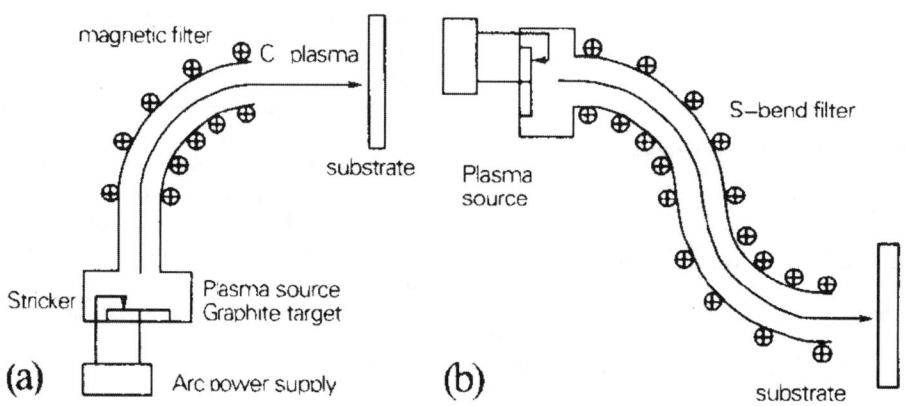

Fig. 3—Schematic of (a) single bend and (b) S-bend FCVA.

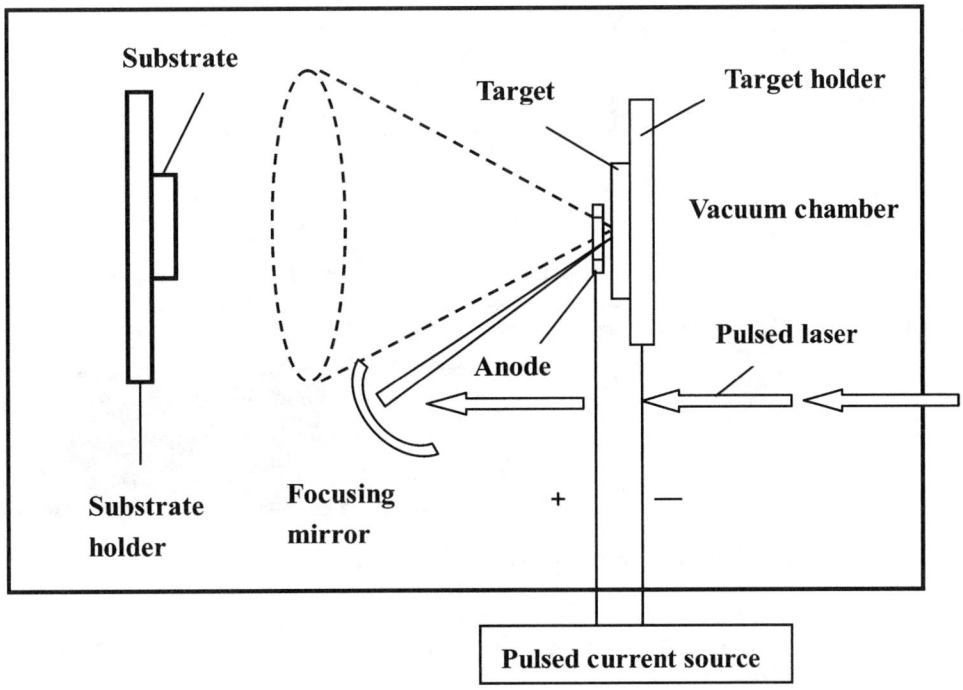

Fig. 4—Schematic diagram of laser-arc deposition system [16].

dently improved compared to that of cathodic vacuum arc deposition.

In plasma enhanced chemical vapor deposition (PECVD), the a-C:H or ta-C:H is formed from a hydrocarbon plasma initiated by an rf power [17–21]. In order to improve the ionization efficiency and to create plasma at lower pressure, an electron cyclotron wave resonance source is commonly used, which can produce high-density plasma of 10^{12} cm^{-3} or over [18,21]. Acetylene and methane are the most popular precursor gases to deposit hydrogen contained DLC coating. The properties of a-C:H and ta-C:H are related with the energy per carbon atom in the plasma. That means the coating deposition from the precursor molecule with multi-carbon atoms, such as acetylene and benzene, needs larger bias to achieve high sp^3 fraction and large hardness as compared with the deposition from methane [13,19,20].

2.2 Tribological Properties of DLC Coatings and Their Modifications

Fricion coefficient is one of the most important properties for tribological applications. For DLC coatings, friction behaviors were extensively studied and various results of friction measurements were reported. However, the differences were very large, ranging from below 0.01 to larger than 0.5 [22–26]. The friction coefficient of DLC coatings is influenced by the structure of the coating, and experimental conditions, such as coupled material, load, sliding velocity, the humidity, and the experimental environment [23,24].

For a-C:H coating, the friction coefficient is strongly influenced by H content in the coating, and by relative humidity (RH). At low humidity, the friction coefficient can be achieved as low as 0.02 [25]. The friction coefficient increases to 0.1~0.15 as the RH rises to 30 % and above. An ultra-low friction coefficient of 0.001~0.005 was reported to be achieved for high hydrogen content a-C:H coating in ultra-high vacuum (UHV) or dry nitrogen [22,26].

As the a-C:H coating slides on a different surface, a transfer layer of a-C:H will be formed on the matched surface. Thus, the contact is practically between two a-C:H surfaces. At the surface of the a-C:H coating with high H content, C atoms are fully hydrogenated into C–H bonds, which makes the a-C:H surface inert and hydrophobic. The interaction between the two a-C:H surfaces is mainly van der Waals force [13]. The weak van der Waals bonds are easily broken under shears, which results in a low friction force and a very small wear rate of the a-C:H coating. However, in a high humidity environment, the transfer layer is difficult to form on the matched surface, so that van der Waals bonds no longer form and the friction coefficient is much higher. In the case of two a-C:H coating surfaces in high humidity, the water molecules are adsorbed on the surface and form van der Waals bond with H atoms. The bond strength between H atoms and water molecules is larger than the strength of H–H van der Waals bonds [27]. As a result, the friction force is larger than that in a low humidity environment.

The friction mechanisms of lower hydrogen content a-C:H and hydrogen free ta-C coatings are quite different from that of a-C:H coatings with high H content. For ta-C coating, there is a free bond for each surface C atom. And the free bonds are passivated by the adsorbed molecules such as water, hydrogen, and oxygen in air. In the case of sliding between two contacted ta-C coatings in UHV or inert gases, the adsorbed layer will be removed. Thus, the free bonds on the two surfaces form strong C–C covalent bonds, which result in a large friction coefficient of larger than 0.5 [26]. When the sliding occurs in ambient air, the worn adsorbed layer can be supplemented instantly, and the friction coefficient reduces to a range of 0.1–0.2 [26,28].

In the case of ta-C coatings sliding on different materials in humid ambient, the top layer of a ta-C coating in the con-

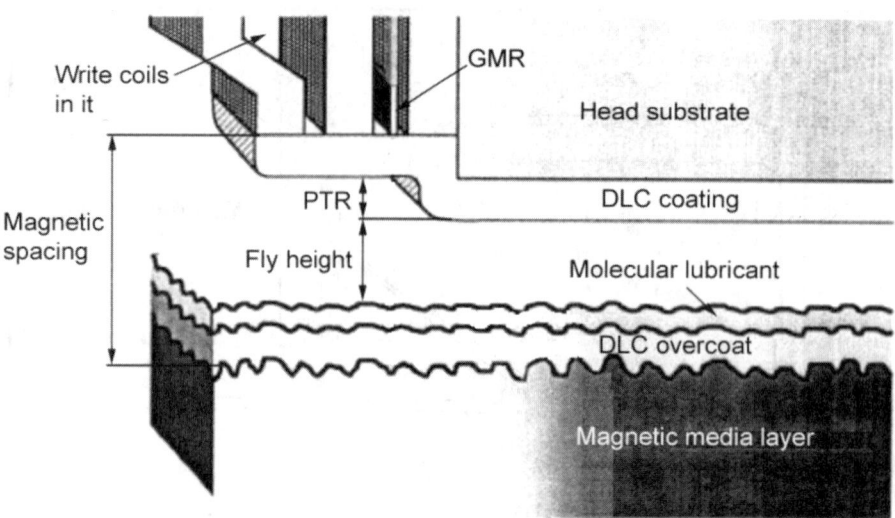

Fig. 5—Schematic of the read/write head-magnetic disk interface.

tact area is transformed to a graphitic layer due to the contact stress and the wear. The graphitic layer can be transferred to the matched surface and acts as a solid lubricant, and the friction coefficient is about 0.1 [28]. Adsorbed water between the graphitic layers can lower the friction slightly, due to the passivation of the dangling C bonds by water molecules. Thus, the friction coefficient decreases slightly with the increase of humidity [27,28].

The properties of DLCs can be modified by incorporating dopants, such as silicon, fluorine, and various metals. Incorporation of silicon and fluorine into DLC coating can significantly decrease the surface energy [29]. Silicon doped DLCs show a reduced friction coefficient (<0.1) in humid ambient air as compared with the conventional undoped DLCs [30–32]. However, the Si doped DLCs cannot provide lower friction coefficient under higher contact pressure (>1 GPa) than undoped DLCs [32]. The tribological properties of fluorine doped DLCs are similar to the silicon doped coatings. In addition, highly fluorinated DLCs appear to be soft with very low wear resistance [32].

Metal contained DLC coatings with various metals, such as Ti, Nb, Ta, Cr, Mo, W, Ru, Fe, Co, Ni, Al, Cu, Au, and Ag, have been reported [32–36]. The metals in the coatings are in the form of nanocrystallites of pure metal or metal carbide depending on the nature and concentration of the metal. The incorporation of the metal can reduce the compressive stress as compared with undoped DLCs [32], which is helpful to enhance the adhesion strength of the coatings on the substrate. The metal doped DLC coatings show a friction coefficient in the range of 0.1–0.2 in ambient air conditions, and a slight dependence on the humidity and metal concentration [32].

Recently, rare earth element of La_2O_3 was doped into DLC coatings by the method of unbalanced magnetron sputtering. No obvious difference in the structure of the coating was found as compared with the conventional DLC coatings, except the La_2O_3 formed nano-crystalline grains and embended into the DLC coating as the content of La_2O_3 increased to 10% [37].

2.3 Applications of DLC Coatings

The low friction coefficient and wear resistance of DLC coatings make them an ideal anti-wear and self lubrication coating for bearings, gears, seals, and blades [38]. The transparency for infrared ray (IR) makes DLCs useful as antireflective and scratch resistance coatings for IR optics [38]. The low deposition temperature also allows DLC coatings to be applied as an anti-wear coating on polymeric material products such as polycarbonate sunglass lenses [39]. DLC coatings are also suitable for biological applications due to their biocompatibility, chemical inertness, and liquid impermeability. They can be deposited on artificial biocomponents made from metals or polymers to improve their compatibility with body tissues [40,41], and also can be used on artificial joints and heart valves, to improve their tribological properties and protect them against corrosion by the body fluid. In addition, the low dielectric constant and negative electron affinity of DLC coatings provide application potentials in ultra-large scale integrated circuits [42] and field emission displays [38].

2.4 Ultra-thin DLC Coatings in Magnetic Disk Driver

One of the most widespread uses of DLC coatings is as protection coating on magnetic head and disk surfaces in hard disk drivers. As shown in Fig. 5, the magnetic disk consists of an Al–Mg alloy or glass substrate, a magnetic layer of cobalt-based alloy thin film, and an ultra-thin DLC overcoat. On the carbon overcoat, one-two monolayers of perfluoropolyether such as ZDOL or Fomblin are used as the molecular lubricant. The read/write head consists of a giant magnetoresistive (GMR) read element and write coil embedded in the Al_2O_3 substrate, and is also protected by an ultra-thin DLC coating. The read/write head is fixed on a slider with an aerodynamic bearing, and flies above the rotating disk. The primary role of the ultra-thin DLC coating is to protect the read/write head from corrosion by the lubricants. On disk, the DLC coating separates the magnetic layer from the ambient, and provides a surface for the molecular lubricant to adhere and move on. Thus, the DLC coatings must be continuous,

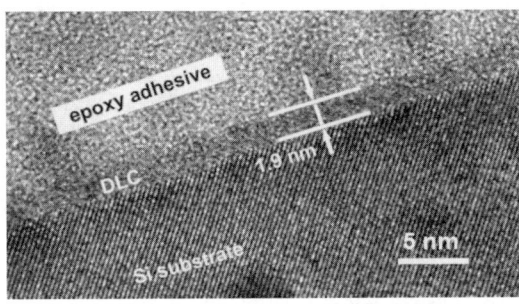

Fig. 6—Bright field high resolution TEM image of 2 nm thick ta-C coating deposited on Si substrate by FCVA.

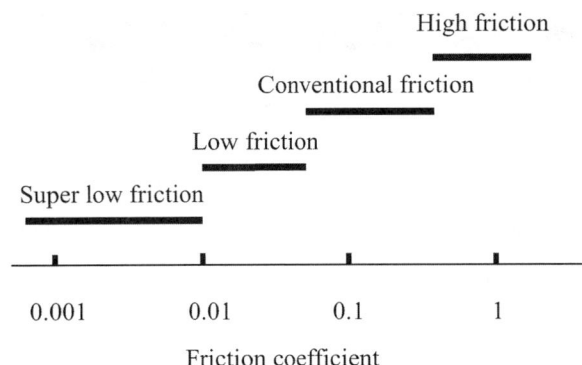

Fig. 7—Classification of friction coefficient.

defect free, and dense enough to act as a corrosion barrier and diffusion barrier.

A decisive factor for the storage density is the magnetic spacing that consists of the fly height, the pole tip recession (PTR), the lubricant thickness, and the thickness of the DLC coatings on both head and disk surface. Larger storage density means a smaller magnetic spacing. A direct method to decrease the magnetic spacing is to decrease the thickness of the DLC protect coatings on both the head and the disk. Initially, sputtered DLC coatings with a thickness of 7 nm were used for drivers with an area density of 10 Gbits/in.^{2} [43–45]. Later, PECVD was used to deposit a-C:H coatings with a thickness of no more than 5 nm for the drivers with an area density of 20–30 Gbits/in.^{2} [45,46]. Recently, a-CNx coating [43,47] and Si doped coating [48] were introduced to substitute a-C:H. In order to achieve a storage density of 200 Gbits/in.^{2}, two requirements must be satisfied. First, the surface of the head and the disk must possess an atomic smooth surface. Second, the magnetic spacing must not exceed 10 nm. Subtracting fly height of 5 nm, PTR and molecular lubricant of 1 nm, a space of only 4 nm is left for the two DLC protective coatings on the head and disk surface, i.e., 2 nm for each coating. In order to synthesize such a thin and smooth DLC coating, it must grow layer by layer to avoid formation of islands. Thus, a deposition method with a high fraction of energetic carbon ions is needed. Traditional deposition methods, such as magnetron sputtering, and PECVD, cannot satisfy the requirement. Due to the nearly 100 % ions fraction, FCVA is a promising technique to deposit the ultra-thin DLC coating with a thickness of 2 nm or even less. Recent research indicates that a continuous DLC coating with a thickness of 2 nm can be synthesized by FCVA as shown in Fig. 6.

2.5 Near Frictionless Carbon Coating

Friction coefficient is a most commonly used factor indicating friction behavior of material pairs. In practical application, friction coefficient could generally be divided into four classes: high friction, conventional friction, low friction, and super low friction. Figure 7 gives a rough classification of friction coefficient. For most of the materials conventionally used in engineering applications, friction coefficient usually varies between 0.05 and 0.5 without additional lubrication. Super low friction generally implies that the friction coefficient is below 0.01.

It is recognized that friction property of a given material is not intrinsic but remarkably influenced by extrinsic conditions. In other words, friction behavior of materials is usually environment dependent. Researchers revealed that some materials like hydrogenated carbon films (a-C:H) exhibited super low friction under specific environments [49,50].

At the end of last century, a near frictionless carbon (NFC) coating was reported, which is practically hydrogen contained DLC film grown on steel and sapphire substrates using a plasma enhanced chemical vapor deposition (PECVD) system [50]. By using a ball on a disk tribo-meter, a super low friction coefficient of 0.001–0.003 between the films coated on both the ball and the disk was achieved [50]. A mechanistic model was proposed that carbon atoms on the surface are partially di-hydrogenated, resulting in the chemical inertness of the surface. Consequently, adhesive interaction becomes weak and super low friction is achieved [22].

It should be stressed that environment conditions strongly influence the friction behavior of NFC coatings. In Erdemir's work, super-low friction coefficient was attained only in vacuum or an inert gas environment [22,51]. When oxygen and moisture are presented, friction coefficient dramatically increased [22,52,53].

The works on super low friction coatings in the past decade brought a hope to approach the dream of eliminating friction by surface coating technology. However, no further important progress has been achieved recently. There are still great difficulties either in theoretical explanation, or in experimental.

3 CNx Films

3.1 Prediction of β-C_3N_4

Friction and wear properties of materials usually correlate to their hardness. For an ideal solid, hardness is generally considered to be correlated with its bulk modulus (a different opinion considers that there is clearer dependence between the hardness of a material and its shear modulus, rather than its bulk modulus [54,55]), which in turn depends on the nature of the chemical bonding within the solid [1]. Mostly, materials with low ionicity and short covalent bonds have large bulk modulus and high hardness [1,56]. Based on calculation, Liu and Cohen [1,56] predicted that the covalent carbon-nitrogen solid such as β-C_3N_4, would have a bulk modulus comparable or greater than diamond, the hardest known material up to now. The hypothetical β-C_3N_4 adopts

TABLE 1—Techniques extensively used for characterization of CNx films.

Items	Analysis Techniques
Structure	X-ray photoelectron spectroscopy (XPS)
	Raman spectroscopy
Nitrogen content	Rutherford backscattering spectroscopy (RBS)
Chemical composition	Auger electron spectroscopy (AES)
	Electron energy-loss spectroscopy
Bonding configuration	Fourier-transform infrared spectrometry (FTIR)
	X-ray photoelectron spectroscopy (XPS)
Crystalline feature	X-ray diffraction (XRD)
Microstructure	Electron diffraction
	Reflection high-energy electron diffraction (RHEED)
Surface morphology	Atomic force microscopy (AFM)
Surface roughness	
Hardness	Nano-indentation
Adhesive strength	Scratch test
Optical property	US-VIS Spectrometry

the known structure of β-Si_3N_4 with C substituted for Si [1]. Each sp^3-hybridized carbon atom is bonded to four nitrogen atoms and each sp^2-hybridized nitrogen atom is bonded to three carbon atoms. The calculated average C–N bond length is 0.147 nm and the theoretical value of its bulk modulus reaches to 427 GPa [1]. The work of Liu and Cohen brought an upsurge in the research of the novel C-N materials.

Attracted by the exciting prospect due to the ideal structure of the covalent carbon-nitrogen solids, a number of groups attempted to synthesis this "harder than diamond" material in the laboratory [57]. Shortly after the publication of Liu and Cohen's work, Niu et al. [58] reported their work on synthesis of the covalent solid carbon nitride. They prepared C–N thin films on Si(100) substrate using pulsed laser ablation method, and claimed having obtained β-C_3N_4 crystallite in the film. A half year later, Yu et al. [59] presented their experimental results that the β-C_3N_4 phase was synthesized on Si and Ge wafers by rf diode sputtering of a pure graphite target with pure N_2. The β-C_3N_4 particles with typical dimensions of approximately 0.5 μm are embedded in a 1-μm thick layer of a C-N polymer [59]. The subsequent researchers also reported that they have prepared thin films containing β-C_3N_4 or having nitrogen content near to the stoichiometric value [60–63]. However, though great efforts have been made, up to the present, pure phase of crystalline β-C_3N_4 is still not obtained. Mostly, nonstoichiometric amorphous carbon nitride (CNx) films with or without crystallite in the films were obtained yielding a hardness value between several GPa to about 40 GPa, far lower than expected.

3.2 Preparation and Characterization of CNx Films

Up to the present, a number of conventional film preparation methods like PVD, CVD, electro-chemical deposition, etc., have been reported to be used in synthesis of CNx films. Muhl et al. [57] reviewed the works performed worldwide, before the year 1998, on the methods and results of preparing carbon nitride films. They divided the preparation techniques into several sections including atmospheric-pressure chemical processes, ion-beam deposition, laser techniques, chemical vapor deposition, and reactive sputtering [57]. The methods used in succeeding research work basically did not go beyond the range summarized by Muhl et al.

For CNx films, the ratio of N/(N + C) (content of nitrogen within CNx film) is one of the most important parameter, which determines the chemical structure of the films. For the correct C_3N_4 stoichiometry, the ideal value of the N/(N + C) ratio is about 0.57. The nitrogen content in a film strongly depends on the deposition methods and the corresponding parameters. Moreover, post-heat treatment (frequently annealing) to the amorphous film could also remarkably alter the content of the nitrogen. Annealing of an as-deposit CNx film usually induces nitrogen loss, while the extent is related to the temperature. Loosely bonded nitrogen can be easily expelled under a relatively low temperature (<200°C) [64]. At a relatively high temperature (550°C to 800°C), the decrease of nitrogen content is frequently accompanied with the degradation of the film. In this process, nitrogen atoms bonded to sp^3 hybridized carbon are preferentially removed and those bonded to sp^2 hybridized carbon are relatively stable, resulting in the graphitization of the film [64–67].

The properties of a CNx film strongly depend on the structure of the film. Consequently, the emphasis of characterization for CNx film is naturally put on the structure, bonding configuration, crystalline feature, as well as nitrogen content. The techniques extensively used for the characterization of CNx film are listed in Table 1.

It should be noted that precise structure characterization of CNx films still remains somewhat difficult. Therefore, in order to obtain a convincing information, it is suggested to apply at least two independent methods for studying the structure of CNx film and moreover, the results derived should be mutually consistent [57].

3.3 Properties of CNx Films Related to Tribological Applications

The research work on carbon nitride materials was first motivated due to its hypothetical super-hardness. The tribological properties of the CNx films are, of course, the emphasis of the succeeding research works. Although the expected structure and hardness have not been achieved up to the present, the potential interesting properties still stimulate many approaches on the tribological behavior of CNx films. Figure 8 gives the results of hardness measurement reported by the

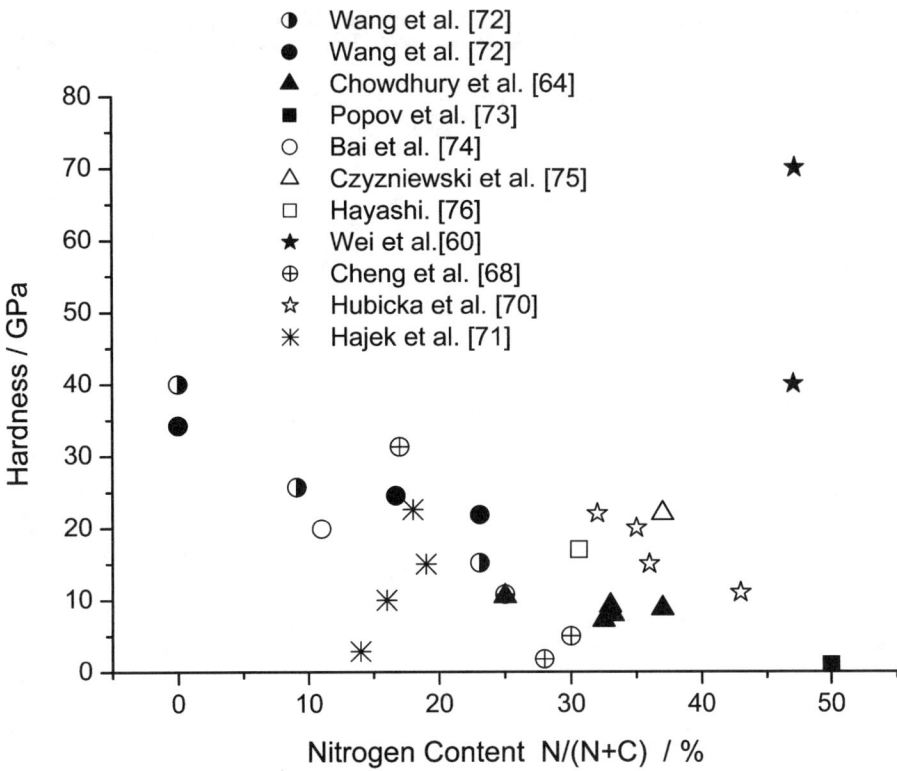

Fig. 8—Hardness versus nitrogen content, from Refs. [60,64,68,70–76].

selected works. From the figure, it can be seen that the hardness has a tendency of decreasing with the increase of the nitrogen content except for the work of Wei et al. [60], who claimed that they obtained hardness values between 40 and 70 GPa with the N content of 0.47. Theoretically, large nitrogen content in the film is expected to be beneficial to the formation of β-C_3N_4, as the consequent, the hardness should increase with the nitrogen content. However, the results shown in Fig. 8 have given an opposite tendency, indicating that the hardness is shown to be not evidently dependent on the total nitrogen concentration. It was further pointed out that the hardness of carbon nitride films is mainly determined by the amount of $C \equiv N$ triple bonding [66,68,69].

Table 2 lists the results of the friction coefficient derived by different researchers under different test conditions [71,73,75–80]. From the table, it can be seen that the CNx films prepared by different techniques did not demonstrate surprising frictional behavior.

Wettability of CNx film is another interesting character for tribological applications. There have been only a few approaches on the solid/liquid wettability of CNx films [81–85]. Figure 9 illustrates the results of the contact angle of water on CNx films by different researchers. Some of the results showed that increase of nitrogen content in the film led to the decrease of the contact angle of water on CNx films [81–83]. However, an opposite trend was obtained indicating that the contact angle increased slightly with the nitrogen content [85].

Though considerable effort has been made on synthesizing β-C_3N_4, investigations on the current CNx films did not show unusual tribological properties. However, attempts are still being made to prepare the expected films, not limited in obtaining β-C_3N_4. Both theoretical and experimental results suggest that, other than β-C_3N_4, there might be structural phase of carbon nitrides, which may exhibit large values for bulk moduli and hardness [86,87]. Teter and Hemley [87] predicted that, using principle calculations, the bulk modulus of a cubic form of C_3N_4 may exceed that of both diamond and β-C_3N_4. Another direction in CNx research is incorporation of some element for film enhancement. An example is the incorporation of silicon (Si) into CNx matrix [88], which was proved to be effective in increasing oxidation resistance of the film at elevated temperature. The role of Si is suggested to catalyze the chemical reactions in the plasma and eventually leads to the formation of crystalline C–N network [89,90].

4 Multilayer Films

In 1970, Koehler [91] presented a concept for the design of strong solids prepared by epitaxial crystal growth which consist of alternate layers of crystals A and B, as shown in Fig. 10. Such multilayer materials are nowadays called heterostructures or superlattices. In such materials, A and B should have similar lattice parameters and thermal expansions, but different elastic constants. Under applied stress, dislocations would form in B layers (supposing $E_A > E_B$), and move towards the A/B interface. Elastic strain induced in the A layer would cause a repulsing force that would hinder the dislocation from crossing the interface. Thus, a large external stress will be required to drive dislocations from B into A, and the strength of the superlattice is enhanced strongly. Later, the prediction of Koehler was verified by Lehoczky [92,93] who deposited Al/Cu and Al/Ag laminate sys-

TABLE 2—Friction coefficient of CNx films—Data selected from published articles.							
Preparation Method	Nitrogen Content N/(N+C)	Micro-hardness GPa	Friction Coefficient	Counterpart Material	Environment	Substrate Material
Inductively coupled plasma chemical vapor deposition (Popov [73])	0.5	0.5-1	0.6	Steel ball	Air	Si(100)
Reactive magnetron sputtering (Czyzniewski [75])	0.37	22	0.45	Bearing steel ball	Air RH20~30 %	Steel
Electron beam evaporation (Hayashi [76][a])	0.306	15	0.2~0.25	Steel ball	Air	Si(100)
			0.2~0.5		1.3×10^{-4} Pa	
		20	0.2~0.3		Air	TiN(200)
			0.2~0.4		1.3×10^{-4} Pa	
		14				TiN(111)
DC magnetron sputtering (Fernández [77])	0.2~0.3	...	0.15~0.2	Steel ball	Air	Si
Electrochemical Deposition (Yan [78])	0.074	...	0.08	Al_2O_3 ball	Air RH40~45 %	Si(100)
Pulsed vacuum arc deposition (Koskinen [79])	0~0.34	...	0.25~0.3	Steel ball	Air RH50 %	High chromium tool steel
			0.11~0.25	Al_2O_3 ball		Si(100)
DC magnetron sputtering (Hajek [71][a])	0.138~0.193	2.2~22.6	0.1~0.55	Al_2O_3 ball, Steel ball Diamond coated Si_3N_4 ball	Air RH40 %	Si(100)
Ion beam assisted deposition (Tokoroyama [80])	...	23	<0.01	Si_3N_4 ball	Dry N_2	Si(100)

[a]The values of friction coefficient are estimated according to the figures presented.

tems. Although the layers in the laminate systems are polycrystalline, when the layer thickness is in the range of 20–70 nm, the coatings still show an enhanced strength about 4.2 times greater than that expected from the rule of mixture given by the equation:

$$H(A_xB_y) = [xH(A) + yH(B)]/(x + y) \quad (1)$$

where H is the hardness, x and y are the volume fraction of A and B. Many researchers have been devoted to the synthesis of various multilayer systems including metal/metal superlattices, metal nitride/metal superlattices, metal nitride/metal nitride superlattices, metal nitride/CNx multilayers, and metal carbide/metal multilayers. In many cases, the strength of the multilayer systems was improved when the modulation wavelength decreased to about 5–7 nm.

4.1 Deposition of Multilayer Films

In order to achieve the enhanced strength, atomically sharp A/B interfaces are needed for the mechanism proposed by Koehler. The multilayer films should be deposited under a relative low temperature in order to avoid the interdiffusion between the layers. Thus, PVD techniques such as evaporation and sputtering deposition are the reasonable methods to fabricate multilayer films. There are several ways to achieve the alternate deposition of the composing layers A and B: fixed substrate position mode, repeated substrate position mode, and single source mode. In the fixed substrate position mode, two sources are aimed at a fixed substrate position and a mechanical shutter is used to modulate the fluxes, as shown in Fig. 11(a). Such systems can be operated without the shutter, in which the two sources work alternately. In the repeated substrate position mode (Fig. 11(b)),

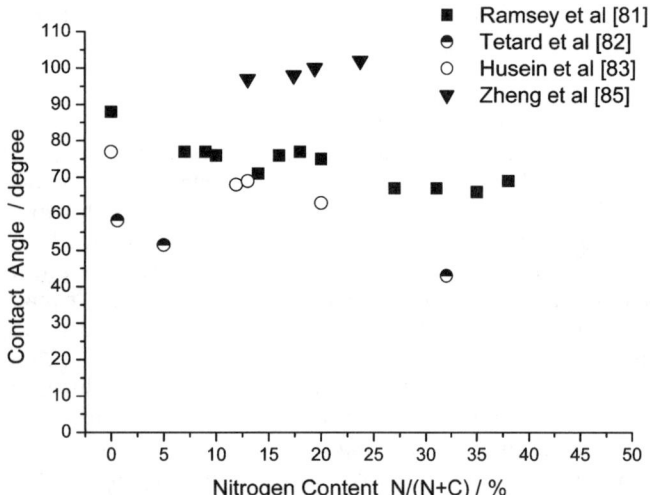

Fig. 9—Contact angle (to water) versus nitrogen content, from Refs. [81–83,85].

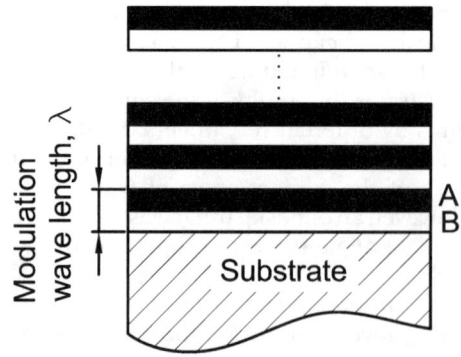

Fig. 10—Schematic of the structure of superlattice film.

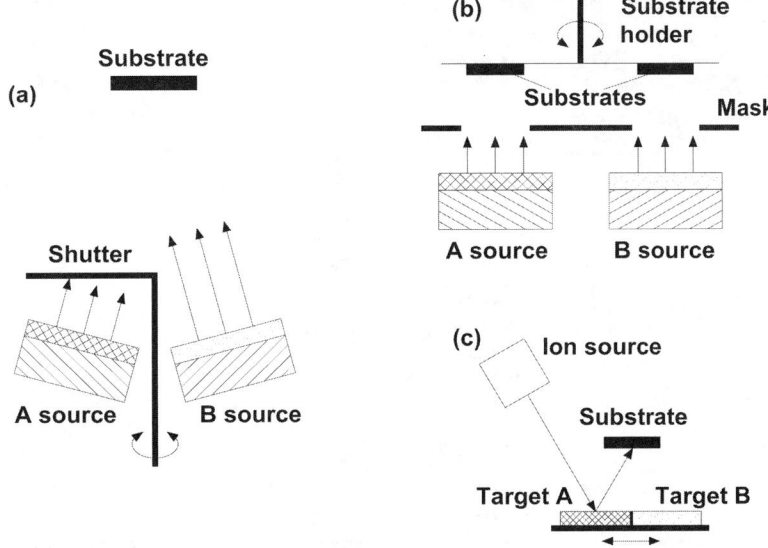

Fig. 11—Schematics of various deposition methods of multilayer films.

the substrate is positioned towards two sources alternately. A mask is placed in front of the substrate holder in order to avoid the mixing of the species from the two fluxes. In the single source mode, two different targets A and B are moved towards one sputtering beam (e.g., Ar^+ beam) or evaporation beam (e.g., laser beam) alternately, as shown in Fig. 11(c). The substrate can rotate in order to obtain uniform coating over a large area. In addition, a single source can be used to deposit the metal/metal compound multilayers, e.g., Ti/TiN [94] and Cr/CrN [95], by modulating the reactive gas flow during reactive sputtering or evaporation.

4.2 Microstructure of Multilayer Films

The microstructures of multilayer films are determined by the constituents, the modulation wavelength, the deposition conditions, and also the substrate crystalline orientation. In multilayers fabricated from two miscible face-centered-cubic (fcc) constituents with small lattice mismatch, such as Cu/Ni multilayer [96], single crystal epitaxial multilayers can be formed by either magnetron sputtering or e-beam evaporation. In immiscible fcc/bcc (body-centered-cubic) multilayer systems, room temperature sputtering deposition often results in the formation of columnar polycrystalline grain structure in the multilayers with a strong textured orientation relationship, such as Cu/Nb multilayer coating [97]. In addition, there are some multilayer systems without epitaxial growth, consisting of crystal/amorphous constituents, amorphous/amorphous constituents, such as TiN/CNx [98] and BN/CNx [99], respectively. Such multilayer systems do not possess the superlattice structure, but they still show improvement in mechanical or tribological properties.

4.3 Mechanical Properties of Multilayer Films

Two requirements should be satisfied in order to achieve the enhancement in strength of the superlattices: modulation wavelength (λ) in the scale of nanometre and sharp interfaces between layers. There is an optimum λ for a superlattice film to achieve the maximum strength, as shown in Fig. 12 [100–102]. The strength of the superlattice film will decreases as the modulation wavelength is larger than the optimum λ due to the increase in the dislocation activities. On the other hand, the strength also decreases as the modulation wavelength is less than the optimum λ. And the strength is similar to that given by the rule of mixture [Eq (1)] as the modulation wavelength decreases to about 2 nm due to the diffusion between layer A and layer B and the disappearance of the sharp interfaces. Table 3 lists the maximum hardness and the optimum λ of several typical superlattice or multilayer films. In addition, the strength of the multilayer is also influenced by the layer thickness ratio t_A/t_B. Normally, the coating with the layer thickness ratio of about 1 shows the largest strength [107].

The mechanism of the strength enhancement can be explained on the basis of dislocation activity. For a coating with large modulation wavelength (such as several tens of

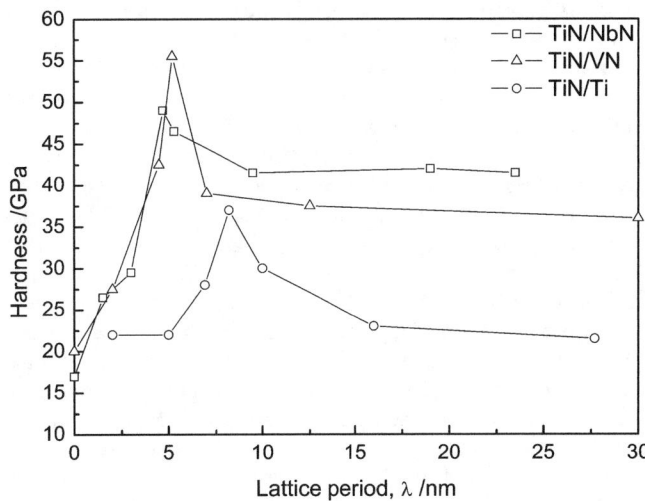

Fig. 12—Hardness of TiN/NbN superlattice after Shinn et al. [100], TiN/VN superlattices after Helmersson et al. [101], and TiN/Ti superlattice after Shih et al. [102] as functions of the superlattice period.

Materials (A/B)	Maximum Hardness (GPa)	Modulation Wave Length λ (nm)	Reference
Cu/Ni	3.9	5	[103]
Cu/Cr	7.1	2.5-5	[103]
Cu/Nb	7.3	1.2	[103]
TiN/NbN	49	4.6	[100]
TiN/VN	56	5.2	[101]
VN/VNbN	22	3-10	[104]
TiN/Ti	37	16	[102]
WN/W	35	38	[102]
HfN/Hf	50	8	[102]
TiN/Pt	22.3	8.2	[105]
CrN/Cr	27.5	11	[95]
NbN/Mo	33	2	[106]
NbN/W	29	2-3.5	[106]
TiC/Cr	29.7	9	[107]
TiC/Mo	28.9	10.4	[107]
TiC/Fe	27.5	10.2	[107]

TABLE 3—Maximum hardness and modulation wavelength of several multilayer systems.

nanometres), the dislocation multiplication source could operate. The operation of Frank-Read sources within soft layers B results in the stress concentrations at dislocation pileups, which can drive dislocations into the A layers. The critical stress required for the operation of a Frank-Read source is given by the equation [108]:

$$\sigma_{FB} = Gb/l \qquad (2)$$

where G is the shear modulus, b is the Burgers vector, and l is the distance between the dislocation pinning sites. For a Frank-Read source confined in a B layer, l is less than $t_B/\sin\theta$ (t_B is the thickness of one single B layer, and θ is the angle between the A/B interfaces and the dislocation glide plane), and the minimum stress required for dislocation source operation is given by the following equation:

$$\sigma_{FB} = G_B b_B \sin\theta/t_B \qquad (3)$$

Hence, the strength of the superlattice increases with the decrease in the single layer thickness.

For superlattices with small modulation wavelength of several nanometres, the dislocation multiplication cannot occur, and the dislocation activity is demonstrated by the movement of individual dislocations from B layer into A layer by stress. The critical shear stress to move a dislocation from B layer into A layer ($\sigma_{A/B}$) can be given by the Lehoczky theory equation [108] as shown in Fig. 13. Figure 13 also gives the normalized $\sigma_{A/B}$ as function of t_B/b. It can be seen that there is no strength enhancement as $t_B \leq 4b$, which corresponds to very small layer thickness (<1 nm), and the disappearance of interfaces due to the diffusion between layer A and layer B. The $\sigma_{A/B}$ increases rapidly with the increase of t_B as t_B larger than $4b$, which indicates the strength of the multilayer system is enhanced strongly. It also can be seen that the $\sigma_{A/B}$ almost reaches the asymptotic value as the t_B is equal to $15b$ which corresponds to the layer thickness by several nanometres. With further increase of t_B ($t_B \gg 4b$), the Lehoczky theory is invalidated and the strength enhancement should attribute to the mechanism described in the previous paragraph.

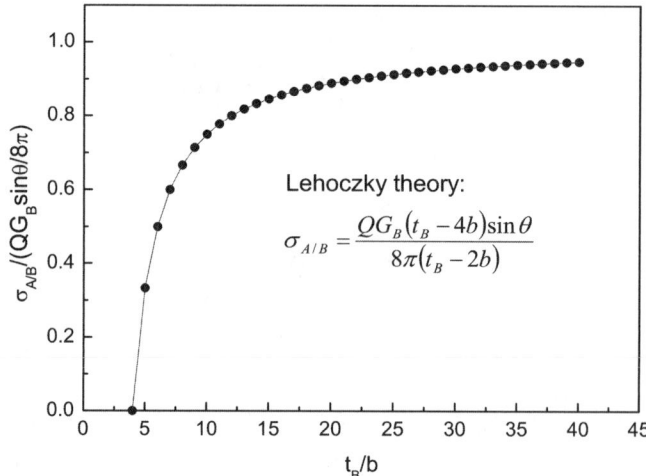

Fig. 13—Normalized $\sigma_{A/B}$ as function of t_B/b, $\sigma_{A/B}$ is the critical shear stress to move a dislocation from the B layer into the A layer, $Q=(G_A-G_B)/(G_A+G_B)$, G_A and G_B are the shear moduli of A and B, b is the Burgers vector, t_B is the thickness of one single B layer, and θ is the angle between the A/B interfaces and the dislocation glide plane.

Lehoczky theory:
$$\sigma_{A/B} = \frac{QG_B(t_B - 4b)\sin\theta}{8\pi(t_B - 2b)}$$

4.4 Tribological Applications of Multilayer Films

Multilayer films are widely used for protection from abrasion and fatigue wear in tribological applications due to their improved mechanical properties. Many multilayer systems showed improved tribological performance as compared with single component films. And the wear rate of many multilayer films shows modulation wavelength dependence in a similar way of that of strength. For example, Holleck [109] reported improvement in wear resistance of TiC/TiB$_2$ multilayer coating under an optimized modulation wavelength of 20 nm. He attributed this to the crack deflection and the stress relaxation mechanism. Similar dependence was found in multilayer films, such as CrN/Cr [95,110], TiN/Ti [111], CNx/BNx [99,112], and WS$_2$/MoS$_2$ [113], etc. Holmberg et al. [114] described the deformation process of the multilayer films consisting of alternate hard layer and soft layer in the following way. The multilayer coating will bend under the normal load and the layers will slide over each other, i.e., a shear zone exists at the interface which permits the harder layers (more brittle) to deflect without fracture. Hence, a lower stress is generated in the layers as compared with a thicker single component film with the same bending radius as shown in Fig. 14. And a larger stress is needed for the crack generation and propaga-

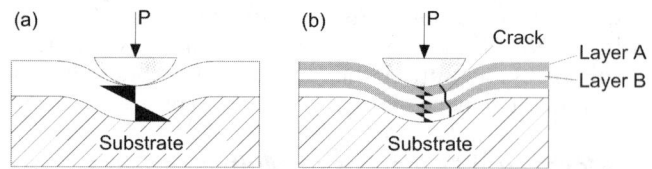

Fig. 14—Schematics of shear stress in (a) single constituent thick coating and in (b) multilayer coating. The shear stress in each layer of multilayer is less than that in the single constituent thick coating due to the layer slide at the interface of the multilayer coating. In addition, crack deflection in the multilayer coating was illustrated schematically.

tion in the multilayer films. Such an explanation is suitable for the fatigue wear of the multilayer films.

In case of abrasion, a plowing effect is the main wear type. Berger [110] gave an explanation for the improvement in the abrasive wear resistance of the CrN/Cr multilayer film. He suggested that there is a supported zone in the soft Cr layer close to the Cr/CrN interface. In the supported zone, the abrasive penetration is reduced owing to the higher hardness of the underlying CrN layer. And thus the supported zone possesses a higher abrasive wear resistance as compared with the single component Cr or CrN film. Miyake [99,112] investigated the abrasive microwear behavior of CNx/BN multilayer films using AFM with a diamond tip scanning on the film. The results show that the wear resistance increased at the interface between the CNx and BN layers, coinciding with the mechanism proposed by Berger.

In conclusion, one should choose an appropriate multilayer system for different application purposes. For the case of fatigue wear, multilayer films consisting of two hard materials with different shear modulus, such as DLC/WC multilayer film [115], would satisfy the requirement for wear resistance. While for abrasive wear, multilayer films consisting of hard ceramic layers and soft metal layers, such as TiN/Ti and CrN/Cr [116,117] multilayer films are more competent.

5 Superhard Nanocomposite Coatings

Superhard materials implies the materials with Vickers hardness larger than 40 GPa. There are two kinds of superhard materials: one is the intrinsic superhard materials, another is nanostructured superhard coatings. Diamond is considered to be the hardest intrinsic material with a hardness of 70–100 GPa. Synthetic c-BN is another intrinsic superhard material with a hardness of about 48 GPa. As introduced in Section 2, ta-C coatings with the sp^3 fraction of larger than 90 % show a superhardness of 60–70 GPa. A typical nanostructured superhard coating is the heterostructures or superlattices as introduced in Section 4. For example, TiN/VN superlattice coating can achieve a superhardness of 56 GPa as the lattice period is 5.2 nm [101].

In 1995, Veprek et al. [118,119] presented the concept for the design of superhard nanocomposite coatings, which consist of a few nanometre small crystallites of a hard transition metal nitride "glued" together by about one-monolayer-thin layer of amorphous nonmetallic, covalent nitride such as Si_3N_4 or BN, as schematically shown in Fig. 15. The mechanical properties of such nanocomposite coatings are strongly enhanced due to the absence of dislocation activity in the nano-crystallites and the strong boundaries "glued" by the amorphous phase. The superhard nanocomposite coating shows an enhanced hardness approaching that of diamond, and a higher oxidation resistance as compared to diamond and c-BN, which makes it a promising candidate for wear and oxidation protection.

5.1 Preparation of Superhard Nanocomposite Coatings

Unlike the heterostructures whose periodic structure must be accurately controlled, the formation of nanocomposite structure is self-organized based upon thermodynamically driven spinodal phase segregation [118–121]. For the CVD

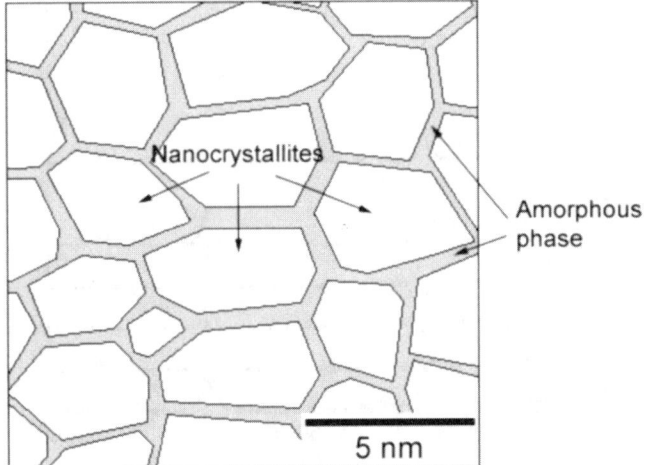

Fig. 15—Schematic of the microstructure of the superhard nanocomposite coating.

method, two conditions are necessary for preparing superhard nanocomposite [121]. One is the sufficiently high activity of nitrogen (partial pressure >1 Pa) that provides the necessary thermodynamic driving force for spinodal decomposition. Another is the deposition temperature of 500–600°C that assures the diffusion-rate-controlled phase segregation to proceed sufficiently fast during deposition [121].

The superhard nanocomposite coatings can be deposited by either CVD or PVD. Veprek et al. [118–123] deposited the superhard TiN/Si_3N_4 nanocomposite coating with maximum hardness of 50 GPa by the PECVD method using $TiCl_4$, SiH_4, and N_2 as precursor gases. Later, they also deposited W_2N/Si_3N_4, VN/Si_3N_4, and TiN/BN nanocomposite coatings with hardness larger than 50 GPa [124–126]. Deposition of TiN/Si_3N_4 nanocomposite coating using reactive magnetron sputtering methods was reported [127–129]. The nanocomposite coating was also deposited by the method of vacuum arc cathodic evaporation from a combined cathodic target [130]. In our former work [131], the IBAD method was used to deposit TiN/Si_3N_4 nanocomposite coatings, in which a titanium target and a silicon target were sputtered by two Ar^+ beams, and a N_2^+ beam was employed to bombard the growing coating simultaneously.

Some metal nitride/soft metal composite coatings, such as ZrN/Cu, ZrN/Ni, and ZrN/Y, were deposited by unbalanced magnetron sputtering at negative bias, and were claimed to be superhard [132–134]. However, the hardness of these coatings showed that they were thermally unstable, and decreased from larger than 40 GPa to about 15 GPa after annealing at the temperature of 600°C [135]. Veprek et al. [121] attributed the superhardness of such coatings to the synergistic effect of ion bombardment during their deposition involving a decrease of crystallite size, densification of the grain boundaries, formation of Frenkel pairs and other point defects, and built-in biaxial compressive stress. And the ion-induced effects will anneal out at a temperature of above 600°C.

In addition, the impurities, especially the oxygen, can strongly decrease the hardness of nanocomposite coatings. That should be the primary reason for the lower hardness

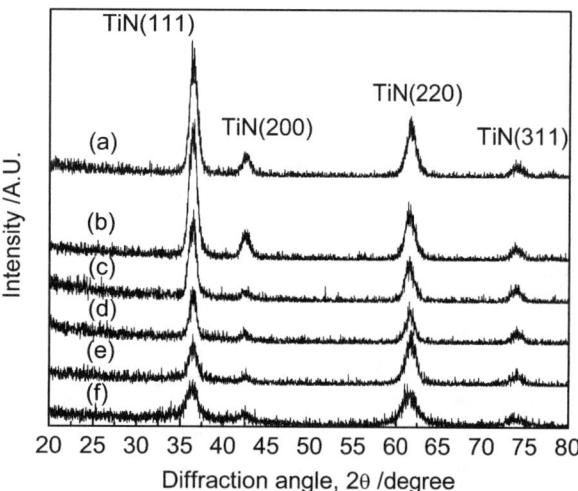

Fig. 16—XRD partens of the TiN/Si$_3$N$_4$ nanocomposite coatings with different Si content deposited by reactive magnetron sputtering. (a) 0, H=26.0 GPa, (b) 2.3 at. %, H=31.9 GPa, (c) 5.7 at. %, H=43.2 GPa, (d) 7.4 at. %, H=43.3 GPa, (e) 10.8 at. %, H=47.1 GPa, (f) 17.2 at. %, H=40.3 GPa.

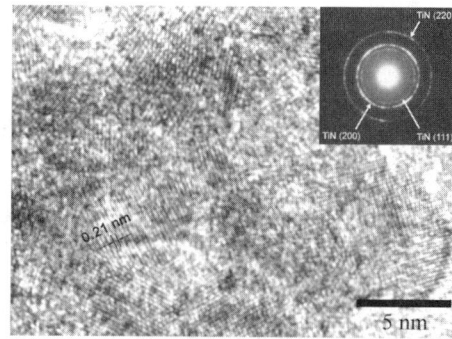

Fig. 18—High resolution TEM plan-view image of TiN/Si$_3$N$_4$ nanocomposite coating with Si content of 10.8 at. % and hardness of 42 GPa. The coating was deposited on NaCl substrate for about 50 nm thick and then was floated off onto a mesh. The crystallites were confirmed to be TiN by the interplanar distance of 0.21 nm, which is the TiN (200) interplanar distance. The gain size of the TiN crystallites is less than 5 nm.

(<40 GPa) of the TiN/Si$_3$N$_4$ coatings reported by several authors [127,136–139]. A superhardness of nanocomposite coatings is accompanied by an oxygen content of less than 0.2 at. % [121]. In order to minimize the oxygen content in the coating, a base pressure of less than 1×10^{-4} Pa in the deposition chamber and a deposition rate of larger than 1 nm/s should be satisfied. Baking the chamber at a temperature a little higher than the deposition temperature before deposition procedure is advantageous for outgassing the water adsorbed on the chamber wall that is the primary source of oxygen impurities in the coating.

The TiN/Si$_3$N$_4$ coating is the first synthesized superhard nanocomposite coating, and has been widely studied by many groups. We will focus on the characteristics of the TiN/Si$_3$N$_4$ nanocomposite coating in the following section, and the properties of other nanocomposite coatings are similar to it.

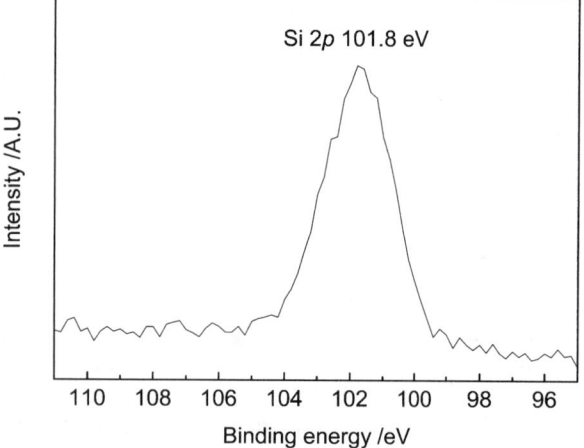

Fig. 17—XPS spectrum of Si 2p peak for the TiN/Si$_3$N$_4$ nanocomposite coating with the optimum Si content of 10.8 at. % and the maximum hardness of 47.1 GPa.

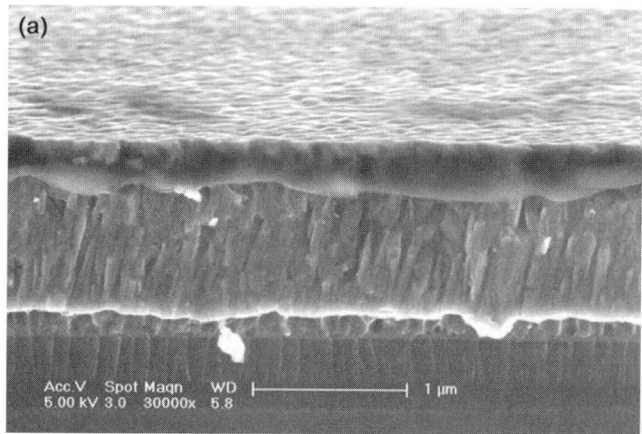

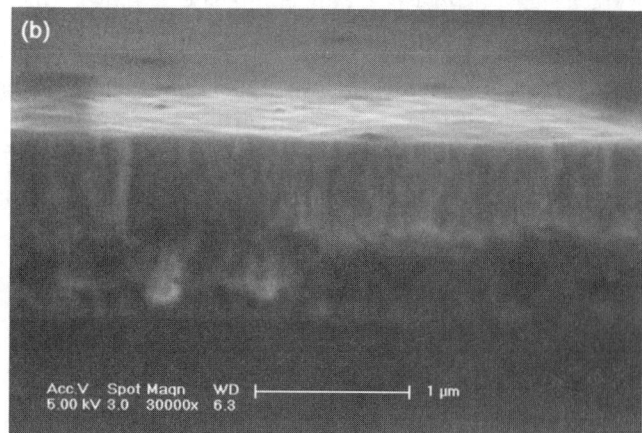

Fig. 19—Cross-sectional morphologies of (a) TiN coating with hardness of 26 GPa, and (b) TiN/Si$_3$N$_4$ coating with optimum Si content of 10.8 at. % and hardness of 47.1 GPa deposited by reactive magnetron sputtering.

TABLE 4—The maximum hardness and optimum Si content of several MeN/Si$_3$N$_4$ (Me=Ti, W) coatings prepared by different authors.

Coating	Maximum Hardness (GPa)	Optimum Si Content (at.%)	Deposition Method	Authors	Reference
TiN/Si$_3$N$_4$	≥50	8~10	PECVD	Veprek et al.	[118–126]
TiN/Si$_3$N$_4$	42	10.8	IBAD	Zhang et al.	[131,143]
TiN/Si$_3$N$_4$	47.1	10.8	Reactive magnetron sputtering	Zhang et al.	[145,146]
TiN/Si$_3$N$_4$	39~43	4~10	Reactive magnetron sputtering	Rebouta et al.	[128]
TiN/Si$_3$N$_4$	38	5	Reactive magnetron sputtering	Diserens et al.	[127]
TiN/Si$_3$N$_4$	36	4	Reactive magnetron sputtering	Hu et al.	[137]
TiN/Si$_3$N$_4$	38	11	Reactive magnetron sputtering	Kim et al.	[138]
TiN/Si$_3$N$_4$	35	9	Reactive magnetron sputtering	Jiang et al.	[139]
TiN/Si$_3$N$_4$/TiSi$_2$	≥80	5.4	PECVD	Veprek et al.	[142]
W$_2$N/Si$_3$N$_4$	50	7~8	PECVD	Veprek et al.	[124]

5.2 Characterization of TiN/Si$_3$N$_4$ Superhard Nanocomposite Coatings

As mentioned in the foregoing section, the TiN/Si$_3$N$_4$ coating was spinodal segregated into TiN nanocrystallites and amorphous Si$_3$N$_4$ phase. The structure information of TiN crystallites can be derived from the X-ray diffraction (XRD) spectrum as shown in Fig. 16. The grain size of the TiN crystallites can be calculated from the Scherrer formula [140]. The Si bonding structure in the coating can be investigated by means of X-ray photoelectron spectroscopy. If the Si was fully nitridized into Si$_3$N$_4$, a symmetric characteristic peak of the Si 2p electrons in Si$_3$N$_4$ should be found at the binding energy range of 101.6–102.2 eV [141], as shown in Fig. 17. High resolution transmission electron microscope (HR-TEM) can provide direct observation of the nanocomposite coatings due to its high resolution of 0.1 nm. Figure 18 is a typical HRTEM image of a nanocomposite TiN/Si$_3$N$_4$ coating with hardness of 42 GPa. Several nanocrystallites with size of less than 5 nm can be distinguished in the figure, where crystal lattice plane are seen. The crystallites are confirmed to be TiN (200) phase according to the interplanar distance of 2.12 angstroms. Because of random orientation of the crystallites, only the crystallites whose lattice plains are parallel to the electron beam can appear lattice resolution. A small tilting of the sample in the microscope results in the vanishing of these lattice images and the appearance of others in areas where the micrograph in Fig. 18 appears structureless. Thus, one cannot claim that the amorphous-like region in Fig. 18 is the amorphous Si$_3$N$_4$ phase. In fact, if the coating truly forms the structure that one-monolayer-thin layer of amorphous Si$_3$N$_4$ surrounds the TiN nanocrystallites described by Veprek et al. [118,119], it is difficult to distinguish such a thin amorphous layer from the HRTEM image due to the resolution limitation.

The appearance of amorphous Si$_3$N$_4$ phase changes the coating morphology from a columnar morphology of TiN coating to a more dense and isotropic morphology, as shown in Fig. 19. The vanishment of the columnar structure is important to the excellent mechanical properties of the nanocomposite coating, due to the absence of the weaker bonds between the columns.

As mentioned in the foregoing sections, the nanocomposite coatings will achieve the largest hardness as the nanocrystallites are fully covered by a one-monolayer-thin layer of amorphous phase. Hence, there must be an optimum Si content in the coating with the maximum hardness. It can be calculated that for a nanocomposite TiN/Si$_3$N$_4$ coating with a TiN grain size of 4 nm, the volume fraction of the Si$_3$N$_4$ phase is about 25 vol. %, i.e., about 10 at. % Si content in the coating. This has been confirmed by many authors. The TiN/Si$_3$N$_4$ coatings show a maximum hardness as the Si content is in the range of 8~10 at. % [118–128,131,136–139,142,143], as listed in Table 4. Notice that the coatings of Rebouta, Diserens, Hu, Kim, and Jiang possess hardness less than 40 GPa. It should attribute to either the dissatisfaction of the deposition conditions described in Section 5.2 or the high impurity content in the coating, especially the oxygen impurity. And the TiN/Si$_3$N$_4$/TiSi$_2$ coating deposited by Veprek et al. [121,142] possesses an ultra-high hardness of 80~105 GPa, which is as high as the hardness of diamond. However, the ultra-hardness decreases to about 50 GPa after exposing it to air for a period of six to eight months. Veprek et al. attribute it to the chemical attack on the TiSi$_2$ phase by the polar water molecules [121].

Veprek et al. [120,121,144] present the explanation of the superhardness for the nanocomposite coating on the basis of the absence of dislocation activity in a few-nanometre-

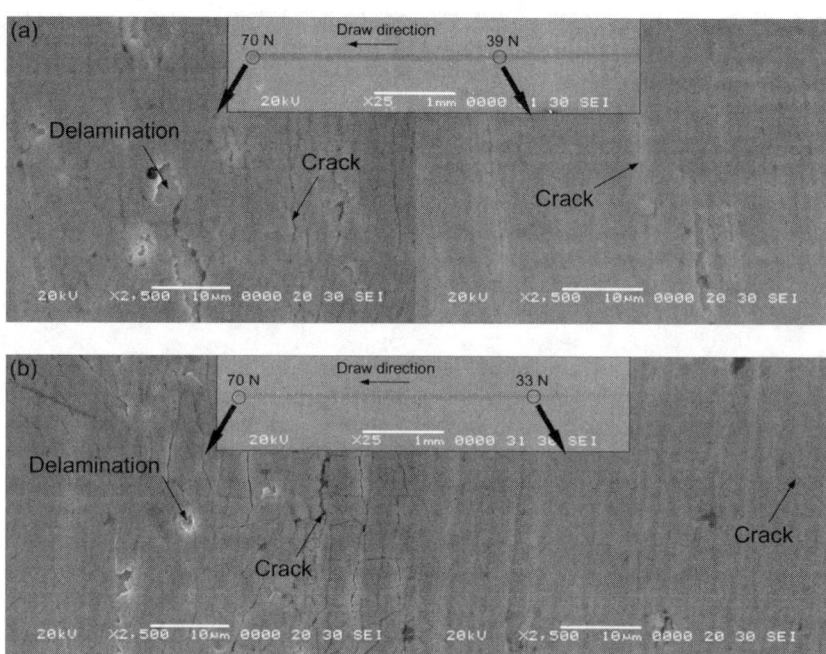

Fig. 20—Cracks in the scratch tracks on (a) TiN coating with hardness of 26 GPa, and (b) TiN/Si$_3$N$_4$ coating with optimum Si content of 10.8 at. % and hardness of 47.1 GPa deposited by reactive magnetron sputtering.

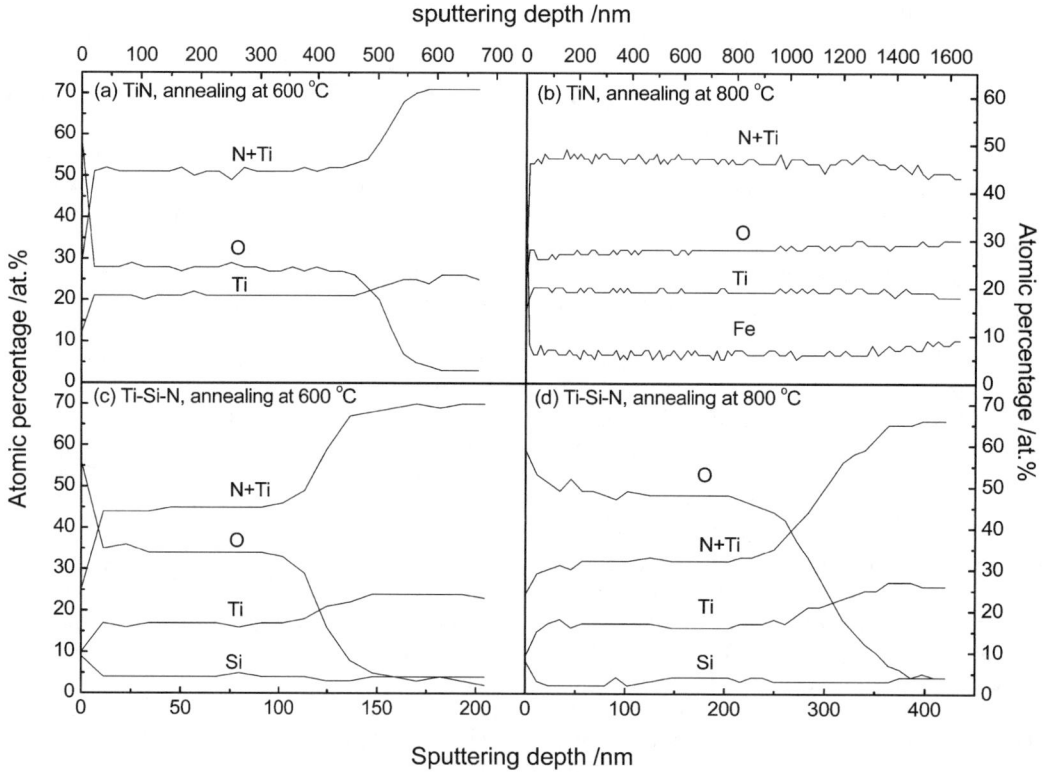

Fig. 21—AES depth profiles of the TiN coatings (a and b) and the TiN/Si$_3$N$_4$ coating with optimum Si content of 10.8 at. % and hardness of 47.1 GPa (c and d) annealed at the temperature of 600 or 800°C in ambient atmosphere. The oxidation depth of the coatings is the sputtering depth where the oxygen atomic percentage reaches the minimum level.

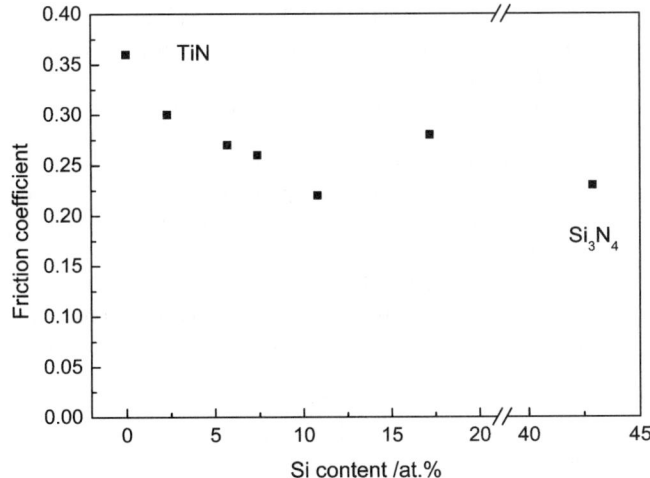

Fig. 22—Friction coefficients between WC ball and TiN/Si$_3$N$_4$ nanocomposite coatings as function of the Si content. The coatings were deposited by reactive magnetron sputtering. The friction coefficients of the TiN/Si$_3$N$_4$ coatings were obtained under the load of 20 N. In the case of the TiN coating and the Si$_3$N$_4$ coating, the load is 5 N, because the two coatings will fail and peel off from the substrate under the load of 20 N.

small crystallites, and a very small stress concentration factor for ≤1-nm-small nanocracks. Therefore, a large stress is needed to initiate and propagate a crack in such systems. In addition, the thermodynamically driven phase segregation results in a strong interface between the TiN nanocrystallites and the Si$_3$N$_4$ phase, which suppress the grain boundary sliding. Thus, the strength of the nanocomposite coating approaches the ideal decohesion strength of the materials given by the equation

$$\sigma_C \approx (E_Y \cdot \gamma_S / a_0)^{0.5} \qquad (4)$$

where E_Y is Young's modulus, γ_S is surface energy, and a_0 is interatomic bond distance [121].

Besides the superhardness, the TiN/Si$_3$N$_4$ nanocom-

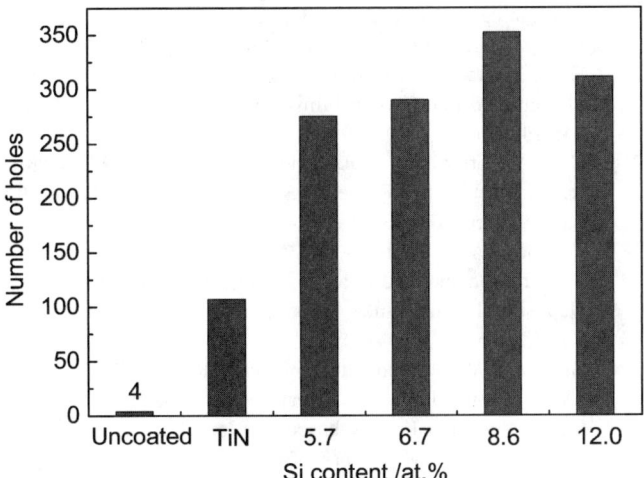

Fig. 23—The cutting life of the uncoated drill and the drills deposited with TiN coating and TiN/Si$_3$N$_4$ nanocomposite coatings drilling holes on quenched AISI 420 stainless steel. The coatings were deposited by reactive magnetron sputtering.

posite coatings also show high Young's modulus of larger than 350 GPa [118,121,127,128,139,145], high elastic recovery of larger than 80 % [122,123,145] at the optimum Si content. The fracture toughness of the coatings was evaluated by Veprek et al. [121] by means of making indentations on the coating using a Vickers indentor with a load of 1,000 mN. No radial cracks were found at the corners of the indentations, which mean a high fracture toughness of the coating. In our recent work [145,146], scratch tests were employed to evaluate the fracture toughness by using a Rockwell-C indentor to make scratches on the coating. If fracture toughness failure occurs, cracks can be found in the scratch track, as shown in Fig. 20. For the TiN/Si$_3$N$_4$ coating with the hardness of 47.1 GPa, toughness failure did not occur until the load increased to 33 N. However, the fracture toughness of the TiN/Si$_3$N$_4$ coating is a little less than that the TiN coating that failed at the load of 39 N.

The strong and dense Si$_3$N$_4$ interface can hinder oxygen diffusion along the grain boundaries, which provide a much higher oxidization resistance of the TiN/Si$_3$N$_4$ coating with the optimum Si content and superhardness as compared with the TiN coating [146–148]. Our recent work indicates that for the coatings annealed at 600°C in air, the oxidation depth of superhard TiN/Si$_3$N$_4$ nanocomposite coating is a quarter of that of the TiN coating, as shown in Fig. 21 [146]. In the case of annealing at 800°C, the oxidation depth of the superhard TiN/Si$_3$N$_4$ nanocomposite coating is less than 400 nm, whereas the TiN coating was fully oxidized and diffusion Fe from the steel substrate even was found in the coating.

In addition, the oxidation behavior of the Ti-Si-N coating is different from that of the Ti-Al-N coating. In the latter case, the Al ions could easily diffuse to the surface and caused a rich aluminum oxide layer to form near the surface, which played a barrier role against further oxidation [149]. Conversely, no significant migration of Si can be found in the oxidizing Ti-Si-N coatings (see Fig. 21(c) and 21(d)). This is attributed to the very low solubility of Si in a TiN crystal lattice, which inhibits the Si ions from penetrating the TiN crystal lattice towards the surface. In addition, the Si atoms cannot easily escape from the Si$_3$N$_4$ phase because of the strong Si–N covalent bonds.

The tribological properties of nanomocomposite coatings were seldom reported because more attention was attracted to their superhardness and the mechanism of strength enhancement. Our recent research indicates that the TiN/Si$_3$N$_4$ nanocomposite coatings show a low wear rate in the magnitude of 10^{-6} mm^3/N·m as sliding against a tungsten carbide (WC) ball [145]. The friction coefficient between the TiN/Si$_3$N$_4$ coating and WC ball in ambient atmosphere is about 0.2~0.3, and shows a little dependence on the Si content in the coating, as shown in Fig. 22. The TiN/Si$_3$N$_4$ coatings show better wear resistance than the TiN coating, especially under a larger load. For example, in our experiments under the load of 20 N, the TiN coating failed rapidly only after a sliding distance of 6 m, which should be attributed to the high temperature induced by the friction and the lower oxidation resistance of the TiN coating. In the case of the TiN/Si$_3$N$_4$ coatings, the coatings display an excellent wear protection due to their superhardness and high oxidation resistance.

The super-hardness, superior oxidation resistance, and the high wear resistance make the superhard nanocomposite TiN/Si_3N_4 coating a promising protection coating under extreme conditions such as large load, high temperature, and oxidation environment, e.g., high speed cutting on difficult-to-machine materials such as aluminum alloy, titanium alloy, and quenched steel. In our recent work [145,146], the drills deposited with the TiN/Si_3N_4 coating with the optimum Si content show a more than 80 times cutting life as compared with the uncoated drills when drilling holes on quenched AISI 420 stainless steel, as shown in Fig. 23.

References

[1] Liu, A. Y. and Cohen, M. L., "Prediction of New Compressibility Solids," *Science*, Vol. 245, 1989, pp. 841–842.

[2] Aisenberg, S. and Chabot, R., "Ion-beam Deposition of Thin Films of Diamond-like Carbon," *J. Appl. Phys.*, Vol. 42, 1971, pp. 2953–2958.

[3] Kaufmann, H. R., "Technology of Ion Beam Sources Used in Sputtering," *J. Vac. Sci. Technol.*, Vol. 15, 1978, pp. 272–276.

[4] Monaghan, D. P., Teer, D. G., Laing, K. C., Efeoglu, I., and Arnell, R. D., "Deposition of Graded Alloy Nitride Films by Closed Field Unbalanced Magnetron Sputtering," *Surf. Coat. Technol.*, Vol. 59, 1993, pp. 21–25.

[5] Qi, J., Luo, J. B., Wen, S. Z., Wang, J., and Li, W. Z., "Mechanical and Tribological Properties of Non-hydrogenated DLC Films Synthesized by IBAD," *Surf. Coat. Technol.*, Vol. 128/129, 2000, pp. 324–328.

[6] Coll, B. F. and Chhowalla, M., "Modelization of Reaction-Kinetics of Nitrogen and Titanium During Tin Arc Deposition," *Surf. Coat. Technol.*, Vol. 68, 1994, pp. 131–140.

[7] Chhowalla, M., Robertson, J., Chen, C. W., Silva, S. R. P., and Amaratunga, G. A. J., "Influence of Ion Energy and Substrate Temperature on the Optical and Electronic Properties of Tetrahedral Amorphous Carbon (ta-C) Films," *J. Appl. Phys.*, Vol. 81, 1997, pp. 39–145.

[8] Brown, I. G., "Cathodic Arc Deposition of Films," *Ann. Rev. Mater. Sci.*, Vol. 28, 1998, pp. 243–269.

[9] Boxman, R. L., Zhitomirsky, V., Alterkop, B., Gidalevich, E., Beilis, O., Keidar, M., and Goldsmith, S., "Recent Progress in Filtered Vacuum Arc Deposition," *Surf. Coat. Technol.*, Vol. 86/87, 1996, pp. 243–253.

[10] Fallon, P. J., Veerasamy, V. S., Davis, C. A., Robertson, J., Amaratunga, G. A. J., Milne, W. I., and Koskinen, J., "Properties of Filtered-Ion-Beam-Deposited Diamond-like Carbon as a Function of Ion Energy," *Phys. Rev. B*, Vol. 48, 1993, pp. 4777–4782.

[11] Shi, X., Flynn, D., Tay, B. K., Prawer, S., Nugent, K. W., Silva, S. R. P., Lifshitz, Y., and Milne, W. I., "Mechanical Properties and Raman Spectra of Tetrahedral Amorphous Carbon Films with High sp^3 Fraction Deposited Using a Filtered Cathodic Arc," *Philos. Mag. B*, Vol. 76, 1997, pp. 351–361.

[12] Polo, M. C., Andujar, J. L., Robertson, J., and Milne, W. I., "Preparation of Tetrahedral Amorphous Carbon Films by Filtered Cathodic Vacuum Arc Deposition," *Diamond Relat. Mater.*, Vol. 9, 2000, pp. 663–667.

[13] Robertson, J., "Diamond-like Amorphous Carbon," *Materials Sci. Eng. R*, Vol. 37, 2002, pp. 129–281.

[14] Friedmann, T. A., Sullivan, J. P., Knapp, J. A., Tallant, D. R., Follstaedt, D. M., Medlin, D. L., and Mirkarimi, P. B., "Thick Stress-free Amorphous-Tetrahedral Carbon Films with Hardness Near that of Diamond," *Appl. Phys. Lett.*, Vol. 71, 1997, pp. 3820–3822.

[15] Scheibe, H. J. and Schultrich, B., "DLC Film Deposition by Laser-arc and Study of Properties," *Thin Solid Films*, Vol. 246, 2005, pp. 92–102.

[16] Jiang, N. N., Shao, T. M., and Chen, D. R., "Pulsed Laser Arc Deposition of Diamond-like Carbon Films," *J. Inorg. Mater.*, 2005, Vol. 20, No. 1, pp. 187–192.

[17] Catherine, U., and Couderc, C., "Electrical Characteristics and Growth Kinetics in Discharges Used for Plasma Deposition of Amorphous Carbon," *Thin Solid Films*, Vol. 144, 1986, pp. 265–280.

[18] Qi, J., Lai, K. H., Lee, C. S., Bello, I., Lee, S. T., Luo, J. B., and Wen, S. Z., "Mechanical Properties of a-C:H Multilayer Films," *Diamond Relat. Mater.*, Vol. 10, 2001, pp. 1833–1838.

[19] Koidl, P., Wild, Ch., Dischler, B., Wagner, J., and Ramsteiner, M., "Plasma Deposition, Properties and Structure of Amorphous Hydrogenated Carbon Films," *Mater. Sci. Forum*, Vol. 52, 1990, pp. 41–70.

[20] Jiang, X., Zou, J. W., Reichelt, K., and Grunberg, P., "The Study of Mechanical Properties of a-C:H Films by Brillouin Scattering and Ultralow Load Indentation," *J. Appl. Phys.*, Vol. 66, 1989, pp. 4729–4735.

[21] Weiler, M., Lang, K., Li, E., and Robertson, J., "Deposition of Tetrahedral Hydrogenated Amorphous Carbon Using a Novel Electron Cyclotron Wave Resonance Reactor," *Appl. Phys. Lett.*, Vol. 72, 1998, pp. 1314–1316.

[22] Erdemir, A., "Design Criteria for Superlubricity in Carbon Films and Related Microstructures," *Tribol. Int.*, Vol. 37, 2004, pp. 577–583.

[23] Gangopadhyay, A., "Mechanical and Tribological Properties of Amorphous Carbon Films," *Tribol. Lett.*, Vol. 5, 1998, pp. 25–39.

[24] Tay, B. K., Sheeja, D., Choong, Y. S., Lau, S. P., and Shi, X., "Pin-on-disk Characterization of Amorphous Carbon Films Prepared by Filtered Cathodic Vacuum Arc Technique," *Diamond Relat. Mater.*, Vol. 9, 2000, pp. 819–824.

[25] Enke, K., "Some New Results on the Fabrication of and the Mechanical, Electrical and Optical Properties of I-Carbon Layers," *Thin Solid Films*, Vol. 80, 1981, pp. 227–234.

[26] Erdemir, A., "The Role of Hydrogen in Tribological Properties of Diamond-like Carbon Films," *Surf. Coat. Technol.*, Vol. 146/147, 2001, pp. 292–297.

[27] Grill, A., "Tribology of Diamond-like Carbon and Related Materials: An Updated Review," *Surf. Coat. Technol.*, Vol. 94/95, 1997, pp. 507–513.

[28] Voevodin, A. A., Phelps, A. W., Zabinski, J. S., and Donley, M. S., "Friction Induced Phase Transformation of Pulsed Laser Deposited Diamond-like Carbon," *Diamond Relat. Mater.*, Vol. 5, 1996, pp. 1264–1269.

[29] Grischke, M., Bewilogua, K., Trojan, K., and Dimigen, H., "Application-oriented Modifications of Deposition Processes for Diamond-like Carbon-based Coatings," *Surf. Coat. Technol.*, Vol. 74/75, 1996, pp. 739–745.

[30] Oguri, K. and Ariai, T., "Friction Coefficient of Si-C, Ti-C and Ge-C Coatings with Excess Carbon Formed by Plasma-assisted Chemical Vapour Deposition," *Thin Solid Films*, Vol. 208, 1992, pp. 158–160.

[31] Goranchev, B., Reichelt, K., Chevallier, J., Hornshoj, P., Dimi-

[31] gen, H., and Hubsch, H., "R. F. Reactive Sputter Deposition of Hydrogenated Amorphous Silicon Carbide Films," *Thin Solid Films*, Vol. 139, 1986, pp. 275–285.

[32] Donnet, C., "Recent Progress on the Tribology of Doped Diamond-like and Carbon Alloy Coatings: A Review," *Surf. Coat. Technol.*, Vol. 100/101, 1998, pp. 180–186.

[33] Dimigen, H., Hubsch, H., and Memming, R., "Tribological and Electrical Properties of Metal-containing Hydrogenated Carbon Films," *Appl. Phys. Lett.*, Vol. 36, 1987, pp. 1056–1058.

[34] Dimigen, H. and Klages, C. P., "Microstructure and Wear Behavior of Metal-containing Diamond-like Coatings," *Surf. Coat. Technol.*, Vol. 49, 1991, pp. 543–547.

[35] Klaffke, D., Wasche, R., and Czichos, H., "Wear Behaviour of I-Carbon Coatings," *Wear*, Vol. 153, 1992, pp. 149–162.

[36] Sjostrom, H., Hultman, L., Sundgren, J. E., and Wallenberg, L. R., "Microstructure of Amorphous C:H and Metal-containing C:H Films Deposited on Steel Substrates," *Thin Solid Films*, Vol. 232, 1993, pp. 169–179.

[37] Zhenyu Zhang, Xinchun Lu, Jianbin Luo, Yang Liu, Chenhui Zhang, "Preparation and Characterization of La_2O_3 Doped Diamond-Like Carbon Nanofilms (I): Structure Analysis," *Diamond Relat. Mater.*, Vol. 16, No. 11, 2007, pp. 1905–1911.

[38] Grill, A., "Diamond-like Carbon: State of the Art," *Diamond Relat. Mater.*, Vol. 8, 1999, pp. 428–434.

[39] Kimock, F. M. and Knapp, B. J., "Commercial Applications of Ion Beam Deposited Diamond-like Carbon (DLC) Coatings," *Surf. Coat. Technol.*, Vol. 56, 1993, pp. 273–279.

[40] Evans, A. C., Franks, J., and Revell, P. J., "Diamond-like Carbon Applied to Bioengineering Materials," *Surf. Coat. Technol.*, Vol. 47, 1991, pp. 662–667.

[41] Thomson, L. A., Law, C. F., and Rushton, N., "Biocompatibility of Diamond-like Carbon Coating," *Biomaterials*, Vol. 12, 1991, pp. 37–40.

[42] Grill, A., Patel, V., and Jahnes, C., "Novel Low K Dielectrics Based on Diamond-like Carbon Materials," *J. Electrochem. Soc.*, Vol. 145, 1998, pp. 1649–1653.

[43] Robertson, J., "Requirements of Ultrathin Carbon Coatings for Magnetic Storage Technology," *Tribol. Int.*, Vol. 36, 2003, pp. 405–415.

[44] Robertson, J., "Ultrathin Carbon Coatings for Magnetic Storage Technology," *Thin Solid Films*, Vol. 383, 2001, pp. 81–88.

[45] Goglia, P. R., Berkowitz, J., Hoehn, J., Xidis, A., and Stover, L., "Diamond-like Carbon Applications in High Density Hard Disc Recording Heads," *Diamond Relat. Mater.*, Vol. 10, 2001, pp. 271–277.

[46] Fung, M. K., Chan, W. C., Lai, K. H., Bello, I., Lee, C. S., Wong, N. B., and Lee, S. T., "Deposition of Ultra-thin Diamond-like Carbon Protective Coating on Magnetic Disks by Electron Cyclotron Resonance Plasma Technique," *J. Non-Cryst. Solids*, Vol. 254, 1999, pp. 167–173.

[47] Cutiongco, E. C., Li, D., and Chung, Y. W., "Tribological Behavior of Amorphous Carbon Nitride Overcoats for Magnetic Thin-film Rigid Disks," *J. Tribol.*, Vol. 118, 1996, pp. 543–548.

[48] Papakonstantinou, P., Zhao, J. F., Richardot, A., McAdams, E. T., and McLaughlin, J. A., "Evaluation of Corrosion Performance of Ultra-thin Si-DLC Overcoats with Electrochemical Impedance Spectroscopy," *Diamond Relat. Mater.*, Vol. 11, 2002, pp. 1124–1129.

[49] Chamberlain, G., "Carbon Coating Reported to be Near-frictionless," *Des. News*, Vol. 54, No. 1, 1998, p. 15.

[50] Erdemir, A., Eryilmaz, O. L., and Fenske, G., "Synthesis of Diamond-like Carbon Films with Superlow Friction and Wear Properties," *J. Vac. Sci. Technol. A*, Vol. 18, No. 4, 2000, pp. 1987–1992.

[51] Erdemir, A., Nilufer, I. B., Eryilmaz, O., Beschliesser, L. M., and Fenske, G. R., "Friction and Wear Performance of Diamond-like Carbon Films Grown in Various Source Gas Plasmas," *Surf. Coat. Technol.*, Vol. 121, 1999, pp. 589–593.

[52] Heimberg, J. A., Wahl, K. J., Singer, I. L., and Erdemir, A., "Friction Behavior of Diamond-like Carbon Coatings: Time and Speed Effects," *Appl. Phys. Lett.*, Vol. 78, 2001, pp. 2449–2451.

[53] Eryilmaz, O. L., and Erdemir, A., "Surface Analytical Investigation of Nearly-Frictionless Carbon Films After Tests in Dry and Humid Nitrogen," *Surf. Coat. Technol.*, Vol. 201, No. 16–17, 2007, pp. 7401–7407.

[54] Teter, D. M., Computation Alchemy: "The Search for New Superhard Materials," *MRS Bull.*, Vol. 23, No. 1, 1998, pp. 22–27.

[55] Haines, J., Leger, J. M., and Bocquillon, G., "Synthesis and Design of Superhard Materials," *Annu. Rev. Mater. Res.*, Vol. 31, 2001, pp. 1–23.

[56] Cohen, M. L., "Calculation of Bulk Moduli of Diamond and Zinc-blended Solids," *Phys. Rev. B*, Vol. 32, 1985, pp. 7988–7991.

[57] Muhl, S. and Méndez, J. M., "A Review of the Preparation of Carbon Nitride Films," *Diamond Relat. Mater.*, Vol. 8, 1999, pp. 1809–1830.

[58] Niu, C. M., Yuan, Z. L., and Charles, M. L., "Experimental Realization of the Covalent Solid Carbon Nitride," *Science*, Vol. 261, 1993, pp. 334–337.

[59] Yu, K. M., Cohen, M. L., Haller, E. E., Hansen, W. L., Liu, A. Y., and Wu, I. C., "Observation of Crystalline C_3N_4," *Phys. Rev. B*, Vol. 49, No. 7, 1994, pp. 5034–5037.

[60] Aixiang Wei, Dihu Chen, Ning Ke, Shaoqi Peng, S. P. Wong, "Characteristics of Carbon Nitride Films Prepared by Magnetic Filtered Plasma Stream," *Thin Solid Films*, Vol. 323, No. 1-2, 1998, pp. 217–221.

[61] Yen, T.-Y. and Chou, C.-P., "Growth and Characterization of Carbon Nitride Thin Films Prepared by Arc-Plasma Jet Chemical," *Appl. Phys. Lett.*, Vol. 67, No. 19, 1995, pp. 2801–2803.

[62] Fu, Y. Q., Wei, J., Yan, B. B., and Loh, N. L., "Characterization and Tribological Evaluation of Duplex Treatment by Depositing Carbon Nitride Films on Plasma Nitrided Ti-6Al-4V," *J. Mater. Sci.*, Vol. 35, No. 9, 2000, pp. 2215–2227.

[63] Gonzalez, P., Soto, R., Leon, B., and Perez, A. M., "Comparison of Pulsed Laser Deposition CNx Films Grown from Organic and Inorganic Targets," *J. Vac. Sci. Technol. A*, Vol. 19, No. 2, 2001, pp. 499–502.

[64] Chowdhury, A. K. M. S., Monclus, M., Cameron, D. C., Gilvarry, J., Murphy, M. J., Barradas, N. P., and Hashmi, M. S. J., "The Composition and Bonding Structure of CNx Films and Their Influence on the Mechanical Properties," *Thin Solid Films*, Vol. 308-309, 1997, pp. 130–134.

[65] McCulloch, D. G. and Merchant, A. R., "The Effect of Annealing on the Structure of Cathodic Arc Deposited Amorphous Carbon Nitride Films," *Thin Solid Films*, Vol. 290, 1996, pp. 99–102.

[66] Aoi, Y., Ono, K., Sakurada, K., Kamijo, E., Sasaki, M., and Sakayama, K., "Effects of Heat Treatment on Structure of Amorphous CNx Thin Films by Pulsed Laser Deposition," *Thin Solid Films*, Vol. 389, No. 1-2, 2001, pp. 62–67.

[67] Wan, L. and Egerton, R. F., "Preparation and Characterization of Carbon Nitride Thin Films," *Thin Solid Films*, Vol. 279, No. 1-2, 1996, pp. 34–42.

[68] Cheng, D. G., Lu, F. X., Yang, L. Y., Song, B., and Tong, Y. M., "Preparation and Characterization of Magnetron Sputtered CNx Films," *Journal of University of Science and Technology Beijing*, Vol. 19, No. 1, 1997, pp. 100–104.

[69] Ming, Y. and Kramer, D. J., "Properties of Carbon Nitride Films Deposited With and Without Electron Resonance Plasma Assistance," *Thin Solid Films*, Vol. 382, 2001, pp. 4–12.

[70] Hubička, Z., Šícha, M., Pajasová, L., Soukup, L., Jastrabík, L., Chvostová, D., and Wagner, T., "CNx Coatings Deposited by Pulsed RF Supersonic Plasma Jet: Hardness, Nitrogenation and Optical Properties," *Surf. Coat. Technol.*, Vol. 142-144, 2001, pp. 681–687.

[71] Hajek, V., Rusnak, K., Vlcek, J., Martinu, L., and Hawthrone, H. M., "Tribological Study of CNx Films Prepared by Reactive dc Magnetron Sputtering," *Wear*, Vol. 213, 1997, pp. 80–89.

[72] Wang, X., Martin, P. J., and Kinder, T. J., "Optical and Mechanical Properties of Carbon Nitride Films Prepared by Ion-Assisted Arc Deposition and Magnetron Sputtering," *Thin Solid Films*, Vol. 256, No. 1-2, 1995, pp. 148–154.

[73] Popov, C., Zambov, L. M., Plass, M. F., and Kulisch, W., "Optical, Electrical and Mechanical Properties of Nitrogen-rich Carbon Nitride Films Deposited by Inductively Coupled Plasma Chemical Vapor Deposition," *Thin Solid Films*, Vol. 377-378, 2000, pp. 156–162.

[74] Mingwu, Bai, Koji, Kato, Noritsugu, Umehara, Yoshihiko, Miyake, Xu, Junguo, and Hiromitsu, Tokisue, "Dependence of Microstructure and Nanomechanical Properties of Amorphous Carbon Nitride Thin Films on Vacuum Annealing," *Thin Solid Films*, Vol. 376, 2000, pp. 170–178.

[75] Czyzniewski, A., Precht, W., Pancielejko, M., Myslinski, P., and Walkowiak, W., "Structure, Composition and Tribological Properties of Carbon Nitride Films," *Thin Solid Films*, Vol. 317, No. 1-2, 1998, pp. 384–387.

[76] Toshiyuki, Hayashi, Akihito, Matsumuro, Mutsuo, Muramatsu, Masao, Kohzaki, and Katsumi, Yamaguchi, "Wear Resistance of Carbon Nitride Thin Films Formed by Ion Beam Assisted Deposition," *Thin Solid Films*, Vol. 376, No. 1-2, 2000, pp. 152–158.

[77] Fernández, A., Fernández, R. C., and Sánchez, L. J. C., "Preparation, Microstructural Characterisation and Tribological Behaviour of CNx Coatings," *Surf. Coat. Technol.*, Vol. 163-164, 2003, pp. 527–534.

[78] Yan, X. B., Xu, T., Chen, G., Yang, S. R., Liu, H. W., and Xue, Q. J., "Preparation and Characterization of Electrochemically Deposited Carbon Nitride Films on Silicon Substrate," *J. Phys. D*, Vol. 37, No. 6, 2004, pp. 907–913.

[79] Koskinen, J., Hirvonen, J. P., Levoska, J., and Torri, P., "Tribological Characterization of Carbon-Nitrogen Coatings Deposited by Using Vacuum Arc Discharge," *Diamond Relat. Mater.*, Vol. 5, No. 6–8, 1996, pp. 669–673.

[80] Tokoroyama, T., Umehara Fuwa, Y., and Nakamura, T., "Effect of the Shear Strength of Coating of CNx on Friction Sliding Against Si_3N_4 Pin in Nitrogen," *Proceedings of WTC2005*, Washington, DC, September 2005, WTC2005-63419.

[81] Ramsey, M. E., Poindexter, E., Pelt, J. S., Marin, J., and Durbin, S. M., "Hydrophobic CNx Thin Film Growth by Inductively-coupled RF Plasma Enhanced Pulsed Laser Deposition," *Thin Solid Films*, Vol. 360, No. 1–2, 2000, pp. 82–88.

[82] Tétard, F., Djemia, P., Besland, M. P., Tessier, P. Y., and Angleraud, B., "Characterizations of CNx Thin Films Made by Ionized Physical Vapor Deposition," *Thin Solid Films*, Vol. 482, No. 1–2, 2005, pp. 192–196.

[83] Husein, F., Imad, Zhou, YuanZhong, Li, Fan, Ryne, C., Allen, Chung Chan, Jacob, I., Kleiman, Yu, Gudimenko, and Clark, V., Cooper, "Synthesis of Carbon Nitride Thin Film by Vacuum Arcs," *Mater. Sci. Eng., A*, Vol. 209, 1996, pp. 10–15.

[84] Tessier, P. Y., Pichon, L., Villechaise, P., Linez, P., Angleraud, B., Mubumbila, N., Fouquet, V., Straboni, A., Milhet, X., and Hildebrand, H. F., "Carbon Nitride Thin Films as Protective Coatings for Biomaterials: Synthesis, Mechanical and Biocompatibility Characterizations," *Diamond Relat. Mater.*, Vol. 12, 2003, pp. 1066–1069.

[85] Zheng, C. L., Cui, F. Z., Meng, B., Ge, J., Liu, D. P., and Lee, I. S., "Hemocompatibility of C-N Films Fabricated by Ion Beam Assisted Deposition," *Surf. Coat. Technol.*, Vol. 193, No. 1–3, 2005, pp. 361–365.

[86] Côté, M. and Cohen, M. L., "Carbon Nitride Compounds with 1:1 Stoichiometry," *Phys. Rev. B*, Vol. 55, No. 9 1997, pp. 5684–5687.

[87] Teter, D. M. and Hemley, R. J., "Low-Compressibility Carbon Nitrides," *Science*, Vol. 271, 1996, pp. 53–55.

[88] Riedel, R., Kleebe, H. J., Schonfelder, H., and Aldinger, F., "A Covalent Micro-Nanocomposite Resistant to High-Temperature Oxidation," *Nature*, Vol. 374, No. 6522, 1995, pp. 526–528.

[89] Chen, L. C., Chen, C. K., Wei, S. L., Bhusari, D. M., Chen, K. H., Chen, Y. F., Jong, Y. C., and Huang, Y. S., "Crystalline Silicon Carbon Nitride: A Wide Band Gap Semiconductor," *Applied Physica Letters*, Vol. 72, No. 19, 1998, pp. 2463–2465.

[90] Chen, L. C., Chen, K. H., Wei, S. L., Kichambare, P. D., Wu, J. J., Lu, T. R., and Kuo, C. T., "Crystalline SiCN: A hard material rivals to cubic BN," *Thin Solid Films*, Vol. 355–356, 1999, pp. 112–116.

[91] Koehler, J. S., "Attempt to Design a Strong Solid," *Phys. Rev. B*, Vol. 2, 1970, pp. 547–551.

[92] Lehoczky, S. L., "Strength Enhancement in Thin-Layered Al-Cu Laminates," *J. Appl. Phys.*, Vol. 49, No. 11, 1978, pp. 5479–5485.

[93] Lehoczky, S. L., "Retardation of Dislocation Generation and Motion in Thin-Layered Metal Laminates," *Phys. Rev. Lett.*, Vol. 41, No. 26, 1978, pp. 1814–1818.

[94] Bull, S. J. and Jones, A. M., "Multilayer Coatings for Improved Performance," *Surf. Coat. Technol.*, Vol. 78, 1996, pp. 173–184.

[95] Martinez, E., Romero, J., Lousa, A., and Esteve, J., "Wear Behavior of Nanometric CrN/Cr Multilayers," *Surf. Coat. Technol.*, Vol. 163/164, 2003, pp. 571–577.

[96] Tsakalakos, T. and Hilliard, J. E., "Elastic Modulus in Composition-Modulated Copper-Nickel Foils," *J. Appl. Phys.*, Vol. 54, 1983, pp. 734–737.

[97] Kueny, A., Grimsditch, M., Miyano, K., Banerjee, I., Falco, C. M., and Schuller, I. K., "Anomalous Behavior of Surface Acoustic Waves in Cu/Nb Superlattices," *Phys. Rev. Lett.*,

Vol. 48, No. 3, 1982, pp. 166–170.

[98] Jensen, H., Sobota, J., and Sorensen, G., "A Study of Film Growth and Tribological Characterization of Nanostructured C-N/TiNx Multilayer coatings," *Surf. Coat. Technol.*, Vol. 94/95, 1997, pp. 174–178.

[99] Miyake, S., "Tribology of Carbon Nitride and Boron Nitride Nanoperiod Multilayer Films and Its Application to Nanoscale Processing," *Thin Solid Films*, Vol. 493, 2005, pp. 160–169.

[100] Shinn, M., Hultman, L., and Barnett, S. A., "Growth, Structure, and Microhardness of Epitaxial TiN/NbN Superlattices," *J. Mater. Res.*, Vol. 7, No. 4, 1992, pp. 901–911.

[101] Helmersson, U., Todorova, S., Barnett, S. A., Sundgren, J. E., Markert, L. C., and Greene, J. E., "Growth of Single-Crystal TiN/VN Strained-Layer Superlattices With Extremely High Mechanical Hardness," *J. Appl. Phys.*, Vol. 62, No. 2, 1987, pp. 481–484.

[102] Shih, K. K. and Dove, D. B., "Ti/Ti-N Hf/Hf-N and W/W-N Multilayer Films With High Mechanical Hardness," *Appl. Phys. Lett.*, Vol. 61, No. 6, 1992, pp. 654–656.

[103] Misra, A., Verdier, M., Lu, Y. C., Kung, H., Mitchell, T. E., Nastasi, M. A., and Embury, J. D., "Structure and Mechanical Properties of Cu-X (X=Nb,Cr,Ni) Nanolayered Composites," *Scripta Mater.*, Vol. 39, 1998, pp. 555–560.

[104] Shinn, M. and Barnett, S. A., "Effect of Superlattice Layer Elastic Moduli on Hardness," *Appl. Phys. Lett.*, Vol. 64, 1994, pp. 61–63.

[105] He, J. L., Wang, J., Li, W. Z., and Li, H. D., "Simulation of Nacre with TiN/Pt Multilayers and a Study of Their Mechanical Properties," *Mater. Sci. Eng., B*, Vol. 49, 1997, pp. 128–134.

[106] Madan, A., Wang, Y. Y., Barnett, S. A., Engstrom, C., Ljungcrantz, H., Hultman, L., and Grimsditch, M., "Enhanced Mechanical Hardness in Epitaxial Nonisostructural Mo/N and W/N Superlattices," *J. Appl. Phys.*, Vol. 84, 1998, pp. 776–785.

[107] Wang, J., Li, W. Z., and Li, H. D., "Mechanical Properties of TiC/Metal Multilayers Synthesized by Ion Beam Sputtering Technique," *J. Vac. Sci. Technol. B*, Vol. 19, No. 1, 2001, pp. 250–254.

[108] Barnett, S. A., "Deposition and Mechanical Properties of Superlattice Thin Films," *Physics of Thin Films, Vol. 17, Mechanic and Dielectric Properties*, M. H. Francombe and J. L. Vossen, Eds., Academic Press, Inc., 1993, pp. 2–77.

[109] Holleck, H., Lahres, M., and Woll, P., "Multilayer Coatings—Influence of Fabrication Parameters on Constitution and Properties," *Surf. Coat. Technol.*, Vol. 41, 1990, pp. 179–190.

[110] Berger, M., Wiklundl, U., Eriksson, M., Wngqvist, H., and Jacobson, S., "The Multilayer Effect in Abrasion—Optimising the Combination of Hard and Tough Phases," *Surf. Coat. Technol.*, Vol. 116/119, 1999, pp. 1138–1144.

[111] Leyland, A. and Matthews, A., "Thick, Ti/TiN Multilayered Coatings for Abrasive and Erosive Wear Resistance," *Surf. Coat. Technol.*, Vol. 70, 1994, pp. 19–25.

[112] Miyake, S., "Improvement of Mechanical Properties of Nanometer Period Multilayer Films at Interfaces of Each Layer," *J. Vac. Sci. Technol. B*, Vol. 21, No. 2, 2003, pp. 785–789.

[113] Miyake, S., Sekine, Y., Noshiro, J., and Watanabe, S., "Low-Friction and Long-Life Solid Lubricant Films Structured of Nanoperiod Tungsten Disulfide and Molybdenum Disulfide Multilayer," *Jpn. J. Appl. Phys.*, Vol. 43, 2004, pp. 4338–4343.

[114] Holmber, K., Matthews, A., and Ronkainen, H., "Coatings Tribology—Contact Mechanisms and Surface Design," *Tribol. Int.*, Vol. 31, 1998, pp. 107–120.

[115] Matthews, A. and Eskildsen, S. S., "Engineering Applications for Diamond-Like Carbon," *Diamond Relat. Mater.*, Vol. 3, 1994, pp. 902–911.

[116] Leyland, A., Sudin, M. B., James, A. S., Kalantary, M. R., Wells, P. B., Housden, J., Garside, B., and Matthews, A., "TiN and CrN PVD Coatings on Electroless Nickel Coated Steel Substrates," *Surf. Coat. Technol.*, Vol. 60, 1993, pp. 474–479.

[117] Sudin, M. B., Leyland, A., James, A. S., Matthews, A., Housden, J., and Garside, B., "Substrate Surface Finish Effects in Duplex Coatings of PAPVD TiN and CrN with Electroless Nickel-Phosphorus Interlayers," *Surf. Coat. Technol.*, Vol. 81, 1996, pp. 215–224.

[118] Veprek, S. and Reiprich, S., "A Concept for the Design of Novel Superhard Coatings," *Thin Solid Films*, Vol. 268, 1995, pp. 64–71.

[119] Veprek, S., Reiprich, S., and Li, S., "Superhard Nanocrystalline Composite Materials: The TiN/Si_3N_4 System," *Appl. Phys. Lett.*, Vol. 66, 1995, pp. 2640–2642.

[120] Veprek, S., "The Search for Novel, Superhard Materials," *J. Vac. Sci. Technol. A*, Vol. 17, 1999, pp. 2401–2420.

[121] Veprek, S., Heijman, M. G. J., Karvankova, P., and Prochazka, J., "Different Approaches to Superhard Coatings and Nanocomposites," *Thin Solid Films*, Vol. 476, 2005, pp. 1–29.

[122] Veprek, S., "Conventional and New Approaches Towards the Design of Novel Superhard Materials," *Surf. Coat. Technol.*, Vol. 97, 1997, pp. 15–22.

[123] Veprek, S. and Argon, A. S., "Mechanical Properties of Superhard Nanocomposites," *Surf. Coat. Technol.*, Vol. 146/147, 2001, pp. 175–182.

[124] Veprek, S., Haussmann, M., and Reiprich, S., "Superhard Nanocrystalline W_2N/ Amorphous Si_3N_4 Composite Materials," *J. Vac. Sci. Technol. A*, Vol. 14, 1996, pp. 46–51.

[125] Veprek, S., Haussmann, M., Reiprich, S., Li, S., and Dian, J., "Novel Thermodynamically Stable and Oxidation Resistant Superhard Coating Materials," *Surf. Coat. Technol.*, Vol. 86/87, 1996, pp. 394–401.

[126] Veprek, S., Nesladek, P., Niederhofer, A., Glatz, F., Jilek, M., and Sima, M., "Recent Progress in the Superhard Nanocrystalline Composites: Towards Their Industrialization and Understanding of the Origin of the Superhardness," *Surf. Coat. Technol.*, Vol. 108/109, 1998, pp. 138–147.

[127] Diserens, M., Patscheider, J., and Levy, F., "Mechanical Properties and Oxidation Resistance of Nanocomposite TiN-SiN_x Physical-Vapor-Deposited Thin Films," *Surf. Coat. Technol.*, Vol. 120/121, 1999, pp. 158–165.

[128] Rebouta, L., Tavares, C. J., Aimo, R., Wang, Z., Pischow, K., Alves, E., Rojas, T. C., and Odriozola, J. A., "Hard Nanocomposite Ti-Si-N Coatings Prepared by DC Reactive Magnetron Sputtering," *Surf. Coat. Technol.*, Vol. 133/134, 2000, pp. 234–239.

[129] Vaz, F., Rebouta, L., Ph., Goudeau, Giratdeau, T., Pacaud, J., Riviere, J. P., and Traverse, A., "Structural Transitions in Hard Si-Based Tin Coating: The Effect of Bias Voltage and Temperature," *Surf. Coat. Technol.*, Vol. 146/147, 2001, pp. 274–279.

[130] Holubar, P., Jilek, M., and Sima, M., "Present and Possible Future Applications of Superhard Nanocomposite Coatings," *Surf. Coat. Technol.*, Vol. 133/134, 2000, pp. 145–151.

[131] Zhang, C. H., Luo, J. B., Li, W. Z., and Chen, D. R., "Mechani-

cal Properties of Nanocomposite TiN/Si$_3$N$_4$ Films Synthesized by Ion Beam Assisted Deposition (IBAD)," *J. Tribol.*, Vol. 125, 2003, pp. 445–447.

[132] Musil, J., Zeman, P., Hruby, H., and Mayrhofer, P. H., "ZrN/Cu Nanocomposite Film—A Novel Superhard Material," *Surf. Coat. Technol.*, Vol. 120/121, 1999, pp. 179–183.

[133] Musil, J., Karvankova, P., and Kasl, J., "Hard and Superhard Zr-Ni-N Nanocomposite Films," *Surf. Coat. Technol.*, Vol. 139, 2001, pp. 101–109.

[134] Musil, J. and Polakova, H., "Hard Nanocomposite Zr-Y-N Coatings, Correlation Between Hardness and Structure," *Surf. Coat. Technol.*, Vol. 127, 2000, pp. 99–106.

[135] Karvankova, P., Mannling, H. D., Eggs, C., and Veprek, S., "Thermal Stability of ZrN-Ni and CrN-Ni Superhard Nanocomposite Coatings," *Surf. Coat. Technol.*, Vol. 146/147, 2001, pp. 280–285.

[136] Hauert, R., Patscheider, J., Knoblauch, L., and Diserens, M., "New Coatings by Nanostructuring," *Adv. Mater.*, Vol. 11, 1999, pp. 175–177.

[137] Hu, X. P., Li, G. Y., Dai, J. W., Ding, Z. M., and Gu, M. Y., "Influences of Si Content and Substrate Temperature on Ti-Si-N Nanocomposite Films," *J. Shang Hai Jiao Tong Univ.*, Vol. 37, 2003, pp. 252–256 (in Chinese).

[138] Kim, S. H., Kim, J. K., and Kim, K. H., "Influence of Deposition Conditions on the Microstructure and Mechanical Properties of Ti-Si-N Films by DC Reactive Magnetron Sputtering," *Thin Solid Films*, Vol. 420/421, 2002, pp. 360–365.

[139] Jiang, N., Shen, Y. G., Mai, Y. W., Chan, T., and Tung, S. C., "Nanocomposite Ti-Si-N Films Deposited by Reactive Unbalanced Magnetron Sputtering at Room Temperature," *Mater. Sci. Eng., B*, Vol. 106, 2004, pp. 163–171.

[140] Klug, H. P. and Alexander, L. E., *X-Ray Diffraction Procedures*, Wiley, New York, 1974, p. 687.

[141] Taylor, J. A., Lancaster, G. M., Ignatiev, A., and Rabalais, J. W., "Interactions of Ion Beams with Surfaces. Reactions of Nitrogen with Silicon and Its Oxides," *J. Chem. Phys.*, Vol. 68, 1978, pp. 1776–1784.

[142] Veprek, S., Niederhofer, A., Moto, K., Bolom, T., Mannling, H. D., Nesladek, P., Dollinger, G., and Bergmaier, A., "Composition, Nanostructure and Origin of the Ultrahardness in nc-TiN/a-Si$_3$N$_4$/a- and nc-TiSi$_2$ Nanocomposites with Hv =80 to ≥105 GPa," *Surf. Coat. Technol.*, Vol. 133/134, 2000, pp. 152–159.

[143] Zhang, C. H., Liu, Z. J., Li, K. Y., Shen, Y. G., and Luo, J. B., "Microstructure, Surface Morphology, and Mechanical Properties of Nanocrystalline TiN/Amorphous Si$_3$N$_4$ Composite Films Synthesized by Ion Beam Assisted Deposition," *J. Appl. Phys.*, Vol. 95, 2004, pp. 1460–1467.

[144] Veprek, S., and Argon, A. S., "Towards the Understanding of Mechanical Properties of Super- and Ultrahard Nacocomposites," *J. Vac. Sci. Technol. B*, Vol. 20, 2002, pp. 650–664.

[145] Zhang, C. H., "Studies on the Properties and Applications of TiN/Si$_3$N$_4$ Nancomposite Coatings," Ph.D. thesis, Tsinghua University, Beijing, China, 2003.

[146] Zhang, C. H., Lu, X. C., Wang, H., Luo, J. B., Shen, Y. G., and Li, K. Y., "Microstructure, Mechanical Properties, and Oxidation Resistance of Nanocomposite Ti-Si-N Coatings," *Appl. Surf. Sci.*, Vol. 252, 2006, pp. 6141–6153.

[147] Choi, J. B., Cho, K., Kim, Y., Kim, K. H., and Song, P. K., "Microstructure Effect on the High Temperature Oxidation Resistance of Ti-Si-N Coating Layers," *Jpn. J. Appl. Phys.*, Vol. 42, 2003, pp. 6556–6559.

[148] Kim, K. H. and Park, B. H., "Mechanical Properties and Oxidation Behavior of Ti-Si-N Films Prepared by Plasma-Assisted CVD," *Chem. Vap. Dep.*, Vol. 5, 1999, pp. 275–279.

[149] Huang, C. T. and Duh, J. G., "Stress and Oxidation Behaviours of R.F.-Sputtered (Ti, Al)N Films," *Surf. Coat. Technol.*, Vol. 81, 1996, pp. 164–171.

9

Friction and Adhesion

Yuanzhong Hu[1]

1 Introduction

STUDY ON ADHESION AND FRICTION IN MICROscopic scale has received great attention in recent decades. The development of Micro Mechanical Electrical System (MEMS) and nanotechnology, for instance, requires a better understanding of the interfacial phenomena which significantly affect performance of micro and nano-devices. Meanwhile the inventions of new scientific instruments, such as the Scanning Tunnel Microscope (STM), Atomic Force Microscope (ATM), Surface Force Apparatus (SFA), Quartz Crystal Microbalance (QCM), etc., and rapid progress of computer simulation technology allows scientists to explore and resolve the secrets of adhesion and friction in more efficient ways than ever before.

This chapter intends to give an introduction to the fundamental studies in the area of atomic-scale adhesion and friction. The emphasis will be focused on molecular origin of friction and connection between adhesion and friction. The chapter was written based on experiences from the present author in studying the fundamental of friction for years. Section 2 describes the surface forces that are responsible for origin of adhesion, and presents an atomic scale analysis by the present author to show how mechanical instability occurs in a process of approach/separation. Section 3 discusses the wearless friction models, both in atomic and asperity levels, which interpret the origin of friction in terms of instable atomic motion and energy dissipation. Section 4 provides the author's own view on interrelations between friction and adhesion, and the role of adhesion hysteresis in particular. Section 5 compares static friction with stick-slip transition to show the similarity and difference between the two events, which provides important information for the understanding of static friction on the basis of the principle of energy optimization. Finally, a summary is given in Section 6 with an expectation that the energy approach presented in this chapter has to be combined with a nonequilibrium thermodynamic model in order to provide a satisfactory solution to the mystery of friction.

2 Physics and Dynamics of Adhesion

2.1 Surface Forces and Adhesion

When two solid bodies have been pressed together under applied load, a normal force is generated at the contact surfaces due to repulsive interaction between atoms. The normal force gradually decreases if the solids in contact are separated along the direction normal to the contact surfaces. In many cases, however, the contact holds even if the normal force has reached zero, which means that to pull the two surfaces apart an additional tensile force, usually defined as negative in value, has to be applied. This phenomenon, known as "pull-off," is a manifestation of adhesion.

Figure 1 shows the results obtained by Qian et al. [1] in a process when AFM probe approaches and then separates from a SiO_2 substrate. The normal force required for separating the probe-substrate contact reads 33 nN. From a thermodynamic point of view, adhesion is in fact a state of the system at the energy minimum when the contact pairs interact with each other through interface, and additional work has to be applied to change the state of the system.

Surface energy or surface tension, γ, has been an important parameter widely used for characterizing adhesion. It is defined as half of the work needed to separate two bodies of unit area from contact with each other to an infinite distance, as schematically shown in Fig. 2. If the contact pairs are of the same material, the surface energy is identical to the cohesive energy.

Several models have been proposed to predict adhesion force—the maximum force required to pull off the surfaces. Among these, the JKR theory is one receiving the greatest attention [2], which says that for an elastic spherical body in contact with a semi-infinite plane, the adhesion force can be estimated by

$$F = -3\pi R \gamma \quad (1)$$

where R is the radius of the sphere and γ stands for surface energy at the contact interface. Other adhesion theories, such as the Bradley and DMT models, were proposed, too, which aroused a debate as for which one was more accurate. Further studies on adhesion mechanics had revealed that these models could be all valid but in different ranges of the parameter μ, corresponding to the ratio of the elastic deformation to the action range of the adhesion forces [3]. Nevertheless, the relationship proposed by Eq. (1) agrees quite well with experiments conducted on AFM and SFA [4,5].

The emphasis of this chapter is on the dynamic nature of adhesion and its interrelations to friction, but before discussing the dynamics a brief introduction to the surface forces responsible for creation of adhesion is given in the following.

2.1.1 van der Waals Forces

The van der Waals forces result from several sources among which the dispersion forces make the most important contribution. The origin of the dispersion forces may be understood as follows: A dipole may appear instantaneously on a

---

[1] *State Key Laboratory of Tribology, Tsinghua University, Beijing, China.*

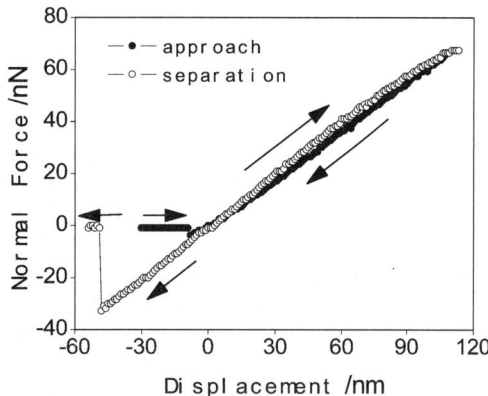

Fig. 1—Results from an experiment of approach/separation between AFM probe and SiO$_2$ substrate (from Ref. [1]).

nonpolar atom. This dipole moment generates an electric field that polarizes a nearby neutral atom, inducing a dipole moment in it. Interactions between these two dipoles give rise to an instantaneous attractive force between the two atoms.

The van der Waals forces are present universally, regardless of the species and polarity of the interacting atoms or molecules. The forces can be attractive or repulsive, but mostly attractive and long-range, effective from a distance longer than 10 nm down to the equilibrium interatomic distance (about 0.2 nm).

The interaction energy can be written as $w(r) = -C/r^6$ where C is a constant and r denotes atomic spacing. The van der Waals forces between two macroscopic objects can be obtained by integrating atomic interactions. Taking two planar surfaces in contact as an example, the integration gives the adhesive pressure as $P = A/6\pi D^3$ where A is the Hamaker constant. With a typical value $A = 10^{-19}$ J, the pressure for the contact at a distance of $D \approx 0.2$ nm is estimated as 700 MPa, and the pressure is reduced by a factor of about 10^5 when the distance increases to $D = 10$ nm. This indicates that the van der Waals interaction between macroscopic solids can be very large, and it plays an important role in a host of adhesion phenomena, not only for the magnitude of force, but also for its persistence in most circumstances.

2.1.2 Electrical Forces

In simple systems such as nonpolar films wetting on surfaces, the van der Waals forces play a dominant role, but in more complex systems long-range electrostatic forces are also involved.

Electrical or Coulomb force results from the interactions between charged particles. For two charges q_1 and q_2

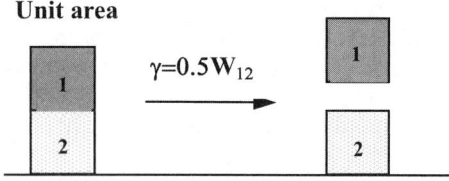

Fig. 2—Surface energy is related to the work needed to separate two solids in contact.

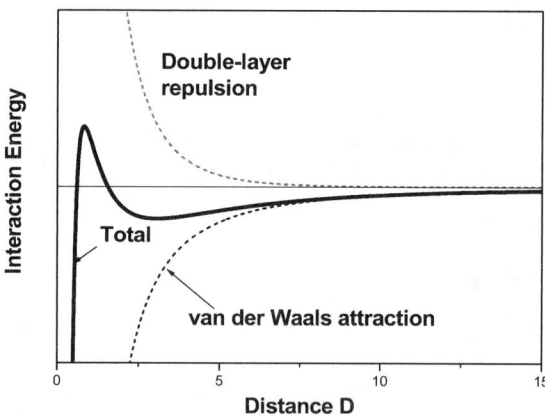

Fig. 3—Schematic energy versus distance profiles of DLVO interaction.

separated at a distance r, the force is given by [6]

$$F = \frac{q_1 \cdot q_2}{4\pi\varepsilon_0 \varepsilon r^2} \quad (2)$$

where ε is the dielectric constant of the medium. The force can be repulsive or attractive depending on like or unlike charges. From the inverse-square distance dependence, it is natural to assume the Coulomb force to be long-range, similar to that of gravitational force. For two isolated ions (e.g., Na$^+$ and Cl$^-$) in contact, the binding energy due to the Coulomb interaction is estimated as in the order of 200 kT, close to the strength of a covalent bond, which is the strongest physical interaction to be considered here.

It has been noted that the electrostatic forces between two parallel flat surfaces uniformly charged with density σ are distance independent if they are interacting in the air or vacuum. When the two surfaces are charged oppositely (positive and negative, respectively), the intensity of the electrical field between them is σ/ε_0, and there would be no electrical field in the gap between similarly charged surfaces.

The situations would be totally different when the two surfaces are put in electrolyte solutions. This is because of formation of the electrical double layers due to the existence of ions in the gap between solid surfaces. The electrical double layers interact with each other, which gives rise to a repulsive pressure between the two planar surfaces as

$$P = \frac{2\sigma^2 e^{-D/\lambda}}{\varepsilon \cdot \varepsilon_0} \quad (3)$$

where λ is known as the Debye length and D is the distance between the surfaces.

Hence, for two similarly charged surfaces in electrolyte, interactions are determined by both electrostatic double-layer and van der Waals forces. The consequent phenomena have been described quantitatively by the DLVO theory [6], named after Derjaguin and Landau, and Verwey and Overbeek. The interaction energy, due to combined actions of double-layer and van der Waals forces are schematically given in Fig. 3 as a function of distance D, from which one can see that the interplay of double-layer and van der Waals forces may affect the stability of a particle suspension system.

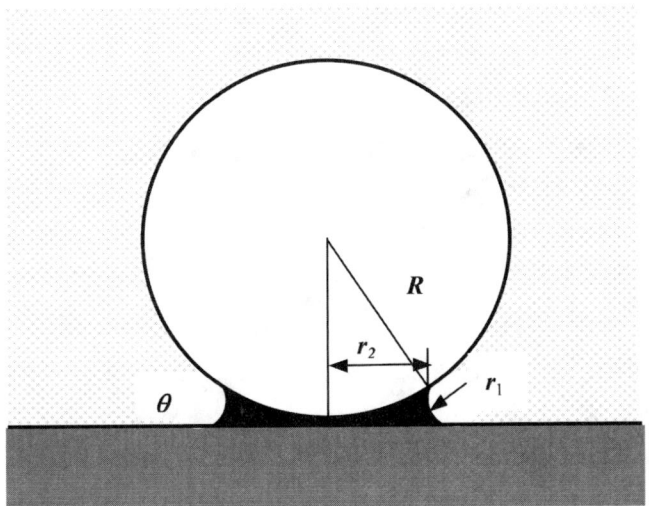

Fig. 4—Liquid bridge and meniscus formed around the contact spot between a microscopic sphere and a solid plane.

2.1.3 Capillary Force

For solid contacts in vapor atmosphere, liquid would condense from the vapor into cracks and pores formed between the contacting surfaces. As a result, a small liquid bridge appears around the contact spot and a meniscus with the curvature of $(1/r_1+1/r_2)$ forms at the solid-liquid-vapor interface, as illustrated in Fig. 4 for a microscopic sphere in contact with a solid plane.

The surface tension acting on the meniscus would pull the sphere toward the plane and give rise to an attractive pressure P over the contact region, which can be calculated in terms of the *Laplace* equation.

$$P = \gamma_L \left(\frac{1}{r_1} + \frac{1}{r_2} \right) \qquad (4)$$

where γ_L denotes surface tension of the liquid, r_1 is the Kelvin radius, and r_2 the radius of liquid pool. The adhesion force under the condition of $r_1 \ll r_2$ is then obtained, according to the geometry shown in Fig. 4.

$$F = 4\pi R \gamma_L \cos\theta \qquad (5)$$

where θ is the contact angle. When adhesion force from solid-solid contact inside the liquid radius is included, the final result is written as

$$F = 4\pi R (\gamma_L \cos\theta + \gamma_{SL}) = 4\pi R \gamma_{SV} \qquad (6)$$

where γ_{SL} and γ_{SV} represent the solid surface energies in liquid and vapor, respectively. Since $\gamma_L \cos\theta$ is often much greater than γ_{SL}, the adhesion force is dominated by surface tension of condensed liquid.

2.1.4 Chemical and Hydrogen Bond Forces

Chemical bonds formed in contact junctions are responsible for the adhesion in some cases. Our concerns here are the forces associated with covalent and hydrogen bonds. Covalent bonding refers to the interaction between atoms that share common electrons. Covalent interactions are short range, operating over a distance in a range of 0.1–0.2 nm, and are very strong with bond strength around 100–300 kT (200–800 kJ mol^{-1}). Hydrogen bond, originally found between water molecules, could also be an important source of adhesion on the surfaces where considerable hydroxide groups exist. The strength of most hydrogen bonds lies between 10 and 40 kJ mol^{-1}, weaker than covalent bonds but stronger than a typical van der Waals "bond" ($\sim$1 kJ mol^{-1}).

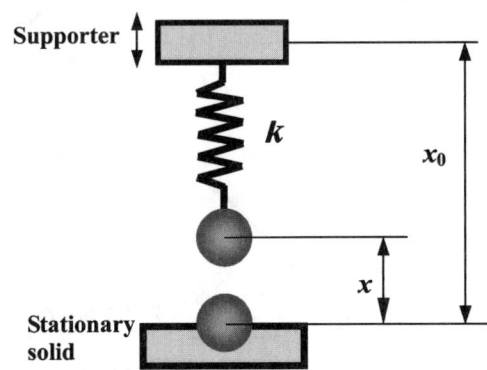

Fig. 5—A system of two atoms for modeling the dynamics of adhesion.

2.2 Dynamic Analyses of Adhesion at Micro and Macroscopic Scales

The studies on adhesion are mostly concerned on predictions and measurements of adhesion forces, but this section is written from a different standpoint. The author intends to present a dynamic analysis of adhesion which has been recently published [7], with the emphasis on the mechanism of energy dissipation. When two solids are brought into contact, or inversely separated apart by applied forces, the process will never go smoothly enough—the surfaces will always jump into and out of contact, no matter how slowly the forces are applied. We will show later that this is originated from the inherent mechanical instability of the system in which two solid bodies of certain stiffness interact through a distance dependent on potential energy.

Let us start from a simple system that consists of only two atoms, as sketched in Fig. 5. The lower atom has been fixed on a stationary solid surface while the upper atom is connected via a spring to the supporter that can move up and down in a normal direction. The two atoms stand for the top layer atoms on the moving and stationary surfaces, and the spring constant k characterizes the interaction between the surface atom and substrate. The position of the moving atom and the supporter are denoted by x and x_0, respectively, corresponding to the distances to the stationary surface. Suppose the supporter moves so slowly that the system can be considered to be in a quasi-static process that the system always remains in the equilibrium state. Hence, the position of the moving atom, x, can be determined through energy minimization, i.e., $dU/dx=0$, where U is the total energy of the system. The interactions between the atoms are assumed to follow the Lennard-Jones (LJ) potential.

$$V(x) = \varepsilon \left[\left(\frac{x}{\sigma} \right)^{-12} - \left(\frac{x}{\sigma} \right)^{-6} \right] \qquad (7)$$

where ε stands for the strength of the potential, and σ is a characteristic distance where the attractive and repulsive interactions cancel each other so that $V(x)=0$. For the system

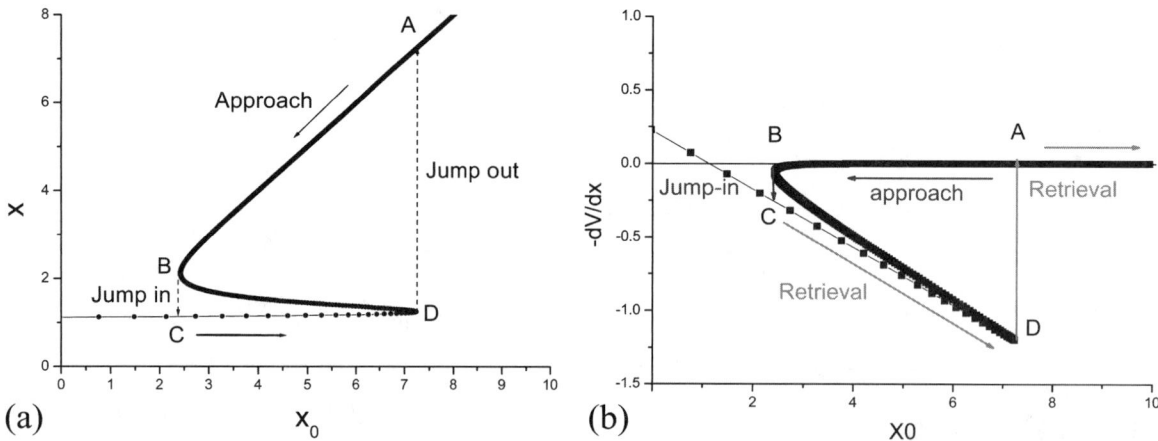

Fig. 6—(a) Position of the upper atoms, and (b) acting force on the atom, plotted as a function of supporter position x_0.

shown in Fig. 5, the total energy can be written as the sum of LJ potential and the deformation energy:

$$U(x) = V(x) + \frac{k}{2}(x_0 - x)^2 \qquad (8)$$

Substitution of energy expression of the system into the condition $dU/dx = 0$ gives rise to the equation that determines the equilibrium position of the moving atom.

$$x_0 = x + \frac{\varepsilon}{k}[c_1 x^{-7} - c_2 x^{-13}] \qquad (9)$$

The force acting on the moving atom can be calculated by the derivative $-dV/dx$. The upper atom is going through an approach/separation process as the supporter moves relatively to the stationary solid. The trajectory of the moving atom and the interacting forces during this process can be obtained from the solutions presented above. They are plotted in Figs. 6(a) and 6(b) as a function of x_0.

Figure 6(a) gives the position of the upper atom under the equilibrium condition $dU/dx = 0$. When approaching, both the supporter and upper atom move down, as illustrated by the trajectory line AB in Fig. 6(a), while the acting force is nearly zero, as can be seen in Fig. 6(b). At position B, however, the moving atom reaches a point of bifurcation where it has to choose between two possible equilibrium positions, B and C. Since the system would possess a lower energy at position C, the atom jumps into contact with the stationary surface, accompanied by a small change in force. This phenomenon is known as the "*jump-in*." From this moment, if the supporter is pulled upward, the system goes into the process of separation, during which the supporter moves from C to D while the upper atom keeps in contact with the lower surface, demonstrated by the x coordinate remaining almost constant. In the meantime, the adhesion force undergoes largely a linear increase. The adhesion force reaches maximum when arriving at position D, and the atom once again meets the point of bifurcation. At this moment the atom has to jump from D to A where the energy level is much lower, while the acting force suddenly goes back to zero. The sudden separation is referred to as the "*pull-off*" by investigators. The mechanical energy accumulated during the process, corresponding to the area surrounded by A, B, C, and D, dissipates through atomic vibration and phonon emission.

The method described above for the atomic system can be extended to a macroscopic system shown in Fig. 7 where a spherical body is connected via the spring k to a supporter in relative motion with respect to a stationary plane.

The total potential energy of the system takes a similar form to Eq (8) but with $V(x)$ denoting intersurface potential, which can be calculated through integrating the atomic potential over the interacting macroscopic bodies. When LJ potential is for characterizing atomic interactions, the potential energy between a macroscopic sphere and a semi-infinite solid is described by

$$V(x) = 2\pi R \gamma \left[A\left(\frac{x}{\sigma}\right)^{-7} - B\left(\frac{x}{\sigma}\right)^{-1} \right] \qquad (10)$$

where γ denotes the surface energy between interacting solid surfaces and R is the radius of the sphere. The position of the spherical body and the acting force given by $-dV/dx$ are plotted in Fig. 8 as a function of x_0.

The mechanical instability, jump-in and pull-off phenomenon, can also be observed in a macroscopic system, and both the trajectory and force curves exhibit similar patterns to those in Fig. 6. As a comparison, Fig. 9 shows a force curve obtained from SFA experiments of mica surface separation in dry air [8]. The pattern of the force variation, the

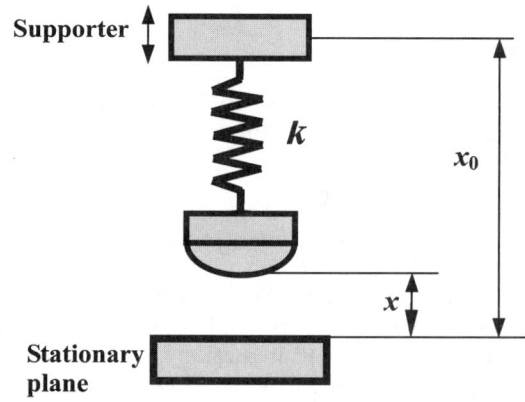

Fig. 7—A macroscopic system for modeling the dynamics of adhesion.

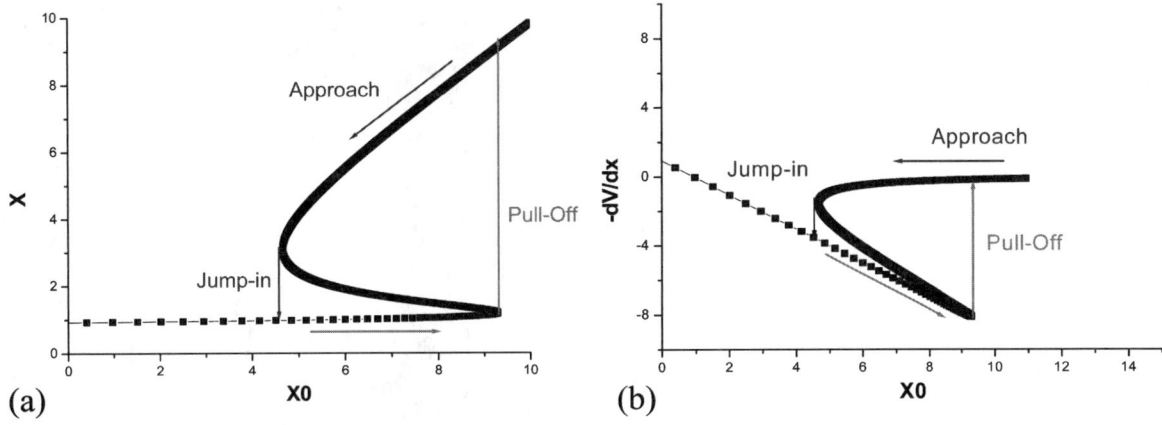

Fig. 8—(a) Position of the macroscopic body, and (b) the acting force, plotted as a function of supporter position, x_0.

jump-out of contact or pull-off, and the subsequent vibrations are consistent with model predictions.

The dynamical adhesion process described in this section is also referred to as *adhesion hysteresis*. We will come back to this subject in later sections when discussing the correlation between adhesion and friction.

3 Models of Wearless Friction and Energy Dissipation

3.1 A Brief Review

Scientific studies of friction can be traced back to several hundreds years ago when the pioneers, Leonardo da Vinci (1452–1519), Amontons (1699), and Coulomb (1785), established the law of friction that "friction is proportional to the normal load and independent of the nominal area of contact," which are still being taught today in schools. Since then, scientists and engineers have been trying to answer two fundamental questions: where friction comes from and why it exhibits such a behavior as described above. Impressive progress has been made but the mystery of friction has not been resolved yet. In an attempt to interpret the origin of friction in terms of surface topography, Coulomb proposed a model [9], as illustrated in Fig. 10, that the asperity on one surface in sliding could climb up along the slope of another asperity at the opposite surface, so that friction would be proportional to the work required for climbing up the hill. This model, though providing a very illustrative view of energy consumption, fails to explain the relationship among the friction, load, and nominal contact area.

An impressive progress in the fundamental study of friction was made more than half a century ago when Bowden and Tabor proposed that friction resulted from shear of adhesive junctions at the real contact area, which took up only a tiny portion of nominal contact zone and was proportional to the load [10], as schematically shown in Fig. 11. The model presents a satisfied explanation as to why friction is proportional to the load and independent of nominal contact area.

The model proposed by Bowden and Tabor has been regarded as the most successful one for presenting a simple and logical theory capable of explaining the Amontons friction law. However, suspicions concerning the two fundamental assumptions in the model were gradually aroused over past years. Friction has been attributed, in Bowden and Tabor's model, to the adhesion between asperities in contact and torn-off of the adhesive junctions when the shear stress exceeds a critical value. This implies that plastic flow and surface destruction may occur at the moment of slip, and that friction is dominated by the shear strength of the adhesive conjunctions, which is material dependent.

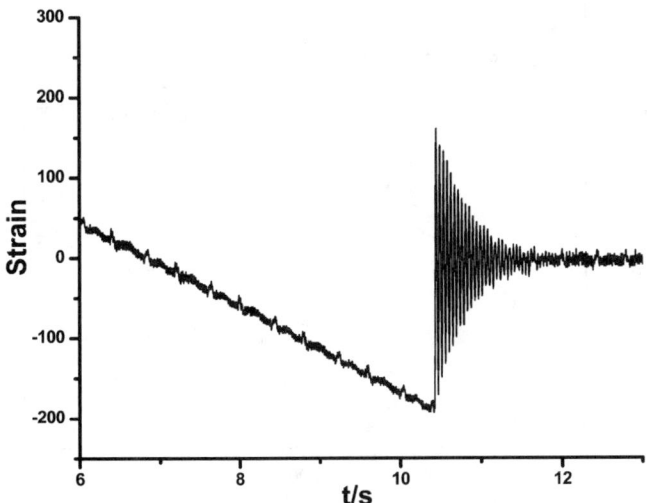

Fig. 9—Forces between mica surfaces in a process of separation, measured on SFA. Showing a linear increase in adhesion force, followed by a sudden separation and vibration [7].

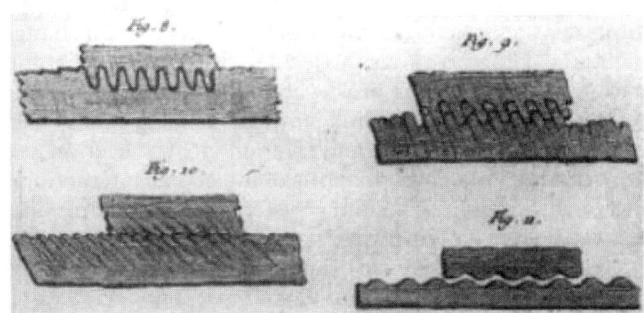

Fig. 10—Coulomb's model for the creation of friction (from Ref. [9]).

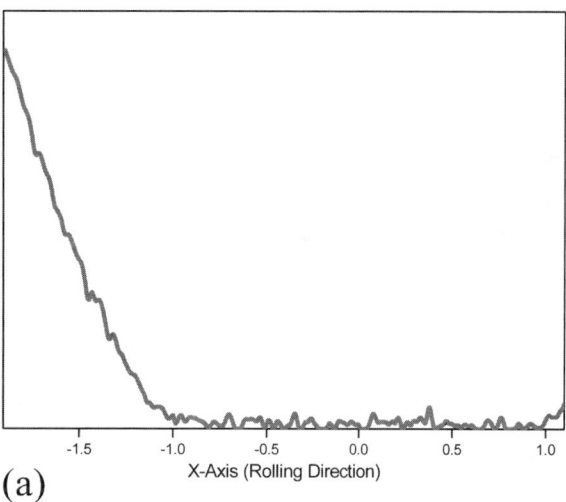 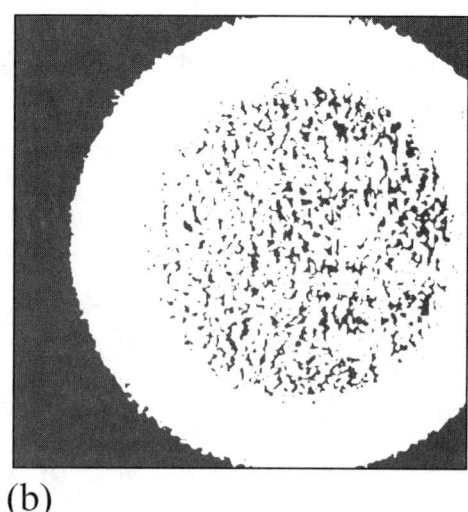

Fig. 11—Nominal contact zone and real contact areas between rough surfaces in contact, (a) film thickness profile along the central line of contact, (b) a contour plot of the contact geometry where the white circular area and gray spots inside the circle correspond to the nominal and real contact area, respectively.

It is true that there are general correlations between friction and structural damages of surface materials, such as breaking of covalent bonds, dislocations, crack propagation, plastic flow, plow, and eventual material losses. Undoubtedly, material damages would cause energy dissipation and friction, but the role has been somewhat overestimated. In fact, the energy losses due to friction are mostly converted into heat, and only a small portion contributes to the material damage and wear. Otherwise, 10 % of world's energy consumed by friction would have caused most machines to be out of operation in a very short time if frictional work had been mostly converted into wear. It has been frequently observed in practices that friction exists or remains quite high sometimes, even when the wear rate is very small or nearly zero. Experiments on SFA reported by Israelachivili et al. [11] reveal that the sliding friction between smooth mica films is much higher than that after the mica surfaces are damaged due to wear (Fig. 12). The evidence indicates that it is unnecessary to presume friction to be a direct result of plastic deformation or wear, and there exists a *wearless friction* whose mechanism has not been fully understood.

Furthermore, introducing the concept of *shear strength* implies that friction is dominated by the properties of interfacial materials. This could lead to a misunderstanding on the complicity of friction. It is difficult to explain, for example, why a slight modification of lubricant, a small change in chemical compositions, molecular structure and conformations, would result in drastic changes in frictional behavior. Researchers start to think more about the origin of friction in microscopic scale, and try to model the friction at a more fundamental level, in terms of molecular interactions, interfacial instability of sliding system, and energy dissipation.

The studies on friction originated from sliding-induced instability of interfacial atoms and consequent energy dissipation are reviewed in this section, including the author's work to extend the models of wearless friction. This type of friction has been addressed in the literature by different names, such as "interfacial friction" or "atomic-scale friction;" and here we prefer the term of "wearless" to emphasize that it is a kind of friction without involving any material damage or wear.

3.2 An Atomic-Scale Model of Wearless Friction

Since the idea that all matters are composed of atoms and molecules is widely accepted, it has been a long intention to understand friction in terms of atomic or molecular interactions. One of the models proposed by Tomlinson in 1929 [12], known as the independent oscillator model, is shown in Fig. 13, in which a spring-oscillator system translates over a corrugating potential. Each oscillator, standing for a surface atom, is connected to the solid substrate via a spring of stiffness k, and the amplitude of the potential corrugation is λ.

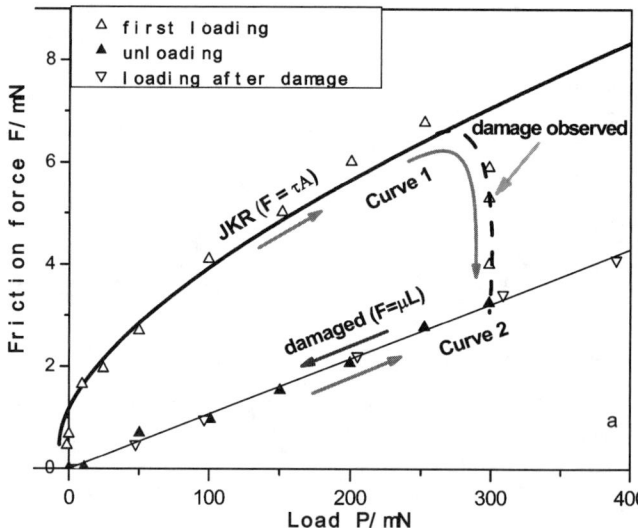

Fig. 12—Friction force versus normal load, measured on SFA, illustrating the transition in frictional performances before (curve 1) and after (curve 2) mica films are damaged.

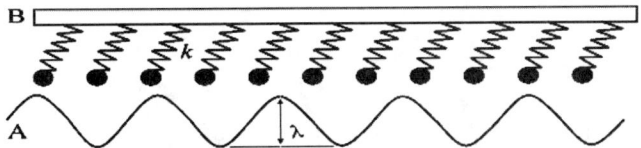

Fig. 13—Tomlinson or independent oscillator model of friction.

The system could be energy conservative if the atoms moved smoothly over the potential field. In that case, an atom, when traveling over one period of the potential, would experience a symmetrically distributed lateral force so that its time average and the net work done by the force would be zero. In reality, however, this is not going to happen that way. The author will demonstrate in the following how the system becomes unstable which inevitably leads to energy dissipation and friction.

The assumption of independent oscillators allows us to study a simplified system containing only one atom, as illustrated in Fig. 14 where x and x_0 denote, respectively, the coordinates of the atom and the support block (substrate). The dynamic analysis for the system in tangential sliding is similar to that of adhesion, as described in the previous section. For a given potential V and spring stiffness k, the total energy of the system is again written as

$$U = V + \frac{1}{2}k(x_0 - x)^2 \qquad (11)$$

Assume the system in sliding always remains at equilibrium state so that the atom's position x can be determined by solving the equation of $dU/dx = 0$, i.e.,

$$\frac{dU}{dx} = V' + k(x - x_0) = 0 \qquad (12)$$

and the lateral force acting on the moving atom is determined by the derivative of the potential, $-dV/dx$. For a sinusoidal potential, $V = \lambda \sin 2\pi x$, the position x and the lateral force F are plotted in Fig. 15 as a function of x_0, the traveling distance of the support.

It can be seen from Fig. 15(a) that the atom moves in a stick-slip way. In forward motion, for example, it is a "stick phase" from A to B during which the atom stays in a metastable state with little change in position as the support travels forward. Meanwhile, the lateral force gradually climbs up in the same period, leading to an accumulation of elastic energy, as illustrated in Fig. 15(b). When reaching the point B where a saddle-node bifurcation appears, the metastable state annihilates and the atom jumps suddenly from B to C, accompanied by a rapid drop of the lateral force, which is the "slip phase." In sweeping over the distance from A to B, the energy has been accumulated due to the work done by the lateral force, corresponding to the dotted area in Fig. 15(b). The sudden slip from B to C results in atomic vibration so that a part of the accumulated energy would dissipate into the environment through phonon emissions. Meanwhile, the rapid drop of the lateral force breaks the symmetry in the force distribution, which gives rise to a nonzero average friction force. In this way, the model successfully reveals how atomic interactions in a sliding system can cause interfacial instability that will eventually lead to energy dissipation and friction.

Stability of the atomic system depends on the spring stiffness and the potential corrugation, or more specifically, depends on the ratio of k/λ. The system would become more stable if the stiffness increases or the potential corrugation decreases, which means less energy loss and lower friction.

3.3 Microscopic Model of Wearless Friction—Sliding Between Elastic Bodies

The model discussed in the previous section can be extended to macroscopic scale for the friction between elastic bodies, following the approach suggested by Caroli [13]. Assume two asperities on the opposite surfaces form a pinning center, which can be either repulsive or attractive depending on the nature of the interactions, as sketched in Fig. 16. The pinning centers distribute on the surfaces in sliding, and are assumed to be independent of each other, i.e., interactions between the neighboring asperities on the same surface have been ignored, so the study of friction on sliding surfaces can be carried out by examining the behavior of one pinning center.

The two asperities in relative sliding are separated from each other at a nominal distance ρ. One of the asperities is assumed rigid while the other one deforms elastically with a deflection u. The total energy of the pinning center can be written as

$$U = V(\rho + u) + ku^2 \qquad (13)$$

where $V(\rho+u)$ denotes the potential energy between the interacting asperities and k is the stiffness of the elastic asperity. Let $\eta = \rho + u$, Eq (13) is rewritten as

$$U = V(\eta) + k(\eta - \rho)^2 \qquad (14)$$

Following the approach similar to that of the atomic-scale model, the evolution of the system state and the lateral force on the asperity can be determined in terms of $dU/d\eta = 0$. If we chose $V(\eta) = \pm V_0 \cos(\pi\eta)$ as the potential function for the repulsive and attractive pinning center, respectively, the lateral force $F = -dV/d\eta$ can be plotted as a function of the traveling distance ρ, as shown in Fig. 17.

The changes of lateral force F in forward and backward motions follow the curve 1 and 2, respectively. It can be observed that there is one saddle-node bifurcation for the repulsive pinning center, but two bifurcations for the attractive pinning center. This suggests that the interfacial instability results from different mechanisms. On one hand, the asperity suddenly looses contact as it slides over a repulsive pinning center, but in the attractive case, on the other hand, the

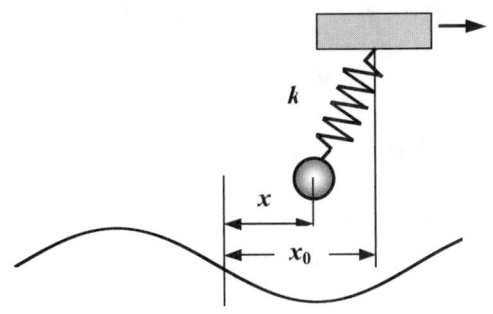

Fig. 14—An atom moves long a corrugating potential.

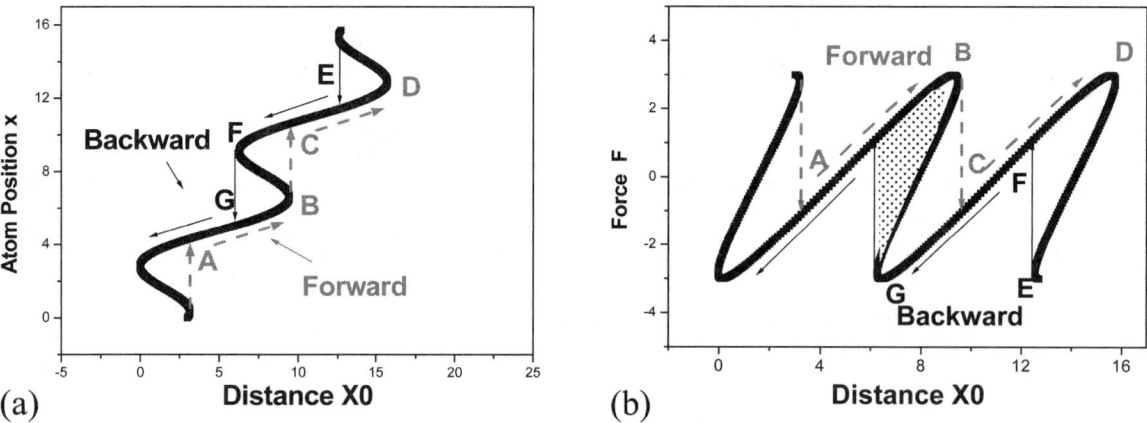

Fig. 15—Position and force of the moving atom as a function of traveling distance x_0, (a) position x, (b) acting force F with the dotted area corresponding the net work done by the force.

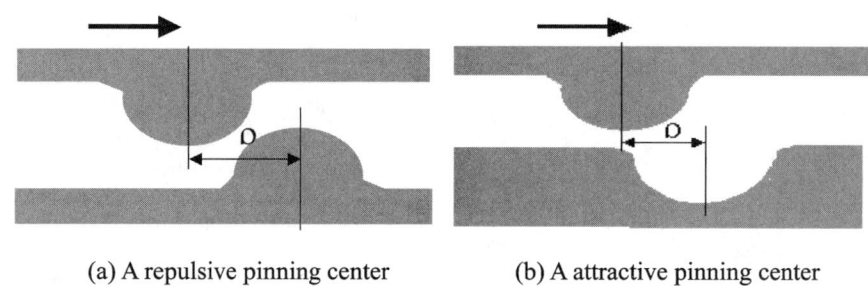

(a) A repulsive pinning center   (b) A attractive pinning center

Fig. 16—A pair of asperities in relative sliding and interacting with each other via a pinning center.

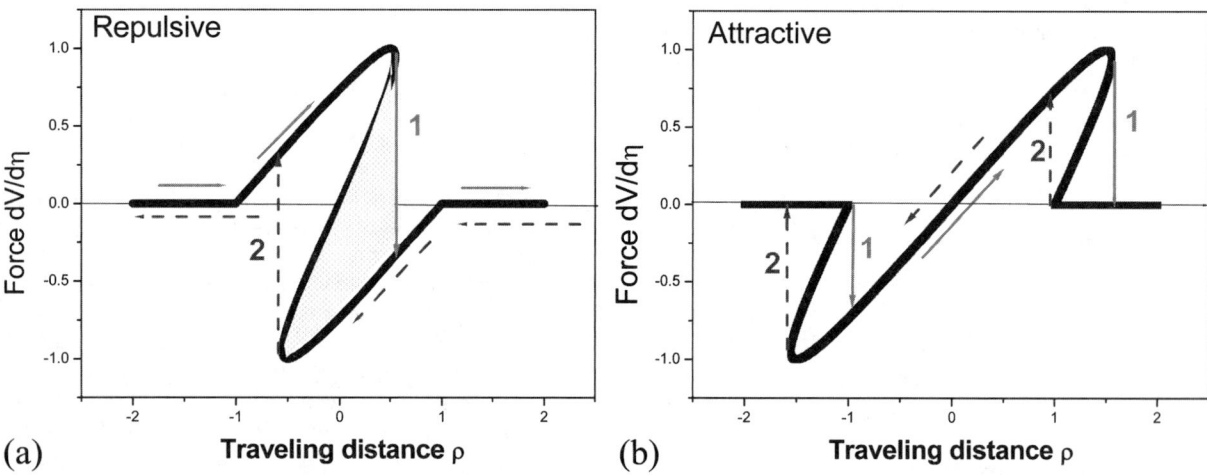

Fig. 17—Lateral force acting on the asperity versus the traveling distance r, (a) for repulsive pinning center, the dotted area corresponds to the net work done by the force, (b) for attractive pinning center.

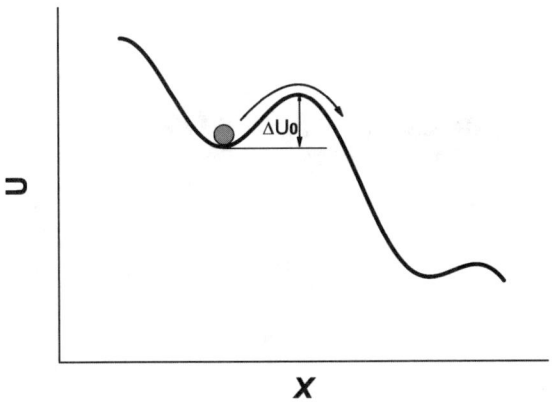

Fig. 18—A sketch for energy curve and barrier.

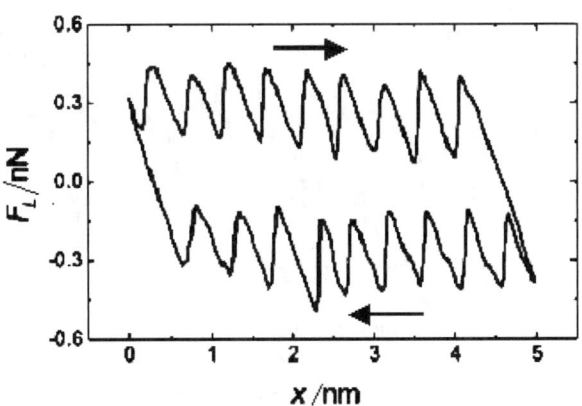

Fig. 19—Force curve in a AFM experiment as the probe travels along the surface of NaCl crystal (from Ref. [14]).

process consists of two successive events, jump-in and pull-off. As the asperity sweeps over the asperity, the work done by the force F is

$$W = \int_{-\infty}^{+\infty} F(\rho) d\rho \qquad (15)$$

The finite network, as marked by the dotted area in Fig. 17, will lead to energy loss that dissipates through elastic wave.

The friction from the repulsive pinning center is of particular interest because it is contrary to the common belief that friction must result from attractive interactions between sliding surfaces. The results presented in Fig. 17(a) demonstrate that friction can be created by purely repulsive interactions. What really matters is the instability of the sliding body and energy dissipation, rather than the attractive or repulsive nature of interactions. This may also shed a light on the efforts to explore the correlation between friction and adhesion.

3.4 Noise Activation

The occurrence of atomic jump or slip implies that for an atom translating over a corrugated potential, there may be more than one metastable position where the energy remains the minimum, which allows the atom jump from one equilibrium position to another if the energy barrier between the two states approaches zero. The energy distribution for such a system is sketched in Fig. 18 where the total system energy U, as defined in Eq (11), is composed of a periodic potential and energy due to elastic deformation of the spring. The energy barrier, denoted as ΔU_0, prevents the atom from going to the position with lower energy. As the energy curve is biased by the driven force, however, the barrier gradually disappears and the atom starts to jump at the time when ΔU_0 reaches zero.

It has been recognized that the behavior of atomic friction, such as stick-slip, creep, and velocity dependence, can be understood in terms of the energy structure of multi-stable states and noise activated motion. Noises like thermal activities may cause the atom to jump even before ΔU_0 becomes zero, but the time when the atom is activated depends on sliding velocity in such a way that for a given energy barrier, ΔU_0, the probability of activation increases with decreasing velocity. It has been demonstrated [14] that the mechanism of noise activation leads to "the velocity strengthening," i.e., the friction force increases logarithmically with the velocity.

3.5 Observations of Instability in Atomic Motion

Evidences for atomic jump or stick-slip induced by instability can be found in both experiments and computer simulations. When an AFM probe travels along the smooth surface of an NaCl crystal [15], the lateral forces fluctuate in a zigzag way, as illustrated in Fig. 19, which results from probe vibration. Considering the smoothness of the surface, the vibration originates most likely from the changes in surface energy instead of surface roughness. If we regard the AFM probe and cantilever as the oscillator and spring sketched in Fig. 14, and the crystal substrate as an equivalent of the corrugated potential, the mechanism of probe vibration can be well interpreted in terms of the model presented in Section 3.2.

The instability similar to that of the oscillator-spring system can be observed on sliding of self-assembly monolayers (SAMs). Figure 20 shows a system where two opposite SAMs, composed of chain molecules chemisorbed to gold (111) surface, are in relative sliding. Molecular dynamics simulations were performed to study friction behavior between the sliding SAMs and examine the molecule motion during the process of sliding [16]. It is found from simulations that stick-slip motion persists when two *commensurate* monolayers move over each other. In stick phase the methyl tail groups are pinned together due to van der Waals interactions while the chain molecules are pulled forward along the sliding direction, which leads to a gradual increase in lateral

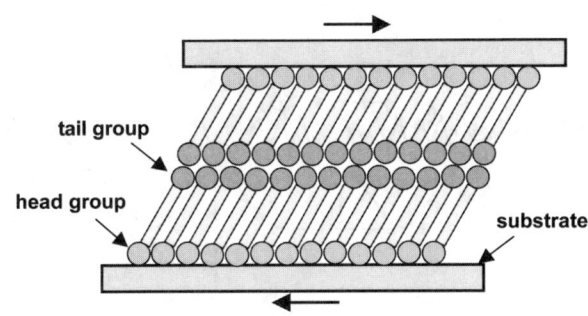

Fig. 20—Two self-assembly monolayers in relative sliding.

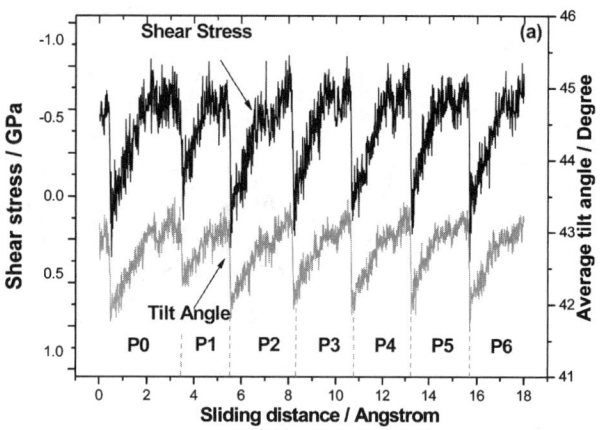
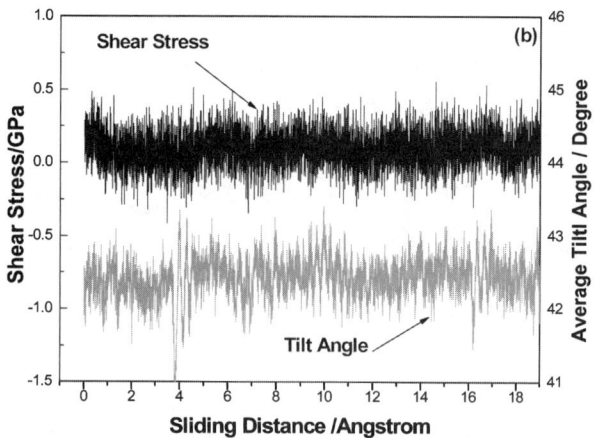

Fig. 21—Changes in lateral force and tilt angle for SAMs in sliding, (a) results for commensurate monolayers, (b) results for incommensurate case.

force and molecule tilt angle. As elastic energy accumulates up to a critical point, the sticking tail groups are suddenly pulled apart, and molecules swing backward rapidly, accompanied by the quick falls in both lateral force and tilt angle. This has been confirmed by simulation results given in Fig. 21(a), where the lateral force and tilt angle fluctuate in the periodic patterns.

The example demonstrates that the instability and consequent energy dissipation, similar to those in the Tomlinson model, do exist in a real molecule system. Keep in mind, however, that it is observed only in a commensurate system in which the lattice constants of two monolayers are in a ratio of rational value. For incommensurate sliding, the situation is totally different. Results shown in Fig. 21(b) were obtained under the same conditions as those in Fig. 21(a), but from an incommensurate system. The lateral force and tilt angle in Fig. 21(b) fluctuate randomly and no stick-slip motion is observed. In addition, the average lateral force is found much smaller, about one-fifth of the commensurate one.

The comparison reveals that in addition to the activity of individual atoms, the next-neighbor interaction and collective atomic motion must play an important role in creating friction. This mechanism can be investigated more efficiently via the Frenkel-Kontorova model.

3.6 Frenkel-Kontorova Model

The Frenkel-Kontorova (FK) model [17] investigates static and dynamic properties of a harmonic chain subjected to a periodic potential with strength b, as sketched in Fig. 22. The system can be considered as a simple model of wearless friction for adsorbed atoms on substrate. The chain contains N particles connected via springs of stiffness k, and a uniform lateral force F is applied on each particle. The equation of motion in dimensionless form can be written as

$$\ddot{x}_j + \gamma \dot{x} = x_{j+1} - 2x_j + x_{j-1} - b \sin x_j + F, \quad j = 1, 2, \ldots, N \tag{16}$$

where γ is a phenomenological damping coefficient, and the potential amplitude b has been normalized. Periodic boundary conditions are adopted.

$$x_{j+N} = x_j + 2\pi L, \quad L \text{ integer} \tag{17}$$

The system is regarded as commensurate if there is a rational value for the ratio $\alpha = a/2\pi = L/N$, where a denotes the mean space of the particles, 2π is the potential period, and L indicates the number of potential periods within the calculation window. By using different combinations of L and N, one may be able to alter the commensurability of the system, which will become incommensurate when α approaches an irrational value. The golden mean, $\alpha = (\sqrt{5}-1)/2 \approx 144/233$, for example, is usually chosen to characterize a typical incommensurate case.

In stationary states, particles stay at the equilibrium with the minimum energy, whose positions can be determined through solving Eq (16) with the left terms being set as zero. By increasing the driving force F gradually, the system evolves into metastable states, and the chain will eventually start to slide. The driving force F_s at the moment of sliding is defined as static friction. It has been found that static friction depends strongly on the ratio α and the potential strength b. In incommensurate cases, there is a critical strength b_c^S, below which the static friction disappears, i.e., the chain starts to slide if an infinitely small force F is applied. The critical strength drops to zero for the commensurate systems, which means at a finite value of b the static friction always exists and grows as the interaction strength increases.

Kinetic friction in the quasistatic limit of $v \to 0$ exhibits a similar dependence on α and b. Friction disappears if the strength b is less than a critical value b_c^K, but above this threshold it is the instability similar to that discussed in the

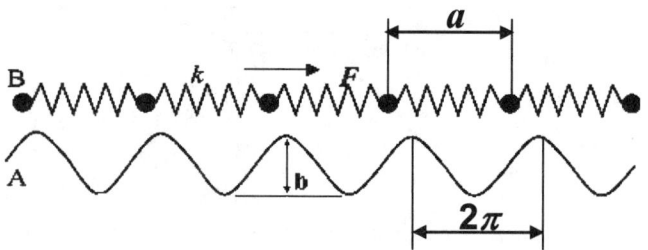

Fig. 22—The Frendel-Kontorova model driven by a uniform force F.

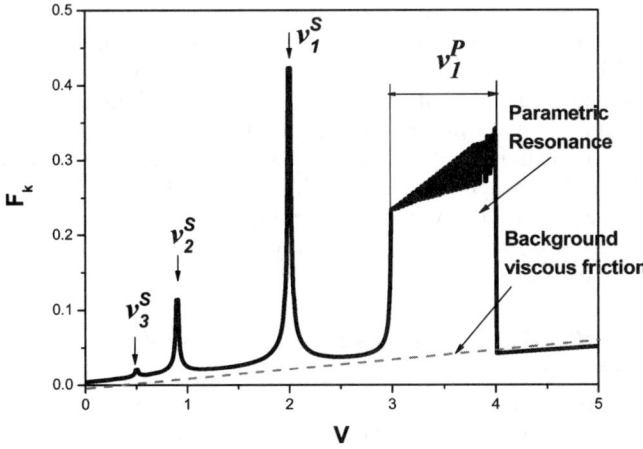

Fig. 23—A schematic force curve plotted as a function of sliding velocity. A viscous friction forms the background of the force curve upon which the frictions from superharmonic and parametric resonance are superposed.

Tomlinson model that results in a finite kinetic friction.

At finite velocity kinetic friction behaves quite differently in the sense that the commensurability plays a less significant role. Besides, the system shows rich dynamic properties since Eq (16) may lead to periodic, quasi-periodic, or chaotic solutions, depending on damping coefficient γ and interaction strength b. Based on numerical results of an incommensurate case [18,19], we outline a force curve of F in Fig. 23 as a function of v, in hopes of gaining a better understanding of dynamic behavior in the F-K model.

Periodic solutions exist in both overdamped and underdamped conditions, which give rise to a kinetic friction increasing linearly with the velocity. This viscous-like friction forms the background of the force curve, as indicated by a dashed line in Fig. 23. For underdamped sliding, however, *superharmonic resonance* appears at specific velocities, denoted by v_i^S in Fig. 23, which leads to resonance peaks in F. As γ further decreases, the periodic solutions may become unstable, and reorganize themselves into quasi-periodic solutions in a certain range of velocities, corresponding to the region marked by v^P in Fig. 23. This is another type of instability known as the *parametric resonance*, which will result in a large kinetic friction. If v is much larger than the maximum resonance velocity, all particles are traveling with velocity $v = F/\gamma$ plus a tiny oscillation. This can be called solid-sliding state. For even smaller values of the damping constant γ or for larger values of the strength of the potential b, the quasi-periodic solutions will transit into a severely unstable state, leading to chaotic solutions. We call the chaotic sliding state the fluid sliding, which also gives rise to a very high friction.

In summary, large kinetic friction develops at the times when the superharmonic and parametric resonance takes place, or when sliding becomes chaotic. Considering the resonant or chaotic states are all unstable, it can be speculated that large kinetic friction is attributed generally to the system instability, otherwise the chain slides following the law of viscous friction $F = \gamma v$. However, a more significant message carried by the FK model is that there exists a new type of source of friction, in different nature from that in the Tomlinson model. The kinetic solutions indicate that friction can result from superharmonic, parametric resonance and chaotic sliding, even for an incommensurate system where friction is supposed to be very low, if predicted in terms of the Tomlinson model. This reveals an origin of friction associated with dynamic response of an atomic system in sliding, which requires experimental validation.

The FK model accounts for the effects that have been ignored in the Tomlinson model, resulting from the interactions of neighboring atoms. For a more realistic friction model of solid bodies in relative sliding, the particles in the harmonic chain have to be connected to a substrate. This motivates the idea of combining the two models into a new system, as schematically shown in Fig. 24, which is known as the Frenkel-Kontorova-Tomlinson model. Static and dynamic behavior of the combined system can be studied through a similar approach presented in this section.

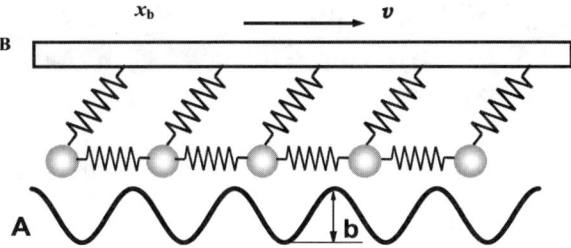

Fig. 24—The Frenkel-Kontorova-Tomlinson model.

3.7 Energy Dissipation—Phononic or Electronic Damping

It has been demonstrated in previous sections that sliding of atomic systems may induce instability and result in vibration. The vibration may or may not lead to energy dissipation, depending on the architecture and thermodynamic nature of the system. During the vibration of a nonlinear driven oscillator, for example, energy may transit alternatively into and out of the system [20]. For the systems with many degrees of freedom, however, occurrence of the instability and vibration are generally regarded as the initiation of energy dissipation.

The intensity of the energy dissipation can be estimated by the damping force acting on a particle i vibrating in a velocity v_i, which follows a linear relation [21].

$$\vec{F}_{fi} = -M\eta\vec{v}_i \quad (18)$$

where M denotes the mass of the particle and η is the damping coefficient that results in atomic scale friction. The average friction F_{fa} of an atomic system in the period of T can be obtained through balancing the work done by the system.

$$\vec{F}_{fa} \cdot \vec{v} = T^{-1}\int_0^T \sum_i \vec{F}_{fi} \cdot \vec{v}_i \quad (19)$$

For insulating surfaces, the friction η can be only due to phonon emission into the substrate, but on metal surfaces damping to vibration may result from both phononic and electronic excitations so that $\eta = \eta_{el} + \eta_{ph}$. The damping coefficient is assumed to be in the form of a diagonal matrix.

TABLE 1—Electronic damping deduced from surface resistivity data [20].

Molecules	$(\eta_\parallel)_{el}$ (S^{-1})
CO	3×10^9
C_2H_4	1.4×10^9
Xe	$\sim 3 \times 10^8$
C_6H_6	7×10^8
C_6H_{12}	6×10^8
C_2H_6	3×10^8

$$\eta = \begin{bmatrix} \eta_\square & & \\ & \eta_\square & \\ & & \eta_\perp \end{bmatrix} \quad (20)$$

It has been found that damping to the motion normal to the surface is usually much larger than that in the parallel direction.

For molecules adsorbed on a solid surface, the damping coefficient η is closely related to energy loss spectra of adsorbate vibrations, which can be deduced experimentally from infrared spectroscopic and inelastic helium scattering measurements. Moreover, for molecules adsorbed on metals the electronic contribution to the parameter η can be deduced from measuring the changes in surface resistivity. Electronic damping, $(\eta_\square)_{el}$, deduced from the surface resistivity data in several different physisorption systems is summarized in Table 1, while phononic damping, $(\eta_\perp)_{ph}$, deduced from the He-atom scattering measurement for hydrocarbons physisorbed on Cu(100) is given in Table 2.

Theoretical analysis indicates that the phononic damping depends strongly on resonance frequency of molecule vibrations. The experimental values of $(\eta_\perp)_{ph}$ in Table 2 are found much larger than the contributions from electronic damping, which is mainly due to the higher resonance frequency of perpendicular vibrations of hydrocarbons on Cu(100).

In summary, wearless or atomic scale friction can be modeled by a simple system in which a group of spring-connected oscillators slide with respect to a corrugated potential. Depending on potential corrugation and spring stiffness, the system may transit into a state of instability, which leads to vibrations. It is has been revealed that mechanical energy of the atomic or molecular vibrations dissipates through phonon emission or electron excitation, but the dependence of energy dissipation on the structure and dynamical response of the system needs further investigation.

4 Correlations Between Adhesion and Friction

Adhesion and friction result from the interactions between the surfaces in contact or relative motion. It is therefore expected from physical intuition that there should be some sort of connection between the two phenomena. In Bowden and Tabor's model, for example, friction is assumed to originate from adhesion at real contact areas. Based on a general belief that strong adhesion leads to higher friction, attempts were made to reduce friction by lowering surface energy, but with only limited success. In this section, the author provides a personal view for the correlations between adhesion and friction, which is important for understanding the origin of friction.

4.1 Adhesion Hysteresis Due to Mechanical Instability

The strength of adhesion between two solids in van der Waals interactions, or the normal force required to separate the two surfaces, is related directly to surface energy, or to surface tension and contact angle if meniscus force dominates. The separation, however, is most unlikely to take place in a smooth and continuous fashion no matter how slowly the force is applied. On the contrary, the surfaces will always suddenly jump apart, leading to energy dissipation. The mechanism to the instability in adhesion has been discussed throughout Section 2. From a thermodynamic point of view, if we separate two surfaces and then bring them back into contact again, the process of approach/separation will be irreversible in the sense that the work expensed during separation will be larger than that in reverse course, i.e., $\Delta W = W_s - W_a > 0$. This is the phenomenon referred to as *adhesion hysteresis* [10].

Adhesion strength and hysteresis are two important features, which are related to each other but from different origins. The adhesion strength, or the force F, can be evaluated by integrating the atomic interactions along contacting surfaces while hysteresis results from the system instability, which depends on stiffness of the system and distance dependence of the interactions. When two macroscopic bodies, interacting each other via the L-J potential, are brought into contact and then separated, similar to the case we discussed in Section 2 (Fig. 7), the force curve is given in Fig. 25(*a*) as a function of the separation distance x_0. Point C on the curve defines the pull-off force while the possible energy loss in the approach/separation cycle corresponds to the area surrounded by points A, B, C, and D. The question is whether or not it is possible for two solids in strong adhesion to yield less energy loss during the forward and backward motions. The answer can be found in Fig. 25(*b*) where the force curve is obtained under the same surface energy but for a system with increased stiffness. In comparison with Fig. 25(*a*), the pull-off force remains almost the same but the energy loss, the area marked by A, B, C, and D, looks much smaller. Similar results can be obtained if we replace the potential with the one in less fluctuation. The example demonstrates that the adhesion strength is dominated by surface energy, but the energy dissipation in the adhesion hysteresis is determined by system instability, which depends on system stiffness and potential corrugation.

4.2 Connection Between Adhesion Hysteresis and Friction

From the discussions on dynamic adhesion and wearless friction, one may realize that the two events originate from

TABLE 2—Phononic damping deduced from He-atom scattering measurement [20].

Molecules	$(\eta_\perp)_{ph}$ (S^{-1})
n-Hexane	2.1×10^{12}
n-Octane	1.8×10^{12}
n-Decane	2.9×10^{12}
Cyclohexane	3.3×10^{12}

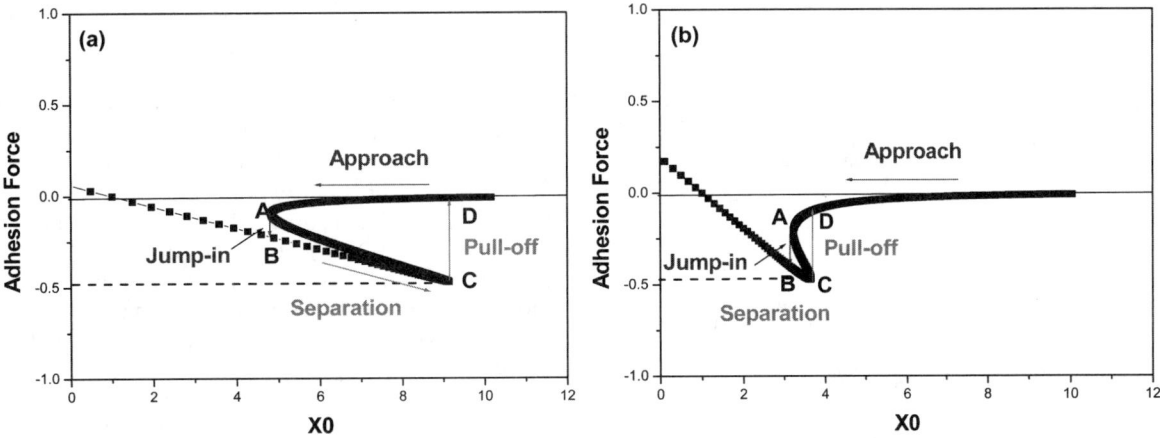

Fig. 25—Adhesion force versus separation distance x_0, calculated based on the system in Figs. 7–9, (a) for the system with a low stiffness, (b) for the system with the same surface energy but increased stiffness.

the same mechanism; that is, for two solids in relative motion along the normal or tangential directions, the system may evolve to a point of bifurcation and a state of instability, resulting in vibration and energy dissipation. It is natural to expect there is a connection between adhesion and friction, but as we have proved in previous sections that as long as energy dissipation is concerned, what really matters is the adhesion hysteresis. Consequently, we can reasonably speculate it is the adhesion hysteresis, instead of the adhesive strength, that will give rise to a direct impact on the performance of friction. This expectation was confirmed in SFA experiments [22], as shown in Fig. 26, where two mica surfaces covered by monolayers were brought slowly into contact and then separated. For the experiment conducted in inert air, the surface energy in advancing, γ_A, was found smaller than that in receding motion, γ_R, so that a hysteresis cycle was observed. The hysteresis in this experiment attributes mainly to the molecular relaxations or rearrangements in monolayers, namely the *chemical hysteresis*, which enhances the adhesion during separation [10]. If decane vapor was introduced into the system, however, the hysteresis disappeared, i.e., $\gamma_A = \gamma_B$. In correspondence, friction coefficient observed in the reversible advancing/receding was much lower than that measured in the process where hysteresis presented.

The experimental results in Fig. 26 have a further implication that the nominal value of surface energy does not associate directly with the friction level. The FFM measurements of friction force between gold surfaces and Si tips

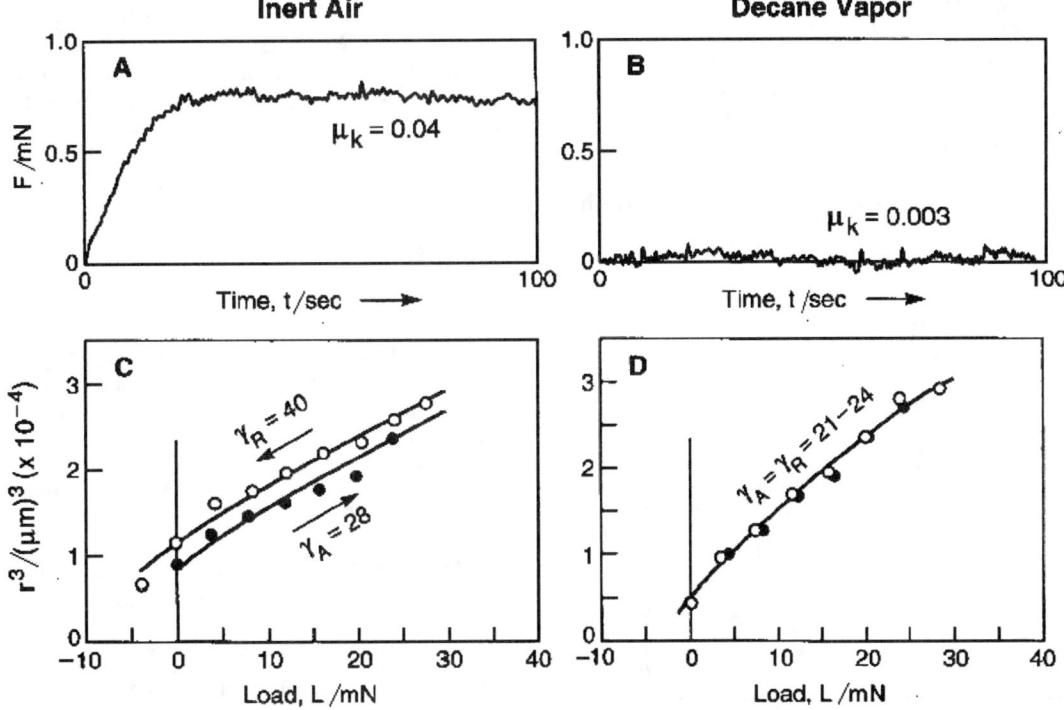

Fig. 26—Connections between adhesion hysteresis and friction: on the left the hysteresis results in a relatively large friction, on the right the hysteresis disappears due to introducing decane vapor, which leads to a smaller friction (from Ref. [21]).

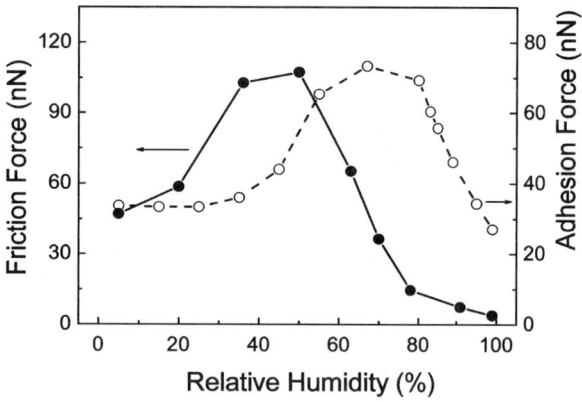

Fig. 27—Friction and adhesion forces measured on AFM show different dependences on relative humidity (from Ref. [23]).

provide another example [23]. The samples were immersed in a liquid such as ethanol to eliminate the adhesion force ($\gamma < 2$ mJ/m^2), but relatively high friction coefficients $\mu = 0.4 \sim 0.5$ were reported, which is believed to result from the sources independent of adhesion energy. In an AFM experiment of a silicon nitride probe on hydrophilic SiO$_2$ substrate [24], Qian et al. observed that as the relative humidity increased both friction and adhesion forces exhibited a similar pattern of rising/dropping, as illustrated in Fig. 27, but friction peaks at lower humidity, revealing that the dependences of friction and adhesion on the humidity are different. All the evidences presented above suggest that the relation between friction and adhesion would be more intricate, which requires a systemic investigation.

4.3 A Model to Relate Friction and Adhesion

Israelachivil et al. [25] proposed a phenomenal model for describing the interrelations between friction and adhesion. Consider the system shown in Fig. 28, where a spherical molecule slides over a corrugated solid surface. The scenario is somehow like pushing the wheel of a cart over a road paved with cobblestones, so it is also known as the cobblestone model.

The molecule initially sits in potential energy minimum. A lateral force F is required to overcome attractive interaction and to initiate the motion. While translating a distance Δd along the surface, the molecule has to be raised in normal direction due to existence of atomic roughness. As a result, the separation between the molecule and the substrate increases by a small amount, ΔD. The work done by the lateral force, F, during the translating is $W_F = F \times \Delta d$, meanwhile the increase in system energy, corresponding to the work needed to move the molecule upward from surface, can be estimated as $\Delta E_{up} \approx 4\gamma A(\Delta D/D_0)$ where γ is the surface tension, A is the area of contact, and D_0 denotes the mean space between molecule and surface. When reaching a critical point, the molecule will turn down toward substrate. If the molecule moves in a conservative manner, there will be no energy dissipation, which gives rise to a zero friction. In reality, however, a part of energy accumulated in upward motion will dissipate during the downward period, due to the impact with substrate or lattice vibrations as discussed in previous sections. Define the net energy loss in an upward/downward cycle as

$$\Delta E = \varepsilon \cdot \Delta E_{up} \quad (21)$$

where ε stands for a dissipation coefficient, which indicates a fraction of the energy ΔE_{up} are dissipated during the upward/downward cycle. The lateral force F averaged in such a period can be determined in terms of energy balance between input work and change of system energy, i.e., $W_F = \Delta E$, which gives the critical shear stress as

$$\tau_C = \frac{F}{A} = \frac{4\gamma\varepsilon \cdot \Delta D}{D_0 \cdot \Delta d} \quad (22)$$

At first sight, Eq (22) seems to suggest that friction is directly related to the surface energy γ but a further examination indicates what really matters is the product of γ and ε. The physical interpretation of the term $\gamma\varepsilon$ is the energy loss of the system during an upward/downward cycle, so that it should be in the same value as the difference between γ_R and γ_A, the surface energy in receding and advancing. In other wards, it reflects the effect of adhesion hysteresis. If we define

$$\gamma\varepsilon = \Delta\gamma = (\gamma_R - \gamma_A) \quad (23)$$

Eq (22) can be rewritten as

$$\tau_C = \frac{\Delta\gamma}{\sigma} \quad (24)$$

where $\sigma = D_0 \Delta d/4\Delta D$ is a characteristic length of molecule motion. Assume that molecule displacements in normal and sliding directions are in the same level, i.e., $\Delta d \approx \Delta D$, the above equation becomes

$$\tau_C \approx \frac{4\Delta\gamma}{D_0} \quad (25)$$

Two conclusions can be drawn from Eqs (24) and (25). First, friction is in direct proportion to the energy difference $\Delta\gamma$ in the adhesion hysteresis, which is consistent with the experimental results presented in the last section. Secondary, an increase in D_0, namely the mean separation or film thickness, will lead to decrease in friction, which agrees well with the observation in boundary lubrication. If the coefficient ε somehow remains fixed, it can be deduced from Eq

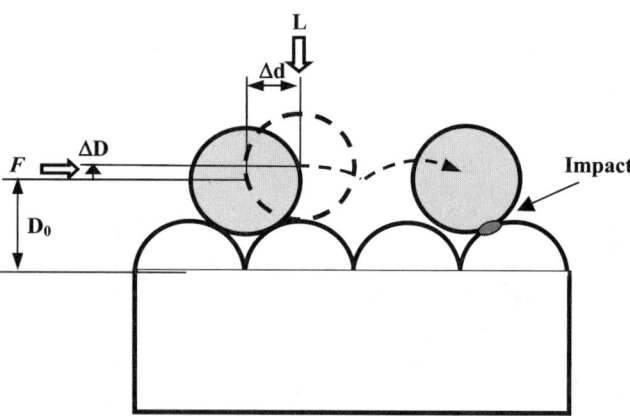

Fig. 28—The cobblestone model of friction, a spherical molecule moving on atomic roughness (reproduced after Ref. [24]).

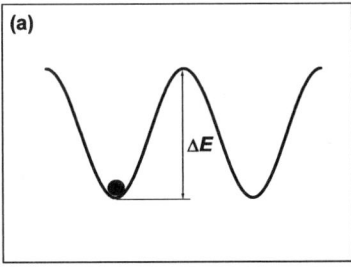 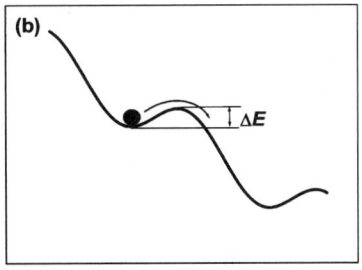 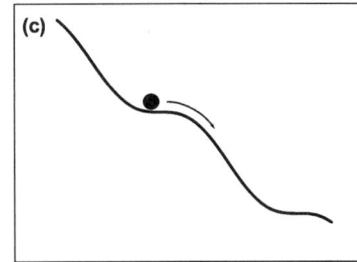

Fig. 29—A diagram of the energy barrier, (a) for a system in static contact, (b) and (c) variations of the barrier with applied lateral force.

(22) that friction would be in direct proportion to the surface energy γ. This explains why in many experiments stronger adhesion usually corresponds to a higher friction.

In the above discussions the only normal force on the sliding molecule results from the attractive interactions. When externally applied load or pressure is to be considered, the energy balance has to be modified as

$$F \times \Delta d = \Delta E + F_{ext} \times \Delta D \quad (26)$$

where ΔE is the change in system energy defined in Eq (21), the second term in the right hand represents the work done by external force during the upward motion of the molecule. The shear stress is

$$\tau_C = \frac{\Delta \gamma}{\sigma} + \frac{\Delta D}{\Delta d} \cdot \frac{F_{ext}}{A} = C_1 + C_2 P_{ext} \quad (27)$$

Equation (27) provides important information on the nature of wearless friction, which deserves a further discussion.

The wearless friction consists of two components, namely the adhesion term C_1 and the pressure term $C_2 P_{ext}$. Similar expressions for the shear stress, e.g., $\tau_C = \tau_0 + \mu P$, were also derived by other investigators [26,27], where τ_0 represents the contribution from adhesion and μ is referred to as the friction coefficient.

The adhesion term is a primary contribution in most cases of atomic-scale friction since a typical external pressure is in the order of 10 MPa while the internal van der Waals pressure is estimated as 1 GPa when using a typical Hamaker constant. The adhesion term should be also considered as a function of the applied normal load because it is inversely proportional to the mean space D_0, which is dependent on the external pressure.

If the surfaces are damaged during sliding so that wear debris and multi-asperity contacts are involved in the process, the mechanism of friction will be substantially different from what we discussed for wearless friction.

In summary, sliding can be regarded as a process during which interfacial atoms would experience a series of stick-slip motions, similar to the jump in and out in the adhesion case, and it is the energy loss in this approach/separation cycle that determines the level of friction.

5 The Nature of Static Friction

5.1 Molecular Origin of Static Friction

Figure 29 schematically shows the energy barrier of a system in contact and its change with applied lateral force. For two surfaces in static contact, interfacial atoms or molecules are pinned in potential energy minima (Fig. 29(a)). If a lateral force is applied on one of the surfaces in an attempt to initiate motion, the force biases the potential energy curve so that the energy barrier, preventing the trapped atoms from escaping the energy minima, decreases with increasing force (Fig. 29(b)). Eventually, the atoms will start to move once the energy barrier drops to zero (Fig. 29(c)).

From a microscopic point of view, the change of system state from rest to sliding is a process of pinning/depinning transition. The interfacial atoms in a sliding system may not depin at the same time. When a part of atoms are depinned, the system will reorganize itself into a new state that requires a larger force to depin. Therefore, static friction is defined as the largest force when all the pinning states disappear.

Studies based on the Frenkel-Kontorova model reveal that static friction depends on the strength of interactions and structural commensurability between the surfaces in contact. For surfaces in incommensurate contact, there is a critical strength, b_c, below which the depinning force becomes zero and static friction disappears, i.e., the chain starts to slide if an infinitely small force F is applied (cf. Section 3). This is understandable from the energetic point of view that the interfacial atoms in an incommensurate system can hardly settle in any potential minimum, or the energy barrier, which prevents the object from moving, can be almost zero.

Solid contacts are incommensurate in most cases, except for two crystals with the same lattice constant in perfect alignment. That is to say, a commensurate contact will become incommensurate if one of the objects is turned by a certain angle. This is illustrated in Fig. 30, where open and solid circles represent the top-layer atoms at the upper and lower solids, respectively. The left sector shows two surfaces in commensurate contact while the right one shows the same solids in contact but with the upper surface turned by 90 degrees. Since the lattice period on the two surfaces, when measured in the x direction, are $5\sqrt{3}$ Å and 5 Å, respectively, which gives a ratio of irrational value, the contact becomes incommensurate.

In reality, static friction is always observed regardless of whether the surfaces in contact are commensurate or not. This raises a new question as to why the model illustrated in Fig. 29 fails to provide a satisfactory explanation for the origin of static friction.

Another mechanism of static friction suggests that when two surfaces are pressed together under a normal load, the atoms or molecules at the interface will rearrange themselves to minimize the energy and to form localized junctions called *cold welding*, which is often observed in contacts

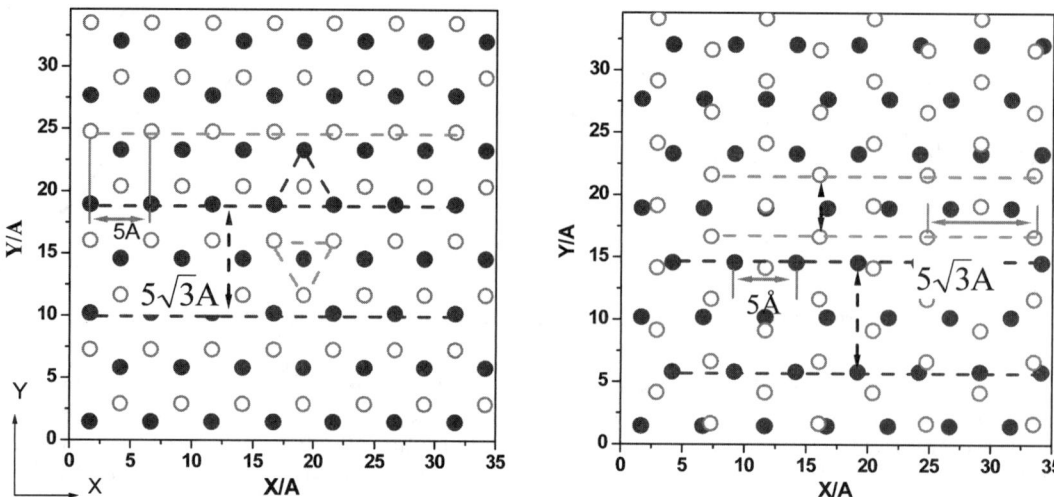

Fig. 30—Change in commensurability of two surface in contact, on the left is a commensurate contact, on the right the contact becomes incommensurate when the upper body being turned by 90 degrees, the open and solid circles denote the surface atoms of the upper and lower bodies, respectively.

between clean metal surfaces in vacuum. Cold-welded junctions provide one explanation for static friction, but rupture of the junctions will inevitably cause structural damage and wear. This brings us back to the old question, raised earlier in this chapter, which was how to explain the static friction that occurs without wear.

It has been proposed recently [28] that static friction may result from the molecules of a third medium, such as adsorbed monolayers or liquid lubricant confined between the surfaces. The confined molecules can easily adjust or rearrange themselves to form localized structures that are conformal to both adjacent surfaces, so that they stay at the energy minimum. A finite lateral force is required to initiate motion because the energy barrier created by the substrate-medium system has to be overcome, which gives rise to a static friction depending on the interfacial substances. The model is consistent with the results of computer simulations [29], meanwhile it successfully explains the sensitivity of friction to surface film or contamination.

Another possibility associates with the thermodynamic nature of the system. It has been recognized that the energy barrier diminishes for the incommensurate contacts because the potential energies from two surfaces in contact are combined together and canceled with each other. However, local and instantaneous energy barriers randomly distributed over time and space may appear due to thermodynamic fluctuations. It is thus possible that friction may result from a nonzero time and spatial average of the instantaneous energy barriers.

5.2 Static Friction and Stick-Slip Motion

From the point of view of system dynamics, the transition from rest to sliding observed in static friction originates from the same mechanism as the stick-slip transition in kinetic friction, which is schematically shown in Fig. 31. The surfaces at rest are in stable equilibrium where interfacial atoms sit in energy minima. As lateral force on one of the surfaces increases (loading), the system experiences a similar process as to what happens in the stick phase that the surface starts to slide when the energy barrier disappears. The initiation of surface motion in this sense is a result of system instability, the same as the occurrence of slip. The similarity allows us to employ the models in studying wearless kinetic friction, to examine the performances of static friction, for example, the creation of creeps.

In the experiments of static friction where two solids were nominally at rest, a small displacement between the surfaces, or a *creep*, was observed, prior to rapid slip that occurred at $F = F_s$. The creep length d_0 was reported to be in a range of 1 μm [30,31]. The formation of creep was conventionally interpreted as a result of pulling and breaking the adhesive junctions. However, the presumption of regarding plastic deformation or cold-welding as the sole cause of static friction is fundamentally inaccurate. What is to be emphasized here is that the criterion for two surfaces under shearing to hold together without macroscopic motion, or to slide with each other, simply depends on the favorable energy conditions of the system. In other words, the occur-

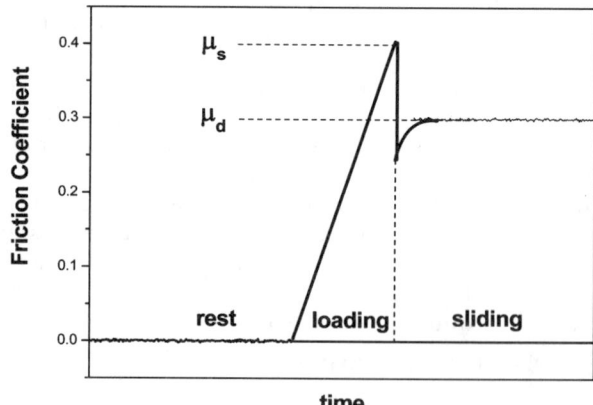

Fig. 31—A schematic plot of friction versus time, illustrating the climbing of the friction coefficient followed by a sudden drop at the time of slip.

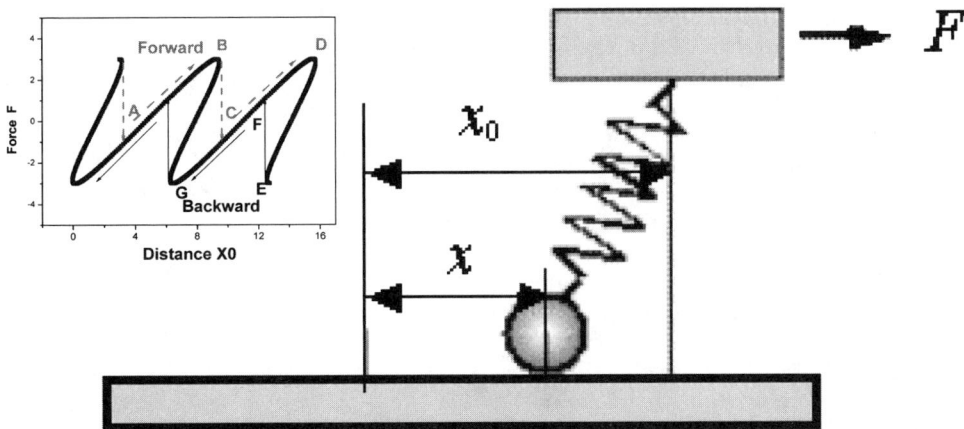

Fig. 32—Static friction and creep in an atomic system: the creep distance is defined as the maximum value of (x_0-x), inserted panel shows the variation of lateral force.

rence of stick-slip does not have to be interpreted in terms of the formation or rupture of adhesive junctions, but is all about which state is more favorable for the system to realize the energy minimization.

Although a system under shear may not exhibit a macroscopic motion during the stick process, there will be a small relative displacement between the sticking surface atoms and the substrate. This can be demonstrated by an example shown in Fig. 32 where the oscillator stands for a surface atom that sticks at position x, and the supporting block represents the substrate where force F is applied. The coordinate difference between the oscillator and the support is ($x_0 - x$) that increases with the growing force. At the moment right before the atom starts to jump, the value of ($x_0 - x$) reaches the maximum, which corresponds to the creep length of static friction. The creep length in this case is comparable to the period of the potential function of the opposite surface, which has been demonstrated by the force curve given in the inserted panel of Fig. 32.

A similar process can be observed at the asperity level, as shown in Fig. 33, where a lateral force F pulls the upper solid forward by a distance, u, while the asperity attached to the solid body remains in contact with the lower asperity. The value of u at the moment when the asperity is suddenly pulled out of contact gives rise to creep length of static friction. By referring to the force curve shown in the inserted panel of Fig. 33, the creep distance for this system is estimated to be similar with the asperity dimension in the sliding direction, which is in agreement with the measured creep length, ~ 1 μm, as reported in Ref. [30].

So far we have compared the static friction with the stick-slip transition. In both cases the system has to choose between the states of rest and motion, depending on which one is more favorable to the energy minimization. On the other hand, the differences between the two processes deserve a discussion, too. In stick-slip, when the moving surface slides in an average velocity V, there is a characteristic time, $t_c = d_0/V$, that defines how long the two surfaces can

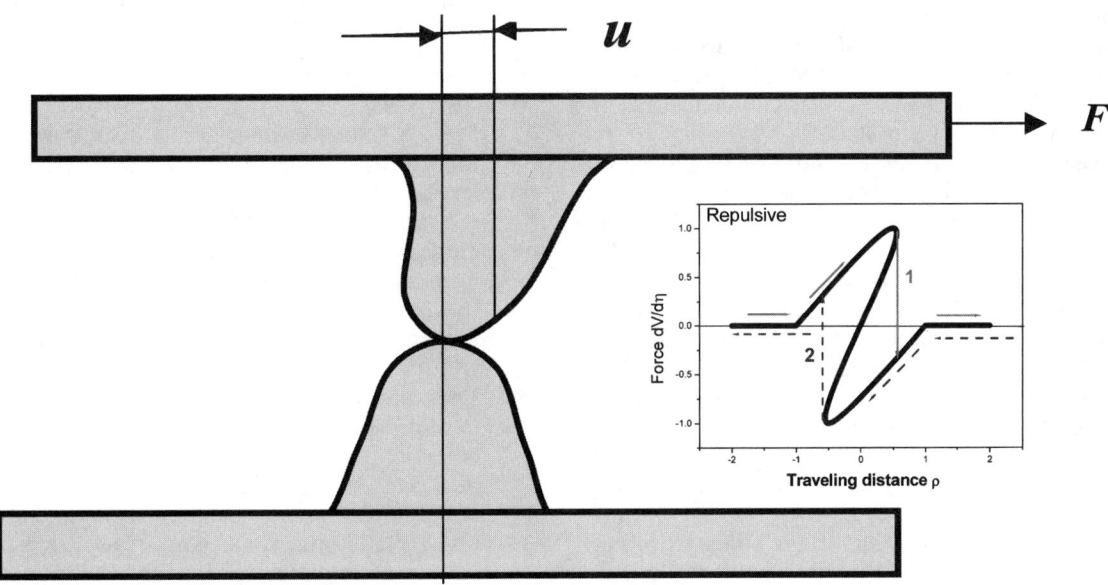

Fig. 33—Static friction between asperities, inserted panel shows variation of lateral force.

hold together, or the "age" of the contact. This is the time needed for the interfacial molecules to relax and to reestablish the contact. For this reason, stick-slip will disappear under a large sliding velocity because time is too short to resume the contact population. In static friction, on the other hand, the surfaces usually are set at rest for a long time before being pulled apart so that the contact age in most cases can be calculated from the time of $-\infty$. This would lead to two consequences. First, static friction, the maximum force needed to initiate sliding, is usually higher than the force peak appeared in stick-slip, mainly due to the longer contact time. Secondly, instability in static friction occurs at the beginning of sliding, possibly followed by a steady sliding, while the stick-slip will proceed in a periodic manner provided that the dynamical set up of the system remains unchanged. Nevertheless, one has to keep in mind that the contact time in some cases is an important factor to be accounted for in the study of static friction.

5.3 Static Friction in Lubricated Contacts

The mechanisms of static friction and stick-slip motion, as discussed in the last section, are supposed to be a good description of dry friction. Another case, perhaps more general in engineering practices, to be addressed in this section is lubricated sliding where liquid lubricant, consisting of a few molecule layers, is confined between two solid walls. Both experimental and theoretical studies indicate, as we have discussed in Chapter 5, that there are substantial changes in rheology of the confined lubricant, and the liquid may transit practically to a solid-like state when film thickness becomes molecularly thin [32,33].

As a lateral force drags one solid surface along the x direction, the solidified thin films experience an increasing shear stress until reaching a certain limit σ_a when the lubricant yields and starts to flow again like a liquid. The solid surface pulled by the lateral force then starts to move forward with a certain sliding velocity. If the solid slows down, the stress will decrease correspondingly but the lubricating film may remain at liquid state until it meets the second critical value, σ_c, when the lubricant returns to solid state. In this model, the static friction, or transition from rest to motion is interpreted as a result of yield of solidified lubricating films, and stick-slip is regarded as a process of periodic transition between melting and freezing phases [24].

On the basis of experimental observations and computer simulations, Persson [34] proposed a stress-velocity relation, $\sigma=f(v)$, as schematically reproduced in Fig. 34, which describes how shear stress changes with sliding velocity. Assume the system begins from a stationary state with both sliding velocity and shear stress being zero. After a lateral force is applied on one surface and pulls it along the sliding direction, the shear stress on the confined lubricant rises up but the surface remains stationary, i.e., the shear stress increases along a vertical line of V=0 in Fig. 34. When the growing shear stress goes to a critical value, σ_a, the solidified lubricant yields to the stress and starts to flow. As a result, the velocity of the solid surface increases rapidly from 0 to v_a. For a further increase of velocity, the stress will go up along the oblique line BE, following the viscous law of fluid traction. If the sliding slows down, however, the lubricant does not immediately resume the solid state, but remains as liq-

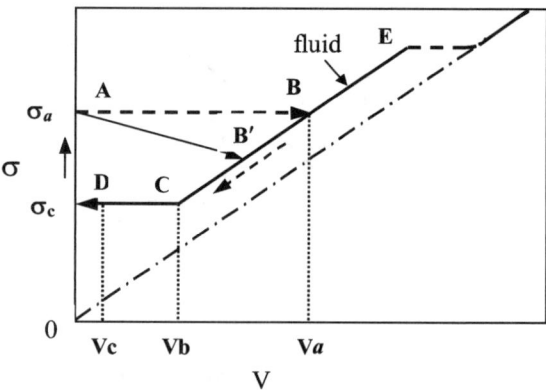

Fig. 34—Shear stress on boundary films versus sliding velocity (reproduced after Ref. [33]).

uid so that the stress decreases along the line BC until it meets the point C where the sliding velocity and corresponding stress are denoted as v_b and σ_c, respectively. As the velocity decreases continuously, the lubricant will go through a state known as "granular fluid" in the range of $v_c < v < v_b$, where the shear stress keeps almost constant. At the point D the lubricant returns completely to solid state.

At the beginning of sliding, the system is accelerated because the driven force must excess the resistance from lubricating film. For this reason, the system actually jumps from A to the point B′, instead of B, to gain a shear stress lower than the critical value σ_a. This phenomenon, so called "velocity-weakening" has been regarded widely in the literatures as the cause for instability and stick-slip motion in lubricated systems.

Finally, it deserves to be mentioned that considerable numbers of models of static friction based on continuum mechanics and asperity contact were proposed in the literature. For instance, the friction at individual asperity was calculated, and the total force of friction was then obtained through a statistical sum-up [35]. In the majority of such models, however, the friction on individual asperity was estimated in terms of a phenomenal shear stress without involving the origin of friction.

6 Summary

1. For solid surfaces interacting in air, the adhesion forces mainly result from van der Waals interaction and capillary force, but the effects of electrostatic forces due to the formation of an electrical double-layer have to be included for analyzing adhesion in solutions. Besides, adhesion has to be studied as a dynamic process in which the approach and separation of two surfaces are always accompanied by unstable motions, jump in and out, attributing to the instability of sliding system.

2. The conventional interpretation for the origin of friction in terms of material damage or wear is not satisfactorily convincing for it greatly underestimates the energy loss in sliding. We have demonstrated in this chapter that a wearless friction may originate from unstable motion of interfacial atoms, molecules, or asperities, resulting in vibrations that are consequently damped through phonon emission or electron excitation, and finally dissipated into heat.

3. An important message carried in this chapter is that friction and adhesion are connected by the similarity in the dynamic feature that they both are accompanied by unstable motion of interfacial atoms, i.e., jump in and out of contact, and it is the energy loss in approach/separation cycle that determines the magnitude of friction.
4. In static friction, the change of state from rest to motion is caused by the same mechanism as the stick-slip transition. The creation of static friction is in fact a matter of choice of system state for a more stable and favorable energy condition, and thus does not have to be interpreted in terms of plastic deformation and shear of materials at adhesive junctions.
5. While the models of wearless friction succeed in understanding how friction results from the mechanical instability and energy dissipation, they fail to explain the friction observed for sliding between incommensurate surfaces in which the system is predicted to be stable and frictionless that is in conflict with reality. Attempts have been made to settle this contradiction, for example, to interpret the friction in terms of contributions from the third medium confined between the surfaces.

The Amontons' friction law was extracted from experiential observations, but it has exhibited a nature of universality, that is, the law holds in a great majority of friction tests, regardless of the system set-up, material, and operation conditions. Most friction models, on the other hand, are system dependent, each with specific assumptions of their own. A universal experimental law requires a fundamental theory to provide more convincing explanations for widely diverging phenomena of friction. New concepts, such as dynamically originated friction, revealed by the Frenkel-Kontorova model, and the thermodynamic origin of friction as suggested by Gao [36], began to emerge in recent years. As a prospect remark, friction has to be regarded as a result of great numbers of dynamic and chaotic events that occurred in the interface region, and it is the nonequilibrium thermodynamic nature of the system that gives rise to the highly defined property of friction.

References

[1] Xiao, X. D. and Qian, L. M., "Investigation of Humidity-Dependent Capillary Force," *Langmuir*, Vol. 16, No. 21, 2000, pp. 8153–8158.

[2] Johnson, K. L., Kendall, K., and Roberts, A. D., "Surface Energy and the Contact of Elastic Solids," *Proc. R. Soc. London, Ser. A*, Vol. 324, 1971, pp. 301–313.

[3] Tabor, D., "Surface Forces and Surface Interactions," *J. Coll. Interface Sci.*, Vol. 58, 1976, pp. 1–13.

[4] Meyer, E., Lüthi, R., Howald, L., Bammerlin, M., Guggisberg, M., and Güntherodt, H. J., "Site-Specific Friction Force Spectroscopy," *J. Vac. Sci. Technol. B*, Vol. 14, 1996, pp. 1285–1288.

[5] Homola, A. M., Israekachivili, J. N., Gee, M. L., and McGuigan, P. M., "Measurements of and Relation Between the Adhesion and Friction of Two Surfaces Separated by Molecularly Thin Liquid Films," *Trans. ASME, J. Tribol.*, Vol. 111, 1989, pp. 675–682.

[6] Israelachvili, J. N., *Intermolecular and Surface Forces*, 2nd ed., Academic Press, 1992, London, pp. 312–337.

[7] Hu, Y. Z., Zhang, T., Ma, T. B., and Wang, H., *Computational Material Science*, Vol. 38, 2006, pp. 98–104.

[8] Yang, D., Wang, H., Hu, Y. Z., and Zhang, L., "Redesign of the Device for Measuring Normal Force and Experimental Verification," Proceedings, *4th China International Symposium on Tribology*, Xian, China, 2004, pp. 240–246.

[9] Dowson, D., *History of Tribology*, Longmans, London, 1979.

[10] Bowden, F. P. and Tabor, D., *Friction and Lubrication of Solids*, Oxford University Press, 1964, pp. 52–86.

[11] Israelachvili, J. N., "Adhesion, Friction and Lubrication of Molecularly Smooth Surfaces," *Fundamental of Friction*, I. L. Singer and H. M. Pollock, Eds., Kluwer Academic Publishers, 1991, pp. 351–385.

[12] Tomlinson, G. A., "A Molecular Model of Friction," *Phil. Mag.*, Vol., 7, 1929, pp. 905–939.

[13] Caroli, C., "Dry Friction as a Hysteretic Elastic Response," *Physics of Friction*, B. N. J. Persson and E. Tosatti, Eds., Kluwer Academic, 1996, pp. 27–49.

[14] Gnecco, E., Bennewitz, R., Gyalog, T., and Meyer, E., "Friction Experiments on the Nanometer Scale," *J. Phys. Condens. Matter*, Vol. 13, 2001, pp. R619–R642.

[15] Gnecco, E., Bennewitz, R., Gyalog, T., Loppacher, C., Bammerlin, M., Meyer, E., and Güntherodt, H.-J., "Velocity Dependence of Atomic Friction," *Phys. Rev. Lett.*, Vol. 84, 2000, pp. 1172–1175.

[16] Zhang, T., Wang, H., and Hu, Y. Z., "Atomic Stick-Slip Friction Between Commensurate Self-Assembled Monolayers," *Tribology Letters*, Vol. 14, No. 2, 2003, pp. 69–76.

[17] Strunz, T. and Elmer, F. J., "On the Sliding Dynamics of the Frenkel-Kontorova Model," *Physics of Friction*, B. N. J. Persson and E. Tosatti, Eds., Kluwer Academic Publishers, 1996, pp. 149–161.

[18] Weiss, M. and Elmer, F. J., "A Simple Model for Wearless Friction: The Frenkel-Kontorova-Tomlinson Model," *Physics of Friction*, B. N. J. Persson and E. Tosatti, Eds., Kluwer Academic Publishers, 1996, pp. 163–178.

[19] Weiss, M. and Elmer, F. J., "Dry Friction in the Frenkel-Kontorova-Tomlinson Model: Dynamical Properties," *Zeitschrift fur Physik B—Condensed Matter*, Vol. 104, No. 1, 1997, pp. 55–69.

[20] Sokoloff, J. B., "Theory of the Contributions to Sliding Friction from Electronic Excitation in the Microbalance Experiment," *Phys. Rev. B*, Vol. 52, No. 7, 1995, pp. 5318–5322.

[21] Persson, B. N. J., and Volokitin, A. I., "Electronic and Phononic Friction," *Physics of Friction*, B. N. J. Persson and E. Tosatti, Eds., Kluwer Academic Publishers, 1996, pp. 253–264.

[22] Chen, Y. L., Gee, M. L., Helm, C. A., Israelachvili, J. N., and McGuiggan, P. M., "Effects of Humidity on the Structure and Adhesion of Amphiphilic Monolayers on Mica," *J. Phys. Chem.*, Vol. 93, 1989, pp. 7057–7059.

[23] Ruths, M., "Boundary Friction of Aromatic Self-Assembled Monolayers: Comparison of Systems with One or Both Sliding Surfaces Covered with a Thiol Monolayer," *Langmuir*, Vol. 19, 2003, pp. 67–88.

[24] Qian, L. M., Tian, F., and Xiao, X. D., "Tribological Properties of Self-Assembled Monolayers and Their Substrates Under Various Humid Environments," *Tribol. Lett.*, Vol. 15, No. 3, 2003, pp. 169–176.

[25] Israelachvili, J. N., "Microtribology and Microrheology of Molecularly Thin Liquid Film," *Modern Tribology Hand-*

book, B. Bhushan, Ed., CRC Press LLC, New York, 2001, pp. 568–610.

[26] Derjaguin, B. V., Muller, V. M., and Toporov, Y. P., "Effect of Contact Deformation on the Adhesion of Particles," *J. Coll. Interface Sci.*, Vol. 67, 1975, pp. 314–326.

[27] Briscoe, B. J. and Evans, D. C., "The Shear Properties of Langmuir-Blodgett Layers," *Proc. R. Soc. London, Ser. A*, Vol. A380, 1982, pp. 389–407.

[28] He, G., Müser, M. H., and Robbins, M. O., "Adsorbed Layers and the Origin of Static Friction," *Science*, Vol. 284, 1999, pp. 1650–1652.

[29] He, G. and Robbins, M. O., "Simulations of the Kinetic Friction Duet to Adsorbed Surface Layers," *Tribology Letters*, Vol. 10, No. 1–2, 2001, pp. 7–14.

[30] Baumberger, T., "Dry Frictions Dynamics at Low Velocities," *Physics of Friction*, B. N. J. Persson and E. Tosatti, Eds., Kluwer Academic Publishers, 1996, pp. 1–26.

[31] Baumberger, T., Caroli, C., Perrin, B., and Ronsin, O., "Nonlinear Analysis of the Stick-Slip Bifurcation in the Creep-Controlled Regime of Dry Friction," *Phys. Rev. E*, Vol. 51, 1995, pp. 4005–4010.

[32] Granick, S., "Motions and Relaxations of Confined Liquids," *Science*, Vol. 253, 1991, pp. 1374–1379.

[33] Hu, Y. Z., Wang, H., and Guo, Y., "Simulation of Lubricant Rheology in Thin Film Lubrication, Part I: Simulation of Poiseuille Flow," *Wear*, Vol. 196, 1996, pp. 243–248.

[34] Persson, B. N. J., "Theory of Friction: The Role of Elasticity in Boundary Lubrication," *Phys. Rev. E*, Vol. 50, No. 7, 1994, pp. 4771–4786.

[35] Kogut, L. and Etsion, I., "A Static Friction Model for Elastic-Plastic Contacting Rough Surfaces," *Trans. ASME, J. Tribol.*, Vol. 126, 2004, pp. 34–40.

[36] Gao, J. B., Luedtke, W. D., Gourdon, D., Ruths, M., Israelachivili, J. N., and Landam, U., "Friction Forces and Amontons; Law: From the Molecular to the Macroscopic Scale," *J. Phys. Chem. B*, Vol. B108, 2004, pp. 3410–3425.

10

Microscale Friction and Wear/Scratch

Xinchun Lu[1] and Jianbin Luo[1]

1 Introduction

MICROTRIBOLOGY IS A KEY SCIENCE AND TECH-nology for micro-electromechanical systems (MEMS), high density magnetic recording system, micro-robot, etc. Also, it is important for developing new micro devices and for understanding the origins of friction and wear [1,2]. The ultimate goal of microtribology research is to create practical zero-wear micro devices. Microtribology consists of microfriction, microwear, and microlubrication, while microfriction and wear plays an important role in microtribology.

1.1 Origin of Microscale Friction and Wear

The word "Microtribology" was first developed in the late 1980s [2,3]. From the last decade, microtribology has become a foremost area in tribology. The original background of microtribology is from the development of high density magnetic storage systems. As we know, the near field technique is used for reading and writing in magnetic storage. The closer the head to the magnetic media, the higher the storage density will be achieved. In the early 1990s, the magnetic recording head sliders having very small mass (less than 10 mg) were beginning to be used [2]. Using photolithography technologies, micro head assemblies having an effective mass of less than 1 mg have been developed for contact magnetic recording. Therefore, zero wear rate was required in such a small size with the tiny mass tribological system. In the late 1990s, the gap between head and media was over 30 nanometers for the areal density 10 Gb/in.^{2} of hard disk drive (HDD), in 2002 it was about 10 nanometers for the areal density 60 Gb/in.^{2}. Nowadays it is about 8 nanometers for the density of over 100 Gb/in.^{2}. In such a tiny gap, wear is strictly controlled to ensure the magnetic media is being protected during its using life. The using life for a hard disk drive is generally three to five years, which means the wear rate of HDD should be zero during one million kilometers sliding.

Also, other micromechanical components having submillimeter dimensions such as micromotors, microgear trains, micropumps and valves, which are known as a MEMS, have also been fabricated using photolithography technologies. The micromechanical components also have small mass. The common-sense view that the role of a bearing is to support a load no longer holds for such systems; the wear on sliding surfaces is primarily the result of surface interaction forces rather than the load.

With the rapid progress of microtribology, in 1991, Granick first pointed out the conception of "Nano-tribology" or "molecular-tribology" [4]. To promote microtribology communication, Japan established the Society of Microtribology in 1990, subsequently in 1992 and 1993, two international microtribology proceedings were conducted in Japan and Poland, respectively. Furthermore, in 1993, the *Journal of Wear* published a special issue about "microtribology" and the Society of Material Research of America published a special issue on "nanotribology." Then, since the early 1990s the topic of microtribology has appeared in a lot of important international proceedings, and microtribology has attracted more and more attention of scientists.

1.2 Definition of Microwear

Microwear is the surface damage or removal of material from one or both of two solid surfaces in a sliding, rolling, or impact motion relative to one another [5]. Microwear damage precedes actual loss of material, and it may also occur independently. The definition of microwear is generally based on loss of material from one or both of the mating surfaces. Strictly, the microwear, as in the case of friction, is not an inherent material property; it depends on the operating conditions and surface conditions. Microwear rate does not necessarily relate to friction. Microwear resistance of a material pair is generally classified based on a wear coefficient, a nondimensional parameter, or wear volume per unit load per unit sliding distance. Between the limit of microwear and nanowear, it is not very clear. In general, wear, between the surface and subsurface at the range of 100 nm–100 μm, can be believed as microwear.

1.3 Contents of this Chapter

Although the traditional theories of lubrication, friction, and wear, which have been developed for over 100 years, formed the important basis for general engineering design and material development, there are still many problems left unresolved when the theories are applied to tribological micromechanisms and active controlling of lubrication, friction, and wear. Especially, the theories of macro-tribology are no longer suitable for explaining tribological behavior and mechanisms which occur in micromachines or super precision instruments because of the ultra light load and nanometer scale. As it is known, many kinds of contact surfaces with relative movement consist of thin films deposited on bulk materials, especially in micromachines or super precision instruments. However, the properties of thin films are often different from that of bulk materials. Therefore, it is very difficult to use the traditional experimental methods

---

[1] *State Key Laboratory of Tribology, Tsinghua University, Beijing, China.*

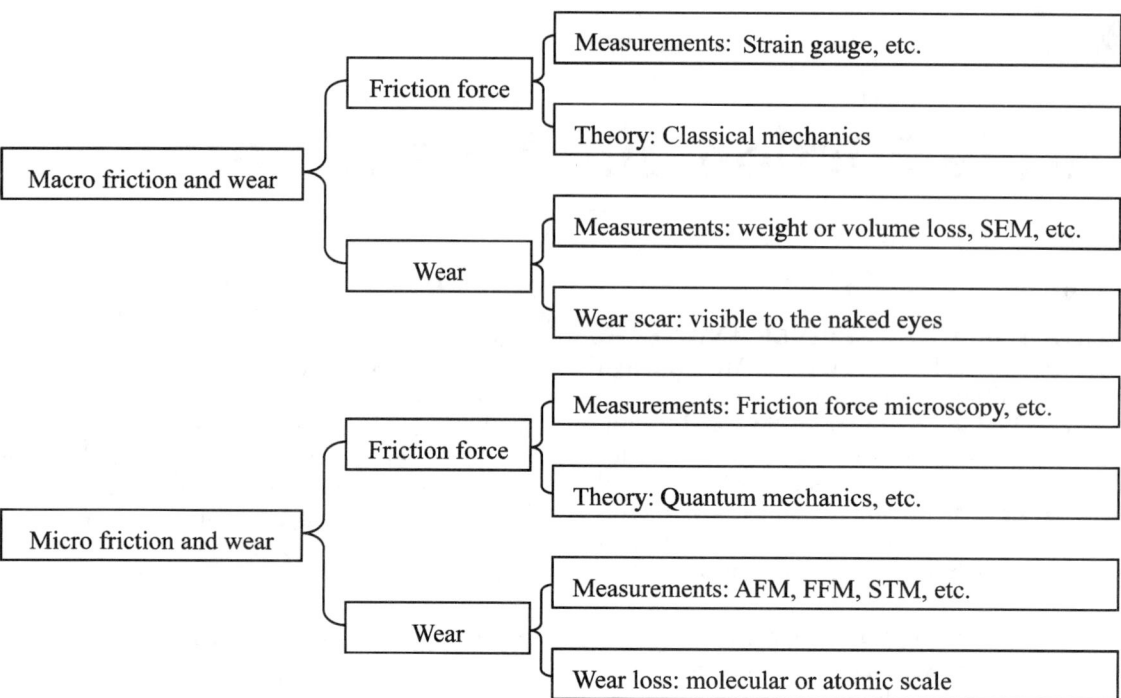

Fig. 1—The brief comparison between macro and micro scale friction and wear.

and theories to study and explain the tribological behavior of these thin films for special use.

In order to study microscale friction and wear, scientists have developed the friction force microscope (FFM), nanoindentation and nanoscratch tester which serve as excellent tools in microtribological research [1,6–9]. In this chapter, we first compare the differences between macro and micro friction and wear, and then introduce some results of our research group on microscale friction and wear of ordered films, thin solid films, and multilayers.

2 Differences Between Macro and Micro/Nano Friction and Wear

2.1 General Differences

In many cases of traditional tribology, friction and wear are regarded as the results of surface failure of bulk materials, the solid surface has severe wear loss under high load. Therefore, the mechanical properties of bulk material are important in traditional friction and wear. However, in microscale friction and wear, the applied load on the interactional surface is light and the contact area is also under millimeter or even micrometer scale, such as the slider of the magnetic head whose mass is less than 10 mg and the size is in micrometer scale. Under this situation, the physical and chemical properties of the interactional surface are more important than the mechanical properties of bulk material. Figure 1 shows the general differences between macro and micro scale friction and wear.

2.2 Measurement for Microscale Friction Using Friction Force Microscopy

The FFM based on the atomic force microscopy (AFM) is the most available tool to study the feature of microscale friction and wear of material surface with high resolution [6]. So the FFM has received more attention and has been widely used in micro/nanotribology research since it was developed.

Figure 2 shows the brief principle of a laser-detected FFM. A sample is put on a piezoelectrical tube (PZT), which scans X, Y plane and controls the feedback of Z axis. The laser beam from a diode is focused on the mirror of the free end of a cantilever with lens, and the reflected beam falls on the center of a position-sensitive detector (PSD), a four-quadrant photodiode. When the sample contacts with the tip and relatively moves under the control of a computer, the reflected beam deflects and changes the position on the PSD due to the twist and deflection of the cantilever caused by the changes of surface roughness, friction force, and adhesive force between the sample and the tip. The extension and re-

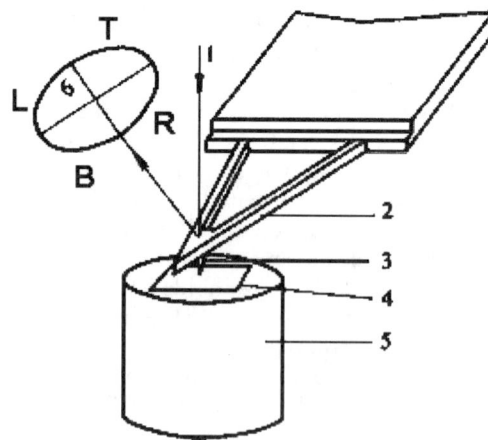

Fig. 2—Brief principle of a laser-detected FFM. 1, laser beam; 2, cantilever; 3, tip; 4, Sample; 5, piezoelectrical tube; 6, position-sensitive detector.

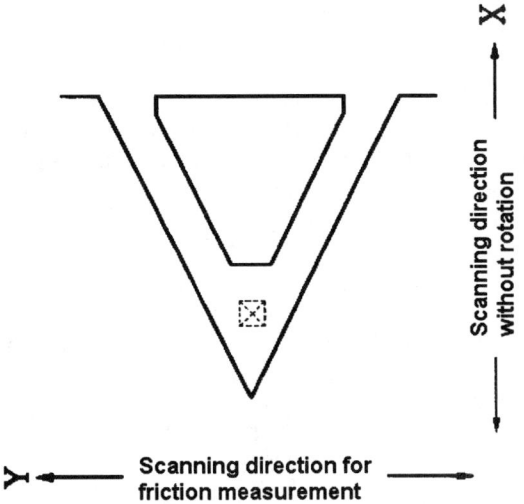

Fig. 3—A schematic diagram of the scanning direction of FFM.

traction of the PZT is feedback controlled by the electrical signals from the vertical direction of the PSD. Thereby the surface morphology of the sample is obtained according to the movements of the PZT. Meanwhile, the image of lateral force in the scanning area is obtained by the electrical signals from the horizontal direction of the PSD caused by the twist of the cantilever.

Figure 3 shows two scanning directions of a tip. For the microscale friction test, the surface morphology as well as the friction force image is generally obtained by scanning along the Y direction with the constant height mode. The normal force and friction signals are recorded by upper and lower, left and right segments of the PSD, respectively.

The friction coefficient can be easily obtained in traditional tribological test. However, in the microscale friction test, the relative moving status of the tip is greatly affected by the surface morphology, and the friction coefficient should be obtained by theoretical analysis according to the experimental condition. A model shown in Fig. 4 is set up to analyze the relative moving status of the tip.

Assuming that the applied load on the sample surface is F_L, the friction force is F_f, the force perpendicular to F_f is F_n, the force applied to the tip by the fixed end of cantilever is F_e. For state A shown in Fig. 4, an equilibrium formula of the torsion moment can be presented as:

$$k\alpha = (F_n \sin\theta \cdot h \cos\alpha + F_f \cos\theta \cdot h \cos\alpha) + (F_n \cos\theta \cdot h \sin\alpha - F_f \sin\theta \cdot h \sin\alpha) \quad (1)$$

Equation (1) can be expressed as

$$\frac{k\alpha}{h} = F_n \sin(\theta + \alpha) - F_f \cos(\theta + \alpha) \quad (2)$$

where k is the spring constant of cantilever, α is the twist angle of cantilever, and θ is the angle between the horizontal and tangential line on the contact spot. Another formula of force equilibrium can be expressed as

$$F_L = F_n \cos\theta - F_f \sin\theta \quad (3)$$

Here the friction coefficient is defined as $\mu = F_f/F_n$. According to Eqs (2) and (3) it can be expressed as

$$\mu = \frac{k\alpha \cos\theta - F_L \cdot h \sin(\theta + \alpha)}{k\alpha \sin\theta - F_L \cdot h \cos(\theta + \alpha)} \quad (4)$$

For state B, the friction coefficient can be expressed as

$$\mu = -\frac{k\alpha \cos\theta + F_L \cdot h \sin(\theta + \alpha)}{k\alpha \sin\theta + F_L \cdot h \cos(\theta + \alpha)} \quad (5)$$

For state C, the friction coefficient can be expressed as

$$\mu = \frac{k\alpha - F_L \cdot h \sin\alpha}{F_L \cdot h \cos\alpha} \quad (6)$$

Therefore, for a rough surface, although it is smooth macroscopically, the friction coefficient is greatly affected by its surface morphology. For a smooth surface on atomic scale, θ can be neglected and the friction coefficient is only related to the twist angle of the cantilever.

3 Calibration of the Friction Force Obtained by FFM

Mate [6] first obtained the friction force signal as well as the normal load signal by modifying an AFM in 1987. The modified AFM was called FFM or LFM (lateral force microscopy). The friction force signal was obtained by detecting the tor-

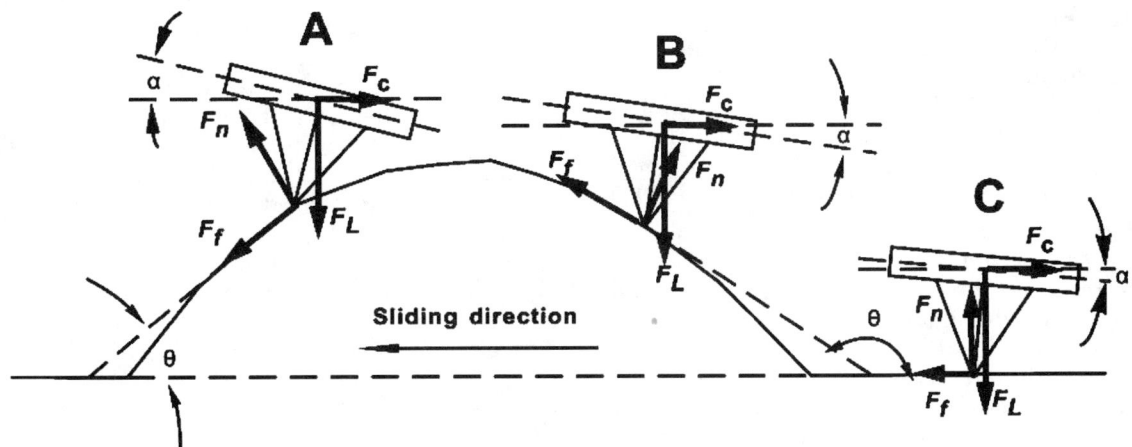

Fig. 4—The status of probe during scanning.

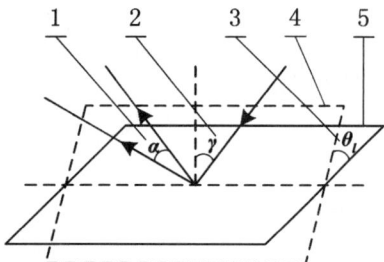

Fig. 5—The relation between the rotated angle of reflection beam (α) and the torsional angle of cantilever (θ_l), (1-rotated angle of the reflection beam, α; 2-incident angle, γ; 3-torsional angle of the cantilever, θ_l; 4-reflection surface before torsion of the cantilever; 5-reflection surface after torsion of the cantilever).

sion of cantilever along the axis of X direction (Fig. 3). However, even for a commercial FFM, the measurement and the calibration of real friction force is an unsolved problem. We developed a friction force microscope employing laser beam deflection for force detection in 1996 [9]. The following is how to measure the lateral force and calibrate the friction force based on the design of this FFM.

3.1 Measurement of Lateral Force

The lateral force is obtained directly from the FFM. So the lateral force is needed to be quantitatively calibrated first and then the friction force can be calculated. As shown in Fig. 2, the lateral force can be expressed by

$$F_l = \frac{K_l \times \theta_l}{h_t} \quad (7)$$

where K_l is the torsional stiffness of the cantilever, θ_l is the torsional angle, and h_t is the height of the tip (about 2.856 μm). The torsional stiffness of the cantilever is calculated theoretically, while the torsional angle is measured experimentally.

3.1.1 The Calculation of Torsional Stiffness of Cantilever

The cantilevers we used are microfabricated silicon nitride (Si_3N_4) triangular beams with integrated square pyramidal tips made of plasma enhanced chemical vapor deposition (PECVD) using photolithographic techniques [10]. The cantilever stiffness in the longitudinal direction (Z-direction) K_z is 0.38 N/m, provided by the producer. The torsional stiffness K_l of the tip is 5.120×10^{-9} Nm/rad, which is calculated by the general finite element software ALGOR with 2,000 quadrilateral shell elements.

3.1.2 The Measurement of Torsional Angle

There is a corresponding relationship between the lateral voltage (V_l) of four quadrant position detectors and the torsional angle. So θ_l can be obtained from V_l.

Figure 5 shows the laser beam path reflected by the torsional cantilever. The incident angle is γ on the cantilever surface before the cantilever torsion. When the torsional angle of the cantilever is θ_l, reflection ray turned an angle α. Their relationship can be expressed by

$$\sin \theta_l = \frac{\sin(0.5\alpha)}{\cos \gamma} \quad (8)$$

The incident angle is a constant for this equipment, $\gamma = 15°$. The torsional angle θ_l, and the rotated angle α of the reflection ray are both infinitesimals. The formula can be simplified as

$$\theta_l = 0.518\alpha \quad (9)$$

Transverse moving head of four-quadrant position detectors (Fig. 2), precise measuring lateral displacement which corresponds to lateral voltage V_l when it varies at the linear range of ± 10 V, thus we can compute the rotated angle α of the reflection ray:

$$\alpha = \frac{\delta S_{max} \times V_l}{\delta V_{max} \times L} \quad (10)$$

where δS_{max} is lateral displacement corresponding to the lateral voltage $\delta S_{max} = 20$ V. The value of δS_{max} is about 0.125 mm obtained experimentally. The optical distance of reflection ray L, is about 33 mm.

Through measurement of the lateral-voltage variation of the position detectors, the torsional angle θ_l of the cantilever can be given by

$$\theta_l = \frac{0.518 \times \delta S_{max} \times V_l}{\delta S_{max} \times L} = \frac{0.518 \times 0.125 \text{ mm} \times V_l}{20V \times 33 \text{ mm}}$$
$$= 9.811 \times 10^{-5} \times V_l (\text{rad}) \quad (11)$$

Then, the corresponding relationship between lateral voltage of the four-quadrant position detectors (V_l) and the torsional angle of the cantilever (θ_l). Thus we can obtain the variation of torsional angle θ_l through reading the variation of lateral voltage (V_l) from the front panel.

3.2 Calibration of Friction Force

The friction force can be obtained by the product of the load and the friction coefficient. We carried on the experiment to measure the lateral force on the single crystal silicon surface without any precleaning. The procedures are shown as follow: first, we approached the tip to the sample surface till the work status. At that time the load is zero. We recorded the Z-direction voltage of the corresponding piezoelectric tube and the lateral output voltage of the detector with four quadrants, which acted as the lateral voltage of the detector with four quadrants before scanning (V_{ls}^0) under the load zero. Then, we let the tip scan laterally on the sample surface, and we took the lateral voltage of the detector with four quadrants, which acted as the scanning lateral voltage of the detector with four quadrants (V_{lk}^0) under the load zero. After the scanning, we adjusted and then recorded the Z-voltage of the piezoelectric tube under the static load. The variety of the lateral output voltage before and after loads reflected the variety of the lateral force signals that arose by the sample topography. Similarly, we recorded the lateral voltage of the detector with four quadrants that acted as V_{ls}^1 under the corresponding loads. And then we took the lateral output voltages of the detector with four quadrants that acted as V_{lk}^1 under those loads when the tip is scanning laterally on the sample surface. We repeated the above experimental procedures by the continued loadings and obtained the data listed

TABLE 1—The lateral force measurements on the single crystal silicon surface.

Exp. No.	Z-voltage of Piezoelectric Ceramic	V_{ls} (V)	V_{lk} (V)	$V_{lk}-V_{ls}$ (V)
1-0	−6.0	2.98	2.90	−0.08
1-1	5.1	1.87	1.78	−0.09
1-2	15.7	0.81	0.72	−0.11
1-3	25.7	−0.22	−0.34	−0.12
1-4	35.9	−1.44	−1.56	−0.12
1-5	45.9	−2.45	−2.58	−0.13
1-6	56.0	−3.54	−3.68	−0.14
1-7	66.1	−4.51	−4.67	−0.16

Note: V_{ls} is the lateral voltage of the detector with four quadrants before scanning; V_{lk} is the lateral voltage of the detector with four quadrants while scanning; and $V_{lk}-V_{ls}$ is the difference between the lateral output voltages of the detector with four quadrants.

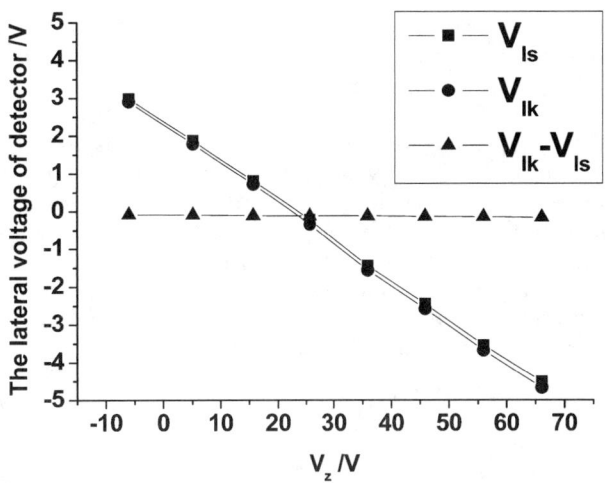

Fig. 6—The curves of the calibrations for the friction forces.

in Table 1. Here, V_{ls} were the static lateral output voltages under the loads, which reflected the lateral force signals that resulted from the sample surface topography. And V_{lk} were the lateral output voltages while scanning, which were the integrated reflections of the lateral force endured by tips while scanning, including the lateral force from the surface topography, the friction force, the adhesion, and other factors, which are dominated by the surface topography and the friction force. The difference between the lateral output voltages of the detector with four quadrants ($V_{lk}-V_{ls}$) mainly reflected the varieties of the lateral force signals that resulted from the friction force. We can deduct the micro-friction coefficient of the sample surface and calculate the friction force of the tips under those loads.

In Table 1, the varieties of Z-voltage of the piezoelectric tube correspond to the magnitudes of the loads. Because the increasing and decreasing of the voltages on the Z direction of the piezoelectric tube after the approach of the tip to the sample surface, resulted in the extension and retraction of the piezoelectric tube in Z direction, which is presented as the augmentation and reduction of the cantilever deformation in Z direction. Sequentially, the varieties of the loads on the sample surface arose. It is listed below with the relationship between the varieties of the Z-voltage of the piezoelectric ceramic (δV_z) and the load (δW):

$$\delta W = \delta V_z \times P \times K_z \qquad (12)$$

Here, P is the piezoelectric constant of the piezoelectric tube, K_z is the lateral stiffness of the cantilever.

We established the coordinate system and plotted the curves using the Z-voltage of the piezoelectric ceramic as the X axis and the lateral voltage of the detector with four quadrants (V_{ls}, V_{lk} and $V_{lk}-V_{ls}$) as the Y axis, as shown in Fig. 6. We fit linearly the varieties of the lateral voltages in the detector with four quadrants ($V_{lk}-V_{ls}$) to the varieties of the Z-voltages of the piezoelectric tube and obtained the slopes of the lines, which are attributed to the friction coefficients of the sample surfaces.

To the single crystal Si, the calculations of the friction coefficient are listed below using the data in Fig. 6:

$$V_{lk} - V_{ls} = -0.08782 - 0.99101 V_z \qquad (13)$$

then,

$$\delta(V_{lk} - V_{ls}) = -0.00101 \delta V_z \qquad (14)$$

The friction coefficient can be calculated as follows:

$$f = \frac{\delta F_f}{\delta W} = \frac{K_l \times \delta \theta_l}{\delta V \times P \times K_z \times h_t} \qquad (15)$$

$$f = \frac{5.120 \times 10^{-9} \text{ Nm/rad} \times 9.811 \times 10^{-5} \text{ rad}/v \times 0.00101}{7.5 \times 10^{-9} \text{ m}/v \times 0.38 \text{ N/m} \times 2.856 \times 10^{-6} \text{ m}}$$

$$\cong 0.06 \qquad (16)$$

Follow the above procedures, we obtained the friction coefficient of the unclean single crystal Si surface is about 0.06, which agrees very well with the result obtained under the same experimental condition by Bhushan et al. It is validated with the feasibility of the calibration method. Based on the friction coefficient, we can easily obtain the friction forces under the loads.

4 Microscale Friction and Wear of Thin Solid Films

Thin solid films have a wide range and play the most important roles on surface engineering. With the deposition technology development, more and more thin solid films are used in micro devices, precision instruments, IC technology, etc., for different purposes. In the tribology research area, the tribological properties of thin solid films are of the first concern, especially at micrometer scale. Here we introduce some results obtained by using FFM on microscale friction and wear of gold films and PTFE multilayers.

4.1 Microscale Friction and Wear of Gold Film

The microscale friction and wear behavior of thin film of gold (Au) which was prepared in a vacuum sputtering apparatus was investigated. The substrate is Si(100) wafer. The film thickness is about 800 nm. For comparison, the microscale friction and wear of the substrate was also studied.

The tests were carried out in an atomic force microscope and friction force microscope developed by the authors' laboratory [9]; the spring constant of Si_3N_4 cantilever used in this research was 0.06 N/m (A) and 0.38 N/m (B), respectively. The scanning direction of the tip is shown in Fig. 3.

During the micro friction test cantilever A was used. The surface morphology as well as the friction force image was obtained by scanning along the Y and −Y direction simultaneously in the constant height mode. The normal force and the friction signals were, respectively, recorded by upper and lower, left and right segments of the quadrant position sensitive diode. The real friction force image was calculated by following equations [8,11]: Y direction,

$$F_Y(i,j) = f_0(i,j) + f(i,j) \quad (17)$$

−Y direction,

$$F_{-Y}(i,j) = f_0(i,j) - f(i,j) \quad (18)$$

Therefore,

$$f(i,j) = (F_Y(i,j) - F_{-Y}(i,j))/2 \quad (19)$$

Here, $F_Y(i,j)$, $F_{-Y}(i,j)$ is the signal of friction force obtained by scanning along the direction of Y and −Y, respectively. $f(i,j)$ is the signal of real friction force. $f_0(i,j)$ is the signal produced by other lateral forces except friction force. Hence, the average of $f(i,j)$ is

$$f = \frac{1}{N^2} \sum_{i=0, j=0}^{N-1} f(i,j) \quad (20)$$

where n is the number of acquisition data points along the X and Y direction.

During the microscale wear test cantilever B was used. First the probe scanned for a set number of times in an area along the X direction, and then the worn surface morphology was measured in a larger area. The worn depth can be calculated by measuring the difference between the worn area and the initial unworn area.

As it is known, during the micro friction force measurement, it is difficult to obtain the real and precise friction force due to the small size of cantilever as well as the slight difference between cantilevers. According to the principle of FFM, the friction force signal can be used as the representative of the real friction force. Therefore, a friction coefficient factor is introduced in order to compare the relative micro friction characteristics among the samples. Corresponding to the relationship between the friction force and the friction

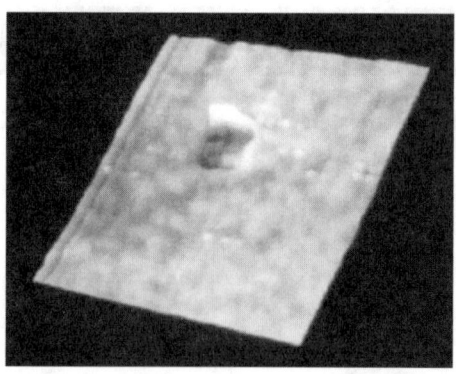

Fig. 8—Morphology of Au film on Si wafer after micro wear. (50 nN, 20 cycles, scanning area 2 μm by 2 μm.)

force signal, the slope of micro friction force signal versus load is referred to as the friction coefficient factor, which is the equivalent of friction coefficient, under the micro friction test [12].

Figure 7 is the dependence of micro friction force signal of Au film and Si wafer on load. It can be observed that the micro friction force increases with load. Figure 8 is the morphology of Au film on Si wafer after the micro wear test under 50 nN. There is also a relative deep wear scar on the film. Figure 9 shows the micro wear scar on Si wafer after the micro wear test. The wear scar is shallow even under a load of 110 nN. It is obvious that the Si wafer gives good wear resistance under such experimental conditions.

For Au film, because it is soft and has lower shear strength than silicon, it is normal that Au film has a higher micro friction coefficient factor than silicon, and can be easily worn under the current experimental condition.

4.2 Microscale Friction and Wear of PTFE/Si_3N_4 Multilayers

The PTFE/Si_3N_4 multilayers were prepared by Ar^+ ion beam alternatively sputtering pure PTFE and Si_3N_4 ceramic target in a polyfunctional beam assisted deposition system. Si (100) wafer is the substrate. The mutlilayer film has eleven layers with alternatively PTFE and Si_3N_4 layers, the inmost and

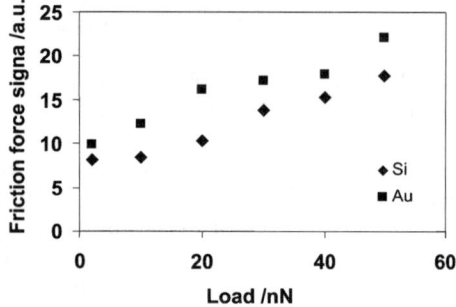

Fig. 7—Dependence of friction force signal of Au film and Si wafer on load.

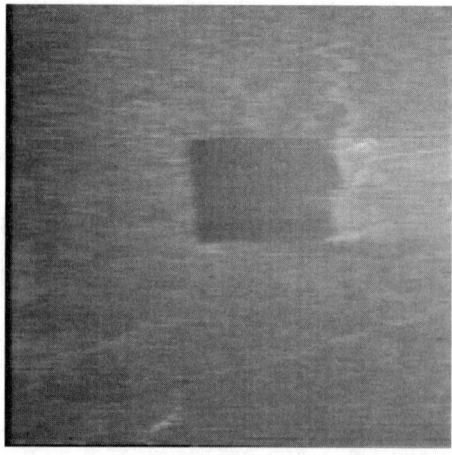

Fig. 9—Morphology of Si wafter after micro wear. (110 nN, 50 cycles, scanning area 2 μm by 2 μm.)

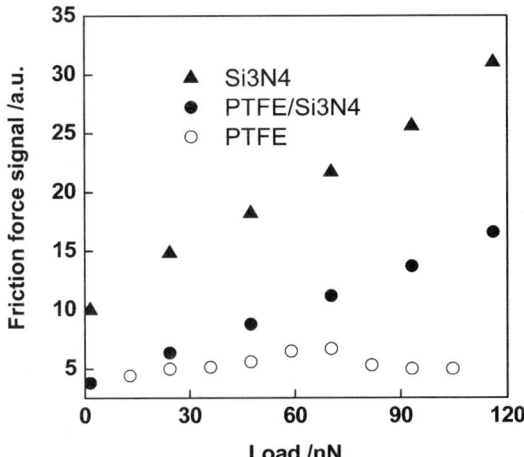

Fig. 10—Dependence of friction force of PTFE, Si$_3$N$_4$, and PTFE/Si$_3$N$_4$ film on load.

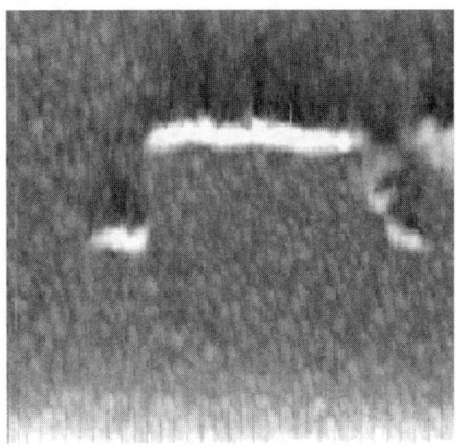

Fig. 12—Morphology of PTFE/Si$_3$N$_4$ film after 116 nN load and ten cycles of scanning. (Scanning area 2 μm by 2 μm.)

outmost layers are Si$_3$N$_4$ film. According to the deposition rate of PTFE and Si$_3$N$_4$, the sputtering time is five minutes for each PTFE layer, and ten minutes for each Si$_3$N$_4$ layer. The thickness of each PTFE layer is about 80 nm, almost the same as that of Si$_3$N$_4$ layer.

Pure PTFE and pure Si$_3$N$_4$ film were also prepared using the same sputtering parameters. The sputtering time is one hour for pure PTFE film, and two hours for Si$_3$N$_4$. The film thickness of pure PTFE or pure Si$_3$N$_4$ is almost equal to that of PTFE/Si$_3$N$_4$ multilayers. It has been shown by Wang et al. [13] that PTFE in the two kinds of samples is in crystalline state.

Figure 10 shows the effect of load on the micro friction force signal of pure PTFE, Si$_3$N$_4$, and PTFE/Si$_3$N$_4$ multilayers. It can be seen that the micro friction force signals of Si$_3$N$_4$ film and PTFE/Si$_3$N$_4$ multilayers vary almost lineally with load. As for pure PTFE film, when the load is less than 70 nN, the micro friction force signal increases linearly with the load. When the load is greater than 70 nN, the micro friction force signal doesn't increase with the load, the friction force is more or less constant.

From Fig. 10, it can be calculated that the friction coefficient factor of pure PTFE film is about 0.057 which is very small, the friction coefficient factor of PTFE/Si$_3$N$_4$ multilayers is about 0.115 which locates between the pure PTFE and Si$_3$N$_4$ film whose friction coefficient factor is about 0.171.

For PTFE film and PTFE/Si$_3$N$_4$ multilayers, through the observation of the surface morphology after the micro wear test, it is found that there is obvious worn mark and projection in the edge of worn mark when the load is above 70 nN (Figs. 11 and 12). As to the Si$_3$N$_4$ film, there is no worn mark observed even when the maximum load is used in the micro friction test.

After the micro wear tests, the dependence of worn depth of PTFE and PTFE/Si$_3$N$_4$ film on load is shown in Fig. 13. The worn depth of both PTFE and PTFE/Si$_3$N$_4$ film is in the nanometer scale. It can be seen that the worn depth increases linearly with load. However, the worn depth of PTFE/Si$_3$N$_4$ multilayers is about one-tenth of PTFE film at the same load. All these results demonstrate that the wear resistance of PTFE/Si$_3$N$_4$ multilayers is greatly improved after micro-assembling of soft and hard layers.

Because PTFE has low adhesion, high lubricity, and low friction coefficient [14], it is not difficult to understand PTFE film has a lower friction coefficient factor and higher wear rate than PTFE/Si$_3$N$_4$ multilayers under the current experi-

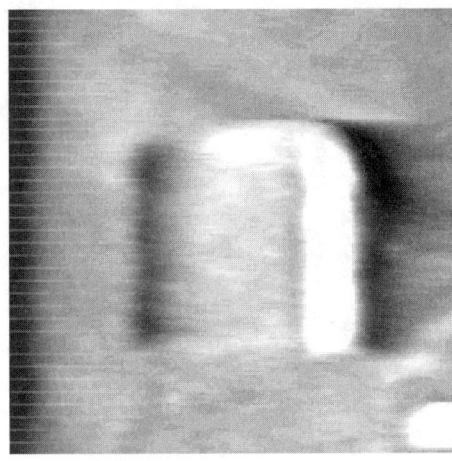

Fig. 11—Worn morphology of PTFE under 110 nN load and 50 cycles scanning. (Scanning area 2 μm by 2 μm.)

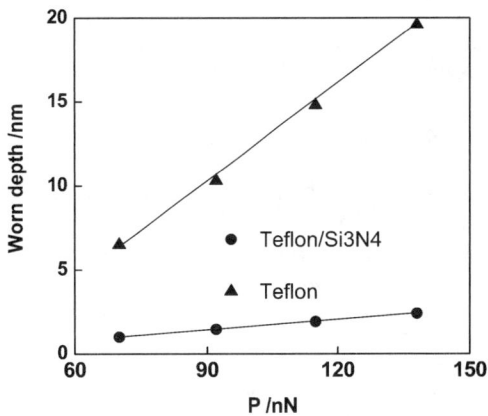

Fig. 13—Dependence of worn depth of PTFE and PTFE/Si$_3$N$_4$ film on load. (20 cycles, scanning area 2 μm by 2 μm.)

mental condition. For PTFE/Si_3N_4, because a little PTFE is exposed on the surface of multilayers, the friction coefficient factor is smaller than Si_3N_4 but larger than PTFE, and the worn mark can be found on the PTFE/Si_3N_4 multilayers.

During the friction and wear tests of PTFE film, two zones can be classified according to the load. One is the load below 70 nN, the friction force which was created in friction and wear tests is too small to make the PTFE film to shear. Within this zone the friction force increases linearly with the load, and there are no transfer of atoms and no worn marks. The second zone is when the load is above 70 nN, and the friction force created in the friction and wear tests is large enough to force the PTFE molecular atom to slip. So there were obvious worn mark and projection in the film, and the friction force stayed almost constant with load.

Through micro-assembling of PTFE and Si_3N_4 layers, the PTFE layer in PTFE/Si_3N_4 multilayers is located between two Si_3N_4 layers. Because of the interaction between PTFE and Si_3N_4 film, the mobility of PTFE film is limited by the two Si_3N_4 layers. The PTFE molecules cannot slip easily, therefore the shear stress required to slip PTFE increases, so the multilayers film has high wear resistance.

4.3 Closure

Au film can be worn under a high load, but Si wafer is difficult to be worn under the experimental condition.

There are two zones in the friction and wear tests of PTFE film. When the load was less than 70 nN, the micro friction force increases with the load. When the load is greater than 70 nN, the friction force of PTFE film is almost constant, and there is obvious worn mark in the PTFE film.

PTFE/Si_3N_4 multilayers not only have the property of PTFE's low friction coefficient but also have the property of Si_3N_4's high wear resistance.

5 Microscale Friction and Wear of Modified Molecular Films

The organized molecular films have potential applications in biology, electronics, optical devices, and MEMS, etc. [15]. These films are generally fabricated in processes such as the molecular deposition (MD) method, Langmuir-Blodgett (L-B) method, and self-assembling techniques. In these applications, tribological properties of macromolecules are the main focus. These properties are important in the understanding of the physical and chemical mechanisms on the surface and interface under micro/nanometer scales. This will lead to wider applications in the future.

5.1 Microscale Friction and Wear of a Triacetic Acid Molecular Film

A single Langmuir-Blodgett trough rinsed with deionized distilled water was employed in the preparation of L-B films. Two layers of L-B film consisting of triacetic acid and $CrCl_2$ was prepared on Si (100) substrate which was treated with hydrophilic solution and then thoroughly cleaned before the L-B film making.

Figure 14 is the dependence of micro friction force signal of the L-B film on load. It can be observed that the micro friction force keeps almost the same. Figure 15 shows micro wear scar of the L-B film. The L-B film was worn by the tip under the load less than 20 nN because the L-B film con-

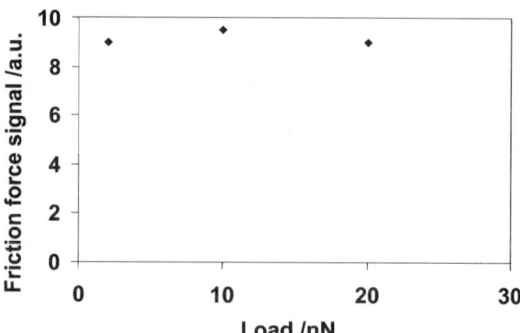

Fig. 14—Dependence of friction force signal of a triacetic acid L-B film on load.

sisted of an organic molecule and was soft. The load is light, but the wear scar is large. The depth of wear scar can be obtained according to the cross section of the worn area. Because the adsorbed force between the terminal and the substrate was limited, the L-B film was seriously worn under a load of 20 nN.

The L-B film studied consists of two-layer organic molecules. The first layer of the L-B film is adsorbed on silicon substrate by the polarization terminals of the molecules. During the micro friction test, the probe contacted with the polarization terminal of the second layer. As a result, there was a special attractive force between the polarization terminal of the second layer and the probe. Therefore, the L-B film does not have the function of reducing friction force under the current experimental condition.

5.2 Microscale Friction and Wear of C_{60} Surface-Modified Molecular Films

In order to improve the tribological properties of molecular films, molecular surface modification is the first choice to make an approach. A Diblock polymer polystyrene–poly(ethylene)oxide (PS-PEO) thin-films were studied in our previous research because of its interesting structure (one

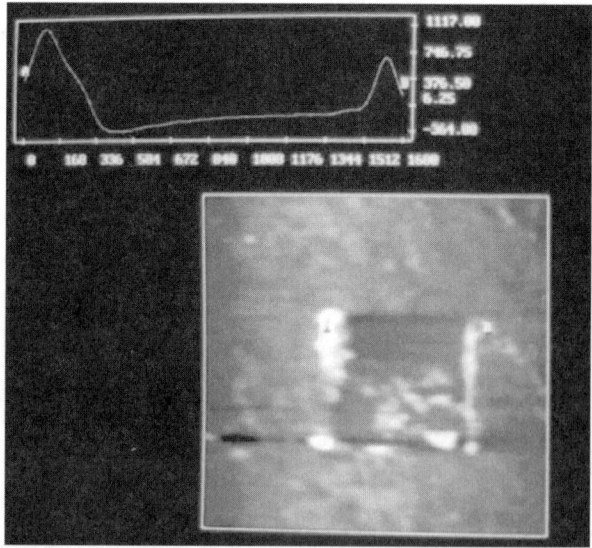

Fig. 15—Morphology of a triacetic acid L-B film after micro wear. (20 nN, ten cycles, scanning area 6 μm by 6 μm.)

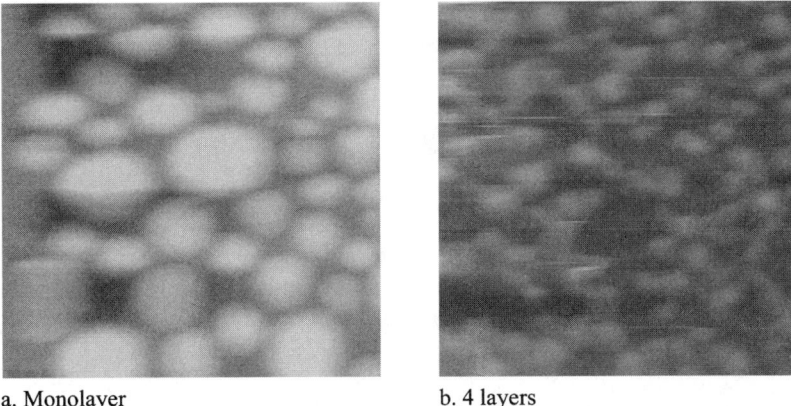

Fig. 16—Topography of star-shaped C_{60}-Pst L-B films (scan range: 2 μm), (a) monolayer, (b) four layers.

block is soft, and the other is rigid). The unique structure provides promising tribological behavior, such as low friction [16]. In order to enhance mechanical strength, another type of macromolecular polymer, C_{60}-Pst (C_{60}-polystyrene) is investigated. The C_{60}-Pst has different forms of molecules, such as mono-substituted, multi-substitutes (star-shaped structure) and hyperbranched (dendrimer structure), etc. Each of these types has unique physical and chemical properties.

5.2.1 C_{60}-Pst Star-Shaped Molecules

C_{60} surface-modified macromolecules, C_{60}-Pst (C_{60}–polystyrene), were synthesized by using the radical-initiated polymerization reaction method. Benzoyl peroxide was used as the initiator. The molecules of the product using this synthesis method are star-shaped. The C_{60} particles are multi-substituted, different from those of mono-substituted and hyperbranched.

The Langmuir-Blodgett technique was used in making L-B films of the macromolecules on a mica surface. A single Langmuir trough rinsed with deionized distilled water was employed in the preparation of L-B films. The floating film was compressed at a rate of 10 cm^2/min. The C_{60}-Pst star-shaped polymer molecules were transferred onto the mica chips by the vertical dipping method at room temperature.

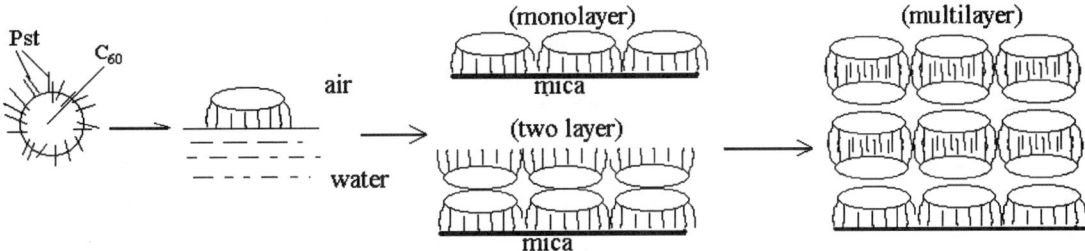

Fig. 17—Schematic model for the star-shaped C_{60}-Pst L-B film of monolayer. (a) Forward direction, (b) reverse direction.

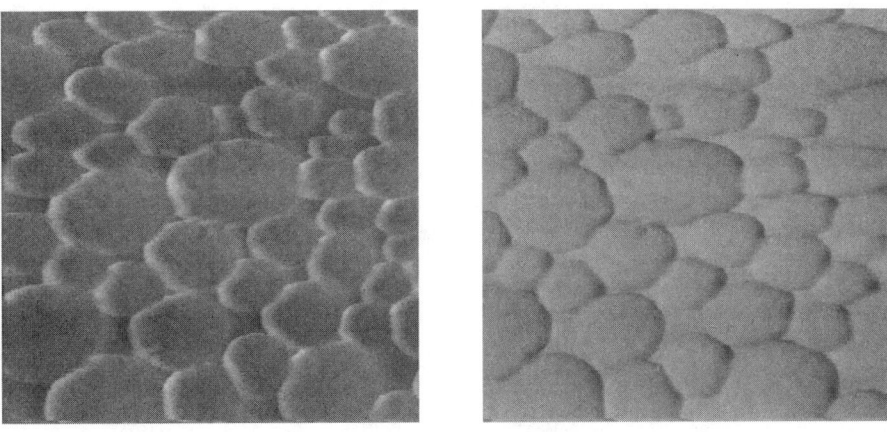

Fig. 18—Friction force images of L-B monolayer film, normal force: 1 nN (scan range: 2 μm). (a) Monolayer L-B film, (b) four-layer L-B films.

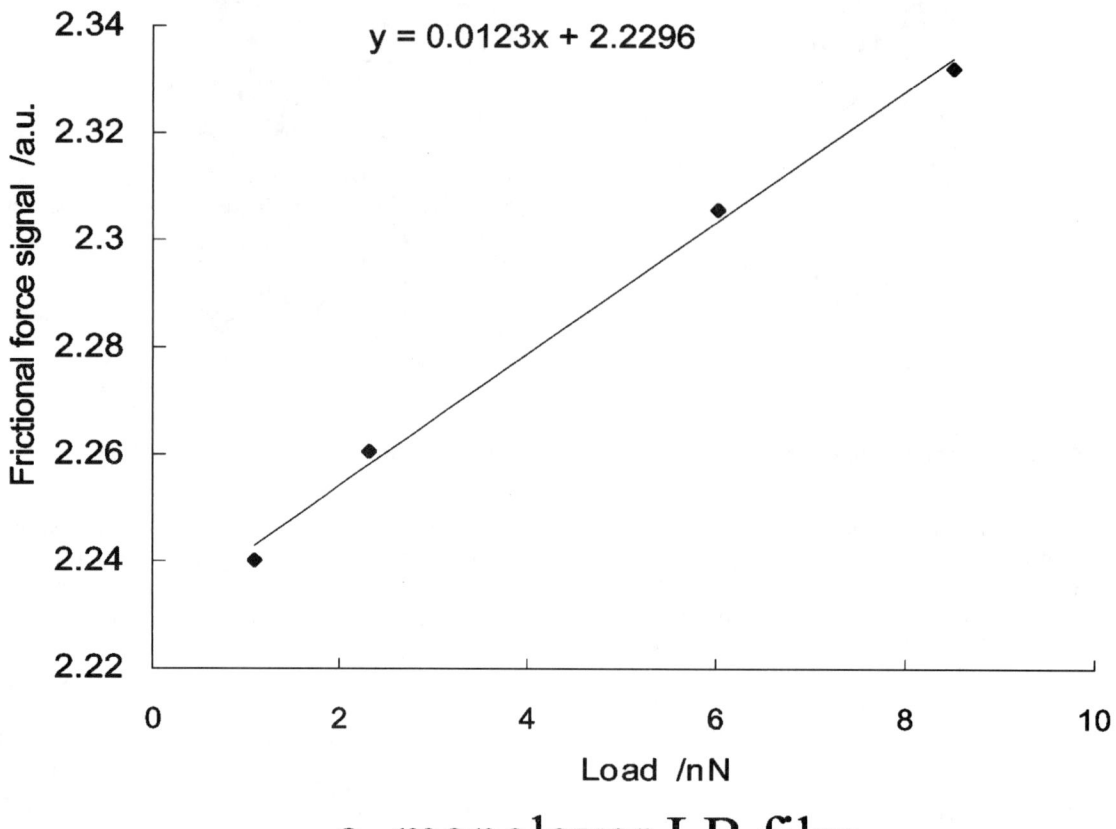

a. monolayer LB film

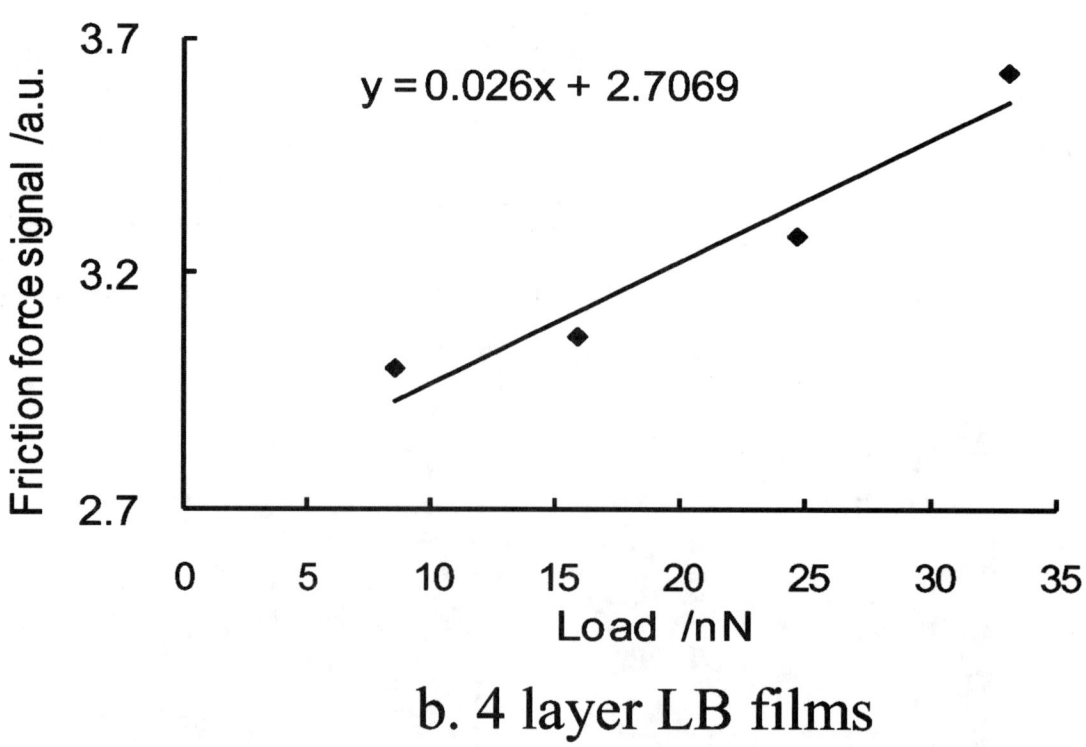

b. 4 layer LB films

Fig. 19—Diagram of friction versus load of mono- and multilayer L-B films of star-shaped C_{60}-Pst polymer.

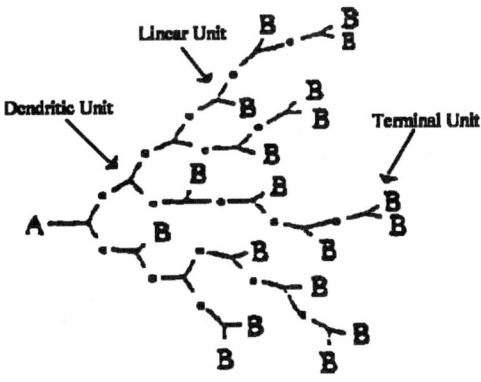

Fig. 20—Schematic illustration of hyperbranched polymer.

The nanotribological properties of this highly ordered thin films were investigated with an AFM/FFM. A commercial silicon nitride cantilever with tip was used in the measurements. The images were taken in the contact mode. The friction force images and topography of all measurements here are original without being processed.

AFM is used in the surface analysis. Figure 16 is the AFM topography of the monolayer and the multilayer L-B films. It shows that the monolayer L-B film is well packed and highly ordered on the mica surface. The surface of the monolayer film (shown in Fig. 16(a)) has a higher packing density than that of the four-layer L-B film (shown in Fig. 16(b)). This is because the molecules form the different structures in the monolayer film from those in four-layer

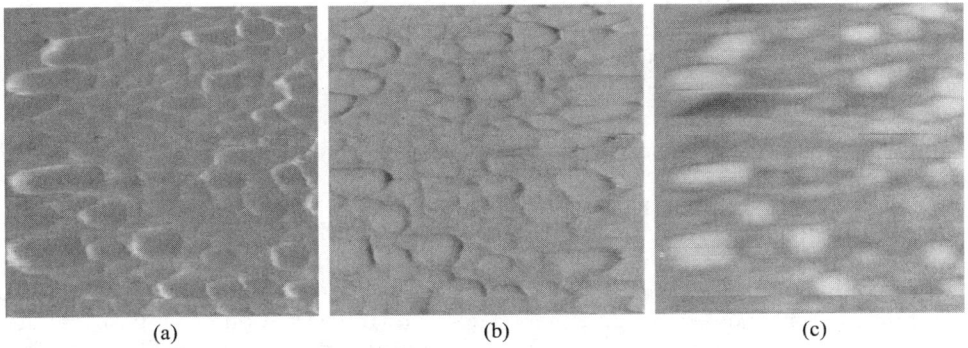

Fig. 21—Friction force and topography images of the L-B films of C_{60}-Pst hyperbranched polymer under load 2 nN using AFM/FFM, image (a) and (b) are friction force images and image (c) is the topography.

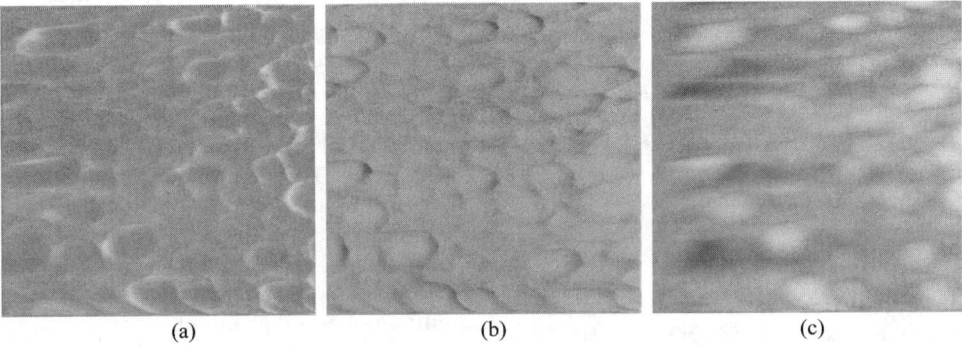

Fig. 22—Friction force and topography images of the L-B films of C_{60}-Pst hyperbranched polymer under load 42 nN using AFM/FFM, image (a) and (b) are friction force images and image (c) is the topography.

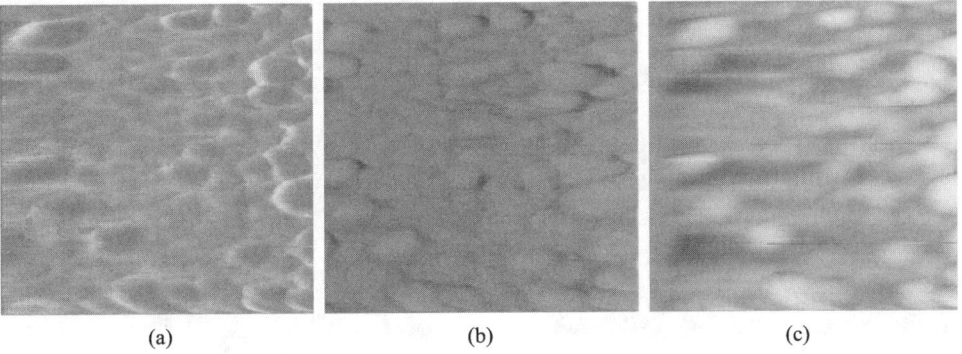

Fig. 23—Friction force and topography images of the L-B films of C_{60}-Pst hyperbranched polymer under load 87 nN using AFM/FFM, image (a) and (b) are friction force images and image (c) is the topography.

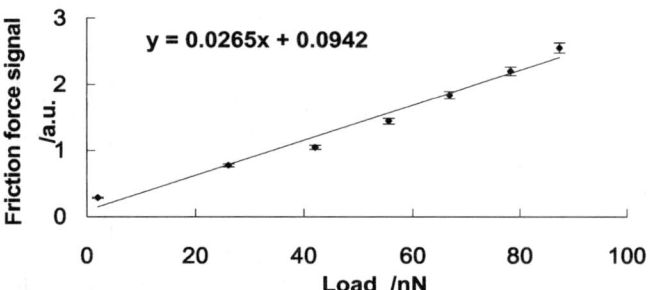

Fig. 24—Diagram of friction versus Load of monolayer C_{60}-Pst polymer molecules L-B film (point A, scan range: 2 μm).

film. The surface of the monolayer film is composed with bigger groups of molecules compared with the four-layer L-B films. Unlike other structures of normal L-B films, the structure of the star-shaped C_{60}-Pst L-B film is special. The structure of the monolayer and multilayer L-B films can be clearly seen in Fig. 17. C_{60} particles are modified with Pst polymers with hydrophilic end-groups. When the monolayer film is formed, it is Pst with polar end group of the star-shaped C_{60}-Pst molecules attaching to the mica surface because they are all hydrophilic. Therefore the surface of the monolayer film is formed with C_{60} molecules. When the second layer of the film is formed, the surface is the Pst molecules with polar end groups. This phenomenon will repeat

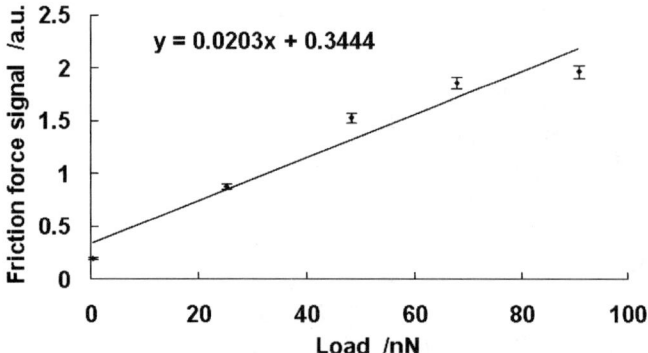

Fig. 25—Diagram of friction versus load of monolayer C_{60}-Pst polymer molecules L-B film (point B, scan range: 1 μm).

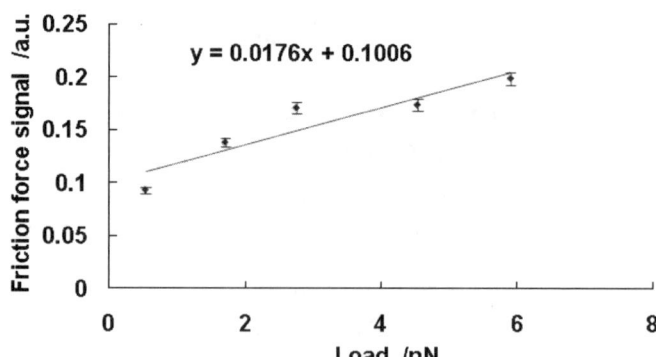

Fig. 27—Diagram of friction versus load of multilayer L-B film of C_{60}-Pst hyperbranched polymer molecules (two layers).

every two layers in making the films. In odd layer films, like the monolayer film shown in Fig. 16(a), the surface is formed with the highly packed C_{60} molecules. In even layer films, like the four-layer film shown in Fig. 16(b), the surface is formed with polar end groups of Pst molecules. The measurement results using AFM show that the L-B films possess a well-ordered layer structure and highly oriented.

Nanotribological properties of L-B films are tested by using FFM. The friction force images are shown in Fig. 18. Figure 18(a) is the forward friction force image and Fig. 18(b) is the reverse one. Comparing with the topography image of the monolayer L-B film as shown in Fig. 16(a), the three images are excellently matched with each other. The almost-perfect correspondence of these three images indicates that the surface topography plays a very important role in friction properties of the thin films. Figure 19 is the diagrams of friction versus load of L-B films. The trend of the friction force versus load curves is linear basically. Two friction factors (friction force signal and normal force) are given: 0.012, 0.026 for Figs. 19(a) and 19(b), respectively, which could express the relative friction coefficient (friction force/load). The experimental results show that the L-B films perform a stable friction characteristic in the friction test. The surface of a four-layer L-B film is smooth with Pst chains outside. The surface of a monolayer L-B film with the end group of C_{60} outside is smoother than that of four-layer film with a polar end group of Pst molecules outside. Its friction

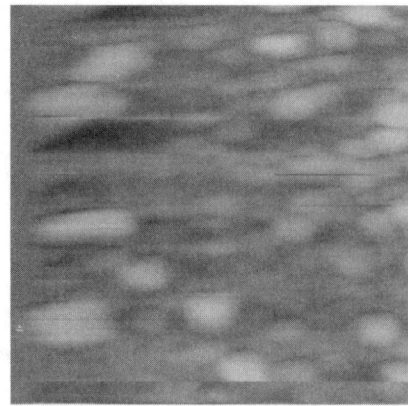

a. scan range: 2μm

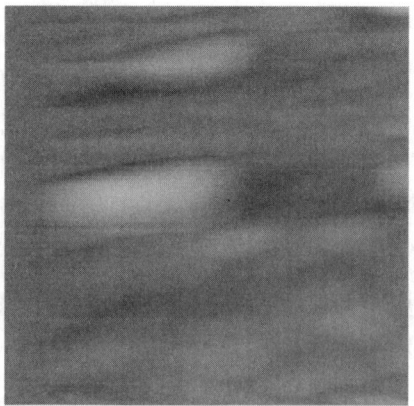

b. scan range: 1μm

Fig. 26—AFM topography of monolayer L-B film for two different areas [(a) point A, (b) point B]; (a) scan range: 2 μm (b) scan range: 1 μm.

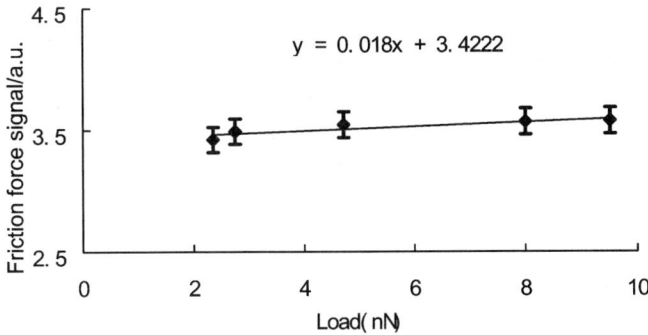

Fig. 28—Diagram of friction versus load of multilayer L-B film of C_{60}-Pst hyperbranched polymer molecules (four layers).

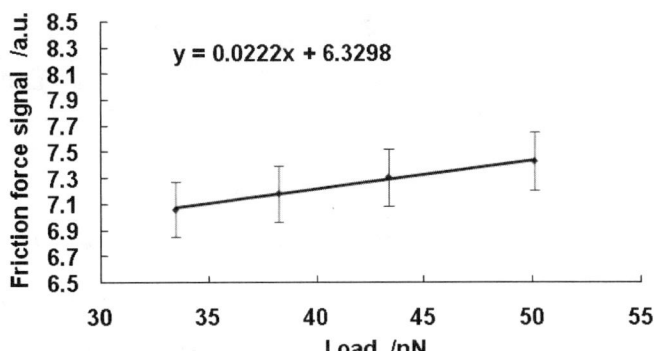

Fig. 30—Diagram of friction versus load of multilayer L-B film of C_{60}-Pst hyperbranched polymer molecules (eight layers).

factor is lower than that of four-layer L-B film as well. This indicates microtribological properties of C_{60} for a low friction application. The scanning experiments have been conducted repeatedly on the surface of L-B films. The L-B films are fixed on the substrate firmly. No obvious wear has been found on the L-B films.

The results show that the C_{60}-Pst star-shaped polymer L-B films are firmly fixed on the surface of the substrate, even after the surface was scanned for many times. The topography, high density, order and preferred orientation of the films are dominating factors in friction. The C_{60}-Pst film could play a significant role in microtribological applications.

5.2.2 C_{60}-Pst Hyperbranched Molecules

The molecular structure of a C_{60}-Pst hyperbranched polymer can be described as in Fig. 20. C_{60}-Pst hyperbranched polymer molecules were transferred onto mica chips by the vertical dipping method of L-B film techniques as introduced above.

Figures 21–23 show the AFM/FFM images of friction force and topography on a monolayer L-B film of C_{60}-Pst polymer in the same scan range under different normal forces. The first image of each experiment is the image of friction force. The second is also the image of friction force but done in the reverse direction. The third is the topography image. The scan range is 2 μm. Experiments were first done in lighter normal force. The normal force from Fig. 21 to Fig. 23 is increased gradually.

Experimental results on the topography show that the L-B film is compact and highly ordered. The images of friction force match with the topography fundamentally. There is little difference between the two friction force images due to shape difference of the topography in the opposite direction. The images shown in Fig. 21 under lighter normal force are clear. With increasing normal force and more times of scanning, the images are not as clear as those at the beginning of the experiments. Blurring of the images can be observed in Fig. 23, even though the films are fixed on the mica surface and the surface of the L-B film has not obviously been worn. This shows that the C_{60}-Pst hyperbranched polymer molecules are immobilized on the surface of the substrate. Only a little transfer can be seen on the surface the L-B film after scanning for many times.

Figure 24 is a diagram of friction versus load derived from the friction force images of the above figures. Figure 25 is also a diagram of friction versus load done in the same monolayer L-B film of C_{60}-Pst polymer but in a different scan range.

The Y-axis represents the magnitude of the friction signal force and the X-axis is the load. The slope of the trend line is defined as the friction factor (friction force signal/load) which is used to express the relative friction coefficient (friction force/load). Experiments that have been done in the same monolayer L-B film but different scan ranges give similar results as shown in Fig. 24 and Fig. 25. The friction factors of this monolayer L-B film, 0.0265 and 0.0203, are similar. The topographies of these two areas are shown in Fig. 26.

Figures 27–30 are diagrams of friction force versus load for different multilayer L-B films of the C_{60}-Pst hyperbranched polymer molecules obtained using AFM/FFM. The

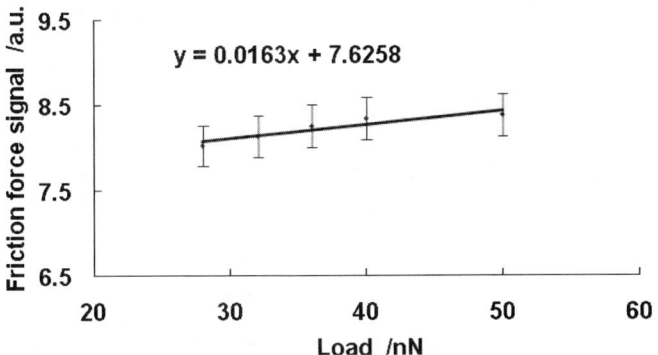

Fig. 29—Diagram of friction versus load of multilayer L-B film of C_{60}-Pst hyperbranched polymer molecules (six layers).

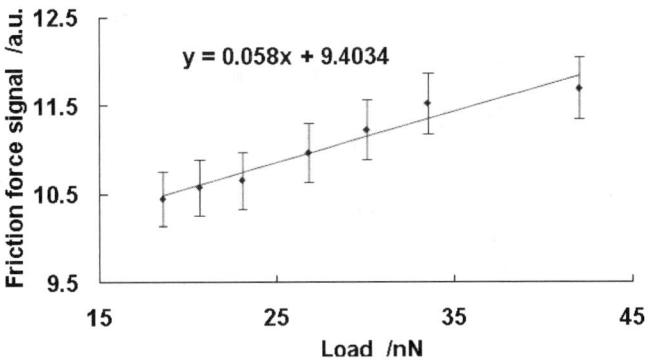

Fig. 31—Diagram of friction versus load of multilayer L-B film of C_{60}-Pst hyperbranched polymer molecules (six layers).

TABLE 2—H:DLC multilayers thickness and process (Ar/CH₄ ratio: 2/1, gas flow: 36 sccm).

Samples	Process and Layers	Periodic Number	Film Thickness (Å)
1	Substrate/A	...	10,000
2	Substrate/B	...	10,000
3	Substrate/B(3 Å)/A(2 Å)/B/A...B/A	2000	10,000
4	Substrate/B(4 Å)/A(4 Å)/B/A...B/A	1250	10,000
5	Substrate/B(4 Å)/A(10 Å)/B/A...B/A	714	9,996
6	Substrate/B(500 Å)/C(500 Å)/B/C...B/C, C=B(3 Å)/A(2 Å)/B/A...B/A	1010	10,000
7	Substrate/B1(1000 Å)/B2(4000 Å)/B1/B2	2	10,000

TABLE 3—Comparison of hardness between measurement and calculation.

Hardness (GPa)	Sample 3				Sample 4				Sample 5			
	H_A	R_A	H_B	R_B	H_A	R_A	H_B	R_B	H_A	R_A	H_B	R_B
	13.66	0.4	5.25	0.6	13.66	0.5	5.25	0.5	13.66	0.71	5.25	0.29
Measured			7.32				9.36				12.5	
Calculated			8.61				9.45				11.27	

friction factors of multilayer L-B films of C_{60}-Pst hyperbranched polymer molecules are similar according to these diagrams. There is no big difference among the results of the friction factors. Most of these multilayer L-B film friction factors are a little lower than the one for the monolayer L-B film. The similarity of the experimental results shows that the L-B films have a rather stable friction characteristic in the friction test using AFM/FFM.

Though most experiments get similar results, some special results are also obtained in the measurements, e.g., the result shown in Fig. 31. The surface roughness of the point in the experiments of Fig. 31 is a little larger than for most other parts of the film. This shows that the topography of the L-B film plays an important role in the friction properties obtained by using FFM measurement. Therefore having the well-prepared and highly ordered L-B film is one important factor for friction measurements. However, results like the one shown in Fig. 31 can still be accepted. Because there are so many factors that can affect the experiment results at the nanometer scale, more experiments need to be done.

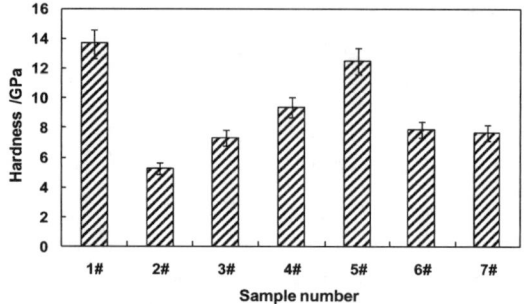

Fig. 32—The hardness of all samples.

5.2.3 Closure

It has been shown that surface-modified C_{60}-Pst polymer was synthesized and was transferred onto mica chips by using the vertical dipping method of L-B film techniques. Though the surface was scanned many times, only a small amount of transfer could be found on the surface of the L-B films. L-B films of C_{60}-Pst polymer, even with only one layer, showed good microtribological performance. The topography and the order of the films are the most important factors influencing the microscale friction and wear.

6 Microscale Friction and Scratch of Multilayers

Multilayer coatings or films have been proven to be successful in various applications, especially for wear protection. In multilayer systems, the intrinsic stress can be effectively reduced by designing interface number and composite materials in terms of the application and process technique. Therefore, the multilayer technique by which a specific functional composite is able to approach received more attention for many years [17,18]. Here we introduce the results on microscale friction and scratch of DLC/DLC and Fe-N/TiN multilayers.

6.1 Microscale Friction and Scratch of DLC Multilayers

Amorphous carbon coatings, often called diamond-like carbon (DLC) coatings, have a wide range of uses including optical, electronic, biomedical, and tribological applications [19,20]. One successful example is the DLC protective thin film for heads and magnetic media in the hard disk drive system [21]. Recently, researchers have been pursuing DLC as a functional surface coating for MEMS. One of the main functions of DLC coatings in MEMS is their low friction and high resistance to wear in their role as protective overcoats. In

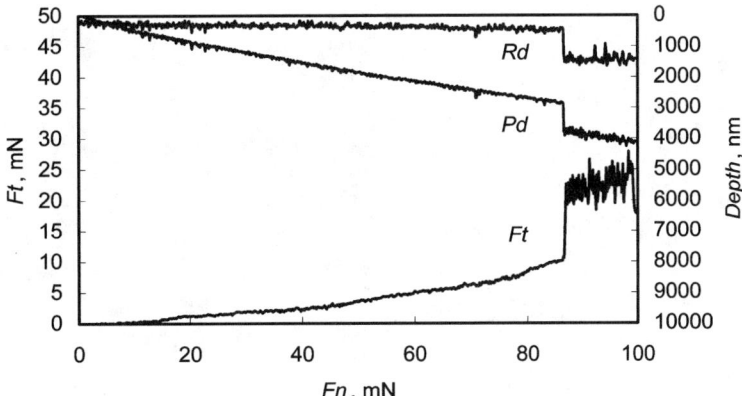

Fig. 33—Typical scratch curve of Sample 4, Fn is the normal load, Ft is the measured tangential force, Pd is the penetration depth, Rd is the residual depth. The critical load L_c is 86.63 mN.

this case, the micromechanical properties of the DLC coatings are important to the whole system.

6.1.1 Samples and Test Procedures

H:DLC coatings on a single crystal silicon (100) wafer were prepared in an rf (13.56 MHz) plasma enhanced CVD (PE-CVD) with a reactive rf magnetron sputtering coating system. The vacuum chamber of the system consists of two coating zones: Zone A is rf PE-CVD to obtain H:DLC called Layer A, and Zone B is reactive rf magnetron sputtering to obtain H:DLC called Layer B. A rotating water-cooled substrate holder biased by rf allows automated program-controlled substrate placement in either Zone A and B during the H:DLC deposition. Therefore, the H:DLC coatings could be monolayer A or B, or multilayers A/B/A/B..., depending on the deposition process. The thickness of each layer was strictly controlled by the deposition time. The deposition rate which was derived based on thick film thickness divided by time is about 1 Å per second. Table 2 shows the thickness and the process of the samples. For all the samples, the substrate bias during the deposition was set at −150 V(dc), the rf power was kept at 700 W, except that for Layer B2 in Sample 7 was 500 W. The total gas flow was kept constant at 36 sccm at Ar/CH_4 ratio of 2/1 under a processing pressure of 0.5 Pa.

A nano hardness tester (CSEM) with a Berkovich indenter tip (diamond, radius 100 nm) was used. For all samples, the maximum load was 5 mN, loading and unloading speed was 10 mN/min. The data of hardness and elastic modulus are the average from ten repeated indentation tests.

A nano scratch tester (CSEM) was employed to carry out the scratch test. A Rockwell diamond tip with a radius of 2 μm was used to draw at a constant speed 3 mm/min across the coating/substrate system under progressive loading of 130 mN maximum at a fixed rate 130 mN/min. The total length of the scratch scar is 3 mm. The critical load (L_c) here is defined as the smallest load at which a recognizable failure occurs. The failure can be observed both by the built-in sensors and by the optical microscope.

The same nano scratch tester was used to carry out the friction tests. The Rockwell diamond tip (radius 2 μm) was used to draw at a constant speed 3 mm/min across the sample surface under a constant load of 20 mN for which no scratches occurred for all the samples. Feedback circuitry in the tester ensures the applied load is kept constant over the sample surface. The sliding distance is 3 mm. The friction coefficient is defined normally as the ratio of the friction force and the applied load.

6.1.2 Nanoindentation

Figure 32 shows the hardness of all samples. The monolayer Samples 1 and 2 have the highest and the lowest hardness, respectively. The multilayers have a higher hardness than the monolayer B, Sample 2. Among the multilayer samples, Sample 5 showed the highest hardness values, but the values are still lower than those obtained from the monolayer A film, Sample 1.

Under current experimental condition, the hardness increases mainly with the thickness of Layer A (Samples 3–5). It is understandable because Layer A is harder than Layer B. The hardness can be simply calculated by the following equation.

$$H = H_A \times R_A + H_B \times R_B \qquad (21)$$

where H_A is the hardness of film A (Sample 1), H_B is the hardness of film B (Sample 2), R_A and R_B are the ratios of Layers A and B of the total multilayers. Table 3 shows the comparison between measured and calculated hardness. The data and the trend are close to each other. This also indicates that the

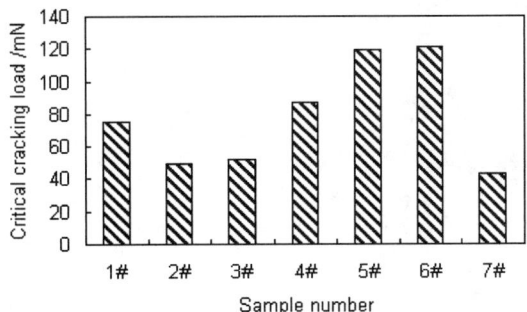

Fig. 34—Critical load of all samples. (a) Sample 1, L_c=76.26 mN; (b) Sample 2, L_c=50.49 mN; (c) Sample 3, L_c=54.59 mN; (d) Sample 4, L_c=86.63 mN; (e) Sample 5, L_c=119.03 mN; (f) Sample 6, L_c=121.44 mN; (g) Sample 7, L_c=43.25 mN.

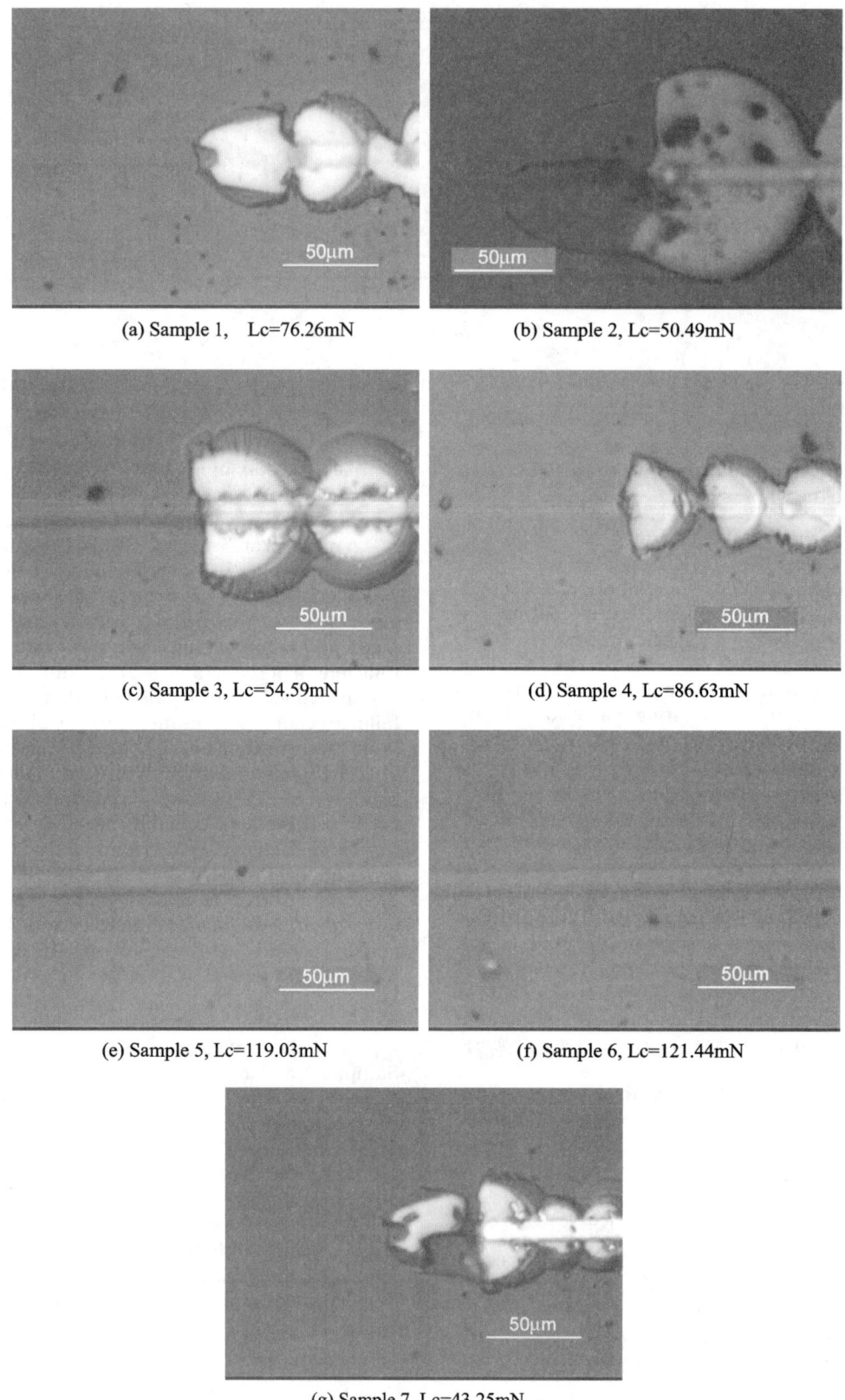

Fig. 35—The typical optical microscopic images (500×) of the first crack in the scratch test.

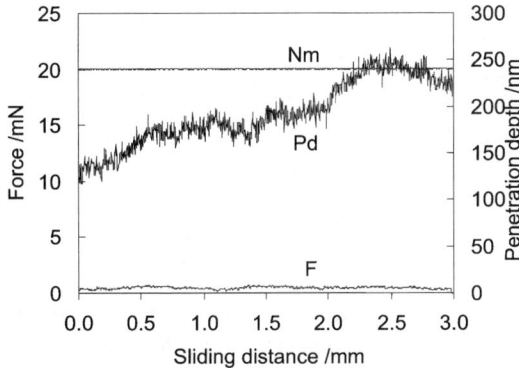

Fig. 36—Friction test of Sample 2, F is the friction force, Pd is the penetration depth, Nm is the measure normal load (20 mN) during the friction test.

hard Layer A plays the role to improve the hardness of the coatings.

The interfaces here also have contribution to the hardness. For Samples 3 and 4, there are no obvious interfaces between Layers A and B because the individual layer thickness is so thin that it is hard to form sharp interfaces. It is more like a mixed structure according to the process. For Sample 5, the interfaces are possibly formed due to the increase of Layer A. So the hardness of Samples 3 and 4 is still much lower than monolayer A, but the hardness of Sample 5 is close to the monolayer A.

For Sample 6 R_A is 0.2, but the hardness is at the same level as Sample 3. From the deposition process (refer to Table 2) we can regard Sample 2 and Sample 3 as individual layers in Sample 6. There should be obvious interfaces between Layer B and Layer C, and the interfaces have contributed to the hardness.

For Sample 7, the hardness and the elastic modulus are

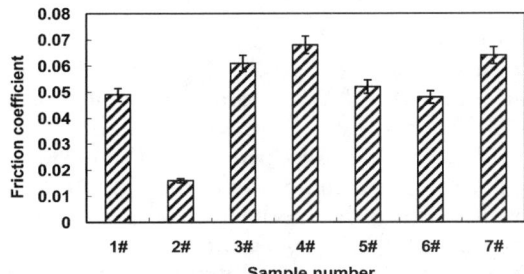

Fig. 37—Friction coefficient of all samples.

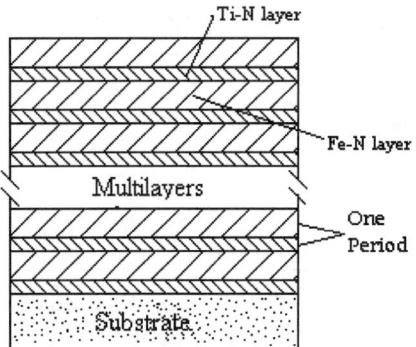

Fig. 38—Schematic diagram of multilayers.

higher than that of Sample 2; this is because Sample 7 was prepared in Zone B of the deposition system with different rf power, and has interfaces in Sample 7 while Sample 2 is only a monolayer. The interfaces here have contributed to the hardness of Sample 7.

6.1.3 Microscale Scratch

Figure 33 shows a typical scratch test curve of Sample 4. Both the penetration depth and the residual depth as well as the tangential force can be obtained from this curve. The critical load can be found from the transition stages plotted in the three curves. The critical load (L_c) of Sample 4 is 86.63 mN.

Figure 34 shows the critical load of all the samples. For the monolayer samples, Sample 1 has a higher critical load than Sample 2. The multilayers Samples 4, 5, and 6 have higher critical loads than monolayer Samples 1 and 2. Samples 5 and 6 have excellent scratch resistant properties. Only extremely small cracks are found in the scratch tracks of Samples 5 and 6. Therefore, there is no sudden change found in the force and penetration depth curves. Sample 7 has the lowest critical load, similar to the monolayer Sample 2.

Figure 35 shows the optical microscopic images of the first crack point on the sample surface. The scratch scar of monolayer Sample 1 has the feature of brittleness. However, there is an obvious crack along the scratch scar of Sample 2 before the coating delamination. This indicates that monolayer Sample 2 has the feature of ductility, and the adhesion between the film and the substrate is poor. However, there is no obvious crack before the delamination in the scratch scars of other samples. The feature of multilayer Samples 3 and 4 is different from monolayer Samples 1 and 2. There are no obvious cracks in the scratch scars of Samples 5 and 6, except several small cracks along the edge of the scars. These

TABLE 4—Thickness of each layer of Fe-N/Ti-N multilayers.

Sample Number	Outermost Layer	Thickness of Fe-N Layer (nm)	Thickness of Ti-N Layer (nm)	Periodic Number	Total thickness (nm)
1	Fe-N	5	2	70	490
2	Fe-N	10	2	40	480
3	Fe-N	20	2	22	484
4	Fe-N	30	2	15	480
5	Fe-N	40	2	12	504
6	Fe-N	450	...	...	450
7	Ti-N	...	450	...	450

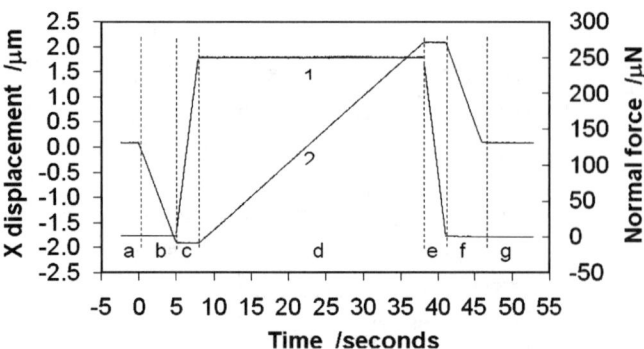

Fig. 39—Loading-unloading curve and displacement of the tip for microscratch test. 1-applied normal force; 2-X direction displacement.

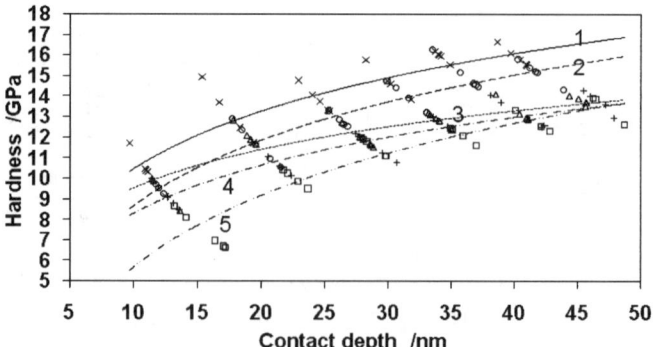

Fig. 41—Hardness of Fe-N/Ti-N multilayers versus contact depth. 1-Fe-N(5 nm)/Ti-N(2 nm); 2-Fe-N(10 nm)/Ti-N(2 nm); 3-Fe-N(20 nm)/Ti-N(2 nm); 4-Fe-N(30 nm)/Ti-N(2 nm); 5-Fe-N(40 nm)/Ti-N(2 nm).

cracks are so small that they were not detected by the built-in force sensors of the nano scratch tester. There are cracks and delaminations at the first crack point of Sample 7.

Hard layer and soft layer combined together can reduce the intrinsic stress of the whole coating [17,18,22–27]. Samples 4, 5, and 6 have higher critical load than that of monolayer A and B. For Samples 5 and 6, no obvious crack occurs during the scratch test. Sample 5 has the highest hardness and reduced elastic modulus among the multilayer samples, and the interfaces in Sample 5 also contribute to scratch resistance. So it has the best micromechanical properties here.

Sample 6 has the highest critical cracking load among the samples. The main reason is that there are layers A(2 Å)/B(3 Å)/…A/B and the hardness of the whole multilayer is not too high. Previous study on multilayers showed that the interfaces of multilayers have a great contribution to the multilayers' scratch resistance [28].

The critical cracking load of Sample 7 is at almost the same level as that of Sample 2. This is because Samples 7 and 2 were made in Zone B and have similar properties. However, there is a difference from cracks between Samples 2 and 7 (refer to Fig. 35). This is mainly due to the fact that Sample 7 is a multilayer structure.

6.1.4 Microscale Friction

Figure 36 shows the friction test result of Sample 2, F is the friction force, Pd is the depth of the sliding tip penetrating into the sample surface, Nm is the measure normal load (20 mN) during the friction test. The averaged friction coefficient of Sample 2 is 0.016.

The constant applied load is only 20 mN so there were no crack formations occurring during the friction tests of all the samples though there is penetration depth on the sample surface. The friction coefficient reflects the surface feature of H:DLC samples. Figure 37 shows the friction coefficient of all samples. Sample 2 has the lowest friction coefficient compared to other samples which have the friction coefficient in a range of 0.049–0.065.

The friction coefficient of Sample 2 is quite different from the other samples; this can be attributed to the surface difference. The previous research shows that the friction coefficient of DLC is related to the deposition parameter [29]. In this study, in order to evaluate the surface properties in the same condition, we designed Layer A as the outermost layer for all the multilayer samples. Among all the samples, only the surface of Sample 2 is from Layer B.

6.1.5 Closure

H:DLC coatings were prepared by rf PE-CVD and rf magnetron sputtering with different process conditions. H:DLC monolayer obtained by rf PE-CVD has a higher hardness than H:DLC monolayer obtained by rf magnetron sputtering, but the critical cracking load of the scratch test shows the opposite result. The mechanical properties including hardness, elastic modulus, and critical load under scratch of H:DLC coating can be improved by depositing H:DLC multi-

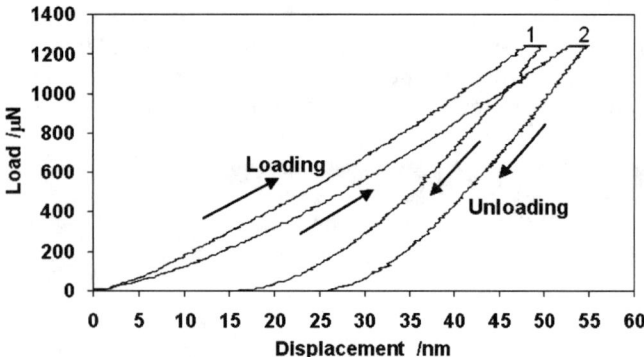

Fig. 40—Typical load-displacement curves during a loading/unloading cycle. Applied load 1250 μN, 1-Fe-N(5 nm)/Ti-N(2 nm), 2-Fe-N(20 nm)/Ti-N(2 nm).

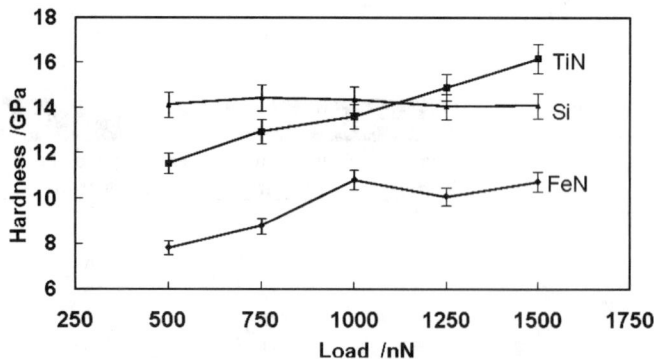

Fig. 42—The dependence of hardness of single layer on load.

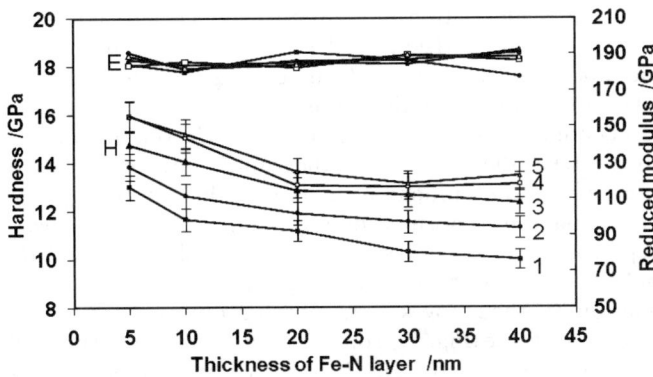

Fig. 43—Hardness and reduced modulus of Fe-N/Ti-N multilayers versus thickness of Fe-N layer. Load: 1–500 µN; 2–750 µN; 3–1,000 µN; 4–1,250 µN; 5–1,500 µN.

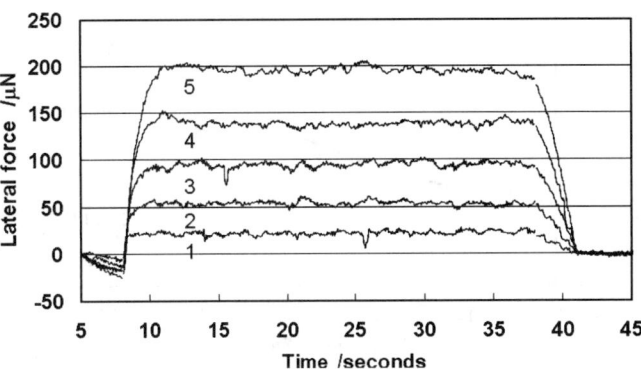

Fig. 45—Measurement of lateral force for different normal forces, Fe-N(5 nm)/Ti-N(2 nm) 1–250 µN; 2–500 µN; 3–750 µN; 4–1,000 µN; 5–1,250 µN.

layers first under rf magnetron sputtering and then under rf PE-CVD with a thin individual layer. Thick individual layers do not improve the mechanical properties of the coatings. Interfaces have contributions to both hardness and anti-scratch resistance.

6.2 Microscale Friction and Scratch of Fe-N/TiN Multilayers

Scientists have found that using Fe-N thin film would be the most prospective approach to enhance the capabilities of the write head of HDD because the Fe-N thin film has high saturation magnetization and low coerciveness [30–33]. Fe-N multilayers were developed in order to improve the magnetic properties of Fe-N films. For these kinds of multilayers, the mechanical properties will change with the number of layers, the thickness of Fe-N films, etc. On the other hand, in the process of designing and applying this write head material, its micromechanical properties that are of concern.

6.2.1 Samples and Test Procedures

The Fe-N/Ti-N nano-multilayers were prepared by using the magnetron-sputtering technique [34]. Si (111) wafers are used as the substrate. The multilayers have a total thickness of about 500 nm with alternately Fe-N and Ti-N layers (shown in Fig. 38). The Fe-N layer was the outermost layer and the Ti-N layer was the innermost layer. The thickness of each layer was strictly controlled by the sputtering time. Table 4 shows the thickness of each layer of the samples. Because the multilayer sample was supposed to be used as the magnetic write head, the thickness of the nonferromagnetic Ti-N layer was not changed.

Nanoindentations were carried out by using a commercial AFM (AutoProbe CP Research, Park Scientific Instruments) equipped with a commercial capacitance transducer (TriboScope, Hysitron) with a three-sided pyramidal dia-

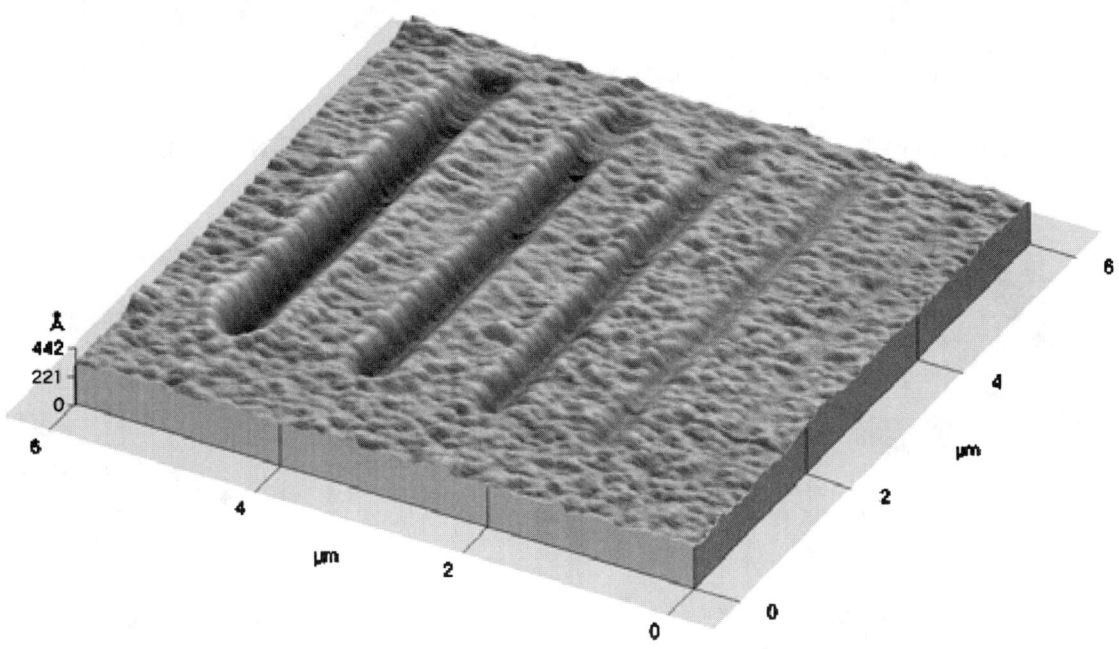

Fig. 44—Image of microscratch scars on Fe-N(5 nm)/Ti-N(2 nm). Load for right to left is 250 µN, 500 µN, 750 µN, 1,000 µN, 1,250 µN, respectively.

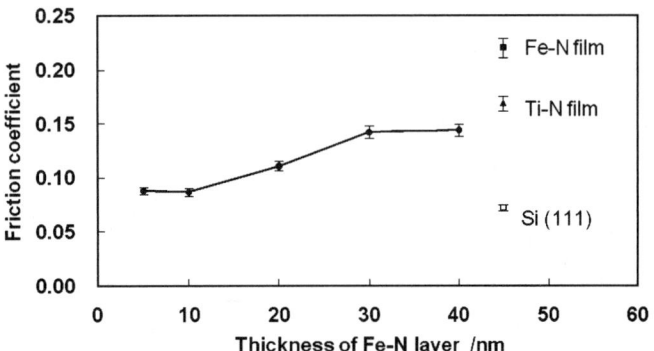

Fig. 46—Average friction coefficient (from the 15th second to the 35th second) of the multilayers versus thickness of Fe-N layers. The average friction coefficient of Fe-N, Ti-N film and Si (111) wafer under the same condition is also shown in this figure. Load, 250 μN. (a) Fe-N(40 nm)/Ti-N(2 nm) (b) Fe-N film, 450 nm (c) Ti-N film, 450 nm (d) Si(111) wafer.

mond tip (Berkowich tip with a radius of about 100 nm) under ambient condition. At least five indentations were made for each datum point. The microhardness and the reduced elastic modulus of the samples were evaluated from the loading-unloading curves by the method described by Oliver et al. [35].

Microscale friction and scratch tests were carried out by using the above AFM equipped with another commercial capacitance transducer (TriboScanner, Hysitron) with a conical diamond tip (1 μm tip radius 90° cone angle) under ambient condition. Friction coefficient could be obtained at the same time as the tip scratched the sample surface. Figure 39 is a real loading-unloading curve of the scratching test whose maximum normal force is 250 μN. There are seven steps for each single-pass scratch. First, the diamond tip makes contact with the sample surface (Fig. 39, Zone a) at position "0," then moves to position "–2 μm" without changing contact force (Fig. 39, Zone b); then loads applied normal force in three seconds (Fig. 39, Zone c), and moves to position "2 μm" (Fig. 39, Zone d) with the applied normal force. This step takes about 30 seconds (from the 8th second to the 38th second) and makes a scar on the sample surface. After that, the tip unloads the normal force (Fig. 39, Zone e) and moves to "0" position (Fig. 39, Zones f, g). The data from the 15th to the 35th second are used to calculate the average friction coefficient. The image of the microscratch scar could be obtained by using the AFM with the diamond tip after the microscratch test.

6.2.2 Nanoindentation

Typical load-displacement curves, with maximum applied load of 1,250 μN are shown in Fig. 40. It can be observed that

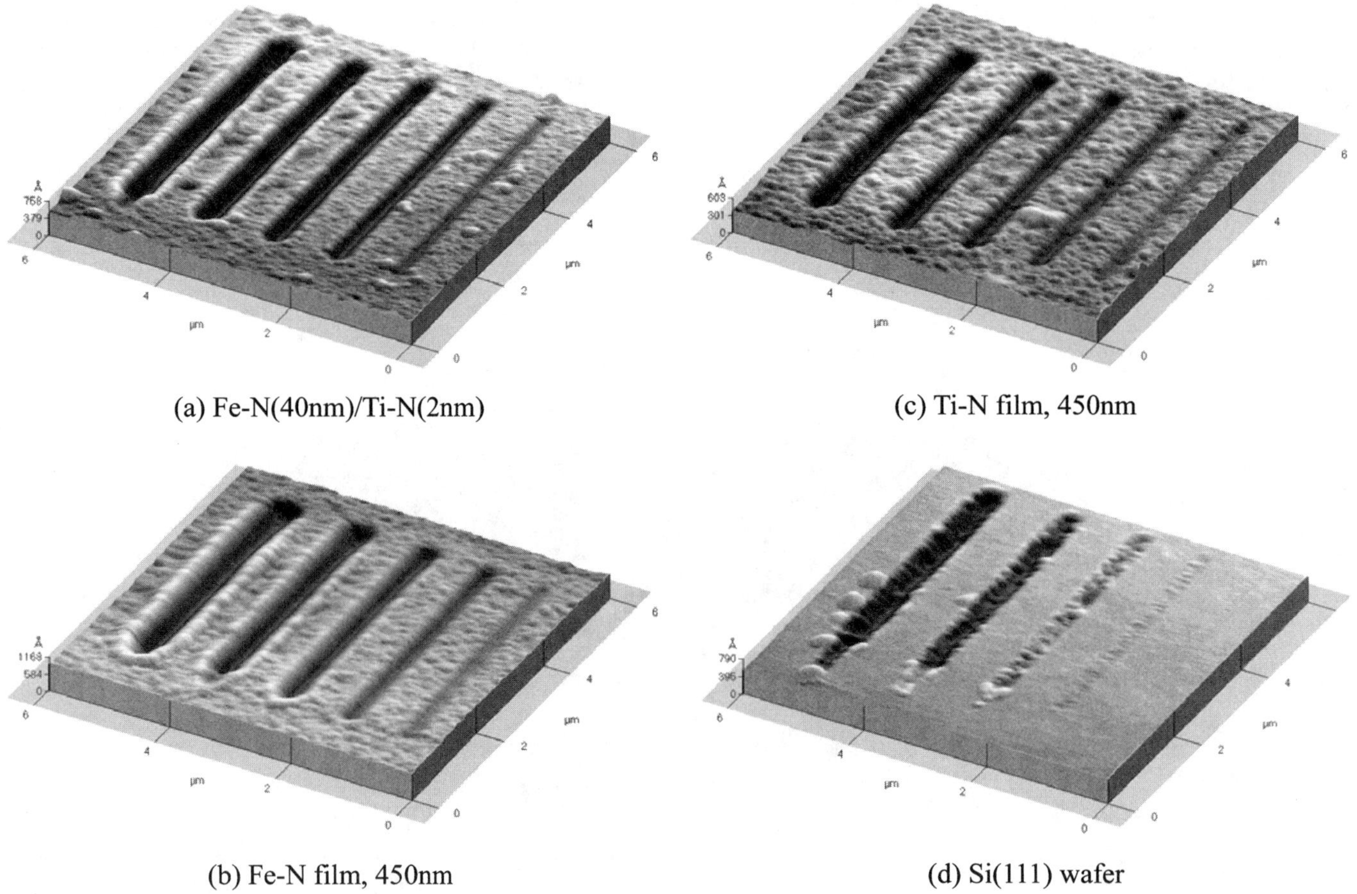

(a) Fe-N(40nm)/Ti-N(2nm)

(c) Ti-N film, 450nm

(b) Fe-N film, 450nm

(d) Si(111) wafer

Fig. 47—Images of scratching scar of Fe-N/Ti-N multilayers, Fe-N and Ti-N film as well as Si (111) wafer. (Load, 250 μN, 500 μN, 750 μN, 1,000 μN, 1,250 μN from right to left, respectively).

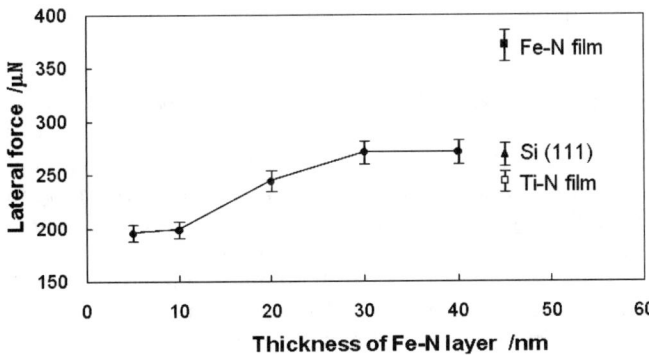

Fig. 48—Average lateral force (from the 15th second to the 35th second) of the multilayers versus thickness of Fe-N layers. Load, 1,250 μN.

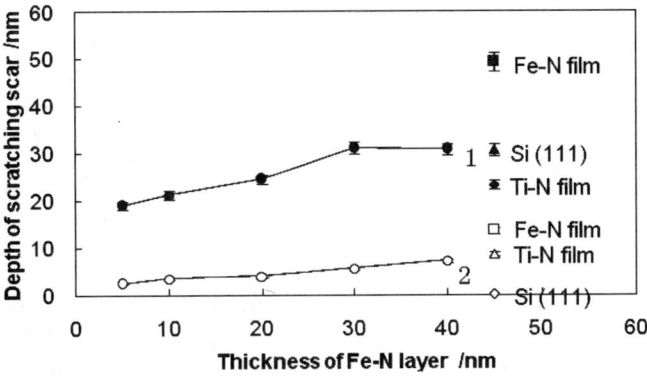

Fig. 49—Depth of scratching scar versus thickness of Fe-N layer, (1–250 μN; 2–1,250 μN).

the maximum indentation in the Fe-N(20 nm)/Ti-N(2 nm) multilayer is deeper than that in the Fe-N(5 nm)/Ti-N(2 nm) multilayer.

Figure 41 shows the dependence of hardness of multilayer on contact depth. The contact depth, h_c, is estimated from the load-displacement data using [35,36]:

$$h_c = h_{max} - \varepsilon \frac{P_{max}}{S} \qquad (22)$$

where P_{max} is the maximum indentation load, $S = dP/dh$ is the contact stiffness, and ε is a constant that depends on the indenter geometry, where $\varepsilon = 0.75$ for the Berkovich tip. According to the data obtained, trend lines were made to analyze the relationship between the hardness and the contact depth. Sample 1 has the highest hardness for all the contact depths. It shows the hardness of all samples increases with the contact depth, but the trend of hardness versus contact depth of Samples 3, 4, and 5 is different from Samples 1 and 2. Lines 3, 4, and 5 begin to merge when the contact depth increases. It appears that the thickness of Fe-N layer has less effect on the hardness if it is thicker than 20 nm.

Figure 42 shows the dependence of average hardness of Fe-N(450 nm) and Ti-N(450 nm) single layer as well as Si (111) wafer on applied normal force. The hardness of Si (111) wafer almost remeains constant, but under current experimental condition, the hardness of both the Fe-N and Ti-N layer increases with the normal force. This is probably because the thick layers are not homogeneous. The Fe-N layer has the lowest hardness for all the applied normal forces.

Figure 43 shows the hardness and reduced modulus of Fe-N/Ti-N multilayers versus the thickness of the Fe-N layer. It is found that the hardness of the multilayers decreases with the thickness of the Fe-N layer, but increases with the applied normal force. Under higher normal force, the hardness does not change much if the thickness of the Fe-N layer is thicker than 20 nm. The reduced modulus of the multilayers with different Fe-N layer thickness is always within the range of 170–190 GPa. There is no obvious relationship between the reduced modulus and the thickness of the Fe-N layer.

6.2.3 Microscale Friction

Figure 44 shows the AFM image of microscratch scars on Fe-N(5 nm)/Ti-N(2 nm) with different applied normal force. Under 250 μN and 500 μN, the average depth of the scratch scar is 2.5 nm and 6.7 nm, respectively. The tip slid on the sample surface under 250 μN but plowed into the sample surface under 500 μN. Figure 45 shows curves of lateral force versus time under different normal load. Here the lateral force includes friction force and plowing force. According to the scratch scars in Fig. 44, when the applied normal force is larger than 250 μN, the lateral force includes a large portion of plowing force. Therefore, the lateral force under normal load 250 μN is used for the calculation of the friction coefficient between the tip and the sample surface.

Figure 46 shows the average friction coefficient (data obtained from the 15th second to the 35th second) of the multilayers versus thickness of Fe-N layers under normal force 250 μN. The friction coefficient increases with the thickness of the Fe-N layers if the thickness of the Fe-N layers is more than 10 nm. For comparison, the friction coefficient of Fe-N, Ti-N film and Si (111) wafer under 250 μN is also shown in Fig. 46. The friction coefficient of the multilayer is lower than the friction coefficient of Fe-N and Ti-N films but more than the one of the Si (111) wafer.

6.2.4 Microscale Scratch

Figure 47 shows the typical microscratch scars of a part of the samples under different normal forces. The difference can be found from the scar under 250 μN. Only a shallow scratch scar is visible on the surface of Fe-N (5 nm)/Ti-N (2 nm) (Fig. 46), but the scars are deeper for other samples except the single crystal silicon wafer (Fig. 47). If the normal force was over 250 μN, the diamond tip penetrated into and plowed the sample surface. Figure 48 shows the relationship between the lateral force and the thickness of the Fe-N layer under 1,250 μN. It is found that the lateral force increases with the thickness of the Fe-N layer. For comparison, the lateral force of Fe-N, Ti-N, film and Si (111) wafer under 1,250 μN is also shown in Fig. 48. The lateral force of the multilayer is less than the Fe-N and Si wafer but more than the Ti-N film. Figure 49 shows the depth of the microscratch scar versus thickness of the Fe-N layer under 250 μN and 1,250 μN. It indicates that the scar depth increases with the thickness of the Fe-N layer, but the scar depth remains constant under high normal force if the thickness of the Fe-N layer is more than 30 nm. For comparison, the microscratch scar depth of Fe-N, Ti-N film, and Si (111) wafer is also shown in Fig. 49. The Fe-N single layer has the deepest microscratch scar under both high load (1,250 μN) and low load (250 μN). The Ti-N single layer has the shallowest scar under high load but the Si (111) wafer has the shallowest scar under low load. Here the microscratch scar depth can be used to represent wear loss. It is found that microscratch resistance of the Fe-N/Ti-N multilayer is much better than Fe-N film.

The friction coefficient is lower for the multilayers than for the Fe-N single layer. This is because the multilayers have a smaller grain size than the Fe-N single layer [37]. For multilayers, the forces applied on the tip are complex during the scratching process. The reason why the lateral force increases with the thickness of the Fe-N layer is mainly because the scratch scar increases with the thickness of the Fe-N layer (Fig. 49). It is the same reason why the lateral force of the Fe-N single layer is larger than that of Fe-N multilayers.

6.2.5 Closure

The hardness of the Fe-N/Ti-N multilayers decreases with the thickness of the Fe-N layer, but increases with the applied normal force. Among the investigated samples, the Fe-N(5 nm)/Ti-N(2 nm) multilayer has the highest hardness. Under current experimental condition, its hardness value is about 50 % more than that of Fe-N.

The friction coefficient of the Fe-N/Ti-N multilayers increases with the thickness of the Fe-N layer if the thickness of the Fe-N layer is more than 10 nm. It is lower than that of Fe-N film.

For the same applied normal force, the microscratch scar depth of the Fe-N/Ti-N multilayer increases with the thickness of the Fe-N layer, but the scar depth does not change much under high normal force if the thickness of the Fe-N layer is more than 30 nm. The microscratch resistance of the Fe-N/Ti-N multilayer is much better than the one of pure Fe-N film.

7 Summary

As is known, microscale friction and wear is important in microtribology. However, it is not easy to get real friction force on micro/nano scale during the tests. The surface morphology at nanometer scale, the scanning direction of the FFM, etc., have significant effects on friction force measurement. Even nowadays for commercial SPM we are not quite sure if the friction force we get is a real one or not.

This chapter introduces several research results on micro friction and wear obtained by the authors.

First we analyzed the states of a tip scanning along an ascent and descent surface on nanometer scale, and then we calibrated the lateral force obtained by the FFM we modified. It may be helpful to understand how to measure the true lateral force by an FFM.

Second, the properties of micro/nano friction and wear/scratch of several representative films are introduced. These films include from organic molecular films, solid films, to multilayers. The experiments were designed reasonably to understand the behaviors of micro/nano friction and wear of the films. The sample preparation methods were also described in detail. It is useful for further study of micro/nano friction and wear/scratch of thin films.

The mechanism of micro/nano friction and wear/scratch is still not well known. There are several questions that need to be answered: (a) How are the materials removed at very low load as nano Newton scale. Is it similar to that at macro scale? Can we explain the material loss according to the traditional wear mechanisms? (b) What is the connection between micro and macro friction and wear?

Therefore, further research on micro/nano friction and wear/scratch is still needed although we tribologists have been studying this subject for over 20 years.

References

[1] Bhushan, B., Israelachvili, J. N., and Landman, U., "Nanotribology: Friction, Wear and Lubrication at Atomic Scale," *Nature*, Vol. 374, 1995, pp. 607–616.

[2] Kaneko, R., Umemura, S., Hirano, A. M., Miyamoto, Y., and Fukui, T. S., "Recent Progress in Microtribology," *Wear*, Vol. 200, 1996, pp. 296–304.

[3] Kaneko, R., Oguchi, S., Miyamoto, T., Andoh, Y., and Miyake, S., "Micro-Tribology for Magnetic Recording," *Tribology and Mechanics of Magnetic Storage Systems*, Vol. VII, STLE Special Publication SP-29, 1990, pp. 31–34.

[4] Xue, Q. and Zhang, J., "On the Advances in Microtribology Research," *Tribology*, Vol. 14, No. 4, 1994, pp. 360–369.

[5] Bhushan, B., *Introduction to Tribology*, Wiley, New York, 2002.

[6] Mate, G. M., McClelland, R., Erlandsson, R. et al., "Atomic-Scale Friction of a Tungsten Tip on a Graphite Surface," *Phys. Rev. Lett.*, Vol. 59, No. 17, 1987, pp. 1942–1945.

[7] Mayer, G. and Amer, N. M., "Simultaneous Measurement of Lateral and Normal Forces with an Optical-Beam-Deflection Atomic Force Microscope," *Appl. Phys. Lett.*, Vol. 57, 1990, pp. 2089–2091.

[8] Ruan, J. and Bhushan, B., "Atomic-Scale Friction Measurements Using Friction Force Microscopy: Part I. General Principles and New Measurement Techniques," *ASME J. Tribol.*, Vol. 116, 1994, pp. 378–388.

[9] Lu, X. C., Wen, S. Z., Meng, Y. G., Huang, P., Dai, C. C., Huang, G. Z., Wang, P. S., and Bai, C. L., "A Friction Force Microscope Employing Laser Beam Deflection for Force Detection," *Chinese Science Bulletin*, Vol. 41, 1996, pp. 1873–1876.

[10] Bhushan, B., *Micro/Nano Tribology*, CRC Press, 1999.

[11] Wang, J. H., Lu, X. C., and Wen, S. Z., "Surface Morphology and Microtribological Behaviour of Magnetic Video Tapes," *Chinese Journal of Materials Research*, Vol. 11, 1997, pp. 351–356.

[12] Lu, X. C., Shi, B., Li, L. K. Y., Luo, J. B., Wang, J. H., and Li, H. D., "Investigation on Microtribological Behaviour of Thin Films Using Friction Force Microscope," *Surface and Coatings Technology*, Vol. 128-129, 2000, pp. 341–345.

[13] Wang, J. H., Wang, L. D., Lu, X. C., Wen, S. Z., and Li, H. D., "Surface and Microtribological Behaviour of Teflon/Si_3N_4 Micro-Assembling Film," *Acta Metallurgica Sinica*, Vol. 34, 1998, pp. 655–660.

[14] Bhushan, B. and Gupta, B. K., *Handbook of Tribology*, McGraw-Hill Inc., New York, 1991.

[15] Ulman, A., *An Introduction to Ultrathin Organic Films*, Academic Press, Boston, MA, 1991.

[16] Shi, B., Lu, X. C., Luo, J. B., Huang, L., and Hong, H., *Thin-*

ning Films and Tribological Interfaces, D. Dowson et al., Eds., Elsevier Science, B. V.: Amsterdam, 2000, p. 705.

[17] Bull, S. J., and Jones, A. M., "Multilayer Coatings for Improved Performance," *Surface and Coatings Technology*, Vol. 78, 1997, pp. 173–184.

[18] Strondl, C., van der Kolk, G. J., Hurkmans, T., Fleischer, W., Trinh, T., Carvalho, N. M., and de Hosson, J. Th. M., "Properties and Characterization of Multilayers of Carbides and Diamond-Like Carbon," *Surface and Coatings Technology*, Vol. 142-144, 2001, pp. 707–713.

[19] Sundararajan, S., and Bhushan, B., "Micro/Nanotribology of Ultra-thin Hard Amorphous Carbon Coatings Using Atomic Force/Friction Force Microscopy," *Wear*, Vol. 225-229, 1999, pp. 678–689.

[20] Lettington, A. H., "Application of Diamond-Like Carbon Thin Films," *Carbon*, Vol. 36, 1998, pp. 555–560.

[21] Bhushan, B., "Chemical, Mechanical and Tribological Characterization of Ultra-Thin and Hard Amorphous Carbon Coatings as thin as 3.5 nm: Recent Developments," *Diamond Relat. Mater.*, Vol. 8, 1999, pp. 1985–2015.

[22] Ziegele, H., Scheibe, H. J., and Schultrich, B., "DLC and Metallic Nanometer Multilayers Deposited by Laser-Arc," *Surface and Coatings Technology*, Vol. 97, 1998, pp. 385–390.

[23] Delplancke-Ogletree, M. P. and Monteiro, O. R., "Wear Behavior of Diamond-Like Carbon/Metal Carbide Multilayers," *Surface and Coatings Technology*, Vol. 108-109, 1998, pp. 484–488.

[24] Delplancke-Ogletree, M. P., Monteiro, O. R., and Brown, I. G., "Preparation of TiC and TiC/DLC Multilayers by Metal Plasma Immersion Ion Implantation and Deposition: Relationship Between Composition, Microstructure and Wear Properties," *Materials Research Society Symposia Proceedings*, Vol. 438, 1997, p. 639.

[25] Deng, J. G. and Braun, M., "Residual Stress and Microhardness of DLC Multilayer Coatings," *Diamond Relat. Mater.*, Vol. 5, 1996, pp. 478–482.

[26] Voevodin, A. A., Walck, S. D., and Zabinski, J. S., "Architecture of Multilayer Nanocomposite Coatings with Super-Hard Diamond-Like Carbon Layers for Wear Protechtion at High Contact Load," *Wear*, Vol. 203-204, 1997, pp. 516–527.

[27] Dekempeneer, E., Van Acker, K., Vercammen, K., Meneve, J., Neerinck, D., Eufinger, S., Pappaert, W., Sercu, M., and Smeets, J., "Abrasion Resistant Low Friction Diamond-Like Multilayers," *Surface and Coatings Technology*, Vol. 142-144, 2001, pp. 669–673.

[28] Lu, X. C., Shi, B., Li, L. K. Y., Luo, J. B., and Mou, J. I., "Nanoindentation and Nanotribological Behavior of Fe-N/Ti-N Multilayers with Different Thickness of Fe-N Layers," *Wear*, Vol. 247, 2001, pp. 15–23.

[29] Santos, L. V., Trava-Airoldi, V. J., Corat, E. J., Iha, K., Massi, M., Prioli, R., and Landers, R., "Friction Coefficient Measurementsby LFM on DLC Films as Function of Sputtering Deposition Parameters," *Diamond Relat. Mater.*, Vol. 11, 2002, pp. 1135–1138.

[30] Kryder, M. H., Messner, W., and Carley, L. R., "Approaches to 10 Gbit/in.^{2} Recording," *J. Appl. Phys.*, Vol. 79, 1996, pp. 4485–4488.

[31] Kryder, M. H., Wang, S., and Rook, K., "FeAlN/SiO$_2$ and FeAlN/Al$_2$O$_3$ Multilayers for Thin-Film Recording Heads," *J. Appl. Phys.*, Vol. 73, 1993, p. 6212.

[32] Sin, K. and Wang, S. X., "FeN/AlN Multilayer Films for High Moment Thin Film Recording Heads," *IEEE Trans. Magn.*, Vol. 32, 1996, pp. 3509–3511.

[33] Wang, Z. J., Wen, L. S., Chang, X. R. et al., "Oscillatory Magnetic Interlayer Exchange Coupling in Fe-N/TiN Multilayers," *Appl. Phys. Lett.*, Vol. 68, 1996, pp. 2887–2889.

[34] Wang, Z. J., "Microstructure and Magnetic Properties of Fe-N/TiN Nano-Multilayer," Ph.D. Thesis, University of Science and Technology Beijing, 1996.

[35] Oliver, W. C. and Pharr, G. M., "An Improved Technique for Determining Hardness and Elastic Modulus Using Load and Displacement Sensing Indentation Experiments," *J. Mater. Res.*, Vol. 7, 1992, p. 1564.

[36] Pharr, G. M., "Measurement of Mechanical Properties by Ultra-Low Load Indentation," *Mater. Sci. Eng., A*, Vol. A253, 1998, pp. 151–159.

[37] Lu, X. C., Shi, B., Luo, J. B., Li, L. K. Y., Chang, X. R., and Tian, Z. Z., "Microstructure and Properties of Fe-N/Ti-N Magnetic Multilayers," *Surf. Interface Anal.*, Vol. 32, 2001, pp. 66–69.

11

Tribology in Magnetic Recording System

Jianbin Luo,[1] Weiming Lee,[2] and Yuanzhong Hu[1]

1 Introduction

A HARD DISK DRIVER (HDD) AS A HIGH SPEED DIGItal recording system has been a main part of the computer. It also has been widely used as a mobile data storing set in almost all electronic devices, including a video recorder, camera, etc. The recording density of a hard disk driver has been increasing at a high rate of 100 % per year in the past ten years. It is much faster than the rate of the Moore's law for silicon devices ($\sim$50 %) [1]. It is expected that the recording density will increase to 1,000 Gbit/in.^{2}, and the fly height will decrease to about 3 nm in the next several years [2,3]. There are three major challenges that tribologists are facing today. The first is how to make solid protective coatings, i.e., diamond-like carbon (DLC) layer, with a thickness of about 1 nm without any micro-pinholes; the second is how to make a lubricant film about 1 nm on the surface of a disk or head to minimize the wear, friction, and erosion; and the third is how to control the vibration of the magnetic head and its impacting on the surface of a disk.

The biggest challenge to produce an ultra-thin (about 1 nm) overcoat is to make the coating free of pin-holes while maintaining the durability and tribological properties. In an HDD system, pin-holes can cause much more contaminants from all sources, such as outgas compounds from polymeric foam components, pressure sensitive adhesives, ionic residues from improperly cleaned components and ambient pollutants, which can be detrimental to the tribology and durability of the HDD. Therefore, efforts have been made mainly on the improvement of carbon film [4–7].

A stable lubrication is very important to the slider/disk interface as the demand of HDD life increases. The lubricant films need to have a strong adhesion and bonding with the carbon surface in order to be effective in reducing friction and wear of the hard disk interface [8]. Organic films with one or a few monolayers have been used for lubrication of the HDD. Perfluoropolyethers (PFPEs) lubricant is one of the synthetic lubricants that are widely applied due to its excellent performances, such as chemical inertness, oxidation stability, lower vapor pressure, and good lubrication properties [9]. Generally, lubricant is deposited on surfaces of a hard carbon overcoat of magnetic recording media. The lubricant and carbon films protect the underlying soft magnetic media from mechanical damage caused by intermittent contacts with the slider. During operation of an HDD, contacts between slider and lubricated film will result in the loss of lubricant in the contact region. Unless being continuously replenished, the film thickness of PFPE in the depleted contact zone will decrease with the increase of contacts, resulting in the loss of the interface life [10].

As recording density approaches 1 Tb/in.^{2} in the near future, the flying height will be so small that it will fall into the categories of contact recording regime [11–13]. Sliding contacts between the surface of the head and disk will be much more frequent than ever before, and will happen throughout almost the whole operation. Good surface mobility of lubricant film will ensure the lube reflowing and cover the area where the lubricant molecules are depleted after head-disk interaction.

The relative mobility of thin PFPE films has attracted considerable attention. The spreading behavior of perfluoropolyalkylethers (PFPAE) on silicon surfaces was studied by O'Connor et al. [14]. They found the surface diffusion coefficients did not obey a simple linear mixing rule, and spreading characteristics of thin PFPE films are related to molecular weight, film thickness, chain-end functionality, temperature, and humidity. In 2003, Deoras et al. [15] made an analysis of the spreading and mobility of PFPE using a surface reflectance analyzer. Sinha et al. [16] investigated the wear and friction properties of ultra-thin (2 nm) PFPE films on glass substrate magnetic hard disks.

The work of Karis et al. [17] indicates that PFPE is gradually removed in the sliding process. The removal rate is proportional to a parameter containing PFPE bulk viscosity, degree of polymerization, temperature, and structure scaling coefficient. Lubricant additives that could stabilize the head-disk interface and minimize the decomposition of lubricants become more anticipant. Partially fluorinated hexaphenoxy cyclotriphosphazene (X-1P) has been suggested by Pevettie et al. [18] as an additive to Z-dol to cover the catalytically active surface of the head, and thereby prevent decomposition of the lubricant Z-dol.

In HDD, high rotational speed and short spacing between a head and a disk are known to increase tribological problems in the head/disk interface (HDI) [19]. To solve the problems, the ramp load/unload (L/UL) operation has been used instead of the contact-start-stop (CSS) operation [20,21]. Although the L/UL technology eliminates the stiction and wear problems associated with the contact-start-stop technology [22], it may cause slider-disk contact problems during the operation. It is supposed that the slider vibration after unloading is responsible for a part of the physical disk damage [23]. Therefore, research on how to im-

---

[1] State Key Laboratory of Tribology, Tsinghua University, Beijing, China.
[2] R&D Director, Kaifa Magnetics Engineering Department, Kaifa Magnetics Ltd., Shenzhen, China.

prove tribological properties of sliders at slider-disk contacts is necessary.

This chapter introduces three kinds of surface organic modification films on a magnetic head that we have studied. These are polyfluoroalkylmethacrylate films, X-1P films, and self-assembled monolayers (SAMs). It also reviews the works of surface lube on a hard disk surface. In the last, the challenges on the development of a magnetic recording system are discussed.

2 Surface Modification Films on Magnetic Head

2.1 Stiction and Friction Properties of Polyfluoroalkylmethacrylate Films

2.1.1 Experimental Details

In order to improve the stiction and friction properties of the HDI of a hard disk drive, the effect of polyfluoroalkyl-methacrylate (PFAM) films on the surface properties of a slider has been tested by coating the film on the surface using the dipping method. The thickness, contact angle, and surface topography of the thin films at different concentrations were measured by Jiang et al. [24] and Yang et al. [25] using a time-of-flight secondary ion mass spectrometer (TOF-SIMS), a contact angle measurement, and an atomic force microscopy (AFM), respectively. All experiments were carried out in a class 100 clean room at the temperature of $20\pm2°$ and relative humilities of $60 \sim 70\%$. The sliders were cleaned, after being coated with DLC film, with ultra-clean cotton buds under a microscope to remove particles brought from the environment. They were then dipped into PFAM solutions in different concentrations for 5 min. Finally all the samples were annealed in an oven at 120° for 30 min. The thickness of the PFAM film was measured by a TOF-SIMS. The surface topography was detected by AFM. The hydrophobic wet ability was measured using a VCA2500XE contact angle system, (the data were reproducible within 0.1° for a given sample). An HDI reliability test system was used for the stiction/friction measurement in the contact-start-stop (CSS) operation. The system measured the acoustic emission signal (AE signal) to predict stiction and friction of the head/disk interface and to infer the integrity of the data written on the disk and the anticipated usable life of the driver.

2.1.2 The Influence of the PFAM Concentration on the Film Thickness, the Water Contact Angle, and the Surface Topography

The apparatus for the PFAM film coating on the slider surface is shown in Fig. 1(a). The film thickness was measured by the TOF-SIMS as shown in Fig. 1(b). It used a pulsed primary Ga$^+$ ion beam to impact the surface of the PFAM film with an inset energy of 15 keV, an extractor current of 2 μA, beam current of 600 pA, a pulse width of 17.5 ns, and a frequency of 10 kHz, respectively. The positive TOF-SIMS spectra on the slider surface is shown in Fig. 2 where the peaks at m/z 31, 50, 69, 100, and 131 in Fig. 2(a) correspond to the positive secondary ion fragments of CF^+, CF_2^+, $C_2F_4^+$, and $C_3F_5^+$, respectively. The peak at m/z 469 apparent in Fig. 2(b) corresponds to the ion $C_{12}H_7F_{15}O_2H^+$ which is the characteristic ion of PFAM molecules. Therefore, the positive TOF-SIMS spectra demonstrates the existence of PFAM film [24,25]. The thickness of the PFAM film can be determined

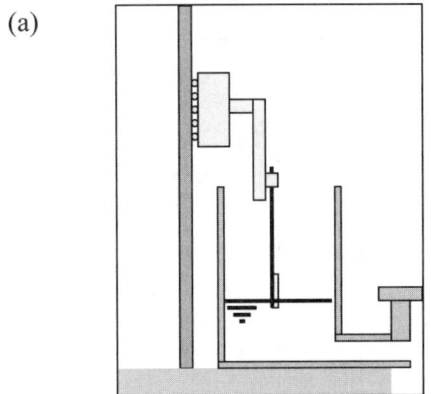

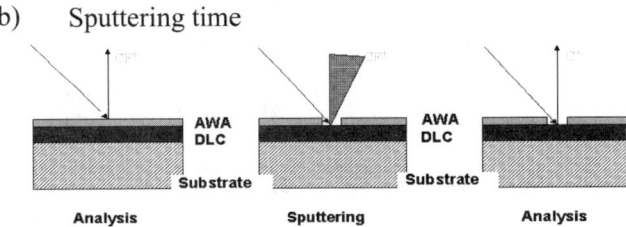

Fig. 1—The formation and measurement of molecular films: (a) dipping of magnetic head immersing in solution, and (b) the process of TOF-SIMS to measure the AWA film.

by TOF-SIMS when it is thinner than 2.0 nm as shown in Fig. 3 [24,25]. It can be seen that the film thickness increases with the number of PFAM molecules at the concentration of 500 ppm. The film thickness does not change with the position of the slider surface, indicating the uniformity as shown in Fig. 4. However, after 20,000 times of CSS tests, much more PFAM molecules were lost between position 2 to position 5 than at other positions [Fig. 4(b)]. The thickness of PFAM films also changed with the concentrations of the PFAM solutions as shown in Fig. 5 [26]. The film thickness sharply increased with the increase of the concentrations, when they were below 200 ppm. Above 200 ppm, the thickness became less sensitive to the concentration.

The values of the water contact angle of the PFAM film on the slider surface also increased with the concentration of PFAM solutions (Fig. 5 [26]). The water contact angle increased up to 121° at 200 ppm from the initial values of 88.4° without the PFAM film, indicating that the PFAM had covered the slider surface well. For N-Hexane, the contact angle was only about 48°, but the contact area was much larger than that on the slider surface without the PFAM film as shown in Fig. 6.

Hu et al. [26] have discussed the influence of the concentration of PFAM solution on a film's surface topography as shown in Fig. 7. At the concentration of 50 ppm, the surface was very smooth. The surface roughness was 0.269 nm which was very close to the bare surface of the slider [Fig. 7(a)]. At the concentration of 500 ppm, a thin and uniform film was formed on the slider surface. The roughness was 0.32 nm as shown in Fig. 7(b), indicating that much more PFAM molecules had been adsorbed on the surface, and the surface coverage greatly increased. As the concentration was raised to 1,000 ppm as shown in Fig. 7(c), the surface became quite rough. The surface roughness Ra was 1.911 nm

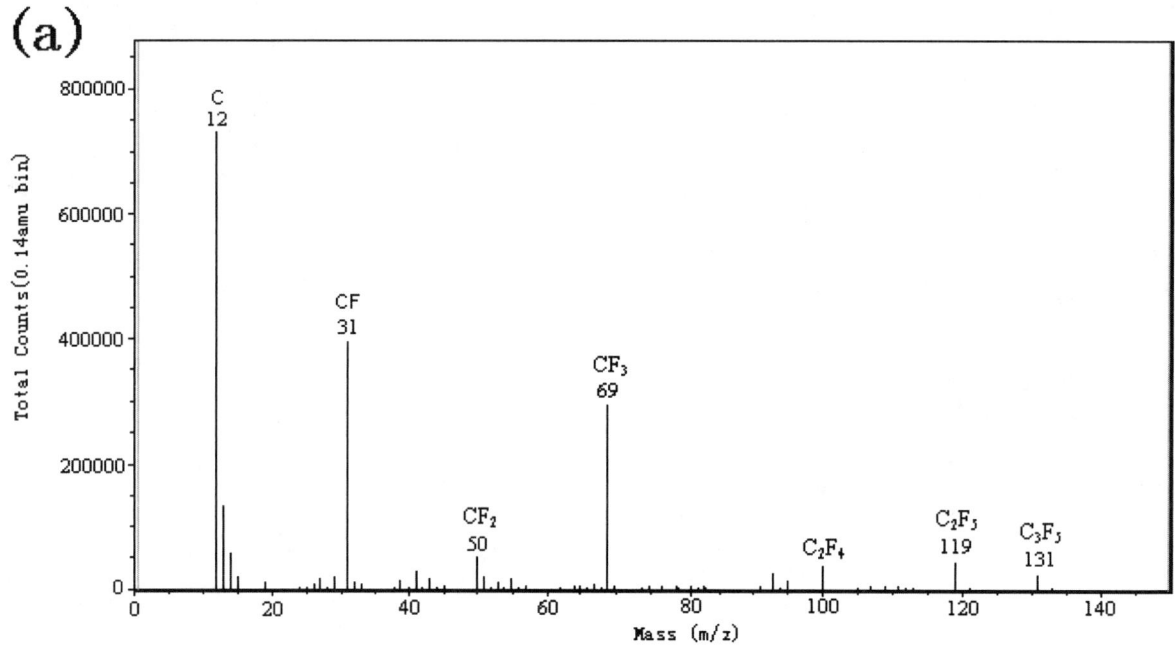

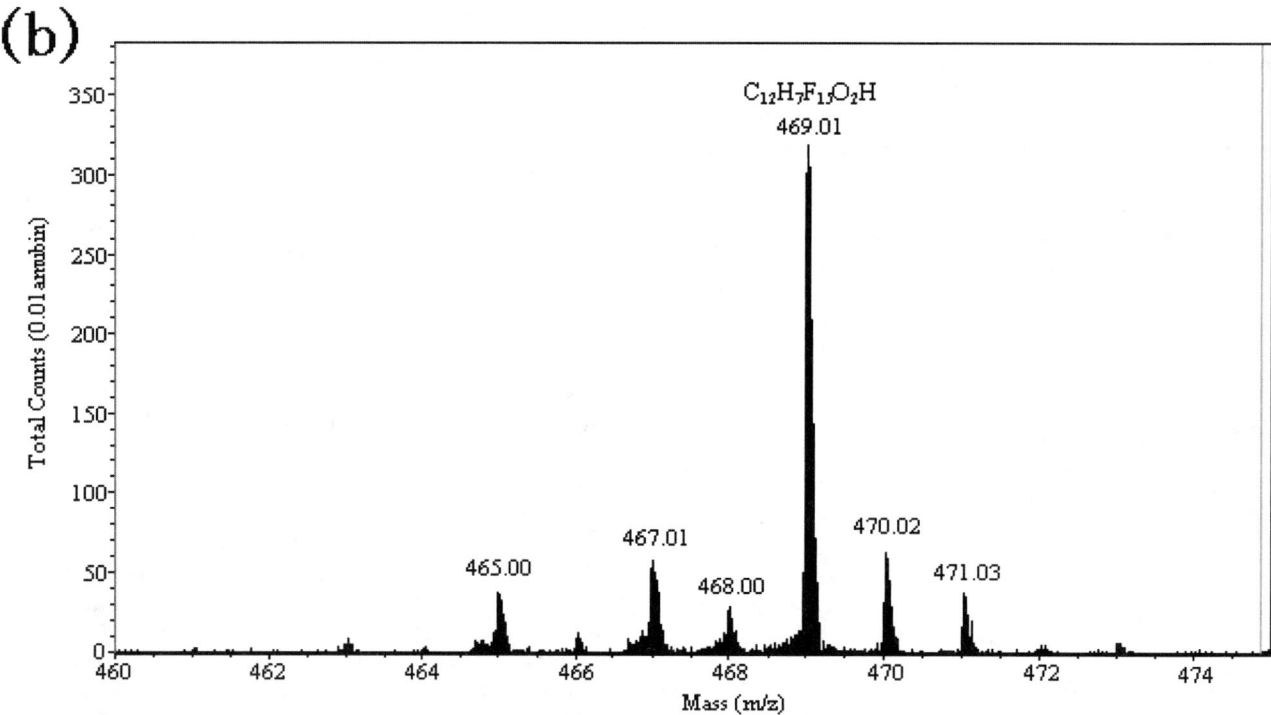

Fig. 2—Positions of the positive TOF-SIMS spectra of PFAM thin film on slider, showing the most characteristic fragments: (a) positive, m/z=0–150; (b) positive, m/z=460–475.

and the maximum height was 50.977 nm. It could be considered that molecules in the solution aggregated prior to be adsorbed on the slider surface when the concentration of PFAM was too high. Then the molecular group adsorbed on the surface, causing a much rougher surface. Doudevski [27] observed a similar phenomenon in his work about the island nucleation and growth of SAMs. Therefore, it can be considered that the ideal surface topography of the PFAM film was formed around 500 ppm.

2.1.3 The Influence of the PFAM Concentration on the Stiction and Friction Properties in the CSS Test

Figure 8 shows the stiction of the sliders at the different stages of the CSS test from the initial cycle to the final cycle, with the PFAM films prepared at different concentrations [26]. There was a 2 h parking after every 5,000 CSS and a 24 h parking at the end of the final 20,000 CSS. Without PFAM film, the stiction increased with the increase of the

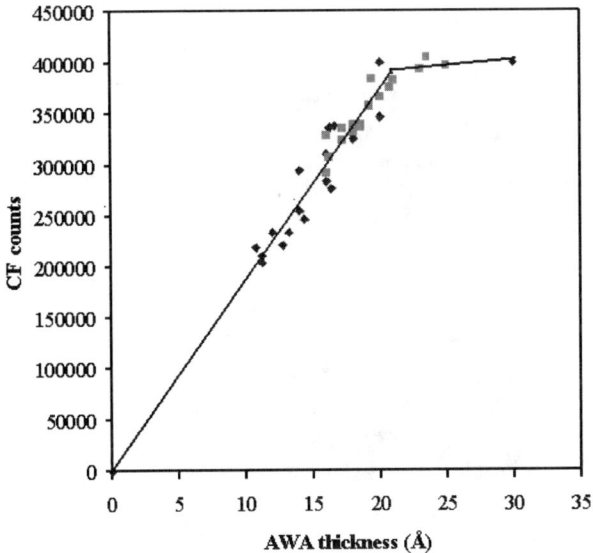

Fig. 3—Relation between the PFAM film thickness and CF number.

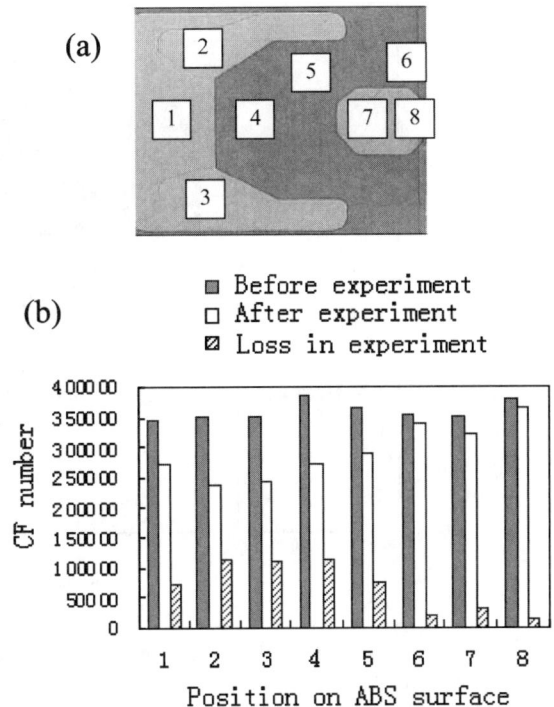

Fig. 4—The number of CF$^+$ at different positions of the slider surface, (a) air bearing surface (ABS) of magnetic head, (b) CF numbers on different positions of ABS.

CSS number or the testing time. At 5 K CSS after a 2 h stop, the stiction increased slightly with the number of flying cycles. At the last cycle after the 24 h stop, the stiction increased to 5.712 g [Fig. 8(a)]. When the PFAM film was formed on the slider surface, even if the concentration was only 50 ppm, the stiction became stable in the whole CSS process [Fig. 8(b)]. In the case of 500 ppm, the stiction value became much lower and stable than that in other concentrations [Fig. 8(c)]. At the concentration of 1,000 ppm, the stiction in every 5 K CSS remained quite small. However, it increased to 6.484 g after the 24 h stop, which was much larger than that in other conditions.

To evaluate wear properties of the PFAM film, the friction after 20 K flying cycles was measured by Yang et al. [28] and Hu et al. [26] as shown in Fig. 9. Each cycle of the takeoff and landing process was performed within 2 min. The existence of the PFAM film has no influence on the normal takeoff and landing of the slider. In the case of 50 ppm and 500 ppm, the PFAM film maintained a low friction at the last cycle in a CSS test. The maximum friction value of the latter is less than 2 g, which is about one-third of that of the bare slider surface and the surface modified at the concentration of 1,000 ppm.

For a further investigation of head/disk working conditions, the profiles of acoustic emission (AE) signal recording the output of an ultra-low noise preamplifier from the contact between head and disk have been measured and shown in Fig. 10. The AE signals normally appear at the takeoff and landing process. It should be zero in the flying process to make sure that there is no contact between head and disk. As shown on the figure, at the 20,000th single flying cycle, the AE signal was zero in the flying process when the PFAM film concentrations were at 50 ppm and 500 ppm [Figs. 10(b) and 10(c)]. It means that the flying balance has been kept in these conditions. However, for the bare surface and the surface with the PFAM film at 1,000 ppm, AE signals were detected

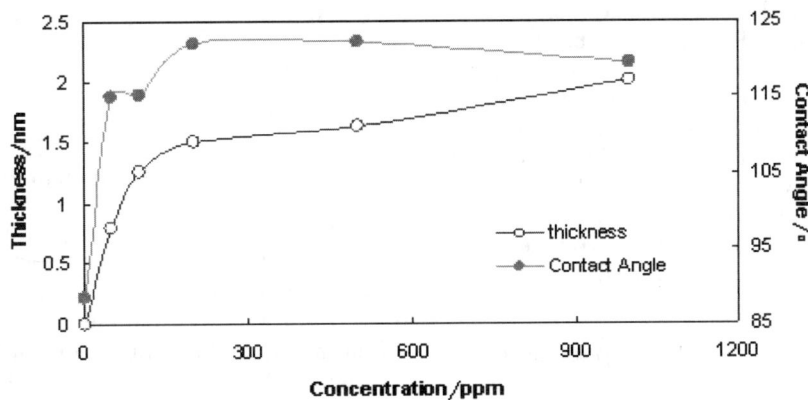

Fig. 5—The thickness and water-contact angle of PFAM thin film as a function of solution concentration.

	Without AWA Coating	With AWA Coating
DI Water	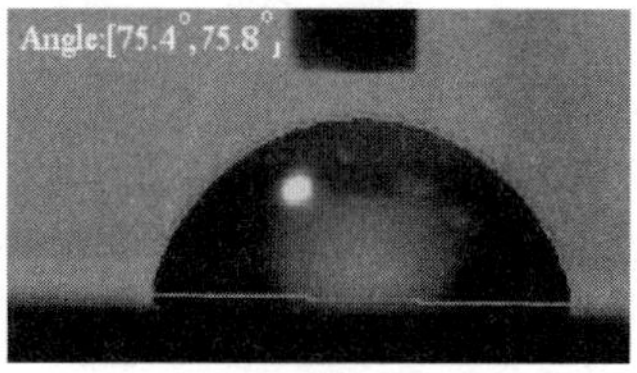 Contact Angle=75.4°~ 75.8°	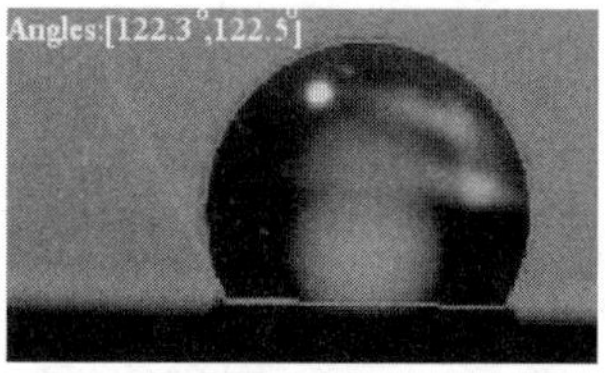 Contact Angle =122.3°~ 122.5°
N-Hexane	Completely wetted	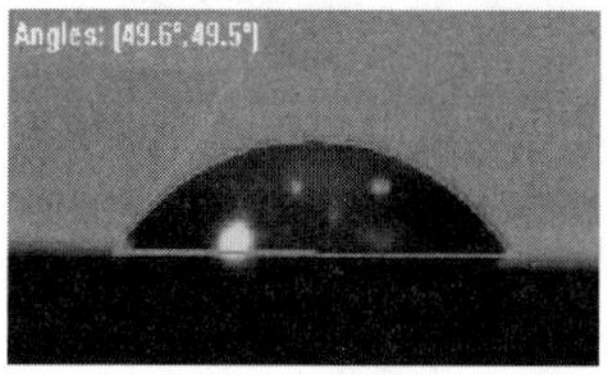 Contact Angle =49.5°~ 49.9°

Fig. 6—Images of the contact angles.

in the flying process. It means that there are some contacts between the head and the disk, or, the flying balance is damaged. The possible reasons are that there are contaminants on the surfaces of the sliders or disks from the environment, or, there are wear remains of lubricants and the PFAM films on the slider surfaces. The water contact angle value of the bare slider was 88.4°. It is much easier for the bare surface to adsorb contaminants from the environment, or to pick up lubricant from the disk than the surface with the PFAM film as shown in Fig. 11 [28]. In the case of 1,000 ppm, the surface roughness of the PFAM film is too high, causing the head/disk contacts to easily take place in the flying process.

It can be concluded that the concentrations of the PFAM solution is an important factor for the PFAM film formed on the slider surface to affect the stiction and friction in the CSS tests. If the concentration is controlled around 500 ppm, an ideal surface topography, good hydrophobic nature, a preferred film thickness, and better frictional and anti-wear properties can be obtained.

2.2 Tribological Characteristics of the X1-P Film on the Surfaces of HDD Heads

Partially fluorinated X-1P has been used for a number of years as an additive in the inert lubricant PFPE film on the surface of a magnetic hard disk to enhance start/stop durability of PFPE lubricants [29,30]. Recently it has been used as a vapor lubricated film on the surface of the disks [31]. In order to avoid the PFPE being catalyzed to decomposition by the slider material Al_2O_3 (refer to Section 3.4), X1-P was also examined as a protective film on the surface of the magnetic heads [25,32]. The results of CSS tests indicate that the thermal stability of the lubricant was greatly improved in the presence of X-1P, and the thickness of X-1P film on the slider surface has an important influence on HDD lubrication properties.

2.2.1 X-1P Action in TGA and FTIR Spectroscopy Tests

X-1P is a compound with a chemical structure of $(p\text{-}FC_6H_4O)_n\text{-}N_3P_3\text{-}(m\text{-}OC_6H_4CF_3)_{6-n}$, where $n = 0, 1, 2,$ …6. X-1P species in the experiment is $n = 2$. Its chemical diagram is illustrated in the following:

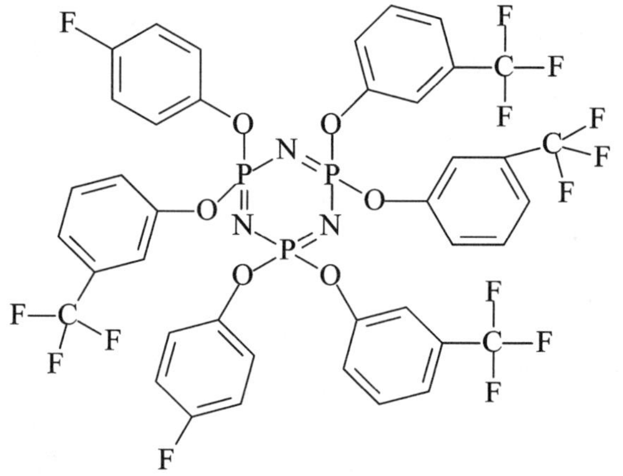

Diagram of X-1P

The PFPE liquid used in the experiment has a typical backbone group "$-CF_2$" as shown in the diagram of PFPE.

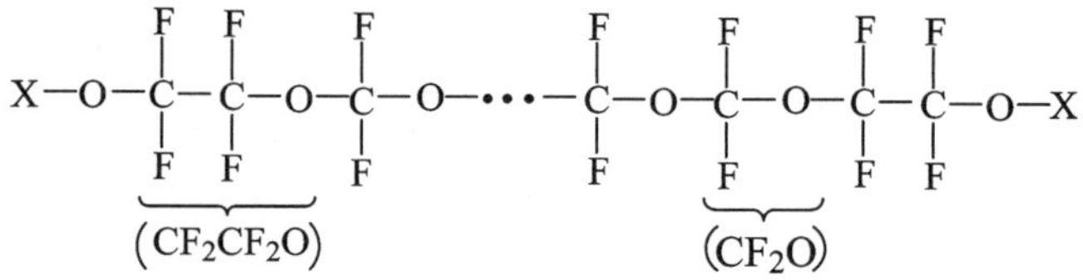

Diagram of PFPE

A thermogravimetric apparatus (TGA) was used to verify the weight loss of the chemical materials of different samples which are PFPE (Sample 1), PFPE + Al_2O_3 (Sample 2), PFPE + X-1P (Sample 3), and PFPE + Al_2O_3 + X-1P (Sample 4) at the temperature of 220°C, respectively. The weight of PFPE of Sample 1 was 500 mg; for Sample 2, PFPE was 500 mg plus 200 mg of Al_2O_3; for Sample 3, PFPE was 500 mg and X-1P was 100 mg; for Sample 4, the weights of

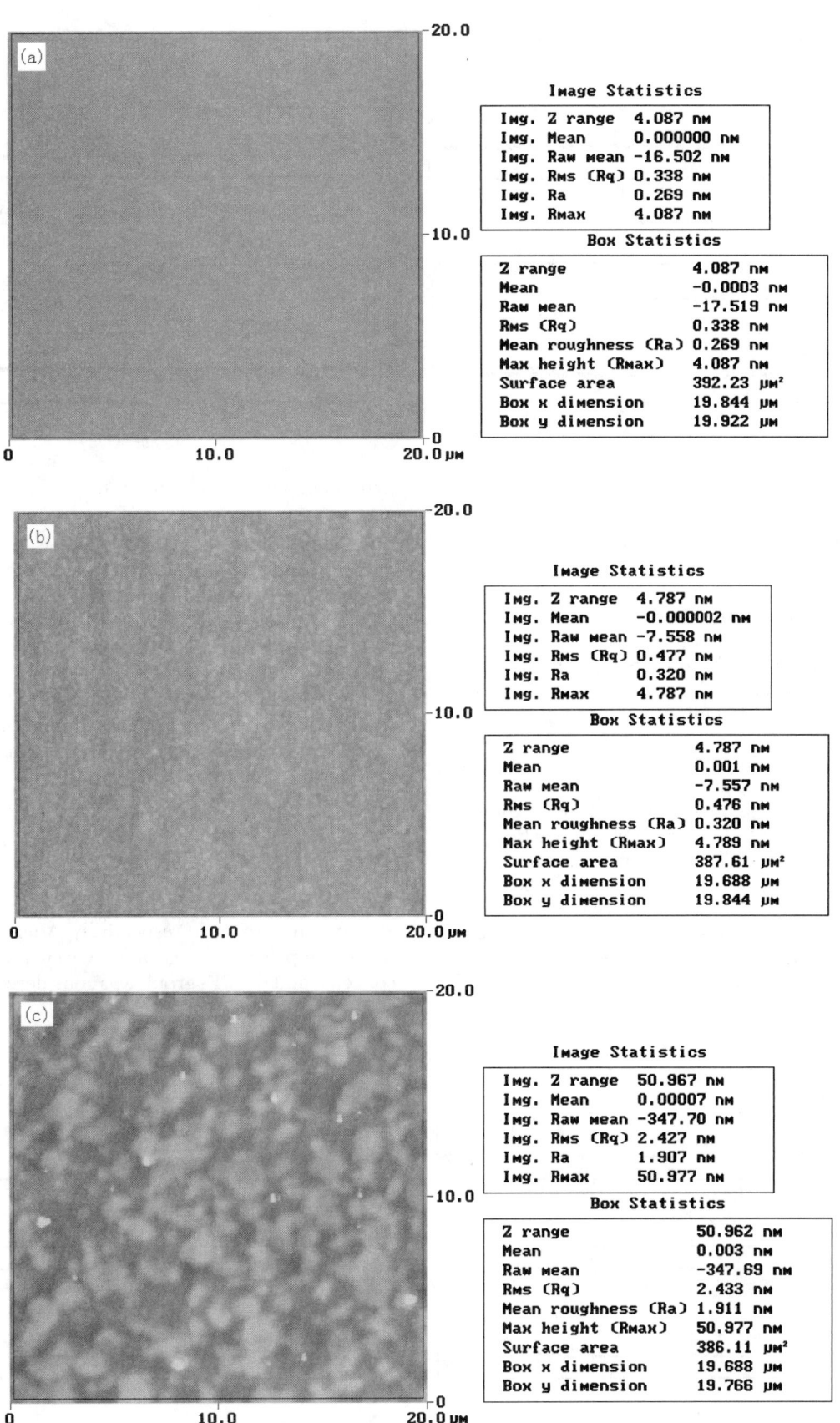

Fig. 7—AFM images of PFAM films in different concentrations: (a) 50 ppm; (b) 500 ppm; (c) 1,000 ppm.

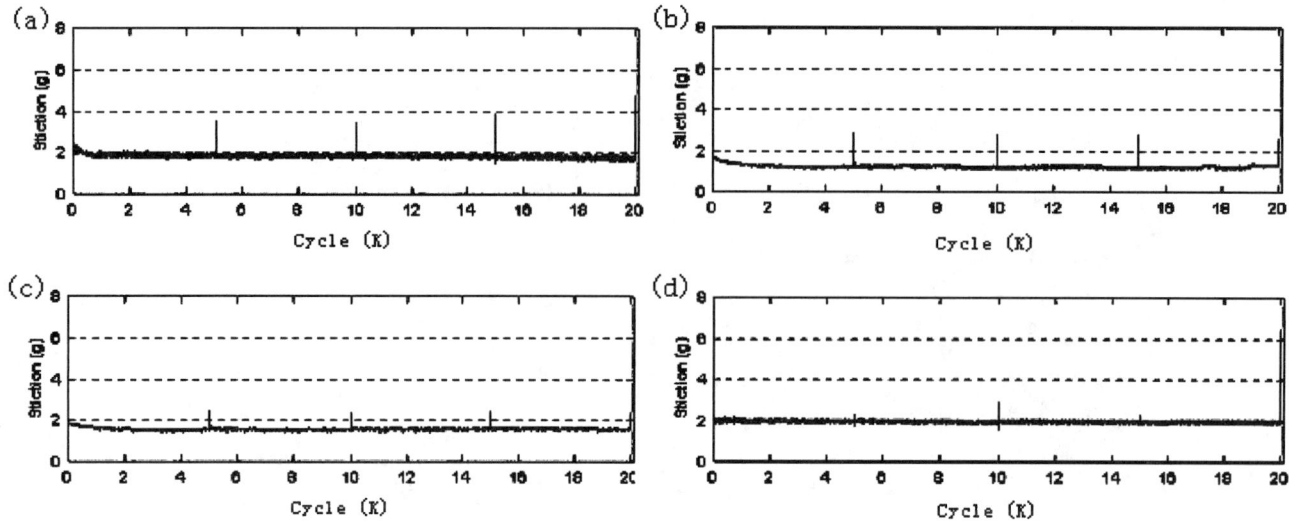

Fig. 8—Stiction profiles of (a) bare slider and PFAM thin film in different concentrations: (b) 50 ppm; (c) 500 ppm; (d) 1,000 ppm.

PFPE, Al$_2$O$_3$, and X-1P were 500 mg, 200 mg, and 100 mg, respectively. The materials in Samples 2 to 4 were sufficiently mixed to ensure that the liquid PFPE and X-1P completely wet the alumina powders. These specimens were put in a closed space filled with inert nitrogen. The flow rate of the nitrogen gas was kept at 20 milliliters per minute. The environmental temperature of the system was set at 220°C and the duration time was 250 minutes per sample for each individual operation procedure. PFPE used in the experiment was Z-dol and the alumina was in ultra-fine powders with chemical analytic grade purity.

As shown in Fig. 12, the weight loss of PFPE for Sample 1 was very small during the whole 250 minutes reaction process, except for an apparent loss in the first 20 minutes because of evaporation. For Sample 3, the weight loss was similar to that of Sample 1, so that the reaction between PFPE and X-1P was very weak. However, for Sample 2, PFPE+Al$_2$O$_3$, the reaction rate was expected to increase greatly with time because the weight loss grew very fast with time. This indicates that Al$_2$O$_3$ which is the basic material of an HDD head, induced the catalytic decomposition of PFPE. When X-1P was added into Sample 2, that is, PFPE+Al$_2$O$_3$+X-1P (Sample 4), the weight loss became very smaller or the reaction between PFPE and Al$_2$O$_3$ had been reduced greatly by X-1P.

A Fourier transform infrared (FTIR) spectrometer, widely used to investigate characteristic changes of possible functional groups for many years [33], was employed in the present test. The samples were classified into three groups, PFPE, PFPE+Al$_2$O$_3$, and PFPE+Al$_2$O$_3$+X-1P. The weight ratios for the latter two were 10+10 and 9+10:1, respectively. Samples were heated to 220°C in an oven to provide a favorable environment for thermo-induced decomposition. They were taken out after heating for 15 min, 45 min, 75 min, and 105 min, respectively. After a chemical treatment, samples were measured by FTIR. Infrared data were collected and the CF$_3$ group was considered as an indication

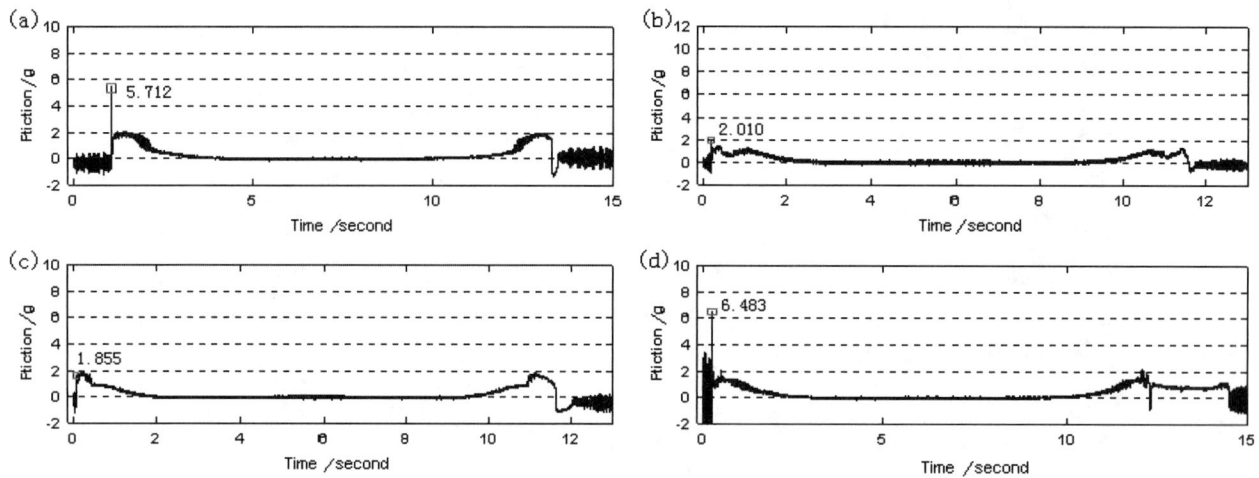

Fig. 9—Friction profiles at 20 K cycles of (a) bare slider and PFAM thin film in different concentrations: (b) 50 ppm, (c) 500 ppm, and (d) 1,000 ppm.

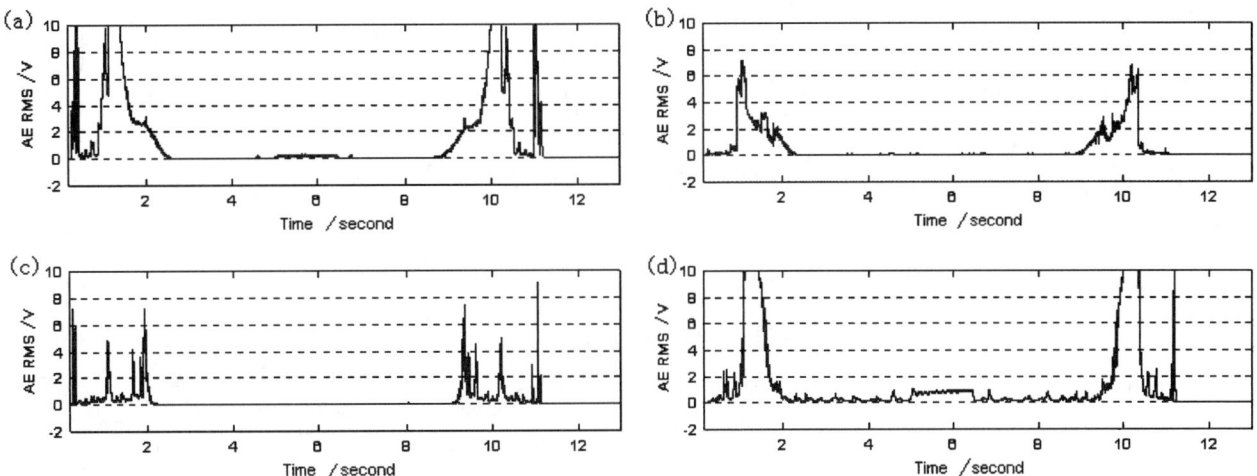

Fig. 10—AE signal profiles at 20 K cycles of (a) bare slider and PFAM thin film in different concentrations: (b) 50 ppm, (c) 500 ppm, and (d) 1,000 ppm.

of a new species generated due to thermal catalytic formation.

In order to reduce the differences of the initial weights, the ratio between the content of CF_3 and CF_2 was used. Based on the obtained infrared absorption curve, relative area ratios of CF_3/CF_2 absorption peaks were calculated and are shown in Fig. 13 [25]. With PFPE only, the content of CF_3 increased slightly with the heating time. With the mixture of PFPE and Al_2O_3, the content of CF_3 increased greatly. This indicates that many groups of CF_2 in PFPE disappeared in the heating process; that is, the reaction between PFPE and Al_2O_3 had taken place. Consequently, PFPE was decomposed. However, when X-1P was added into the mixture of PFPE and Al_2O_3, the ratio between the group CF_3 and CF_2 increased more slowly than the former. As a result, the reaction between PFPE and Al_2O_3 had been greatly reduced by X-1P.

2.2.2 Measurement of X-1P Films on the Surfaces of HDD Heads

X-1P is useful to reduce the reaction between PFPE and Al_2O_3. In this test, a usual dipping method was used to prepare X-1P films on the surfaces of the head. The X-1P solution was diluted by 1-methoxynonafluorobutane. The thickness of the X-1P film was controlled by adjusting the solution concentration and the time of the samples being immersed into the solution. Yang et al. [25,32] used a concentration of 1,000 ppm and an immersing time of 5 min, and all heads were cleaned by washing with detergent and deionized (DI) water prior to deposition in order to remove organic contaminants adsorbed from the environment. As shown in Fig. 14, molecular ion (m/z = 1,002) of X-1P was found in the spectroscopy of TOF-SIMS. It proved the existence of X-1P film on the slide. The ion spikes of m/z = 1,052, 1,102 corresponds to m-trifluoromethyl-phenoxy. X-1P film was uniform on the slider surface because the colors of the CF_3 (imaging of TOF-SIMS), on different micro-areas (100 μm by 100 μm) were homogeneous and very close to each other as shown in Fig. 15 where a, b, c, and d were from different positions of the air bearing surface of the HDD head.

2.2.3 Effects of the X1-P Film Thickness on the Stiction

The CSS test was used to evaluate the tribological properties of X1-P film. Aluminum substrate disks with two zones characterized by different surface roughness were used in the CSS test. The smooth zone with a roughness Ra of 1.5 nm allowed the recording head to fly without contact with the disk surface in the writing and reading process. The landing zone was a rougher annular region with discrete bumps produced by a laser near the center of the disk where the head resided when the disk stopped as shown in Fig. 16. The laser

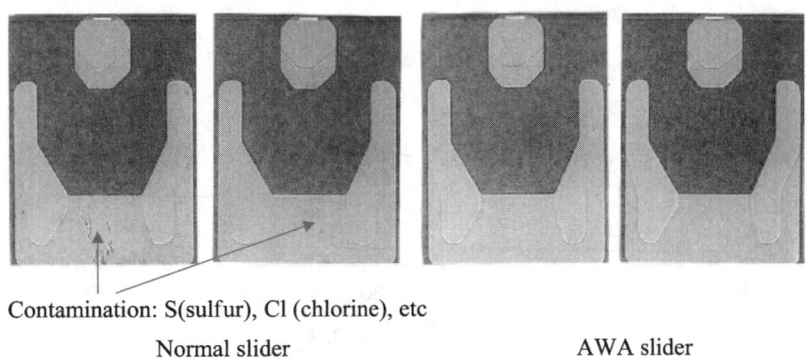

Contamination: S(sulfur), Cl (chlorine), etc

Normal slider   AWA slider

Fig. 11—The slider surfaces after a 20 K CSS test.

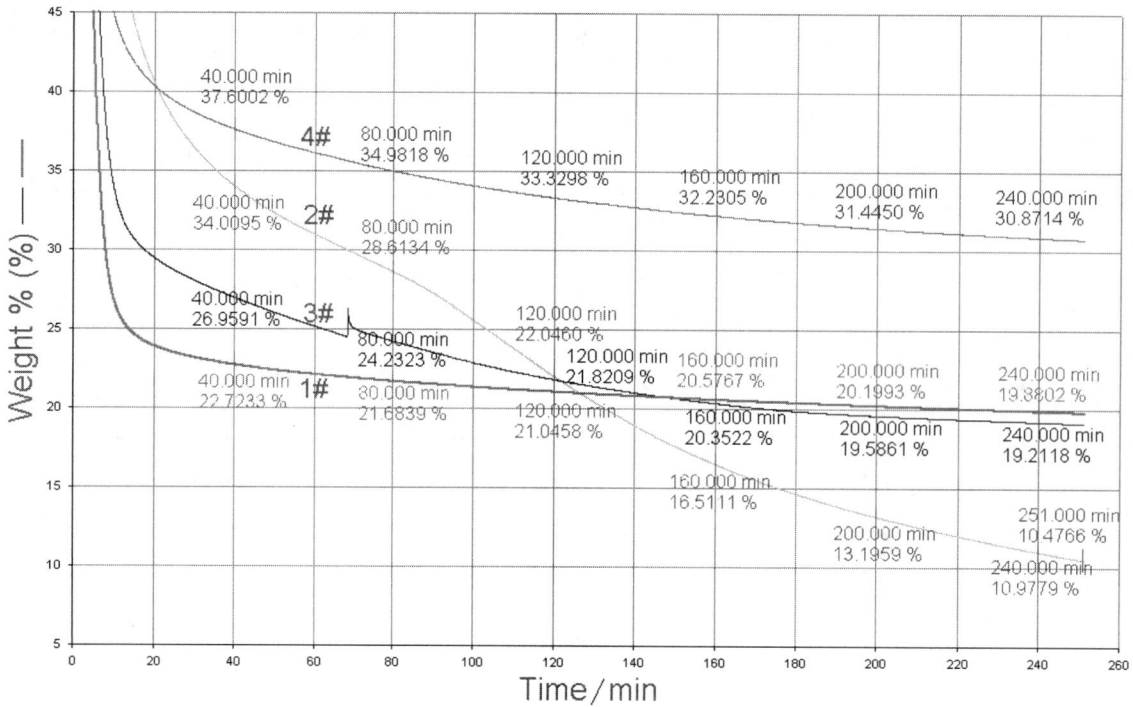

Fig. 12—TGA results for specimens at 220°C: 1–PFPE, 2–PFPE+Al$_2$O$_3$, 3–PFPE+X1-P, 4–PFPE+Al$_2$O$_3$+X1-P.

texture bumps in the landing zone are of 11.2 nm high, 59.8 nm in depth, and 10.313 μm in diameter to reduce the stiction between the disk and the head. The disk rotated at 5,400 r/min. The preload on the head was 2.56 g. The resolution of the strain-gage sensor for measuring the friction force is less than 0.03 g. The ambient temperature was 20°C and the humidity was 50 %. Five pieces of heads with different thicknesses of X-1P film measured by TOF-SIMS were tested by CSS as shown in Fig. 17. The stiction for the surface without X-1P was a little larger than that with X-1P at a thickness of 1.23 nm and 1.35 nm in the immediate CSS process. They were nearly the same after the 2 hr or 24 hr parking as shown in Fig. 17. As the X-1P film is thicker than 1.83 nm [Figs. 17(d) and 17(e)], the stiction in the immediate CSS process becomes much larger than that in the (a), (b), and (c). However, the stiction would converge to similar values for different thicknesses of X-1P film after the 2 hr or 24 hr parking. Therefore, the X-1P film with the thickness around 1.3 nm is favorable to the decrease of stiction in the immediate CSS process.

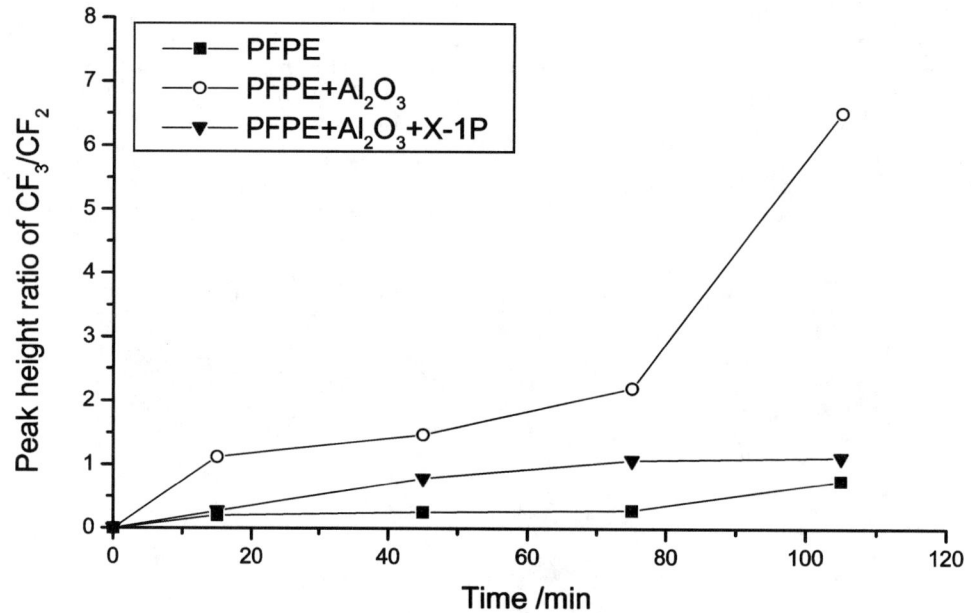

Fig. 13—Computation result of IR adsorption peaks for CF$_3$/CF$_2$ group.

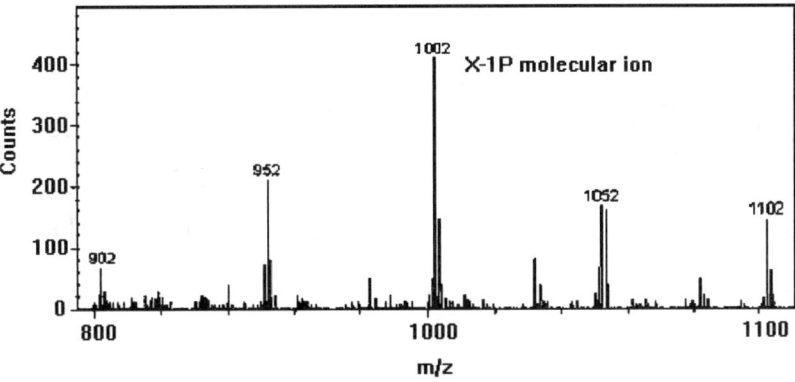

Fig. 14—Dipping-draining method.

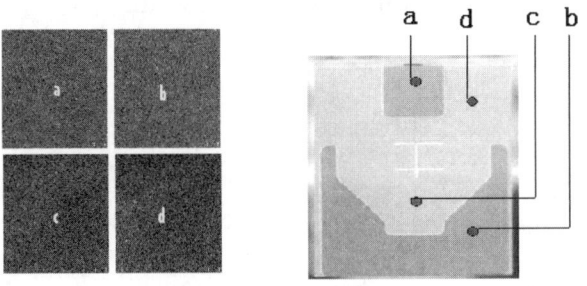

Fig. 15—CF_3 imaging on surface of HDD head in the area 100 by 100 μm.

Surfaces of heads were checked after the CSS tests by an optical inspection. For the heads with X-1P films about 1.3 nm thick, both clear surfaces of the head and disk in the landing zone were obtained after the CSS test as shown in Figs. 18(a) and 18(b). Nevertheless, there were some liquid droplets of PFPE on the surfaces of heads with X-1P films at 2.53 nm thick as shown in Figs. 18(c) and 18(d). It also can be seen that there were footprints on the surfaces of disks where the head had parked for 24 h as shown in Figs. 18(e) and 18(f). It indicates that the thick X-1P film will result in the dewetting of PFPE.

2.2.4 Summary

The results from TGA and FTIR tests indicate that PFPE will be more stable when X-1P is in the presence of the system

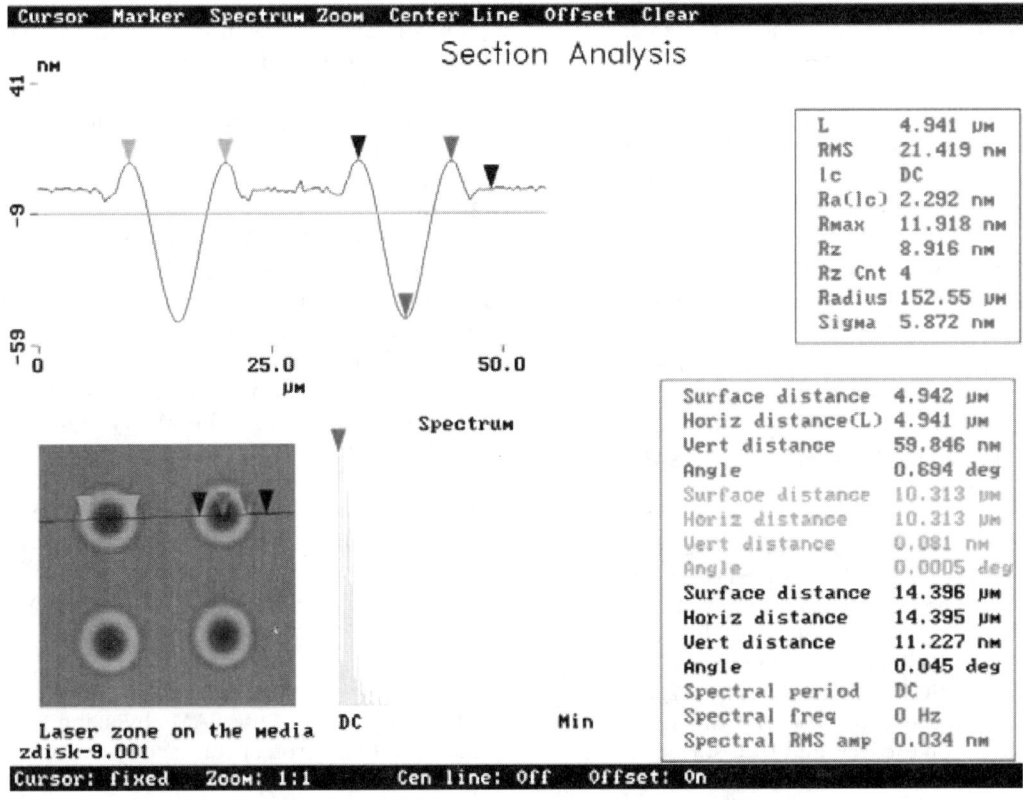

Fig. 16—The topograph of the landing zone of HDD disks.

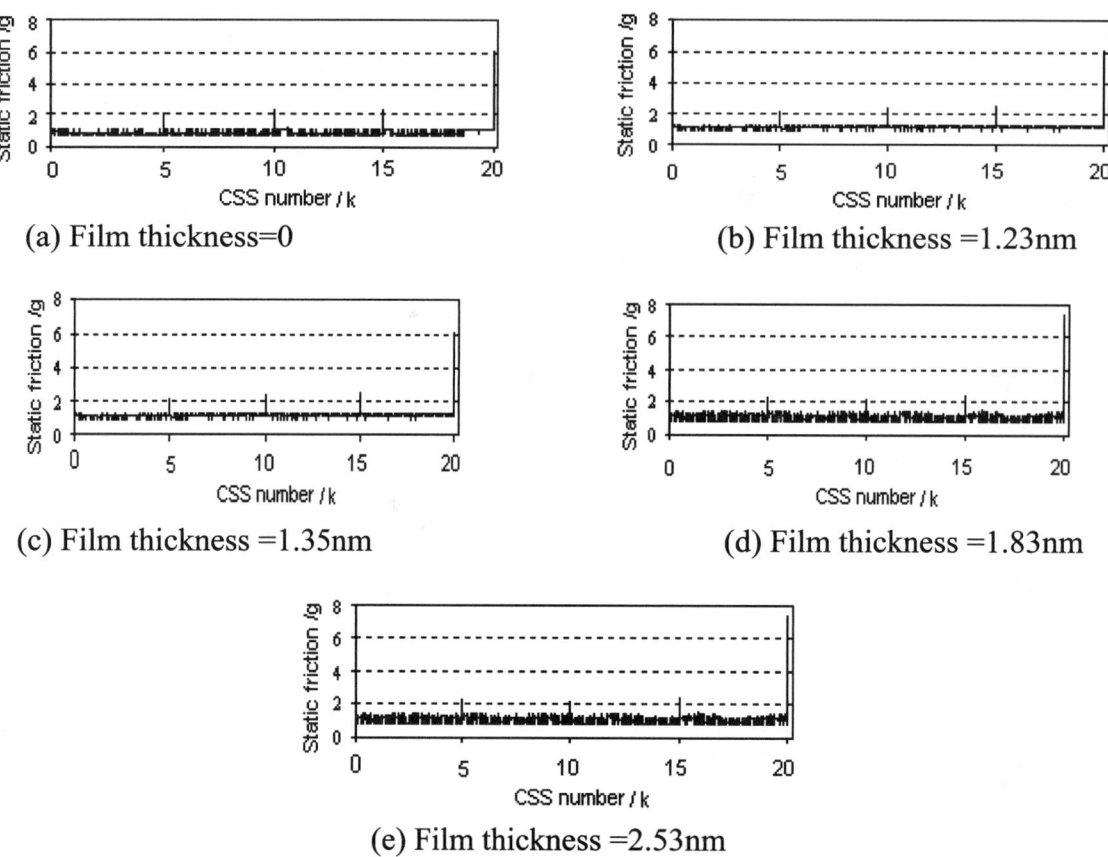

Fig. 17—The surface of landing zone after CSS test with thin X-1P film on head surface.

even if Al_2O_3 exists. If the X-1P film on the surface of an HDD head is thicker than 2 nm, it will induce a larger stiction in the immediate CSS process and make the lubricant PFPE dewetted. Therefore, the thinner X-1P films are preferred for they not only reduce the reaction between PFPE and Al_2O_3 but also improve lubrication properties of the HDD head surface.

2.3 Stiction and Friction Properties of Self-Assembled Monolayer on the Magnetic Head Surface

2.3.1 Characteristics of SAMs on the DLC Surface

In order to study adhesive behavior of organic films on the surface of magnetic heads, different kinds of SAMs have been tested as an overcoat on a diamond-like carbon (DLC) film which is coated on the surfaces of magnetic heads.[26] The SAMs are of 1H,1H,2H,2H-perfluorodecyltriethoxy-silane ($CF_3(CF2)_7(CH_2)_2Si(OCH_2CH_3)_3$, FTE), and three kinds of fluoroalkylsilane (FAS) with a different molecular chain length, that is, trichlorosilane ($CF_3(CH)_2SiCl_3$, FPTS), 1H,1H,2H,2H-perfluorooctyltrichlorosilane ($CF_3(CF_2)_5(CH_2)_2SiCl_3$, FOTS), and 1H,1H,2H,2H-perfluorodecyltrichlorosilane ($CF_3(CF_2)_7(CH_2)_2SiCl_3$, FDTS).

The reaction between the FTE and DLC surface is schematically shown in Fig. 19. Hu et al. [26,34], Ryan et al. [35], and Choi et al. [36], assumed that the Si-X (X=Cl, OR, and NR) groups on the silane molecule would change into Si-OH groups in a hydrolysis process, and a Langmuir-like monolayer was formed on the hydrophilic surface with the presence of a thin water film under ideal conditions. A cross-linked surface assembly was then formed by condensation reactions between the –OH groups on the silane molecules and those on the oxide surface or on neighboring silanes [34–36].

The FTE SAMs have a good hydrophobic property. Chio et al. [36] have compared the variation of contact angles with immersing time in a neat FTE and a 100 mM FTE solution. The contact angles of water and hexadecane increased to about 110° and 73° from the initial value 76° and 36°, respectively, after 24 h immersion. Their works also indicate that the adsorption rate in 100 mM FTE solution is slightly faster than that in neat FTE.

From different solutions, Hu et al. [26,34] made SAMs over DLC films on magnetic heads. The samples were pulled out at different immersing times, cleared using an ultrasonic cleaner in octane solution, and then washed by DI water. After that, the samples dried by blowing nitrogen were annealed in a cleaning box for 30 min at a temperature of 120°C.

The X-ray photoelectron spectra were obtained to examine C_{1s} peaks of the SAMs by using PHI Quantum 2000 X-ray photoelectron spectroscopy (XPS) with a Al K_α X-ray (1,468.6 eV) anode source operated at 15 kV. The diameter of the X-ray beam was 100 μm. A 1.5 nm thick DLC film was coated on a 1.5 nm thick intermediate Si layer which was

CHAPTER 11 ■ TRIBOLOGY IN MAGNETIC RECORDING SYSTEM

Fig. 18—Head and media surface status for the heads with thicker X-1P overcoat after CSS test.

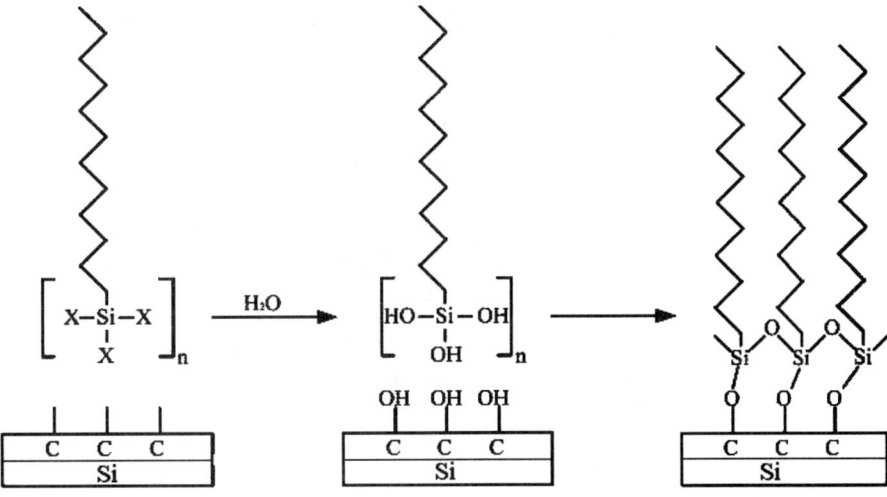

Fig. 19—Scheme of the formation of the FTE SAM on a hydroxylated DLC surface.

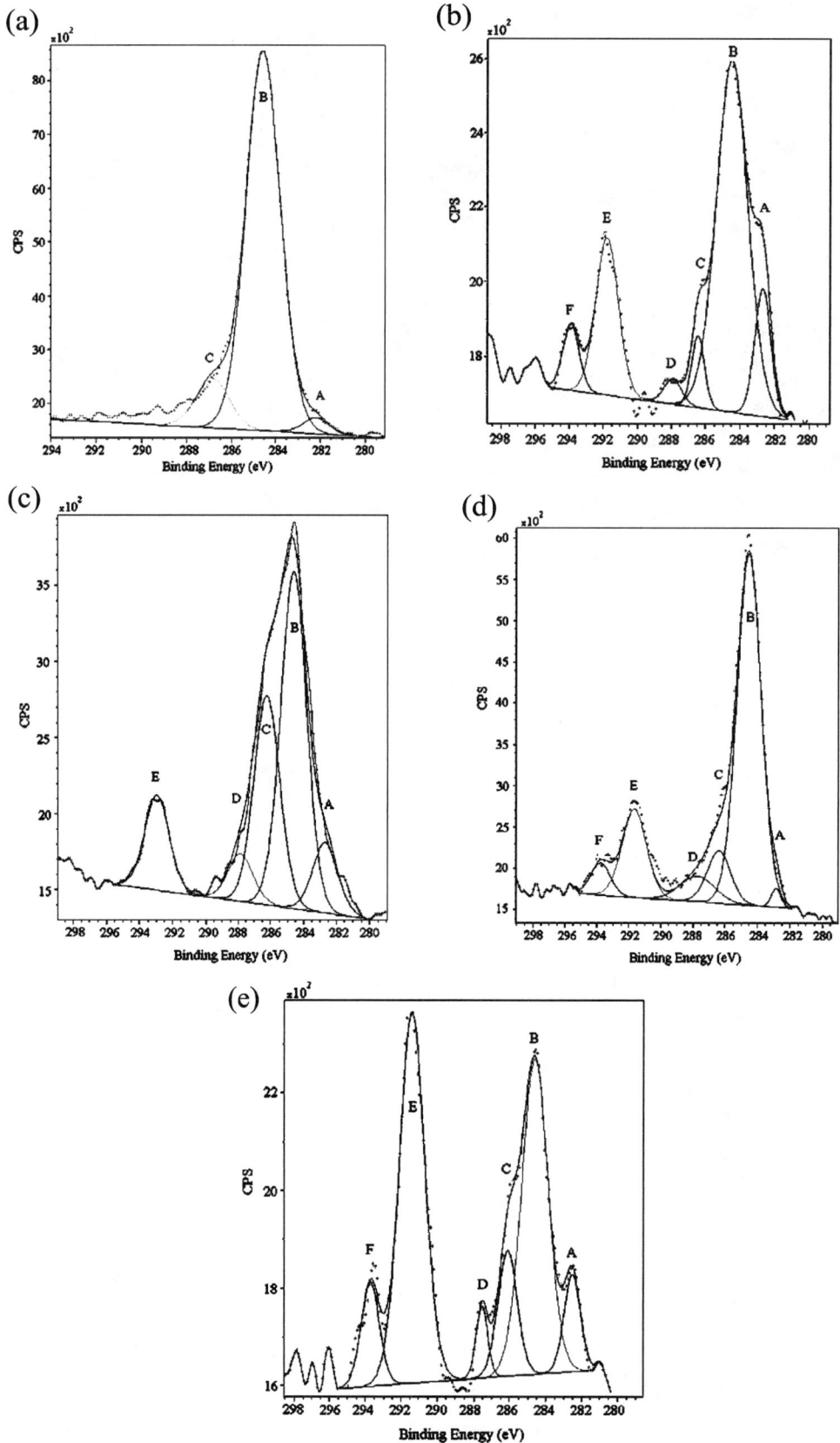

Fig. 20—X-ray photoelectron spectra of a C_{1s} peak of (a) 1-DLC film and the Si layer on the surface of magnetic head, (b) the FTE SAMs on the DLC film, (c) the FPTS SAM on the DLC film, (d) the FOTS SAM on the DLC film, and (e) the FDTS SAM on the DLC film.

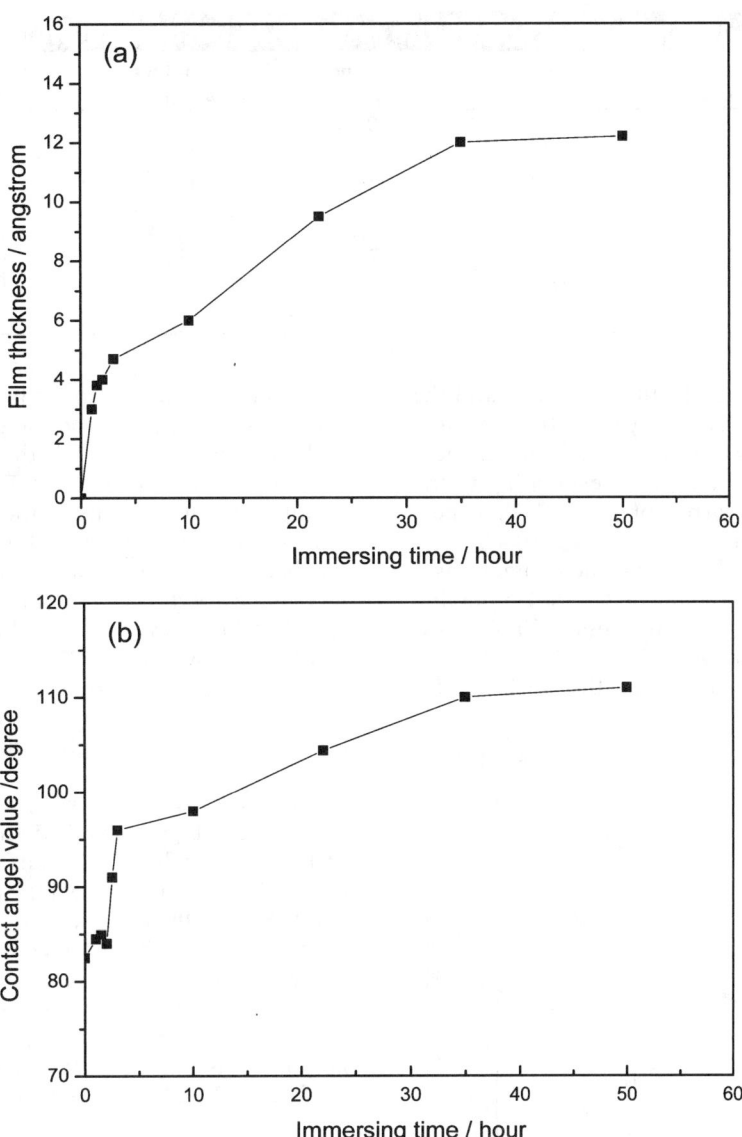

Fig. 21—The thickness and water contact angle of FTE SAMs on the DLC surfaces as a function of the immersing time in a 10 mM FTE solution at temperature of 20°C: (a) film thickness of FTE SAM, and (b) water contact angle.

coated on the surface of the slider (Al_2O_3). SAMs were formed on the surface of the DLC film. Figure 20 shows XPS survey spectra of the C_{1s} peaks of (a) a DLC film plus a intermediate layer of Si on the surface of a magnetic head, (b) a SAM of PTE on the surface of DLC film, (c) a SAM of FPTS on the surface of DLC film, (d) a SAM of FOTS on the surface of DLC film, and (e) a SAM of FDTS on the surface of DLC film. As shown in Fig. 20(a), the C_{1s} peaks of the DLC film and Si layer coated on the slider surface are at the banding energy of 282.188 eV (A), 284.552 eV (B), and 286.825 eV (C), corresponding to the bonds C–Si, C–C, and C–O, respectively. They are attributed to the DLC and Si layers. For the SAM of FTE [Fig. 20(b)], the new peak at about 287.9 eV (D) corresponding to the bond C–O–Si appears, which indicates that the SAMs have been formed on the DLC surface, and peaks at 291.8 eV (E) and 293.9 eV (F) corresponding to the band $-CF_2$ and $-CF_3$ of FTE molecules, respectively, are also detected [34,35]. For the SAMs of FPTS, FOTS, and FDTS, the peak D, E, F also can be observed as shown in Figs. 20(c)–20(e).

The solution concentration also has an influence on the formation of the SAMs. The maps of X-ray photoelectron spectra show that the C(1s) peaks for the samples in 1 mM and 5 mM FTE solutions with 24 h immersing are similar to those in Fig. 20(a), which indicates that there was hardly any SAMs on the DLC surface. The thickness of FTE SAMs and the water contact angle value increased with the immersing time. Hu et al. [26,34] obtained FTE SAMs with a thickness of 1.2 nm and a water contact angle value of about 110° by immersing a sample in a 10 mM FTE solution for 35 hours as shown in Figs. 21(a) and 21(b).

In order to reduce the immersing time and to examine the influence of the chain length of the adsorbed molecules on the formation of SAMs, Hu et al. [26,34] has investigated the tribological properties of the SAMs of FPTS, FOTS, and FDTS, which have a different molecular chain length as shown in Table 1.

TABLE 1—The properties of the FATS SAM on the magnetic head.

No.	Reagent	Chain Length	Immersing Time	Thickness (nm)	Contact Angle (°)	Roughness (nm)	Description of SAMs
1	FPTS	3 C	1 h	0.46	94.0	0.177	Monolayer
2	FOTS	8 C	1 h	0.8	107.6	0.125	Monolayer
3	FOTS	8 C	24 h	1.0	111.7	0.439	Defective double layers
4	FDTS	10 C	1 h	1.2	110.4	0.144	Monolayer
5	FDTS	10 C	12 h	2.3	130.7	0.123	Double layers
6	FDTS	10 C	48 h	3.7	123.9	0.541	Defective multilayers

For the three sorts of molecules, the film thickness and the water contact angle value of the SAMs grow with increasing chain length. In the immersing time of 1 h, the film thickness obtained is 0.48 nm, 0.8 nm, 1.2 nm, corresponding to the molecules with the number of carbon of 3, 8, and 10, respectively, and the contact angle value is 94°, 107.6°, 110.4°.

The variations of film thickness and the contact angle value of FDTS SAM are also closely related to the immersing time as shown in Fig. 22. The film thickness of FDTS SAM increases to about 20 Å within 2 h, which is close to the molecular chain length of FDTS, and then the thickness increases very slowly up to the immersing time within 12 h. If the immersing time is too long, the thickness of FDTS film will increase to about 37 Å, that is, another layer is formed. The water contact angle of FDTS SAM increases fast to about 110° within 60 min, and then it changes very slowly with the immersing time. However, once the second layer has been formed, the contact angle will decrease. The contact angle for the tested samples with the longest chain length, FDTS SAM, reaches 130.7° with the immersing time of 12 h as shown in Fig. 22(b). However, the long immersing time will result in a rough surface topography and a thicker film as shown in Fig. 23, which may be harmful for it will reduce the effective fly height of the magnetic head.

2.3.2 Tribological Properties of the SAMs

As shown in Figs. 24 and 25 the longer chain length of the $-CF$ group and the smaller surface roughness lead to considerably improved tribological properties of the magnetic head. Especially, the stiction of the magnetic head coated with the FDTS SAM obtained in an immersing time of 12 h, decreased to 2.207 g in CSS test. However, the number of film layers hardly has an influence on stiction or cleanliness of the surface. Besides, the more layers the film has, the larger the distance is between the disk and head. Therefore, a uniform FATS SAM is favorable to magnetic heads.

2.3.3 Other Properties of SAMs

Figure 26 gives the corrosion percentage of magnetic heads immersing for 4 min in a 25°C oxalic acid solution with a concentration of 0.05 mol/L. The Sample 0 is the head without any SAMs, and the samples from number 1 to 6 are the heads with SAMs listed in Table 1. For the heads without SAMs and those with FPTS SAMs but in shorter molecular chain length, the corrosion rate is 64 % and 60 %, respectively. However for the heads with FDTS SAMs, the corrosion rate is reduced to around 40 %. The experimental results indicate that the longer molecular chain length and the more layers of SAMs contribute largely to enhancing the corrosion resistance of magnetic heads in oxalic acid. And the corrosion percentage of the magnetic head coated with the FDTS SAM at 12 h decreases to 41 %.

The electron charge on the magnetic head was measured by a Guzik instrument. The parameter R^2 is chosen to judge the level of the surface electron charge of magnetic heads. When the R^2 value is close to 1, the electron charge on the magnetic head is close to zero. As shown in Fig. 27, with increasing the immersing time, the FOTS and FDTS SAMs have enhanced the ability of magnetic heads to resist electron charge adsorption, resulting in an R^2 value of about

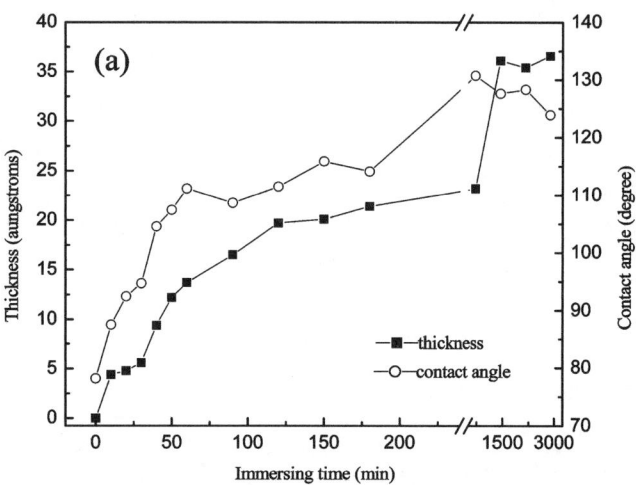

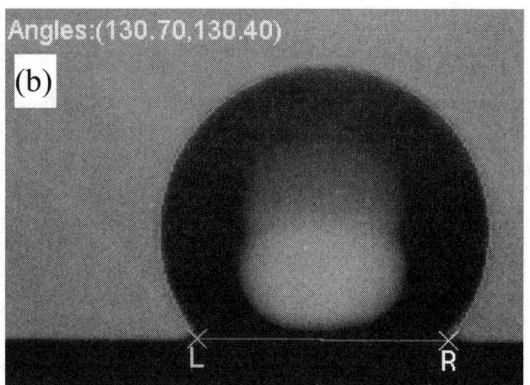

Fig. 22—Contact angle and film thickness of FDTE SAMs on the DLC surfaces as a function of the immersing time in a 10 mM FDTE solution at the temperature of 60°C: (a) contact angle value and film thickness, (b) water contact angle of the FDTS SAM with an immersing time of 12 h.

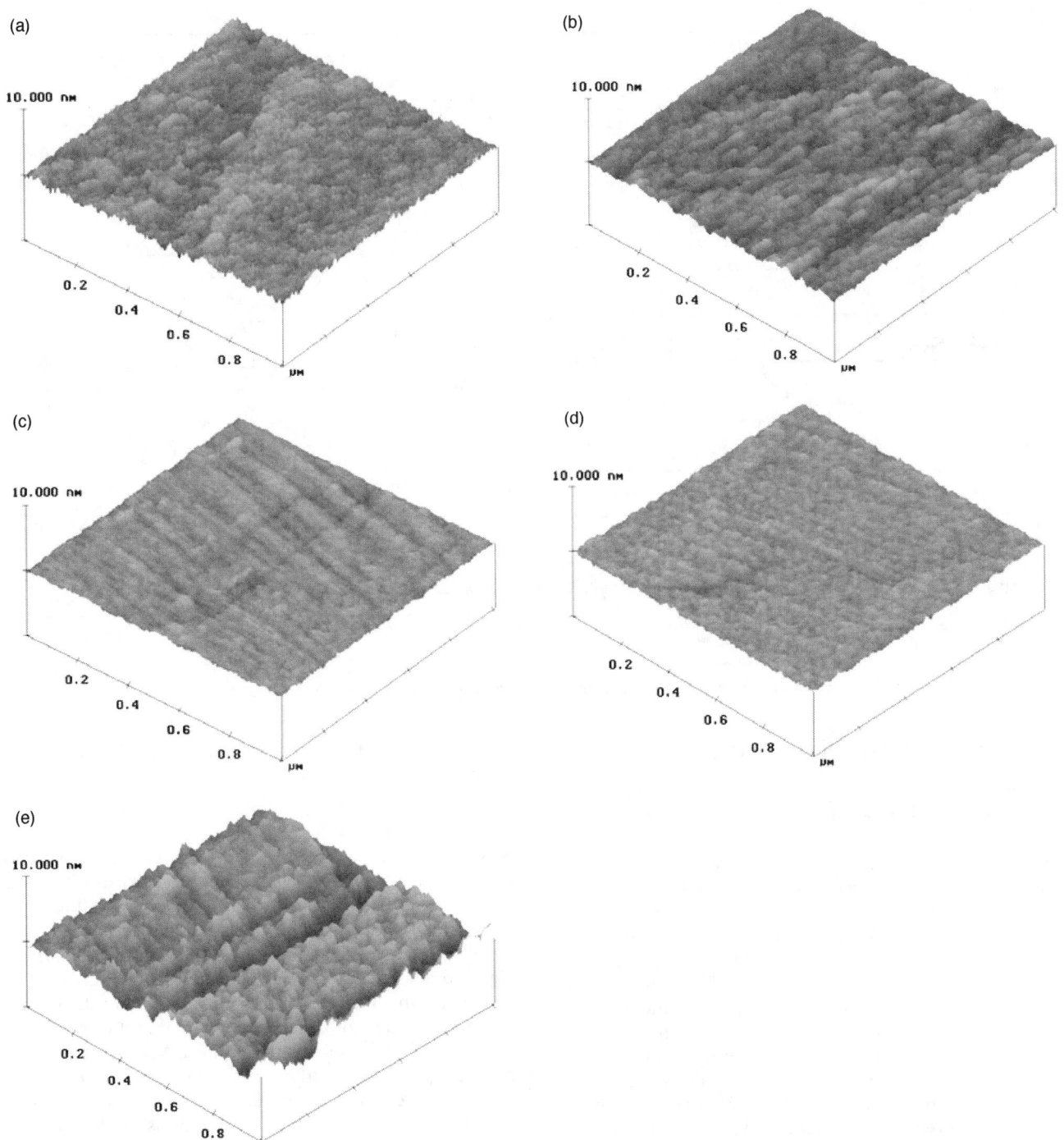

Fig. 23—The AFM surface topography of FDTE SAMs in a 10 mM FDTE solution at the temperature of 60°C with an immersing time of (a) 10 min, (b) 30 min, (c) 50 min, (d) 2 h, (e) 48 h.

0.97. The property is hardly influenced by the molecular chain length, the film layers, and the surface roughness, but the uniformity of a FATS SAM was important considering that R^2 decreases when the immersing time was less than 50 min.

2.3.4 Summary

The FAS SAMs have been grafted on magnetic heads by a robust covalent bond. The thickness, morphology, and the tribological properties of the FAS SAMs are influenced by the concentration of the solution and the immersing time. The FAS SAMs led to a considerable improvement in the magnetic head performances, such as tribological properties, corrosion-resistant property, and electron charge adsorption-resistant property. The longer the CF_2 chain is, and the smoother SAM surface is, the better the properties of the magnetic heads will be.

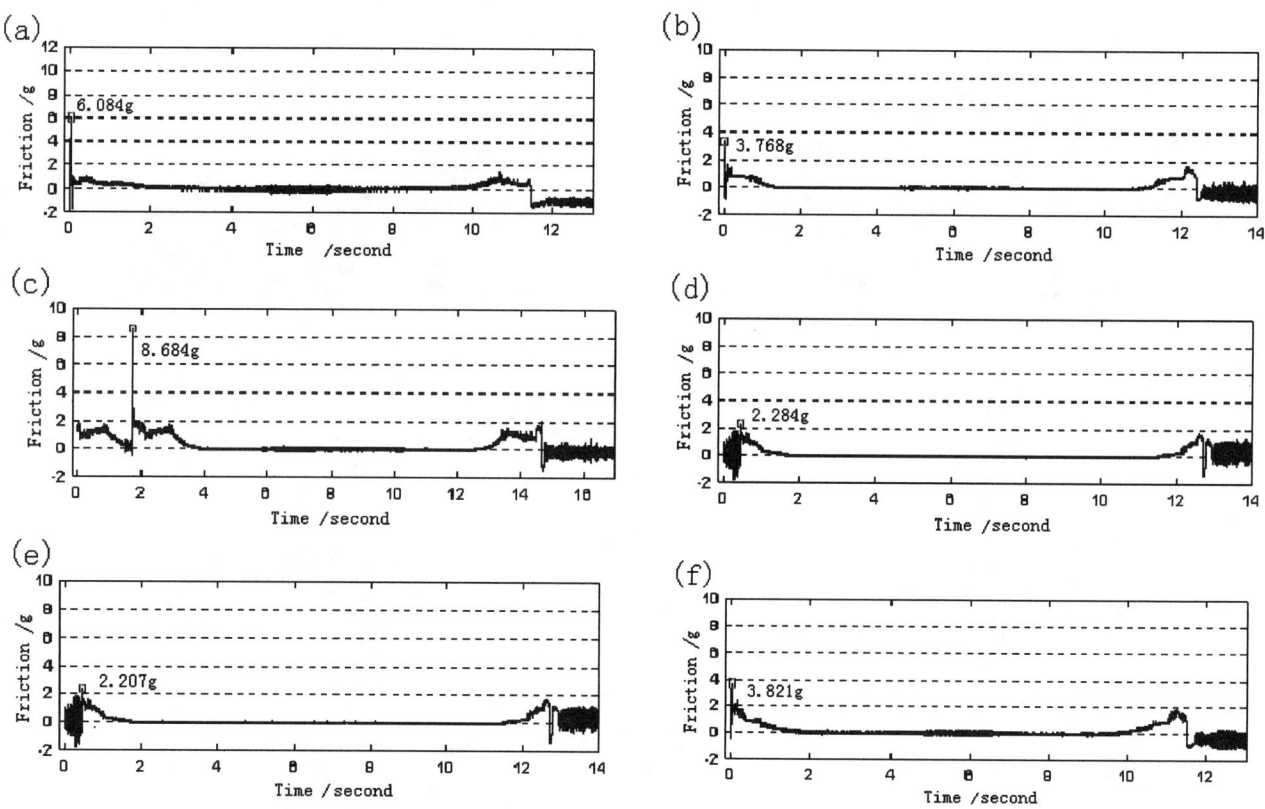

Fig. 24—The friction at the 20,001th CSS cycle of sample (a) No. 1, (b) No. 2, (c) No. 3, (d) No. 4, (e) No. 5, (f) No. 6.

3 Lubricants on Hard Disk Surface

A commercial hard disk is composed of several layers, including the substrate, underlayer, magnetic layer, carbon overcoat, and nanoscale lubricanting film [37]. Due to the rapid increase in recording density, the space between the disk and magnetic head, or the so-called fly height, has been reduced to a few nanometers, resulting in more frequent head-disk contacts [38]. In case of contact, the lubricant may be depleted from the contacted region, which may lead to disk failure if the lubricant is unable to reflow into the depleted contacted zone within the time scale of contacts [39]. Thereby, the mobility of nanoscale lubricating films plays a significant role in the HDI design. The stability of lubricating films is also of great importance for the HDI design since an unstable lubricant film would lead to an uneven surface morphology, which directly deteriorates the flying stability of a magnetic head [40]. Moreover, the degradation of lubricant has been a research subject receiving great attention since it concerns the durability of the lubricant and the service life of the hard disk driver (HDD) itself. The discussions in this section are therefore focused on the subjects of the nanostructure, mobility, and stability of nanoscale lubricating films, as well as the lubricant degradation.

At the present, perfluoropolyether or PFPE, a random copolymer with a linear principal chain structure, has been widely used in HDD as the lubricant. Its chemical structure can be described by $X-[(OCF_2CF_2)_p-(OCF_2)_q]-O-X$ ($p/q \cong 2/3$), with the average molecular weight ranging from 2,000 to 4,000 g/mol. Here, the symbol $-X$ denotes the end-bead (eb), corresponding to $-CF_3$ (nonfunctional) in PFPE Z-15, or $-CF_2CH_2OH$ (functional) in PFPE Z-DOL, respectively.

3.1 Structure of PFPE Films

Novotny et al. [41] used p-polarized reflection and modulated polarization infrared spectroscopy to examine the conformation of 1–1,000 nm thick liquid polyperfluoropropylene oxide (PPFPO) on various solid surfaces, such as gold, silver, and silica surfaces. They found that the peak frequencies and relative intensities in the vibration spectra from thin polymer films were different from those from the bulk, suggesting that the molecular arrangement in the polymer films deviated from the bulk conformation. A two-layer model has been proposed where the films are composed of interfacial and bulk layers. The interfacial layer, with a thickness of 1–2 monolayers, has the molecular chains preferentially extended along the surface while the second layer above exhibits a normal bulk polymer conformation.

Mate and Novotny [42] studied the conformation of 0.5–13 nm thick Z-15 on a clean Si (100) surface by means of AFM and XPS. They found that the height for PFPE molecules to extend above a solid surface was no more than 1.5–2.5 nm, which was considerably less than the diameter of gyration of the lubricant molecules ranging between 3.2–7.3 nm. The measured height corresponds to a few molecular diameters of linear polymer chains whose cross-sectional diameter is estimated as 0.6–0.7 nm. The experimental results imply that molecules on a solid surface have an extended, flat conformation. Furthermore, they brought forward a model, as shown in Fig. 28, which illustrates two

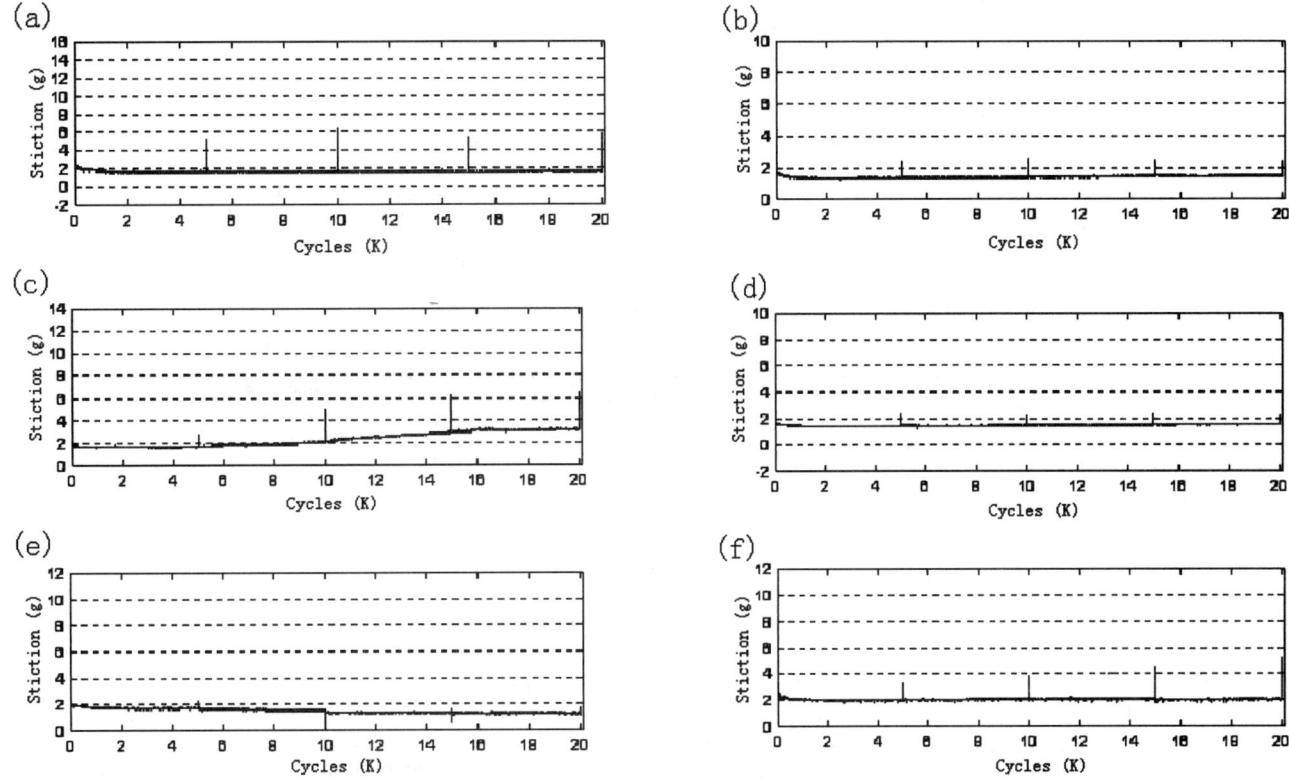

Fig. 25—The stiction during a 20,000 CSS cycle of sample (a) No. 1, (b) No. 2, (c) No. 3, (d) No. 4, (e) No. 5, (f) No. 6.

extreme conditions for a nonfunctional linear molecule on a solid surface. At one extreme, when a strong attraction exists between the molecule and the solid, the molecules lie flat on the surface, reaching no more than their chain diameter δ above the surface. At the other extreme, when the molecule/solid attraction is weak, the molecules adopt a conformation close to that of the molecules in the bulk state with the microscopic thickness equal to about the diameter of gyration $2R_g$.

Spreading of PFPE films on amorphous carbon surfaces

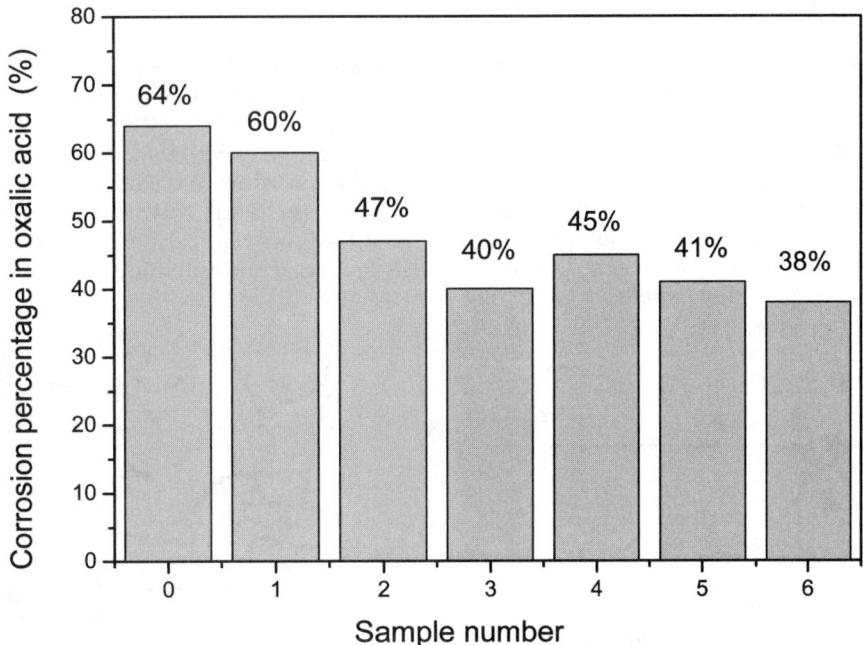

Fig. 26—Corrosion test of the magnetic heads immersing in oxalic acid solution with a concentration of 0.05 mol/L for 4 min at a temperature of 25°C.

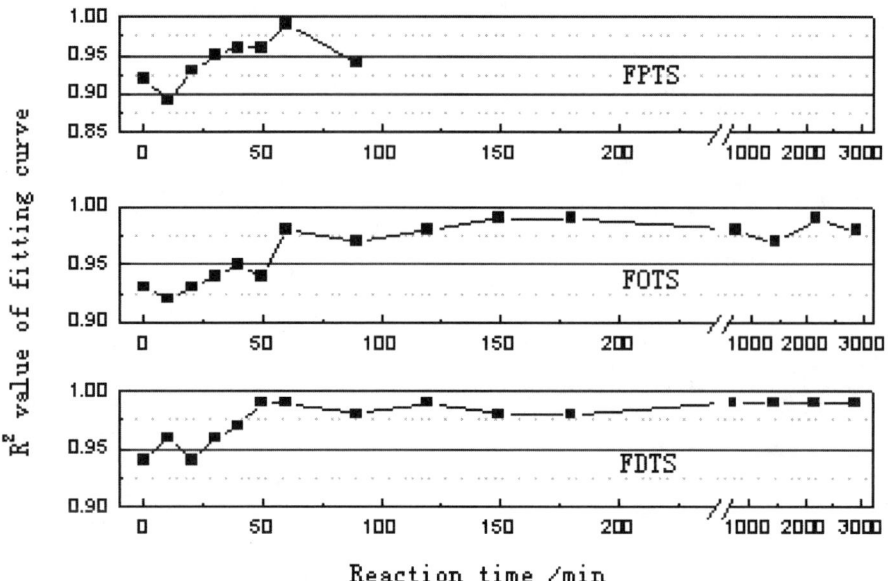

Fig. 27—The R² value as functions of immersing time for a FATS SAM on the magnetic head.

was studied by Ma et al. [43] using SEM (scanning microellipsometry). They found that Z-dol molecules in spreading behaved differently from Z–15 molecules. A complex layering develops during the spreading of Z-dol molecules, with the layer thickness in the order of gyration radius. However, the layers do not appear in equal thickness since the second layer is nearly twice thick as the first layer. That interesting phenomenon is considered to be related to molecular conformation on amorphous carbon surfaces. Based on the observations, a possible molecular conformation of PFPE films is therefore proposed. In the first layer adjacent to the solid, the Z-dol molecules have a coil conformation with the endbeads anchored to the carbon surface via strong attractive interactions, resulting from dangling bonds and surface functionalities. Above the first layer, the Z-dol molecules form a dimmer structure, most likely through hydrogen bonding, forming a layer with a thickness nearly double that of the first layer.

It is worth to note that the conformation model of Z-dol is speculated upon based on the observations of spreading, but detailed molecular arrangements are difficult to know owing to the limitation of instruments. Computer simulations such as the Monte Carlo (MC) and molecular dynamics (MD) were also performed in expecting to detect such information and reach a deep insight to the molecular conformation. Guo [44] and Li [45] carried out MD simulations using a coarse-grained bead-spring model to study the conformation of functional PFPE films. Results show that molecules near the surface exhibit a cluster structure in oblate form, but it recovers a spherical shape as the distance from the surface increases. The density profile of the functional endbeads for the PFPE molecules shows a characteristic oscillation as a function of the distance from the surface, revealing a layering structure, as illustrated in Fig. 29, which originates from the endbeads-endbeads and endbeads-surface couplings. The thickness of the two neighboring layers is approximately $4R_g$, which attributes to the dimmer structure formed from significant endbeads coupling between layers.

3.2 Lubricant Mobility

As a crucial factor that dominates the behavior of lubricant flow, the mobility of PFPE molecules has been studied extensively in both experiments and simulations, through observing the spreading of the lubricant on solid substrates. Investigators, including Novotny [46], O'Connor et al. [47], Min et al. [48], and Ma et al. [49], in collaboration with IBM scientists, carried out systemic experimental studies on spreading

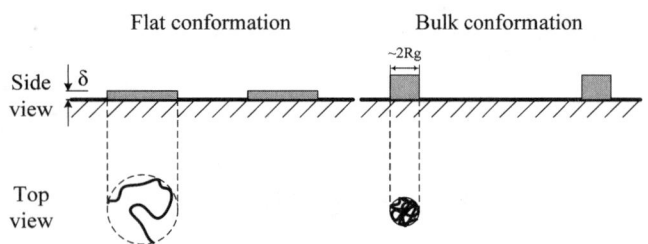

Fig. 28—Schematic representation of two extreme polymer conformations at the surface of the solid at low surface coverage: δ is the cross-sectional diameter of the polymer chain, and R is the radius of gyration of the molecule in the bulk [42].

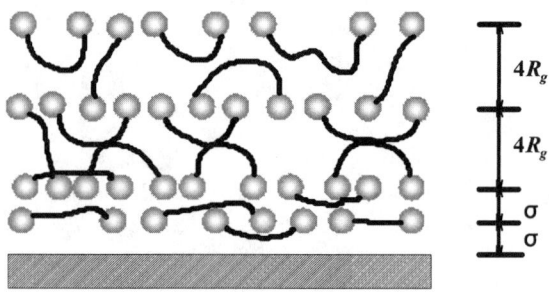

Fig. 29—Sketch for the structure of functional PFPE molecules [45], where Rg is the radius of gyration of PFPE molecules and σ is the diameter of PFPE molecules.

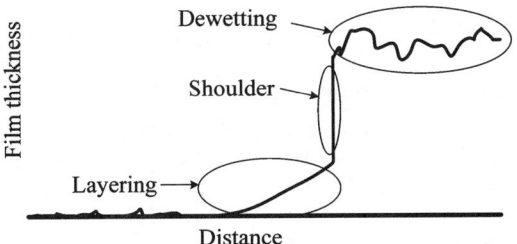

Fig. 30—SME thickness profiles for PFPE Z-dol on silica wafers [48].

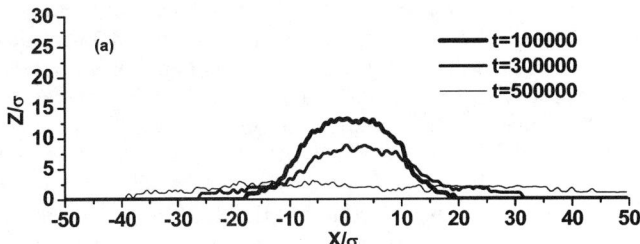

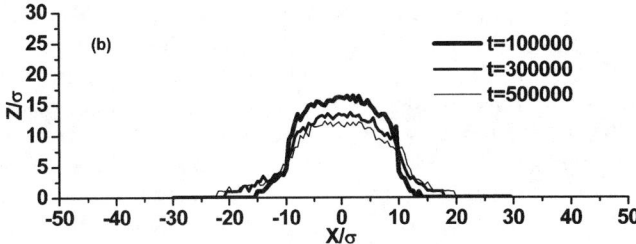

Fig. 31—Thickness profiles at time t=100,000, 300,000, and 500,000 MD steps, where X is in the direction of PFPE spreading, Z is in the direction of PFPE falling down; (a) nonfunctional PFPE; (b) functional PFPE [51].

behavior of PFPE films on silica and carbon surfaces. From the observations using SEM and XPS, it is found that the spreading of nonfunctional (Z-15) and functional (Z-dol) PFPE exhibits distinctly different thickness profiles and mobility performances. Spreading of Z-15 occurs mainly via diffusion-like movement of the fast moving front or foot of gradually decreasing thickness, and displays a greater mobility than Z-dol. The spreading profile for Z-dol, on the other hand, shows a characteristic shoulder by forming an apparent monomolecular "anchored" layer, which separates out from the initial film layer with a sharp boundary (Fig. 30).

The experiments further reveal that the spreading of functional PFPE proceeds in a layering fashion, which is apparently different from that of nonfunctional PFPE. It is reported [49] that in spreading of Z-dol multiple layers develop from the liquid front, but each layer exhibits a different thickness. The thickness of the first layer is close to the diameter of gyration of the polymer in bulk and the second layer is nearly twice as thick as the first layer. No such layering was observed in the case of Z-15. Instead, the liquid front evolves smoothly with time, leading to a gradual and diffusive profile.

The process of spreading of liquid lubricant on solid substrate is driven by hydrodynamic flow and surface diffusion. For PFPE films in nanometer scales, the speed of spreading can be characterized by the surface diffusion coefficient, D, which is dependent on film thickness h. Experiments show that the surface diffusion coefficient D increases as the film thickness decreases down to 1 nm, but below which it becomes independent of PFPE film thickness [46]. The Matano interface method was employed to extract the thickness-dependent diffusion coefficient $D(h)$, directly from the spreading thickness profiles [49]. It is found that $D(h)$ is not a monotonic function of film thickness. As the thickness of Z-dol film increases, the diffusion coefficient $D(h)$ increases initially, reaches a maximum, and then drops abruptly to nearly zero, during which subsequent maxima are also observed, manifesting the layering effect on diffusion coefficients. The peak $D(h)$ values, corresponding to each successive maximum, become lower with increasing thickness, suggesting that the layering effect diminishes as the film approaches bulk properties. For nonfunctional lubricant Z-15, $D(h)$ initially increases, too, as film thickness rises, but it reaches its maximum value at a lower thickness of 1 nm. Thereafter, $D(h)$ decreases monotonically with the increasing film thickness, following approximately a relationship of $D(h) \sim 1/h$.

In addition to the end-bead functionality, other factors, such as molecular weight, temperature and humidity have been studied as to their effects on spreading behavior. The height of the layers in spreading profiles of Z-dol, for example, is found to increase in a power law, $\sim Mn^{0.6}$, where Mn is the mean molecular weight [49]. It is reported that the apparent surface diffusion coefficients of both Z-15 and Z-dol increase with decreasing molecular weight. The mobility of lubricant Z-15 is not significantly affected by humidity, whereas the PFPE with functional endbeads exhibits a dramatic increase in mobility with growing relative humidity [47,48].

Theoretically, the diffusion coefficient can be described as a function of the disjoining pressure Π, the effective viscosity of lubricant, η, and the friction between lubricant and solid surfaces. In relatively thick films, an expression derived from hydrodynamics applies to the diffusion coefficients.

$$D(h) = \int_0^h \frac{(h-z)^2}{\eta(z)} \left(\frac{\partial \Pi}{\partial h}\right) dz \qquad (1)$$

In ultra-thin films, on the other hand, the mechanism of spreading is dominated by surface diffusion so that the diffusion coefficients can be written as

$$D_M(h) = -\frac{V_M}{\gamma} h \frac{\partial \Pi}{\partial h} \qquad (2)$$

where γ denotes the friction coefficient between lubricant molecule and solid surface and V_M is the molecular volume.

It can be seen from the above expressions that the driving force for the spreading process is the disjoining pressure, which is determined in turn by surface energy and its variations with film thickness. Attempts were made by Tyndall et al. [50], to measure the surface energy of Z-dol films on amorphous carbon as a function of film thickness. Results show that the polar component of the surface energy exhibits oscillations as a function of PFPE film thickness, which explains the origins of the complex spreading profiles.

For a better understanding of the molecule mechanism of spreading, the present authors performed MD simulations [51] based on a coarse-grained bead-spring model, in which a droplet of nonfunctional or functional PFPE was allowed to spread on a solid surface and the spreading profiles at different times were recorded. Typical results of the spreading profiles are presented in Fig. 31. We found that for nonfunctional PFPE, molecules in the fluid interface region of the droplet would flow down to form a cap structure on the substrate. Simultaneously molecules at the liquid front will diffuse along the substrate to form a precursor film while the molecules above the diffusion layer will move downward to fill the cavities created by the diffusion. For functional films, on the other hand, the endbeads-endbeads interaction may sharpen the spreading profiles to form a vertical step and slow down the whole spreading process, while the endbeads-surface interaction may weaken the process of molecule diffusion and cavity filling in the precursor film.

3.3 Film Stability

Thermodynamic instability of PFPE films, or the dewetting in particular, has drawn increasing attention recently for its great influence on surface morphology of the film and consequent head/disk contacts. The evidence of dewetting was observed in the experiments [48], as indicated by the fluctuations in spreading profiles shown in Fig. 30. The onset of dewetting was found to be relating to the film thickness, approximately in dimension of the radius of gyration. The film thickness dependence was confirmed in the experiments by Kim et al. [52], using optical microscopy, AFM, and ellipsometry to observe dewetting of Z-dol films on silicon substrates. They found that no dewetting is observed in the thickness range of 0–2.4 nm which corresponds to a monolayer thickness.

The dependence on film thickness is attributed to the dewetting nucleation, which occurs in the 2.5–4.5 nm thickness range via the formation of randomly distributed droplets rather than the formation of holes. When the initial film thickness exceeds 4.5 nm, dewetting is trigged via nucleation of holes instead of droplets, and for film thickness above 10 nm, dewetting develops slowly via hole nucleation at defects. The different dewetting processes observed for different initial film thicknesses can be explained in terms of the variation of disjoining pressure and the inability of the polymer to spread on its own monolayer.

When one takes surface roughness of PFPE films as an indicator of dewetting, it has to be done very carefully. Toney et al. [53] measured the roughness of PFPE films on silicon substrates and carbon overcoats, as the film thickness increased from partial coverage to complete coverage, and finally to a thickness when dewetting occurred. They found that the rms roughness of the PFPE–air interface grew slowly from about 0.2 to 0.4 nm as the film thickness increased from 0.5 to 3.3 nm. This roughness is attributed primarily to thermally excited capillary waves, which is different from Kim's explanation for dewetting induced fluctuations. They reported that the PFPE polymer was in fact able to smooth the surface in such a way that the rms roughness decreased from 0.9 to 0.4 nm for PFPE on a rougher surface of amorphous hydrogenated carbon. The surface smoothing is believed due to the surface tension of the liquid polymer pulling the polymer–air surface flat over the carbon surface.

Izumisawa et al. [54] investigated the thermodynamic stability of Z-dol films via experiments and MC simulations. They predicted that PFPE films became less stable for lower molecular weights and lower surface polarity. For the stable films, no dewetting is observed, but a rougher surface morphology appears as the molecular weight decreases. In unstable conditions, films locally change their thickness to become thermodynamically more stable, causing a rough surface. MD simulations based on a coarse-grained, bead-spring model were performed to study surface morphology of PFPE films on solid surfaces as a function of end-group functionality and polymer length [55]. Results show that a rougher surface morphology is observed for strong endbead functionalities and shorter polymer lengths.

3.4 Mechanisms of PFPE Degradation

Another important issue associated with PFPE application in HDD is the mechanisms of lubricant degradation, which have received much attention in recent years for the concern of lubricant durability and the reliability of HDD operation. In general, lubricant decomposition would take place under high temperature, which is the most considered mechanism for lubricant degradation. It is reported that the thermal decomposition temperature of PFPEs is above 350°C, but sliding experiments showed that the decomposition occurred at a much lower temperature than that in the static thermal process [56]. The sliding may trigger a transient interfacial temperature high enough to cause lubricant decomposition, but this is not expected in the case of low speed sliding, or there are other explanations. It is recognized that the decomposition may be catalyzed by the presence of metal oxide, such as Al_2O_3 commonly used for making magnetic heads. Kasai et al. [57] reported that Lewis acid sites on Al_2O_3 are catalytic and cause PFPE lubricant to decompose at temperatures as low as 200°C. To prevent this, an additive known as XP-1 has been introduced to the disk-head interfaces and has been proven to be effective to slow down the reactions between PFPE and Al_2O_3. Refer to Section 2.2 for further information.

Another possible mechanism for degradation associates with tribochemical reactions at disk-slider interface activated by low-energy electron (exoelectron) emission. Exoelectron emission occurs when surface materials experience plastic deformation, abrasion, fatigue cracking, or phase transitions. As suggested by Kajdas [58], the electron emission would ionize lubricant molecules to form anions which react in turn with the surface already positively charged due to loss of electrons. The evidence for this suggested mechanism has been obtained from studies using mass spectrometry and FTIR spectroscopy [59]. The wear fragments generated during sliding were collected and compared with the products from electron decomposition of the PFPE lubricants. The similarity between the two groups of fragments suggests that the decomposition of the PFPE lubricants is an electron-mediated process. Moreover, UV illustration has been employed to improve the bonding strength of the PFPE on the disk surface, but at the same time it is found that low-energy electrons created by the illustration accelerate the decomposition of the PFPE lubricants, which also confirms the contributions of electron emission to the lubricant degrada-

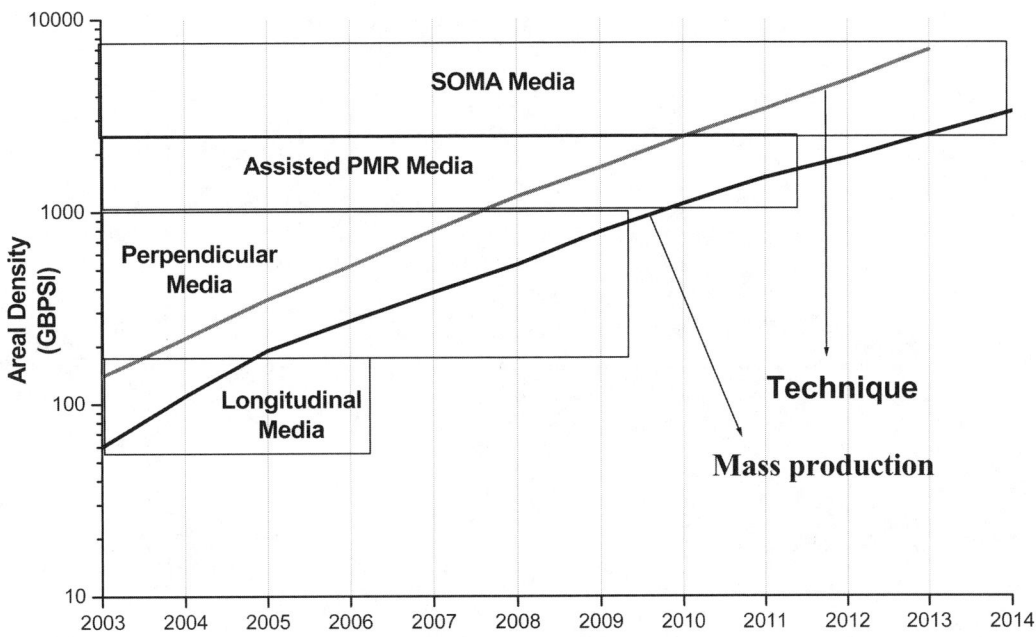

Fig. 32—Roadmap of HDD development.

tion [56]. In addition, the PFPE lubricants certainly experience mechanical shear stress during sliding which can result in polymer scission and lubricant decomposition at a relatively low temperature.

In summary, the degradation of the PFPE lubricants is a complex process involving several mechanisms, including thermal decomposition, catalytic decomposition, tribochemical reactions activated by exoelectron emission, and mechanical scission, which comes into the play simultaneously.

4 Challenges from Developments of Magnetic Recording System

4.1 Overview

The HDD industry is experiencing a significant upturn. The expansion of emerging consumer electronics (CE) markets has fueled the industry's strong recovery while the traditional computing markets continue to grow, yielding the optimistic outlook in the industry's forecast. The areal density has increased exponentially in the past years, while the disk size has been reduced to 0.85 in. in diameter, the rotating rate has increased towards 20,000 r/min, and the maximum internal data rate is more than 100 Mbytes/s. In addition to impacting the HDD demand, the CE markets also have led HDD in the directions of robust mobility and high capacity.

Due to new opportunities and continued computer miniaturizations (e.g., laptops replacing desktops and desktops getting smaller), the HDD is in transition to smaller form factors. The 2.5 in. and smaller disk sizes are rapidly growing. In 2004, a quarter of HDD shipments were 2.5 in. or smaller. By 2008 these smaller form factor drives could count for about one-half of HDD shipments.

New applications in consumer electronics require more storage capacities than the traditional computing section, leading HDD to a high demand on higher capacity drives. As shown in Fig. 32, the areal density roadmap in mass production is hoping to increase from 120 Gbits/in.^{2} in 2005 to more than 2 Tbits/in.^{2} in 2013 (about half of that are already in laboratory demo). In order to face the challenge, the following questions need to be answered: (1) how to improve the writing capability at high frequency and the reading signal quality of magnetic heads with reducing track width; (2) how to successfully implement perpendicular magnetic recording (PMR) technology into production; (3) how to reduce grain size of the magnetic media to less than 10 nm [60] and the surface roughness to about 0.1 nm to improve its quality; (4) how to overcome the problems in HDI and tribology.

Projections for the requirements of disk drivers storing data at densities in excess of 1 Tbits/in.^{2} indicate that the head-disk spacing will be as small as 35 Å [61,62]. The proximity of the read-write head to the disk surface will lead to more frequent high-speed contacts causing probable damage to the surface. When the flying height (FH) is down to about 2 nm, the challenges for flying abilities are how to get high stability of heads, how to get two-dimensional stability of heads, how to solve the problem of the interface dynamics to improve the slider design, and how to improve the anti-wear capacity of the slider and disk with more frequent contacts between them.

Nano-metrology is also a big problem for 1 Tbites/in.^{2} HDD. For the absolute FH testing at 2–3 nm, an advanced optical FH testing at the foundation of the present optical method needs to be developed and the repeatability of 0.15 nm is needed to feed the urgent need of flying height testing technology. Other techniques may be developed as well to realize the repeatability of 0.1 nm for 1 Tbites/in.^{2} The surface and film characterization measuring techniques for the carbon wear and corrosion protective property are also to be developed.

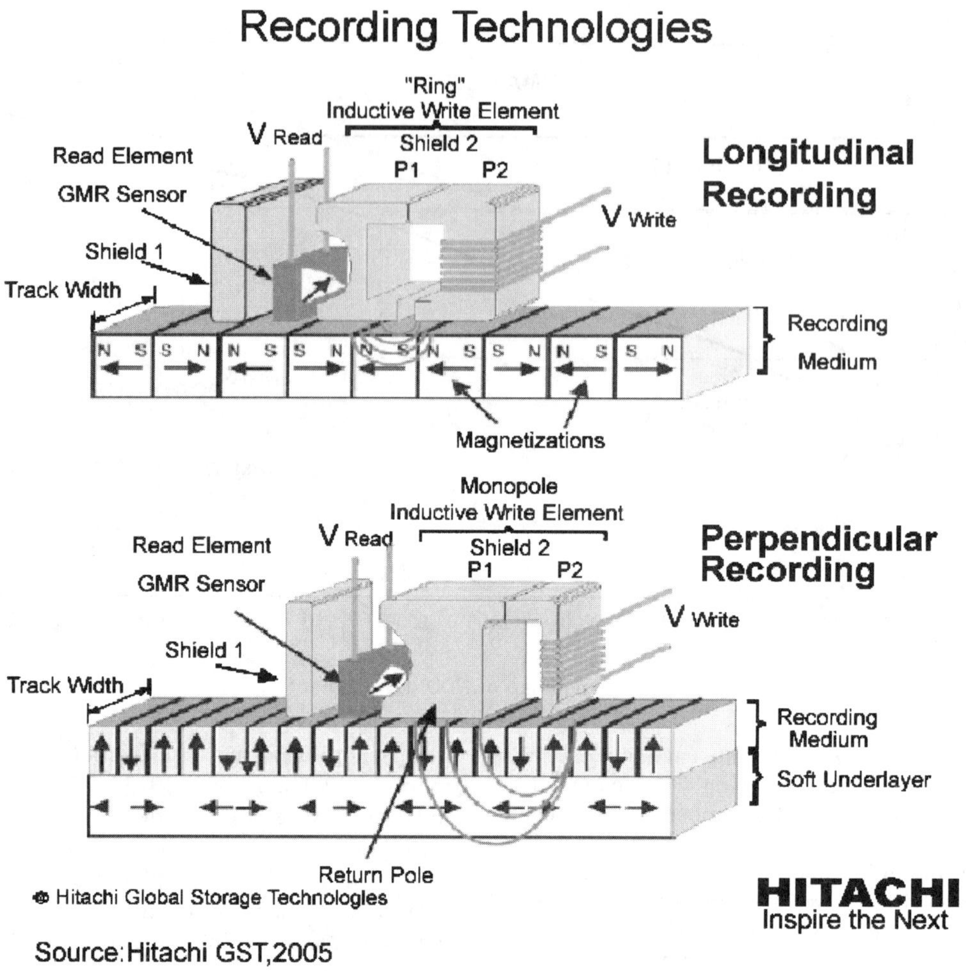

Fig. 33—Magnetic recording technologies; longitudinal versus perpendicular recording.

4.2 Perpendicular Magnetic Recording (PMR)

Magnetic recording head technology has realized many revolutions in the past decade which are responsible for the growth of the HDD capacity. It had a transition from thin film inductive heads to magnetoresistive (MR) and giant MR (GMR) heads. The continuation of improvement of the media alloy sputter process is pushing the current media longitudinal technology to the physical limit. PMR, therefore, as it has been the subject of laboratory research for many years, has been deployed for the fulfillment of capacity requirement of robust mobility and high capacity of the HDD industry. While head technology advance requires a complete change in recording architecture and new manufacturing processes, the transition from longitudinal magnetic recording (LMR) to PMR will be difficult as a media manufacturing process overhauls but no change is required in the system architecture.

In today's longitudinal recording method for media, magnetization of the media is parallel to the surface as shown in Fig. 33. The write head generates a field that follows the media surface across the head gap and returns through the back portion of the head to complete a magnetic circuit. The north/south polarity of the bits is parallel to the media surface. Alternatively, in perpendicular recording, a write head is designed in the way that the field is generated through a single pole element, it passes normal to the disk surface, and returns through a broad collector. The return path in the media is a magnetically soft underlayer, which conducts the magnetic flux.

In the perpendicular recording, the magnetic polarity is perpendicular to the media surface and can be read with a GMR sensor similar to the sensor used in LMR. In this case, the media is effectively situated within the head gap, potentially relaxing many limits on both head and media design experienced in longitudinal recording.

Current products using LMR have an areal density as high as 160 Gb/in.^{2} Within two years, it is approaching 200 G/in.^{2}, which is believed to be the practical limits. Some PMR based products will be released with similar or slightly higher areal density than LMR; the establishment of these early products will provide a baseline for the future PMR based products. It is thought that the limit for PMR is about 1 Tbit/in.^{2}, which is about five times that in the longitudinal recording technology. The former should support the market needs in the next ten years. Within five years, PMR will dominant the industry, virtually all products will apply the technology to achieve an areal density of four to five times that today. Within ten years, PMR (or hybrid methods including PMR) will reach the areal density of ten times today's capabilities.

TABLE 2—Roadmap of HDD technology.

Year	2003	2004	2005	2006	2007
Areal density (Gb/in.^{2})	200	300	500	800	1000
Fly Height (nm)		3-5		2.5	2.0
Pole-Tip Recess Control		<1 nm		<0.6 nm	<0.4 nm
Carbon Overcoat Thickness (nm)	4	3	~2.5	~2	~1
Average Roughness (Ra)			<0.20 nm		<0.15 nm

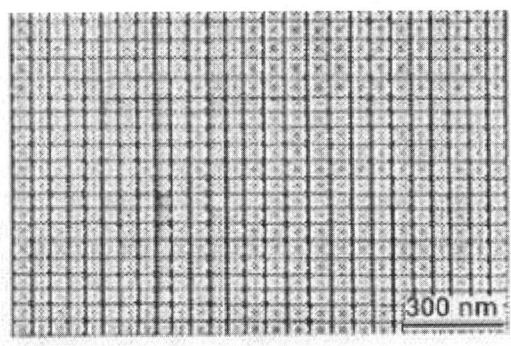

Fig. 34—Bit pattern media (BPM) or self-ordered magnetic arrays.

The optimal media is thicker in perpendicular recording, which allows smaller diameter alloy grains with the same volume relative to LMR, leading the way for smaller bits to be written without a serious thermal stability problem arising from the superparamagnetic effect. PMR can yield recording tracks that are both narrower and better defined as compared with the longitudinal recording. This enables higher track densities.

4.3 PMR and Beyond

4.3.1 Heat Assisted Magnetic Recording (HAMR)

With the increasing areal density, the magnetic medium will be approaching the limits imposed by superparamagnetism. A new recording technique, HAMR, sometimes called optically-assisted magnetic recording, has been proposed as one of the methods to surpass the superparamagnetic limit. This technique, with a type of patterned or self-ordered magnetic array, is viewed as a possible candidate for the next generation of magnetic recording media technology to meet the areal density requirement. It combines the conventional magnetic recording technology of HDD and the optical recording technology used in CD rewritable drives to create a promising recording technique that could increase areal storage density beyond 1 Tbits/in.^{2}. In HARM, a tiny laser light spot focuses onto a small region of the disk surface, and then rapidly heats the high coercivity magnetic media to an elevated temperature (>400°C), thus temporarily lowering its coercivity to the point that writing is possible [60]. The effect is that when the media is heated, the coercivity or the field required to write on the media is reduced, making it possible to write on the high-coercivity media in spite of the limited fields that can be produced by recording heads. This requires that the lubricant must withstand periodic rapid laser heating to temperatures in excess of 400°C.

With the increasing areal density, many problems in overcoat and surface technology need to be solved. The first

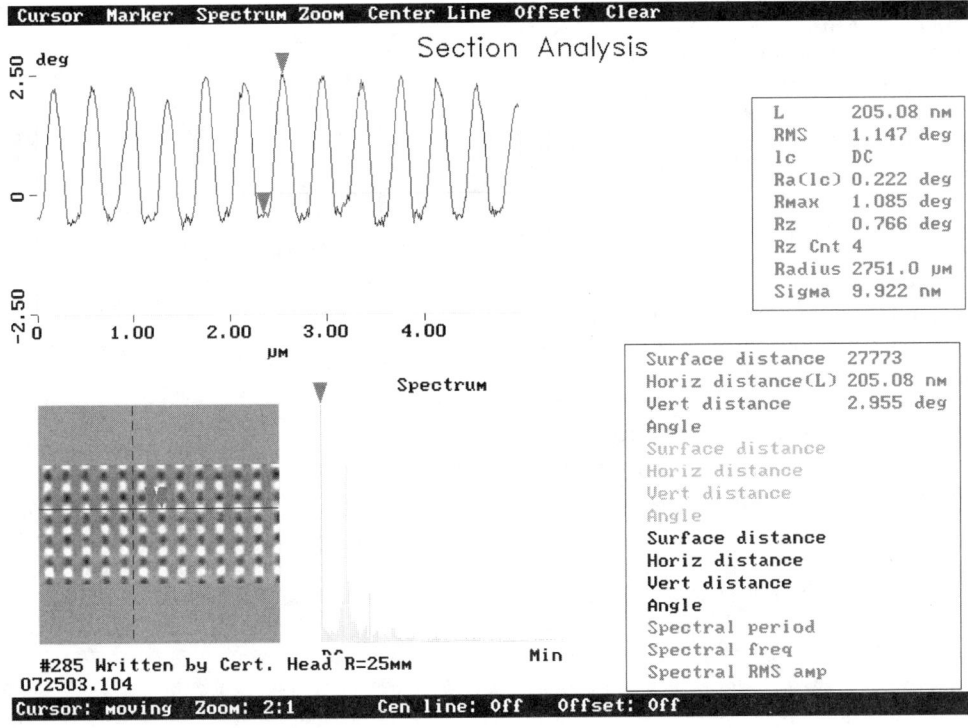

Fig. 35—AFM and MFM of a discrete track media.

problem is that the thickness of the overcoat should be reduced to about 1 nm as shown Table 2. It is a great challenge to find such a new technique as to make this super-thin overcoat without pinholes while maintaining the durability and tribological properties. In addition, an overcoatless media is also a good option to reduce the head-disk space, but many more details need to be worked out. Another problem is the surface roughness control (about 0.15 nm) after the sputtering process and prior to carbon overcoat deposition as shown in Table 2. One solution is to use an ion/plasma beam to polish and mill the surface at grazing incidence angle with existing equipment and optimized processes. Another approach is to find a new technique to improve the surface roughness.

4.3.2 SOMA+HAMR

Beyond perpendicular recording, technologists have already begun to explore alternative methods of recording technology.

Modeling of perpendicular magnetic recording suggests that extensions to about 1 Tbit/in.^{2} should be possible. It is envisioned that higher areal densities are possible by using assisted assembly process. Technologists are also working on bit-patterned media (BPM), or sometimes called "self-ordered magnetic arrays (SOMA)" (see Fig. 34). A typical bit of information is made up of about 100 grains of material. As a result, a large gain in bit density would be achieved. Writing on such self-organized magnetic array media will require temporal heating and cooling in a magnetic field. A combination of SOMA and HAMR may once lead to single particle per bit recording, with ultimate density capability of 50 Tbit/in.^{2}

4.3.3 Discrete Track Recording (DTR)

Various methods for producing DTR media have been studied by many different researchers. One of the demonstrated processes employs nano-imprint lithography (NIL) techniques to pattern either the substrate of the media or the sputtered magnetic layer of the media. NIL provides a potential cost-effective manufacturing process for patterning the large area of a disk surface needed to increase areal density growth (Fig. 35). DTR media provides the potential for improving recording performance due to the inherent inter-track isolation, resulting in improved signal-to-noise ratio. Also, edge track effects, such as overwrite and transition curvature that normally would degrade error rate performance, can be reduced or eliminated by properly designing the magnetic write width of the recording head greater than the pre-patterned data area of a DTR medium. These improvements in the recording performance offer the opportunity to extend the useful life of longitudinal recording as well as to provide a path to sustain the areal density growth of perpendicular media as it is to be introduced in the near future.

4.3.4 Probe Storage

Probe storage may be one of the unusual methods of the proposed recording since it does not involve the use of disks at all. Rather, probe storage technology could be implemented in something like the size of a typical semiconductor chip. It works like a scanning microscope, except that there is an array of these microscopes or probes that reads and writes the data. Each probe addresses an array of bits of information, and the probes write and read in parallel. Several media candidates are under consideration, including magnetic media. And, unlike typical silicon chips, they will not lose memory once power is turned off. It is estimated that about ten gigabytes of information will be able to be stored on a centimeter-sized chip.

References

[1] Robertson, J., "Requirements of Ultrathin Carbon Coatings for Magnetic Storage Technology," *Tribol. Int.*, Vol. 36, 2003, pp. 405–415.

[2] Bogy, D., "Dynamic Flying at sub 5 nm," Invited speaker at TISD, 2003.

[3] Demczyk, B., Liu, J., Chao, Y., and Zhang, S. Y., "Lubrication Effects on Head-Disk Spacing Loss," *Tribol. Int.*, Vol. 38, 2005, pp. 562–565.

[4] Hyodo, H., Yamamoto, T., and Toyoguchi, T., "Properties of Tetrahedral Amorphous Carbon Film by Filtered Cathodic Arc Deposition for Disk Overcoat," *IEEE Trans. Magn.*, Vol. 37, 2001, pp. 1789–1791.

[5] Chiba, T., Ariake, J., Honda, N., and Ouchi, K., "A Highly Durable Structure of Carbon Protective Layer," *IEEE Trans. Magn.*, Vol. 38, 2002, pp. 2129–2131.

[6] Yamamoto, T., Hyodo, H., Tsuchitani, S., and Kaneko, R., "Ultrathin Amorphous Carbon Overcoats by Filtered Cathodic Arc Deposition," *IEEE Trans. Magn.*, Vol. 39, 2003, pp. 2201–2204.

[7] Song, Y. G., Chu, X., and Yang, M. M., "Ultrathin Ion-Beam Carbon as an Overcoat for the Magnetic Recording Medium," *IEEE Trans. Magn.*, Vol. 39, 2003, pp. 594–598.

[8] Klaus, E. E. and Bhushan, B., "Lubricants in Magnetic Media-Review," *Tribol. Mech. Magn Storage Syst.*, Vol. 2, 1985, pp. 7–15.

[9] Carre, D. J. and Markowitz, J. A., "The Reaction of Perfluoropolyalkylether Oil and FeF3, A1F3, and A1C13 at Elevated Temperature," *ASME Trans.*, Vol. 28, 1985, pp. 40–46.

[10] Talke, F. E., "On Tribological Problems in Magnetic Disk Recording Technology," *Wear*, Vol. 190, 1995, pp. 232–238.

[11] Luo, J. B. and Yang, M. C., "Surface Modification of Computer Magnetic Head," *Sino-German Symposium on Micro Systems and Nano Technology*, 7–9 Sept., Braunschweig, Germany, 2001.

[12] Yang, M. C., Luo, J. B., Wen, S. Z., et al., "Failure Characterization at Head/Write Interface of Hard Disk Drive," *Science in China*, Vol. 44 (Supp), 2001, pp. 407–411.

[13] Novotny, V. J., Karis, T. E., and Johnson, N. W., "Lubricant Removal, Degradation, and Recovery of Particulate Magnetic Recording Media," *ASME J. Tribol.*, Vol. 114, 1992, pp. 61–67.

[14] O'Connor, T. M., Back, Y. R., Jhon, M. S., et al., "Surface Diffusion of Thin Perfluoropolyalkylether Films," *J. Appl. Phys.*, Vol. 79, 1996, pp. 5788–5790.

[15] Deoras, S. K., Chun, S. W., Vurens, G., et al., "Spreading and Mobility Analysis of PFPE Lubricants Using a Surface Reflectance Analyzer," *Tribol. Int.*, Vol. 36, 2003, pp. 241–246.

[16] Sinha, S. K., Kawaguchi, M., Kato, T. et al., "Wear Durability Studies of Ultra-Thin Perfluoropolyether Lubricant on Magnetic Hard Disks," *Tribol. Int.*, Vol. 36, 2003, pp. 217–225.

[17] Karis, T. E. and Jhon, M. S., "The Relationship Between PFPE Molecular Rheology and Tribology," *Tribology Letters*, Vol. 5, 1998, pp. 283–286.

[18] Perettie, D. J., Morgan, T. A., and Kar, K. K., "X-1P as a Dual

[19] Karis, T. E., "Tribochemistry in Contact Recording," *Tribol. Lett.*, Vol. 10, 2001, pp. 149–162. Purpose Lubricant for Pseudo-Contact Recording," *Insight*, Vol. 9, 1996, pp. 3–6.

[19] Karis, T. E., "Tribochemistry in Contact Recording," *Tribol. Lett.*, Vol. 10, 2001, pp. 149–162.

[20] Suk, M. and Gills, D. R., "Comparison of Friction Measurement Between Load/Unload Ramps and Suspension Lift Tabs Using Strain Gage and Actuator Current," *IEEE Trans. Magn.*, Vol. 36, 2000, pp. 2721–2723.

[21] Mitsuhiro, S. and Hiroshi, T., "Catastrophic Damage of Magnetic Recording Disk Caused by Slider-Disk Impact During Loading/Unloading," *IEEE Trans. Magn.*, Vol. 39, 2003, pp. 893–897.

[22] Albrecht, T. R. and Sai, F., "Load/Unload Technology for Disk Drives," *IEEE Trans. Magn.*, Vol. 35, 1999, pp. 857–862.

[23] Suk, M. and Gillis, D., "Effect of Slider Burnish on Disk Damage During Dynamic Load/Unload," *ASME J. Tribol.*, Vol. 120, 1998, pp. 332–338.

[24] Jiang, Z. C., Liu, Y. W., and Chung, W. F., "TOF-SIMS Analysis: Application to Ultra-Thin AWA Film on Magnetic Head," *Sci. China, Ser. A: Math., Phys., Astron.*, Vol. 44, 2001, pp. 393–399.

[25] Yang, M. C., "Study on Nano-Tribology at Head/Disk Interface of Computer Hard Disk Drives," Ph.D. Thesis, Tsinghua University, Beijing, China, 2001.

[26] Hu, X. L., Zhang, C. H., Luo, J. B., and Wen, S. Z., "Formation and Tribology Properties of Polyfluoroalkylmethacrylate Film on the Magnetic Head Surface," *Chin. Sci. Bull.*, Vol. 50, 2005, pp. 2385–2390.

[27] Doudevski, I. and Daniel, K. S., "Concentration Dependence of Self-Assembled Monolayer Island Nucleation and Growth," *J. Am. Chem. Soc.*, Vol. 123, 2001, pp. 6867–6872.

[28] Yang, M. C., Luo, J. B., Wen, S. Z., et al., "Failure Characterization at Head/Write Interface of Hard Disk Drive," *Science in China*, Vol. 44 (Supp.), 2001, pp. 407–411.

[29] Chen, C. Y., Bogy, D. B., Chen, T., and Bhatia, C. S., "Effect of the Additive X1-P on the Tribological Performance and Migration Behavior of PFPE Lubricant at the Head-Disk Interface," *IEEE Trans.*, Vol. 36, 2000, pp. 2708–2710.

[30] Perettie, D. J., "The Effect of Phosphazene Additives to Passivate and Stabilize Lubricants at the Head-Disk Interface," *Tribol. Int.*, Vol. 36, 2003, pp. 489–491.

[31] Zhou, L., DeKoven, B. M., Chen, S., Chao, J., and Talke, F. E., "Tribology of X1-P Vapor Lubricated Hard Disks," *Tribol. Int.*, Vol. 38, 2005, pp. 574–577.

[32] Luo, J. B., Yang, M. C., Zhang, C. H., et al., "Study on the Cyclotriphosphazene Film on Magnetic Head Surface," *Tribol. Int.*, Vol. 37, 2004, pp. 585–590.

[33] Briggs, D., in *Practical Surface Analysis*, 2nd ed., 1992.

[34] Hu, X. L., "Studies on the Formation and Properties of Ultra-Organic Thin Films on the Magnetic Head," Ph.D. Thesis, Tsinghua University, Beijing, China, 2005.

[35] Ryen, C. M. and Zhu, X. Y., "Two-Step Approach to the Formation of Organic Monolayers on the Silicon Oxide Surface," *Langmuir*, Vol. 17, 2001, pp. 5576–5580.

[36] Choi, J., Ishida, T., Kato, T., and Fujisawa, S., "Self-Assembled Monolayer on Diamond-Like Carbon Surface: Formation and Friction Measurements," *Tribol. Int.*, Vol. 36, 2003, pp. 285–290.

[37] Wang, S. X. and Taratorin, A. M., *Magnetic Information Storage Technology*, Academic Press, San Diego, CA, YEAR, pp. X.

[38] Guo, G., Li, L., Hsia, Y. T., and Jhon, M. S., "Stability Analysis of Ultrathin Lubricant Films Via Surface Energy Measurements and Molecular Dynamics Simulations," *J. Appl. Phys.*, Vol. 97, 2005, 10P302.

[39] Tyndall, G. W. and Karis, T. E., "Spreading Profiles of Molecularly Thin Perfluoropolyether Films," *Tribol. Trans.*, Vol. 42, 1999, pp. 463–470.

[40] Ogata, S., Mitsuya, Y., Zhang, H. D., and Fukuzawa, K., "Molecular Dynamics Simulation for Analysis of Surface Morphology of Lubricant Films with Functional End Groups," *IEEE Trans. Magn.*, Vol. 41, 2005, pp. 3013–3015.

[41] Novotny, V. J., Hussia, I., Turlet, J. M., and Philpott, M. R., "Liquid Polymer Conformation on Solid Surfaces," *J. Chem. Phys.*, Vol. 90, 1989, pp. 5861–5868.

[42] Mate, C. M. and Novotny, V. J., "Molecular Conformation and Disjoining Pressure of Polymeric Liquid Films," *J. Chem. Phys.*, Vol. 94, 1991, pp. 8420–8427.

[43] Ma, X., Gui, J., Smoliar, L., Grannen, K., Marchon, B., Jhon, M. S., and Bauer, C. L., "Spreading of Perfluoropolyalkylether Films on Amorphous Carbon Surfaces," *J. Chem. Phys.*, Vol. 110, 1999, pp. 3129–3137.

[44] Guo, Q., Izumisawa, S., Phillips, D. M., and Jhon, M. S., "Surface Morphology and Molecular Conformation for Ultrathin Lubricant Films with Functional End Groups," *J. Appl. Phys.*, Vol. 93, 2003, pp. 8707–8709.

[45] Li, X., Hu, Y. Z., and Wang, H., "A Molecular Dynamics Study on Lubricant Perfluoropolyether in Hard Disk Driver," *Acta Phys. Sin.*, Vol. 54, 2005, pp. 3787–3792.

[46] Novotny, V. J., "Migration of Liquid Polymers on Solid Surfaces," *J. Chem. Phys.*, Vol. 92, 1990, pp. 3189–3196.

[47] O'Connor, T. M., Jhon, M. S., Bauer, C. L., Min, B. G., Yoon, D. Y., and Karis, T. E., "Surface Diffusion and Flow Activation Energies of Perfluoropoly-Alkylethers," *Tribol. Lett.*, Vol. 1, 1995, pp. 219–223.

[48] Min, B. G., Choi, J. W., Brown, H. R., Yoon, D. Y., O'Connor, T. M., and Jhon, M. S., "Spreading Characteristics of Thin Liquid Films of Perfluoropolyalkylethers on Solid Surfaces: Effects of Chain-End Functionality and Humidity," *Tribol. Lett.*, Vol. 1, 1995, pp. 225–232.

[49] Ma, X., Gui, J., Smoliar, L., Grannen, K., Marchon, B., Bauer, C. L., and Jhon, M. S., "Complex Terraced Spreading of Perfluoropolyalkylether Films on Carbon Surfaces," *Phys. Rev. E*, Vol. 59, 1999, pp. 722–727.

[50] Tyndall, G. W. and Karis, T. E., "Spreading Profiles of Molecularly Thin Perfluoropolyether Films," *Tribol. Trans.*, Vol. 42, 1999, pp. 463–470.

[51] Li, X., Hu, Y. Z., and Wang, H., "Modeling of Lubricant Spreading on a Solid Substrate," *J. Appl. Phys.*, Vol. 99, 2006, pp. 024905-1--5.

[52] Kim, H. I., Mate, C. M., Hannibal, K. A., and Perry, S. S., "How Disjoining Pressure Drives the Dewetting of a Polymer Film on a Silicon Surface," *Phys. Rev. Lett.*, Vol. 82, 1999, pp. 3496–3499.

[53] Toney, M. F., Mate, C. M., and Leach, K. A., "Roughness of Molecularly Thin Perfluoropolyether Polymer Films," *Appl. Phys. Lett.*, Vol. 77, 2000, pp. 3296–3298.

[54] Izumisawa, S. and Jhon, M. S., "Stability Analysis and Molecular Simulation of Nanoscale Lubricant Films with Chain-End Functional Groups," *J. Appl. Phys.*, 2002, Vol. 91, 2002, pp. 7583–7585.

[55] Ogata, S., Mitsuya, Y., Zhang, H. D., and Fukuzawa, K., "Molecular Dynamics Simulation for Analysis of Surface Morphol-

ogy of Lubricant Films with Functional End Groups," *IEEE Trans. Magn.*, Vol. 41, 2005, pp. 3013–3015.

[56] Zhao, X., Bhushan, B., and Kajdas, C., "Lubrication Studies of Head-Disk Interfaces in a Controlled Environment, Part 2: Degradation Mechanisms of Perfluoropolyether Lubricants," *Proc. Inst. Mech. Eng., Part J: J. Eng. Tribol.*, Vol. 214, 2000, pp. 547–559.

[57] Kasai, P. H., Tang, W. T., and Wheeler, P., "Degradation of Perfluoropolyethers Catalyzed by Aluminum Oxide," *Appl. Surf. Sci.*, Vol. 51, 1991, pp. 201–211.

[58] Kajdas, C., "Tribochemistry," *Surface Modification and Mechanisms*, G. E. Totten and H. Liang, Eds., Marcel Dekker, New York, 2004, pp. 99–104.

[59] Vurens, G., Zehringer, R., and Saperstein, D., "The Decomposition Mechanism of Perfluoropolyether Lubricants During Wear," *Surface Science Investigation in Tribology*, ACS Symposium Series, Vol. 48, Y. W. Chung, A. M. Homola, and G. B. Street, Eds., American Chemical Society, Washington, DC, 1992, pp. 169–180.

[60] Lim, M. S. and Gellman, A. J., "Kinetics of Laser Induced Desorption and Decomposition of Forblin Zdol on Carbon Overcoats," *Tribol. Int.*, Vol. 38, 2005, pp. 554–561.

[61] Zhang, B. and Nakajima, A., "Possibility of Surface Force Effect in Slider Air Bearings of 100 Gbit/in.^{2} Hard Disks," *Tribol. Int.*, Vol. 36, 2003, pp. 291–296.

[62] Tyndall, G. W. and Waltman, R. J., "Thermodynamics of Confined Perfluoropolyether Film on Amorphous Carbon Surface Determined from the Time-Dependent Evaporation Kinetics," *J. Phys. Chem. B*, Vol. 104, 2000, pp. 7085–7095.

12

Tribology in Ultra-Smooth Surface Polishing

Jianbin Luo,[1] Xinchun Lu,[1] Guoshun Pan,[1] and Jin Xu[1]

1 Introduction

ULTRA-SMOOTH SURFACES HAVE BEEN WIDELY used in many areas, e.g., large scale integration (LSI), computer hard disk driver (HDD), optic lenses, connectors of optic fibers, and so on. These surfaces should be of very low roughness and waviness, very high flatness, and very few defects. In order to achieve these requirements, many kinds of planarization techniques have been developed in the past 15 years, e.g., ultra-fine diamond (UFD) powder polishing [1] which has been used in the manufacturing of magnetic heads, chemical mechanical polishing (CMP) [2] which is recognized as the best method of achieving global planarization in ultra-smooth surface fabrication and also widely used in ultra-large scale integration fabrication, electric chemical polishing (ECP) and electric chemical mechanical polishing (ECMP) [3] which has also been used in global planarization of LSI fabrication, abrasive-free polishing [4], magnetofluid polishing [5], ultrasonic polishing [6], etc. These polishing techniques can be divided into three groups, one is the mechanical process, including UFD powder polishing, magnetofluid polishing, and ultrasonic polishing, where the polished surface cannot be used under erosion circumstance; another is the chemical process, including ECP, where chemical dissolving takes a major role; and the other is the chemical-mechanical process, including CMP and abrasive-free polishing, where the balance of chemical effect and mechanical removal is very important. During these polishing processes, tribology is one of the important factors. Good lubrication during the polishing process is the key factor to get an ultra-smooth surface, as severe wear related to the material removal rate (MRR) will cause scratches on the surface. Therefore, a better understanding of tribological behaviors between slurry, pad, and polished materials is helpful for the improvements of a polished surface.

In this chapter, an introduction of experimental and theoretical studies on nanoparticles collision has been made, as nanoparticles impact on an ultra-smooth surface always occur in the ultra-smooth surface manufacturing. Then the development of CMP technology is introduced. And at last, the polishing of magnetic head surface is discussed.

2 Nanoparticles Impact

2.1 Surface Observation

Polished surface is usually observed by optical interference techniques, e.g., Chapman MP2000+, WYKO, and Zygo. As shown in Table 1, different parameters of a polished surface, e.g., the surface waviness, roughness, flatness, dub-off, roll-off, etc., can be gotten by HDI instruments or Chapman MP2000+ (Fig. 1) [7]. However, a different measurement profile gives different surface data even with the same surface. As shown in Fig. 2, the surface roughness Ra is about 0.12 nm measured by AFM, while it is only about 0.06 nm measured by Chapman MP2000+. The data from these two tools have a difference of about two times because of the different resolution in X.Y. and the different scanning range. Therefore, to discuss the data difference among different surfaces, the measuring techniques or instruments must be pointed out.

Global planeness and large scale scratches are usually evaluated by HDI instruments as shown in Fig. 3(a) [8], which is a surface reflectance analyzer to measure flatness, waviness, roughness of a surface, and observe scratches (Fig. 3(b)), pits (Fig. 3(c)), particles (Fig. 3(d)) on a global surface. These surface defects can also be observed by SEM, TEM, and AFM. Shapes of slurry particles can be observed by SEM and TEM, and their movement in liquid by the fluorometry technique as shown in Chapter 2.

2.2 Experimental Observation of Nanoparticles Collision with Solid Surface

What will occur when nanoparticles collide with a solid surface? The answer to this question is important for a general understanding of collision between nanoparticles and solid surface, which is essential to control and prevent surface damage or defects generated by nanoparticles collision during the polishing process in devices manufacturing [9]. In recent years, a great deal of the simulations and tests of collision between energetic clusters and surfaces are performed to explore the interactions between energetic clusters and solid surfaces [10–12] because of their importance in surface cleaning, surface metallization, and ultra-precision machining. Xu et al. [13] designed a collision tester, as shown in Fig. 4, from which, solid surface after collision can be gotten and then measured. A cylindrical liquid jet, deionized water, or slurry with nanoparticles, impacts on the solid surface at a speed of 50 m/s, with an incidence angle of 45°. The solid specimen is Si (100) wafer with a surface roughness of 5 Å. Particles in the slurry are silica with a mean diameter about 60 nm (Fig. 5). Particle concentration in the slurry is maintained at 5 wt % and pH value of the slurry is 11. The impacting time is set as 3, 10, 30, and 60 min. All solid specimens are ultrasoniclly cleaned with DI water before the tests. To eliminate the adhesion effect of particles on a solid surface, silicon surfaces are ultrasonically cleaned with 0.5 % HF so-

---

[1] *State Key Laboratory of Tribology, Tsinghua University, Beijing, China.*

TABLE 1—Measuring parameters.

	MRR (nm/min)	Flatness (μm)		FSWa (Å)		Duboff (Å)		Wa (Å)		Ra[a] (Å)	
		A	B	A	B	A	B	A	B	A	B
QH-1#	112	0.23	0.21	4.54	4.21	148	209	0.53	0.57	0.44	0.44
QH-2#	84	0.26	0.25	4.23	3.75	163	199	0.50	0.50	0.42	0.42
Control level	>76	7	7	8	8	500	500	0.8	0.8	0.8	0.8

[a]Wa and Ra are tested by Chapman MP2000+.

lution for 20 s, and then with DI water. The tests are conducted at room temperature of 23 °C.

The collision process can be captured by a high speed video camera as shown in Fig. 6 [14]. The slurry is about 50 mm apart away from the solid surface at 0 s (Fig. 6(a)), and reaches the surface at 0.018 s (Fig. 6(b)). Then the slurry reflects at an angle as same as the incidence angle (Fig. 6(c)). As time goes, the reflected liquid beam is divided into two beams, one is in the reflected direction and another is parallel to the solid surface as shown in Fig. 6(d). When time reaches 0.068 s, most of the reflected slurry moves along the solid surface.

Figure 7 [13] shows AFM morphologies of the damaged surface after collision. After the surface undergoes three minutes of impacting, damage is observed on the impacted surface as shown in Fig. 7(a). After undergoing impacting for ten minutes, many pits appear on the surface (Fig. 7(b)), and their sizes are in the same order of that of the particles, and their depths are generally less than 2 nm. When the impacting time is kept longer to 60 minutes, the maximum depth increases to about 8 nm (Fig. 7(d)). Compared with the TEM image of a solid surface impacted with DI water without particles (Fig. 8(a)), many significant elongated damage regions can be observed on the surface impacted with slurry (Fig. 8(b)), and dimension of the damage region is in nanoscale similar to the size of the particles.

The TEM micrograph of a collision region is shown in Fig. 9 [13]. From the differences along the symmetry axis, the damaged region can be divided into two parts, i.e., an elliptical area at central and its adjacent annular area. Six positions (denoted as I, II, III, IV, V, and VI) are selected for high resolution TEM analysis as shown in Fig. 10 [13], and a significant change in lattice structure at different location in the region is found. Location I is on the boundary of the damage region, and some changes in the atomic array direction can be seen in Fig. 10(a). It becomes severe in Location II. Considerable atomic disorder in the central area occurs, as Location III in Fig. 10(c), where the amorphous phase is also formed. The disorder grade reduces gradually in the direction from Location III to IV (Figs. 10(c) and 10(d)). In the Location V, the central region of the atomic pile, the region with both crystalline pocket and amorphous is observed as shown in Fig. 10(e). In another boundary region (Location VI), a clear boundary between the damaged and undamaged region can be seen [Fig. 10(f)].

Central damage region, corresponding to the site where a particle first contacts with the surface, undergoes the greatest stress, which leads to crystal lattice deformation.

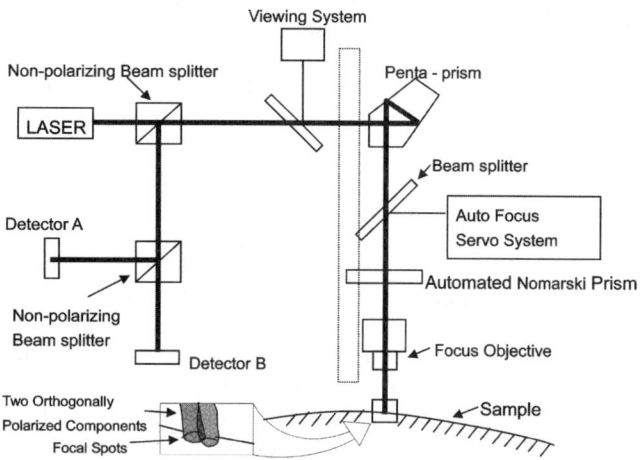

Fig. 1—Profile measurement technique of Champper 2000+. A surface measurement is made with a linearly polarized laser beam that passes to translation stage which contains a penta-prism. The beam then passes through a Nomarski prism which shears the beam into two orthogonally polarized beam components. They recombine at the Nomarski prism. The polarization state of the recombined beam includes the phase information from the two reflected beams. The beam then passes to the nonpolarizing beam splitter which directs the beam to a polarizing beam splitter. This polarizing beam splitter splits the two reflected components to detectors A and B, respectively. The surface height difference at the two focal spots is directly related to the phase difference between the two reflected beams, and is proportional to the voltage difference between the two detectors. Each measurement point yields the local surface slope [7].

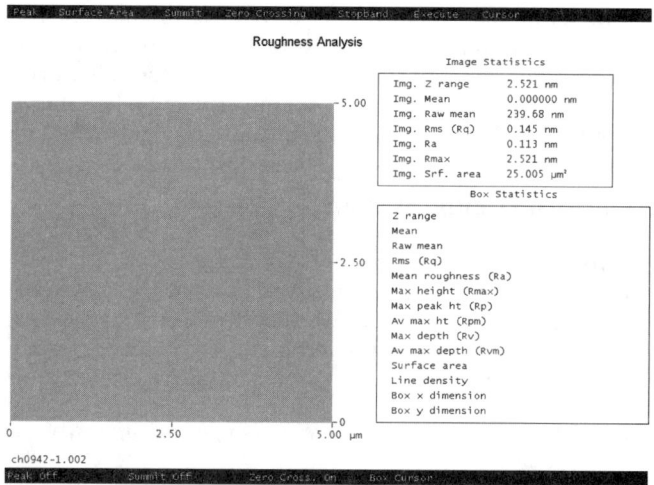

Fig. 2—AFM measurement.

Due to particles extrusion, crystal lattice deformation expands to the adjacent area, though the deformation strength reduces gradually (Figs. 10(a)–10(d)). On the other hand, after impacting, the particle may retain to plow the surface for a short distance to exhaust the kinetic energy of the particle. As a result, parts of the free atoms break apart from the substrate and pile up as atom clusters before the particle. The observation is consistent with results of molecular dynamics simulation of the nanometric cutting of silicon [15] and collision of the nanoparticle with the solid surface [16].

Figure 11(a) [13] shows the cross section of HRTEM image of the tested specimen. Amorphous silicon area occurs and thickness of the deformation layer is about 15 nm. The microprobe result (Fig. 11(b)) is sufficient to rule out the presence of oxygen in this area. Johansson et al. [17] also reported amorphous material on mechanically polished surfaces of silicon (100) [18]. Explanation of such phenomenon is reformation generated by particles indenting and heat by the impact of the abrasive particles. The heat causes a local melting and then is followed by a rapid quenching of the liquid. The amorphous state should be due to the very high compressive stress and temperature. Presence of crystal grains in the deformation layer is also observed. The result is well correlated with the results of molecular dynamics analysis in the nano-indentation of silicon monocrystals [16,19].

The damaged region after impacting for 30 min can also be observed by AFM as shown in Fig. 12 [13]. The maximum depth of the groove is approximately 3 nm. A clear mark of extrusion edge of the impacted region can also be observed. However, no damage region can be found on the surface impacted with DI water (without particle) for 30 minutes.

When particle impacts with a solid surface, the atoms of the surface layer undergo crystal lattice deformation, and then form an atom pileup on the outlet of the impacted region. With the increase of the collision time, more craters present on the solid surface, and amorphous transition of silicon and a few crystal grains can be found in the subsurface.

2.3 Molecular Dynamics Simulation (MDS) on Collision of a Nanoparticle

Molecular dynamics simulation (MDS) is a powerful tool for the processing mechanism study of silicon surface fabrication. When a particle impacts with a solid surface, what will happen? Depending on the interaction between cluster and surface, behaviors of the cluster fall into several categories including implantation [20,21], deposition [22,23], repulsion [24], and emission [25]. Owing to limitations of computer time, the cluster that can be simulated has a diameter of only a few nanometres with a small cohesive energy, which induces the cluster to fragment after collision.

In order to explain how the trajectory of the nanoparticle and the damage region of the surface are formed, the

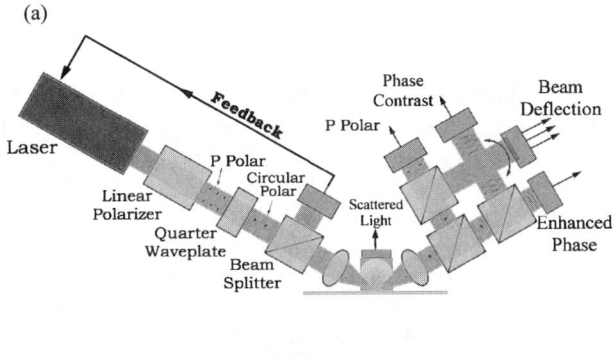
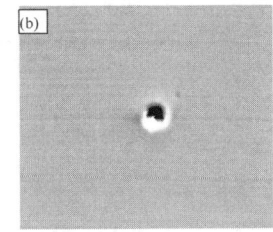

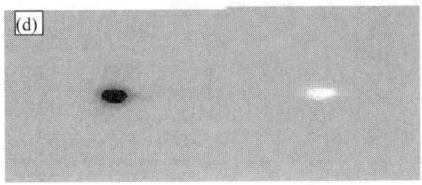

Fig. 3—Measurement of surface by HDI surface reflectance analyzer. In electromagnetic radiation (light), the polarization direction is defined as the direction of the electric field vector. The incident polarization of the light can be controlled. The instrument uses a variety of detectors to analyze the reflected polarization state of the light. (U.S. Patent 6,134,011). (a) Plane of the disk: The SRA uses a fixed 60 degree (from the surface normal) angle of incidence. The plane of incidence is the same as the paper plane; (b) Pit on a surface detected by reflected light channels of HDI instrument; (c) Scratches on disk surface measured by HDI surface reflectance analyzer; (d) Particles on the surface of disk detected by reflected light (black spot) and by scattered light (white spot) [8].

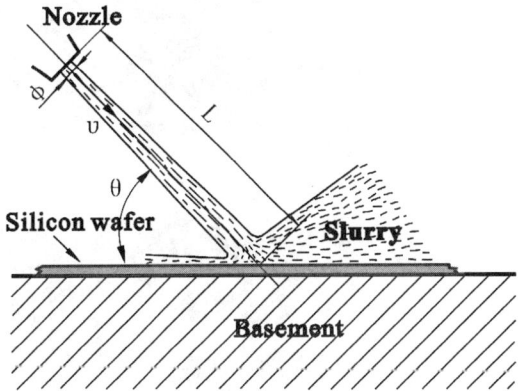

Fig. 4—Profile of a cylindrical liquid jet containing deionized water or a slurry with nano-particles impacting on a surface of a silicon wafer with an incident angle θ at a speed v. (L=100 mm, Φ=2 mm, θ=45°.)

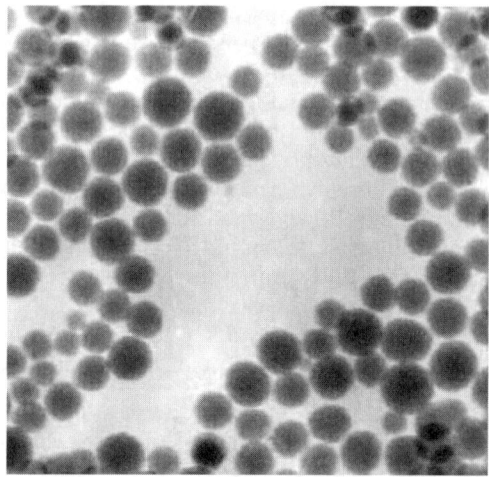

Fig. 5—TEM image of silica nano-particle.

collision processes with different incident angles and different incident velocities are presented by Duan et al. [16] as shown in Fig. 13. They made the simulation of nanoparticle impacting with solid surface. In their simulation, high hardness nanoparticle and low cohesive energy between nanoparticle and surface are achieved by setting the parameters of Lennard-Jones potential, and then the phenomena of collision and recoil of a nanoparticle with a silicon surface is created.

The interaction between particle and surface and the interaction among atoms in the particle are modeled by the Lennard-Jones potential [26]. The parameters of the Lennard-Jones potential are set as follows: $\varepsilon_{pp} = 0.86$ eV, $\sigma_{pp} = 2.27$ Å, $\varepsilon_{ps} = 0.43$ eV, $\sigma_{ps} = 3.0$ Å. The Tersoff potential [27], a classical model capable of describing a wide range of silicon structure, is employed for the interaction between silicon atoms of the surface. The particle prepared by annealing simulation from 5,000 K to 50 K, is composed of 864 atoms with cohesive energy of 5.77 eV/atom and diameter of 24 Å. The silicon surface consists of 45,760 silicon atoms. The crystal orientations of [100], [010], [001] are set as x, y, z coordinate axes, respectively. So there are 40 atom layers in the z direction with a thickness of 54.3 Å. Before collision, the whole system undergoes a relaxation of 5,000 fs at 300 K.

2.3.1 Typical Collision Process

2.3.1.1 The Particle with an Incident Angle of 0°

The process of a particle with a 2 nm diameter impacting with a silicon wafer surface with an incident angle of 0° at a speed of 3,000 m/s is shown in Fig. 14 [16]. When the particle impacts into the surface, a compress region near the particle will be formed, and the contact region of the Si surface layer will vibrate or change between the compressible state and tensile state. From 800 fs to 1,700 fs (Figs. 14(a)–14(c)), compression of the contact region becomes heavier, and the potential energy of Si surface layer becomes highest, and then it will release energy. The released elastic

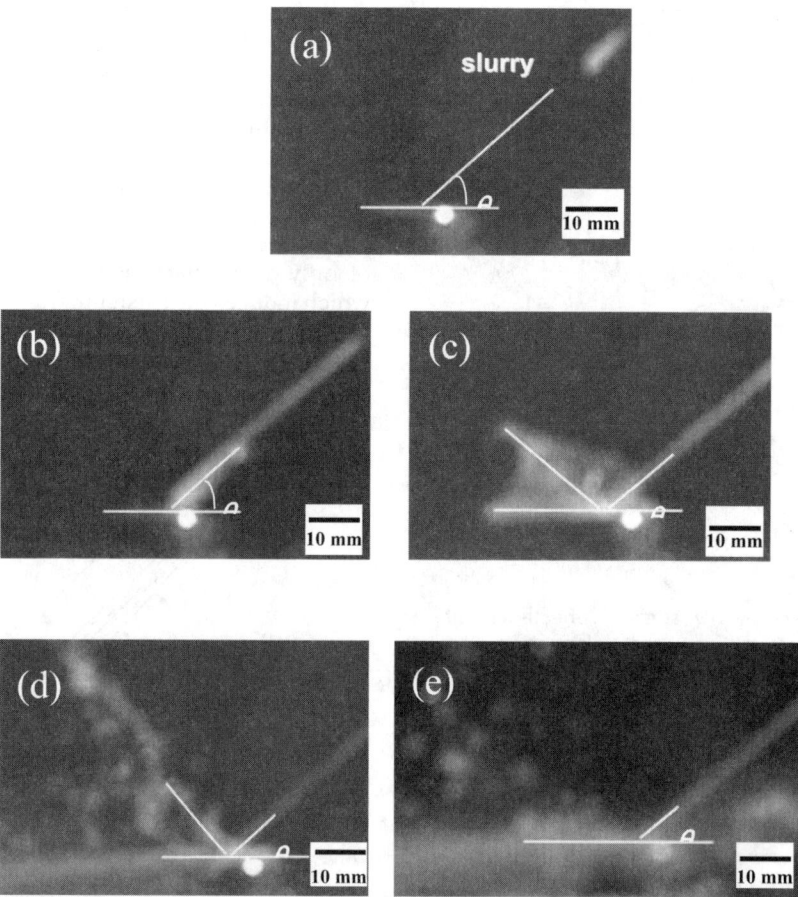

Fig. 6—Images of fluid beam impacting on solid surface at different time (a) 0 s; (b) 0.018 s; (c) 0.032 s; (d) 0.050 s; (e) 0.068 s.

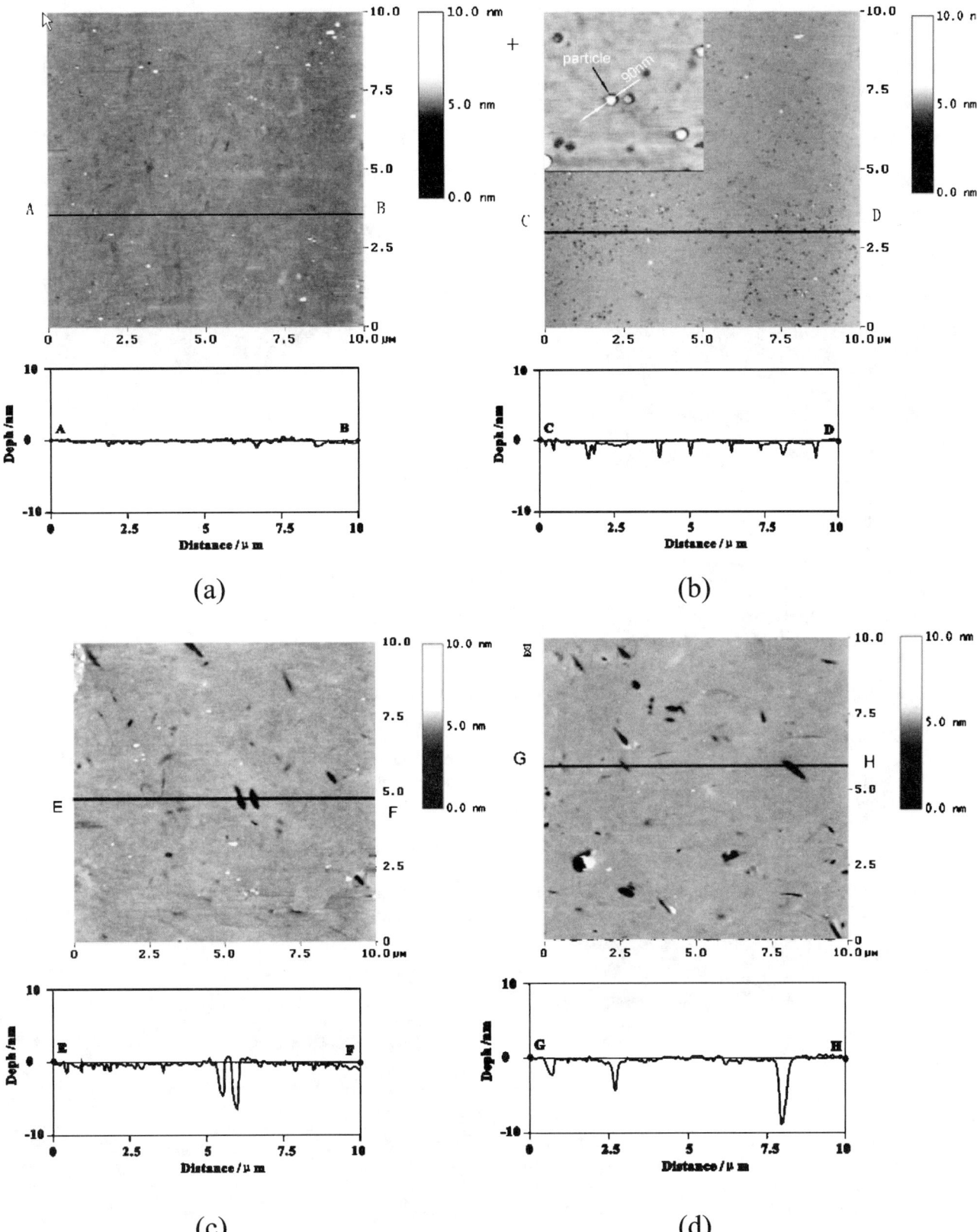

Fig. 7—AFM images of the damaged wafers undergoing different collision conditions (a) 3 min, (b) 10 min, (c) 30 min, (d) 60 min.

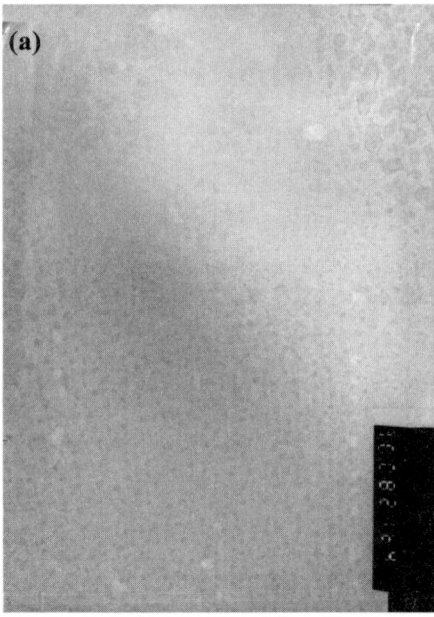

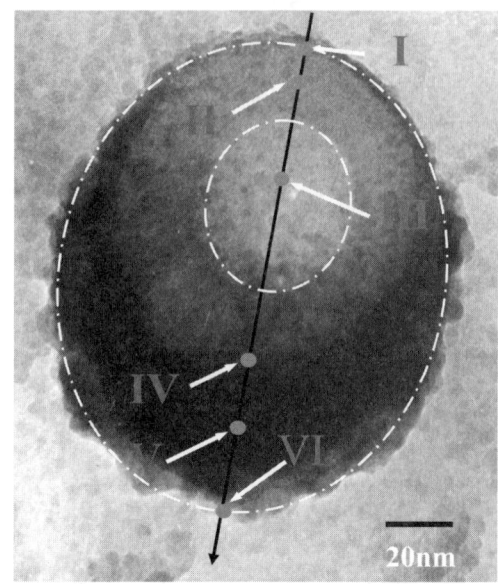

Fig. 9—TEM image of wafer undergoing collision for three minutes.

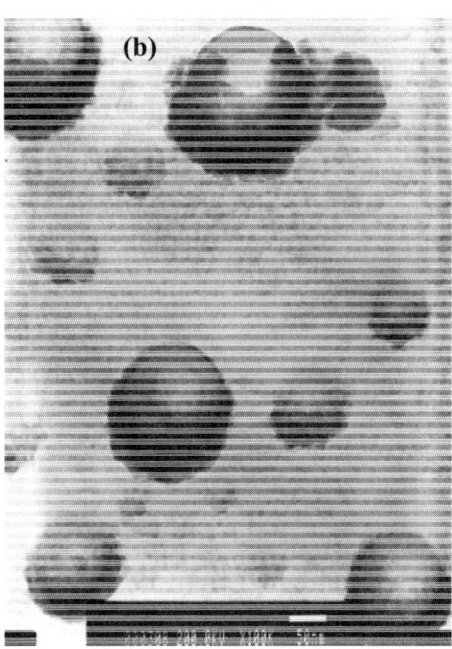

Fig. 8—TEM images of wafer surfaces undergoing different collisional conditions, (a) without particles, (b) with particles.

deformation energy of the surface acts on the lower right part of the particle, and then push the particle move away in the opposite direction as shown in Figs. 14(d)–14(g). When the particle leaves the Si surface, the contact region of the surface layer undergoes a relaxation or vibration process, and balances step by step (Figs. 14(h) and 14(i)). The variations of the position and the speed of the particle in the vertical direction are shown in Fig. 15 [16]. When the particle reaches the deepest position at 1,800 fs, the speed of the particle becomes zero. Then it moves away from the Si surface forced by the vibration of the Si surface layer. Before it parts from the surface, its speed decreases a little because of the adsorbed force between the particle and the Si surface in the contact region. After the particle leaves the surface, it moves at a constant speed.

2.3.1.2 The Particle with an Incident Angle of 45° and 60°

The cross section of the collision region that the particle impacts with the Si surface with an incident angle of 45° at a speed of 2,100 m/s is shown in Fig. 16 [28]. As the particle impacts into the Si surface layer, the contact region of the Si surface layer transforms from crystal into amorphous phase immediately. The area of the depressed region and the thickness of the amorphous layer increase with the penetration depth of the particle (Figs. 16(a)–16(c)). After it reaches the deepest position, the particle then moves both upwards and rightwards, and some silicon atoms ahead of the particle are extruded out and result in a pileup of atoms. Then the released elastic deformation energy of the Si surface pushes

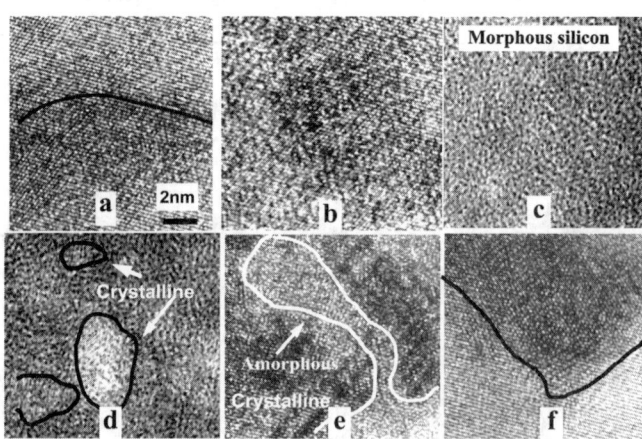

Fig. 10—HTEM images of different positions on wafer undergoing collision for three minutes, (a) Position I; (b) Position II; (c) Position III; (d) Position IV; (e) Position V; (f) Position VI.

CHAPTER 12 ■ TRIBOLOGY IN ULTRA-SMOOTH POLISHING

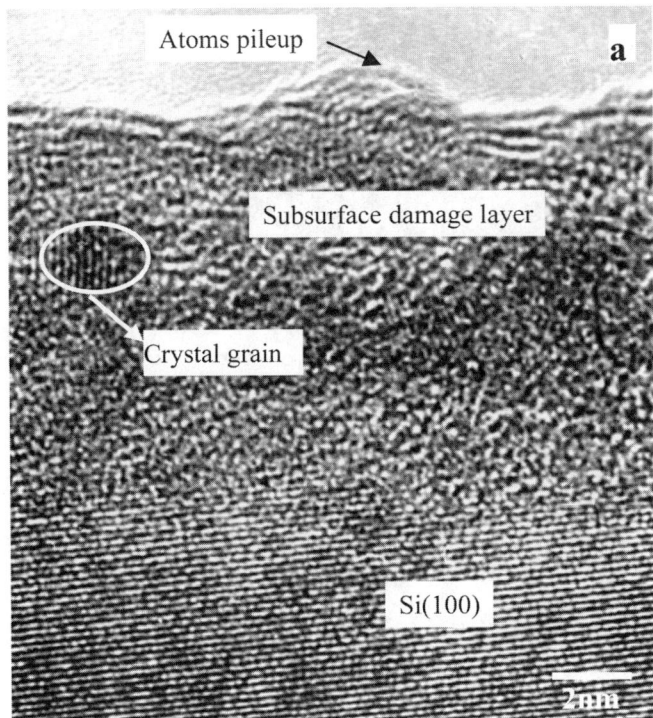

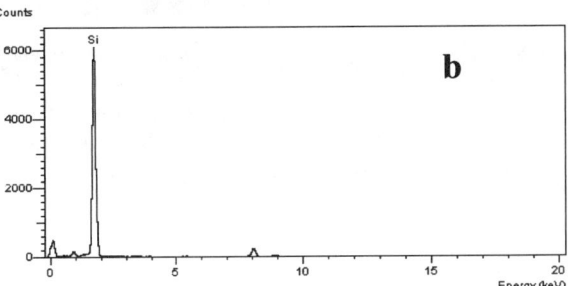

Fig. 11—Observation of subsurface damage. (a) Cross section HTEM images of surface undergoing collision for ten minutes. (b) Electron diffraction pattern from an amorphous area.

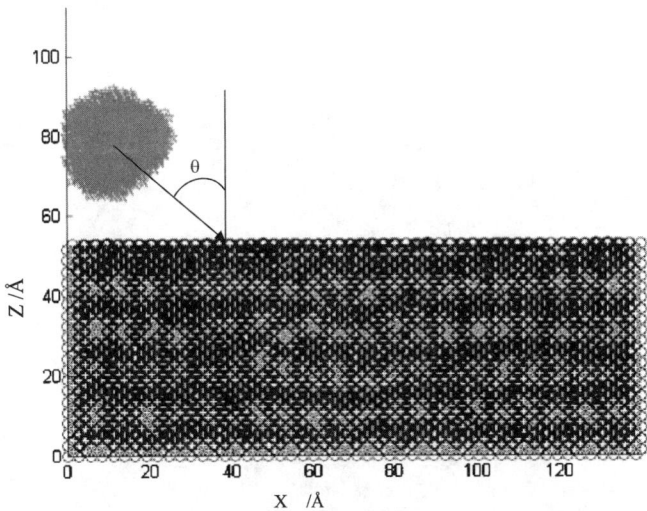

Fig. 13—The particle impacting on a surface at a incident angle of θ.

the particle to move away from the SI surface (Figs. 16(c)–16(f)).

Looking from the top side as shown in Fig. 17(a), it can be seen that the arrangement of atoms in the contact region is much different from that in the outside region. It becomes amorphous in central contact region of the Si surface, and the height of the pileup of atoms at the outlet of the impacted region is different at the different position. It is about 3 nm from the top of the pileup to the deepest position of such an impacted region, as shown in Fig. 17 [28].

Figure 18 [28] shows the variation of the particle speed and the potential energy of the silicon wafer in the collision process. The dashed line means the speed of the particle in the vertical direction and the black one indicates the variation of potential energy of the silicon disk. When the particle penetrates into the wafer surface, its vertical speed becomes lower and lower. Once the particle reaches the deepest position, the speed of the particle becomes zero and the potential energy of the silicon wafer increases to the highest one, and

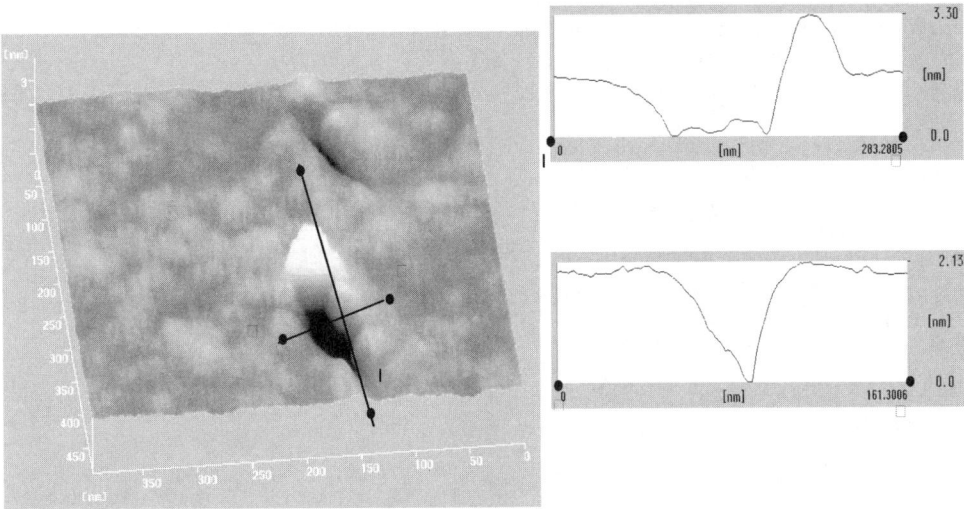

Fig. 12—AFM image of collided surface undergoing 30 minutes.

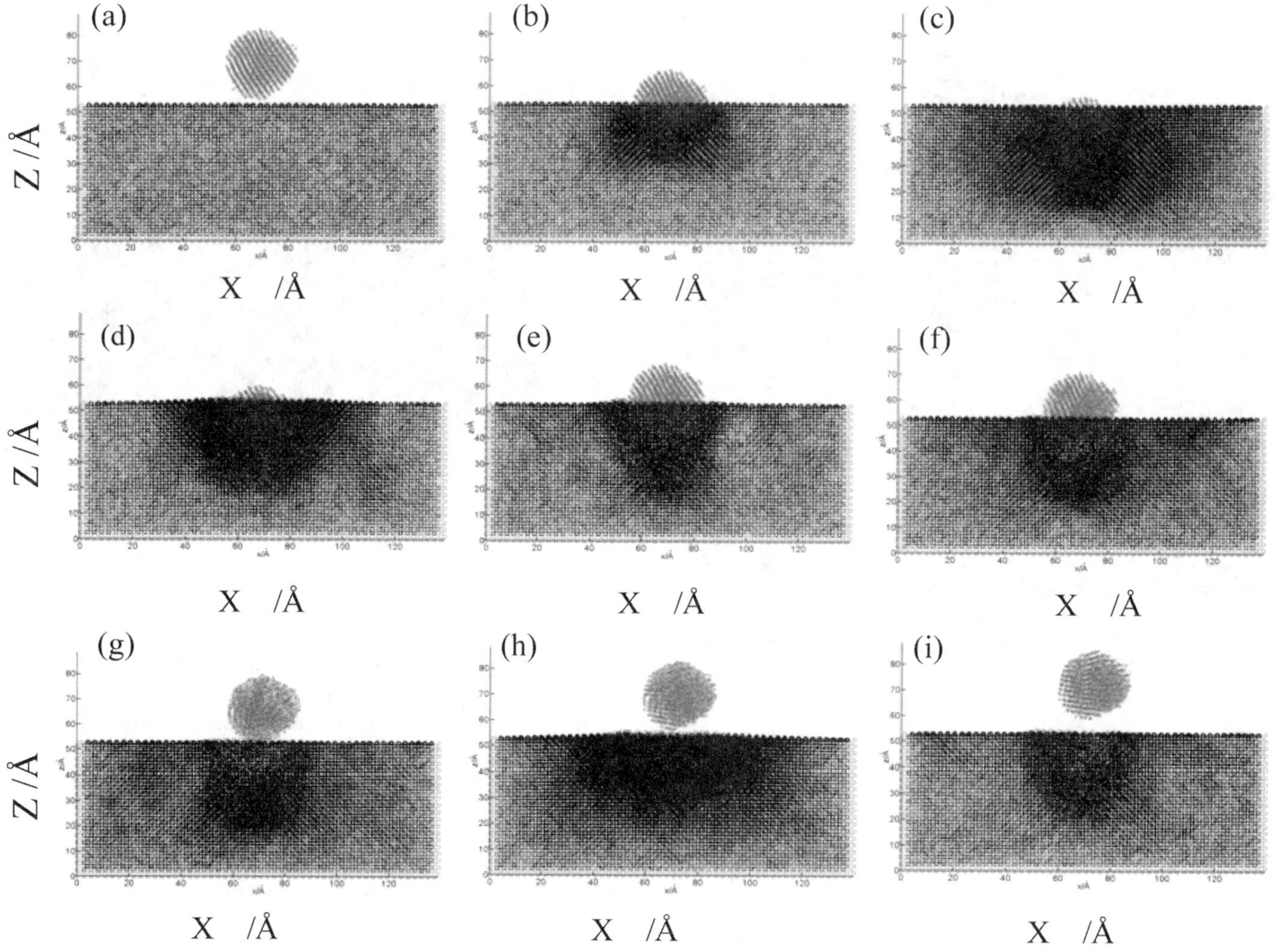

Fig. 14—Particle position and variation of the solid surface in the collision process a incident angle of 0° at the time: (a) 200 fs; (b) 800 fs; (c) 1,700 fs; (d) 2,600 fs; (e) 4,100 fs; (f) 4,500 fs; (g) 7,000 fs; (h) 8,200 fs; (i) 9,000 fs.

so the contact region of the silicon wafer changes into the compressed state. The surface will release elastic deformation energy and push the particle move away from the substrate, and the particle speed in the vertical direction changes into positive. When the speed becomes highest, the surface will change from the compressed state into the tensile state which will pull the particle to go back. Therefore, the speed of the particle will decrease a little. After the particle leaves the surface, its speed keeps constant, and the silicon surface vibrates and gets balanced later.

When the incident angle increases to 60° as shown in Fig. 19, the smaller initial speed in vertical direction v_z leads to a shallower depth of impacted pit. When v_z decreases to zero, v_x still has a large value of 1,500 m/s, resulting in a long scratch on the surface, and a shape of the flatten arc has been left as shown in Figs. 19(c)–19(f).

2.3.2 Effect of the Incident Angle

The impacted regions have different shapes with different incident angles from 0° to 75° as shown in Fig. 20. As incident angle increases, the shape of the impacted region becomes flatter and longer. Until the incident angle increases to 75°, the depressed region can hardly be observed on the sur-

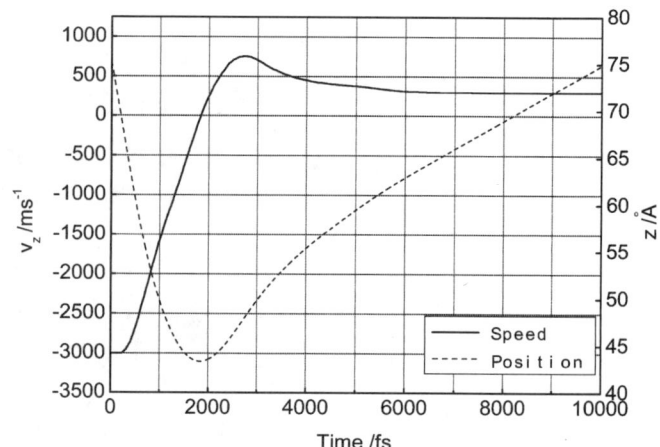

Fig. 15—Speed of the particle in the vertical direction at different position with different time. The diameter of the particle is 2 nm with an incident angle of 0° at an incident speed of 3,000 m/s.

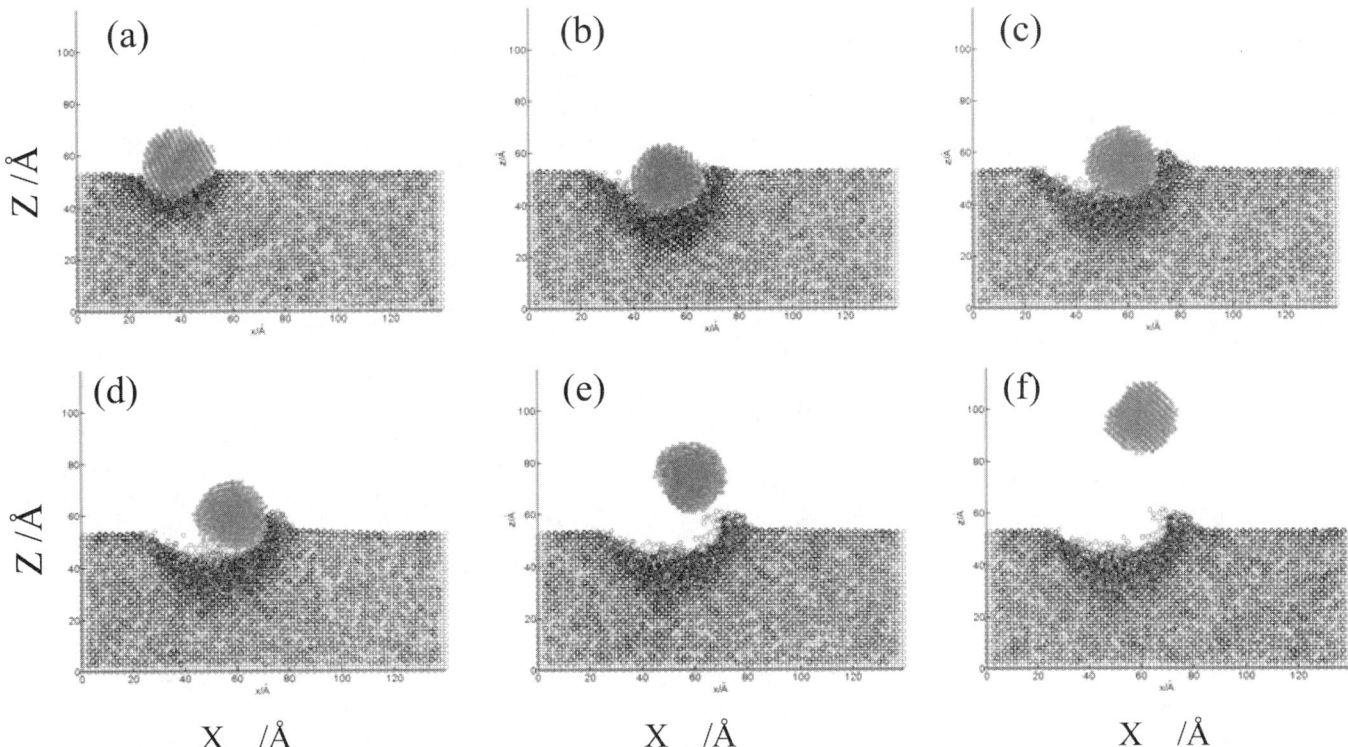

Fig. 16—Snapshots of the nanoparitcle-surface collision with 45° incident angle at an incident speed of 2,100 m/s at different times of (a) 900 fs; (b) 1,900 fs; (c) 3,100 fs; (d) 3,600 fs; (e) 5,700 fs; (f) 9,000 fs.

face. There are hill-shapes of atom pileups at the rim of the impacted region when the incident angle is in the range from 15° to 60°. For the incident angle 0° and that larger than 75°, the penetration depth is so small that the atoms of the surface can't be extruded out to form pileup.

2.3.3 Conclusions

The incident angle decides the position of the particle to contact with the impacted surface when the particle reaches the deepest position, and it is the contact part of the particle where the released elastic deformation energy of the surface acts on, these leading to the recoil angle sensitive to the incident angle in the collision process. The shapes of the impacted region change from a deep scoop to a flatten arc with the incident angle changing from 0° to 75°. Some silicon atoms on the surface have been extruded out by the incident particle, and form a pileup at the rim of the impacted region. Amorphous phase transformation and plastic flow inside the amorphous zone is the main deformation in the silicon surface. There is no dislocation and micro-crack in the silicon surface after the collision of a nanoparticle.

3 Chemical Mechanical Polishing (CMP)

3.1 Introduction

According to the international technology roadmap for semiconductors, chips with a wafer diameter of 450 mm and a feature size of 0.05 μm by 2011 will serve to decrease manufacturing costs [29]. In the integrated circuit (IC) manufacture process as shown in Fig. 21 [30], dielectric stacks have been formed by the ion etching on the coating of dielectric material formed on the silicon surface (Fig. 21(a)).

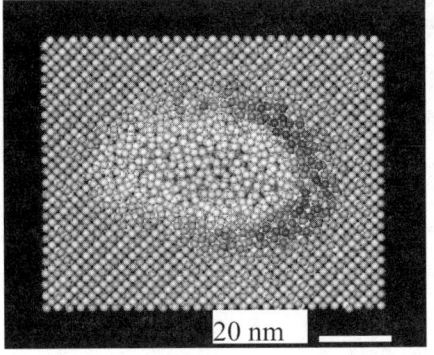
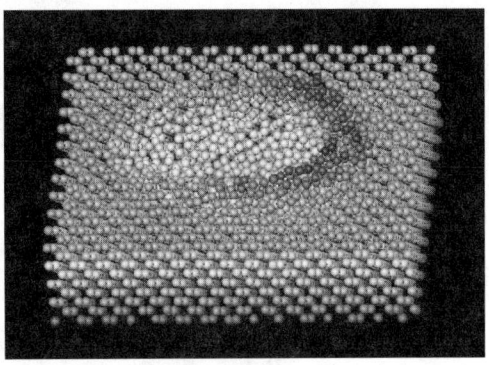

Fig. 17—Incident angle: 45°, Incident energy: 5 eV/atom.

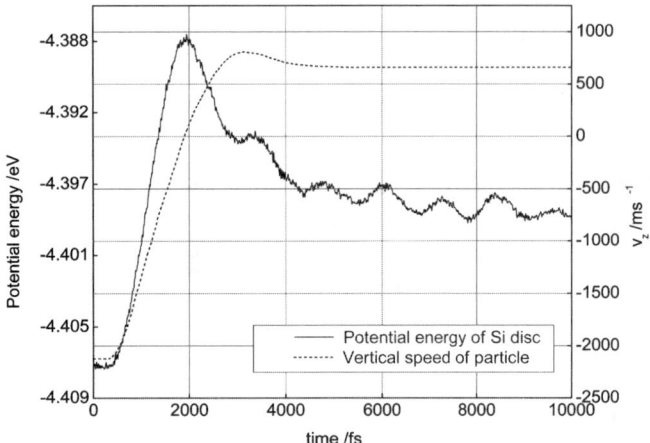

Fig. 18—Potential energy of Si disk and particle speed in vertical direction during the collision process with an incident angle of 45° at an incident speed of 2200 m/s.

surfaces. So global planarization technology is needed. The corresponding wafer surface quality is also expected to meet the unprecedented requirement. For example, surfaces and subsurfaces of wafers must be damage-free after polishing, and the requirements for surface quality of a computer hard disk become much higher with the increase of storage density. These requirements are close to the limits of the present manufacturing technology.

The technique of planarization has become the key factor to meet such requirements. Chemical Mechanical Polishing, also often referred to as Chemical Mechanical Planarization (CMP), is recognized as the best method to achieve global planarization in super-precision surface fabrication. From the technology point view, the initial work to develop CMP into semiconductor fabrication was done by IBM, where an expertise including understanding of machine, pads, and slurries was used for the silicon wafer fabrication [31,32]. It has also been a key technology for facilitating the development of high density multilevel interconnected circuit [33–35], such as silicon, dielectric layer, and metal layers [36,37]. With the development of these high-performance electrical products, CMP has to be improved to meet even more stringent requirements, and to face many new challenges.

There are many types of CMP equipment from different companies for different usages. As shown in Fig. 22, equipment (a) for the planarization of the copper layer in IC pro-

Then a very thin barrier layer and a copper seed layer are formed (Figs. 21(b) and 21(c)>). In order to conduct the electric current, good conductive material, e.g., copper, will be coated on the surface of the copper seed layer which forms a rough surface as shown in Fig. 21(d). Since multilayer's introduction into IC production, the surface coated with copper must be very smooth, clean, and bare of dielectric stacks

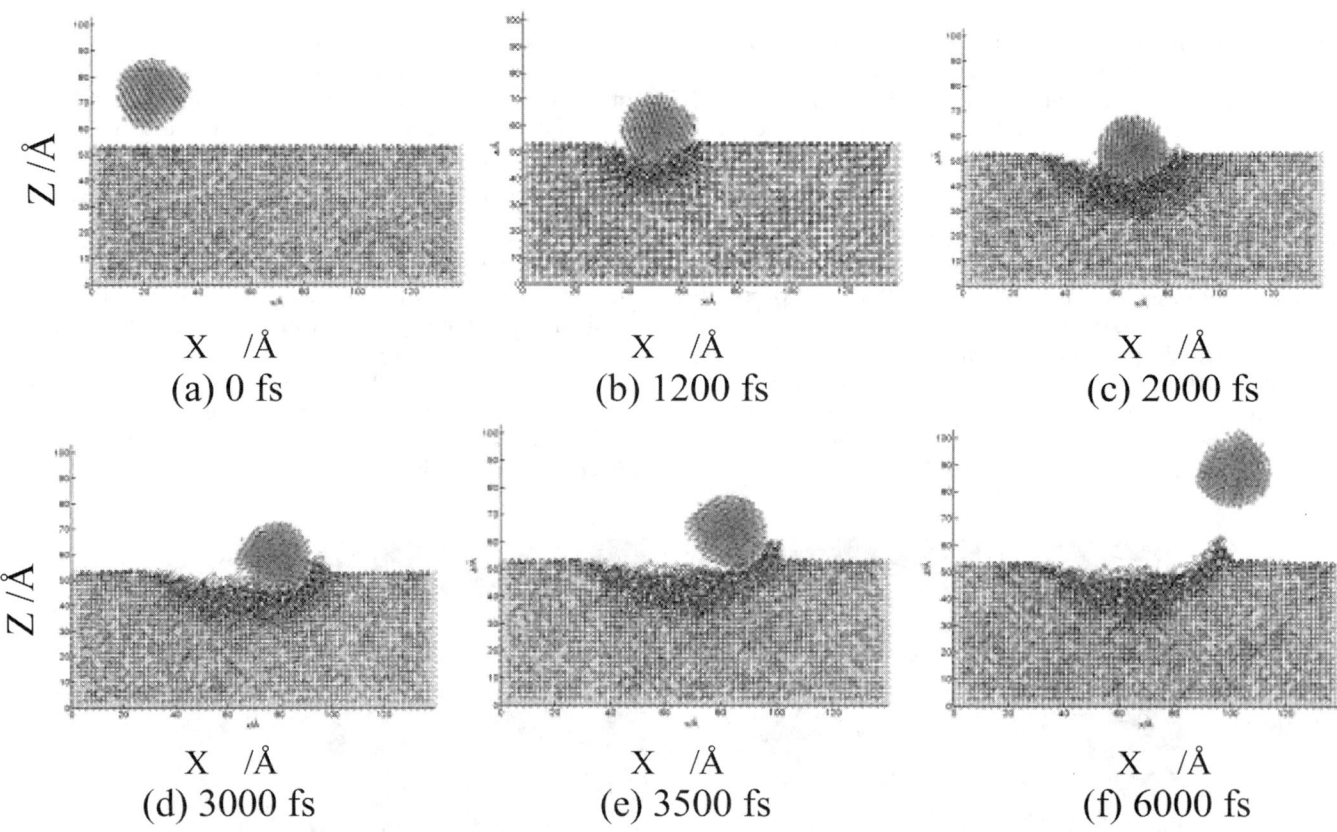

Fig. 19—Snapshots of the nanoparitcle-surface collision with the incident angle of 60° at the time: (a) 0 fs; (b) 1,200 fs; (c) 2,000 fs; (d) 3,000 fs; (e) 3,500 fs; (f) 6,000 fs.

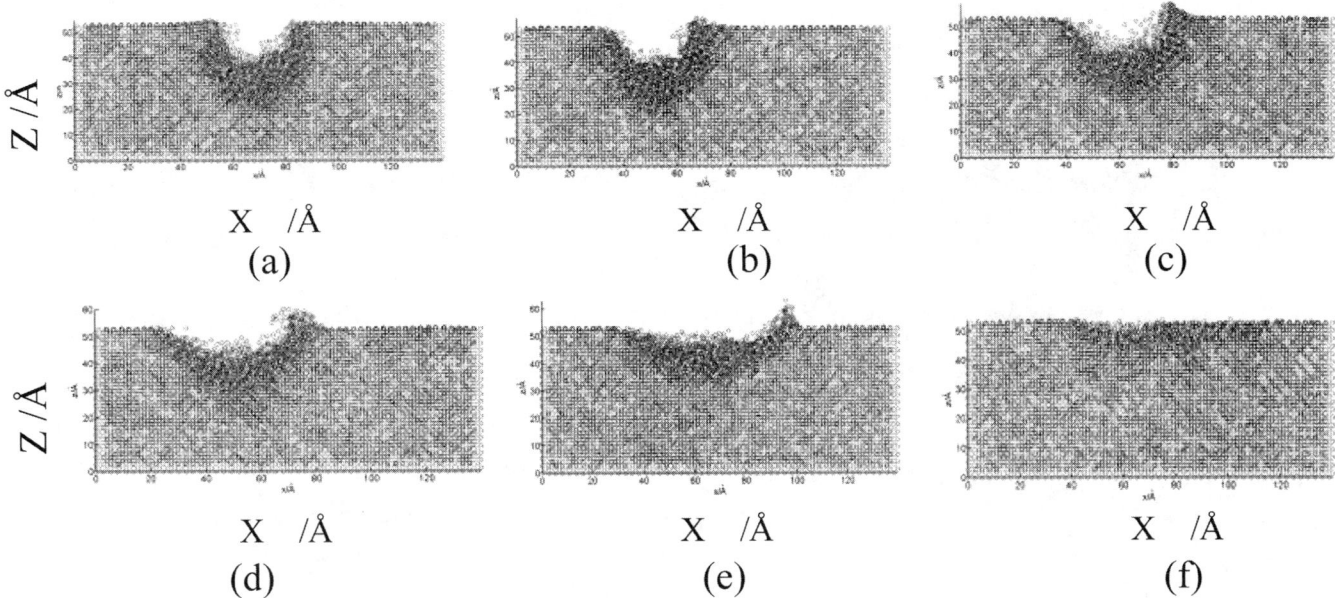

Fig. 20—Configurations of the silicon surface after a nano-particle impacting with the incident angle of (a) 0°, (b) 15°, (c) 30°, (d) 45°, (e) 60°, (f) 75°.

duction, (b) for Si wafer, and (c) for the polishing of both sides of HDD substrate disk. However, the typical polishing tool is shown in Fig. 23(a). The polishing pad is circular which is adhered on the surface of a platen that can rotate round its own axis. The wafer placed in a carrier face down is forced down against the pad, and it also can rotate round its own axis. The force applied through the carrier onto the wafer is generally 0.5–10 psi for different materials. In the polishing process, the slurry is dispensed from a tube in front of the wafer, and is pulled under the wafer by the pad surface, which usually has many small hollow spherical pores exposed on the surface to contain the slurry as it rotates with the platen. Therefore, there is a slurry layer between the wafer and the pad surface as shown in Fig. 23(b). The carrier will vibrate due to the speed difference between the two sides, side close to the axis central of the platen and side far away from the axis.

Conditioning, an important technology in the CMP process, is to maintain the asperity structures on the pad surface, which force the abrasive particles against the wafer. A conditioner is usually used to move a hard abrading surface, often a matrix with embedded diamond points, across the pad surface to roughen it. An inadequate roughened pad surface results in a very low polish rate [31].

3.2 CMP Slurry

The effect of abrasive particle size on material removal rate and surface finish has been widely reported [38–42]. Xie [39] observed that the polishing rate increased with the increase of particle size, while Bielmann [40] claimed that a decreased alumina particle size led to higher tungsten removal rate and the local roughness of the polished W surfaces was insensitive to alumina particle size. Zhou et al. [41] found that neither small (10 nm) nor large (140 nm) but 80 nm SiO_2 particle exhibited the highest material removal rate and the best surface finish during wafer CMP. However Liu's results [34] indicated that small size (ranging from 15 to 20 nm) silica instead of large size (ranging from 50 to 70 nm) silica as abrasives could effectively reduce the roughness and the damaged layer thickness during silicon

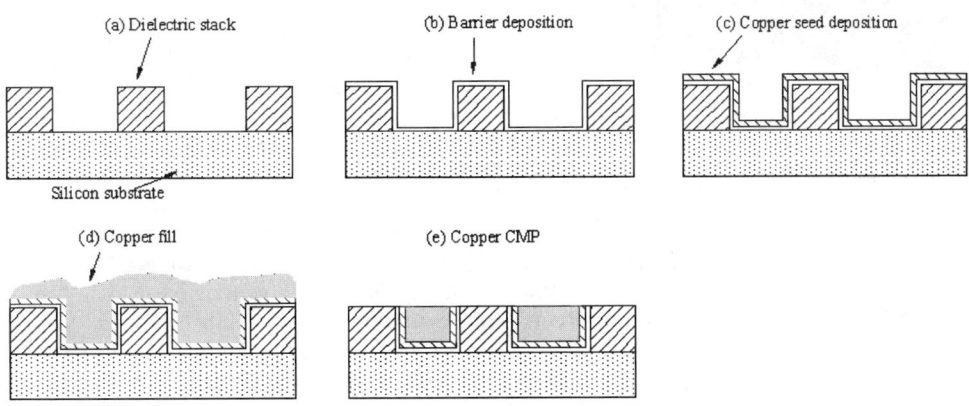

Fig. 21—Production process of integrated circuit (IC).

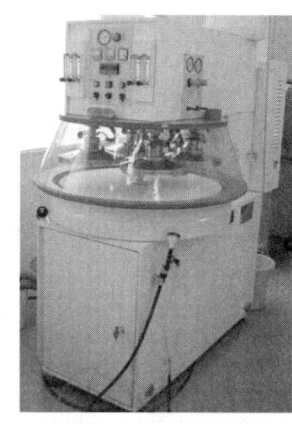

Fig. 22—The photograph of SPEEDFAM–16B–4M polisher.

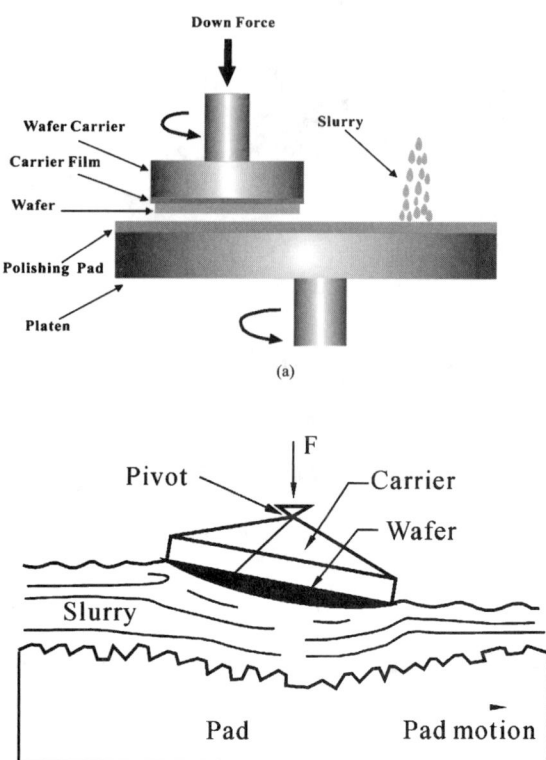

Fig. 23—Sketch of CMP tester (a) polish rig, (b) contact area of carrier, slurry, and polish pad.

substrate CMP. Basim's results [42] showed that the surface damage increased with the increasing of SiO_2 size, and the presence of coarser particles tended to create critical defects on the oxide film and change the polishing mechanism. As of yet there appears to be no consensus on the influence of particle size on the polishing performances. Lei and Luo [43] investigated the effect of the particle sizes ranging from 10 nm to 160 nm on the removal rate and polished surface waviness and roughness. From their results, slurry with smaller particles can reach lower surface roughness but also smaller material removal rate.

3.3 CMP Pads

The polishing pad, as another consumable material, also has a dominating effect in the CMP process, which is usually made of a matrix of cast polyurethane foam with filler material to control hardness of polyurethane impregnated felts. The pad carries the slurry on top of it, executes the polishing action, and transmits the normal and shear forces for polishing, thereby playing a very crucial role in process optimization [44–46].

Pads used in IC manufacturing can be grouped into four classes as shown in Table 2 [44,47]: Class I, felts and polymer impregnated felts; Class II, microporous synthetic leathers; Class III, filled polymer films; Class IV, unfilled textured polymer films. Different fabrication processes lead to different mechanical properties of the pad as shown in Table 3. The mechanical, physical, and chemical properties of polyurethane material can be permanently or temporarily altered if the pad is heated beyond a particular limit [48]. The tests about the thermomechanical analysis, dynamic mechanical analysis, and thermogravimeteric analysis have been performed for material chrarcterization of the polishing pad by Lu et al. [49], Li et al. [50], and Powell [51]. The effect of pad texture on tribological and kinetic properties during the polishing process has been studied by Philipossian and Olsen [52]. Real time monitoring of friction coefficient was also

	Class I	Class II	Class III	Class IV
Major structural characteristic	Felted fiber with polymer binder	High-porosity film on substrate	Solid urethane sheet with filer (voids, SiO_2, CeO_2, etc.)	Solid polymer sheet with surface texture
Subcategories	Spun bond nonwovens	Free-standing thin films; felt substrates	Foams; oxide filled	
Hardness	Medium to high	Low	High	Very high
Compressibility	Medium to high	High	Low	Very low
Slurry carrying capacity	Low to high	Very high	Low	Minimal
Bulk microstructure	Continuous channels Between fibers	Complex foam to vertically oriented channels	Closed cell to open cell foam	None
Polymer types employed	Urethanes; polyolefins (fiber phase)	Urethanes; polyolefins (fiber phase)	Urethanes	Various
Representative commercial trade names	Pellon™, Suba™	Politex™	IC 1000	OXP3000
Application(s)	Si stock polish; tungsten damascene CMP	Si final polish; CMP; post-CMP buff	Si stock; ILD CMP; metal dual damascene	ILD CMP; shallow trench Isolation; metal dual damascene

TABLE 2—Summary of polishing pad classes [44,47].

done by them to estimate the normal shear forces originating during a particular CMP process. In their experiment, the Stribeck curve shows that mechanism of polishing for the K-grooved pad remains in boundary lubrication throughout the range of parameters studied. Flat and XY-grooved pads begin in boundary lubrication and then migrate to partial lubrication as the Sommerfeld number increases. Perforated pads begin in boundary lubrication and then transit to partial lubrication at a higher value of the Sommerfeld number. Thus different pad surface texture showed different material removal mechanism, and then different tribological properties, and different removal rate [44].

The chemical and mechanical properties have a great influence on the polish quality. Ma et al. [53] have observed that, as shown in Fig. 24, the waviness of polished surface decreases with polish time in the first 15 hours, and then becomes stable. Pad A as shown in Fig. 25(a) can be used stably for more than 65 hours. However, for pad B as shown in Fig. 25(b), it just can be used for only 25 hours and then became unstable.

3.4 CMP in IC Fabrication
3.4.1.1 Silicon

Silicon wafer has been extensively used in the semiconductor industry. CMP of silicon is one of the key technologies to obtain a smooth, defect-free, and high reflecting silicon surfaces in microelectronic device patterning. Silicon surface qualities have a direct effect on physical properties, such as breakdown point, interface state, and minority carrier lifetime, etc. Cook et al. [54] considered the chemical processes involved in the polishing of glass and extended it to the polishing of silicon wafer. They presented the chemical process which occurs by the interaction of the silicon layer and the polishing particle with water and other chemicals in solution, and the removal process as a traveling indenter plowing across the silicon wafer surface. With examinations through TEM, it is believed that an alternative mechanism of material removal during CMP could be plastic cutting/shearing in the outermost Si surface layer, or simply removal of the newly formed oxide [55]. According to Ref. [56], abrasive size is the key factor to affect the removal mechanism. Yield mechanism for large abrasive size ($>2~\mu m$) appears to be brittle in nature, and the main mode of material removal is fracture. Whereas dislocation networks and slip planes along the {111} direction seem to be the mechanism by which the material undergoes plastic deformation, and the transition from brittle to ductile yield occurs for alumina slurries at a particle size below 0.3 μm. However, completely damage-free polishing occurs at a particle size of 50 nm by HRTEM images (Fig. 26).

The atomic-scale removal mechanism during chemical mechanical polishing of silicon had been investigated by Graf et al. [57] in combination with examinations through XPS and high resolution electron energy loss spectroscopy. They proposed that abrasive particles carry a high local OH^- concentration which will transfer to the wafer surface and catalyze the corrosive action of water resulting in the cleavage of silicon backbonds, and therefore facilitates the removal mechanism. Liu et al. [34] presented that CMP is a complicated multiphase reaction process. It mainly includes the following two dynamics processes. First, the active component in polishing slurry reacts with the atoms of silicon wafer. The second step is the process of resultants desorption induced by mechanical action between abrasive particles and Si wafer. Resultants come to slurry and are carried away immediately. The balance and synergetic effects of these two

TABLE 3—Physical properties of the polishing pad.

Thickness (mm)	Weight (g/m^2)	Density (g/m^3)	Compressibility (%)	Elasticity (%)	Nap Layer Thickness (μm)	Pore Size (μm) min	max
1.0	375	0.44	8.4	84.7	600	10	50

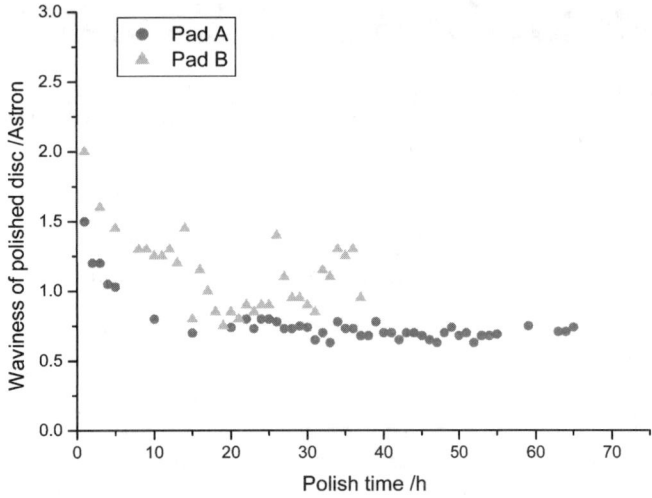

Fig. 24—The waviness of polished surface with polish time using different pads.

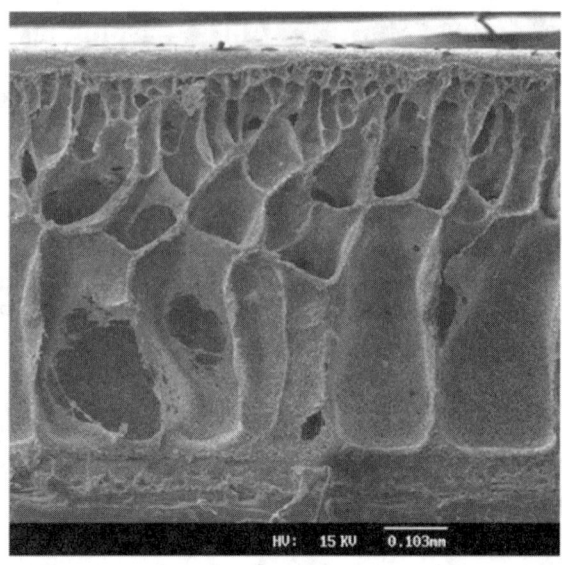

(a) The cross section of polish pad A

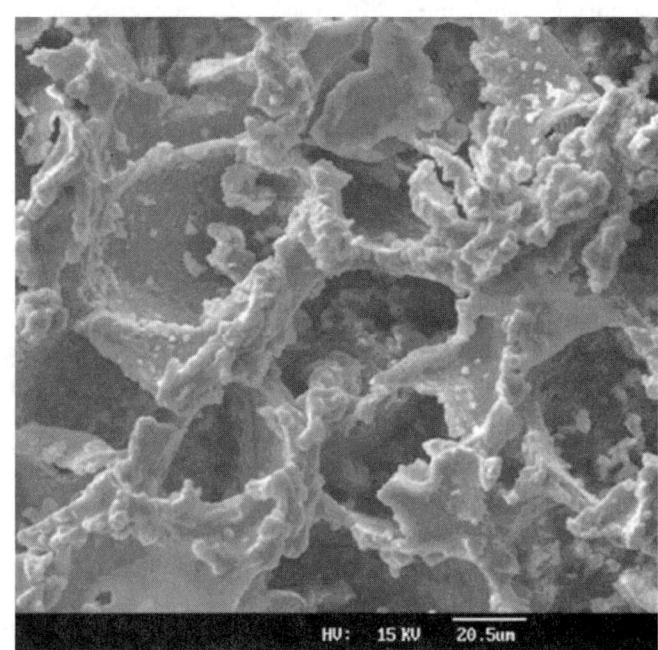

(b) The surface of polish pad B

Fig. 25—SEM photos of polish pads: (a) the cross section of polish pad A, (b) the surface of polish pad B.

steps decide the total removal rate and surface finishing degree.

3.4.2.2 Dielectric Layer

CMP is increasingly applied in the planarization of silica which is used widely as dielectric material in microelectronic devices [58–61]. Particle size also has a significant influence on the material removal rate during SiO_2 CMP, similar to that of Si wafer. Mahajan [58] considered four different particle sizes, namely, 0.2, 0.5, 1.0, and 1.5 μm. The results indicate that as particle size increases, there is a transition in the mechanism of material removal from surface area-based mechanism to indentation-based mechanism as shown in Fig. 27 [62]. The proposition is reasonable in explaining the variation of the removal rate with the weight concentration of particle in CMP. However, when oxide CMP is carried out with slurry with particle size below 100 nm, existing viewpoints can be classified into the following two: Tomozawa et al. [59,60] proposed that two inter-related processes can lead to the wear of the oxide layer during CMP. One process is heating whose source is the friction between the oxide wafer and abrasives. The second contributing process is the hydration of the oxide that occurs during plastic deformation. Plastic deformation is assisted by friction heating and the removal of the resulting softer hydrated surface layer takes place by the plowing action of the abrasive particles. On the other hand, Hoshino et al. [61] proposed the SiO_2 layer is first reacted with abrasive particles and a multiple number of chemical bondings are formed on the surface. Then mechanical tearing of Si–O–Si bonds leads to the SiO_2 removal as a lump, and the lump is released from the particles downstream.

Currently, there is a trend of low dielectric constant (low-k) interlevel dielectrics materials to replace SiO_2 for better mechanical character, thermal stability, and thermal conductivity [37,63,64]. The lower the k value is, the softer the material is, and therefore, there will be a big difference between the elastic modulus of metal and that of the low-k material. The dehiscence between the surfaces of copper and low-k material, the deformation and the rupture of copper wire will take place during CMP as shown in Fig. 28 [65].

Successful CMP of low-k material ($k<2.2$), such as organic polymers (SILK), organosilicate glasses (OSG), and porous media has become a very important problem in IC fabrication. Extensive studies of slurry chemistry, abrasive particles, CMP pressures, and new CMP tools are bursting to be done. Experimental results for dielectric film removal rate, surface roughness, and surface chemical change can also be used to develop a phenomenological understanding of the CMP removal of low-k materials. Borst et al. [37] proposed a removal mechanism model based on phenomenogical progression for SILK CMP. The contents include: (1) mass transport of reactant from the bulk slurry/wafer interface; (2) ab-

Fig. 26—Cross-sectional HRETM images of Si (100) polished using 50 nm diameter abrasive [56].

sorption of reactant to available SILK polymer surface sites; (3) reaction between adsorbed reactant and the SILK polymer surface site to which it is attached; (4) shear-enhanced desorption of weakened altered polymer surface layer; (5) mass transport of polymer product from the slurry wafer interface to the bulk slurry. This mechanism is similar to that of other low-k interlevel dielectrics material, such as OSG and porous media. Wang et al. [65] propose a surface stress free (SSF) polishing method for the removal of Cu layer and the planarization of IC wafer, which will be important for the usage of the soft low-k material.

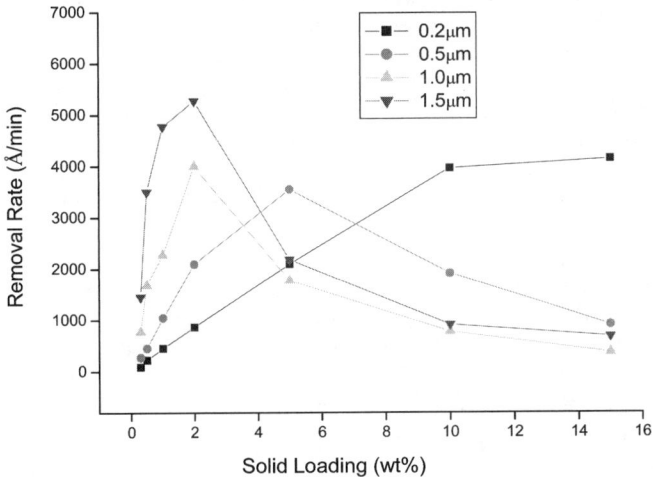

Fig. 27—The effect of particles size on removal rate [62].

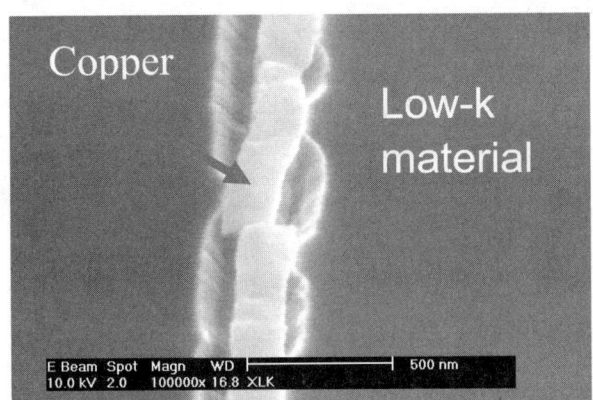

Fig. 28—Polishing for low-k material and copper [65].

3.4.3.3 Metal Material

Metal polishing mechanisms appear to be considerably different from silica polishing. The critical event that determines the polishing process in metal CMP appears not only to be influenced by the crystallographic/microstructure deformation process but also to relate to more complex components of slurry [18]. To better understand the removal mechanism in metal CMP, tungsten is chosen, since both industrial and laboratory CMP data are available for this metal, and its abrasion behavior as a metal is similar to that of other ductile metals which have been studied quite extensively under two- and three-body abrasion [66].

The first mechanism specifically for tungsten CMP was proposed by Kaufman et al. [67]. They thought, first, chemical action dissolves W and forms a very thin passivating film which stops growth as soon as it reaches a thickness of one or a few moleculars later. Second, the film is removed locally by the mechanical action of abrasive particles, which contact with the protrude parts of the wafer surface, and then cause material loss. In recent years, most of the analysis and models for metal CMP are built based on the Kaufman model [68,69]. However, the model is not involved in microscopic structure analysis for the polished surface, but focuses on interpreting macroscopic phenomena happening during CMP [18].

Anantha [70] indicated the texture or the preferred orientation of the grains after CMP is different from the initial texture in W interconnect structures. The deformation of metal oxide film is principally controlled by a stress-induced vacancies migration along a gradient from grain boundaries and atomic diffusion. Once the stress exceeds the threshold, actual material removal occurs. As a result, in a typical metal CMP process, after the creation of the oxide, the abrasive removal is controlled by microstructure in the atomic scale. However, Kneer [71] suggested that transgranular fracture assisted by intergranular corrosion is the dominant material removal mechanism during tungsten CMP. In order to testify the influence of particle plowing and transgranular fracture on material removal during tungsten CMP, Stein [72], in combination with examinations through AFM and TEM, presented that neither intergranular fracture nor mechanical abrasive are likely to be the dominant mechanisms of tungsten removal during CMP. Tungsten is commonly polished using alumina particles slurry. To investigate the interactions of tungsten ions and alumina particles in CMP systems, electrophoretic mobilities of particle were measured as a function of pH value and time in the presence of tungsten ions. The results suggested that metal oxide films on a metal surface are comprised of aggregates of nanosize metal oxide particles, and tungsten ions are strongly chemisorbed by alumina particles and that adsorption may provide a pathway for metal removal during CMP, as reported in Ref. [73].

To develop a high signal speed semiconductor, copper has become the mainstream interconnect metal for the generation silicon devices with the intergration metallization, dual damascene technique. Copper has replaced tungsten in ultra-large scale integration (ULSI) in many devices. There has been a great deal of effort focused on understanding the interaction in copper polishing [74–78]. Similar to tungsten CMP, surface passivation can also occur in copper CMP [74].

However, copper behaves significantly different from tungsten in CMP and it can be polished at low pH value under nonpassivating conditions [75]. Abrasives play an important role in the CMP process, not only in enhancing the polishing rate but also in determining the removal mechanisms. Polishing mechanism of Cu using submicron-sized alumina abrasive is dominated by the wear process in which the material removal depends on the hardness of the abrasive particles and the film being polished [76]. In order to understand the effect of mechanical behavior in CMP, Liang et al. [77] proposed that there are apparently two possible polishing mechanisms during Cu CMP. The first possibility is to assume the formation quickly of the pasivated film on the copper surface at alkaline pH region. The surface energy of oxide film was excited by mechanical interactions between the pad and abrasive particles under the polishing action. Bonding of copper oxide breaks down and therefore it dissolves into the slurry. The second possibility is that the pure copper surface is generated during polishing. The surface energy is excited by mechanical interaction and the copper bonding breaks down. Copper atoms are oxidized immediately in the slurry and Cu_2O remains. Further investigation on the removal mechanism for Cu CMP has been performed by Wei et al. [78]. They presented that the material removal occurs by pulling out the surface fragments bonded to the interacting oxygen atom, that is, each oxygen atom is capable of pulling out two Cu atoms and Cu_2O is formed over the surface which can be easily removed from the surface by particles. It would explain why the removal rate is high at low H_2O_2 concentrations and low at high H_2O_2 concentrations.

Gotkis [79] put forward an available model which gives a more accurate representation for the material removal rate which would be: $RR = k_{chem} \cdot (RR_{mech})_0 + k_{mech} \cdot (RR_{chem})_0$ where k_{chem} and k_{mech} are factors that account for the effects of chemical modification and mechanical activation of polished surfaces, respectively. $(RR_{mech})_0$ and $(RR_{chem})_0$, respectively, account for pure abrasive wear and pure chemical dissolution of surface material in cases where no surface modification and no abrasive activation take place. The change of k_{chem} and k_{mech} for CMP results in different mechanisms of polishing in terms of the effects of physical properties of abrasive particles, pad, and the material being polished. For Cu CMP, two distinct regions of removal are suggested to exist [80]; at low etchant concentrations, where the polishing rate varies with etchant concentration, polishing rates are limited by the dissolution rate of the copper and copper oxides that form the surface layer. At high etchant concentrations, where the polishing rate varies with pressure, the polishing rate is limited by the rate at which surface is abraded. In addition, three regions of material removal are defined in terms of a material removal rate as a function of the abrasive weight concentration during CMP [69]. First a chemically dominant and rapidly increasing region, whose range is determined by the generation rate and hardness of the surface passivation layer; second, a mechanically dominant linear region, where the material removal is proportional to the weight concentration; and third, a mechanical dominant saturation region, where the material removal saturates because the total contact area is fully occupied by the abrasives.

Copper is also well poised to replace aluminum as the mainstream interconnect metal in IC, mainly owing to its lower resistivity and better electromigration resistance. Nevertheless, there are several challenges to use copper metallization in ULSI circuits. Diffusion barriers are required to prevent the diffusion of copper into the silicon and SiO_2 where copper affects the electronic properties of the devices. Tantalum and its nitride have been investigated to develop suitable diffusion barriers as well as an adhesion promotion layer. Since it is necessary to polish the barrier layer after removing the overlaying excess copper, it is essential to understand the removal mechanism of Ta CMP. Similar to the mechanism of copper CMP, Ta CMP also exhibits alternately the formation of passive film on the surface being polished and the abrasive removal film of particles. Among all factors, pH value and ionic strength play crucial roles in determining the polishing mechanism, since variation in pH value and ionic strength of the slurry will result in variations of the surface electrostatic interaction forces among the particles and that between the particles and the tantalum surface [81–84].

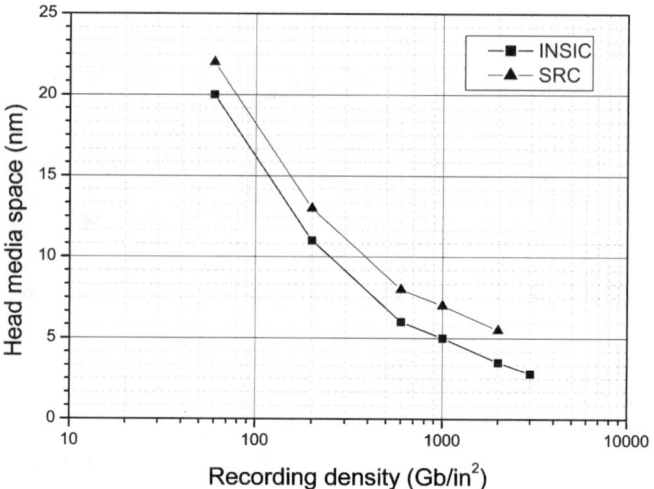

Fig. 29—Head media space with recording density.

3.5 CMP in Computer Magnetic Hard Disk Manufacture

3.5.1 Introduction

With the rapid increase of the capacity of HDD, the areal recording density is growing at an astounding rate of up 100 % annually in recent years [85], which forces magnetic heads to read smaller and weaker signals. To remedy this, scientists are working on technology that enables heads to fly closer to disks to strengthen the output signals [86–89]. When the areal recording density reaches 1,000 $Gb/in.^2$, the flight height should be lowered to approximately 2 to 3 nm [90] as shown in Fig. 29. With heads operating so close to disks, media must be ultra-smooth to avert head crashes [91–94], a kind of damage to read/write heads, which is usually caused by sudden contact with the disk surface. Head crashes can damage the data stored in the disk's magnetic coating [95]. Furthermore, scratches, pits, and other micro defects on the media surface, which may also lead to wrong read/write, must be decreased to a minimum.

To meet these requirements, a local and global planarization technique for the disk substrates becomes neces-

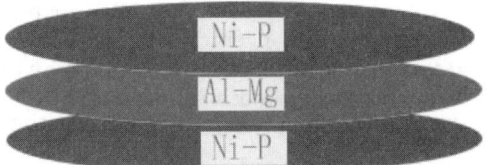

Fig. 30—HDD disk substrate.

sary [96]. The average roughness (Ra) and waviness (Wa), which represents the finer or shorter spatial wavelength features and the longer spatial wavelength features of the surfaces, respectively, have become the most important quality parameters for disk substrates.

Among numerous planarization technologies, CMP provides a global planarization of topography with a low post-planarization slope [97,98]. It can also dramatically reduce topographical variations to a degree not possible with any other planarizing process [97,99,100].

However, compared with a large amount of CMP studies for IC manufacturing, very few researches on CMP of hard disk substrate have been reported to date except some patents on slurries [101–107]. In these patents, the influences of slurry parameters, such as abrasive size, abrasive content, oxidizer content, and pH value on the polishing performances, and the CMP mechanism of disk substrate have seldom been involved.

Lei and Luo [43] have investigated the influence of particles size, pH value, solid concentration, polishing time, etc., on the material removal rate, polished surface roughness, and waviness. Their tests were conducted with a SPEEDFAM-16B-4M CMP equipment as shown in Fig. 22(c). The polishing conditions are as follows: processing pressure of 70 g/cm^2, rotating speed of 25 r/min, processing duration of 8.5 min and slurry supplying rate of 300 mL/min. Workpieces were 95 mm in outside diameter and 50 mm in diameter of the central hole, which is aluminum alloy substrates with NiP coatings as shown in Fig. 30. The coating consists of about 85 wt % nickel and 15 wt % phosphorus elements. The substrates polished in Al_2O_3 slurry have a surface roughness (Ra) of about 6 Å as shown in Fig. 31. The surface morphology and physical properties of the polishing pad are shown in Fig. 32 and Table 3, respectively.

The surface features, especially Wa and Ra, of the above disk substrates were measured to evaluate the polishing effects in different kinds of slurries components. The lower the Ra and Wa values are, the higher level the surface planarization is. Wa and Ra were measured using Chapman MP2000$^+$ surface profiler with the resolution of 0.3 Å for Wa and 0.1 Å for Ra. The measuring wavelength was 80 μm for Ra and 400 μm for Wa. The polished surface topography was measured using a DI D-300 atomic force microscope with a resolution of 0.01 Å in vertical direction.

3.5.2 Effect of Abrasive Size
CMP of HDD substrate in four kinds of slurries, each with a different mean particle diameter of 15, 30, 50, and 160 nm as shown in Fig. 33, was conducted by Lei and Luo [43]. The dependences of Wa, Ra, and material removal rate on the SiO_2 particle size are shown in Figs. 34(a)–34(c), respec-

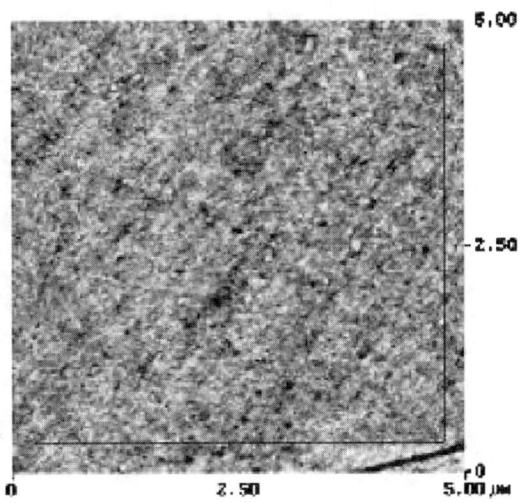

Fig. 31—Surface AFM image for Ni-P coating polished by Al_2O_3 slurry.

tively. It can be seen that the particle size has a strong impact on surface quality and material removal rate (MRR). Wa and Ra of the polished surfaces exhibit the same trend: with increase of the average size of SiO_2 particle from 15 nm to 160 nm, Wa and Ra decreases when the particle size does not exceed 30 nm, and then increase continuously. Wa and Ra reach the minimum at a particle average diameter of 30 nm. However, Pan et al. got a very smooth surface (Ra 0.9 nm under AFM) by using very small SiO_2 particle (about 10 nm in diameter) [108,109]. It should be noticed that small particle size results in lower Wa and Ra of the polished surfaces. This is different from Bielmann's [40] but similar to Zhou's [41] and Liu's [34] results.

Figure 34(c) shows that, with the increasing of SiO_2 particle size, material removal rate increases continuously, which is also different from Bielmann's [40] and Zhou's [41] but similar to Xie's [39] results. This may be due to the stronger mechanical grinding effect, because a larger particle provides stronger impact and grinding on the substrate surfaces. Increasing of the mechanical effect could be dominant during the material removal process when increasing the particle size.

3.5.3 Effect of SiO$_2$ Particle Concentration
The former studies [41,106] have shown that the polishing rate increases with the abrasive content. Taking SiO_2 particle about 30 nm in diameter as an example, the dependences of Wa, Ra, and MRR on the SiO_2 particle content are given in Figs. 35(a)–35(c), respectively [43]. It can be seen that the concentration of solid particles has an obvious effect on surface quality and MRR. With the increasing of the solid concentration, Wa decreases continuously until the SiO_2 particle content reaches 6.2 %, then excessive SiO_2 particle results in increasing of Wa, but the trend of Ra does not seem to keep pace with that of Wa. Higher solid concentration exhibits higher Ra, while it even may be lower at the concentration between 5 % and 6.2 %. The proper size and concentration of solid particles are important to realize the good planarization. It indicates neither strong nor weak but a moderate mechanical effect is needed to match with the chemical effect in the CMP process.

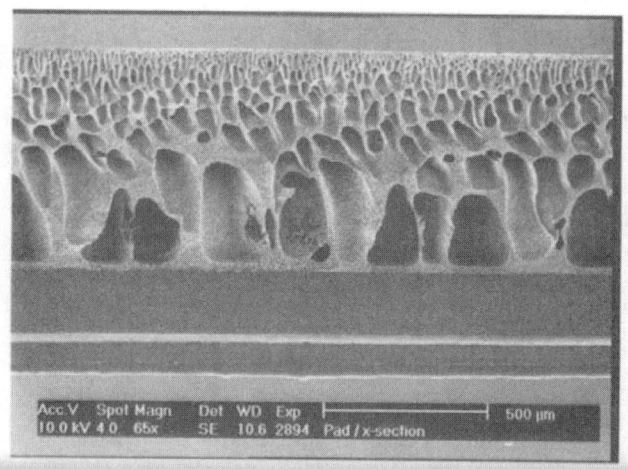

Fig. 32—SEM image of cross section of the polishing pad.

It can be seen from Fig. 35(c) that MRR increases continuously with the solid concentration, which is consentaneous with the former results [41,106]. This may be due to the stronger mechanical grinding effect at the high particle concentration at which particles impact and grind the surfaces.

3.5.4 Effect of Oxidizer Concentration

Oxidizer concentration is also one of the important factors in determining material removal rate, and it was found to increase with the oxidizer concentration [43,106]. With the increase of oxidizer concentration, both Wa and Ra decrease at first and then increase, the optimum concentration is 1 wt %, and relative low oxidizer concentration helps to get lower Wa and Ra values as shown in Figs. 36(a) and 36(b). High concentration of oxidizer may result in excessive corrosive wear, which will lead to the increasing of topographical variations.

MRR increases continuously during the entire testing oxidizer concentration as shown in Fig. 36(c), which is consentaneous with the Stein's results [106], and this may be due to the stronger chemical corrosion effect at high oxidizer concentration. In addition, it should be noticed that varia-

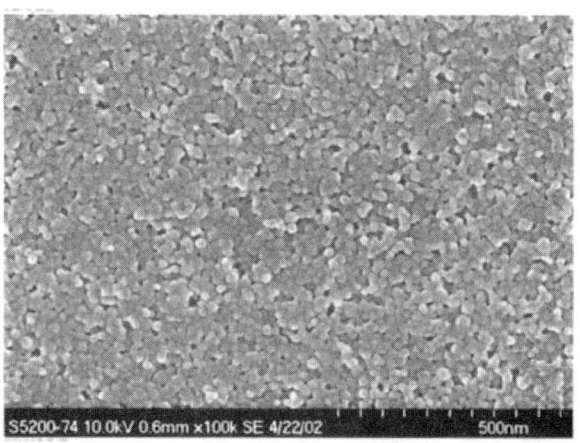

(a) 15 nm SiO_2 particle

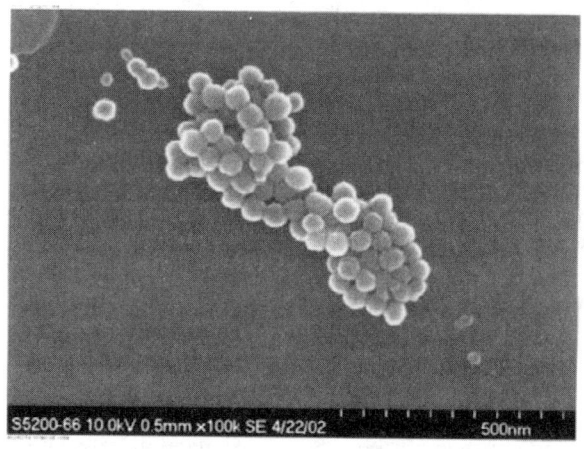

(c) 50 nm SiO_2 particle

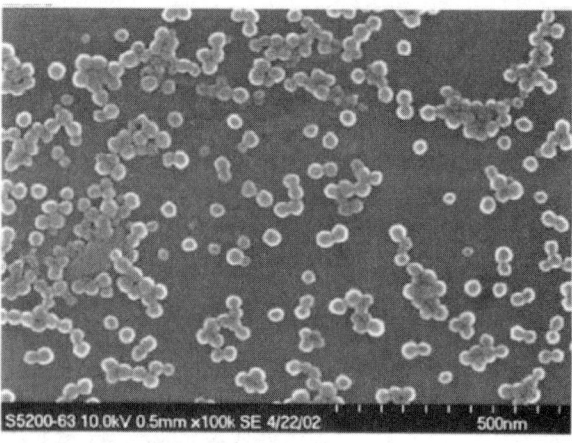

(b) 30 nm SiO_2 particle

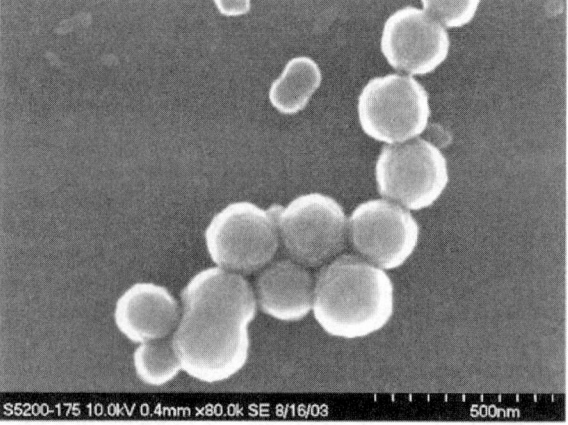

(d) 160 nm SiO_2 particle

Fig. 33—SEM images of SiO_2 particles: (a) average SiO_2 particle size is 15 nm, (b) average SiO_2 particle size is 30 nm, (c) average SiO_2 particle size is 50 nm, (d) average SiO_2 particle size is 160 nm.

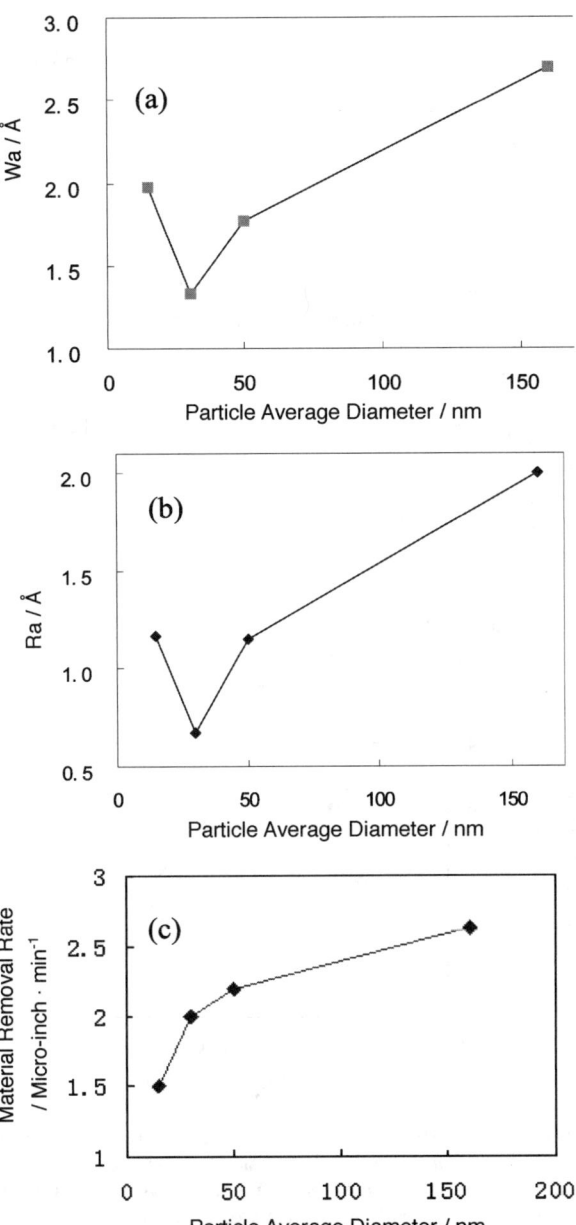

Fig. 34—Effect of the particle size on (a) surface waviness Wa, (b) surface roughness Ra, and (c) material removal rate. The slurry contains 5 wt % SiO_2 particles, 1.5 wt % oxidizer, and 1 wt % lubricant in DI water at pH 1.8.

tions of MRR with SiO_2 size, solid concentration, and oxidizer concentration exhibit the similar trend. It means that increasing any one of the mechanical and chemical effects can enhance MRR.

3.5.5 Effect of Lubricant Additive Concentration

In the CMP process, polishing noise and slight vibrating from CMP equipment are present frequently, which may lead to surface unevenness and the emergence of polishing lines [107]. A kind of carboxylic acid containing more than ten carbons alkyl chain was introduced as lubricant to the slurry [43]. The lubricant is very effective in eliminating the polishing noise. As shown in Figs. 37(a) and 37(b), the addi-

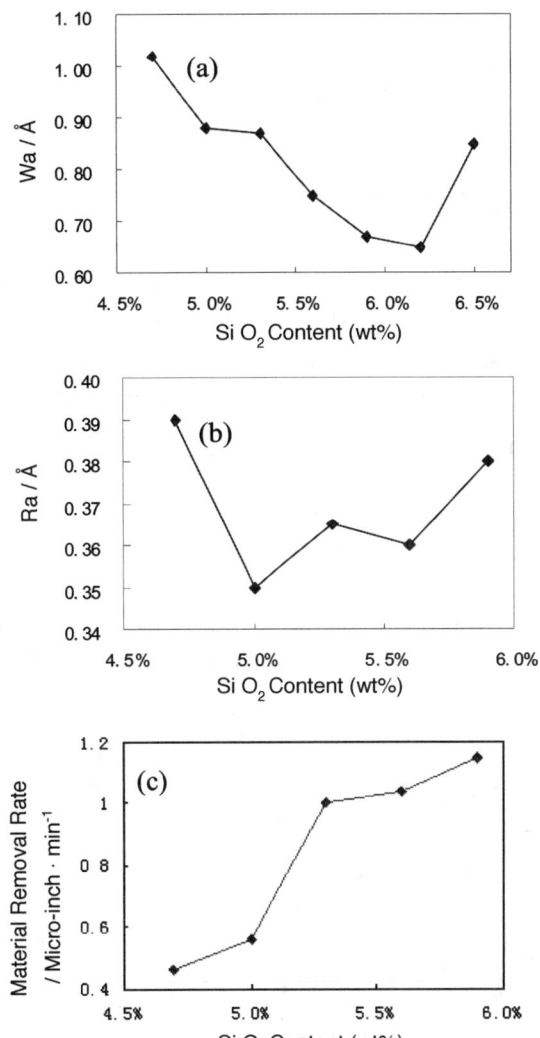

Fig. 35—Effect of the solid concentration of 30 nm SiO_2 particles in slurry on (a) surface waviness Wa, (b) surface roughness Ra, and (c) material removal rate. The slurry contains 1 wt % oxidizer, and 1 wt % lubricant in DI water at pH 1.65.

tion of the lubricant can remarkably decrease the Wa and Ra of the polished surfaces. With the increasing of lubricant content, Wa decreases gradually and reaches the minimum at 2 wt % lubricant, then excessive lubricant leads to the rising of Wa. The change of Ra with lubricant concentration exhibits little difference from that of Wa decreases continuously throughout the testing lubricant concentration, but the decreasing rate is getting smaller with the increasing of lubricant concentration.

Figure 37(c) shows that the addition of the lubricant leads to MRR reduction. It indicates that the lubricant adsorbs on the disk surfaces to prevent them from being removed. It is also noticed that MRR decreases slowly when the lubricant concentration is less than 2 wt %, and then decreases quickly beyond 2 wt %. Therefore, there exists a saturated concentration for the lubricant molecules adsorbing on the disk surfaces.

3.5.6 Effect of the Slurry pH Value

CMP of different materials needs polishing slurries with different pH value, e.g., slurry for oxide CMP has a basic pH

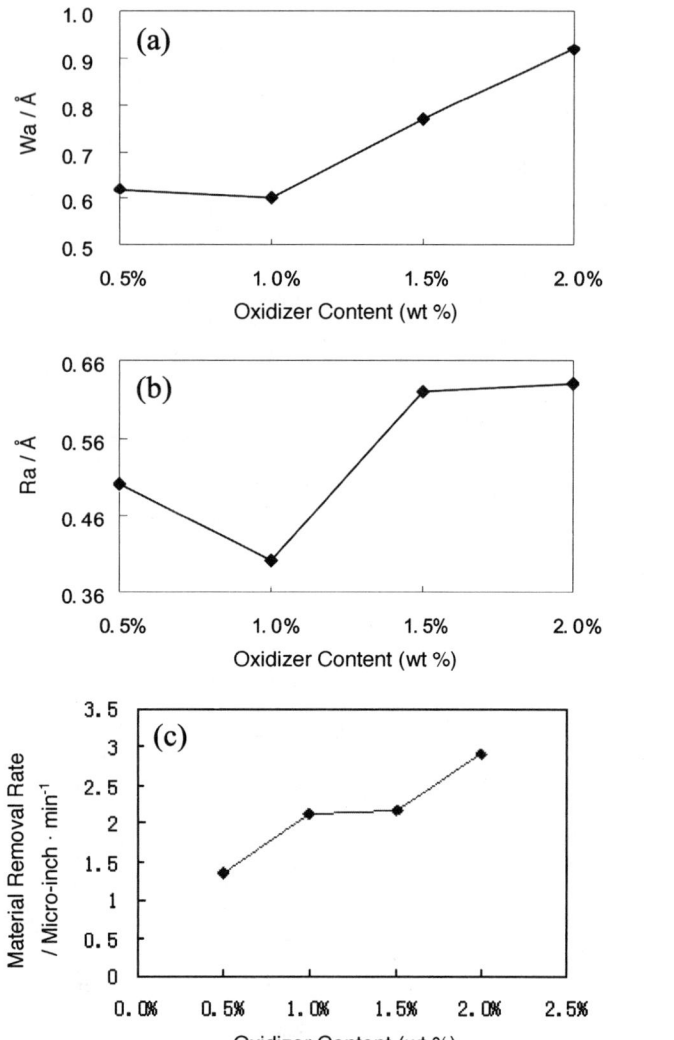

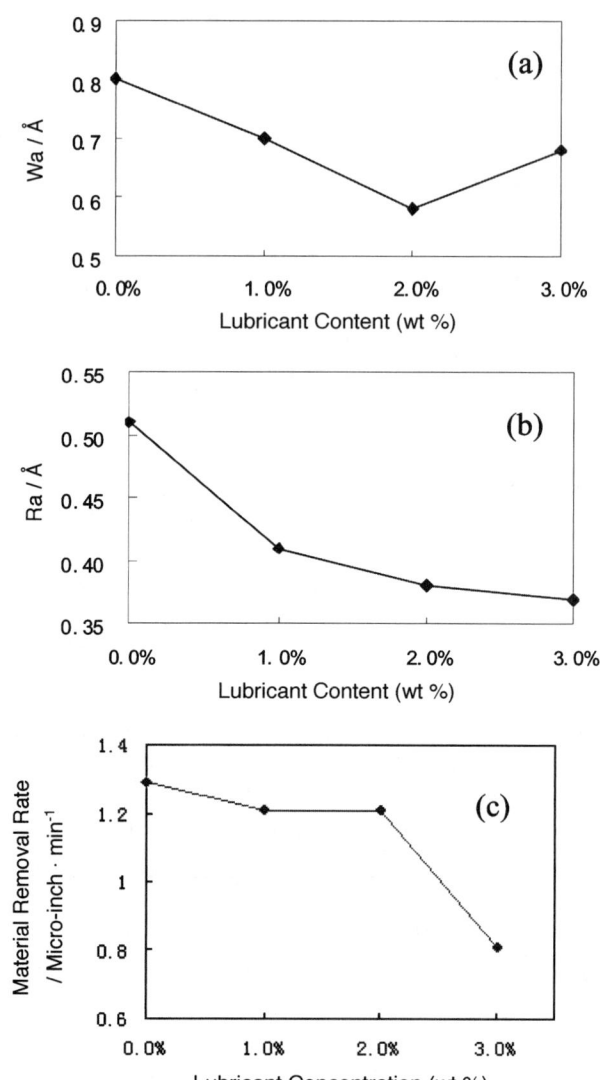

Fig. 36—Effect of the oxidizer concentration on (a) surface waviness Wa, (b) surface roughness Ra, and (c) material removal rate. The slurry contains 6 wt % SiO_2 particles with a diameter of 30 nm and 1 wt % lubricant in DI water at pH 1.9.

Fig. 37—Effect of the lubricant concentration on (a) surface waviness Wa, (b) surface roughness Ra, and (c) material removal rate. The slurry contains 6 wt % SiO_2 particles with a diameter of 30 nm and 1 wt % oxidizer in DI water at pH 1.55.

value more than 10, while that for metal is in the acidic range less than 3 [100]. The pH value of slurry is one important factor influencing the polishing rate [34]. As shown in Fig. 38, with the increasing of pH value, Wa decreases at first and reaches the minimum at pH value of 1.8, and then it begins to increase with the farther increasing of pH value. However, Ra declines slowly when pH value increases from 1.4 to 1.8, then keeps stable when pH value increases further. In other words, higher pH value helps to realize the local planarization, while the global planarization may be more facile to get at the relatively lower pH value. Combining with the trends of Wa and Ra, the suitable pH value of the slurry is around 1.8.

Change of material removal rate with pH value is very interesting. Neither low nor high pH value exhibits the high MRR, which may be attributed to a passivation effect on the disk surfaces under the low pH value and a slow chemical dissolution rate under the high pH value.

3.5.7 Polishing Performances in Different Slurries

Taking the slurry containing 6 wt % SiO_2 particles with a diameter of 30 nm, 1 wt % oxidizer and 2 wt % lubricant at pH 1.8 as an example, the surface features of the disk substrates polished in the slurry are shown in Table 4. For comparison, another kind of SiO_2 slurries with different particle size was conducted under the same conditions. The surfaces after polish with different slurries have different features. Ra and Wa decrease largely after polish in both slurries. Especially, the slurry with small particles exhibits lower Ra, Wa, but also lower MRR [108].

As shown in Fig. 39 [43], it is found that the surface before polishing is uneven and there are a large number of scratches (Fig. 39(a)). After polishing in the slurry I, the surface becomes smoother, and scratches as well as other micro defects could hardly be observed (Fig. 39(b)), by comparison with the surface polished in the slurry II (Fig. 39(c)).

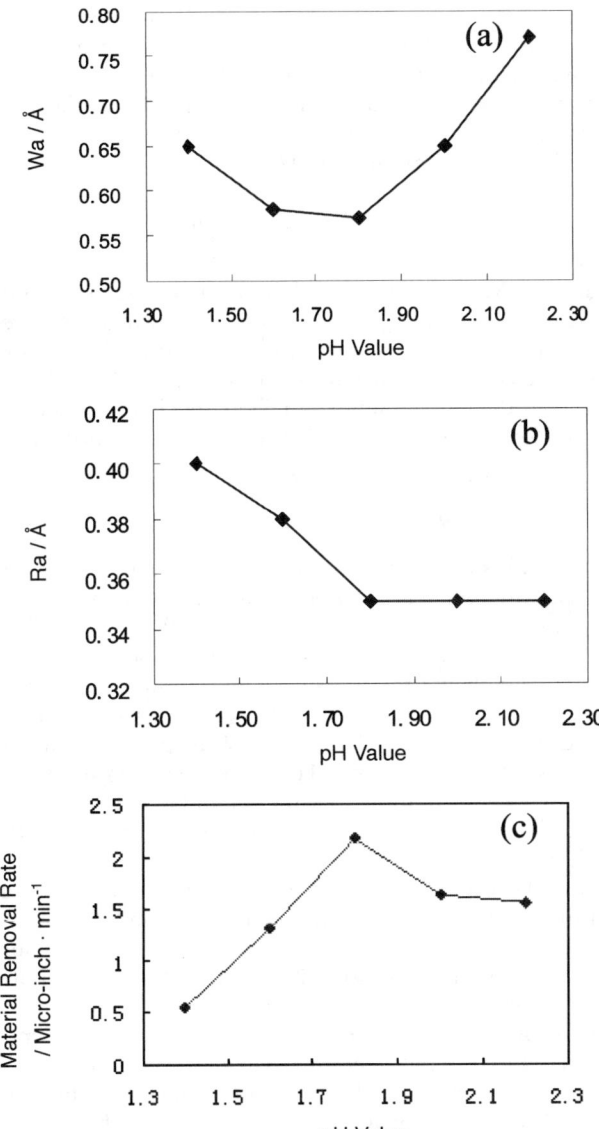

Fig. 38—Effect of pH value of the slurry on (a) Wa, (b) Ra, and (c) material removal amount. The slurry contains 6 wt % SiO$_2$ particles with a diameter of 30 nm, 1 wt % oxidizer and 2 wt % lubricant in DI water.

Luo and Pan [108,109], using a slurry with SiO$_2$ particles about 12 nm in diameter, improved lubricant additives, and a pH value of about 1.5, have got ten a much lower surface roughness and waviness measured by both Chapman MP2000$^+$ and AFM (Fig. 40).

3.5.8 Examination of the Polished Surfaces

In order to investigate chemical changes in the disk surface that has undergone the CMP process, the polished surface was analyzed by AES as shown in Fig. 41. The contents of elements and their deep distribution in the polished surface are shown in Figs. 42(a) and 42(b), respectively. Elements in the polished surface consist of 25.9 wt % nickel, 22.2 wt % phosphorus, 17.7 wt % oxygen and 34.2 wt % carbon. By comparison with that consisting of 84.7 wt % nickel and 15.3 wt % phosphorus in the surface before polishing (as shown in Fig. 41), elements oxygen and carbon were introduced to the surface. The introduction of element oxygen and the reduction in the concentration of element nickel imply that the oxidization reaction occurs in the CMP process. The introduction of element carbon may be due to the adsorption of the lubricant molecules on the disk surface since the lubricant molecule contains a polar group (carboxylic acid) and a long alkyl chain.

As shown in Fig. 42(b), with lasting of the sputtering time, the atomic concentrations of oxygen and carbon decrease rapidly within one minute, and then decrease near to zero. The change of nickel concentration is just the reverse: Ni concentration increases rapidly during the initial one minute and then varies little. It means that the chemical changes of elements Ni and O occurred only within the external layer of the polished surface. The concentration of element phosphorus exhibits little change during the entire sputtering time, which implies that the element P does not join in the oxidization reaction under the testing conditions [43].

3.5.9 Acting Mechanism of CMP for HDD Disk

The CMP process is regarded as a combination of chemical effect, mechanical effect, and hydrodynamic effect [110–116]. Based on contact mechanics, hydrodynamics theories and abrasive wear mechanisms, a great deal of models on material removal mechanisms in CMP have been proposed [110,111,117–121]. Although there is still a lack of a model that is able to describe the entire available CMP process, during which erosion and abrasive wear are agreed to be two basic effects.

At first, the plated NiP layer of aluminum alloy substrate is oxidized to form an oxide film on the surface, which may be softer [122,123] or more friable, and hence is easier to be removed than the NiP layer. Then the oxide film is worn away by SiO$_2$ abrasive particles and is dissolved in the acidic slurry, or both. The protruding region is removed faster than the recessed region [124], resulting in global planarization of the substrate surface. The role of the lubricant is to improve the mobility of the slurry and reduce the excessive friction

Surface Features	Before Polishing	Polished in Slurry Type I	Polished in Slurry Type II
Ra (Å) (A/B)	6.10/5.90	0.52/0.54	0.41/0.45
Wa (Å) (A/B)	6.90/6.65	0.65/0.67	0.52/0.55
MRR (μin/min)	...	2.8	2.5

TABLE 4—The surface features of disk substrates polished in different slurries.

Note: A, B represents the up and down surface of the disks respectively.
Slurry Type I: 6 wt % SiO$_2$ particles with diameters from 10 nm to 60 nm, and pH value of the slurry of 1.6.
Slurry Type II: 6 wt % SiO$_2$ particles with a diameter of 30 nm, 1 wt % oxidizer and 2 wt % lubricant in DI water, and pH value of the slurry is 1.6.

between polishing pad and disks to prevent the emergency of polishing lines.

3.6 Modeling and Simulation in CMP

Numerical simulation and molecular dynamics simulation on the removal action of CMP have been widely studied in recent years. In 1927, Preston [125] presented the mechanical model which relates the removal rate to the down pressure and relative velocity as follows:

$$R = KPV \qquad (1)$$

where R is the polish rate, K is a proportionality constant, P is the applied downward pressure, and V is the liner velocity of the wafer relative to the polishing pad. However, experimental results have shown that more factors in CMP, especially the material of the pad, the hardness and size of abrasives in slurry, have great influence on the material removal rate [107,126]. It is found that a softer and rougher polishing pad will yield a larger material removal rate. This cannot be explained by Preston's equation [110]. Furthermore, experimental results show that the pressure dependence of removal rate for CMP with soft pads satisfies a nonlinear relationship.

In 1991, Warnock [127] described the surface as a set of discrete points. For each point "i," the polish rate is:

$$MRR = K_i A_i / S_i \qquad (2)$$

where K_i is the kinetic factor or horizontal component in the polish removal rate, A_i is an accelerating factor associated with points which protrude above their neighbors, and S_i is a shading factor, describing how the polish rate is decreased by the effect of neighboring points protruding above point "i." His idea is shown in Fig. 43 [127]. The dashed line schematically represents the shape of the polished pad as it is deformed by the topography present. In the "down" region, the S coefficients are large, and the polish rate is low since the pad does not deform into the small openings. For small features projecting above the surroundings, the A coefficients are large and the polish rate is high. On the sloped surface, the K coefficients may be large, depending on the slope. A_i and S_i also vary along the slope. The model gives a reasonable approach for defining the dependence of the polish rate on the wafer shape and takes into account all geometrical cases.

Zhang and Busnaina [128] proposed an equation taking into account the normal stress and shear stress acting on the contact area between abrasive particles and wafer surfaces.

$$MRR = K_p \sqrt{P_0 V} \qquad (3)$$

where K_p represents both chemical effect on mechanical material removal and the effects of consumables. The model is an improved Preston's equation and more close to the experimental results that MRR has a nonlinear relationship with PV.

In 1996, Liu et al. [129] analyzed the wear mechanism based on the rolling kinematics of abrasive particles between the pad and wafer. They summarized that the kinetics of polishing are: (1) material removal rate is dependent on the real contact area between the slurry particle and the wafer surface. The real contact area is related to the applied pressure, the curvature, and Young's modulus of the slurry particles and wafer surface; (2) material removal rate increases linearly with the increasing of applied pressure and relative velocity between pad and wafer surface; (3) material removal rate increases as a result of a reduction in the modulus of slurry particle or film to be polished. They also improved Preston's equation (Eq.(1)) to:

$$MRR = C\left[\frac{1}{E_s} + \frac{1}{E_w}\right]PV \qquad (4)$$

where E_s is Young's modulus of the slurry particles, E_w is that of the wafer surface, P is the applied downward pressure, V is the linear velocity of the wafer relative to the polishing pad, C is a constant assuming fixed slurry, which is related to chemical effect and independent from mechanical factors.

In 1999, Zhao and Shi [130] proposed an equation as:

$$\begin{cases} MRR = K(V)(P^{2/3} - P_{th}^{2/3}) & p \geq p_{th} \\ MRR = 0 & p < p_{th} \end{cases} \qquad (5)$$

where $K(V)$ is a function of the relative velocity V and other CMP parameters. P is the pressure applied to the wafer. P_{th} is a polishing threshold pressure. $P_{th} = (\pi^3 P_c^3 R^2 D_a / 6 E_{pw}^2)$. E_{pw} is related to the elastic modulus (E_p) and poisson ratio (σ_p) of the pad and elastic modulus (E_w) and poisson ratio (σ_w) of wafer. $E_{pw} = (1 - \sigma_p^2 / E_p) + (1 - \sigma_w^2 / E_w)$. P_c is a critical value of pressure. D_a is the asperity density of the pad. R is radius of compressed asperity of pad. They reported that the equation is quite close to that of experimental data. The material removal will occur only when the down pressure is larger than the threshold pressure.

Sundararajan et al. [131] in 1999 calculated the slurry film thickness and hydrodynamic pressure in CMP by solving the Reynolds equation. The abrasive particles undergo rotational and linear motion in the shear flow. This motion of the abrasive particles enhances the dissolution rate of the surface by facilitating the liquid phase convective mass transfer of the dissolved copper species away from the wafer surface. It is proposed that the enhancement in the polish rate is directly proportional to the product of abrasive concentration and the shear stress on the wafer surface. Hence, the ratio of the polish rate with abrasive to the polish rate without abrasive can be written as

$$\frac{MRR \text{ with abrasives}}{MRR \text{ without abrasives}} = 1 + \alpha \tau_w C_A \qquad (6)$$

where α is the proportionality constant, τ_w is the average shear stress on the wafer surface due to slurry, and C_A is the abrasive concentration (wt %) in the slurry. The proportionality constant is the abrasive enhancement factor α which is determined from polish experiments with and without abrasive particles.

Taking into account the enhancement caused by the abrasives, the expression for the average polish rate R_p becomes [131]

$$R_p = \frac{M_{Cu}}{\rho_{Cu}} \times J_s \times (1 + \alpha \tau_w C_A) \qquad (7)$$

where M_{Cu} and ρ_{Cu} are the molecular weight and density of copper, respectively. J_s is the total flux of copper leaving the wafer surface.

The understanding of the removal mechanism is far

CHAPTER 12 ■ TRIBOLOGY IN ULTRA-SMOOTH POLISHING

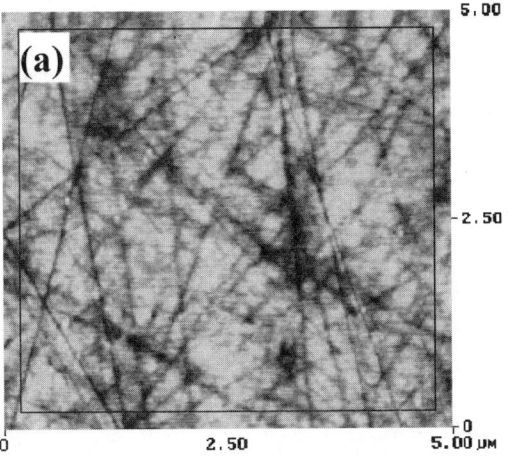

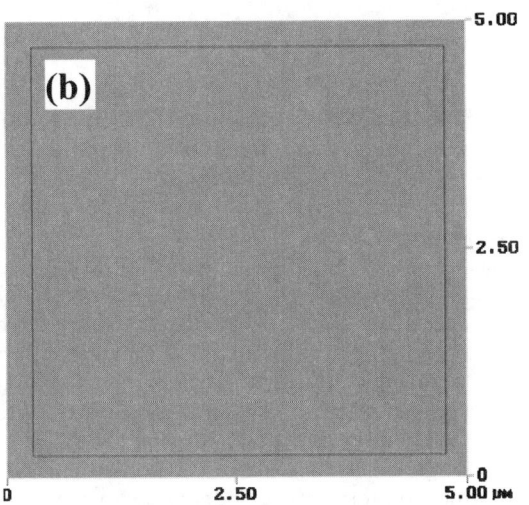

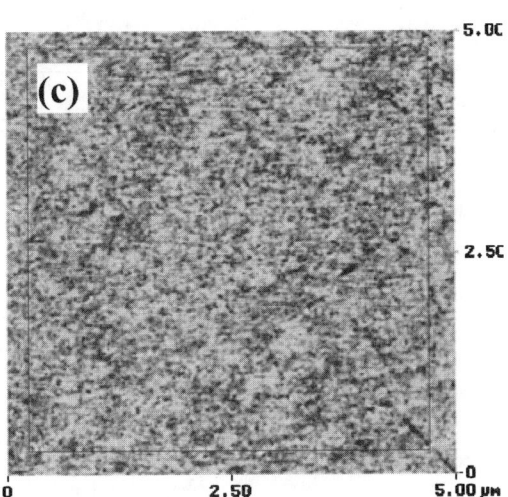

Fig. 39—Surface profiles of disk substrates polished in different slurries: (a) before polishing, (b) polished in the slurry Type I, (c) polished in the slurry Type II.

from completion, but it is clear that the polishing rate and surface quality cannot meet the requirements without abrasive particles during CMP for many materials, such as silicon substrate, dielectric layer, and metal layer. This indicates that the effect of mechanical components on CMP is significant.

In 1999, Luo and Domfeld [110] proposed that there are two typical contact modes in the CMP process, i.e., the hydro-dynamical contact mode and the solid-solid contact mode [110]. When the down pressure applied on the wafer surface is small and the relative velocity of the wafer is large, a thin fluid film with micro-scale thickness will be formed between the wafer and pad surface. The size of the abrasive particles is much smaller than the thickness of the slurry film, and therefore a lot of abrasive particles are inactive. Almost all material removals are due to three-body abrasion. When the down pressure applied on the wafer surface is large and the relative velocity of the wafer is small, the wafer and pad asperity contact each other and both two-body and three-body abrasion occurs, as is described as solid-solid contact mode in Fig. 44 [110]. In the two-body abrasion, the abrasive particles embedded in the pad asperities move to remove materials. Almost all effective material removals happen due to these abrasions. However, the abrasives not embedded in the pad are either inactive or act in three-body abrasion. Compared with the two-body abrasion happening in the wafer-pad contact area, the material removed by three-body abrasion is negligible.

Luo and Domfeld [110] introduced a fitting parameter H_w, a "dynamical" hardness value of the wafer surface to show the chemical effect and mechanical effect on the interface in their model. It reflects the influences of chemicals on the mechanical material removal. It is found that the nonlinear down pressure dependence of material removal rate is related to a probability density function of the abrasive size and the elastic deformation of the pad.

Ahmadi et al. [132] suggested the main wear mechanisms that are expected to occur in the CMP process are abrasive wear, adhesive wear, erosive wear, and corrosive wear. When the wafer and pad move relatively, a particle which is held in the pad may slide or roll against the wafer surface to wear the surface. By using an optical microscope, scanning electron microscope, atomic force microscope, and other surface analysis techniques, Liang et al. [114] found the surface damages resulted from particles sliding or rolling against wafer surfaces. In some conditions, when the particle rolls across the wafer surface, abrasive wear and adhesive wear are expected to be the main wear mechanism [130,132]. Determining whether an abrasive particle is held by a surface or moves against it involves many factors including deformation of particle and substrate, relative velocity, adhesion force, surface hardness, and friction coefficient. In addition, it is also important to determine whether abrasive particles roll or slide on the wafer surface so that the type of wear mechanisms occurring in the CMP process could be identified. Duan and Luo [16] describe a sliding process of a particle on a wafer surface. It must be noted that not all the particles in contact with the wafer contribute to an effective material removal. If an abrasive particle rolls on the wafer surface, its contribution to the removal of material is small because of the adhesion wear being the dominant wear

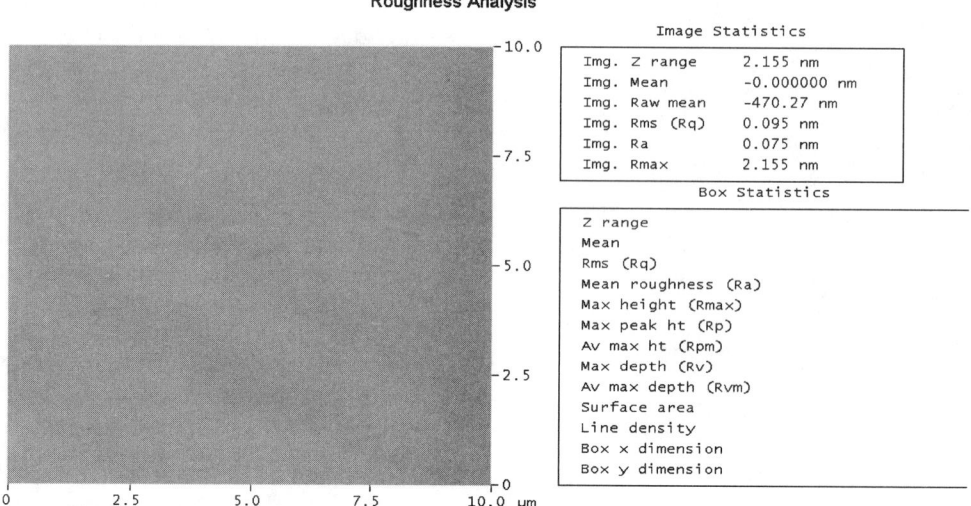

Fig. 40—Polished surface measured by AFM.

mechanism. If the particle is held firmly by the pad, sliding of the particle against the wafer surface occurs, which leads to abrasive wear, considerable structural and chemical changes, and during the process, the interaction between the particle and the wafer surface becomes significant. This sliding condition is determined by the surface friction at particle-pad and particle-wafer interface. In addition to the friction coefficient and contact area, the surface friction is determined by the normal pressure at the contact interfaces. This critical condition for effective removal introduces a polishing threshold pressure in CMP [18,110,130].

The abrasive wear mechanism has been accepted by many researchers. Removal by mechanical abrasion will lead to particles plow marks so that a higher polish rate would also result in rougher surfaces. However, Moon et al. [133], Stein and Cecchi [134] and Bielmann et al. [135] independently reported that few abrasive damages are found in high-quality finished wafer surfaces suggesting that the dominant mechanism of material removal during CMP may not be the mechanical plowing of the slurry particles into the wafer surface. Based on the consideration of the bonding energies of the surface molecules, the mechanism of formation and removal of weakly bonded surface molecular was presented by a number of researchers in the CMP process, such as Hoshino et al. [136], Pietsch et al. [137], Vijayakumar [138], and Zhao [139]. They suggested that chemical reactions convert strongly bonded surface atoms or molecules to weakly bonded molecular species while the mechanical action delivers the energy that is needed to break the weak molecular bonds, thereby removes the high-spot surface materials at the molecular scale. For example, according to Ref. [136], when polishing SiO_2 film by CeO_2 particles, it is believed that the polishing rate is affected by the formation rate of Ce–O–Si bonding and the cleaving rate of Si–O–Si. And the cleaving of Si–O–Si is controlled not only by chemical depolymerization but also by mechanical tearing.

In all of the above discussions, the focus has been entirely on the local condition, such as the pressure exposed on the wafer and relative velocity between pad and wafer. A model is to be proposed to describe the entire available removal process. It is also necessary to characterize some basic interactions, such as action between the individual abrasive particles with the polished surface.

3.7 Summary of CMP

The material removal mechanisms in the CMP process are involved in abrasive action, material corrosion, electrochemistry, and hydrokinetics process. Polishing effects are the sum of a very large number of polishing interactions over a range of operating conditions. To date, it is extremely difficult to clearly separate the key factor associated with a required removal and surface quality during CMP. There is still a lack of enough knowledge about the polishing fundamentals of common material used widely now in the microelectronic industries, such as silicon, SiO_2, tungsten, copper, etc.

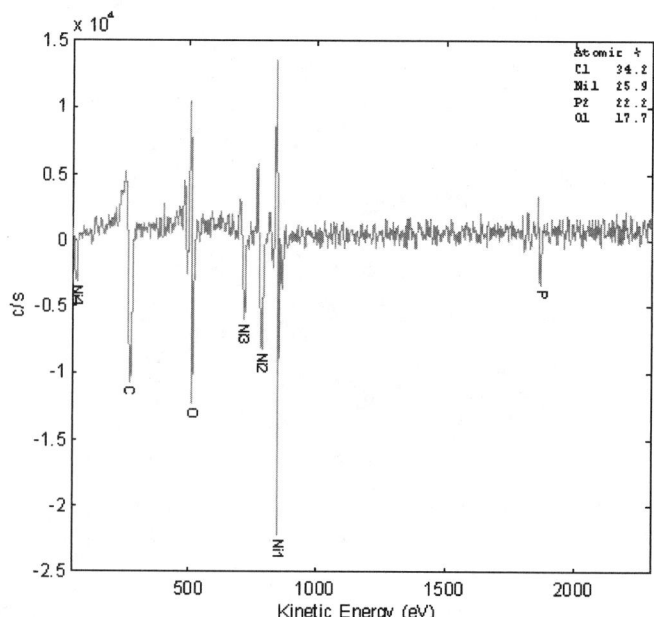

Fig. 41—AES surface survey of elements in the disk surface before polishing.

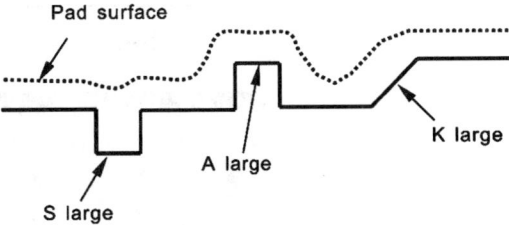

Fig. 43—Cross section of an arbitrary surface being polished [127].

our insight; however, the basic removal mechanisms of CMP are poorly understood. It is helpful to better illustrate the removal mechanism to focus on investigating the follows aspects:

1. To measure an individual particle surface interaction and its material removal effects. Because of the complexity of the polishing system, it is highly desirable to characterize the physical and chemical behavior of individual interactions while other components are fixed. AFM technology can be provided to explore slurry particle interactions with different surfaces in different liquid ambient.

2. To observe the surface damage caused by colliding with slurry particles. It is quite difficult to directly observe the surface modification of the polished materials induced by CMP because of the modification occurring generally in a nano-scale or less. By designing the experiment which can intensify the effects of CMP, the primary goal is to investigate the material removal mechanism by adhesion wear or collision action by analyzing an individual interaction between particles and the surface being polished without a pad during CMP.

3. To test the transport rules of slurry particles in the flow field during CMP. An understanding of the slurry transport and particles motion beneath the wafer plays an important role in revealing the interaction process

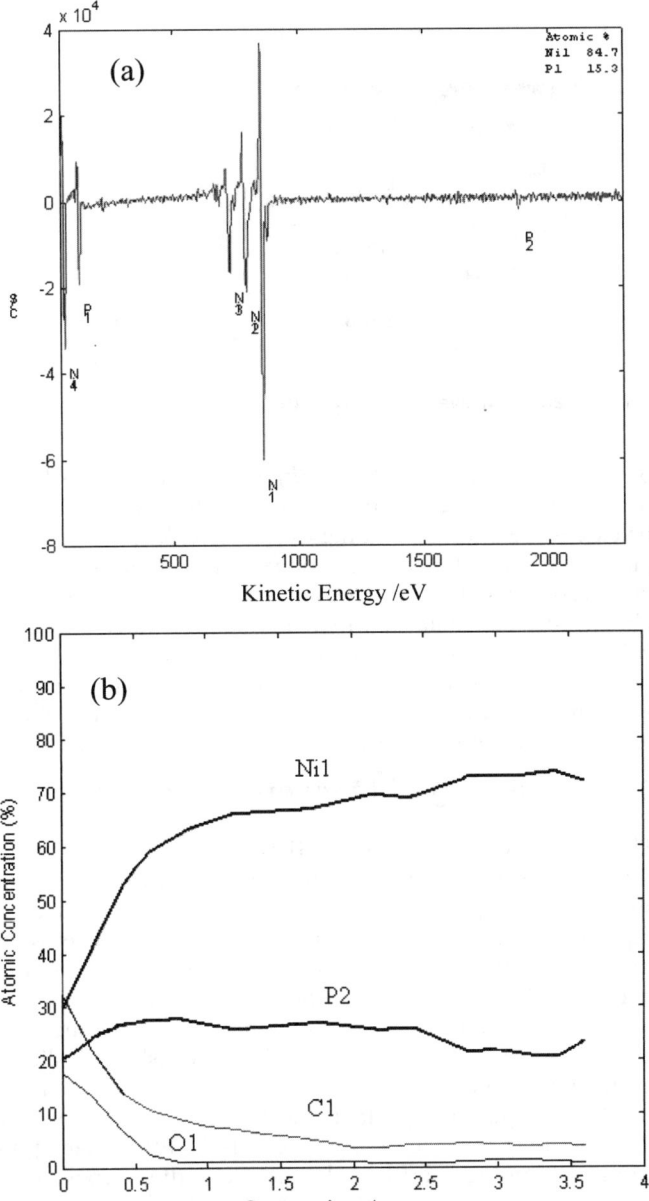

Fig. 42—AES surface survey of elements in the disk substrate surface after polishing. The slurry contains 6 wt % SiO_2 particles with a diameter of 30 nm, 1 wt % oxidizer and 2 wt % lubricant in DI water, and pH value of the slurry is 1.8. (a) Elements in the disk surface, (b) deep distribution of the elements. (The contents of elements and their deep distribution in the polished surface were analyzed by using a PHI 680 auger nanoprobe under determining conditions as follows: ion beam current of 1 μA, ion beam voltage of 2 kV, electron beam current of 10 nA, electron beam voltage of 10 kV and scan area of 20 μm by 20 μm.)

Its widespread application has exceeded the growth of scientific understanding. However, some new materials, such as copper, copper alloy, diffusion barrier, and low-k ILD (polymers, organosilicate glasses), are being introduced into IC manufacturing. The modeling for the materials is now an active area of investigation. The removal mechanisms for the materials appear to be substantially more complex than that for common materials [37]. Recent works have augmented

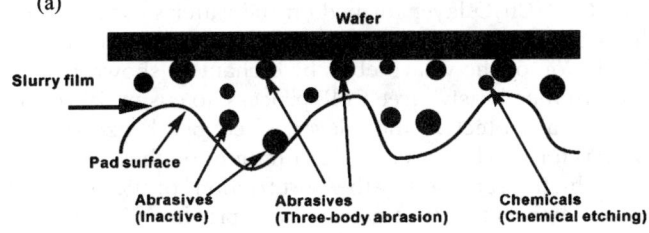
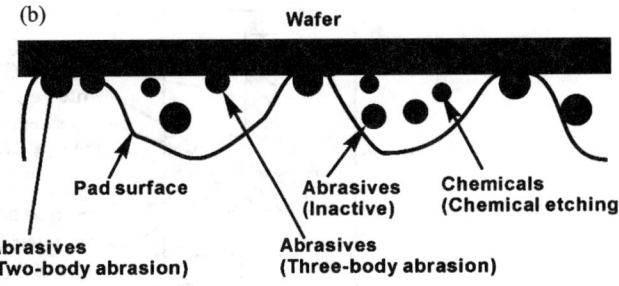

Fig. 44—Schematic of the solid-solid contact in a CMP process [120].

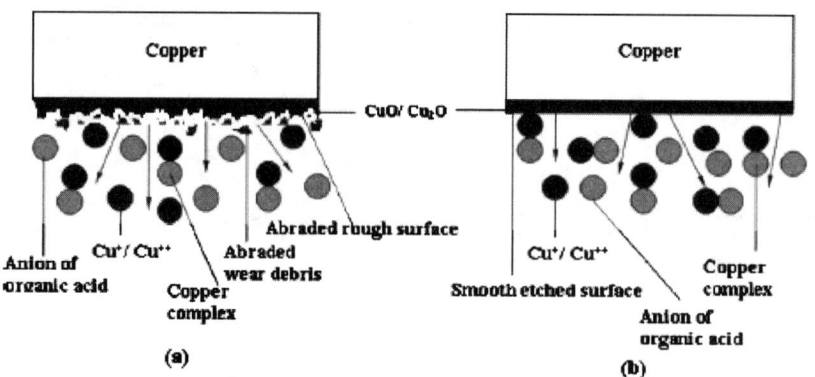

Fig. 45—Mechanism comparison of the normal CMP and abrasive-free CMP [110].

among particles, and between the particle and the surface. For instance, a fluorescence technique can be used, which uses the fluorescence from a few different dyes each fluorescing at different wavelengths to measure mixing motion.

3.8 CMP Technology in the Future

With the development of IC fabrication and HDD manufacturing, there are two challenges for CMP, one is the ultra-low pressure polishing because of low-k material's very low elastic module; another is to get the polished surface particle-free, which becomes a difficult problem since the particles in the slurry become smaller to a get a very smooth surface. In order to fit these challenges, abrasive-free CMP is to be used because it has no abrasive particles in the slurry and requires lower pressure. Abrasive-free CMP is a more chemically active process, which provides lower dishing, erosion, and less or no mechanical damage of low-k materials as compared to conventional abrasive CMP processes where the material is removed by plastic deformation such as abrasive wear, plowing, microcutting, burnishing [140–142]. Balakumar et al. [140] compared the mechanisms of abrasive CMP and Cu abrasive-free CMP as shown in Fig. 45. In abrasive CMP, the thick CuO/Cu_2O layer formed on the wafer surface is mechanically removed by the abrasive particles followed by the dissolution of the wear debris by etchant as shown in Fig. 45(a). In the abrasive-free CMP process, to ensure isotropic etching, a protective monolayer of copper benzotriazole (BTA) is formed by the corrosion inhibitor on the cupric surface, which is removed by the "soft friction" of sliding. Once the protective layer is removed, the cupric layer is expected to be directly etched out, leaving a smooth surface on the top of the wafer as shown in Fig. 45(b) [67,140]. Fang et al. [142] proposed that the Cu surface should be strongly required to be oxide-free during polishing to optimize the pattern in Cu planarization with pressure-independent Cu removal. They used a solution of 1.1 M HNO_3 and 0.1 mM 5Me-BTA as the abrasive-free CMP slurry, which had strong acidity with a pH value about 0.05, to keep the Cu surface oxide-free due to the lack of OH^- to oxidize the Cu surface as Cu oxides. In their experiment, they had gotten a removal rate about 38 nm which is independent on the pressure.

4 The Polishing of Magnetic Head Surface

The areal recording density of HDDs has been increasing at the annual rate of 100 % which achieved the expected data since 1994 [143]. The head-disk interface technology is moving towards deep sub-10 nm mechanical spacing between the head and the disk media, to increase the resolution of the read/write head, and to keep its current annual increasing rate of area density. Such small head-disk spacing will greatly increase the likelihood of slider-disk and slider-particle-disk interaction [144]. The research works on the head-disk interface contact mechanics [145], tribology [144,146,147], surface modification [148,149], surface polishing technology [150], and other recording types [151] have attracted much attention.

The structure of HDD magnetic head is shown in Fig. 46. The image on the left is the planform of ABS which faces the disk during performance. The right image is the cross section. From which we can see more than ten layers of different materials. The function of MR is to read signals from the media. In these magnetically soft materials, the electrical re-

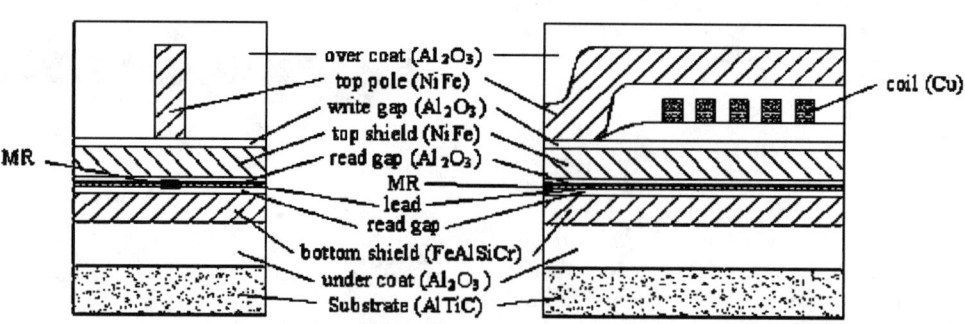

Fig. 46—Structure and material of read-write head.

TABLE 5—IC analysis of slurry (ppb).

Slurry	Powder Size	Anion						
		F	Cl	NO_2	Br	NO_3	PO_4	SO_4
Type I	UFD slurry	N	283.6	13.1	N	72.4	24.7	218.3
Type II	125 nm	N	239.5	59.3	N	230.2	204.2	156.6
Type III	250 nm	N	344.5	33.6	1.0	206.0	18.0	170.7

sistance changes when the material is magnetized. The resistance goes back to its original value when the magnetizing field is turned off. Lead is connected with MR and conducts current to it. It also has some layers that are composed of Ta, Au, and Co-Pt. The layers of MR and lead are all perpendicular to ABS. MRR increases when the slider becomes thinner. Row bars are magnetic heads used in the polishing experiments, which are made up of 46 or 56 sliders juxtaposed together and separated by grooves the depth of which is about 100 μm [152].

Luo et al. [1,153] used a slurry containing ultra-fine diamond (UFD) powders to polish the surface of HDD sliders. The powders are from 3 nm to 18 nm in diameter and 90 % around 5 nm. They are crystal and sphere-like [154]. The pH value of the slurry is kept in the range from 6.0 to 7.5 in order to avoid the corrosion of read-write heads, especially pole areas. A surface-active agent is added into the slurry to decrease the surface tension of the slurry to 22.5 Dyn/cm, and make it spread on the polish plate equably. An antielectrostatic solvent is also added to the slurry to avoid the magnetoresistance (MR) head being destroyed by electrostatic discharge. The anion concentration of the slurry is strictly controlled in ppb level so as to avoid the erosion of magnetic heads as shown in Table 5. The concentration of UFDs in the slurry is 0.4 wt %.

The simplified sketch of the lapping equipment is shown in Fig. 47. Three row bars are attached on the back of the keeper, forming an equilateral triangle, and pressed on the plate tightly under the load of 4 kg. Both the plate and the keeper rotate in the lapping process. The keeper also oscillates along the radial direction of the plate with the speed of 10 cycles per minute. There are lands and grooves on the plate with the width of 30 μm and 70 μm, respectively, as shown in the uper left of Fig. 47. Contrasting to UFD slurry, the diameter of the diamond crocus is 250 nm which is embedded into the lands by a rotating ceramic ring on the plate.

The AFM and SEM images of the sliders polished by the two kinds of slurries are shown in Fig. 48 [1,152], in which (a), (c), and (e) are the AFM and SEM photos of the pole area and the undercoat polished by the slurry Type III with powders around 250 nm in diameter, and (b), (d) and (f) are the same area polished in the slurry Type I, or UFD slurry. The

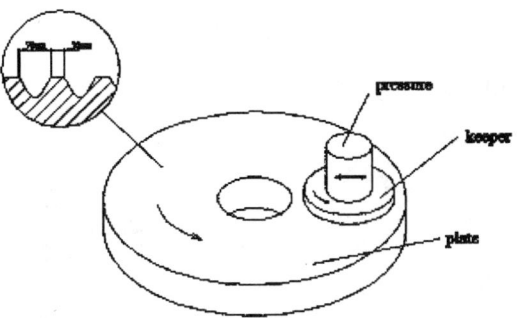

Fig. 47—The polish machine used in experiments.

average roughness Ra of the pole area in Fig. 48(a) is 0.66 nm. However, it is 0.20 nm in Fig. 48(b). The mean roughness Ra of the Al_2O_3 surface are 0.44 nm polished in the slurry Type III as shown in Fig. 48(c), and 0.114 nm polished by the Type I or UFD slurry as shown in Fig. 48(d). It indicates that the average roughness reduced about three times polished by UFD slurry, compared to that polished in slurry Type III. Many micro scratches and black spots can be observed after being polished with slurry Type III as shown in Fig. 48(e) under SEM. These micro scratches cross the MR and lead layer may make them deformed to cause a short circuit with shields. The black spots were confirmed as the element of carbon by Auger electron spectroscopy (AES). The size of the spots is about 100 nm. So they should be the grains of slurry Type III. The surface as shown in Fig. 48(f) polished by UFD slurry almost has no micro scratch and black spot under SEM. Therefore, using UFD slurry in the lapping process greatly contributes to eliminating micro scratches and black spots, reducing roughness and meliorating the surface finish. The grains of UFD slurry are so small that the grains of the slurry are flowing and rolling in the oil film, impacting and brushing the surface in the polish process. Therefore, the micro scratches and the black spots disappear, and the better surfaces have been gotten, but the material remove rate will decrease largely.

After lapping, the sliders will be cleaned, and then a passivation film of diamond-like carbon (DLC) will be deposited on the surfaces of sliders through chemical vapor deposition (CVD) to protect the pole area from chemical-physical corrosion and electrostatic discharge attack. Corrosion in pole areas will result in loss of read/write functions. A corrosion test was taken to examine the ability of the sliders polished by different slurries as shown in Table 6. It can be seen that the MRR change rate of the sliders polished by UFD slurry is much less than that polished by the slurry Type III, that is, the capability of anti-corrosion of the former is much better than that of the latter.

The thickness of DLC coating is about 25 Å or less. The

TABLE 6—MRR change rate of the corrosion test.

Slurry	DLC $pH2-pH1$	RIE $pH4-pH3$	DLC->RIE $pH3-pH2$
Polished by slurry Type III	2.22 %	11.11 %	6.67 %
Polished by UFD slurry	0.89 %	0.45 %	1.79 %

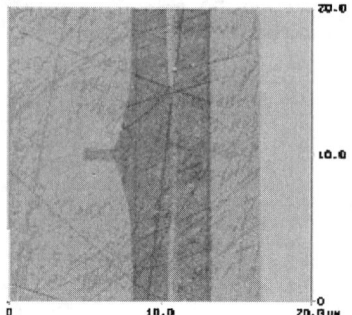

(a) Polished in the type III slurry with powder of 250 nm in diameter

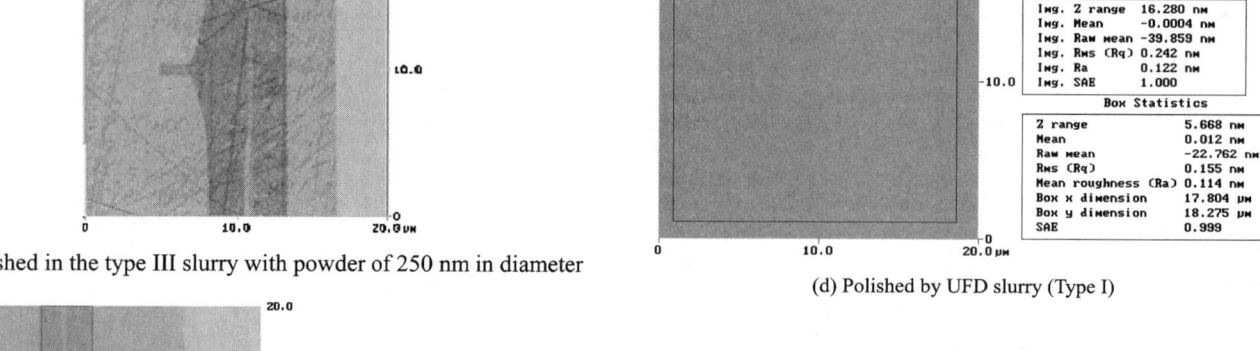

(d) Polished by UFD slurry (Type I)

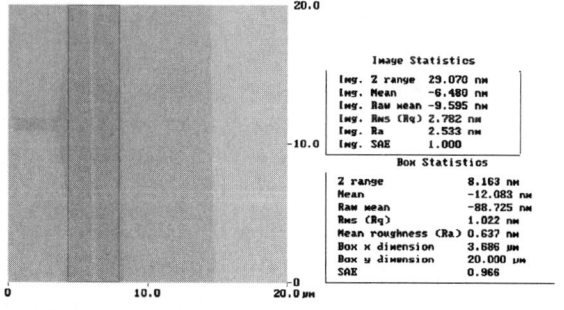

(b) Polished in UFD slurry (Type I)

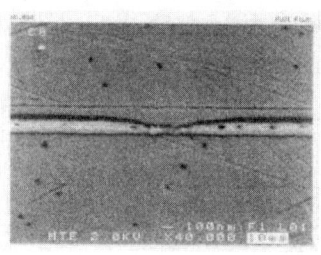

(e) Polished in the type III slurry with powder of 250 nm in diameter (40,000×)

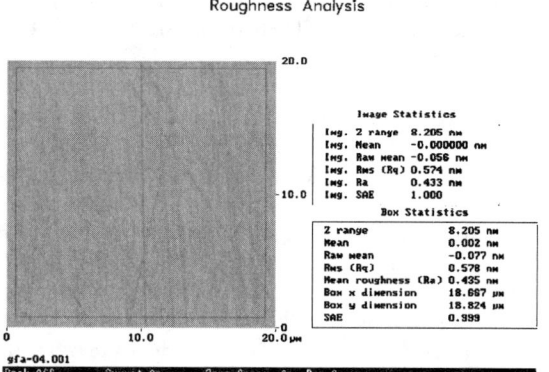

(c) Polished in the type III slurry with powder of 250 nm in diameter

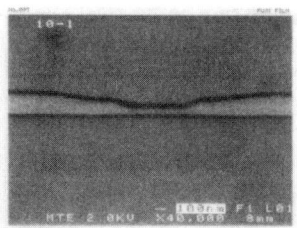

(f) Polished by UFD slurry (40,000×)

Fig. 48—AFM and SEM images of the surfaces polished in two conditions: (a) polished in the Type III slurry with powder of 250 nm in diameter, (b) polished in UFD slurry (Type I), (c) polished in the Type III slurry with powder of 250 nm in diameter, (d) polished by UFD slurry (Type I), (e) polished in the Type III slurry with powder of 250 nm in diameter (40,000×), (f) polished by UFD slurry (40,000×).

roughness of the surfaces polished by the slurry Type III is relatively big, many asperity peaks near the micro scratches are higher than 5 nm, and the black spots embedded in ABS are about 10 nm higher than the surface. Hence the surfaces cannot be entirely covered by the coatings. The places or points without DLC coatings can be corroded easily. The situation is thoroughly different by using UFD slurry in the lapping process. Since there is no micro scratch and black spot, the surfaces can obtain a better protection from DLC film and less corrosion, which enhances the anti-corrosion capability of read/write heads. Moreover, a perfect surface finish enables thinner coatings and the smaller distance between the pole areas and the media.

References

[1] Luo, J. B., Gao, F., Hu, Z. M., and Lei, H., "Surface Finish and Performances of Read/Write Heads by Using Ultra-fine Diamond Slurry in Polishing Process," *Int. J. Nonlinear Sci. Numer. Simul.* Vol. 3, No. 3–4 Sp. Iss., 2002, pp. 449–454.

[2] Fury, M. A., "The Early Days of CMP," *Solid State Technol.*, Vol. 449, No. 5, 1997, pp. 81–86.

[3] Liu, F. Q., Du, T. B., Duboust, A., Tsai, S., and Hsu, W. Y., "Cu Planarization in Electrochemical Mechanical Planarization," *J. Electrochem. Soc.*, Vol. 153, No. 6, 2006, pp. C377–C381.

[4] Fang, J. Y., Huang, P. W., Tsai, M. S., Dai, B. T., Wu, Y. S., and Feng, M. S., "Study on Pressure-Independent Cu Removal in Cu Abrasive-Free Polishing," *Electrochem. Solid St.*, Vol. 9, No. 1, 2006, pp. G13–G16.

[5] Cheng, H. B., Feng, Z. J., Wang, Y. W., and Lei, S. T., "Magnetorheological Finishing of SiC Aspheric Mirrors," *Mater. Manuf. Processes*, Vol. 20, No. 6, 2005, pp. 917–931.

[6] Zhou, Y. and Liang, D. P., "Study on the Theoretic Model of Ultrasonic Polishing," *China Mechanical Engineering*, Vol. 16, No. 8, 2005, pp. 664–666.

[7] http://www.Chapinst.com

[8] http://www.hdi-inst.com

[9] Pai, R. A., Humayun, R., Schulberg, M. T., Sengupta, A., Sun, J. N., and Watkins, J. J., "Mesoporous Silicates Prepared Using Preorganized Templates in Supercritical Fluids," *Science*, Vol. 303, No. 5657, 2004, pp. 507–510.

[10] Hsieh, H., Averback, R. S., Sellers, H., and Flynn, C. P., "Molecular-Dynamics Simulations of Collisions Between Energetic Clusters of Atoms and Metal Substrates," *Phys. Rev. B*, Vol. 45, No. 8, 1992, pp. 4417–4430.

[11] Yamaguchi, Y. and Gspann, J., "Large-Scale Molecular Dynamics Simulations of Cluster Impact and Erosion Processes on a Diamond Surface," *Phys. Rev. B*, Vol. 66, 2002, pp. 155408-1-10.

[12] Kyuno, K., Cahill, D. G., Averback, R. S., Tarus, J., and Nordlund, K., "Surface Defects and Bulk Defect Migration Produced by Ion Bombardment of Si(001)," *Phys. Rev. Lett.*, Vol. 83, 1999, pp. 4788–4791.

[13] Xu, J., Luo, J. B., Lu, X. C., Wang, L. L., Pan, G. S., and Wen, S. Z., "Atomic Scale Deformation in the Solid Surface Induced by Nanoparticle Impacts," *Nanotechnology*, Vol. 16, 2005, pp. 859–864.

[14] Xu, J., Luo, J. B., Zhang, C. H., Zhang, W., and Pan, G. S., "Nano-Deformation of a Ni-P Coating Surface after Nanoparticle Impacts," *Applied Surface Science*, Vol. 252, 2006 (in press).

[15] Komanduri, R., Chandrasekaran, N., and Raff, L. M., "Molecular Dynamics Simulation of the Nanometric Cutting of Silicon," *Philos. Mag. B*, Vol. 81, No. 12, 2001, pp. 1989–2019.

[16] Duan, F. L., Luo, J. B., Wen, S. Z., and Wang, J. X., "Atomistic Structural Change of Silicon Surface Under a Nanoparticle Collision," *Chinese Science Bulletin*, Vol. 50, No. 15, 2005, pp. 1661–1665.

[17] Johansson, S., "Fracture Testing of Silicon Microelements In Situ in a Scanning Electron Microscope," *J. Appl. Phys.*, Vol. 63, 1988, pp. 4799–4808.

[18] Xu, J., Luo, J. B., Lu, X. C., Zhang, C. H., and Pan, G. S., "Progress in Material Removal Mechanisms of Surface Polishing with Ultra Precision," *Chinese Science Bulletin*, Vol. 49, No. 16, 2004, pp. 1687–1693.

[19] Zhang, L. C. and Tanaka, H., "On the Mechanics and Physics in the Nano-indentation of Silicon Monocrystals," *JSME Int. J., Ser. A*, Vol. 42, No. 4, 1999, pp. 546–559.

[20] Diaz, D. L., Rubia, T., and Gilmer, G. H., "Structural Transformations and Defect Production in Ion Implanted Silicon," *Phys. Rev. Lett.*, Vol. 74, No. 13, 1995, pp. 2507–2510.

[21] Ihara, S., Itoh, S., and Kitakami, J., "Mechanisms of Cluster Implantation in Silicon," *Phys. Rev. B*, Vol. 58, No. 16, 1998, pp. 10736–10744.

[22] Biswas, R., Grest, G. S., and Soukoulis, C. M., "Molecular-Dynamics Simulation of Cluster and Atom Deposition on Silicon (111)," *Phys. Rev. B*, Vol. 38, No. 12, 1988, pp. 8154–8162.

[23] Tarus, J. and Nordlund, K., "Molecular Dynamics Study on Si_2O Cluster Deposition on Si (001)," *Nucl. Instr. and Meth. in Phys. Res. B*, Vol. 212, 2003, pp. 281–285.

[24] Takami, S., Suzuki, K., Kubo, M., and Miyamoto, A., "The Fate of a Cluster Colliding onto a Substrate," *J. Res. Nanopart*, Vol. 3, 2001, pp. 213–218.

[25] Cheng, H. P., "Cluster-Surface Collisions: Characteristics of Xe55- and C_2O-Si [111] Surface Bombardment," *J. Chem. Phys.*, Vol. 111, No. 16, 1999, pp. 7583–7592.

[26] Leach, A. R., *Molecular Modeling: Principles and Applications*, England, Longman, 1996.

[27] Tersoff, J., "Modeling Solid-State Chemistry: Interatomic Potentials for Multicomponent Systems," *Phys. Rev. B*, Vol. 39, No. 8, 1989, pp. 5566–5568.

[28] Duan, F. L., Luo, J. B., and Wen, S. Z., "Repulsion Mechanism of Nanoparticle Colliding with Monocrystalline Silicon Surface," *Acta Phys. Sin.*, Vol. 54, No. 6, 2005, pp. 2832–2837, (in Chinese).

[29] http://www.itrs.net/Links/2005ITRS/Home2005.htm

[30] Zant, V. P., *Microchip Fabrication: A Practical Guide to Semiconductor Processing*, 4th ed., McGraw-Hill Companies, Inc., 2000.

[31] Moy, D., Schadt, M., Hu, C. K., et al., *Proceedings, 1989 VMIC Conference*, 1989, p. 26.

[32] Oliver, M. R., *Chemical-Mechanical Planarization of Semiconductor Materials*, Berlin: Springer Series in Material Science, 2004.

[33] Liu, R. C., Pai, C. S., and Martinez, E., "Interconnect Technology Trend for Microelectronics," *Solid-State Electron.*, Vol. 43, 1999, pp. 1003–1009.

[34] Liu, Y. L., Zhang, K. L., Wang, F., and Di, W. G., "Investigation on the Final Polishing and Technique of Silicon Substrate in ULSI," *Microelectron. Eng.*, Vol. 66, 2003, pp. 438–444.

[35] Swetha, T., Arun, K. S., and Ashok, K., "Tribological Issues and Modeling of Removal Rate of Low-k Films in CMP," *J. Electrochem. Soc.*, Vol. 151, 2004, pp. 205–211.

[36] Zhou, C., Shan, L., Hight, J. R., Danyluk, S., Ng, S. H., and Paszkowski, A. J., "Influence of Colloidal Abrasive Size on Material Removal Rate and Surface Finish in SiO_2 Chemical Mechanical Polishing," *Lubr. Eng.*, Vol. 58, No. 8, 2002, pp. 35–41.

[37] Borst, C. L., Gill, W. N., and Gutmann, R. J., "Chemical-Mechanical Polishing of Low Dielectric Constant Polymers and Organosilicate Glasses," Boston: Kluwer Academic Publishers, 2002, pp. 1–5.

[38] Michael, C. P. and Duncan, A. G., "The Importance of Particle Size to the Performance of Abrasive Particles in the CMP Process," *J. Electro. Mater.*, Vol. 25, 1996, pp. 1612–1616.

[39] Xie, Y. and Bhushan, B., "Effects of Particle Size, Polishing Pad and Contact Pressure in Free Abrasive Polishing," *Wear*, Vol. 200, 1996, pp. 281–295.

[40] Bielmann, M., Mahajan, U., and Singh, R. K., "Effect of Particle Size During Tungsten Chemical Mechanical Polishing," *Electrochem. Solid-State Lett.*, Vol. 2, 1999, pp. 401–403.

[41] Zhou, C. H., Shan, L., and Hight, R., "Influence of Colloidal Abrasive Size on Material Removal Rate and Surface Finish in SiO_2 Chemical Mechanical Polishing," *Tribol. Trans.*, Vol. 45, 2002, pp. 232–230.

[42] Basim, G. B., Adler, J. J., Mahajan, U., Singh, R. K., and Moudgil, B. M., "Effect of Particle Size of Chemical Mechanical Polishing Slurries for Enhanced Polishing with Minimal Defects," *J. Elecrochem. Soc.*, Vol. 147, No. 9, 2000, pp. 3523–3528.

[43] Lei, H. and Luo, J. B., "CMP of Hard Disk Substrate Using a Colloidal SiO_2 Slurry: Preliminary Experimental Investiga-

[44] Zantye, P. B., Kumar, A., and Sikder, A. K., "Chemical Mechanical Planarization for Microelectronics Applications," *Materials Science and Engineering R-Reports*, Vol. 45, No. 3–6, 2004, pp. 89–220.

[45] Chen, C. Y., Yu, C. C., Shen, S. H., and Ho, M., "Operational Aspects of Chemical Mechanical Polishing Polish Pad Profile Optimization," *J. Electrochem. Soc.*, Vol. 147, No. 10, 2000, pp. 3922–3930.

[46] Jeng, Y. R., Huang, P. Y., and Pan, W. C., "Tribological Analysis of CMP with Partial Asperity Contact," *J. Electrochem. Soc.*, Vol. 150, No. 10, 2003, pp. G630–G637.

[47] Li, S. H. and Miller, R., "Chemical Mechanical Polishing in Silicon Processing," *Semiconductor Semimet.*, Vol. 63, New York: Academic Press, 2000.

[48] Hepburn, C., *Polyurethane Elastomers*, London: Applied Science, 1982.

[49] Lu, H., Obeng, Y., and Richardson, K. A., "Applicability of Dynamic Mechanical Analysis for CMP Polyurethane Pad Studies," *Mater. Charact.*, Vol. 49, No. 2, 2002, pp. 177–186.

[50] Li, I., Forsthoefel, K. M., Richardson, K. M., Obeng, Y. S., and Easter, W. G., "Dynamic Mechanical Analysis (DMA) of CMP Pad Materials," *Materials Research Society Symposium Proceedings*, Vol. 613, 2000, E7.3.

[51] Powell, P. C., *Engineering with Polymers*, New York: Chapman & Hall, 1983.

[52] Philipossian, A. and Olsen, S., "Effect of Pad Surface Texture and Slurry Abrasive Concentration on Tribological and Kinetic Attributes of ILD CMP," *Materials Research Society Symposium Proceedings*, Vol. 767, 2003, F2.8.

[53] Ma, J. J., "Experimental Research on Chemical Mechanical Polishing of Computer Hard Disk," Master Thesis, Tsinghua University, Beijing, China, 2003.

[54] Cook, L. M., "Chemical Process in Glass Polishing," *J. Non-Cryst. Solids*, Vol. 120, 1990, pp. 152–171.

[55] Stefan, J. S., and Jan, A. S., "Surface Defects in Polished Silicon Studied by Cross-Sectional Transmission Electron Microscopy," *J. Am. Ceram. Soc.*, Vol. 72, No. 7, 1989, pp. 1135–1139.

[56] Kunz, R. R., Clark, H. R., Nitishin, P. M., Rothschild, M., and Ahern, B. S., "High Resolution Studies of Crystalline Damage Induced by Lapping and Single-Point Diamond Machining of Si (100)," *J. Mater. Res.*, Vol. 11, No. 5, 1996, pp. 1228–1237.

[57] Graf, D., Schnegg, A., Schmolke, R. et al., "Morphology and Chemical Composition of Polishing Silicon Wafer Surfaces," *Electrochemical Society Proceedings*, Vol. 96-22, 2000, pp. 186–196.

[58] Mahajan, U., Bielmann, M., and Singh, R. K., *Materials Research Society Symposium Proceedings*, Vol. 566, 2000, pp. 27–32.

[59] Tomozawa, M., "Oxide CMP Mechanisms," *Solid State Technol.*, Vol. 40, No. 7, 1997, pp. 39–53.

[60] Lei, S., "Mechanical Interactions at the Interface of Chemical Mechanical Polishing," Ph.D. Thesis, Georgia Institute of Technology, 2000.

[61] Hoshino, T., Kurata, Y., Terasaki, Y., Terasakiy, Y., and Susa, K., "Mechanism of Polishing of SiO_2 Films by CeO_2 Particles," *J. Non-Cryst. Solids*, Vol. 283, No. 1–3, 2001, pp. 129–136.

[62] Tsai, T. H. and Yen, S. C., "Localized Corrosion Effects and Medications of Acidic and Alkaline Slurries on Copper Chemical Mechanical Polishing," *Applied Surface Science*, Vol. 210, No. 3–4, 2003, pp. 190–205.

[63] Dan, T. and Fury, M., "Chemical Mechanical Polishing of Polymer Films," *J. Electron. Mater.*, Vol. 27, No. 10, 1998, pp. 1088–1094.

[64] Neirynck, J. M., Yang, G. R., Murarka, S. P., et al., "Low Dielectric Constant Materials-Synthesis and Applications in Microelectronics," *Materials Research Society Symposium Proceedings*, Vol. 381, 1995, pp. 229–234.

[65] Wang, D. H., Chiao, S., Afnan, M., Yih, P., and Rehayem, M., "Stress-Free Polishing Advances Copper Integration with Ultralow-k Dielectrics," *Solid State Technol.*, Oct. 2001, pp. 101–104, 106.

[66] Larsen, B. J. and Liang, H., "Probable Role of Abrasion in Chemo-Mechanical Polishing of Tungsten," *Wear*, Vol. 233-235, 1999, pp. 647–654.

[67] Kaufman, F. B., Thompson, D. B., Broabie, R. E., Jaso, M. A., Guthrie, W. L., Pearson, D. J., and Small, M. B., "Chemical-Mechanical Polishing for Fabricating Patterned W Metal Features as Chip Interconnects," *J. Electrochem. Soc.*, Vol. 138, No. 11, 1991, pp. 3460–3465.

[68] Zhao, Y., and Change, L., "A Micro-Contact and Wear Model for Chemical-Mechanical Polishing," *Wear*, Vol. 252, 2002, pp. 220–226.

[69] Luo, J. and Dorfeld, D. A., "Material Removal Regions in Chemical Mechanical Planarization for Sub-micron Integrated Circuit Fabrication: Coupling Effects of Slurry Chemicals, Abrasive Size Distribution and Wafer-Pad Contact Area," *IEEE Trans. Semicond. Manuf.*, Vol. 16, No. 1, 2003, pp. 45–56.

[70] Anantha, R. S. and Wang, J. F., "Microstructural and Surface Phenomena in Metal CMP," *Electrochemical Society Proceedings*, Vol. 22, 1996, pp. 258–266.

[71] Kneer, E. A., Raghunath, C., Raghavan, S., and Jeon, J. S., "Electrochemistry of Chemical Vapor Deposited Tungsten Films with Relevance to Chemical Mechanical Polishing," *J. Electrochem. Soc.*, Vol. 143, No. 12, 1996, pp. 4095–4100.

[72] Stein, D., "Mechanistic, Kinetic and Processing Aspects of Tungsten Chemical Mechanical Polishing," Ph.D. Thesis, University of New Mexico, 1998.

[73] Asare, O. K. and Khan, A., "Chemical-Mechanical Polishing of Tungsten: An Electrophoretic Mobility Investigation of Alumina-Tungstate Interactions," *Electrochemical Society Proceedings*, Vol. 7, 1998, pp. 138–144.

[74] Steigerwald, J., Zirpoli, R., Myrarka, S., et al., "Metal Dishing and Erosion in the Chemical-Mechanical Polishing of Copper Used for Patten Delineation," *Materials Research Society Symposium Proceedings*, ULSI-X, 1995, pp. 55–59.

[75] Oliver, M. R., "CMP Fundamentals and Challenges," *Materials Research Society Symposium Proceedings*, Vol. 566, 2000, pp. 73–79.

[76] Li, Y., Ramarajan, S., Hariharaputhiran, M., Her, Y. S., and Babu, S. V., "Planarization of Cu and Ta Using Silicon and Alumina Abrasives—A Comparison," *Materials Research Society Symposium Proceedings*, Vol. 613, 2000, pp. E2.4.1–E2.4.6.

[77] Liang, H., Martin, J. M., and Brusic, V., "Chemical Wear of Cu CMP," *Materials Research Society Symposium Proceedings*, Vol. 613, 2000, pp. E2.5.1–E2.5.5.

[78] Wei, D., Gotkis, Y., Li, H., et al., "Copper CMP for Dual Damascene Technology: Some Consideration on the Mechanism of

[78] ... Removal," *Materials Research Society Symposium Proceedings*, Vol. 671, 2001, pp. E3.3.1–E3.3.6.

[79] Gotkis, Y., and Kistler, R., "Cu CMP for Dual Damascene Technology: Fundamentals," *Electrochemical Society Proceedings*, Vol. 26, 2000, pp. 253–269.

[80] Steigerwald, J. M., Muraraka, S. P., Ho, J., Gutmann, K. J., and Duquette, D. J., "Mechanism of Copper Removal During Chemical Mechanical Polishing," *J. Vac. Sci. Technol. B*, Vol. 13, No. 6, 1995, pp. 2215–2218.

[81] Hariharaputhiran, M., Li, Y., Ramarajan, S., and Babu, S. V., "Chemical Mechanical Polishing of Ta," *Electrochem. Solid-State Lett.*, Vol. 3, No. 2, 2000, pp. 95–98.

[82] Kuiry, S. C., Sea, L. S., Fei, W. et al., "Effect of pH and H_2O_2 on Ta Chemical Mechanical Planarization," *Journal of the Electrochemical Society*, Vol. 150, No. 1, 2003, pp. c36–c43.

[83] Li, Y. and Babu, S. V., "Chemical Mechanical Polishing of Copper and Tantalum in Potassium Iodate-Based Slurries," *Electrochem. Solid-State Lett.*, Vol. 4, No. 2, 2001, pp. G20–G22.

[84] Ramarajan, S., Li, Y., Hariharaputhiran, M., Her, Y. S., and Babu, S. V., "Effect of pH and Ionic Strength on Chemical Mechanical Polishing of Tantalum," *Electrochem. Solid-State Lett.*, Vol. 3, No. 5, 2000, pp. 232–234.

[85] Han, H., Ryan, F., and McClure, M., "Ultra-thin Tetrahedral Amorphous Carbon Film as Solider Overcoat for High Areal Density Magnetic Recording," *Surface and Coatings Technology*, Vol. 121, 1999, pp. 579–584.

[86] Menon, A. K., "Critical Requirements for 100 Gb/in.^{2} Head Media Interface," *ASME Proceedings of the Symposium on Interface Technology Towards* 100 Gbit/in.^{2}, Orlando, FL, 1999, pp. 1–9.

[87] Bhushan, B., "Chemical, Mechanical and Tribological Characterization of Ultra-thin and Hard Amorphous Carbon Coatings as Thin as 3.5 nm, Recent Development," *Diamond Relat. Mater.*, Vol. 8, 1999, pp. 1985–2015.

[88] Zhao, Z. and Bhushan, B., "Studies of Fly Stiction," *Proceedings of the Institution of Mechanical Engineers, Part J: Journal of Engineering Tribology*, Vol. 215, 2001, pp. 63–76.

[89] Xu, J. and Tsuchiyama, R., "Ultra-Low-Flying-Height Design from the Viewpoint of Contact Vibration," *Tribol. Int.*, Vol. 36, 2003, pp. 459–466.

[90] Demczyk, B., Liu, J., Chao, Y., and Zhang, S. Y., "Lubrication Effects on Head-Disk Spacing Loss," *Tribol. Int.*, Vol. 38, No. 6–7, 2005, pp. 562–565.

[91] Li, Y. and Bhushan, B., "Wear and Friction Studies of Contact Recording Interface with Microfabricated Heads," *Wear*, Vol. 202, 1996, pp. 60–67.

[92] Bhushan, B., Gupta, B. K., and Azarian, M. H., "Nanoindentation, Microsratch, Friction and Wear Studies of Coating for Contact Recording Applications," *Wear*, Vol. 181-183, 1995, pp. 743–758.

[93] Bhushan, B., *Tribology and Mechanics of Magnetic Storage Devices*, 2nd ed., New York: Springer-Verlag, 1996.

[94] Poon, C. Y. and Bhhushan, B., "Surface Roughness Analysis of Glass-Ceramic Substrates and Finished Magnetic Disks, and Ni-P Coated Al-Mg and Glass Substrates," *Wear*, Vol. 190, 1995, pp. 89–109.

[95] IDEMA (The International Disk Drive Equipment and Materials Association), *Disk Drive Technology*, IDEMA, 2000, pp. 7–9.

[96] Horn, M., "Antireflection Layers and Planarization for Microlithography," *Solid State Technol.*, Vol. 31, 1991, pp. 57–62.

[97] Ali, I., "Chemical-Mechanical Polishing of Interlayer Dielectric: A Review," *Solid State Technol.*, Vol. 34, 1994, pp. 63–70.

[98] Rahul, J., Janos, F., and Huang, C. K., "Chemical-Mechanical Polishing: Process Manufacturability," *Solid State Technol.*, Vol. 34, 1994, pp. 71–75.

[99] Skidmore, K., "Techniques for Planarizing Device Topography," *Semicond. Int.*, 1988, p. 115.

[100] Farid, M. and Masood, H., "Manufacturability of the CMP Process," *Thin Solid Films*, Vol. 270, 1995, pp. 612–615.

[101] Atsugi, Takeshi, Shibata, "Abrasives Composition, Substrate and Process for Producing the Same, and Magnetic Recording Medium and Process for Producing the Same," *USP 5, 868, 604*, 1999.

[102] Streinz, C. and Neville, C., "Composition and Method for Polishing Rigid disks," *USP 6, 015, 506*, 2000.

[103] Hagihara, Toshiya, Naito, "Polishing Composition," *USP, 6, 454, 820*, 2002.

[104] Fang, M. M., "Method for Polishing a Memory or Rigid Disk with an Oxidized Halide-Containing Polishing System," *USP 6, 468, 137*, 2002.

[105] Fang, M. M., "Method of Polishing a Memory or Rigid Disk with an Ammonia- and/or Halide-Containing Composition," *USP 6, 461, 227*, 2002.

[106] Stein, D. J., Hetherington, D. L., and Cecchi, J. L., "Investigation of the Kinetics of Tungsten Chemical Polishing in Potassium Iodate-Based Slurries I: Role of Alumina and Potassium Iodate," *Journal of the Electrochemical Society*, Vol. 146, 1999, pp. 376–381.

[107] Lei, H. and Luo, J. B., "Chemical Mechanical Polishing of Computer Hard Disk Substrate in Colloidal SiO_2 Slurry," *Int. J. Nonlinear Sci. Numer. Simul.*, Vol. 3, 2002, pp. 455–459.

[108] Luo, J. B., Duan, F. L., Xu, J., and Pan, G. S., "Microwear of the Surface Collided with Nano-Particles," *Forefront of Tribology 2005*, Kobe, Japan, May 2005.

[109] Luo, J. B. and Pan, G. S., "Variations of the Surface Layer during Chemical Mechanical Polish," *Indo-Chinese Workshop on MEMS Devices and Related Technologies*, New Delhi, India, April 5–7, 2006.

[110] Luo, J. F. and Dornfeld, D. A., "Material Removal Mechanism in Chemical Mechanical Polishing: Theory and Modeling," *IEEE Trans. Semicond. Manuf.*, Vol. 14, No. 2, 2001, pp. 112–133.

[111] Cho, C. H., Park, S. S., and Ahn, Y., "Three-Dimensional Wafer Scale Hydrodynamic Modeling for Chemical Mechanical Polishing," *Thin Solid Films*, Vol. 389, 2001, pp. 254–260.

[112] Park, S. S., Cho, C. H., and Ahn, Y., "Hydrodynamic Analysis of Chemical Mechanical Polishing Process," *Tribol. Int.*, Vol. 33, 2000, pp. 723–730.

[113] Larsen-Basse, J. and Liang, H., "Probable Role of Abrasion in Chemomechanical Polishing of Tungsten," *Wear*, Vol. 233-235, 1999, pp. 647–654.

[114] Liang, H., Kaufman, F., Sevilla, R., and Anjur, S., "Wear Phenomenon in Chemical Mechanical Polishing," *Wear*, Vol. 211, 1997, pp. 271–279.

[115] Grover, G. S., Liang, H., Ganeshkumar, S., and Fortino, W., "Effect of Slurry Viscosity Modification on Oxide and Tungsten CMP," *Wear*, Vol. 214, 1998, pp. 10–13.

[116] Zhou, C. H., Shan, L., Hight, J. R., Ng, S. H., and Danyluk, S., "Fluid Pressure and Its Effects on Chemical Mechanical Polishing," *Wear*, Vol. 253, 2002, pp. 430–437.

[117] Zhao, Y. W. and Chang, L., "A Micro-Contact, and Wear Model for Chemical-Mechanical Polishing of Silicon Wafers," *Wear*, Vol. 252, 2002, pp. 220–226.

[118] Zhao, Y. W., Chang, L., and Kim, S. H., "A Mathematical Model for Chemical-Mechanical Polishing Based on Formation and Removal of Weakly Bonded Molecular Species," *Wear*, Vol. 254, 2003, pp. 332–339.

[119] Nanz, G. and Camilletti, L. E., "Modeling of Chemical-Mechanical Polishing: A Review," *IEEE Trans. Semicond. Manuf.*, Vol. 8, 1995, pp. 382–389.

[120] Seok, J., Sukam, C., Kim, A. T., Tichy, J. A., and Cale, T. S., "Multiscale Material Removal Modeling of Chemical Mechanical Polishing," *Wear*, Vol. 254, No. 3–4, 2003, pp. 307–320.

[121] Warnock, J., "A Two-Dimensional Process Model for Chemomechanical Polishing Planarization," *J. Electrochem. Soc.*, Vol. 138, 1991, pp. 2398–2402.

[122] Hsu, J., Chiu, S., Wang, Y., et al., "The Removal Selectivity of Titanium and Aluminum in Chemical Planarization," *J. Electrochem. Soc.*, Vol. 149, 2002, pp. 204–208.

[123] Steigerward, J. M., Muraka, S. P., and Gutmann, R. J., *Chemical Mechanical Planarization of Microelectronic Materials*, New York: Wiley, 1997.

[124] Liang, H. and Xu, G. H., "Lubricating Behavior in Chemical-Mechanical Polishing of Copper," *Scr. Mater.*, Vol. 46, 2002, pp. 343–347.

[125] Preston, F., "The Theory and Design of Plate Glass Polishing Machines," *Society of Glass Technology—Journal*, Vol. 11, No. 42, 1927, pp. 214–256.

[126] Pohl, M. C. and Griffiths, D. A., "The Importance of Particle Size to the Performance of Abrasive Particles in the CMP Processes," *J. Electron. Mater.*, Vol. 25, 1996, pp. 1612–1616.

[127] Warnock, "A Two-Dimensional Process Model for Chemimechanical Polish Planarization," *Journal of the Electrochemical Society*, Vol. 138, No. 8, 1991, pp. 2398–2402.

[128] Zhang, F., Busnaina, A. A., Feng, J., and Fury, M. A., "Particle Adhesion Force in CMP and Subsequent Cleaning Processes," *Proceedings, 4th International Chemical-Mechanical Planarization for ULSI Multilevel Interconnection Conference*, Santa Clara, CA, Feb. 11–12, 1999, pp. 61–64.

[129] Liu, C., Dai, B., Tseng, W., et al., "Modeling of the Wear Mechanism During Chemical-Mechanical Polishing," *Journal of the Electrochemical Society*, Vol. 143, No. 2, 1996, pp. 716–721.

[130] Zhao, B. and Shi, F. G., "Chemical Mechanical Polishing: Threshold Pressure and Mechanism," *Electrochem. Solid-State Lett.*, Vol. 2, No. 3, 1999, pp. 145–147.

[131] Sundararajan, S. and Thakurta, D., "Two-Dimensional Wafer-Scale Chemical-Mechanical Planarization Models Based on Lubrication Theory and Mass Transport," *Journal of the Electrochemical Society*, Vol. 146, No. 2, 1999, pp. 761–766.

[132] Ahmadi, G. and Xia, X., "A Model for Mechanical Wear and Abrasive Particle Adhesion During the Chemical Mechanical Polishing Process," *Journal of the Electrochemical Society*, Vol. 148, No. 3, 2001, pp. G99–G109.

[133] Moon, Y., "Mechanical Aspect of the Material Removal Mechanism in Chemical Mechanical Polishing," Ph.D. Thesis, University of California, Berkeley, 2002.

[134] Stein, D., "Mechanistic, Kinetic and Processing Aspects of Tungsten Chemical Mechanical Polishing," Ph.D. Thesis, University of New Mexico, 1998.

[135] Bielmann, M., Mahajan, U., and Singh, R. K., "Effect of Particles Size During Tungsten Chemical Mechanical Polishing," *Electrochem. Solid-State Lett.*, Vol. 2, No. 8, 1999, pp. 401–403.

[136] Hoshino, T., Kurata, Y., Terasaki, Y., et al., "Mechanism of Polishing of SiO_2 Films by CeO_2 Particles," *J. Non-Cryst. Solids*, Vol. 283, 2001, pp. 129–136.

[137] Pietsch, G. J., Chabal, Y. J., and Higashi, G. S., "The Atomic-Scale Removal Mechanism During Chemo-Mechanical Polishing of Si (100) and Si (111)," *Surf. Sci.*, Vol. 331-333, 1995, pp. 395–401.

[138] Vijayakumar, A., Du, T., Sundaram, K. B., and Desai, V., "Polishing Mechanism of Tantalum Films by SiO_2 Particles," *Microelectron. Eng.*, Vol. 70, No. 1, 2003, pp. 93–101.

[139] Zhao, Y., Change, L., and Kim, S. H., "A Mathematical Model for Chemical Mechanical Polishing Based on Formation and Removal of Weakly Bonded Molecular Species," *Wear*, Vol. 254, 2003, pp. 332–339.

[140] Balakumar, S., Haque, T., Kumar, A. S., Rahman, M., and Kumar, R., "Wear Phenomena in Abrasive-Free Copper CMP Process," *Journal of the Electrochemical Society*, Vol. 152, No. 11, 2005, pp. G867–G874.

[141] Lai, J., Ph.D. Thesis, Massachusetts Institute of Technology, Cambridge, 2001.

[142] Fang, J. Y., Huang, P. W., Tsai, M. S., et al., "Study on Pressure-Independent Cu Removal in Cu Abrasive-Free Polishing," *Electrochem. Solid-State Lett.*, 9, No. 1, 2006, pp. G13-G16.

[143] Kuroe, A., Muramatsu, S., Fusayasu, H., et al., "Predicted Outputs of High-Frequency Carrier-Type Magnetic Head for 100 Gb/in.^{2} Generation," *J. Magn. Magn. Mater.*, Vol. 235, 2001, pp. 382–387.

[144] Liu, B., Man, Y. J., and Zhang, W., "Slider-Disk Interaction and Tribologically Induced Signal Decay," *J. Magn. Magn. Mater.*, Vol. 239, 2002, pp. 378–384.

[145] Komvopoulos, K., "Head-Disk Interface Contact Mechanism for Ultrahigh Density Magnetic Recording," *Wear*, Vol. 238, 2000, pp. 1–11.

[146] Yang, M. C., Luo, J. B., Wen, S. Z., et al., "Failure Characterization at Head/Write Interface of Hard Disk Drive," *Science in China*, Vol. 44, 2001, pp. 407–411.

[147] Bhushan, B., "Chemical, Mechanical and Tribological Characterization of Ultra-thin and Hard Amorphous Carbon Coatings as thin as 3.5 nm," *Diamond Relat. Mater.*, Vol. 8, 1999, pp. 1985–2015.

[148] Luo, J. B. and Yang, M. C., "Surface Modification of Computer Magnetic Head," *Sino-German Symposium on Micro Systems and Nano Technology*, 7–9 Sept. 2001, Braunschweig, Germany.

[149] Yang, M. C., Luo, J. B., Wen, S. Z., Wang, W., Huang, Q., and Hao, C. W., "Investigation of X-1P Coating on Magnetic Head to Enhance the Stability of Head/Disk Interface," *Science in China*, Vol. 44, 2001, pp. 400–406.

[150] Zhou, C. H., Shan, L., Hight, J. R., et al., "Influence of Colloidal Abrasive Size on Material Removal Rate and Surface Finish in SiO_2 Chemical Mechanical Polishing," *Tribol. Trans.*, Vol. 45, 2002, pp. 232–241.

[151] Wood, R., Sonobe, Y., Jin, Z., and Wilson, B., "Perpendicular

Recording: The Promise and The Problems," *J. Magn. Magn. Mater.*, Vol. 235, 2001, pp. 1–9.

[152] Gao, F., "Application of Ultra-Fine Diamond Slurry in Polishing Magnetic Recording Heads," Master Thesis, Tsinghua University, 2001.

[153] Luo, J. B., Hu, Z. M., Gao, F., and Lu, X. C., "Nano-polish Slurries and Their Preparation Methods," ZL00133674.6, Chinese Patent, 2003.

[154] Shen, M. W., Luo, J. B., and Wen, S. Z., "The Tribological Properties of Oils Added with Diamond Nano-particles," *STLE Tribol. Trans.*, Vol. 44, 2001, pp. 494–498.

Subject Index

A

Adhesion
  and friction, 167–186
  surface force apparatus (SFA), 17–18
Adsorbed monolayers, boundary lubrication, 80–81
AFM. see atomic force microscope (AFM)
Apparatus setup, surface force apparatus (SFA), 14–15
Applications, surface force apparatus (SFA), 17–18
Asperity contact, mixed lubrication at micro-scale, 118
Atomic force microscope (AFM), 19–21
Average flow model, mixed lubrication at micro-scale, 117–118

B

Boundary friction, lubrication, 93–94
Boundary lubrication, 79–95
  adsorbed monolayers, 80–81
  boundary friction, 93–94
  hydrodynamic, 82–83
  surface film formation, 81–82

C

Chemical mechanical polishing (CMP), 4
  ultra-smooth surface polishing, 245–261
CMP. see chemical mechanical polishing (CMP)
CNx films, 151–153
Constant function-based scheme, mixed lubrication at micro-scale, 122
Correlations, friction and adhesion, 178–181
Couple stress, thin film lubrication theoretical model, 76

D

Deterministic mixed lubrication (DML) model, 125–130
Development challenges, magnetic recording system, 231–236
Diamond-like carbon (DLC) coatings, 2
  thin solid coatings, 147–151
DLC. see diamond-like carbon (DLC) coatings
DML. see deterministic mixed lubrication (DML) model

E

EEF. see external electric field (EEF)
EFF. see electric field effect (EFF)
EHL. see elastohydrodynamic lubrication (EHL)
Elastohydrodynamic lubrication (EHL), 37–40
Electric field effect (EFF), thin film lubrication, 45–48
Energy dissipation, friction and adhesion, 171–178

Ensemble average, thin film lubrication theoretical model, 63–65
External electric field (EEF), 55–60

F

FFM. see friction force microscope (FFM)
Fluorometry technique, measuring, 26–28
Flying head air bearing surface, gas lubrication, 111–113
Friction
  and adhesion, 167–186
  correlations, 178–181
  physics and, 167–171
  static, 181–184
  surface force apparatus (SFA), 17–18
  wearless, 171–178
Friction force
  microscale friction and wear/scratch, 189–191
  thin film lubrication experimental study, 43–44
Friction force microscope (FFM), 21
Fringe and shape interference on curves, 41
Frustrated total reflection (FTR) technique, 12–14
FTR. see frustrated total reflection (FTR) technique

G

Gap width on gas flow effect, gas lubrication, 101–103
Gas bubble in liquid film under external electric, thin film lubrication, 55–60
Gas flows
  in microsystems, 113–114
  regimes, gas lubrication, 97–98
Gas lubrication in nano-gap, 96–115
  flying head air bearing surface, 111–113
  gap width on gas flow effect, 101–103
  gas flow regimes, 97–98
  gas flows in microsystems, 113–114
  head-disk, 104–106
  rarefaction, 106–111
  Reynolds equations, 98–101, 103–104
  theory, 97–103
Gas lubrication theory, 3–4
Green's function-based scheme, mixed lubrication at micro-scale, 122

H

Hard disk driver (HDD), 3
  gas lubrication theory, 3–4
HDD. see hard disk driver (HDD)
Head-disk, gas lubrication in nano-gap, 104–106
Hydrodynamic, boundary lubrication, 82–83

I

Interferometry
  thin film colorimetric interferometry (TFCI), 10–12
  wedged spacer layer optical interferometry, 8–9
Ionic liquids, thin film lubrication, 54–55

L

Laser doppler vibrometry (LDV), 30–32
Lateral forces measurement, SFA, 17
Layers
  adsorbed monolayers, boundary lubrication, 80–81
  lubrication and monolayers, 3
  multilayer films, 153–157
  multilayers, 200–208
  scratch, 200–208
  wedged spacer layer optical interferometry, 8–9
LDV. see laser doppler vibrometry (LDV)
Linear interpolation-based scheme, mixed lubrication at micro-scale, 122
Liquid crystal
  film thickness, 45–48
  isoviscosity, 48
  styles, 45
Lubricants
  film failure, 53–54
  on hard disk surface, 226–231
  isoviscosity, 40–41
  viscosity, 40
Lubrication and monolayers, 3

M

Macro scale friction and wear/scratch, 188
Magnetic head surface, ultra-smooth surface polishing, 262–265
Magnetic recording system
  development challenges, 231–236
  tribology, 210–236
Measuring techniques, 7–36
  atomic force microscope (AFM), 19–21
  fluorometry technique, 26–28
  friction force microscope (FFM), 21
  laser doppler vibrometry (LDV), 30–32>
  nanoindentation, 22–25
  nanscratch, 25–26
  optical, 8–14
  Polytec Micro Scanning Vibrometer, 32
  scanning acoustic microscopy (SAM), 28–30
  scanning probe microscope, 18–22
  scanning tunneling microscope (STM), 18–19
  Secondary Ion Mass Spectrometry (SIMS), 32–33
  surface force apparatus (SFA), 14–18
  Time-of-Flight Secondary Ion Mass Spectrometry (TOF-SIMS), 32–33
Micro/nano scale friction and wear/scratch, 188
Micro/nanotribology, 1–2
  applications, 3–4
Microscale friction and wear/scratch, 187–209
  friction force, 189–191
  macro, 188
  micro/nano, 188
  modified molecular films, 194–200
  multilayers, 200–208
  thin solid films, 191–194
Mixed lubrication at micro-scale, 116–144
  asperity contact, 118
  average flow model, 117–118
  constant function-based scheme, 122
  deterministic mixed lubrication (DML) model, 118–121
  Green's function-based scheme, 122
  linear interpolation-based scheme, 122
  numerical methods for surface deformation, 121–125
  performance, 130–144
  Reynolds equations, 121
  rough surface lubrication, 116–117
  statistical approach, 116–118
  stochastic models, 116–117
  thermal analysis, 119–120<
Modified molecular films, microscale friction and wear/scratch, 194–200
Molecule films, 79–95
  boundary friction, 93–94
  ordered, 88–93
  thin film rheology, 83–88
Monolayers, adsorbed, 80–81
Multilayer films, 153–157
Multilayers, microscale friction and wear/scratch, 200–208

N

Nanoindentation, measuring techniques, 22–25
Nanoparticles
  thin film lubrication experimental study, 50–53
  ultra-smooth surface polishing, 237–244
Nanoscratch, measuring techniques, 25–26
Near frictionless coatings (NFC), 2
NFC. see near frictionless coatings (NFC)
Normal force measurement, SFA, 16–17
Numerical methods for surface deformation, 121–125

O

Optical measuring techniques, 8–14
  frustrated total reflection (FTR) technique, 12–14
  relative optical interference intensity method, 9–10
  thin film colorimetric interferometry (TFCI), 10–12
  wedged spacer layer optical interferometry, 8–9
Optical technique for gap measurement, SFA, 15–16

P

Physics, friction and adhesion, 167–171
Polytec Micro Scanning Vibrometer, 32

R

Rarefaction, gas lubrication in nano-gap, 106–111
Relative Optical Interference Intensity Method, 7
Relative optical interference intensity method, 9–10
Reynolds equations
   gas lubrication in nano-gap, 98–101, 103–104
   mixed lubrication at micro-scale, 121
   thin film lubrication theoretical model, 68
Rheology, thin film lubrication theoretical model, 72–74
Rough surface lubrication, 116–117

S

SAM. see scanning acoustic microscopy (SAM)
Sample positioning, SFA, 16–17
Scanning acoustic microscopy (SAM), 28–30
Scanning probe microscope, 18–22
Scanning tunneling microscope (STM), 7, 18–19
Scratch
   macro scale friction and wear/scratch, 188
   microscale friction and wear/scratch, 189–191
   modified molecular films, 194–200
   multilayers, 200–208
   nanoscratch, 25–26
   thin solid films, 191–194
Secondary ion mass spectrometry (SIMS), 32–33
SFA. see surface force apparatus (SFA)
Simulations via micropolar theory, thin film lubrication, 67–72
Slide ratio effect, thin film lubrication, 44–45
Solid surface energy effect, thin film lubrication, 42–43
Spatial average, thin film lubrication, 63–65
Static friction, 181–184
Statistical approach, mixed lubrication, 116–118
STM. see scanning tunneling microscope (STM)
Stochastic models, mixed lubrication, 116–117
Structure, surface force apparatus (SFA), 14–17
Superhard nanocomposite coatings, 2–3
   thin solid coatings, 157–162
Surface coatings, 2–3
Surface film formation, boundary lubrication, 81–82
Surface force apparatus (SFA), 2, 7, 14–18
   adhesion and friction, 17–18
   apparatus setup, 14–15
   applications, 17–18
   lateral forces measurement, 17
   measurement techniques, 14–17
   normal force measurement, 16–17
   optical technique for gap measurement, 15–16
   sample positioning, 16–17
   structure, 14–17
   surface forces, 17
   thin film rheology, 18
Surface forces, 17
Surface modification films on magnetic head tribology, 211–225

T

TFCI. see thin film colorimetric interferometry (TFCI)
Tthermal analysis, mixed lubrication at micro-scale, 119–120
Thin film colorimetric interferometry (TFCI), 10–12
Thin film lubrication experimental study
   advancements, 37
   boundary lubrication, 37–38
   defined, 37–38
   elastohydrodynamic lubrication (EHL), 37–40
   electric field effect, 45–48
   film thickness and influence factors, 39–41
   friction force, 43–44
   fringe and shape interference on curves, 41
   gas bubble in liquid film under external electric, 55–60
   ionic liquids, 54–55
   liquid crystal film thickness, 45–48
   liquid crystal isoviscosity, 48
   liquid crystal styles, 45
   lubricant film failure, 53–54
   lubricant isoviscosity, 40–41
   lubricant viscosity, 40
   lubrication map, 38–39
   lubrication properties, 39–53
   nanoparticles, 50–53
   rolling speed effect, 39–40
   slide ratio effect, 44–45
   solid surface energy effect, 42–43
   thin film time effect, 48–50
   transition form EHL to TFL, 41–42
   ultra-fine diamond powders (UDP), 50–53
Thin film lubrication (TFL), 2, 7
Thin film lubrication theoretical model, 63–78
   couple stress, 76
   ensemble average, 63–65
   Reynolds equation, 68
   rheology, 72–74
   simulations via micropolar theory, 67–72
   spatial average, 63–65
   velocity field of lubricants, 65–67
   viscosity modification, 72–74
Thin film rheology
   molecule films, 83–88
   surface force apparatus (SFA), 18
Thin solid coatings, 147–166
   CNx films, 151–153
   diamond-like carbon (DLC) coatings, 147–151
   multilayer films, 153–157
   superhard nanocomposite coatings, 157–162
Thin solid films, microscale friction and wear/scratch, 191–194
Time effect, thin film lubrication experimental study, 48–50
Time-of-Flight Secondary Ion Mass Spectrometry (TOF-SIMS), 32–33
Transition form EHL to TFL, thin film lubrication, 41–42
Tribology
   defined, 1
   history, 1
   lubricants on hard disk surface, 226–231

magnetic recording system, 210–236
scaledown development, 1
surface modification films on magnetic head, 211–225
ultra-smooth surface polishing, 237–265

U

UDP. see ultra-fine diamond powders (UDP)
Ultra-fine diamond powders (UDP), 50–53
Ultra-smooth surface polishing
  chemical mechanical polishing (CMP), 245–261
  magnetic head surface, 262–265
  nanoparticles impact, 237–244
  tribology, 237–265
Ultra-thin coatings, 3

V

Validation, deterministic mixed lubrication (DML) model, 125–130
Velocity field of lubricants, 65–67
Viscosity
  liquid crystal isoviscosity, 48
  lubricants, 40–41
  modification, thin film lubrication, 72–74

W

Wear/scratch, microscale friction and wear/scratch, 187–209
Wearless friction and energy dissipation, 171–178
Wedged spacer layer optical interferometry, 8–9